Contents

Ceramic Materials and Components for Energy and Environmental Applications

Ceramic Materials and Components for Energy and Environmental Applications

Ceramic Transactions, Volume 210

A Collection of Papers Presented at the 9th International Symposium on Ceramic Materials for Energy and Environmental Applications and the Fourth Laser Ceramics Symposium November 10–14, 2008, Shanghai, China

Edited by
Dongliang Jiang
Yuping Zeng
Mrityunjay Singh
Juergen Heinrich

The American Ceramic Society

WILEY

A John Wiley & Sons, Inc., Publication

Published by John Wiley & Sons, Inc., Hoboken, New Jersey.
Published simultaneously in Canada.

Library of Congress Cataloging-in-Publication Data is available.

ISBN 978-0-470-40842-1

Printed in the United States of America.

10 9 8 7 6 5 4 3 2 1

X. Bioceramics

XI. Laser Ceramics

Preface

The global population growth and tremendous economic development has led to increasing demand for energy from all over the world as well as increasing concern for environment and global warming. The energy efficient and eco-friendly systems and technologies are critically needed for the further global growth and sustainable development. Advanced ceramics are enabling materials for a number of demanding energy efficient and eco-friendly applications in aerospace, power generation, ground transportation, nuclear, and chemical industries. These materials have unique properties such as high strength, high stiffness, long fatigue life, low density, and adaptability to the intended functions. Significant achievements have been made worldwide in the design, development, manufacturing, and application of these materials in recent years and considerable innovative research and technology development is still continuing to address technical and economic challenges.

9th International Conference on Ceramic Materials and Components for Energy and Environmental Applications (9th CMCEE) in Shanghai, China was continuation of series of international conferences held all over the world over the last three decades. The major goal of CMCEE was to bring together academicians, researchers, and end users in various disciplines from all over the world to share knowledge and exchange views leading to industrial applications of these technologies. The current volume contains selected peer reviewed papers from more than 300 presentations from all over the world. The papers in this volume also highlight and emphasize the importance of synergy between advanced materials and component designs. This volume also contains selected papers from 4th International Laser Ceramics symposium which was held during the same time period. We would like to thank organizers and sponsors of this symposium.

We would like to acknowledge the financial support from Chinese Academy of Sciences, Shanghai Municipal Corporation, and Shanghai Institute of Ceramics. Our special thanks to Abhishek Singh from Case Western Reserve University, Cleveland, Ohio for the editing of the manuscripts. We would also like to thank Mr. Greg Geiger, Technical Content Manager of The American Ceramic Society for all the help in the production of this volume. We would like to thank all the contributors and reviewers from all over the world.

Dongliang Jiang
Yuping Zeng
Shanghai Institute of Ceramics, Shanghai, China

Mrityunjay Singh
Ohio Aerospace Institute, Cleveland, USA

Juergen Heinrich
Clausthal University of Technology, Germany

Acknowledgements

9th International Conference on Ceramic Materials and Components for Energy and Environmental Applications (9th CMCEE)

Hosted by:
Shanghai Institute of Ceramics, Chinese Academy of Sciences

Endorsed by:
The Chinese Ceramic Society
The American Ceramic Society
The European Ceramic Society
The Ceramic Society of Japan
The Korean Ceramic Society
The Australian Ceramic Society

Conference Committee:
Conference Chair (Asia):
Prof. Dongliang Jiang
Shanghai Institute of Ceramics,
Chinese Academy of Sciences
China

Co-Chair (America):
Dr. Mrityunjay Singh
Ohio Aerospace Institute
NASA Glenn Research Center
USA

International Advisory Committee

F. Aldinger (Germany)
Longtu Li (China)
José Ferreira (Portugal)
J. Martinez-Fernandez (Spain)
C.X. Ding (China)
S. I. Milieko (Russia)
Ludwig Gauckler (Switzerland)
Dong-Soo Park (Korea)
Jingkun Guo (China)
Mrityunjay Singh (USA)
Jurgen Heinrich (Germany)
Louis Winnubst (The Netherlands)
Zhanping Jin (China)
Paolo Zannini (Italy)
Walter Krenkel (Germany)
M.H. Lewis (U.K)
Jean Baumard (France)
L.M. Manocha (India)
Yibing Cheng (Australia)
G.L. Messing (USA)
Ruiping Gao (China)
Dale E. Niesz (USA)
Peter Greil (Germany)
Yoshio Sakka (Japan)
Stuart Hampshire (Ireland)
Derek Thompson (U.K)
Dongliang Jiang (China)
Koichi Niihara (Japan)
K. Komeya (Japan)
Hasan Mandal (Turkey)
G.N. Babini (Italy)
H.T. Lin (USA)
I-Wei Chen (USA)
S. Mathur (Germany)
M.K. Ferber (USA)
R. Naslain (France)
Takashi Goto (Japan)
Pavol Sajgalik (Slovakia)
Victor Gusarov (Russia)
N. Sobczak (Poland)
S.I. Hirano (Japan)
M. Yoshimura (Japan)
A. Kohyama (Japan)
L.T. Zhang (China)

4th International Laser Ceramics symposium (4th LCS)

Local Organizing Committee

Chair:
Prof. Lidong Chen
Shanghai Institute of Ceramics
Chinese Academy of Sciences

Secretariat:
Prof. Yu-Ping Zeng
Shanghai Institute of Ceramics
Chinese Academy of Sciences

Secretariat assistant:
Mr. Hui Tong
Shanghai Institute of Ceramics
Chinese Academy of Sciences

Hosted by
Shanghai Institute of Ceramics, Chinese Academy of Sciences

Endorsed by
The Chinese Ceramic Society
The American Ceramic Society
The European Ceramic Society
The Ceramic Society of Japan
The Korean Ceramic Society
The Australian Ceramic Society

Conference Chairman:
S. W. Wang, Shanghai Institute of Ceramics, Chinese Academy of Sciences, China

Co-chairs and Program Committee
A. A. Kaminskii, Institute of Crystallography, Russia
K. Ueda, Institute for Laser Science, University of Electro-Communications, Japan
Q. H. Lou, Shanghai Institute of Optics and Fine Mechanics, Chinese Academy of Sciences, China
A. Ikesue, World-Lab. Co. Ltd., Japan
W. Strek, Dept of Excited State Spectroscopy, Poland Academy of Sciences, Poland
V. Lupei, Institute of Atomic Physics, Romania
Bruno Le Garrec, CEA CESTA, France

International Advisory Committee
T. Taira, Laser Research Center for Molecular Science, Institute for Molecular Science, Japan
T. Yanagitani, Konoshima Chem Co Ltd, Takuma, Japan
J. Kawanaka, Osaka University, Japan
R. L. Gentilman, Raytheon Company, USA
T. F. Soules, Lawrence Livermore National Laboratory, USA
M. Dubinskiy, US Army Research Laboratory, USA
G. J. Quarles, II-VI Corp./VLOC, USA
S. B. MIROV, University of Alabama at Birmingham, Birmingham, USA
G. C. Wei, Osram Sylvania Inc., USA
D. L. Jiang, Shanghai Institute of Ceramics, Chinese Academy of Sciences, China
Z. Y. Xu, Institute of Physics, Chinese Academy of Sciences, China
D. Z. Shen, Research Institute of Synthetic Crystals, Beijing, China
D. Y. Tang, School of Electrical & Electronic Engineering, Nanyang Technological University, Singapore
S. N. Bagayev, Institute of Laser Physics, Novosibirsk, Russia
V. B. Kravchenko, Fryazino, FIRE RAS, Russia
T. T. Basiev, Laser Materials and Technology Research Center, Moscow, Russia
G. Boulon, LPCML, CNRS, Lyon, France
M. Mortier, Ecole Nationale Supérieure de Chimie de Paris, France
J. F. Baumard, SPCTS, Limoges, France
Y. Rabinovitch, CILAS, ESTER Technopole, France
H. J. Kong (Korea), Laser Science Research Lab, Korea Advanced Institute of Science and Technology
Witold Łojkowski, Institute of High Pressure Physics, Polish Academy of Sciences, Poland
R. Chaim, Department of Materials Engineering, Israel Institute of Technology, Israel
A. Krell, Fraunhofer Institute for Ceramic Technologies and Systems (IKTS), Germany

Local Organizing Committee

I. Basic Science, Design, Modeling and Simulation

FRACTURE STATISTICS OF SMALL SPECIMENS

Robert Danzer, Peter Supancic

Institut für Struktur- und Funktionskeramik, Montanuniversität Leoben, Peter-Tunner Strasse 5, A-8700 Leoben, Austria
and
Materials Center Leoben, Roseggerstrasse 12, A-8700 Leoben, Austria.
E-mail: isfk@unileoben.ac.at

ABSTRACT

Strength data of brittle materials show significant scatter. Therefore designing with brittle materials has to be made with probabilistic methods. So far this is done using Weibull statistics, which is based on the weakest link hypothesis. It (implicitly) implies a particular type of defect distribution, which can be observed in many (but not in all) ceramic materials. It is shown that for very small specimens the Weibull assumptions claim unrealistic high densities of flaws. Then the flaws will interact, they are not longer statistically independent and the weakest link hypothesis is not valid. Consequently the Weibull distribution predicts too high a strength for very small specimens.

INTRODUCTION

Fracture of ceramics usually initiates from flaws which are randomly distributed in the material. The strength of the specimen then depends on the length of the major flaw, which varies from specimen to specimen. The strength of brittle materials has to be described by statistical means [1 - 3]. It follows from experiments that the failure probability increases with load amplitude and with size of the specimens [1-5]. The first observation is trivial. The second observation follows from the fact that it is more likely to find a major flaw in a large than in a small specimen. Therefore the mean strength of a set of large specimens is smaller than that of small specimens. This size effect of strength is the most prominent and relevant consequence of the statistical behaviour of the strength of brittle materials.

Weibull developed his statistical theory of brittle fracture on the basis of the weakest link hypothesis, i.e. the specimen fails if its weakest element fails [6, 7]. In its simplest form and for an uniaxial homogenous and tensile stress state, σ, and for specimens of the volume, V, the so called Weibull distribution of the probability of failure, F, is given by:

$$F(\sigma,V)=1-exp\left[-\frac{V}{V_0}\left(\frac{\sigma}{\sigma_0}\right)^m\right] \quad .$$

The Weibull modulus, m, describes the scatter of strength data: the distribution is the wider the smaller m is. σ_0 is the characteristic strength and V_0 is the corresponding reference volume. Of course the probability of surviving (the reliability, R) is: $R=1-F$. Freudenthal [8] showed for sparsely distributed flaws, that the probability of failure only depends on the number of destructive flaws, $N_{c,S}$, occurring in a specimen of size and shape, S:

$$F_S(\sigma)=1-exp(-N_{c,S}) \quad .$$

$N_{c,S}$ is the mean number of destructive (critical) flaws in a large set of specimens (i.e. the value of expectation). Jayatilaka et al. [9] showed, that, for brittle and homogeneous materials, the distribution of the strength data is caused by the distribution of sizes (and orientations) of the flaws.

A Weibull distribution of strength will be observed for flaw populations with a monotonically decreasing density of flaw sizes. Danzer et al. [10 - 12] extended these ideas to flaw populations with any size distribution and to specimens with an inhomogeneous flaw population. On the basis of these ideas a direct correlation between the flaw size distribution and the scatter (statistics) of strength data can be defined.

The Weibull distribution is the state of the art statistics in the mechanical design process of ceramic components [1 - 3]. Strength testing and data evaluation are standardised. A sample of at least 30 specimens has to be tested. The range of "measured" failure probabilities increases with the sample size [3, 13] and is - for a sample of 30 specimens - very limited (it is between 1/60 and 59/60). To determine the design stress, the measured data have to be extrapolated with respect to the volume and to the "tolerated" failure probability. This often results in a very large extrapolation span [3].

In this paper the Weibull theory is applied to very small specimens. The analysis follows the ideas presented in [13]. The relationships between flaw population, size of the fracture initiating flaw and strength are discussed. It is shown that a limit for the applicability of the classical fracture statistics (i.e. Weibull statistics based on the weakest link hypothesis) exists for very small specimens (components).

FRACTURE STATISTICS AND DEFECT SIZE DISTRIBUTION

The function $N_{c,S}(\sigma)$ is obtained by integrating the local density, $n_c(\sigma,\vec{r})$, of destructive flaws

$$n_c(\sigma,\vec{r}) = \int_{a_c(\sigma)}^{\infty} g(a,\vec{r})\,da$$

over the volume of the specimen: $N_{c,S} = \int n_c dV$ [3, 8 - 10]. For simplicity, but without loss of generality [8], it is assumed that size and orientation of a flaw are described by a single variable (the flaw size, a). The frequency distribution of the density of flaw sizes, $g(a,\vec{r})$, may depend on the position vector, $\vec{r}$. A local fracture criterion (e.g. the Griffith criterion, [1, 2]) correlates stress amplitude and flaw length: the critical flaw size is the smallest flaw length, which – under the action of the stress – causes failure (the size of the smallest destructive flaw). Since a_c depends on the magnitude of the applied stress, so do the values of n_c and also $N_{c,S}(\sigma)$. For a homogeneous material loaded under uniaxial homogeneous tension the volume integral is trivial. For a flaw population with relative frequencies decreasing with a negative power the flaw size, a,

$$g(a) = g_0 \cdot (a / a_0)^{-r}$$

a Weibull distribution (eq. 1) occurs [9]. This function has only two independent parameters: the exponent ($-r$) and the coefficient ($g_0 \cdot a_0^{\,r}$). Using these assumptions and after some algebra the density of destructive flaws in terms of a critical flaw size is: $n(a_c) = (a_c \cdot g(a_c)) / (r-1)$. The critical flaw size can be defined using the Griffith/Irwin criterion [1 - 3]:

$$a_c = \frac{1}{\pi} \cdot \left(\frac{K_{Ic}}{Y \cdot \sigma} \right)^2 .$$

K_{Ic} is the critical stress intensity factor (the fracture toughness) and Y is a dimensionless geometric factor. Inserting in the above expression analytical equations for the Weibull parameters results: The Weibull modulus is only related to the path of the flaw size distribution:

$$m = 2 \cdot (r-1) .$$

The second parameter in the Weibull statistics is: $V_0 \cdot \sigma_0{}^m = \left((r-1)/g_0 \cdot a_0\right)\left(K_{Ic}/Y\sqrt{\pi \cdot a_0}\right)$[illegible].

In the following a material behaving in the way as described above (eq. 1, eq. 4, etc.) is called "Weibull material".

A Weibull type strength distribution also may arise for inhomogeneous stress and non uniaxial stress states (then the volume has to be replaced by an effective volume, [1 - 3]). If failure is caused by surface flaws, the volume has to be replaced by the surface [1, 3, 12].

Danzer et al. have discussed the influence of other types of flaw populations (e.g. of bimodal distributions) on strength [11, 12]. In these cases the Weibull modulus might depend on the applied load amplitude and on the size of the specimen. Then the determination of a design stress in the usual way may become problematic. A stress and size dependent modulus occurs for materials with an R-curve behaviour [11] and may also be caused by internal stress fields [11].

It should be noted that on the basis of a small sample size, e.g. only 30 specimens, it is not possible to differentiate between a Weibull, a Gaussian, or any other similar distribution functions, as shown by Lu et al. [14] using statistical measures or by Danzer et al. [12] using Monte Carlo simulations. This is caused by the inherent scatter of the data and the difference between sample and true population. The ultimate test for the existence of a Weibull distribution is to test a material on different levels of (effective) volumes.

THE CORRELATION BETWEEN STRENGTH AND FLAW POPULATION

In the following, the relationship between fracture statistics and defect size distribution is discussed for the simple case of tensile tests (uniaxial and homogeneous stress state) on a homogeneous brittle material. The tests are performed on specimens of equal size. It is assumed that the volume of the specimens is: $V = V_0$. The number of tested specimens (the sample size) is X. In each test the load is increased up to the moment of failure. The strength is the stress at the moment of failure. In each sample the strength values of the individual specimens are different, i.e. the strength is distributed.

If data determined in that way are evaluated the specimens are ranked according to their strength, i being the ranking parameter. To estimate the failure probability for an individual specimen an estimation function is used [1, 3, 13]:

$$F_i = (i - 1/2)/X \qquad , \qquad i = 1, 2, ? \;\; X \quad .$$

Inserting eq. 7 into eq. 2 and making a few rearrangements, we get: $N_{c,S}(\sigma_i) = \ln 2X/(2X - 2i + 1)$. In this way the mean number of critical flaws per specimen (volume V_0; stress σ_i) can be read from the ranking number and the sample size. For the weakest specimen $(i = 1)$ of a sample the estimator for the probability of failing is: $F_1 = F(\sigma_1) = 1/2X$. That specimen contains on average $N_{c,S}(\sigma_1) = \ln 2[X/(2X-1)]$ destructive flaws. For the strongest specimen of the sample $(i = X)$ it holds that: $F_X = F(\sigma_X) = (2X-1)/2X$ and $N_{c,S}(\sigma_X) = \ln 2X$.

A special situation occurs if the strength is equal to the characteristic strength (i.e. for $V = V_0$ and $\sigma = \sigma_0$). Then the probability of failure is $F(\sigma_0) = 1 - 1/e$ and $N_{c,S}(\sigma_0) = 1$ and the density of critical flaws is: $n_c(\sigma_0, V_0) = N_{c,S}(\sigma_0)/V_0 = 1/V_0$. If the calculations made for $\sigma = \sigma_0$ and $V = V_0$ are generalized for any stress value σ_i and for specimens of any volume V the equation reads:

$$g(a_{c,i}) = \frac{r-1}{V \cdot a_{c,i}} \cdot ln \frac{2X}{2X - 2i + 1}$$

The use of this equation opens a simple possibility to determine the frequency distribution of flaw sizes in a wide range of parameters by testing specimens of different volume.
In the following these ideas are applied to describe bending test results of a commercial silicon nitride ceramic.

FRACTURE AND FLAW STATISTICS OF A TYPICAL COMMERCIAL SILICON NITRIDE MATERIAL

A typical commercial gas pressure sintered silicon nitride ceramic is used as model material. Its hardness (HV5) is 15.5 ± 0.3 MPa, the fracture toughness (SEVNB, [15, 16]) is 5.0 ± 0.2 MPa√m, the Young's Modulus is 297 ± 2 GPa and the Poisson ratio is 0.27. More details can be found in [19]. A sample of 30 bending specimens was machined out from large discs (diameter 250 mm, thickness 5 mm). The specimen preparation and the tests were done according to EN 843-1 and evaluated according to EN 843-5 [17, 18]. The characteristic strength of this sample is $\sigma_0 = 871$ MPa and the Weibull modulus is $m = 14.1$. The effective volume [1] of the bending specimens is $V_{eff} = V(m+2)/4(m+1)^2 = 8.5 \cdot 10^{-9}$ m^3. This value is used as scaling parameter $V_0 = V$. The Weibull distribution is shown in **Fig. 1.a**. The measured data are nicely distributed around the straight line, which describes the Weibull distribution.

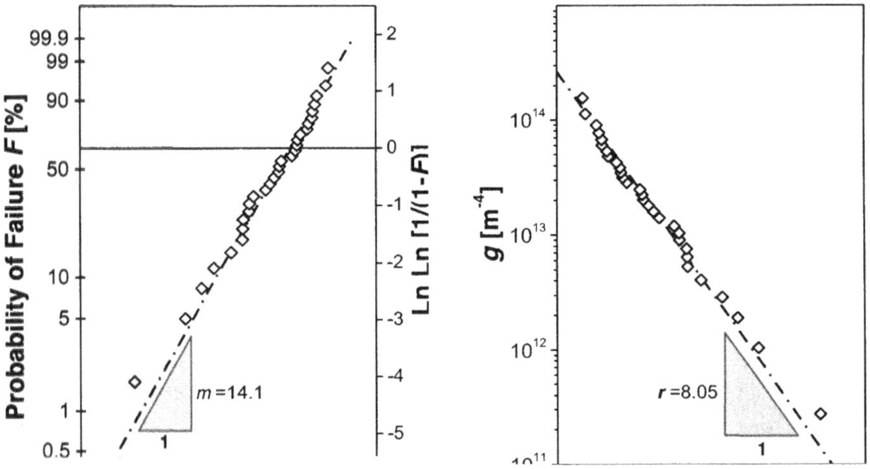

Fig. 1: a) Strength data of a silicon nitride ceramic tested in four point bending (4PB) in a Weibull plot and b) the relative frequency distribution of flaw sizes. The data points were determined by fracture experiments.

By fractographic inspection fracture [20, 21] origins within the volume (inclusions) were found in 3 specimens. In the other 27 specimens no clear evidence for the type of fracture origin was found. It is assumed that all flaws are small surface flaws, i.e. their geometric factor is $Y = 1.12$.

In the following, the strength data are used to determine the frequency distribution function of the defects. Using eq. 5 the size of the critical flaw for the characteristic strength is $a_{c,0}$ = $26 \cdot 10^{-6}$ m. This value is used in the following for the arbitrary scaling parameter: $a_0 = a_{c,0}$. For the exponent r we get using eq. 6: r = 8.05. With these data the factor in eq. 4 is: g_0 = $3.2 \cdot 10^{13}$ m^{-4}, and therefore all parameters in eq. 4 are determined. The relative frequency distribution of flaw sizes is shown in **Fig. 1.b**. The transformation of strength into flaw size data was done using eq. 8. Again the data are nicely distributed around the straight line. This is characteristic for a Weibull material. An extremely strong dependency of the relative frequency on the critical flaw size can clearly be recognized: although the size of the critical flaws in the samples only varies by a factor of around 2.2 the corresponding relative frequency of the flaw sizes varies by a factor of around 500. That means that flaws with a radius of 21 μm are about 500 times more frequent than flaws with a radius of 46 μm.

SIZE EFFECT ON STRENGTH; APPLICATION TO VERY SMALL SPECIMENS

As discussed in the introduction a size effect on strength exists [1 - 3], which – for a Weibull material - is described by:

$$V_1 \cdot \sigma_1{}^m = V_2 \cdot \sigma_2{}^m$$

The probability of failure in a sample of specimens of volume V_2 is equal to that in another sample containing specimens of volume V_1, if the stresses applied to the specimens are related according to eq. 9. Using the Griffith criterion we get [13]:

$$a_{c,1} / a_{c,2} = (V_1 / V_2)^{2/m}$$

For the specimens with volume V_1 the corresponding relative frequency of flaw sizes at $\sigma = \sigma_0$ is $g(a_1) = g_0 \cdot (a_1 / a_0)^{-r}$ and the density of destructive flaws is $n_c(\sigma_{0,1}, V_1) = 1/V_1$. The analogue holds for specimens of volume V_2.

In **Fig. 2** the diameter ($2a_c$) of the critical flaws (for the characteristic strength of the specimens) is plotted versus the (effective) volume in a double logarithmic scale (eq. 10) using the example of the silicon nitride material described above. The slope of the line is: $1/(m/2) = 1/(r-1) = 1/7.05$. The dashed line describes a typical scaling parameter for the volume of the specimen. For simplicity the edge length of a cube with volume V is taken as the characteristic length: $l = V^{1/3}$. For materials with a modulus $m > 6$, there exists a point of intersection between both lines, which is - in the selected example - at a volume of about $V \approx 4.2 \cdot 10^{-17}$ m^3 (this corresponds to the diameter of the critical flaw of about $2a_c \approx 3.4$ μm).

Obviously the assumption made in eq. 4 (the relative frequency of flaws follows an inverse power law) can only approximate the behaviour of materials for large flaws. It fails for very small flaws: the relative frequency goes to infinity if the flaw size goes to zero: $a \rightarrow 0$, $g(a) \rightarrow \infty$ [13]. At the intersection point in **Fig 2**, the density of dangerous flaws gets so high that the volume of the specimens is completely filled with flaws and, left of that point; the "volume of dangerous flaws" even exceeds volume of the specimens. For obvious reasons this is not possible in real materials.

Another inconsistency is caused by the fact, that the derivation of the fracture statistics (eq. 1 and eq. 2) assumes non-interacting flaws [8]. This will only be true in the case of a low flaw density. If fracture statistics are applied to very small specimens made of a Weibull material the density of dangerous flaws gets high and the interaction between flaws cannot be neglected any longer [22]. For that case it can be assumed that interacting flaws link up. This would cause an upper limit for

Table 1: Strength test on specimens of different size.

Data set	4PB	A (B3B-test)	B (B3B-test)	C (B3B-test)	D (B3B-test)
Specimen dimension / mm	45×3.95×2.98	Ø 43; t = 2.484	Ø 20; t= 1.983	Ø 10.8, t = 1.054	Ø 4.7; t = 0.445
Charact. strength / MPa	871	1053	1123	1226	1275
$2a_{c,0}$ / µm	52	36	31	26	24
Weibull modulus / -	14.1 [11 -17]	12.4 [9 – 15]	15.9 [12 – 19]	21.7 [16 – 26]	17.7 [13– 22]
effective volume / 10^{-12} m^3	8500	280	65	4.3	0.6
x_s / µm	2041	654	402	163	84
$x_s / 2a_{c,0}$ / -	39	18	13	6.3	3.5

4PB: four point bending test; B3B: ball on three balls test; Ø : diameter; t: thickness; $2a_{c,0}$: diameter of the critical flaw corresponding to the characteristic strength, l : reference length (defined to be the third root of the effective volume). Numbers in square brackets are limits of the 90 % confidence interval.

the strength, if the distance between the flaws gets too close, say 2 – 3 times their diameter.

Further strength tests (data sets A- D) were made in biaxial bending on specimens of different size. Specimens were cut from the same plates as used for the bending specimens. Tests and results are described in [19]. Key results are summarised in **Table 1**. The data show a significant size effect, i.e. the characteristic strength is much larger for small than for large specimens (**Fig. 3**). The straight line shows the size effect as predicted by eq. 9 based on the bending test data. Although the data sets A and B are in the 90 % confidence interval of the extrapolation, the sets C and D show a significantly lower strength than predicted.

The behaviour of small specimens is discussed in more detail in [13]. A possible reason for this drop of strength is the fact that machining of very small specimens (as is the case of set D) is very demanding and some machining damage cannot be excluded in this case. Additional damage would cause a reduction of strength as observed in **Fig. 3**. Further possible reasons for the (apparent) deviation of the strength of small specimens from the Weibull behaviour are experimental measurement uncertainties, which become large for small specimens and which are not included in the scatter bars shown in **Fig. 3**. The plotted scatter bars refer to the uncertainties due to the sampling procedure (the sample is different from the underlying population, [12]).

Another reason would be the interaction between flaws, as described above. The last line in **Table 1** shows the ratio of the size parameter (corresponding with the effective volume; it is the length of the edge of a cube with the effective volume) divided by the diameter of the critical Griffith flaw for the characteristic strength. This ratio is larger than 10 for the sets 4PB, A and B. Here an interaction seems not to be likely. But for set D the ratio is as small as about 3. Here some overlapping of local stress field and linking of micro defects may become possible. But at present it is not clear if this really happens or not.

Bazant formulated a statistical theory of fracture for quasibrittle materials [5, 23, 24]. He assumed that there exist several hierarchical orders which each can be described by parallel and serial linking of so-called representative volume elements (RVEs). For large specimens (and low probability of failures) the fracture statistics is equal to the Weibull statistics , i.e. if the specimens size is larger than 500 to 1000 times of the size of one RVE. In the actual case this is similar to the diameter of the critical flaw. For smaller specimens the volume effect disappears and the fracture statistics become Gaussian. In that sense Bazant's analysis qualitatively fits to the measured data in Fig. 3 and to the analysis made above.

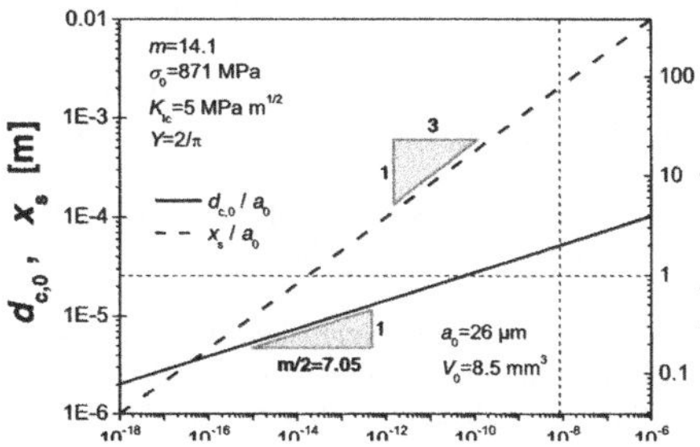

Fig. 2: Diameter of the critical flaw size ($d_c = 2a_c$) versus the volume of the specimen (full line) in a double logarithmic plot. The edge length x_s of a cube having the effective volume is also shown (bold, dashed line).

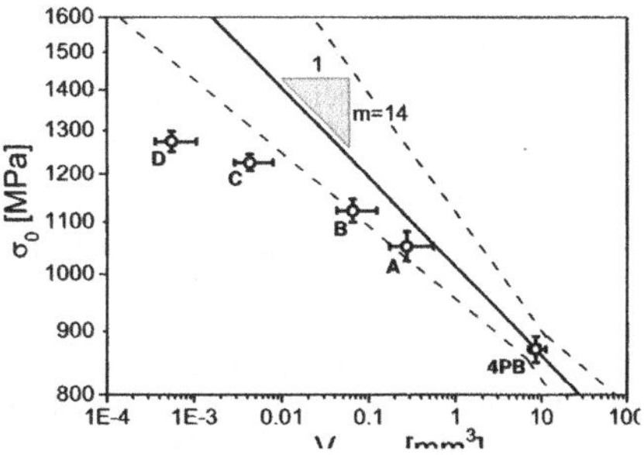

Fig. 3: Characteristic strength versus (effective) volume in a double logarithmic plot. Shown are test results on specimens of different size. The straight line shows the Weibull extrapolation based on the four point bending test results. The dashed lines are the 90 % confidence intervals of the prediction.

CONCLUSIONS

- In brittle ceramic materials there exists a strong correlation between flaw and strength distribution.
- For interacting flaws having a density steeply decreasing with flaw size, the strength is Weibull distributed.
- This implies very low flaw densities for large flaw sizes, and extremely high flaw densities for very small flaw sizes.
- A further and very important consequence of the Weibull distribution is the size effect, i.e. the mean strength decreases with increasing specimen size. This is the most important consequence of fracture statistics for designing with ceramics.
- However, this is not true for very small specimens. Here the flaw densities become so high that interaction between flaws becomes possible. Then Weibull statistics predicts too high a strength; i.e. there exists an upper limit of strength.
- There exist some experimental hints for such a limit but a clear experimental proof is missing.

ACKNOWLEDGEMENT

Financial support by the Austrian Federal Government (in particular from the Bundesministerium für Verkehr, Innovation und Technologie and the Bundesministerium für Wirtschaft und Arbeit) and the Styrian Provincial Government, represented by Österreichische Forschungsförderungsgesellschaft mbH and by Steirische Wirtschaftsförderungsgesellschaft mbH, within the research activities of the K2 Competence Centre on "Integrated Research in Materials, Processing and Product Engineering", operated by the Materials Center Leoben Forschung GmbH in the framework of the Austrian COMET Competence Centre Programme, is gratefully acknowledged.

LITERATURE

1 D. Munz, T. Fett, Ceramics: Mechanical Properties, Failure Behaviour, Materials Selection, Springer Verlag, Berlin/Heidelberg/New York, (1999).
2 J. B. Wachtmann, Mechanical Properties of Ceramics, John Wiley & Sons, New York, (1996).
3 R. Danzer, T. Lube, P. Supancic, and R. Damani, Fracture of Ceramics, Advanced Engineering Materials, **10**, 275-298, (2008).
4 R. Danzer, P. Supancic, J. Pascual, and T. Lube, Fracture Statistics of Ceramics - Weibull Statistics and Deviations from Weibull Statistics, Engineering Fracture Mechanics, **74**, 2919-2932, (2007).
5 Z. P. Bazant, and S. D. Pang, Activation energy based extreme value statistics and size effect in brittle and quasibrittle fracture, Journal of the Mechanics and Physics of Solids, 55, 91-131 (2007).
6 W. Weibull, A Statistical Theory of the Strength of Materials, Ingeniörsvetenskapsakademiens, Handlingar Nr 151, Generalstabens Litografiska Anstalts Förlag, Stockholm, 1 – 45, (1939).
7 W. Weibull, A Statistical Distribution Function of Wide Applicability, Journal of Applied Mechanics, **18**, 293 – 298, (1951).
8 A. M. Freudenthal, Statistical Approach to Brittle Fracture, in H. Liebowitz (ed.) Fracture, Vol II, Academic Press, New York/London, 591 – 619, (1968).
9 A. de S. Jayatilaka, and K. Trustrum, Statistical Approach to Brittle Fracture, J. Mat. Sci., **12**, 1426 – 1430, (1977).
10 R. Danzer, A General Strength Distribution Function for Brittle Materials, J. Eur. Ceram. Soc. **10**, 461 – 472, (1992).
11 R. Danzer, G. Reisner, and H. Schubert, Der Einfluß von Gradienten in der Defektdichte und Festigkeit auf die Bruchstatistik von spröden Werkstoffen, Zeitschrift für Metallkunde, **83**, 508 – 517, (1992).
12 R. Danzer, T. Lube, and P. Supancic, Monte-Carlo Simulations of Strength Distributions of Brittle Materials – Type of Distribution, Specimen- and Sample Size, Zeitschrift für Metallkunde, **92**, 773 - 783, (2001).
13 R. Danzer, Some Notes on the Correlation between Fracture and Defect Statistics: Are Weibull Statistics Valid for Very Small Specimens?, J. Eur. Ceram. Soc., **26**, 3043-3049, (2006).
14 C. Lu, R. Danzer, and F. D. Fischer, Fracture Statistics of Brittle Materials: Weibull or Normal Distribution, Physical Review E, **65**, 1 - 4, (2002).
15 R. Damani, R. Gstrein, and R. Danzer, Critical Notch Root Radius in SENB-S Fracture Toughness Testing, J. Eur. Ceram. Soc., **16**, 695-702, (1996).
16 ISO 23146, Fine ceramics (advanced, advanced technical ceramics) – Test methods for toughness of monolithic ceramics – Single-edge V-notched beam (SEVNB) method, (2008).
17 ISO 843-1, Advanced Technical Ceramics, Monolithic Ceramics, Mechanical Properties at Room Temperature, Part 1: Determination of flexural Strength (1995).
18 ISO 843-5, Advanced Technical Ceramics, Monolithic Ceramics, Mechanical Properties at Room Temperature, Part 5: Statistical Analysis, (1996).
19 W. Harrer, R. Danzer, P. Supancic, and T. Lube, Influence of sample size on the results of B3B tests, Key Engineering Materials, in print, (2009).
20 R. Morrell, Fractography of Brittle Materials, Measurement Good Practice Guide No. **14**, HMSO, National Physical Laboratory, UK, ISSN 1368-6550, (1999).
21 R. Danzer, Mechanical Failure of Advanced Ceramics: The Value of Fractography, Key Engineering Materials, **223**, 1 - 18, (2002).
22 C. Lu, R. Danzer, and F. D. Fischer, Scaling of Fracture Strength in ZnO: Effects of Pore/Grain-Size Interaction and Porosity, J. Eur. Ceram. Soc., **24**, 3643-3651, (2004).

23 Z. P. Bazant, Scaling of quasibrittle fracture: asymptotic analysis, Int. J. Fract., **83**, 19-40, (1997).
24 Z. P. Bazant, Probability distribution of energetic-statistical size effect in quasibrittle Probabilistic Engineering Mechanics, **19**, 307-319, (2004).

STRUCTURE AND PROPERTY OF Ti-Al-C/TiB_2 COMPOSITE CERAMICS

MIN Xin-min[1,2,a], XU Gang[1,b], MEI Bin-Chu[2,c]
[1]Chemistry Department, School of Sciences, Wuhan University of Technology, Wuhan 430070, China;
[2]State Key Laboratory of Advanced Technology for Materials Synthesis and Processing, Wuhan University of Technology, Wuhan 430070, China
mxm@whut.edu.cn, xugang851018@yahoo.com.cn, bcmei@whut.edu.cn

ABSTRACT

The relations between composition, electronic structure, chemical bond and property of composites of Ti_3AlC_2/TiB_2, Ti_2AlC/TiB_2 and related single phases of Ti_3AlC_2, Ti_2AlCas well as TiB_2 are studied by the first principle methods. There are strong ionic and covalent interactions among the interfaces of the composites. These interactions among the interfaces, including whole ionic and covalent bonds of Ti_3AlC_2/TiB_2 are stronger than those of Ti_2AlC/TiB_2. The results are consistent with the experiment, that the mechanical properties of Ti_3AlC_2/TiB_2 are better than that of Ti_2AlC/TiB_2.

Key words: Titanium aluminum carbide; TiB_2; composite ceramic; chemical bond; property

INTRODUCTION

New layered ternary $M_{n+1}AX_n$ compounds (M is transitional metal, A is III or IV group element and X is C or N) are attracting increased interest due to their unique properties. $M_{n+1}AX_n$ compounds combine unusual properties of both metals and ceramics[1-3]. Like metals, they are a good thermal and electrical conductor and are relatively soft. Like ceramics, they are elastically stiff and exhibit excellent high temperature mechanical properties. They are resistant to thermal shock and unusually damage tolerant, and exhibit excellent corrosion resistance. Above all, unlike conventional carbides, they can be machined by using conventional tools without lubricant, which is of great technological importance for their application. Additionally, $M_{n+1}AX_n$ compounds are an exceptional solid lubricant. Unfortunately, a limit to the potential application as a high temperature structural material is the relatively soft and low creep strength of these materials. Incorporation of second phase is an effective way to overcome these weaknesses[4]. There have been some calculations on the single phases of $M_{n+1}AX_n$[5,6], but it is rarely seen for the calculations on the multi-phases of $M_{n+1}AX_n$ composites. Owing to the high hardness, high modulus, excellent chemical stability and approximate thermal expansion coefficient, TiB_2 herein is chosen to produce Ti-Al-C/TiB_2 composites to improve these properties[4]. In this paper, the relations between composition, electronic structure, chemical bond and property of composites of Ti_3AlC_2/TiB_2, Ti_2AlC/TiB_2 and related single phases of Ti_3AlC_2, Ti_2AlCas well as TiB_2 are studied using density function theory and discrete variational method(DFT-DVM)[7], one of the first principle methods.

CALCULATED METHOD AND MODELS

The DFT-DVM method is put forward by Professor Ellis at Northwestern University, U.S.A.[7] to resolve the Kohn-Sham equation. First, a number of discrete sampling points in a three-dimensional grid are chosen. Variation is made to the parameter in the error function to obtain all the minima for the points. Using the multi-dimensional numeral integer method, a self-consistent process is carried out to obtain the energy function, wave function and others. The DFT-DVM method can be used to calculate larger structure models of molecules, clusters and solids, and thus has been widely used in chemistry, physics, material science an so on[7-8].

The space group of Ti_2AlC is D^4_{6h}-$P6_3/mmc$[9], and it has the periodic layered structure. There are atomic layers of $n(Ti):n(Al):n(C) = 2:1:1$ in the z axis direction. In order to consider the interaction between interfaces with different environments and the other atomic planes, different computed models have been designed. There are 11 layers of 71 atoms in model 1 of Ti_2AlC, the central atomic layer is Al, and toward the outside there are Ti, C, Ti, Al and Ti layers, respectively. The upside 6 layers of atoms in Fig. 1 are the underside parts of model 1, namely, the central Al layer and the downside 5 layers of Ti, C, Ti, Al and Ti, respectively (further downward 3 layers are parts of the model of composite system, and will be introduced in the follows). Moreover, the above 5 layers of model 1 can be obtained by the operation of D_{3h} group symmetry. The numbers in Fig. 1 are classes of atom according to the operation of D_{3h} group. Model 2 of Ti_2AlC includes 15 layers of 96 atoms, the central atomic layer is also Al, toward the outside there are Ti, C, Ti, Al, Ti, C and Ti layers in turn. There are differences between model 1 and 2 in the number of atomic layers and the inferior outer atom layer.

Ti_3AlC_2 has the same space group as Ti_2AlC. There are atomic layers of $n(Ti):n(Al):n(C) = 3:1:2$ in the z axis direction. Different computed models have also been designed. There are 13 layers of 81 atoms in model 3 of Ti_3AlC_2, the central atomic layer is Ti, and toward the outside there are C, Ti, Al, Ti, C, and Ti layers, respectively. Model 4 of Ti_3AlC_2 includes 15 layers of 97 atoms, the central atomic layer is Al, toward the outside there are Ti, C, Ti, C, Ti, Al and Ti layers in turn. There are differences between model 3 and 4 in the number of atomic layers and the inferior outer atom layer, too.

TiB_2 is with P6/mm space group, and also has the layer structure. It has Ti and B layers in turn in the z axis direction, such as the underside 4 layers of atoms in Fig. 1. The computed model 5 of TiB_2 includes 3 Ti and 4 B atom layers of 69 atoms.

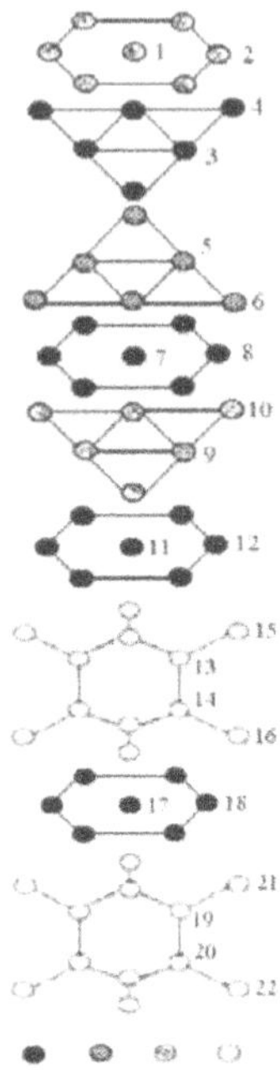

Fig.1 A part atoms of Ti_2AlC and Ti_2AlC/TiB_2 models

Both of Ti_2AlC and TiB_2 have the layer structure, moreover, their crystal plane (001) have the same structure of Ti_6 hexagon with a center Ti(such as Fig. 1), and the difference of Ti-Ti bond length is only about 1%. Therefore, Ti atoms can be made as the outside layer of model of Ti_2AlC phase, and also as the outside layer of model of TiB_2 phase. Namely, the strongest interaction between Ti_2AlC and TiB_2 phases should be in the direction of (001) planes of them. Based on models of Ti_2AlC and TiB_2, model 6 of Ti_2AlC/TiB_2 composite is formed from adding 4 B and 2 Ti atom layers to the upside and underside of model 2, such as Fig. 1. Model 7 of Ti_2AlC/TiB_2 composite is adding 4 B and2 Ti atom layers to the upside and underside of model 3, too. The differences between model 6 and 7 of Ti_2AlC/TiB_2 lie in the interfaces of Ti layers with different environments or with the different inferior interfaces. Similarly, models 8 and 9 of Ti_3AlC_2/TiB_2 composite are adding 4 B and 2 Ti atom layers to the upside and underside of model 4 and 5, respectively. The differences between model 8 and 9 of Ti_3AlC_2/TiB_2 are same as those between model 6 and 7 of Ti_2AlC/TiB_2.

RESULTS AND DISCUSSION

Chemical bonds have an important effect on the property of the materials. Covalent and ionic bonds, the main composition of the chemical bond, are discussed. The ionic bond is much like the Coulomb force. It is inversely proportional to the distance between atoms and directly proportional to the atomic net charge. The net charge is expressed as the difference between the atomic number and the electronic population. The net charge average values of the center atoms of every model are shown in Table 1. For example, the average value of Ti_2AlC is that of model 1 and 2. C and Al are with negative charges, and Ti is with a positive charge in all models, which is consistent with the order of electronegativity of C > Al > Ti. Both the number of copies of Ti atom in Ti_2AlC and Ti_3AlC_2 are 1/2, but the number of copies of C atom, with larger electronegativity, in Ti_2AlC is 1/4, and that in Ti_3AlC_2 is 1/3. Therefore, the net charge of Ti in Ti_2AlC is lower than that in Ti_3AlC_2. Correspondingly, the number of C atoms in Ti_2AlC is less than that in Ti_3AlC_2, C atom in Ti_2AlC obtains more electrons than that in Ti_3AlC_2, and the net charge (refers to absolute value, similarly hereinafter) of C in Ti_2AlC is higher than that in Ti_3AlC_2. For the same reason, the net charge of Al in Ti_2AlC is higher than that in Ti_3AlC_2, too. Generally speaking, the difference of ionic bond strength between Ti_2AlC and Ti_3AlC_2 is not too obvious.

Table 1 Net charges of atom in the models

	Ti($M_{n+1}AX_n$)	Ti(TiB_2)	Ti(interface)	Al	C	B
Ti_2AlC	1.2301			-0.9407	-1.3809	
Ti_3AlC_2	1.3285			-0.9182	-1.2797	
TiB_2		1.6628				-0.7181
Ti_2AlC/TiB_2	1.3841	1.7690	1.4011	-0.7474	-1.5784	-0.9341
Ti_3AlC_2/TiB_2	1.5547	1.7737	1.6104	-0.7092	-1.4453	-1.0712

The covalent bond order average values of the center atoms of every model are shown in Table 2. The strongest bond in Ti_2AlC or Ti_3AlC_2 is Ti-C, the next strongest is Ti-Al and Al-Al, and the weakest is Ti-Ti. There is not much difference of covalent bond strength of Al-Al or Ti-Ti, but the covalent bond of Ti-C or Ti-Al in Ti_3AlC_2 is obviously stronger than that in Ti_2AlC, which is consistent with the result of experiment that the mechanics property of Ti_3AlC_2 is better than that of Ti_2AlC, and Ti_3AlC_2 is easier synthesized than Ti_2AlC[4].

Table 2 Covalent bond orders in the models

	Ti-C	Ti-Al	Al-Al	Ti-Ti($M_{n+1}AX_n$)	Ti-Ti (TiB_2)	Ti-B (TiB_2)	Ti(interface)-C	Ti(interface)-Al
Ti_2AlC	0.2143	0.1427	0.0838	0.0527				
Ti_3AlC_2	0.2227	0.1566	0.0846	0.0506				
TiB_2					0.0330	0.0610		
Ti_2AlC/TiB_2	0.2090	0.1615	0.0852	0.0608	0.0372	0.0628	0.2023	0.1385
Ti_3AlC_2/TiB_2	0.2203	0.1745	0.0856	0.0610	0.0405	0.0645	0.2289	0.1403

In the composite of Ti_2AlC/TiB_2, Ti atoms on the interface connect to C, Al and B of two phases of Ti_2AlC and TiB_2. Therefore, the number of copies of Ti atom in Ti_2AlC/TiB_2 is less than that in Ti_2AlC or TiB_2, and the net charge of Ti in Ti_2AlC/TiB_2 is higher than that in Ti_2AlC or TiB_2 (Table 1). The net charge of Al in Ti_2AlC/TiB_2 is lower than that in single phase, but the net charge of C or B in Ti_2AlC/TiB_2 is higher than that in single phase. Generally speaking, the ionic bond strength of Ti_2AlC/TiB_2 is larger than that of single phase. The net charge of Ti on interface of Ti_2AlC/TiB_2 is lower than that in TiB_2, but is higher than that in Ti_2AlC, so there are some strong ionic interactions on the interface. The net charge of Ti in Ti_3AlC_2/TiB_2 increases more obviously than that in Ti_2AlC/TiB_2, and the ionic bond of Ti_3AlC_2/TiB_2 is stronger than that in Ti_2AlC/TiB_2. The ionic bond on the interface of Ti_3AlC_2/TiB_2 is stronger than that in Ti_2AlC/TiB_2, too.

The covalent bond strength (Table 2) of Ti-C of Ti_2AlC/TiB_2 is less than that in single phase, but that of Ti-Al, Al-Al or Ti-Ti of Ti_2AlC/TiB_2 is larger than that in single phase, and the change extent of Ti-Al is larger than that of Ti-C. Therefore, the covalent bond strength of Ti_2AlC/TiB_2 is larger than that of single phase. The covalent bond strength of Ti-C on interface of Ti_2AlC/TiB_2 is close to that in Ti_2AlC. The covalent bond strength of Ti-Al on interface of Ti_2AlC/TiB_2 is less than that in Ti_2AlC, but is still larger than that of Al-Al or Ti-Ti in single phase or composite, so there are some strong covalent interactions on the interface. The difference between Ti_3AlC_2/TiB_2 and single phase is same as between Ti_2AlC/TiB_2 and single phase. The covalent bond of Ti_3AlC_2/TiB_2 is stronger than that in Ti_2AlC/TiB_2, which is much same as the difference between Ti_3AlC_2 and Ti_2AlC. The covalent bond on interface of Ti_3AlC_2/TiB_2 is also stronger than that in Ti_2AlC/TiB_2, which is consistent with the result of experiment that the mechanics property of Ti_3AlC_2/TiB_2 is better than that of Ti_2AlC/TiB_2[4].

SUMMARY

Composites of Ti_3AlC_2/TiB_2, Ti_2AlC/TiB_2 and related single phases are calculated using the first principle methods. The ionic and covalent bonds of Ti_3AlC_2 are stronger than those of Ti_2AlC, which is consistent with the result of experiment that the mechanics property of Ti_3AlC_2 is better than that of Ti_2AlC, and Ti_3AlC_2 is easier synthesized than Ti_2AlC. The ionic and covalent bonds of Ti_3AlC_2/TiB_2 and Ti_2AlC/TiB_2 composites are stronger than those of the single phases. There are strong ionic and covalent interactions among the interfaces of the composites. The interactions among the interfaces, the whole ionic and covalent bond of Ti_3AlC_2/TiB_2 are stronger than those of Ti_2AlC/TiB_2. The results are consistent with the result of experiment that the mechanics property of Ti_3AlC_2/TiB_2 is better than that of Ti_2AlC/TiB_2.

ACKNOWLEDGMENTS

Thanks for the Subsidization by the Natural Science Foundation of China (No. 50572080), Ministry of Education of China(PCSIRT0644) and Open Fund of the State Key Lab of Theoretical & Computational Chemistry

REFERENCES

[1] M.W. Barsoum: Prog. Solid State Chem. Vol. 28 (2000), p. 201
[2] M.W. Barsoum, H.D. Linh and E.R. Tamer: Allo. Comp. Vol. 350 (2003), p. 303
[3] L. Chaput, G. Hug, P. Pecheur and H. Scherrerl: Phys. Rev. B Vol. 75 (2007) p. 0351071
[4] W.B. Zhou: *Study on fraction and Performance of Ti-Al-C/Ti-B Composites* (Doctor Degree Thesis, Wuhan University of Technology. Wuhan, China 2006).
[5] E. Lofland, J.D. Hettinger and K. Harrell: Applied Phys. Lett. Vol. 84 (2004) p. 508
[6] T. Liao, J.Y. Wang and Y.C. Zhou: Phys. Rev. B Vol. 73 (2006) p. 214109
[7] D.E. Ellis and D. Guenzburger: Adv Quantum Chem. Vol. 34 (1999) p. 51
[8] X.S. Xiao, C.Y. Wang and T.L. Chen. *The Method of Density Function and Discrete Variation Used in Chemistry and Material Physics*. (Science Press, Beijing 1998).
[9] L. Farber, I. Levin and M.W. Barsoum: J. Applied Phys. Vol. 86 (1999) p. 2540

THE EFFECT OF DOPED SINTERING AIDS FOR $Nd(Mg_{0.5}Ti_{0.5})O_3$ MICROWAVE DIELECTRIC CERAMICS PROPERTIES

Kok-Wan Tay[1*], Teng-Yi Huang[2]

[1]Department of Electrical Engineering, Wu-Feng Institute of Technology, Chiayi, Taiwan, R.O.C.
[2]Graduate School of OptoMechatronics and Materials, Wu-Feng Institute of Technology, Chiayi, Taiwan, R.O.C.

ABSTRACT

The research mainly investigated the microwave dielectric properties of $Nd(Mg_{0.5}Ti_{0.5})O_3$ by adding different sintering aids, such as B_2O_3、Bi_2O_3、and V_2O_5 for lowering the sintering temperature. The sintered $Nd(Mg_{0.5}Ti_{0.5})O_3$ ceramics are characterized by X-ray diffraction spectra and scanning electron microscopy (SEM). The sintering temperature of $Nd(Mg_{0.5}Ti_{0.5})O_3$ ceramics with 10 mol% B_2O_3 additions can be effectively reduced from 1500°C to 1325°C, and the dielectric constant (ε_r) value of 26.2, a quality factor (Qxf) value of 61307 (at 9.63GHz), and τ_f value of -45.5ppm/°C. The ε_r 26.8, Qxf 27506 (at 9.87GHz), and τ_f value of -52.6 ppm/°C, respectively, were obtained for 10 mol% Bi_2O_3 –doped $Nd(Mg_{0.5}Ti_{0.5})O_3$ ceramics sintered at 1325°C. The ε_r 24.8, Qxf 15481 (at 10.02 GHz), and τ_f value of -57.8 ppm/°C, respectively, were obtained for 10 mol% V_2O_5 –doped $Nd(Mg_{0.5}Ti_{0.5})O_3$ ceramics sintered at 1375°C. Small values (~ 3.5 ppm/°C) of τ_f are obtained for $Nd(Mg_{0.5}Ti_{0.5})O_3$ ceramics with 10 mol% B_2O_3 additions. Therefore, $Nd(Mg_{0.5}Ti_{0.5})O_3$ with 10 mol% B_2O_3 additions may be suggested for application in microwave communication devices.

Keywords: Sintering、Microwave dielectric ceramic、Dielectric resonator.

1. INTRODUCTION

Due to rapid development in the microwave communication system, satellite broadcasting system, as well as wireless mobile systems, has become more important for the miniaturization of microwave device, such as oscillators, band pass filters, duplexers and global positioning systems (GPS) patch antennas [1, 2]. To miniaturize the devices and for the systems to work with high efficiency and stability, the materials for microwave resonators must be excellent in the following three dielectric characteristics. The first characteristic is a high dielectric constant (ε_r>20). The use of high dielectric constant materials can effectively reduce the size of resonators since the wavelength (λ) in dielectrics is inversely proportional to $\sqrt{\varepsilon_r}$ of the wavelength (λ_0) in vacuum ($\lambda = \lambda_0 / \sqrt{\varepsilon_r}$). The second is a high quality factor (Qxf) value (Q>5000). This is required to achieve high frequency selectivity and stability in microwave transmitters and receiver components. The third is a near zero temperature coefficient of resonant frequency (τ_f) for dielectric resonators and microwave device substrates [3, 4]. Small temperature coefficients of the resonant frequency ensure the stability of the microwave components at different working temperatures. Using two or more compounds with negative and positive temperature coefficients to form a solid solution mixed phases is the most promising method of obtaining a zero temperature coefficient of the resonant frequency. Because most dielectric ceramics with high dielectric constant have positive τ_f value [5], searching for materials with a high dielectric constant, a high Q and a negative τ_f is necessary to achieve this goal.

Low temperature solid-state synthesis is an approach that shows great promise for the synthesis of materials with unusual interesting properties. Usually, three methods are commonly used for

reducing the sintering temperature of dielectric ceramics: low melting-temperature glass addition, chemical processing, and powder with smaller particle sizes [6-9]. The first method using liquid phase glass sintering was found to effectively lower the firing temperature. However, it also decreased the microwave dielectric properties of dielectric resonators, especially quality factor. The chemical process often required a flexible procedure, which was expensive and time consuming. Therefore, the selection of non-glass addition with low melting point is extremely important.
Since B_2O_3, Bi_2O_3, and V_2O_5 is commonly used as a liquid-phase flux and has been shown to accomplish a substantial sintering temperature reduction [10,11], it was selected as a sintering aid in present study. The objective of this study is to develop a new dielectric material which has high dielectric constant, high quality factor and near to zero τ_f by incorporating different amount of B_2O_3, Bi_2O_3, and V_2O_5 added into $Nd(Mg_{0.5}Ti_{0.5})O_3$ ceramics. The resultant microwave dielectric properties were analyzed based upon the densification, the X-ray diffraction patterns and the microstructures of the ceramics.

2. EXPERIMENTAL PROCEDURE

Specimen powders were prepared by a conventional solid-state method. High-purity oxide powders (>99.9%): Nd_2O_3, MgO and TiO_2 were used as raw materials. The powders were weighed according to the composition $Nd(Mg_{0.5}Ti_{0.5})O_3$, and were ground in distilled water for 12h in a balling mill with agate balls. Prepared powders were dried and calcined at 1100°C for 2h in air. The calcined powers were mixed as desired composition $Nd(Mg_{0.5}Ti_{0.5})O_3$ with different sintering aids of 10 mol % B_2O_3, 10 mol % Bi_2O_3, and 10 mol % V_2O_5 additions as sintering aids and re-milled for 12h. The fine powder together with the organic binder was pressed into pellets with dimensions of 11 mm in diameter and 5 mm in thickness was made by pressing at a pressure of $25 kg/cm^3$. These pellets were sintered at temperatures of 1300°C ~ 1375°C for 6 h in air. The heating and cooling rates were both set at 5°C /min.
The microstructure observation of the sintered ceramics surface was performed by means of scanning electron microscopy (SEM, JEOL JSM 6400, Japan). The crystalline phase of sintered ceramics was identified by X-ray diffraction (XRD, RIGAKU D/max 2.B) with CuKa radiation (λ=1.5418Å at 40 kV and 30 mA) and scanned from 20° to 70° with scanning speed of 4°/min. The bulk densities of the sintered pellets were measured by the Archimedes method. The dielectric constant (ε_r) and the quality factor values (Qxf) at microwave frequencies were measured using the Hakki-Coleman dielectric resonator method which had been modified and improved by Courtney [12, 13]. The dielectric resonator was positioned between two brass plates. Microwave dielectric properties of sintered samples were measured by an Anritsu 37347C Network Analyzer. For temperature coefficient of resonant frequency (τ_f), the technique is the same as that of quality factor measurement. The test cavity was placed over a thermostat in the temperature range from 30°C to 80°C. The τ_f value (ppm/°C) can be calculated by noting the change in resonant frequency (f), and is defined by:

$$\tau_f = \frac{f_2 - f_1}{f_1(T_2 - T_1)} \quad (1)$$

Where, f_1 and f_2 represent the resonant frequencies at T_1 and T_2, respectively.

3. RESULTS AND DISCUSSIONS

Figure 1 presents the XRD patterns of $Nd(Mg_{0.5}Ti_{0.5})O_3$ ceramics with (a)10 mol % B_2O_3, (b)10 mol % Bi_2O_3, and (c)10 mol % V_2O_5 additive sintered at different temperatures 1300°C, 1325°C,

1350°C and 1375°C, respectively, for 6hr. Identical XRD patterns of Fig,1(a) and (b) were observed have not change significantly with sintering temperatures in the range of 1300°C ~1375°C. Besides, secondary phases with 10 mol % Bi_2O_3 and B_2O_3 addition are not observed. Fig.1(c) has produce of second phase, but as sintering temperature increased, the second phase became weak.

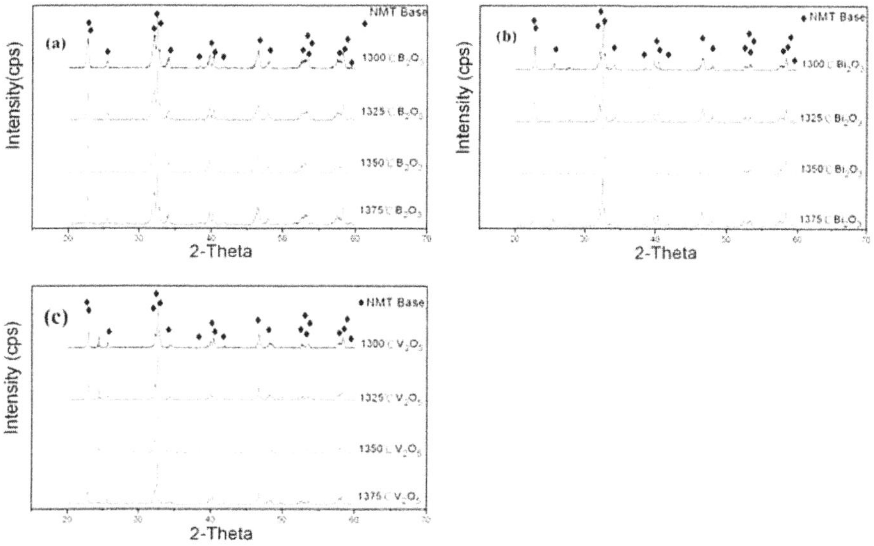

Figure 1. XRD patterns of $Nd(Mg_{0.5}Ti_{0.5})O_3$ with (a) 10 mol % B_2O_3, (b) 10 mol % Bi_2O_3 , and (c)10 mol % V_2O_5 additive sintered for 6 hr.

The SEM micrographs of 10mol% B_2O_3-doped $Nd(Mg_{0.5}Ti_{0.5})O_3$ ceramics with sintering temperatures of 1300°C ~1375°C are shown in figure 2. The porosity decreased with increasing sintering temperature and no pore was observed at temperatures of 1375°C with sintering for 6 hr owing to grain growth uniformly. However, degradation in grain uniformity and abnormal grain growth started to appear for ceramics specimens at sintering temperatures of 1350°C, Moreover, the increase in the grain size was observed at sintering temperatures of 1375°C, which could damage its microwave dielectric properties.

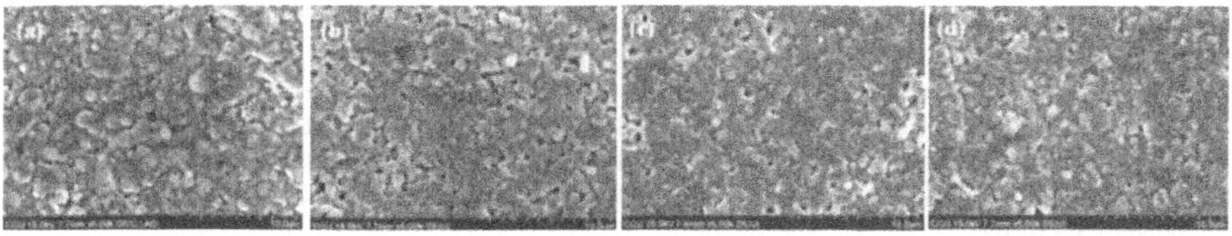

Figure 2. SEM micrographs of 10 mol % B_2O_3–doped $Nd(Mg_{0.5}Ti_{0.5})O_3$ ceramics at different sintering temperatures (a)1300°C, (b)1325°C, (c)1350°C, and (d)1375°C

The SEM micrographs of 10mol% Bi_2O_3 doped-$Nd(Mg_{0.5}Ti_{0.5})O_3$ ceramics with sintering temperatures of 1300°C ~1375°C are shown in figure 3. The porosity decreased with increasing sintering temperature and no pore was observed at temperatures of 1325°C, but non-uniformity grain were observed, and abnormal grain growth started to appear between the sintering temperatures of 1325°C and 1350°C, These may directly affect the microwave dielectric properties.

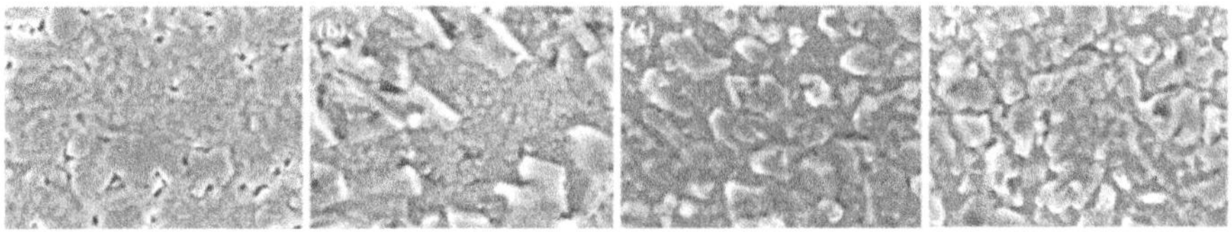

Figure 3. SEM micrographs of 10 mol % Bi_2O_3–doped $Nd(Mg_{0.5}Ti_{0.5})O_3$ ceramics at different sintering temperatures (a)1300°C, (b)1325°C, (c)1350°C, and (d)1375°C

The SEM micrographs of 10mol% V_2O_5-doped $Nd(Mg_{0.5}Ti_{0.5})O_3$ ceramics with sintering temperatures of 1300°C ~1375°C are shown in figure 4. Porous specimens and second phase were also observed at sintering temperatures of 1300°C and 1320°C, as sintering temperature increased to 1350°C and 1375°C, the porous and second phase vanish, and the uniformity grains growth were observed.

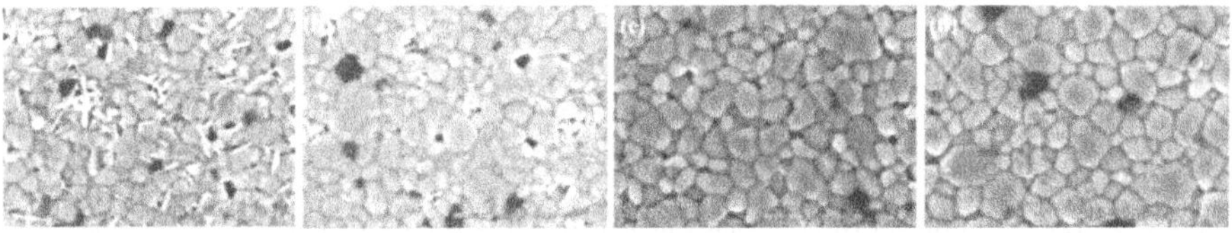

Figure 4. SEM micrographs of 10 mol % V_2O_5–doped $Nd(Mg_{0.5}Ti_{0.5})O_3$ ceramics at different sintering temperatures (a)1300°C, (b)1325°C, (c)1350°C, and (d)1375°C

The density of 10 mol % B_2O_3, Bi_2O_3, and V_2O_5–doped $Nd(Mg_{0.5}Ti_{0.5})O_3$ ceramics at different temperature (1300°C ~1375°C) is shown in figure 5. The density increased with increasing sintering temperature due to dense sample as observed in SEM. The density increased was owing to the decrease in the porosity of the specimen. Moreover, increase the sintering temperature would enhance the grain growth resulting in an increase of the density. At 1375°C, the ceramics with 10 mol % B_2O_3 addition reached the optimal bulk density.

The microwave dielectric characteristics and microstructures of $Nd(Mg_{0.5}Ti_{0.5})O_3$ ceramics were determined by the sintering conditions and the sintering aid. The dielectric constant of $Nd(Mg_{0.5}Ti_{0.5})O_3$ ceramics with doped different B_2O_3, Bi_2O_3, and V_2O_5 additions and sintering temperature (1300°C ~1375°C) were illustrated in figure 6. With sintering temperature was increased to 1325°C, the dielectric constant of B_2O_3 increase slightly than Bi_2O_3, but a higher sintering temperature to 1350°C will cause the crystalline grain size became unequal and degraded its dielectric constant.

The increase in the dielectric constant was attributed to a higher density as well as a lower porosity. With the additive of V_2O_5, the dielectric constant increased linear as the sintering temperature increased, as the crystalline grain is relatively dense. The dielectric constant decreased is because its have second phase and frit when temperature is low.

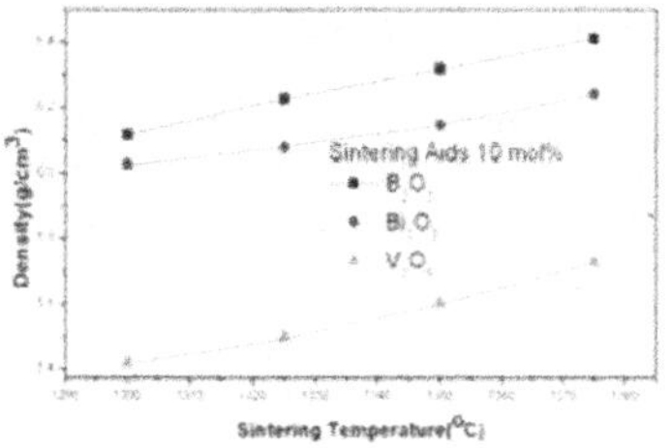

Figure 5. Dependence of sintering temperature of $Nd(Mg_{0.5}Ti_{0.5})O_3$ ceramics on density with B_2O_3, Bi_2O_3, and V_2O_5 additions.

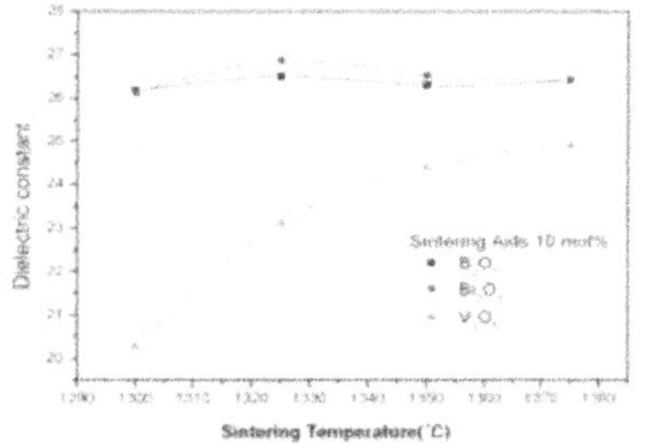

Figure 6. Dependence of sintering temperature of $Nd(Mg_{0.5}Ti_{0.5})O_3$ ceramics on dielectric constant with B_2O_3, Bi_2O_3, and V_2O_5 additions.

Figure 7 shows the quality factor (Qxf) of $Nd(Mg_{0.5}Ti_{0.5})O_3$ ceramics variously sintering temperatures (1300°C ~1375°C) and 10 mol % doped of different B_2O_3, Bi_2O_3, and V_2O_5 additions. For B_2O_3 addition, when sintering temperature increasing to 1325°C, the Qxf increased to a maximum value of 61307 (at 9.63 GHz) and thereafter decreased. Adding 10 mol % Bi_2O_3 is also same like B_2O_3, that its moves towards the trend and have a lower Qxf value compare to B_2O_3.

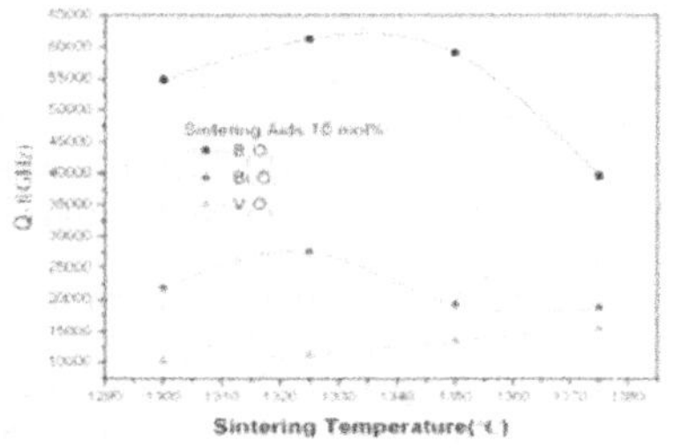

Figure. 7. Dependence of sintering temperature of $Nd(Mg_{0.5}Ti_{0.5})O_3$ ceramics on quality factor (Qxf) with B_2O_3, Bi_2O_3, and V_2O_5 additions.

Add 10 mol % V_2O_5 shown that the Qxf value increased slightly with rising the sintering temperature. As temperature increased will caused the crystalline grain to be dense with each other. The microwave dielectric loss is mainly caused not only by the lattice vibration modes, but also by the pores, second phases, impurities or even the lattice defect.

Figure 8 shows the temperature coefficient of the resonant frequency (τ_f) of $Nd(Mg_{0.5}Ti_{0.5})O_3$ ceramics with different sintering temperature, and B_2O_3, Bi_2O_3, and V_2O_5 additions. All of the τ_f values drift toward positive τf at 1325°C sintering temperature for 10 mol % of B_2O_3, Bi_2O_3, and V_2O_5 additions, the τf is -45.5ppm/°C,-52.6ppm/°C, and -57.8ppm/°C, respectively, thereafter increase to negative τf.

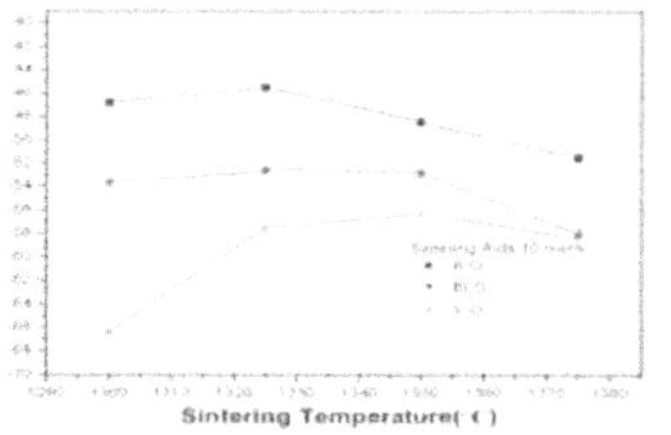

Figure 8. Dependence of sintering temperature of $Nd(Mg_{0.5}Ti_{0.5})O_3$ ceramics on τ_f with B_2O_3, Bi_2O_3, and V_2O_5 additions.

4. CONCLUSIONS

The microwave dielectric properties of $Nd(Mg_{0.5}Ti_{0.5})O_3$ by adding different sintering aids, such as B_2O_3、 Bi_2O_3、 and V_2O_5 for lowering the sintering temperature were investigated. Comparing all this three sintering aids, adding 10 mol % B_2O_3 at 1325°C, can obtain the best characteristic of $Nd(Mg_{0.5}Ti_{0.5})O_3$ ceramics. A large sintering temperature reduction (175/°C) can be achieved, the higher density of 6.1 g/cm^3, dielectric constant value of 26.2, a quality factor (Qxf) value of 61307 (at 9.63GHz), and τ_f value of -45.5ppm/°C. Therefore, $Nd(Mg_{0.5}Ti_{0.5})O_3$ with 10 mol% B_2O_3 additions may be suggested for application in microwave communication devices, which requiring low sintering temperature.

REFERENCES

1. S. Nishigaki, H. Kato, S. Yano, R. Kamimure, Microwave dielectric properties of (Ba, Sr)O-Sm_2O_3-TiO_2 ceramics, Am. Ceram. Soc. Bull. 66 (1987) 1405-1410.
2. K. Wakino, K. Minai, H. Tamura, Microwave characteristics of $(Zr,Sn)TiO_4$ and BaO-PbO-Nd_2O_3-TiO_2 dielectric resonators, J. Am. Ceram. Soc. 67(1984) 278-281.
3. T. Takada, S. F. Wang, S. Yoshikawa, S. J. Yang, R. E. Newnham, Effect of glass additions on BaO-TiO_2-WO_3 microwave ceramics, J. Am. Ceram. Soc. 77 (1994) 1909-1916.
4. G. Kajfezz, P. Guillon, Dielectric Resonators, Artech House, Massachusetts, 1986.
5. C. L. Huang, Y. B. Chen, Microwave properties of B_2O_3-doped $Nd(Mg_{1/2}Ti_{1/2})O_3$–$CaTiO_3$ dielectric resonators at microwave frequency, Mater. Lett. 60 (2) (2006) 198-202.
6. S. I. Hirno, Takashi, Hayashi, A. Hattori, Chemical processing and microwave characteristics of $(Zr,Sn)TiO_4$ microwave dielectrics, J. Am. Ceram. Soc. 74 (1991) 1320-1324.
7. T. Kakada, S.F. Wang, Syoshikawa, S. T. Jang, R. E. Newnham, Effects of glass additions on $(Zr,Sn)TiO_4$ for microwave applications, J. Am. Ceram. Soc. 77 (1994) 2485-2488.
8. T. Kakada, S. F. Wang, Syoshikawa, S. T. Jang, R. E. Newnham, Effect of glass additions on BaO–TiO_2–WO_3 microwave ceramics, J. Am. Ceram. Soc. 77 (1994) 1909-1916.

[9.] V. Tolmer, G. Desqardin, Low-temperature sintering and influence of the process on the dielectric properties of $Ba(Zn_{1/3}Ta_{2/3})O_3$, J. Am. Ceram. Soc. 80 (1997) 1981-1991.
[10]. K. W. Kang, H. T. Kim, M. Lanagan, T. Shrout, Low-temperature sintering and microwave dielectric properties of $CaTi_{1-x}(Fe_{0.5}Nb_{0.5})_xO_3$ ceramics with B_2O_3 addition, Mater. Res. Bull. 41 (2006) 1385-1391.
[11]. R. Umemura, H. Ogawa, A. Yokoi, H.Ohsato, A. Kim, Low-temperature sintering-microwave dielectric property relations in $Ba_3(VO_4)_2$ ceramic, J. Alloys Compd. 424 (2006) 388-393.
[12]. W.E. Courtney, Analysis and evaluation of a method of measuring the complex permittivity and permeability of microwave insulators, IEEE Trans. Microwave Theory Tech. 18 (1970) 476-485.
[13]. B.W. Hakki and P.D. Coleman, A Dielectric Resonator Method of Measuring Inductive Capacities in the Millimeter range, IEEE Trans. Microwave Theory Tech. 8 (1960) 402-410.
*E-mail: kwtay@mail.wfc.edu.tw

MICROWAVE DIELECTRIC PROPERTIES OF (1-x)$(Mg_{0.6}Zn_{0.4})_{0.95}Co_{0.05}TiO_3$-x$SrTiO_3$ CERAMIC SYSTEM

Jun-Jie Wang, Chun-Huy Wang, Ting-Kuei Hsu, and Yi-Hua Liu

Department of Electronic Engineering
Nan Jeon Institute of Technology
178, Chau-Chin Road, Yen-Shui,
Tainan Hsien, Taiwan, 73746

ABSTRACT

The microstructures and the microwave dielectric properties of the (1-x)$(Mg_{0.6}Zn_{0.4})_{0.95}Co_{0.05}TiO_3$-x$SrTiO_3$ ceramic system were investigated. In order to achieve a temperature-stable material, we studied a method of combining a positive temperature coefficient material with a negative one. $SrTiO_3$ has dielectric properties of dielectric constant ε_r ~ 205, $Q \times f$ value ~ 4,200 GHz and a large positive τ_f value ~ 1700 ppm/°C. $(Mg_{0.6}Zn_{0.4})_{0.95}Co_{0.05}TiO_3$ possesses high dielectric constant (ε_r ~ 19.6), high quality factor ($Q \times f$ value ~ 162,000 GHz) and negative τ_f value (-65 ppm/°C). By appropriately adjusting the x value in the (1-x)$(Mg_{0.6}Zn_{0.4})_{0.95}Co_{0.05}TiO_3$-x$SrTiO_3$ ceramic system, a near-zero τ_f value can be obtained. A new microwave dielectric material of 0.96$(Mg_{0.6}Zn_{0.4})_{0.95}Co_{0.05}TiO_3-0.04SrTiO_3$ possesses the excellent dielectric properties of a dielectric constant of 23.5, a $Q \times f$ value of 92,000 GHz and a τ_f value of -2.5 ppm/°C and can be utilized in microwave devices.

INTRODUCTION

The development of microwave dielectric resonators for applications in communication systems, such as cellular phones, wireless local area networks (WLAN), direct broadcasting satellite (DBS) and global positioning systems, has been rapidly progressing in the past decades [1-2]. The unique electrical properties of ceramic dielectric resonators have revolutionized the microwave-based wireless communications industry by reducing the size and cost of filter and oscillator components in circuit systems. The advantage of using dielectric resonators is that it makes the size reduction of microwave components possible. Requirements for these dielectric resonators must be the combined dielectric properties of a high dielectric constant, a low dielectric loss ($Q >$ 5000, where $Q = 1/\tan\delta$) and a near-zero temperature coefficient of resonant frequency (τ_f) [3]. In general, high dielectric constant materials exhibit high dielectric loss (low $Q \times f$ value), while low loss ceramics are usually accompanied by low ε_r value.

$MgTiO_3$-based ceramics have wide applications as dielectrics in resonators, filters and antennas for communication, radar and global positioning systems operating at microwave frequencies. $MgTiO_3$-$CaTiO_3$ ceramics is well known as the material for temperature compensating type capacitor, dielectric resonator and patch antenna. The material is made of a mixture of modified α-Al_2O_3 structured magnesium titanate ($MgTiO_3$: ε_r ~ 17, $Q \times f$ value ~ 160,000 GHz measured at 7 GHz and a τ_f value ~ -50 ppm/°C) [4] and perovskite structured calcium titanate ($CaTiO_3$: ε_r ~ 170, $Q \times f$ value ~ 3,600 GHz (at 7 GHz) and τ_f value ~ 800 ppm/°C) [5]. With the ratio Mg:Ca = 95:5, 0.95$MgTiO_3$-0.05$CaTiO_3$ ceramics gives ε_r ~ 21, $Q \times f$ value ~ 56,000 GHz and a zero τ_f value. However, it required sintering temperatures as high as 1400-1450 °C. Many researchers made effort to study the microstructures and the microwave dielectric properties of 0.95$MgTiO_3$-0.05$CaTiO_3$ ceramics by adding various additives or varying the processing. The dielectric properties of 0.95$MgTiO_3$-0.05$CaTiO_3$ ceramics can be further improved by introducing additions such as Cr, La and B [6-9], although some of the τ_f values were not reported. Huang et al. reported that a new series of microwave dielectric materials with positive temperature coefficient of resonant frequency were added to (Mg, Co, Zn)TiO_3 ceramic system. The experiment results showed that these ceramic systems have excellent microwave dielectric properties [10-13].

In this paper, $(Mg_{0.6}Zn_{0.4})_{0.95}Co_{0.05}TiO_3$ ceramics was investigated to possess dielectric properties with a dielectric constant ~ 19.6, a $Q \times f$ value ~ 162,000 GHz and a τ_f value ~ -65 ppm/°C [14]. In order to achieve the near-zero τ_f value, $SrTiO_3$ was added to $(Mg_{0.6}Zn_{0.4})_{0.95}Co_{0.05}TiO_3$ as a ceramic system of (1-x)$(Mg_{0.6}Zn_{0.4})_{0.95}Co_{0.05}TiO_3$-x$SrTiO_3$. The resultant microwave dielectric properties were analyzed based upon the densification, the X-ray diffraction (XRD) patterns and the microstructures of the ceramics. The correlation between the microstructure and the $Q \times f$ value was also investigated.

EXPERIMENTAL

The starting materials were high-purity oxide powders (>99.9%): $SrCO_3$, ZnO, TiO_2, CoO and MgO. The powders were separately prepared according to the desired stoichiometry $SrTiO_3$ and $(Mg_{0.6}Zn_{0.4})_{0.95}Co_{0.05}TiO_3$, and ground in distilled water for 12h in a ball mill with agate balls. The prepared powders were dried and calcined at 1100 °C for 4h in air. After calcinations, the calcined powders were mixed according to the molar fraction (1-x)$(Mg_{0.6}Zn_{0.4})_{0.95}Co_{0.05}TiO_3$-x$SrTiO_3$ and then re-milled for 12h. The fine powder with 3 wt% of a 10% solution of PVA as a binder (Polyvinyl alcohol 500, Showa, made in Japan) was pressed into pellets with dimensions of 11 mm in diameter and 5 mm in thickness under the pressure of 200 MPa. These pellets were sintered at temperatures of 1200 ~ 1300°C for 4 h in air.

The crystalline phases of the sintered ceramics were identified by XRD using Cu Kα (λ = 0.15406 nm) radiation with a Seimens D5000 diffractometer operated at 40kV and 40mA. The apparent densities of the sintered pellets were measured by the Archimedes method. The dielectric constant (ε_r) and the quality factor values (Q) at microwave frequencies were measured using the Hakki-Coleman dielectric resonator method [15, 16]. A system combining a HP8757D network analyzer and a HP8350B sweep oscillator was employed in the measurement. For temperature coefficient of resonant frequency (τ_f), the technique is the same as that of quality factor measurement. The test cavity is placed over a thermostat and the temperature range used is from 30 to 80°C.

RESULTS AND DISCUSSION

Fig. 1 shows the XRD patterns of 0.96$(Mg_{0.6}Zn_{0.4})_{0.95}Co_{0.05}TiO_3-0.04SrTiO_3$ (hereafter referred to as 96MZCST) ceramics sintered at different temperatures for 4 h. The XRD patterns showed that peaks indicating the presence of $(Mg_{0.6}Zn_{0.4})_{0.95}Co_{0.05}TiO_3$ as the main crystalline phase, in association with $SrTiO_3$ and $(Mg_{0.6}Zn_{0.4})_{0.95}Co_{0.05}Ti_2O_5$ as minor phases. It is understood that crystal structures of $(Mg_{0.6}Zn_{0.4})_{0.95}Co_{0.05}TiO_3$ and $SrTiO_3$ are rhombohedral (ICDD-PDF #01-073-7752) and cubic (ICDD-PDF #00-040-1500), respectively. $(Mg_{0.6}Zn_{0.4})_{0.95}Co_{0.05}Ti_2O_5$, usually formed as an intermediate phase, was identified and difficult to completely eliminate from the sample prepared by mixed oxide route. The formation of $(Mg_{0.6}Zn_{0.4})_{0.95}Co_{0.05}Ti_2O_5$ might lower the $Q \times f$ value of the specimen. The X-ray diffraction patterns of the 96MZCST ceramic system have not change significantly with sintering temperatures in the range 1200 ~ 1300°C

The SEM micrographs of 96MZCST ceramics sintered at different sintering temperatures for 4 h are illustrated in Fig. 2. As the sintering temperature increased, the grain size increased. However, rapid grain growth was observed at temperatures higher than 1250°C, which might degrade the microwave dielectric properties of the ceramics.

Fig. 3 shows the apparent densities of 96MZCST ceramics sintered at different sintering temperatures for 4 h. With increasing sintering temperature, the apparent density was found to increase to a maximum value of 4.24 g/cm^3 at 1250°C and thereafter decreased. Moreover, the degradation of apparent density at temperatures above 1275°C was owing to rapid grain growth.

Fig. 4 shows the dielectric constants of 96MZCST ceramics at different sintering temperatures for 4 h. The relationships between ε_r values and sintering temperatures revealed the same trend with those between densities and sintering temperatures since higher density means lower porosity. The dielectric constant increased with increasing sintering temperature. After reaching maximum at 1250°C, it decreased. A maximum ε_r value of 23.5 was obtained for 96MZCST ceramics sintered at 1250°C for 4 h.

The quality factor values ($Q \times f$) of 96MZCST ceramics at different sintering temperatures for 4 h are demonstrated in Fig. 5. With increasing sintering temperature, the $Q \times f$ value was found to increase to a maximum value and thereafter decreased. A maximum $Q \times f$ value of 92,000 GHz was obtained for 96MZCST ceramics sintered at 1250°C for 4 h. The degradation of $Q \times f$ value was attributed to rapid grain growth resulted in a reduction of density as observed in Figs. 2 and 3. The microwave dielectric loss is mainly caused not only by the lattice vibrational modes, but also by the pores, the second phases, the impurities, or the lattice defect [17]. Apparent density also plays an important role in controlling the dielectric loss, and has been shown for other microwave dielectric materials. Since the $Q \times f$ value of 96MZCST ceramics was consistent with the variation of density, it suggested the dielectric loss of 96MZCST ceramics was mainly controlled by apparent density.

Fig. 6 illustrates the temperature coefficients of resonant frequency (τ_f) of 96MZCST ceramics sintered at 1250°C for 4 h with different x values. The temperature coefficient of resonant frequency is well known to be governed by the composition, the additives, and the second phase of the materials. Increasing $SrTiO_3$ content seemed to make the τ_f value more positive. Since the τ_f values of $(Mg_{0.6}Zn_{0.4})_{0.95}Co_{0.05}TiO_3$ and $SrTiO_3$ are -65 and 1700 ppm/°C [18], respectively, it also implies that zero τ_f can be achieved by increasing the amount of $SrTiO_3$ content. A near-zero τ_f value can be obtained for 96MZCST ceramics sintered at 1250°C for 4 h.

Table 1 demonstrates the microwave dielectric properties of $(1-x)(Mg_{0.6}Zn_{0.4})_{0.95}Co_{0.05}TiO_3$-$xSrTiO_3$ ceramic system sintered at 1250℃ for 4 h. As the x value increased from 0.04 to 0.16, the τ_f values of $(1-x)(Mg_{0.6}Zn_{0.4})_{0.95}Co_{0.05}TiO_3$-$xSrTiO_3$ ceramics varied from -2.5 to 116 ppm/°C. Since the τ_f curves went through zero, it indicates that zero τ_f value can be obtained by appropriately adjusting the x value of $(1-x)(Mg_{0.6}Zn_{0.4})_{0.95}Co_{0.05}TiO_3$-$xSrTiO_3$ ceramics. However, increasing the $SrTiO_3$ content, the $Q \times f$ value would decrease. This is because $SrTiO_3$ ceramics possesses lower $Q \times f$ value of 4,200 GHz.

CONCLUSION

$(1-x)(Mg_{0.6}Zn_{0.4})_{0.95}Co_{0.05}TiO_3$-$xSrTiO_3$ ceramics showed mixed phases of $(Mg_{0.6}Zn_{0.4})_{0.95}Co_{0.05}TiO_3$ as the main phase in association with some minor phases $SrTiO_3$ and $(Mg_{0.6}Zn_{0.4})_{0.95}Co_{0.05}Ti_2O_5$. The existence of $(Mg_{0.6}Zn_{0.4})_{0.95}Co_{0.05}Ti_2O_5$ phase would cause a decrease in the $Q \times f$ value. The microwave dielectric properties are strongly related to the density and the matrix of the specimen. With x = 0.04, a near-zero τ_f value can be obtained for $(1-x)(Mg_{0.6}Zn_{0.4})_{0.95}Co_{0.05}TiO_3$-$xSrTiO_3$ ceramics. A dielectric constant ε_r of 23.5, a $Q \times f$ value of 92,000 GHz and a τ_f value of -2.5 ppm/°C were obtained for 96MZCST ceramics sintered at 1250℃ for 4 h.

REFERENCES

1S. Nishigaki, H. Kato, S. Yano, and R. Kamimure, Microwave Dielectric Properties of $(Ba,Sr)O$-Sm_2O_3-TiO_2 Ceramics, *Am. Ceram. Soc. Bull.*, **66**, 1405-10 (1987).
2K. Wakino, K. Minai, and H. Tamura, Microwave Characteristics of $(Zr,Sn)TiO_4$ and BaO-PbO-Nd_2O_3-TiO_2 Dielectric Resonators, *J. Am. Ceram. Soc.*, **67**, 278-81 (1984).
3T. Kakada, S. F. Wang, S. Yoshikawa, S. J. Jang, and R. E. Newnham, Effect of Glass Additions on BaO-TiO_2-WO_3 Microwave Ceramics, *J. Am. Ceram. Soc.*, **77**, 1909-16 (1994).
4K. Wakino, Recent Development of Dielectric Resonator Materials and Filters in Japan, *Ferroelectrics*, **91**, 69-86 (1989).
5R. C. Kell, A. C. Greenham, and G. C. E. Olds, High-Permittivity Temperature-Stable Ceramic Dielectrics with Low Microwave Loss, *J. Am. Ceram. Soc.*, **56**, 352-4 (1973).
6V. M. Ferreira, F. Azough, J. L. Baptista, and R. Freer, Magnesium Titanate Microwave Dielectric Ceramics, Ferroelectrics, 133, 127-32 (1992).
7V. M. Ferreira, F. Azough, R. Freer, and J. L. Baptista, The Effect of Cr and La on MgTiO3 and MgTiO3-CaTiO3 Microwave Dielectric Ceramics, J. Mater. Res., 12, 3293-9 (1997).
8V. M. Ferreira, J. L. Baptista, S. Kamba, and J. Petzelt, Dielectric Spectroscopy of MgTiO3-based Ceramics in the 109-1014Hz Region, J. Mater. Sci., 28, 5894-900 (1993).
9C. L. Huang and M. H. Weng, Improved High Q Value of MgTiO3-CaTiO3 Microwave Dielectric Ceramics at Low Sintering Temperature, Mater. Res. Bull., 36, 2741-50 (2001).
10C. L. Huang, J. J. Wang, and Y. P. Chang, Dielectric Properties of Low Loss (1-x)(Mg0.95Zn0.05)TiO3
-xSrTiO3 Ceramic System at Microwave Frequency, J. Am. Ceram. Soc., 90, 858-62 (2007).
11C. L. Huang, C. L. Pan, and J. F. Hsu, Microwave Dielectric Properties and Mixture Behavior of (Mg0.95Co0.05)TiO3-Ca0.6La0.8/3TiO3 Ceramic System, J. Alloys Compd., 461, 521-8 (2008).
12Y. B. Chen and C. L. Huang, New Dielectric Material System of x(Mg0.95Zn0.05)TiO3-(1-x)Ca0.8Sm0.4/3TiO3 at Microwave Frequency, Mater. Lett., 62, 2454-7 (2008).
13J. J. Wang, C. L. Huang, and P. H. Li, Microwave Dielectric Properties of (1-x)(Mg0.95Zn0.05)TiO3-xCa0.6La0.8/3TiO3 Ceramic System, Jpn. J. Appl. Phys., 45, 6352-6 (2006).
14H. J. Cha, D. H. Kang, and Y. S. Cho, Optimized microwave dielectric properties of Co- and Ca-substituted Mg0.6Zn0.4TiO3, Mater. Res. Bull., 42, 265-273 (2007).
15B. W. Hakki and P. D. Coleman, A Dielectric Resonator Method of Measuring Inductive Capacities in the Millimeter Range, IEEE Trans. Microwave Theory Tech., 8, 402-10 (1960).
16W. E. Courtney, Analysis and Evaluation of a Method of Measuring the Complex Permittivity and Permeability of Microwave Insulators, IEEE Trans. Microwave Theory Tech., 18, 476-85 (1970).

17B. D. Silverman, Microwave Absorption in Cubic Strontium Titanate, Phys. Rev., 125, 1921-30 (1962).
18P. H. Sun, T. Nakamura, Y. J. Shan, Y. Inaguma, M. Itoh, and T. Kitamura, Dielectric Behavior of (1-x)LaAlO3-xSrTiO3 Solid Solution System at Microwave Frequencies, Jpn. J. Appl. Phys., 37, 5625-9 (1998).

Table 1 Microwave dielectric properties of $(1-x)(Mg_{0.6}Zn_{0.4})_{0.95}Co_{0.05}TiO_3$-$xSrTiO_3$ ceramic system sintered at 1250℃ for 4 h

x value	Apparent density (g/cm^3)	ε_r	$Q \times f$ (GHz)	τ_f (ppm/°C)
0.04	4.24	23.5	92000	-2.5
0.08	4.25	26.9	66000	32.3
0.12	4.26	30.1	51000	78.6
0.16	4.28	37.7	46000	116.3

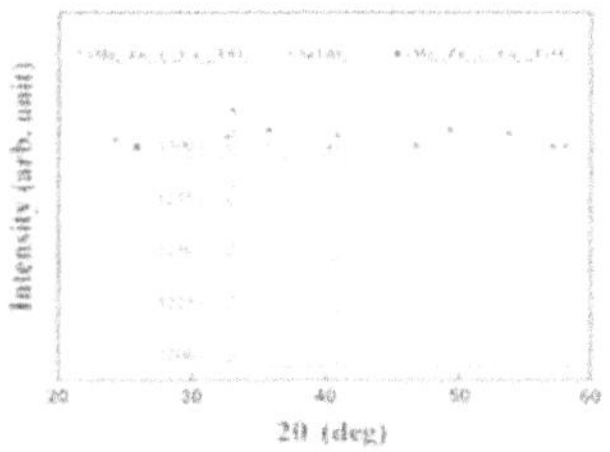

Figure 1. X-ray diffraction patterns of 96MZCST ceramics sintered at different sintering temperatures for 4 h

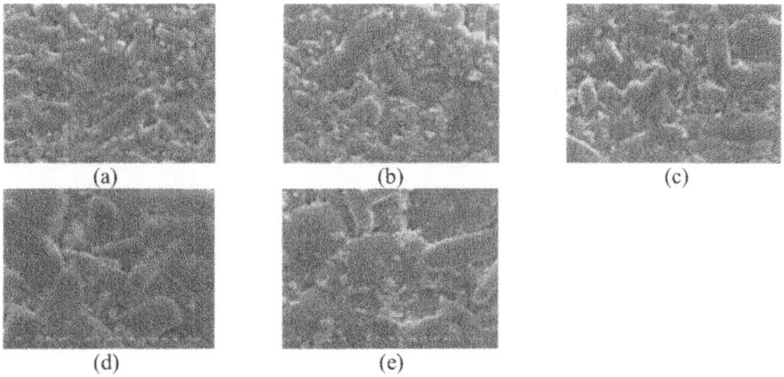

Figure 2. SEM photographs of 96MZCST ceramics sintered at (a) 1200℃ (b) 1225℃ (c) 1250℃ (d) 1275℃ (e) 1300℃ for 4 h

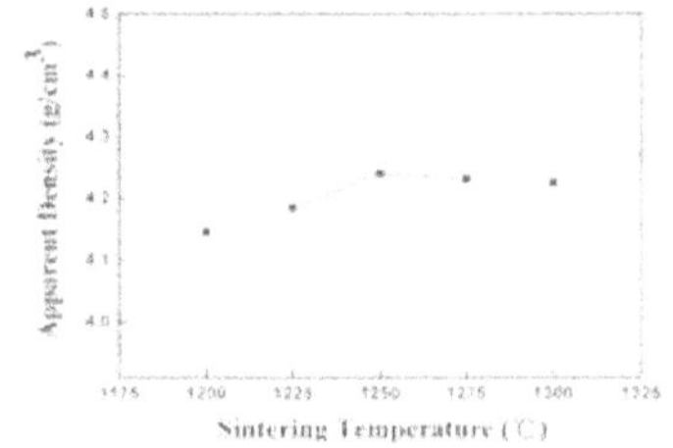

Figure 3. Apparent density of 96MZCST ceramics as a function of its sintering temperature

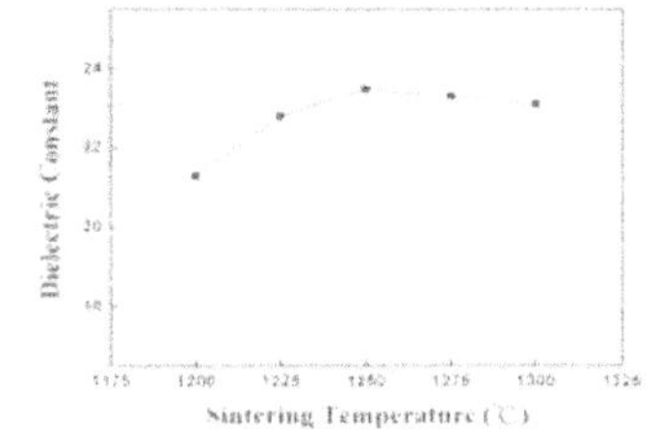

Figure 4. Dielectric constant of 96MZCST ceramics as a function of its sintering temperature

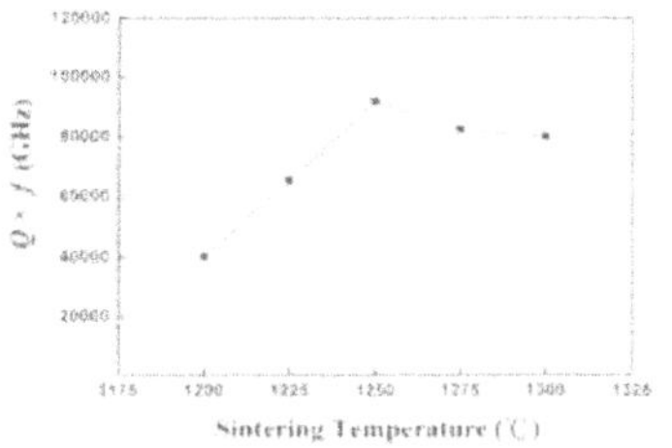

Figure 5. $Q \times f$ value of 96MZCST ceramics as a function of its sintering temperature

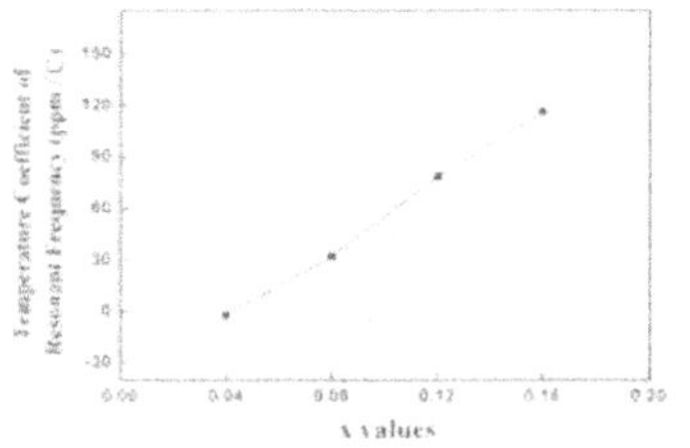

Figure 6. τ_f value of $(1-x)(Mg_{0.6}Zn_{0.4})_{0.95}Co_{0.05}TiO_3$-$xSrTiO_3$ ceramic system sintered at 1250℃ for 4 h with different x values

OXYNITRIDE GLASSES: EFFECTS OF COMPOSITION ON GLASS FORMATION AND PROPERTIES WITH IMPLICATIONS FOR HIGH TEMPERATURE BEHAVIOUR OF SILICON NITRIDE CERAMICS

Stuart Hampshire and Michael J. Pomeroy
Materials and Surface Science Institute
University of Limerick, Limerick, Ireland

ABSTRACT

Oxynitride glasses are silicates or alumino-silicates, containing Mg, Ca, Y or rare earth (RE) ions as modifiers, in which nitrogen atoms substitute for oxygen atoms in the glass network. These glasses are found as intergranular films and at triple point junctions in silicon nitride ceramics and these grain boundary phases affect their fracture behaviour and creep resistance. This paper provides an overview of the preparation of M-SiAlON glasses and outlines the effects of composition on properties. As nitrogen substitutes for oxygen in oxynitride glasses, increases are observed in glass transition and softening temperatures, viscosities, elastic moduli and microhardness. These property changes are related to structure of the glasses and have also been compared with known effects of grain boundary glass chemistry in silicon nitride. The implication for behaviour of silicon nitride ceramics, particularly at high temperatures, is discussed. The paper also outlines new research on oxynitride glasses containing other anions, such as fluorine. The effects of glass chemistry on glass formation and on physical and mechanical properties is presented.

INTRODUCTION

Silicon nitride has become the preferred ceramic for high temperature structural applications because its microstructure gives rise to high toughness and strength[1,2]. Sintering of silicon nitride makes use of additives, for example Y_2O_3 or a rare earth oxide and alumina, which react with a surface layer of silica on the nitride powder particles and some of the nitride itself to form a RE-SiAlON liquid, which on cooling remains either as an amorphous intergranular film (IGF) or as a triple point glass between grains[1,2]. The amount and chemistry of these glass phases determine the properties of these materials, by influencing (i) the glass/grain interfacial debonding conditions[1-3], so affecting fracture behavior or (ii) the glass softening and flow conditions, so affecting creep resistance at high temperature[2,4,5].

In order to understand the nature of these intergranular phases, studies on oxynitride glass formation, structure and properties[4,6-13] have been undertaken which show that oxynitride glasses are effectively silicate or alumino-silicate glasses in which nitrogen atoms partially substitute for oxygen atoms in the SiO_4 tetrahedra within the glass network. Nitrogen incorporation results in higher glass transition temperature (Tg), elastic modulus, viscosity and hardness as a result of the extra cross-linking by 3-coordinated nitrogen within the glass structure. A number of studies on rare earth lanthanide (RE)-Si-Al-O-N glasses have shown that by keeping Si:Al:O:N constant and simply changing the RE cation, density, hardness, Tg, elastic modulus and viscosity all increase with increasing cation field strength (CFS) or decreasing cation radius[13-15]

This paper outlines the effect of composition on oxynitride glass formation and properties that crucially affect the high temperature mechanical behaviour of silicon nitride based ceramics.

EXPERIMENTAL PROCEDURE

The extent of the glass forming regions in various M-Si-Al-O-N systems (M = Mg, Y, Ca, etc.) has been studied previously[6,7,10,11] and represented using the Jänecke prism with compositions expressed in equivalent percent (e/o) of cations and anions instead of atoms or gram-atoms. One equivalent of any element always reacts with one equivalent of any other element or species. For a system containing three types of cations, A, B and C with valencies of v_A, v_B, and v_C, respectively, then: Equivalent conc. of A = (v_A [A])/(v_A [A] + v_B[B] + v_C[C]),

where [A], [B] and [C] are, respectively, the atomic concentrations of A, B and C, in this case, Si^{IV}, Al^{III} and the metal cation, M, with its normal valency.

If the system also contains two types of anions, C and D with valencies v_C and v_D, respectively, then: Equivalent conc. of C = (v_C [C])/(v_C [C] + v_D[D]),

where [C] and [D] are, respectively, the atomic concentrations of C and D, i.e. O^{II} and N^{III}.

Fig. 1 shows glass forming regions in the Y-Si-Al-O-N system which was studied by exploring glass formation as a function of Y:Si:Al ratio on vertical planes in the Jänecke prism representing O:N ratios of 0, 10 and 22 e/o N. The region is seen to expand initially as nitrogen is introduced and then diminishes above approximately 10 e/o N, with glass formation occurring at more Y-rich compositions at higher N contents.

Preparation of glasses involves mixing appropriate quantities of silica, alumina, the modifying oxide and silicon nitride powders by wet ball milling in isopropanol for 24 hours, using sialon milling media, followed by evaporation of the alcohol before pressing into pellets. Batches of 50-60g are melted in boron nitride lined graphite crucibles at ~1700°C for 1h under 0.1MPa nitrogen pressure in a vertical tube furnace, after which the melt is poured into a preheated graphite mould (~900°C). The glass is annealed at a temperature close to the glass transition temperature (Tg) for one hour to remove stresses and slowly cooled.

Bulk densities were measured by the Archimedes principle using distilled water as the working fluid. X-ray analysis was used to confirm that the glasses were totally amorphous. Scanning electron microscopy allowed confirmation of this and assessment of homogeneity.

Differential thermal analysis (DTA) was carried out in order to detect Tg, which is observed as the onset point of the endothermic drift on the DTA curve, corresponding to the beginning of the transition range. Viscosity results were obtained in air between 750 and 1000°C from (1) a compressive creep test on cylinders of 10 mm diameter, (2) three point bending tests on bars of

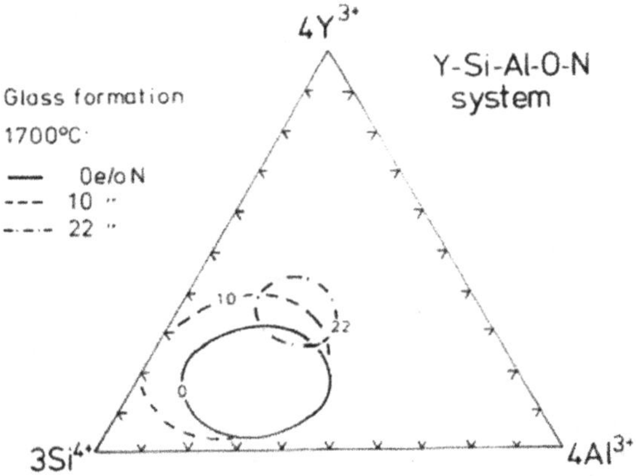

Fig. 1 Y-Si-Al-O-N glass forming regions at 0, 10 and 22 e/o N on cooling from 1700°C[10]

dimensions: 25mm x 4mm (width) x 3mm (height) with a span of 21 mm. Viscosity, η is derived from the relationships between (i) the stress/strain relations in an elastic solid and (ii) those that relate to a viscous fluid:

$$\eta = \sigma / [2(1+\upsilon)\dot{\varepsilon}] \qquad (1)$$

where σ and $\dot{\varepsilon}$ are the applied stress and the creep rate on the outer tensile fibre and υ is Poisson's ratio (taken as 0.5). Results from both types of test show good agreement[4,10-12]. Elastic moduli were determined[16] using an ultrasonic pulse-echo-overlap technique.

RESULTS AND DISCUSSION

EFFECTS OF COMPOSITION ON VISCOSITY

Fig. 2 shows the effect of nitrogen on viscosity – reciprocal temperature relationships for a series of glasses[11] with composition (in e/o) of 28Y:56Si:16Al:(100-x)O:xN (x = 0, 10, 17). It can be seen that, at any temperature close to Tg, viscosity increases by 2 to 3 orders of magnitude simply by replacing 17 e/o oxygen by nitrogen. The increases observed are due to increased cross-linking within the glass structure as 2-coordinated bridging oxygen atoms are replaced by 3-coordinated nitrogen atoms[10]. N may also be 2-coordinated and still act as a bridging ion, as in:

$$\equiv Si - N^{-} - Si \equiv$$

It is also possible that non-bridging nitrogen atoms may also be present, as in:

$$\equiv Si - N^{2-}$$

The glass network contains $(SiO_4)^{4-}$, $(SiO_3N)^{5-}$ and possibly also $(SiO_2N_2)^{6-}$ tetrahedral structural units. It should be noted that the $(SiO_3N)^{5-}$ tetrahedron requires the presence of a cation locally to balance the extra negative charge and this is equivalent to that for an $(AlO_4)^{5-}$ tetrahedron within the network. Therefore, oxynitride glasses containing $(SiO_3N)^{5-}$ tetrahedra can accommodate more cation modifiers in "network dwelling" sites than the equivalent oxide glasses. Raman spectra of oxynitride glasses[17] reveal that, as nitrogen content increases, the proportion of Q^3 species decreases and there is a corresponding increase in the proportion of Q^4 species (Q^n : n = no. of bridging anions joining SiO_4 tetrahedra), confirming that nitrogen increases the crosslinking between individual tetrahedra via the transformation of Q^3 oxide species into Q^4 oxynitride species. Similar trends have been reported for other Y-Si-Al-O-N glasses with different cation ratios[12,18,19]. The effect of fixed Si:Al and Y:Al ratios on properties of glasses with constant O:N ratio[11,12,18] show that as Si:Al ratio increases, Tg and viscosity increase while elastic moduli, hardness and thermal expansion coefficient decrease. With increasing Al:Y ratio, elastic moduli and thermal expansion coefficient decrease while the Tg and viscosity decrease to a minimum (at 16 e/o Al) and then increase with further increase in Al content.

Overall, these effects can be assumed to be related to changes in the density of the glass network and the numbers of non-bridging oxygens as Al changes its coordination. At higher Al:Y ratios, when 4 co-ordinated Al is prevalent, enhancement of the cross-linking of the glass network occurs, caused by the formation of more Al-O-Si linkages as Raman spectroscopy analyses20 would indicate. With constant Y content, an increase in Al:Si ratio allows replacement of Si-O-Si by Al-O-Si bridges and so the Y will act more as a network dwelling ion providing local charge balance. At constant Si, with increasing Al:Y ratio, non-bridging oxygens are replaced by Al-O-Si linkages.

Becher et al.21 have shown using in-situ high-resolution electron microscopy that, in silicon nitride ceramics, debonding at the interface between the grains and the continuous nanometer-thick intergranular film (IGF) or within the IGF is a critical part of the toughening mechanism for these materials and fracture toughness depends on the Y:Al ratio of the IGF, which varies with the yttria:alumina ratio in the fixed total amount of sintering additives. There is also evidence that in silicon nitride grain boundaries, weakening of the amorphous network of the IGF occurs as yttrium levels increase and this is responsible for the observed debonding by both crack propagation along the interface and within the IGF when the sintering additive contains the highest yttria:alumina ratio[21].

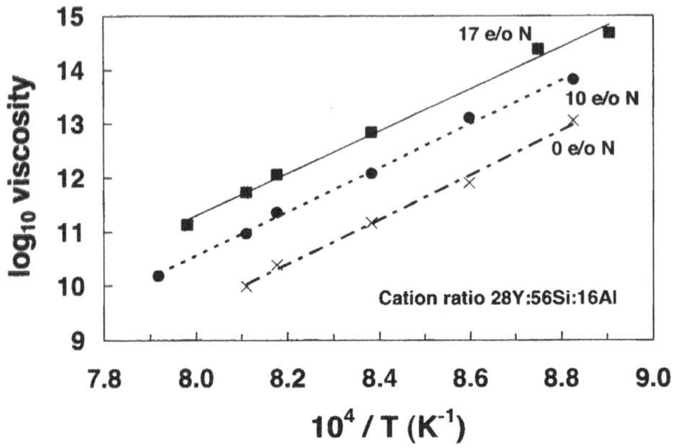

Fig. 2 Viscosity-reciprocal temperature relationships for a series of glasses with composition (in e/o) of 28Y:56Si:16Al:(100-x)O:xN (x=0, 10, 17) (data from Hampshire et al.[11]).

Fig. 3 demonstrates the effects of different rare earth lanthanide cations (RE = Eu, Ce, Sm, and Ho) on viscosity – reciprocal temperature relationships for RE-Si-Al-O-N glasses with fixed cation ratio of 28RE:56Si:16Al[4]. At any temperature (close to Tg), viscosity decreases by ~3 orders of magnitude in the order: Ho>Sm>Ce>Eu. Eu is in the +2 state which is a much larger ion than Eu^{3+}. Viscosity, as with other properties, increases almost linearly with increase in cation field strength of the RE ion[4].

Viscosities of RE-Si-Al-O-N liquids, containing Sm, Ce, Eu, where the ionic radii are larger than that of Y, are less than those of the equivalent Y-Si-Al-O-N liquids and this will have implications for easier densification of silicon nitride ceramics. Viscosity of Y-Si-Al-O-N glasses of the same cation composition and nitrogen content are close to those for Ho glasses. However, there will also be consequences for high temperature properties, particularly creep resistance. Liquids and glasses containing RE cations with ionic radii smaller than Y (Lu, Er, Dy, Yb) have been shown to have higher viscosities than the Y-SiAlON glasses and, in silicon nitride, these RE cations will form grain boundary glasses with higher softening temperatures.

SUMMARY AND IMPLICATIONS FOR SILICON NITRIDE CERAMICS

Modification of grain boundary glass chemistry in silicon nitride has profound effects on properties. Summarising:

1. when oxygen is substituted by nitrogen, there is an increase in viscosity of two to three orders of magnitude.
2. when the Y:Al ratio of the glass is increased, there is a further increase in viscosity of one order of magnitude.

3. by changing the rare earth cation from larger ions such as La or Ce to smaller cations such as Er or Lu, viscosity can be increased by a further two orders of magnitude.

The implications for silicon nitride ceramics are that intergranular glasses containing more N and less Al and smaller RE cations will provide enhanced creep resistance. Overall, a change of five to six orders of magnitude in viscosity can be achieved by careful modification of glass compositions[4,12] as shown schematically in Fig.4. The activation energies for viscous flow increase as cation field strength increases so, at any temperature, glasses with smaller cations have higher viscosities and the activation energies for viscous flow are higher, reaching values of >1300 kJ/mol for Lu-SiAlON glasses. The thickness of IG glass films decreases as RE ion radius decreases so that Lu glass films are thinner than La IG films of the same nominal composition.

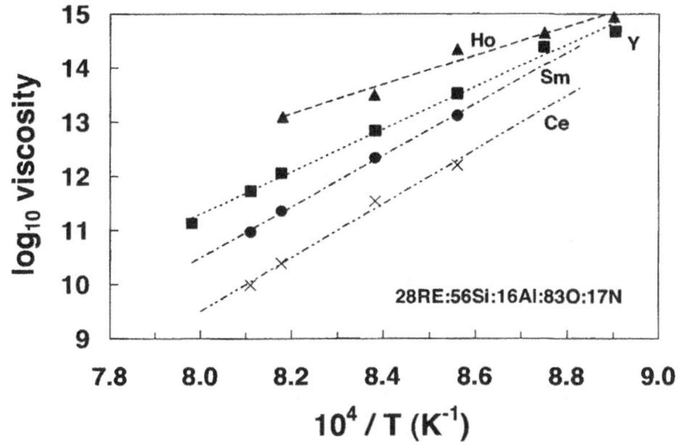

Fig. 3 Viscosity-reciprocal temperature relationships for a series of glasses with composition (in e/o) of 28RE:56Si:16Al:83O:17N (RE = Y, Ho, Sm, Ce)

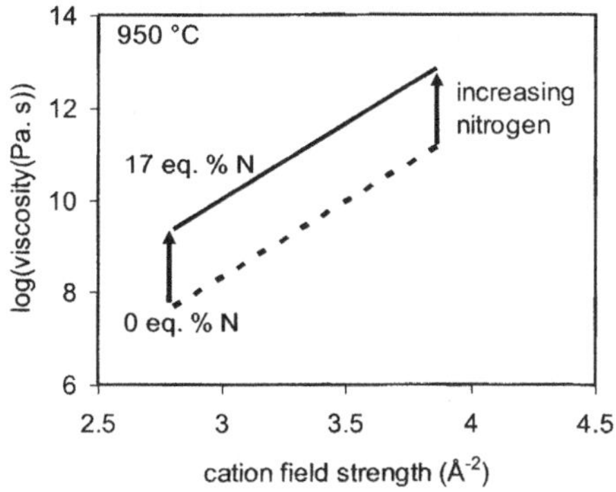

Fig. 4 Combined effects of CFS and N content on viscosity of RE-Si-Al-O-N glasses.

RE ions have different tendencies to segregate to the grain surfaces in silicon nitride[21], changing occupation of adsorption sites on β-Si_3N_4 prismatic plane surfaces. La has a preference for adsorption on the β-Si_3N_4 prismatic sites while Lu ions have much lower tendency for these surfaces, preferring to segregate at the triple points leaving N concentrated in the IGFs. Enhancement of creep resistance of Si_3N_4 sintered with Lu_2O_3[23] may be because Lu_2O_3 produces a more deformation resistant IGF (less Lu, more N) than does larger RE ions. Increasing the viscosity of the IGF relative to the triple points reduces the ability of the material to cavitate during creep, and forces a change in creep mechanism to solution-precipitation[23].

REFERENCES

[1]P. F. Becher, G. S. Painter, N. Shibata, S. B. Waters, H-T. Lin, Effects of rare-earth (RE) intergranular adsorption on the phase transformation, microstructure evolution, and mechanical properties in silicon nitride with RE2O3 + MgO additives: RE=La, Gd, and Lu, *J. Amer. Ceram. Soc.*, **91** [7], 2328-2336, (2008).

[2]S. Hampshire, 2009, Silicon Nitride Ceramics, in: *Advances in Ceramic Materials*, Ed. B Ralph, P. Xiao, Trans Tech Publications, Switzerland, *Mater. Sci. Forum*, **606**, 27-41 (2009).

[3]P. F. Becher, G. S. Painter, N. Shibata, R. L. Satet, M. J. Hoffmann, S. J. Pennycook, Influence of additives on anisotropic grain growth in silicon nitride ceramics, *Mater. Sci. Eng. A –Struct. Mater. Prop. Microstr. Process.*, **422**, 85-91 (2006).

[4]S. Hampshire and M. J. Pomeroy, Effect of composition on viscosities of rare earth oxynitride glasses, *J. Non-cryst. Solids*, **344**, 1-7 (2004).

[5]F Lofaj, Creep mechanism and microstructure evolution in silicon nitride ceramics, *Int. J. Mater. Product Tech.*, **28**, 487-513 (2007).

[6]S. Hampshire and M. J. Pomeroy, Oxynitride Glasses, *Int. J. Appl. Ceram. Tech.*, **5** [2], 155-63 (2008).

[7]S. Hampshire, Oxynitride Glasses, *J. Euro. Ceram. Soc.*, **28** [7], 1475-83 (2008).

[8]S. Hampshire, Oxynitride glasses, their properties and crystallisation - a review, *J. Non-cryst. Solids*, **316**, 64-73 (2003).

[9]R. E. Loehman, Oxynitride Glasses, *J. Non-cryst. Sol.*, **42**, 433-45 (1980).

[10]S. Hampshire, R. A. L. Drew, K. H. Jack, Oxynitride glasses, *Phys. Chem. Glasses*, **26**, 182-6 (1985).

[11]S. Hampshire, E. Nestor, R. Flynn, J –L. Besson, T. Rouxel, H. Lemercier, P. Goursat, M. Sebai, D.P. Thompson, K. Liddell, Yttrium oxynitride glasses: properties and potential for crystallisation to glass-ceramics, *J. Euro. Ceram. Soc.*, **14**, 261-73 (1994).

[12]P. F. Becher and M. K. Ferber, The Temperature Dependent Viscosity of SiREAl-Based Glasses As A Function of N:O and RE:Al Ratios Where RE = La, Gd, Y and Lu, *J. Am. Ceram. Soc.*, **87**, 1274-79 (2004).

[13]R. Ramesh, E. Nestor, M. J. Pomeroy, S. Hampshire, Formation of Ln-Si-Al-O-N Glasses and their Properties, *J. Euro. Ceram. Soc.*, **17**, 1933-9 (1997).

[14]M. Ohashi, K. Nakamura, K. Hirao, S. Kanzaki, S. Hampshire, Formation and Properties of Ln-Si-O-N Glasses (Ln = Lanthanides or Y), *J. Amer. Ceram. Soc.*, **78** [1] 71-76 (1995).

[15]Y. Menke, V. Peltier-Baron, S. Hampshire, Effect of rare-earth cation on properties of SiAlON glasses, *J. Non-cryst. Solids*, **276**, 145-50 (2000).

[16]T. Rouxel, J-L. Besson, D. Fargeot and S. Hampshire, Changes in Elasticity and Viscosity of a SiYAlON Glass during Structural Relaxation in the Transformation range, *J. Non-cryst. Solids*, **175** [1], 44-50 (1994).

[17]E. Dolekcekic, M. J. Pomeroy, S. Hampshire, Structural Characterisation of Er-SiAlON Glasses by Raman Spectroscopy, *J. Euro. Ceram. Soc.*, **27** [2-3], 893-98 (2007).

[18]S. Hampshire, R. A. L. Drew and K. H. Jack, Viscosities, Glass Transition Temperatures and Microhardness of Y-Si-Al-O-N Glasses, *J. Am. Ceram. Soc.*, **67** [3], C46-7 (1984).
[19]Sun, E. Y., Becher, P. F., Hwang, S.-L., Waters, S. B., Pharr, G. M. and Tsui, T. Y., Properties of silicon-aluminum-yttrium oxynitride glasses, *J. Non-Cryst. Sol.*, 1996, **208**, 162-9.
[20]T. Rouxel, J-L. Besson, E. Rzepka and P. Goursat, Raman Spectra of SiYAlON Glasses and Ceramics, *J. Non-cryst. Solids*, **122**, 298-304 (1990).
[21]P. F. Becher, G. S. Painter, M. J. Lance, S. Ii, Y. Ikuhara, Direct observations of debonding of reinforcing grains in silicon nitride ceramics sintered with yttria plus alumina additives, *J. Am. Ceram. Soc.*, **88**, 1222-26 (2005).
[22]N. Shibata, G. S. Painter, R. L. Satet, M. J. Hoffmann, S. J. Pennycook and P. F. Becher, Rare-earth adsorption at intergranular interfaces in silicon nitride ceramics: Subnanometer observations and theory, *Phys. Rev. B* 72 (14), Article No.140101 (2005).
[23]F. Lofaj, S. M. Wiederhorn, G. G. Long, B. J. Hockey, P. R. Jemian, L. Browder, J. Andreason, U. Taffner, Non-cavitation tensile creep in Lu-doped silicon nitride, *J. Euro. Ceram. Soc.*, **22**, 2479-87 (2002).

THE HYDROLYSIS OF ALUMINIUM NITRIDE: A PROBLEM OR AN ADVANTAGE

Kristoffer Krnel and Tomaž Kosmač
Engineering Ceramics Department, Jožef Stefan Institute
Jamova 39, SI-1000 Ljubljana, Slovenia

ABSTRACT

Aluminium nitride is an interesting and useful ceramic material. On the other hand, it is somewhat problematic, mostly due to its reactivity with water. Because of this the important issue in the aqueous processing of AlN powder is the control of the hydrolysis reactions.

In this paper, all aspects of the AlN hydrolysis will be presented. First, the reaction itself and the influence of various parameters on the mechanisms, kinetics and reaction products will be covered. Second, the possibilities of controlling the reaction either to prevent it or to accelerate it will be shown, and a method for the preparation of water-resistant, hydrophilic AlN powder using the dispersion of the powder in a solution of aluminium dihydrogen phosphate will be described. Finally, the exploitation of hydrolysis for the HAS shaping process and for the preparation of nanostructured alumina coatings will be explained.

INTRODUCTION

The hydrolysis of the aluminum nitride powder is well known and has been investigated by many authors. [1,2,3] Bowen et al. proposed the following reaction scheme:[1]

$$AlN + 2H_2O \rightarrow AlOOH_{amorph.} + NH_3 \quad (1)$$
$$NH_3 + H_2O \rightarrow NH_4^+OH^- \quad (2)$$
$$AlOOH + H_2O \rightarrow Al(OH)_3 \quad (3)$$

AlN powder first reacts with water to form amorphous aluminum oxide hydroxide (pseudoboehmite phase, AlOOH), which later crystallizes as aluminum 3-hydroxide (bayerite or gibbsite, $Al(OH)_3$), according to reaction (3).[4,5] The reaction kinetics were described using an unreacted-core model which, for an irreversible reaction, encompasses: (a) mass transfer of the reactant (liquid or gas) through a film surrounding the particle to the surface of the solid, (b) diffusion of the reactant through the product layer to the surface of the unreacted core, and (c) chemical reaction of reactant and solid at the core surface. The slowest step will be the rate-controlling step. In the case of AlN degradation in deionized water, the chemical reaction was proposed to be the rate-controlling step. More recently, Mobley further elaborated and extended Bowen's model of AlN degradation in water.[6] According to his study, the amorphous AlOOH layer that is formed by reaction (1) is dissolved in water, forming different aluminum species which are finally precipitated as crystalline bayerite ($Al(OH)_3$).

However, one of the major problems in the fabrication of AlN ceramics is still the reactivity of AlN powder with water. This means that an important issue in the aqueous processing of AlN powder is the control of the hydrolysis reactions. In the production of ceramics containing AlN as a major or minor constituent (AlN, SiAlONs, SiC (?), etc.) it is necessary to prevent hydrolysis. To do that, non-aqueous powder processing is required, or, alternatively, water-resistant AlN powder should be used.[7] Most of the commercially available water-resistant AlN powders are coated with one of the carboxylic acids. These acids are hydrophobic (i.e., they repel water) and so the powders cannot be dispersed in water without the addition of a hydrophilic surfactant. This in turn requires the addition of an anti-foaming agent to reduce the surface tension, which would otherwise cause extensive foaming of the slurry.[8] On the other hand, AlN powder can be used as a setting agent, e.g., in the Hydrolysis Assisted Solidification (HAS) forming process, where the hydrolysis is exploited for the solidification of the suspension it the closed mold.[9,10] In this process, a few percent of AlN powder is dispersed in the aqueous slurry of the powder to be shaped. After homogenization the suspension is casted into the closed mold, where the hydrolysis is thermally activated. Because of the water consumption, the formation of aluminum hydroxides and the change in the ionic strength of the suspension during the hydrolysis of AlN powder, the viscosity of the host slurry is increased to such extent that a solid body is formed. Another example of the exploitation of the hydrolysis of AlN powder is the preparation of nanostructured alumina coatings on various substrates.

In this work, all aspects of the AlN hydrolysis will be presented. The reactivity of AlN powder in an various aqueous environments was investigated by measuring the pH and the temperature during the hydrolysis of the powder at room and elevated temperatures. The AlN hydrolysis was investigated by measuring the pH of diluted suspensions and by analysis of the reaction products. The results indicate possible solutions for control of the reaction with water in order to exploit it for the HAS shaping process and for the preparation of nanostructured alumina coatings, or to prevent it to enable aqueous AlN powder processing. [11,12,13,14]

EXPERIMENTAL

The AlN powder used in the experimental work was an AlN Grade B powder (H.C. Starck, Berlin, Germany) with a nominal particle size of 1.2 μm, an oxygen content of 2.2 wt.%, and a specific surface area of 3.2 m^2/g. The hydrolysis tests were carried out as follows: a diluted suspension containing 2 wt.% of AlN in water was prepared by stirring and ultrasonication. The pH and the temperature were continuously monitored versus time during the mixing. Powders were also soaked in a solution of silicic acid, phosphoric acid and aluminium dihydrogen phosphate at room and elevated temperatures to evaluate possible permanent protection of the powder. The hydrolysis tests were performed by redispersing these powders in deionized water to check their stability. After the hydrolysis test, the slurries were filtered and washed with 2-propanol to remove the excess water. The cakes were dried at 80°C for 1 hour and then stored in plastic, airtight containers for subsequent analysis.

For the preparation of aluminate coatings, the deionized water was preheated with an electric heater under constant stirring to the desired temperature, a zirconia disc was inserted and then the AlN powder added to the water. The pH and temperature were measured versus time using a combined glass-electrode/Pt 1000 thermometer pH meter (Metrohom 827). In addition, some of the prepared boehmite coatings on zirconia surface were thermally treated in the resistance oven in dry air, at 900 °C, for 1 hour, at a heating rate 10 °C/min.

RESULTS AND DISCUSSION

The results of the measurements of reactivity of AlN powder in water and various aqueous media at room temperature are presented in Figure 1, where the pH is plotted versus time. The results show that there is an incubation time before the start of the AlN hydrolysis reactions. Once this incubation time is over, the pH of the slurry starts to increase, indicating the onset of the reactions. The existence of the incubation time is in agreement with the results of other authors[15,16,17]. One of the suggested reasons for the incubation time is the presence of a thin, hydrated oxide layer on the surface of the AlN particles, which has to be dissolved or penetrated in order that the hydrolysis can start. Our results of the measurement of the reaction rate versus the reaction temperature, showing shortening of the incubation time with increasing reaction temperature, indicate the first reason as the most probable, i.e., the dissolution of the protective oxide scale.[11]

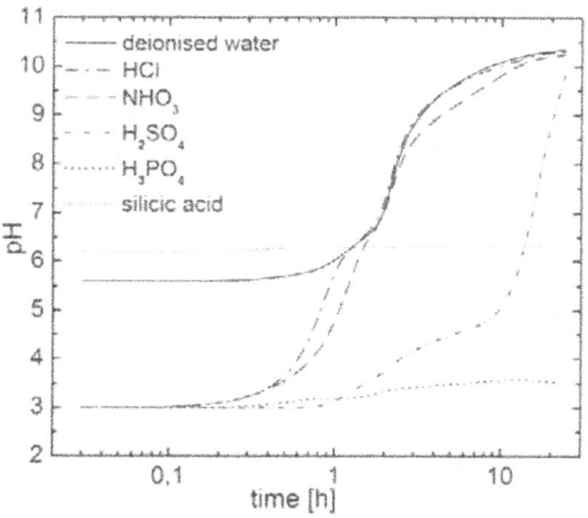

Figure 1: pH versus time for the hydrolysis of a 2 wt.% AlN suspension in deionised water and various aqueous solutions

Also the starting pH has a strong influence on the reaction of AlN with water. [18] At low values of pH, the reaction is much slower than at high values. In an alkaline environment the incubation time is approximately the same as in distilled water, but the reaction is accelerated.[11] At very low pH values (pH = 1), on the other hand, the reaction of AlN powder with water is even prevented. The reason for this is that the reaction of AlN with water is basic-catalyzed, due to the impact of OH^- ions on the Al–N bond. In agreement with that, we did not observe any hydrolysis of AlN powder at very low starting pH (pH~1), regardless of the acid used to adjust the pH. In contrast, in a less acidic environment, i.e., at higher pH values (pH~3), the reaction was fast enough to reveal the influence of different acids on the hydrolysis reaction. The results of the experiments are also plotted in Figure 1. Monoprotonic acids, which are completely dissociated (HCl, HF, HNO_3) and form water-soluble salts with aluminum, did not influence the hydrolysis reactions. In the presence of incompletely dissociated di-protonic H_2SO_4 and H_2CO_3 acids, which form water-soluble salts with aluminum, the reaction was hindered but not prevented. In the presence of phosphoric and silicic acid the hydrolysis was prevented at room temperature, presumably due to the formation of insoluble phosphates or silicates on the powder surface.

A higher starting temperature not only speeds up the reaction of the AlN powder with water, but it also shortens the incubation time as already noted. In addition, the reaction temperature influences the morphology of the reaction product. That is, at temperatures below 60°C, the final product of the hydrolysis are large and elongated $Al(OH)_3$ crystals, whereas at higher temperatures (above 60°C), nano-crystalline AlOOH is formed, which was also confirmed by our further investigation.[19] At around 50°C both products are formed, as shown in Figure 2.

The morphology of the AlOOH observed at elevated temperature and its similarity with the hydroxyapatite coatings prepared by biomimetic deposition using the supersaturated calcium phosphate solutions,[20] gave us the idea for the preparation of the aluminate coatings exploiting the hydrolysis of the AlN powder. In Figure 3a the nano-crystalline AlOOH coating on the zirconia surface prepared by soaking the substrate in the AlN suspension during hydrolysis is presented. The coatings can be further heat-treated at 900 °C, for improved adhesion with the substrate.

During the heat treatment, the transformation of AlOOH to alumina occurs, but morphology of the coating is completely preserved, as visible from Figure 3b.

Figure 2: SEM micrograph of AlN powder after hydrolysis at 50°C, showing large, elongated $Al(OH)_3$ crystals and small, nanometric AlOOH crystals.

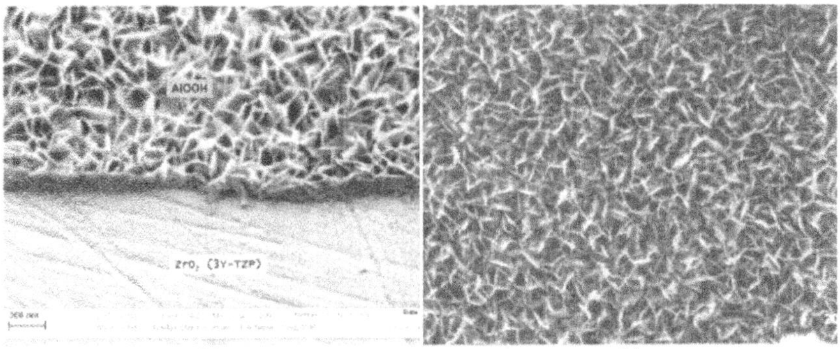

(a) (b)

Figure 3: SEM micrographs of the boehmite coating on the zirconia surface precipitated using AlN powder hydrolysis at 90 °C: (a) after deposition; (b) after heat treatment at 900°C in air.

The results of the pH-time measurements in diluted phosphoric acid show that it is effective in protecting AlN powder at room temperature, whereas the hydrolysis takes place at elevated temperatures. However, as pointed out by Uenishi et al., several inorganic and organic phosphoric acids and/or their compounds are capable of preventing hydrolysis even at higher pH values and at elevated slurry temperatures (e.g., at pH 5.6 and at up to 80°C).[3] When aluminum dihydrogen phosphate was present in the AlN slurry, no hydrolysis was observed even at 70°C. Further experiments indicate that by soaking in aluminum dihydrogen phosphate solution at elevated temperatures (above 50°C), water-resistant AlN powder can be prepared, which will not react with water even after drying and/or redispersion (see Figure 4).[13] The protected AlN powder is hydrophilic, which facilitates aqueous powder processing. It has been suggested that the less-soluble phosphate complexes are formed on the AlN powder surface by chemisorption, as in the case of aluminum protection by phosphoric acid anodization (PAA).[21]

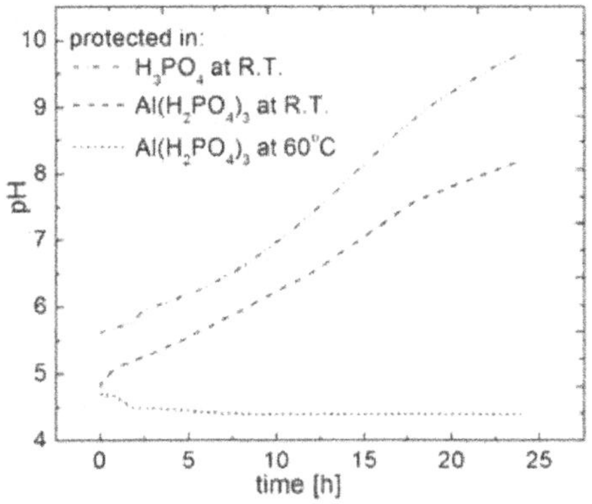

Figure 4: Room-temperature pH-time profiles for diluted suspensions of AlN powder that was first soaked in phosphoric acid and $Al(H_3PO_4)_3$ solutions at room temperature and in $Al(H_2PO_4)_3$ solution at 70°C.

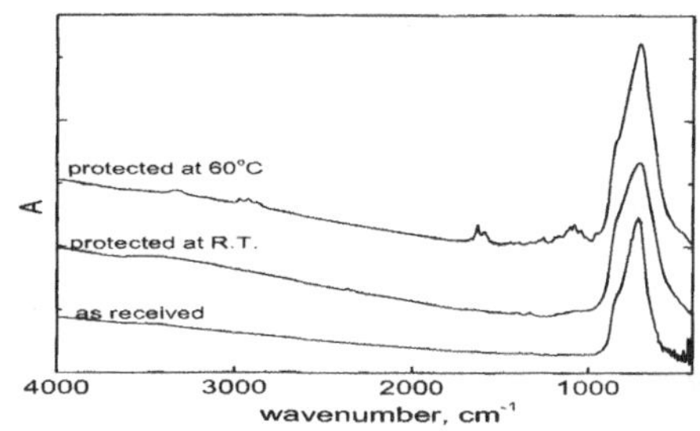

Figure 5: FTIR analysis of the AlN powder before and after soaking in the solution of aluminium dihydrogen phosphate at room temperature and at 60 °C.

To explain the difference between soaking at room temperature and elevated temperature, the powder surfaces were analysed by FTIR after the soaking of the powders in the solution of aluminium dihydrogen phosphate (see Figure 5). The results show that the powder protected at 60 °C shows much stronger signals associated with AlOOH (stretchings around 3400, 2900 and 1600 cm-1), indicating that the surface of this powder is slightly more hydrolysed. It also shows signals indicating the presence of phosphate complexes bonded to aluminium (stretching from 1050 to 1100 cm-1).[22,23] The conclusion is that phosphate anions bond to the hydrolysed surface of the AlN powder, forming phosphates on the surface that prevent the access of water to the AlN core. If the powder is protected at room temperature this phosphate surface layer is too thin, and is partly dissoluted when the powder is redispersed in water. The protection at 60°C first initiates the formation of a thin, freshly hydrolysed surface onto which more phosphate anions can be bonded. This results in a thicker, protective phosphate surface layer that can withstand the redispersion in water and enables aqueous AlN powder processing. Since the phosphate layer on the surface of the AlN powder is thin, the solubility of the complexes formed by the chemisorption of anions onto the AlN powder surface plays an important role in the reactivity of AlN powder with water. At temperatures above 50°C this phosphate layer is somewhat thicker, and the protected powder can withstand the redispersion in water. These results were later confirmed by Olhero et al., who concluded that aluminum dihydrogen phosphate offers the possibility to prepare water-resistant AlN powder that can be used for aqueous colloidal processing and shaping.[24] Our recent results

also show, that the powder protection works also with the nano-sized AlN powder, that can be processed in water, shaped by slip casting and sintered to high density.[25]
The non-reactivity of the AlN powder in the solution of phosphoric acid also indicates possible solutions for the control of the hydrolysis reactions during the HAS forming process that exploits the reaction of the AlN powder with water for the solidification of the aqueous suspensions in closed molds. During HAS forming process, it is important that the hydrolysis is prevented during homogenization and handling of the suspension, but it has to be initiated once the suspension is poured, cast or injected into the mold. The addition of phosphoric acid can be used as an agent for the control of the reactivity, since the reactivity of AlN powder in solution of phosphoric acid is reestablished at temperatures higher than 60 °C, as already shown in figure 4.
As already showed, in the presence of silicic acid, the reaction of AlN powder with water is also prevented, even though the initial pH was higher (pH=5).[12] This was first addressed as a problem when the HAS process was used in the slurry-forming of Si_3N_4 ceramics.[26] In this process, AlN powder was added to an aqueous Si_3N_4 slurry as a setting agent. However, the AlN hydrolysis did not occur and thus the solidification was prevented. The amount of dissolved silica assumed to be present in aqueous Si_3N_4 and SiC slurries[27] should be reflected in the reactivity of the AlN powder in these slurries. In the presence of silicic acid, the reaction was suppressed at both room and elevated temperatures, which was also ascribed to the formation of insoluble silicates. The adsorption of silicate anions onto the powder surface was confirmed by chemical analysis and zeta-potential measurements. Using DRIFT measurements, however, the presence of Si-O bonds on the powder could not be unambiguously confirmed, since the characteristic wavelengths for these bonds are in the region of very strong Al-N stretching frequencies[12]. To further investigate this assumption, the reactivity of AlN powder with water in supernatants obtained from centrifuged Si_3N_4 and SiC slurries were checked. It was found that reactivity depends strongly on the concentration of dissolved silica in these slurries relative to the surface area of the AlN powder in the slurry. Various Si_3N_4 and SiC powders were used, which were fabricated by different production routes and had surfaces oxidized to different degrees. The hydrolysis of AlN did not occur if the concentration of dissolved silica with respect to the AlN-powder surface was high enough (1 mg SiO_2 / m^2 of AlN powder) to form a layer of alumosilicates on the AlN-powder surface.[14]
Aqueous powder processing of Si_3N_4 or SiC suspensions containing unprotected AlN powder as a constituent or sintering additive is therefore possible provided that the concentration of silicic acid in the suspension is high enough with respect to the concentration and specific surface area of the admixed AlN powder. To satisfy this condition, Si_3N_4 powders can be heat-treated in air to increase the silicic acid content in their slurries. On the other hand, if the added AlN powder is to be used as a forming aid in the HAS process, the reactivity of AlN powder with water has to be established. This can be done by removing the silica surface layer, responsible for the presence of silicic acid in the slurry, from the Si_3N_4 or SiC powder by leaching them in a hot alkaline solution.[28] In Figure 5 the results of the measurement of the reactivity of AlN powder in aqueous Si_3N_4 and SiC slurries are presented. The results show that with proper surface treatment of the Si_3N_4 and SiC powders, the AlN hydrolysis in aqueous Si_3N_4 and SiC slurries can be prevented or initiated.

CONCLUSIONS

The results of our extensive research on the hydrolysis of AlN powder improved our understanding of the reactions and mechanisms as well as giving us the possibility to control the hydrolysis reactions by chemical means. Using phosphoric acid, we made it possible to control the start and the speed of the reaction for use in the HAS forming process, which exploits hydrolysis for the solidification of aqueous slurries. On the other hand, with proper surface treatment using aluminum dihydrogen phosphate, water-resistant AlN powder can be prepared, that is hydrophilic, which facilitates aqueous powder processing. It has been suggested that the phosphate complexes with low solubility are formed on the AlN powder surface by chemisorption and prevent the access of water to AlN core. In some cases the hydrolysis can also be prevented "in-situ", as is the case in Si_3N_4 and SiC slurries, which contain dissolved silicic acid.

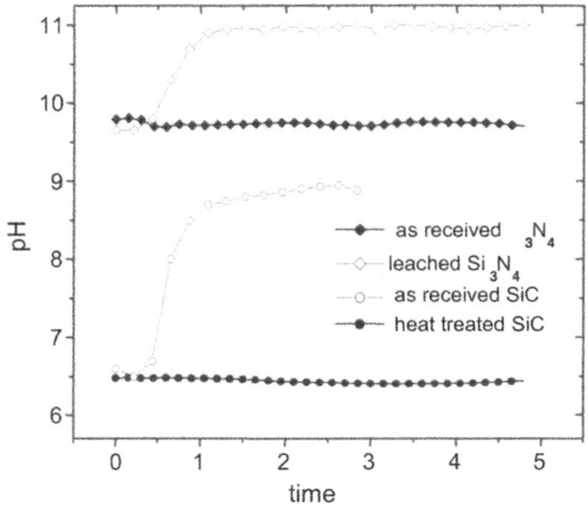

Figure 5: pH–time profiles for UBE E10 Si_3N_4 and Norton FTP-15-NLC SiC slurries containing 5 wt% AlN powder. Powders were surface treated to enable or disable the AlN hydrolysis.

REFERENCES

1 P. Bowen, J.G. Highfield, A. Mocellin, T.A. Ring, "Degradation of Aluminum Nitride Powder in an Aqueous Environment", J. Am. Ceram. Soc., 73 [3] 724-728 (1990).

2 T. Reetz, B. Monch, M. Saupe, "Aluminum Nitride hydrolysis", Ber. DKG, 68 [11-12] 464-465 (1992).

3 M. Uenishi, Y. Hashizume, T. Yokote, "Aluminium Nitride Powder Having Improved Water-Resistance", U.S. Pat. 4,923,689, May 8, 1990.

4 K. Wefers and C. Misra, Oxides and Hydroxides of Aluminum, Technical Paper No. 19 (revised 1987) available from Alcoa, Pittsburg, PA.

5 T. Graziani and A. Belosi, Degradation of Dense AlN Materials in Aqueous Environments, Mater. Chem. Phys., 35 (1993) 43-48.

6 W. M. Mobley, "Colloidal Properties, Processing and Characterization of Aluminum Nitride Suspensions"; Ph.D. Thesis. Alfred University, Alfred, NY, 1996.

7 Groat, E.A., Mroz Jr., J., Aqueous Processing of AlN powders. Ceramic Industry, 1990, march, 34-38.

8 Reed J.S., Introduction to the Principles of Ceramic Processing. John Wiley & Sons, New York, 1988.

9 T. Kosmač, S. Novak, M. Sajko, "Hydrolysis-Assisted Solidification (HAS): A New Setting Concept for Ceramic Net-Shaping", J. Europ. Ceram. Soc., 17 (1997) 427-432.

10 T. Kosmač, S. Novak, K. Krnel, Hydrolysis assisted solidification process and its use in ceramic wet forming : dedicated to professor Dr. Drago Kolar in memory of this brilliant scientist and teacher. Z. Met.kd., 92 (2001) 150-157.

11 K. Krnel, T. Kosmač, Degradation of AlN powder in aqueous environments, J. Mater. Res., 19 (2004) 1157-163.

12 K. Krnel, T. Kosmač, Reactivity of aluminum nitride powder in dilute inorganic acids, J. Am. Ceram. Soc., 83 (2000) 1375-1378.

13 K. Krnel, T. Kosmač, Protection of AlN powder against hydrolysis using aluminum dihydrogen photosphate. J. Eur. Ceram. Soc, 21 (2001) 2075-2079.

14 K. Krnel, T. Kosmač, Reactivity of aluminum nitride powder in aqueous silicon nitride and silicon carbide slurries, J. Am. Ceram. Soc., 85 (2002) 484-486.

15 Görter, H., Gerretsen, J., and Terpstra, R. A., Comparison of the Reactivity of Some Surface Treated AlN Powders with Water, 3rd Euroceramics V1, 615-620, Faenza Editrice Iberica, S.C, 1993.

16 M. Egashira, Y. Shimizu, S. Takasuki, Chemical surface treatments of aluminium nitride powder suppressing its reactivity in water, J. Mater. Sci. Let., 10 (1991) 994-996

17 S. Fukumoto, T. Hookabe, H. Tsubakino, Hydrolysis behyviour of aluminium nitride in various solutions, J. Mater. Sci., 35 (2000) 2743-2748.

18 T. Reetz, B. Monch, and M. Saupe, "Aluminum Nitride Hydrolysis," Ber. Dtsch. Keram. Ges., 68 (1992) 464–65.

19 A. Kocjan, K. Krnel, T. Kosmač, The influence of temperature and time on the AlN powder hydrolysis reaction products, J. Eur. Ceram. Soc., 28 (2008) 1003-1008.

20 S. Beranič, I. Pribošič, T. Kosmač, The formation of an apatite coating on Y-TZP zirconia ceramics. Key eng. mater., 330-332 (2007), 773-776.

21 G.D. Davis, J.S. Ahearn, J.D. Venables, Use of Surface Behaviour Diagrams to Study Hydration/corrosion of Aluminum and Steel Surfaces, J. Vac. Sci. Technol. A, 2 (1984) 763-766 .

22 J.G. Highfield , P. Bowen, Diffuse-Reflectance Fourier Transform Infrared Spectroscopic Studies of the Stability of Aluminum Nitride powder in an Aqueous Environment, Anal. Chem., 61 (1989) 2399.

23 A. I. Omoike, G.W. Vanloon, Removal of phosphorous and organic matter removal by alum during wastewater treatemnt, Wat. Res. 33 (1999) 3617.

24 S.M. Olhero, S. Novak, M. Oliviera, K. Krnel, T. Kosmač, J.M.F. Ferreira, A thermo-chemical surface treatment of AlN powder for the aqueous processing of AlN ceramics. J. mater. res., 19 (2004) 746-751.

25 K. Krnel, T. Kosmač, "The role of chemisorbed anions in the aqueous processing of AlN powder", Z. Met.kd., in press.

26 K Krnel, T. Kosmač, "Use of Hydrolysis Assisted Solidification in Slurry Forming Si3N4 Bodies", Ceramic Processing Science (Ceramic Transactions, vol 83), The American Ceramic Society, Westerville 257-264 (1998).

27 P. Greil, "Processing of Silicon Nitride Ceramics", Materials Science and Engineering, A109 (1989) 27-53.

28 K. Krnel, T. Kosmač, Use of hydrolysis assisted solidification (HAS) in the formation of Si3N4 ceramics, Mater. sci. forum, 413 (2003) 75-80.

PREPARATION AND COMPARISION OF TWO TYPICAL CVD FILMS FROM CH_4 AND C_3H_6 AS CARBON RESOURCES

W. B. Yang, L. T. Zhang*, L. F. Cheng, Y. S. Liu, W. H. Zhang
National Key Laboratory of Thermalstructure Composite Materials, Northwestern Polytechnical University, Xi'an, Shaanxi, 710072, P. R.China
e-mail: wenbinyoung@gmail.com

Abstract

Two boron-carbon coatings were prepared by CVD from BCl_3-CH_4-H_2-Ar and BCl_3-C_3H_6-H_2-Ar precursors with same depositing temperature and total pressure. Microstructure and composition of the prepared coatings were compared by SEM, XRD, and XPS. It was found that the coatings derived from C_3H_6 and CH_4 are both hybrid with different microstructure and composition. The coating derived from C_3H_6 mainly consists of boron doped graphite which decides this coating showing a relatively smooth surface morphology and a finely laminated fracture structure while the coating derived from CH_4 consists dominantly of $B_{51.02}C_{1.82}$ crystals. The main composition of $B_{51.02}C_{1.82}$ shows a distinct crystal orientation surface morphology and a randomly oriented lamellar structure.

Keywords: CVD; boron-carbon material; bonding states; microstructure; XPS;

INTRODUCTION

Non-oxide ceramic matrix composites (CMCs), such as C/C and C/SiC are outstanding high-temperature structural materials in aerospace industries [1-4]. Unfortunately, their applications are badly limited at oxidizing environments due to the poor oxidation resistance of the carbon phase [5]. Boron carbon (B-C) materials [6] are considered promising materials to enhance the oxidation resistance of C/C and C/SiC composites. When introduced into coating [7-9] ⁱterphase [2, 10-11] or matrix [12-13] of C/C and C/SiC, B-C materials are used to prepare self-healing materials (the fluid oxide phase formed in situ by oxidation, filling cracks, slowing down the in-depth diffusion of oxygen) with improved lifetimes.

It was found that the application in different parts of non-oxide CMCs require different compositions and microstructures of the B-C materials [7-13]. Boron doped (substituted) carbon has a boron-to-carbon stoichiometry lower than 0.3 [14-22]. This group of materials usually exhibits a layered structure, which is weak enough to deflect matrix cracks and protect the fibers from matrix crack stress concentrations, so they are suitable for use as interphases [2, 10, 23]. While boron carbide, such as $B_{13}C_2$ (usually known as B_4C), or $B_{51}C$, $B_{50}C_2$, $B_{49}C_3$, $B_{48}C_3$, or B_8C_2, has boron-to-carbon stoichiometries higher than 2 [24-31]. The higher boron content of these materials helps to form more fluid oxide, which makes them more suitable for use in coatings and matrix [7-9, 12-13]. The most popular way to prepare B-C materials for self-healing modification of non-oxide CMCs is chemical vapor deposition (CVD). It is found that the carbon resource has an important influence on the microstructures and compositions of B-C materials. To prepare boron carbide, methane [25-30] (CH_4) was always employed, while, to prepare boron doped carbon, C_2H_4 [14], C_2H_6 [15], C_3H_8 [17], C_6H_6 [16, 19] was always employed.

Preparation and characterization of B-C materials helps to understand their intrinsic properties and guide their applications. While the products differ a lot in microstructure and composition, there is

* Corresponding author. Tel.: +86-29-8848-6068-827; fax: 86-29-8849-4620.
E-mail address: wenbinyoung@gmail.com

lack of systematically comparative study on the microstructure and composition of these two kinds of B-C material.

In this paper, to meet the different requirements in self-healing CMCs, typical boron carbide and boron doped carbon was prepared by CVD from BCl_3-CH_4-H_2-Ar and BCl_3-C_3H_6-H_2-Ar mixture, respectively. Microstructures, phases and chemical bonding characters of the deposits were systematically analyzed and the relationship between microstructures and compositions was discussed.

EXPERIMENTAL

Preparation of specimens

A vertical cold wall CVD furnace was employed to prepare the B-C coatings. Boron trichloride ($BCl_3 \geq 99.99$ vol.% and iron$\leq$10 ppm) was used as the boron source. The carbon source was provided by the methane ($CH_4 \geq 99.95$ vol.%) and propylene ($C_3H_6 \geq 99.95$ vol.%) gas. Hydrogen ($H_2 \geq 99.999$ vol.%) was used as a dilution gas of BCl_3.The deposition parameters are listed in Table 1. T-300 carbon fiber from Toray, Japan was employed as substrate.

Characterizations of the coatings

The surface and fracture section morphologies of two kinds of coatings were observed by SEM (JSM-6700F). Phase identification was carried out by XRD (Rigaku D/MAX-2400 (Cu K_a radiation)) with powder milled from carbon fiber deposited with B-C coatings. To distinguish the carbon component in fiber and deposited coatings, the XRD analysis of carbon fiber that was heat treated at depositing temperature (1273 K) for 20h with argon atmosphere was also preformed. XPS analysis of the as-received coatings was performed with an Axis Ultra spectrometer (AXIS ULTRA, KRATOS ANALYTICAL Ltd.), using monochromatic Al K_α (1486.71 eV) radiation at a power of 150W (10mA, 15kV). To compensate for surface charging effects, binding energies were calibrated use C 1s hydrocarbon peak at 284.8 eV. The bonding states analysis is carried out with the as received surface of the coatings, while the measurement of compositions is carried out after etching with argon ion for 200s.

In this paper, for the convenience of descriptions, the coatings deposited from propylene are referred to as $(B\text{-}C)_P$ and the coatings deposited from methane are referred to as $(B\text{-}C)_m$.

RESULTS AND DISCUSSION

Fig. 1 shows the surface and fracture section morphologies of the $(B\text{-}C)_p$ coating. It is clear that the $(B\text{-}C)_p$ exhibits a relatively smooth surface morphology and a highly organized laminated fracture structure with layers of nanometer distinction which is similar to that of graphite. While the $(B\text{-}C)_m$ coating shows a distinct randomly oriented crystal surface morphology (Fig. 2(a)) and a smooth fracture morphology (Fig. 2(b)) with randomly oriented lamellar structure can be detected.

Fig. 3 shows the XRD patterns of carbon fiber, $(B\text{-}C)_p$ coated carbon fiber, and $(B\text{-}C)_m$ coated carbon fiber. It is shown that $(B\text{-}C)_p$ coating and $(B\text{-}C)_m$ coating are very different in phases. The $(B\text{-}C)_m$ coating shows a series of distinct peaks of $B_{51.02}C_{1.82}$, while the $(B\text{-}C)_p$ coating only shows weak signals of B_4C. It shall also be noticed that the three carbon peaks are quite different according to their d-spacing values listed in Table. 2. The d-spacing value of 3.43Å for carbon fiber coated with $(B\text{-}C)_p$ is the average of carbon fiber and $(B\text{-}C)_p$ coating, which indicated that the carbon phase in $(B\text{-}C)_p$ coating has a d-spacing value less than 3.43Å. For the same reason, the carbon phase in $(B\text{-}C)_m$ coating has a d-spacing value less than 3.39Å, which is close to that of graphite.

The compositions of the $(B\text{-}C)_p$ and $(B\text{-}C)_m$ determined by XPS are listed in Table 3. The atomic percentage of the composition was determined using sensitivity factors of 0.171, 0.314 and 0.733

for B, C and O respectively [32]. The two coatings mainly consisted of boron and carbon with little oxygen. Although both precursor systems had the same B/C ratio, listed in Table 1, the B/C ratio of the $(B\text{-}C)_m$ coating was about 6 times higher than that of the $(B\text{-}C)_p$ coating.

Fig. 4 shows the deconvolutions of XPS B1s spectra the CVD $(B\text{-}C)_p$ and $(B\text{-}C)_m$ coatings, respectively. Both of them exhibit the same 5 bind energy components, which mean that the chemical bonding states of boron atoms in the two coatings were alike. There is no boron (187.0 eV) [14, 20] detected in both of the coatings. The lowest binding energy component at 187.6eV corresponds to $B_{51.02}C_{1.82}$, In the $B_{51.02}C_{1.82}$ crystal, besides boron atoms, there are some carbon atoms surround, the electronegativity of carbon atom is bigger than boron, so the binding energy of $B_{51.02}C_{1.82}$ is higher than that of boron. The binding energy at 188.8 eV was attributed to boron atoms doped graphitic lattices (marked as B_d, the subscript "d" means doped) [14, 20, 33], which can also be understood as boron atoms dissolved by substitution in the graphitic lattices. In B_d, each boron atom is combined with other three carbon atoms, the all-sided intervention of carbon atoms shifts the binding energy of boron atom in B_d to higher direction compare with $B_{51.02}C_{1.82}$. The binding energy at 190.0 eV and 192.0 eV were corresponded to boron species with mixed B–C and B–O bonding, namely boron oxycarbides. The highest binding energy component at 193 eV was due to the presence of B_2O_3, according to the spectrum of the B_2O_3 standard [20]. These partially and completely oxidized boron atoms were induced by the adsorption of oxygen and water occur at the extreme surface of the samples which were not etched for the high resolution analyses as mentioned befor.

Fig. 5 shows the schematic representation of possible local chemical environments of boron atoms doped graphitic lattices with oxygen absorbed. The three covalent bonds formed by the sp2 hybridization of the outmost three electrons are within one plane, and the angles among them are 120 degree, which is similar to the hexagonal net in graphite. Therefore, boron atoms are easy to substitute carbon atoms in a hexagonal net of graphite. The bonding state of boron atoms inside B_d is one boron atom combined with three carbon atoms. The outmost B_d atoms combined with the adsorbent oxygen in varying degrees, in form of BC_2O, BCO_2 and B_2O_3, which are more and more oxidized.

Although the chemical bonding states of boron atoms in the CVD $(B\text{-}C)_p$ and$(B\text{-}C)_m$ coatings are similar, the difference in the contents of species containing boron atoms is obvious. As shown in Table 4, the content of boron atoms involved in boron carbide of $(B\text{-}C)_p$ is much less than that of $(B\text{-}C)_m$. While, the content of boron atoms involved with oxygen in $(B\text{-}C)_p$ is much higher than that of $(B\text{-}C)_m$. It is clear that the content of BC_2O in $(B\text{-}C)_p$ is nearly 5 times higher than that of $(B\text{-}C)_m$. As discussed above, the BC_2O is derived from B_d, so the content of BC_2O could be calculated to that of B_d. This meant that boron atoms in $(B\text{-}C)_p$ are mainly substituted in the graphite (77.3 at.%). While, boron atoms in $(B\text{-}C)_m$ are mainly located in boron carbide (51.0 at.%).

According to the SEM, XRD, and XPS analysis, it was found that both of the prepared coatings are mixtures with different microstructure and composition. $(B\text{-}C)_p$ coating consisted of boron doped graphite, carbon phase and a small amount of B_4C. The main composition of boron doped graphite decided the $(B\text{-}C)_p$ coating showing a graphite like structure. $(B\text{-}C)_m$ coating consisted of $B_{51.02}C_{1.82}$ crystal, boron doped graphite, and carbon phase. The main composition of $B_{51.02}C_{1.82}$ decide $(B\text{-}C)_m$ coating showing a crystal like structure. The typical microstructure of $(B\text{-}C)_p$ makes it a candidate for the self healing mechanical fusion layer of non-oxide CMCs while the $(B\text{-}C)_m$ is more suitable for the use of self healing coating and matrix modification of non-oxide CMCs for its high content of boron.

CONCLUSION

Boron-carbon coatings derived from C_3H_6 and CH_4 are mixture with different microstructure and composition. SEM results showed that $(B\text{-}C)_p$ coating had a relatively smooth surface morphology and a finely laminated fracture structure, while $(B\text{-}C)_m$ coating had a distinct crystal orientation surface morphology and a smooth glassy fracture with randomly oriented lamellar structure. XRD results showed that boron in the $(B\text{-}C)_m$ coating exhibited a series of distinct peaks of $B_{51.02}C_{1.82}$, while that in the $(B\text{-}C)_m$ coating only showed weak signals of B_4C. XPS results showed that both of the $(B\text{-}C)_p$ coatings and $(B\text{-}C)_m$ coatings mainly consisted of boron, carbon and minor amount of oxygen. Despite the same B/C ratio of precursors, the B/C ratio of the $(B\text{-}C)_m$ coatings is about 6 times higher than that of the $(B\text{-}C)_P$ coatings. In $(B\text{-}C)_P$, boron atoms were mainly within the boron doped graphite while in $(B\text{-}C)_m$, boron atoms were mainly located in boron carbide.

ACKNOWLEDGMENTS

The authors acknowledge the support of the National Basic Research Program of China

PICTURES AND TABLES

Fig. 1. SEM photographs of the $(B\text{-}C)_p$ coating (a) surface morphology, (b) fracture section morphology

Fig. 2. SEM photographs of the $(B\text{-}C)_m$ coating (a) surface morphology, (b) fracture section morphology

Fig. 3 XRD patterns showing (a) carbon fiber, (b) $(B\text{-}C)_p$ coated carbon fiber, and (c) $(B\text{-}C)_m$ coated carbon fiber

Fig. 4. Deconvolutions of XPS B1s spectra of the $(B\text{-}C)_p$ and $(B\text{-}C)_m$ coatings

Fig. 5. Schematic representation of possible local chemical environments of boron atoms doped in graphite lattices with oxygen absorbed

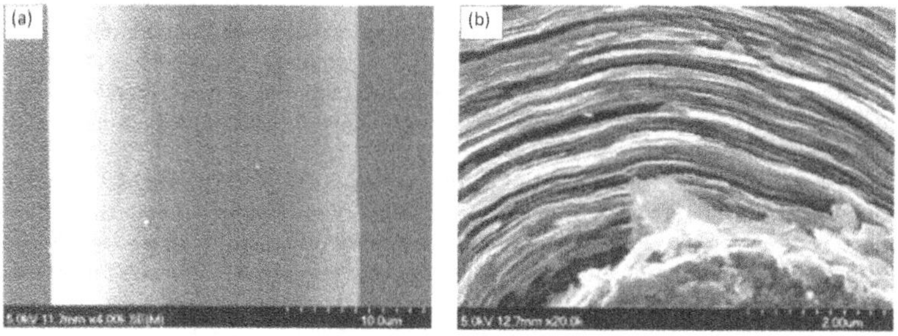

Fig. 1. SEM photographs of the $(B\text{-}C)_p$ coating
(a) surface morphology, (b) fracture section morphology

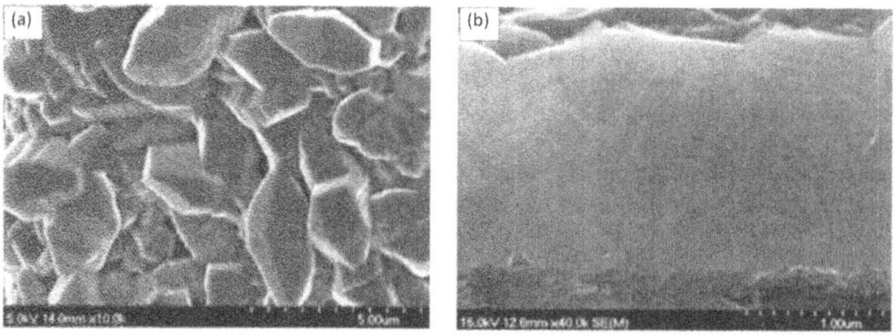

Fig. 2. SEM photographs of the $(B\text{-}C)_m$ coating
(a) surface morphology, (b) fracture section morphology

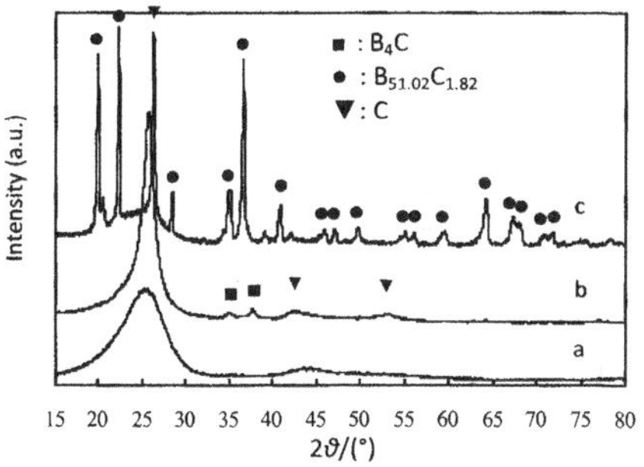

Fig. 3. XRD patterns showing (a) carbon fiber, (b) $(B\text{-}C)_p$ coated carbon fiber, and (c) $(B\text{-}C)_m$ coated carbon fiber

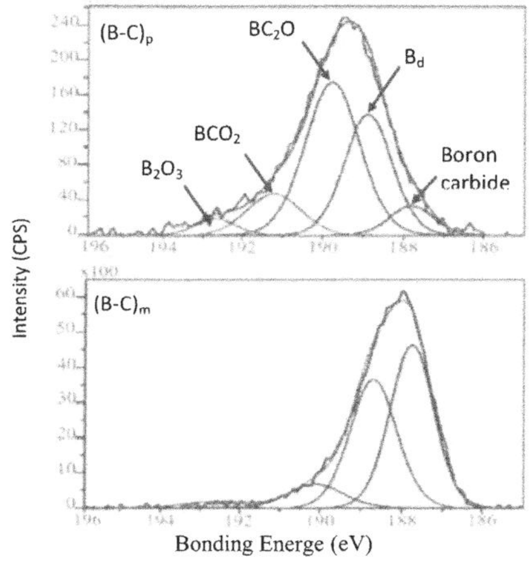

Fig. 4. Deconvolutions of XPS B1s spectra of the $(B\text{-}C)_p$ and $(B\text{-}C)_m$ coatings

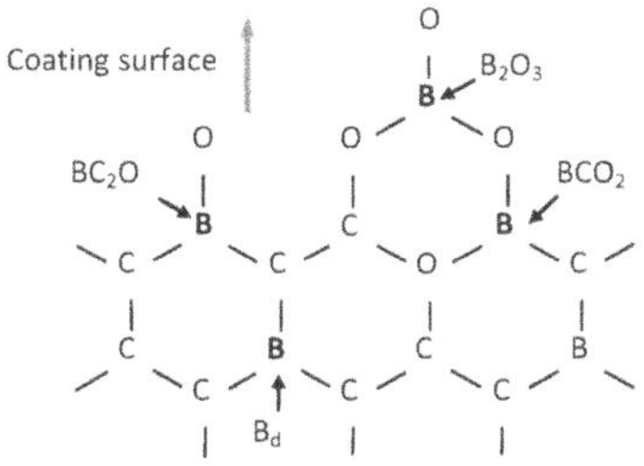

Fig. 5. Schematic representation of possible local chemical environments of boron atoms doped in graphite lattices with oxygen absorbed

Table. 1. The deposition parameters of CVD B-C coatings

BCl_3 (ml/min)	C_3H_6 / CH_4 (ml/min)	H_2 (ml/min)	Ar (ml/min)	Temperature (K)	Time (h)
10	10 / 30	100	100	1273	20

Table. 2. d-spacing values (XRD) of different carbon phases

Carbon phases	C fiber	Pyrolytic Carbon [12]	Carbon fiber coated with $(B\text{-}C)_p$	Carbon fiber coated with $(B\text{-}C)_m$	Graphite [12,21]
d-spacing value (Å)	3.64	3.46	3.43	3.39	3.35

Table. 3. The atomic concentration of the $(B\text{-}C)_p$ and $(B\text{-}C)_m$ coatings

CVD coating	Elements atom content		(at.%)
	B	C	O
$(B\text{-}C)_p$	22.3	75.6	2.1
$(B\text{-}C)_m$	62.8	33.5	3.7

Table. 4. Proportions of different chemical bonding states of boron atoms in $(B\text{-}C)_p$ and $(B\text{-}C)_m$ coatings

	Boron carbide (a.t%)	B_d (a.t%)	BC_2O (a.t%)	BCO_2 (a.t%)	B_2O_3 (a.t%)
$(B\text{-}C)_p$	6.9	31.4	45.9	12.3	3.6
$(B\text{-}C)_m$	51.0	37.5	8.8	1.5	1.3

REFERENCES

[1] *Naslain R., Gutte A., Rebillat F., Pailler R., Langlais F. and Bourrat X.* Boron-bearing Species in Ceramic Matrix Composites for Long-term Aerospace Applications // J. Solid State Chem. 2004. V. 177. No 2. P. 449-456.

[2] *lamouroux F., Bertrand S., Pailler R., Naslain R. and Cataldi M.* Oxidation-resistant Carbon-Fiber-Reinforced Ceramic-Matrix Composites // Compos. Sci. Technol. 1999.V. 59. No 7. P. 1073-108.

[3] *Quemard L., Rebillat F., Guette A., Tawil H. and Louchet-Pouillerie C.* Self-healing Mmechanisms of a SiC Fiber Reinforced Multi-layered Ceramic Matrix Composite in Hhigh Pressure Steam Environments // J. Eur. Ceram. Soc. 2007. V. 27. No 4. P. 2085-2094.

[4] *Rebillat F., Gallet S. L., Guette A., Bourrat X. and Naslain R.*, In: Krenkel W. et al.High Temperature Ceramic Matrix Composites, Wiley-VCH, Weinheim, Germany, (2001) 193-198.

[5] *Naslain R.* Design, Preparation and Properties of Non-oxide CMCs for Application in Engines and Nuclear reactors: an overview // Compos. Sci. Technol. 2004. V. 64. No 4. P. 155-170.

[6] *Schouler M.C., Cheynet M.C., Sestier K., Garden J. and Gadelle P.* New filamentous Deposition

in the Boron-Carbon System // Carbon. 1997. V. 35. No 4. P. 993-1000.

[7] *Schulte-Fischedick J., Schmidt J., Tamme R., Kröner U., Arnold J. and B. Zeiffer.* Oxidation Behavior of C/C–SiC Coated with SiC–B_4C–SiC–Cordierite Oxidation Protection System // Sci. Eng. A. 2004. V. 386. P. 428 - 434.

[8] *Goujard S., Vandenbulcke L., Tawil H.* The Oxidation Behaviour of Two-and Three-Dimensional C/SiC Thermostructural Materials Protected by Chemical-Vapour-Deposition Polylayers Coatings // J. Mat. Sci. 1994. V. 29. P. 6212-6220.

[9] *R'Mili M., Massardier V., Merle P., Vincent H., Vicent C.* The Effect of Thermal Exposure on the Strength Distribution of B_4C Coated Carbon Fibers // Carbon. 1999. V. 37. No 4. P. 129-145.

[10] *Viricelle J.P., Goursat P., Bahloul H.D.* Oxidation Behaviour of a Multi-Layered Ceramic-matrix Composite $(SiC)_f/C/(SiBC)_m$ // Compos. Sci. Technol. 2001. V. 61. No 4. P. 607-614.

[11] *F. Christin.* A Global Approach to Fiber and Architectures and Self Sealing Matrices: From Research to Production // 5th International Conference on High-Temperature Ceramic Matrix Composites. 2004. P. 477-483.

[12] *Naslain R., Pailler R., Bourrat X., Bertrand S., Heurtevent F., Dupel P. and Lamouroux F.* Synthesis of Highly Tailored Ceramic Matrix Composites by Pressure-pulsed CVI // Solid State Ion. 2001. V. 141. P. 541-548.

[13] *Jacques S., Guette A., Langlais F., Naslain R. and Goujard S.J.* Preparation and Characterization of 2D SiC/SiC

[14] Composites with Composition-Graded C(B) Interphase // Eur. Ceram. Soc. 1996. V. 17. P. 1083-1092.

[15] *Charles T.H., Linda E.J., Carl C.* An Investigation of Vapor Deposited Boron Rich Carbon—A Novel Graphite-Like Material–Part I: The Structure of $BC_x(C_6B)$ Thin Films. Carbon. 1999. V. 37. No 2. P. 221–230.

[16] *Jansson U., Carlsson J.O.* Area selective chemical vapor deposition of boron carbide achieved by molecular masking //J. Vac. Sci. Technol. A. 1988. V. 6. No 3. P. 1733-1735.

[17] *Cermignani W., Paulson T.E., Onneby C. and Pantano C.G..* Synthesis and Characterization of Boron-Doped Carbons // Carbon. 1995. V. 33. No 4. P. 367–374.

[18] *Jacques S., Guette A., Bourrat X., Langlais F., Guimon C. and Labrugere C.* LPCVD and Characterization of Boron-Containing Pyrocarbon Materials // Carbon. 1996. V. 34. No 9. P. 1135–1143.

[19] *Kouvetakis J., Kaner R.B., Sattler M.L. and Bartlett N.* A Novel Graphite- Like Materials of Composition BC_3 and Nitrogen-Carbon Graphites // J Chem Commun. 1986. P. 1758–1759.

[20] *Fecko D.L., Jones L.E., Thrower P.A.* The Formation and Oxidation of BC_3, a New Graphite Like Material // Carbon. 1993. V. 31. No 4. P. 637–644.

[21] *Shen C.* Preparations and Characterizations of Novel Graphite-Like Materials and Some High Oxidation State Fluorine Chemistry // Berkeley, CA: University of California, 1992.

[22] *Kouvetakis J., McElfresh M.W and Beach D.B.* Chemical Vapor Deposition of Highly Conductive Boron-Doped Graphite from Triphenyl Boron // Carbon. 1994. V. 32. No 6. P. 1129–1132.

[23] *Jacques S, Guette A, Langlais F, and Naslain R.* C(B) Materials as Interphase in SiC/SiC Model Microcomposites // J. Mater. Sci. 1997. V. 32. No 4. P. 983-988.

[24] *Kerans R.J., Hay R.S., Parthasarathy T.A., Cinibulk M.K.* Interface Design for Oxidation-Resistant Ceramic Composites // J. Am. Ceram. Soc. 2002. V. 85. No 11. P. 2599-2632.

[25] *Sezer A.O., Brand J.I.* Chemical Vapor Deposition of Boron Carbide // Mater. Sci. Eng. B. 2001. V. 70. No 11. P. 191-202

[26] *Tomas S.M, Lackey W.J, More K.L.* Chemical Vapor Deposition of $B_{13}C_2$ from BCl_3-CH_4-H_2-Argon Mixtures // J. Am. Ceram. Soc. 1998. V. 81. No 11. P. 3077-3086.

[27] *Oliveira J.C., Conde O.* Deposition of Boron Carbide by Laser CVD: a Comparison with

Thermodynamic Predictions // Thin Solid Films. 1997. V. 307. No 1. P. 29-37.

[28] *Jansson U., Carlsson J.O., Stridh B., Söderberg S. and Olsson M.* Chemical Vapour Deposition of Boron Carbides I: Phase and Chemical Composition // Thin Solid Films. 1989. V. 172. No 1. P. 81-93.

[29] *Jansson U and Carlsson J.O.* Chemical Vapour Deposition of Boron Carbides in the Temperature Range 1300–1500K and at a Reduced Pressure // Thin Solid Films. 1985. V. 124. No 2. P. 101-107.

[30] *Conde O, Slivestre A.J., Oliveria J.C.* Influence of Carbon Content on the Crystallographic Structure of Boron Carbide Films // Surf. Coat. Technol. 2000. V. 125. No 1. P. 141-146.

[31] *Vincent H., Vincent C. and Berthet M.P.* Boron Carbide Formation From BCl_3-CH_4-H_2 Mixtures on Carbon Substrates and in a Carbon-Fibre Reinforced Al Composite // Carbon. 1996. V. 34, No 9. P. 1041-1055.

[32] *Noyan D.S., Özbelge H.Ö., Sezgi N.A., Doğu T.* Kinetic Studies for Boron Carbide Formation in a Dual Impinging-Jet Reactor // Ind. Eng. Chem. Res. 2001. V. 40. No 3. P. 751-755

[33] http://www.uksaf.org/data/sfactors.html.

[34] *Tan M.L., Zhu J.Q., Han J.C., Gao W., Niu L., Lu J.* Chemical Analysis and Vibrational Properties of Boronated Tetrahedral Amorphous Carbon Films // Diamond Relat. Mater. 2007. V. 16. No 9. P. 1739–1745

KINETIC INVESTIGATION ON THE DEPOSITION OF SIC FROM METHYLTRICHLOROSILANE AND HYDROGEN

Cuiying Lu[1,2], Laifei Cheng [1], Chunnian Zhao [1], Litong Zhang [1], Fang Ye [1]
1.National Key Laboratory of Thermostructural Composite Materials, Northwestern Polytechnical University, Xi'an 710072, China
2. Department of Chemical Industry and Chemistry, Yulin College, Yulin 719000, China

ABSTRACT

Kinetics of chemical vapor deposition of SiC from MTS/H_2 was investigated by magnetic suspension balance in a wide range of processing conditions. The results show that the deposition rate exhibits three deposition behaviours depending on temperature: at lower temperatures (T<1000℃), the deposition rate increases slowly with increasing temperature. The apparent active energy is from 130 to 222kJ/mol; at mediate temperatures (from 1000 to 1100℃) , the deposition rate increases drastically and the apparent active energy is from 100 to 130kJ/mol. However, at higher temperatures (T>1300℃), it decreases quickly. The deposition rate first rises with both of pressure and flow rate and then almost keeps constant. The transformation point lies in 200sccm and from 2 to 3kPa, respectively. Residence time is seen to have a strong positive effect on the deposition rate. Longer residence time corresponding to temperature changing from 1000℃ to 900℃ causes a decrease by 24 times in the deposition rate.

KEYWORDS: Kinetic investigation, MTS, CVD, SiC

INTRODUCTION

Silicon Carbide (SiC) made by chemical vapor deposition (CVD) has good properties such as excellent hardness and chemical resistance at high temperatures. Methyltrichlorosilane (CH_3SiCl_3, MTS) has been the most widely used as source gas. The kinetics of SiC deposition through thermal decomposition of MTS in hydrogen have been studied in a variety of processing conditions. The obtained results showed that the dependence of deposition rate on the processing conditions apparently was as erratic as the appearance of the high and low reactivity value. Some results were consistent with each other but some are different and even opposite. For example, the apparent activation energy reported by researches varied between 60kJ/mol and 410kJ/mol[1-7]and the reported reaction order of MTS was one or zero[2,3,8-10]. The dependence of deposition rate on pressure was equally complex. Besmann's[1,2] results shown that at 1323K the reaction rate did not change appreciably between pressures of 3.3kPa and 1atm, whereas So and Chun[5], studying the deposition of SiC in a cold-wall system, reported a monotonic increase of reactivity as pressure increased between 200 and 500torr. Several other investigations, carried out in hot-wall reactors[3,11], showed complex dependence of reactivity on pressure with the occurrence of maximum and minimum point in the reaction rate vs. pressure diagram depending on temperature and reactive mixture composition. The effect of residence time on the deposition rate was less studied. Little attention was paid to the role of the residence time of the gases in the hot zone of the reactor (shortly called "residence time"), which is inversely proportional to the total gaseous mass flow rate (shortly: "flow rate") in a reactor with laminar flow at constant temperature and pressure. However, theories and experiments show that SiC is formed after a series of gas-phase reactions and surface reactions. So the residence time should play an important role in the deposition rate and composition of the deposit. In addition, a few of results also indicated that the deposition rate and the stoichiometry of the deposit varied significantly with the position in the reactor [12,13]. Clearly, more experimental work is needed to establish the decomposition kinetics of MTS, which are the foundation of modeling and indirect controlling of the deposition of SiC from MTS.

The objective of the paper is to investigate the deposition kinetics of SiC from MTS. It is constructed as follows: in the first section, the dependences of deposition rate on the temperature and pressure are discussed. Then in the second section, attentions are paid to the effects of residence time on the deposition rate.

EXPERIMENTAL DETAILS

MATERIALS

MTS(98%) is produced from JiangSu Meilan Chemical Industry Co. Ltd. The purity of Hydrogen is 99.99% from Messer Gas Products Co. Ltd.

APPARATUS

A schematic of the CVD equipment has been described elsewhere[14]. The reactor is coupled to a sensitive magnetic suspension microbalance (RUBOTHERM, sensitivity±3μg). The hot-wall reactor consists of a vertical alumina tube (1000 mm length, 28 mm in internal diameter) heated in its central part by a electronic resistance furnace using a 220 mm long molybdenum silicide susceptor. The temperature is controlled by a PID temperature controller (Imago 500). The isothermal zone is about 100 mm in length as shown in Fig.1. MTS are fed into the reactor by hydrogen bubbling and the flow rate of hydrogen is controlled by electronic mass flow controllers (D0719FM). The effluent gases are pumped via a cold trap for sampling. The pressure is measured downstream with a variable capacitance sensor (GSR AU606176) and controlled by a throttle valve (S.N.M6207381A, 0 to1atm).

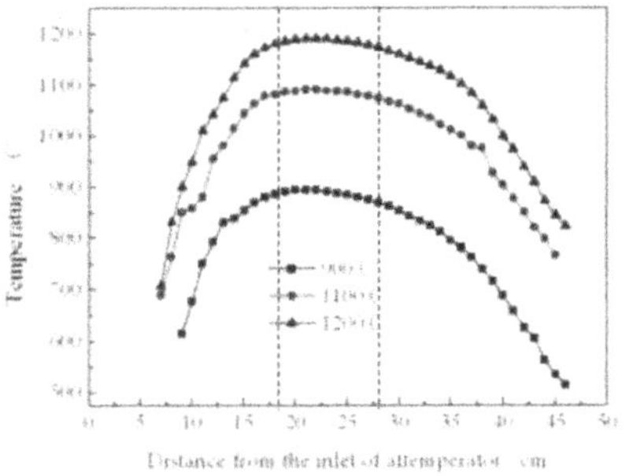

Fig.1Temperature distribution in the reactor at the H_2 flow rate of 88sccm

KINETIC EXPERIMENTS

The growth rate (*r*) has been measured as a function of temperature (from 1173 to 1400K) and residence time (from 0.2 to 5s). The measurements were performed in situ with a microbalance. The purified graphite substrates were hung in the isothermal section by a molybdenum wire, 100μm in diameter, to the microbalance beam. No radial temperature dependence has been detected. The growth rate here is the mass gain measured by means of the microbalance per unit time and substrate area in the steady state.

RESULTS AND DISCUSSION

Effect of temperature on the deposition rate

The dependence of deposition rate on temperature is shown in Fig.2 (r vs. T and ln r vs. $1/T$, with r being the deposition rate and T the absolute temperature in the reactor). The dependence revealed by the results in Fig.2 is qualitatively similar to the report by Loumagne[15]. For both cases, an increase in temperature leads to increase in deposition rate. Three regions are distinguishable on Fig.2, based on the temperature with which the deposition rate changes. In low temperature range(<1000℃), deposition rate has a slight dependence on temperature. By employing linear regression, the activation energy from 130 to 222kJ/mol is calculated, which is the characteristic of deposition process controlled by surface chemical reaction mechanism. From 1100 to 1300℃, deposition rate strongly increases as a function of temperature. The corresponding value of activation energy is from 100 to 130 J/mol. The drastic change of activation energy implies that

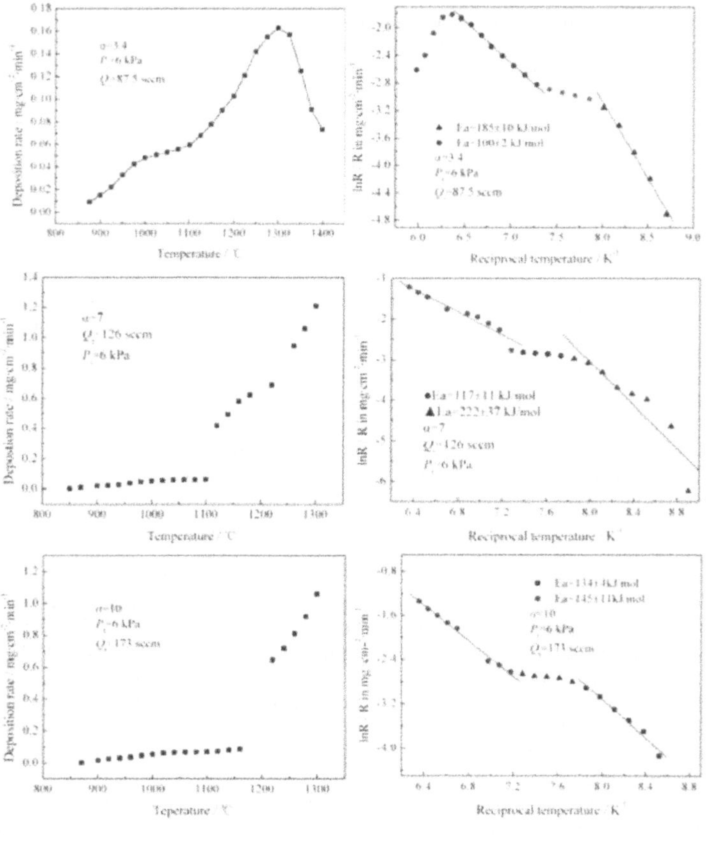

Fig.2 Dependence of deposition rate on temperature

different deposition reaction mechanism occurs corresponding to the two temperature regions. From 1000 to 1100℃, the deposition rate seemly is not largely affected by temperature. The behavior may be related to the transition of the above two deposition mechanisms. The area of the transition region is affected by pressure and the ratio of H_2 to MTS. It will become narrow or even disappear with increasing pressure and the ratio of H_2 to MTS. And the point of the deposition rate change moves to higher temperature. In high temperatures, it is clearly observed the drastic decrease in deposition rate with temperature. On the one hand, higher temperature should lead to higher deposition rate. On the other hand, higher temperature causes the higher reactant depletion on the hot-wall resulting in the decreasing of silicon and carbon containing species adjacent substrate surface, this is, causing the decreasing deposition rate. The two effects of temperature on deposition rate are off-set. So in our work, it is proposed that it is the in-positive effect that mainly leads to the decreasing of deposition rate.

Effects of flow rate and position in the reactor

The effects of substrate location in the reactor and flow rate are discussed in the same section since these two variables are the ones that have the most influence on the residence time of the reactant molecules in the hot zone of the reactor upstream of the substrate. The calculation of residence time is as follows [16]:

$$t = \frac{V_r}{Q_t}\frac{T^0}{T}\frac{P}{P^0} \qquad (1)$$

Where, V_r is the hot zone volume, Q_t is the total flow rate, T and P are temperature and pressure, respectively. At constant temperature and pressure, residence time is adjusted by changing the total flow rate at fix ratio of H_2 /MTS or the position of substrate in the reactor.

Fig.3 presents the effect of flow rate and position of substrate in the reactor on the deposition rate for different combinations of temperature and pressure. The deposition rate on flow rate increases linearly with flow rate, and then reaches a smooth, that is, a decrease in residence time has in general a positive effect on the deposition rate, which is consistent with previous works [16]. The deposition rate decreases drastically as residence time is below 1s while it has weakly dependence for over 1s. Moreover, it should be noted that the deposition rate began to decrease as the residence time is below 0.3s at 1000℃ and 6kPa.

The deposit is rarely formed directly from the gaseous reactant species which are injected into the furnace, but rather from intermediates resulting from homogeneous reactions occurring in the gas phase. Similarly, in that of SiC from MTS/H_2 mixtures, the Si-C bond in the MTS is first broken resulting in free radicals CH_3 and $SiCl_3$. Subsequent homogeneous recombination and or reacting with MTS and/or hydrogen give intermediates such as $SiCl_2$, C_2H_4, and C_2H_2. Finally, in a last step, SiC is formed on the substrate. Longer residence time of reactant in the reactor favors the formation of less active intermediate species. And the concentration of HCl tends to become higher resulting from gas phase reactions and surface reactions, which can significantly decrease the deposition rate [17]. Moreover, an increase in residence time leads to stronger reactant depletion on the reactor wall. Therefore, shorter residence time has a positive effect on deposition rate.

A change in the flow rate not only affects the residence time of the mixture in the reactor, but also the thickness of boundary layer adjacent to the substrate. Large flow rate corresponding to short residence time causes thin boundary layer resulting in high diffusion of reactant in favour to increase the deposition rate.

The position of the substrate in the CVD reactor is another parameter that has a strong effect on the residence time of the reactants. The effect of substrate location on the deposition rate can be seen

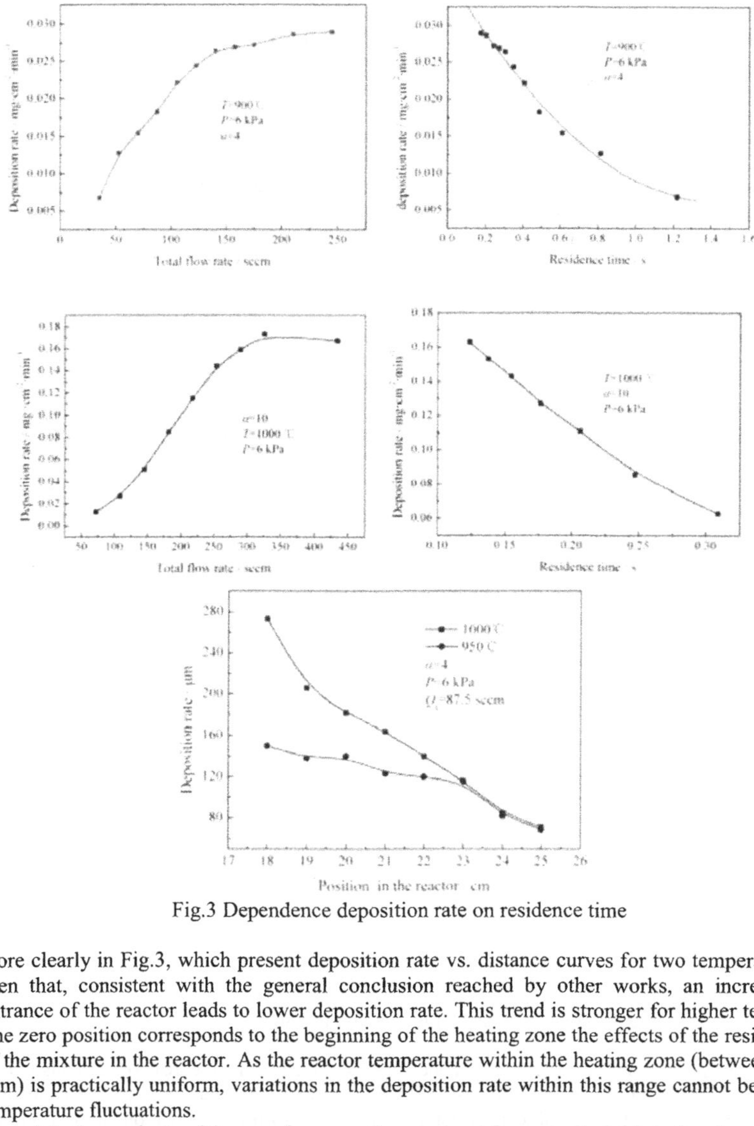

Fig.3 Dependence deposition rate on residence time

more clearly in Fig.3, which present deposition rate vs. distance curves for two temperatures. It is seen that, consistent with the general conclusion reached by other works, an increase in the entrance of the reactor leads to lower deposition rate. This trend is stronger for higher temperature. The zero position corresponds to the beginning of the heating zone the effects of the residence time of the mixture in the reactor. As the reactor temperature within the heating zone (between 0cm and 8cm) is practically uniform, variations in the deposition rate within this range cannot be caused by temperature fluctuations.

Since it is the products of the gas phase reactions and not the reagents fed into the chemical reactor that serve as actual deposition precursors, this observation suggests that the occurrence of any chemical reactions in the entry section of the reactor (before the isothermal hot zone) may have significant effects on the deposition rate profile within the isothermal zone. Silicon carbide films with MTS precursor are deposited through gas phase reactions, mass transfer and surface reactions. The deposition rate at a certain location in the reactor depends on temperature, the kinetic parameters of the heterogeneous reactions, and the concentrations of gas phase species in the vicinity of the surface. The last are in turn influenced by the concentrations in the bulk of the gas phase and the mass transfer resistance from the bulk to the gas phase solid interface. The bulk phase concentrations of the gases are different from those at the entrance of the reactor not only because of the occurrence of gas phase decomposition reactions, but also because of material depletion as a result of solid deposition on the wall. The above discussion points to the conclusion that because of too lower concentration of reactant along the reactor axial, lower deposition rate exhibits.

Effect of pressure on the deposition rate

The effect of total pressure on the deposition rate present in Fig.4 is similar to that of flow rate. An increase in the pressure leads to an increase in the deposition rate. Further increase in the pressure does not affect the rate which reaches a plane. Among different studies on the variation of deposition rate with pressure, the results varied both qualitatively and quantitatively as described in introduction. Besmann[17] explained the pressure dependence by involving two different deposition mechanisms. One mechanism was the direct decomposition of MTS to SiC via Rea.(2) as follows.

$$CH_3SiCl_3 \rightarrow SiC + 3HCl \qquad (2)$$

However, MTS is readily decomposed at elevated temperatures to separate silicon- and carbon-containing gas phase species. A process for which Burgess and Lewis [18] have determined an Arrhenius relationship with first order in MTS concentration, they noted that, in the presence of hydrogen, the dissociation products were likely to be $SiCl_4$ and CH_4, although other species, such as C_2H_2, CH_3, $SiCl_2$ and $SiCl_3$, might be important. Thus, the second SiC deposition mechanism might involve:

$$CH_4 + SiCl_4 \rightarrow SiC + 4HCl \qquad (3)$$

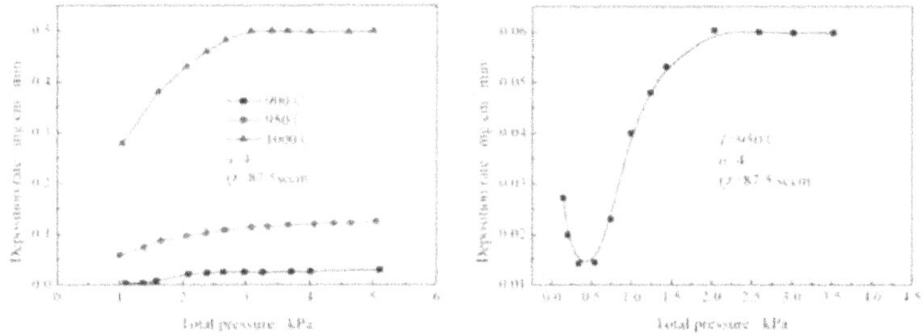

Fig.4 Dependence deposition rate on pressure

A change in pressure not only affects the concentration of MTS in the reactor, but also the residence time of MTS. The positive effect of pressure on deposition rate may be the result of higher concentrations of the actual deposition precursor at substrate deposition sites. Longer residence time leads to the low deposition rate mainly due to the reactant depletion on the hot wall. Both of these occurrences may offset each other, and cause the appearance of a maximum.

At mediate pressures, the SiC deposition proceeds through the first order in MTS concentration. So deposition rate increases linearly as a function of pressure. At lower pressures, it is possible that shorter residence time overwhelms the decrease in overall concentration due to decreasing pressure, resulting in the observed increase in deposition rate.

CONCLUSIONS

The effects of reaction conditions and residence time on the deposition of silicon carbide from mixtures of MTS and hydrogen were investigated using a tubular hot-wall reactor coupled by a magnetic suspension microbalance. The experimental results showed that an increase in temperature had a positive effect on the deposition rate. But the deposition mechanism changed with temperature. The variation of the deposition rates showed a maximum and a minimum with total pressure, which could be explained by two different reaction mechanisms of MTS and the competition between residence time and concentrations of the reactant. Both of the distance into the CVD reactor and the flow rate strongly influenced the reactivity. Longer residence time and farther distance of the substrate in reactor leaded to decreasing deposition rate, mainly duo to the reactant depletion on the wall.

ACKNOWLEDGEMENTS

The authors wish to acknowledge the financial supports of the National Natural Science Funds of China (50820145202).

REFERENCES

[1]T. M. Besmann, B. W. Sheldon, M. D. Kaster, Surface & Coatings Technology, 43:167-175(1990).

[2] T. M. Besmann, M. L. Johnson, Proceedings of 3th International Symposium on Ceramic Material and Components for Engineering, 443-456(1988).

[3] F. Langlais, C. Prebende, B. Tarride, On the Kinetics of the CVD of Si from Si_2Cl_2/H_2 and SiC from CH_3SiCl_3 in a Vertical Tubular Hot-Wall Reactor, J.de Physique, **50**(C5): 93-103(1990).

[4] L. M. Ivanova and A. A. Pletyushkin, Inorganic Materials, **3**:1585-1589(1967).

[5] M. G. So and J. S. Chun, J.Vac.Sci.Technol.A, **6**:5-8(1988).

[6] A.W.C. van Kemenade and C.F.Stemfoot, J.Cryst.Growth,**12**:13-16(1972).

[7] B. J. Choi and D. R. Kim , Growth of Silicon Carbide by Chemical Vapor Deposition. J. Mater. Sci. Lett., **10**: 860-862(1991).

[8] D. Neuschutz, F. Salehomoum, Kinetics of Chemical Vapor Deposition of Sic Between 750 and 850°C at 1Bar Total Pressure. Mat. Res. Soc. Symp. Proc., **250**: 41-46(1992).

[9] D. V. Fedoseev, V. P. Dorokhovich, A. V. Lavrent'ev, et al., Kinetics of Silicon Carbide Crystal Growth, Izv.Akad. Nauk SSSR, Neorgan. Mat.,**12**(10): 1796-1799(1976).

[10] K. Brennfleck, E. Fitzer, G.Schoch et al., in Proc. Ninth Int. Conf. Chem. Vap. Deposition, 649-662(1984).

[11] T. M. Besmann, M. L. Johnson, Third International Symposium of Ceramic & compounents for Engines,443-456(1988).

[12] F. Loumagne, F. Langlais, R. Naslain, Experimental Kinetic Study of The Chemical Vapor Deposition of SiC-Based Ceramics from CH_3SiCl_3/H_2 Gas Precursor. Journal of Crystal Growth, **155(3-4)**:198-204(1995).

[13] A. Josiek, F. Langlais, Residence-Time Dependent Kinetics of CVD Growth of SiC in the MTS/H_2 System, Journal of Crystal Growth, **160(3-4)**:253-260(1996).

[14] Lu Cuiying, Cheng Laifei, Zhang Litong, Xu Yongdong, Zhao Chunnian; In-situ kinetic investigation on carbon deposition from propylene, Chinese Journal of material science and engineering, 26(6):843-846(2008)

[15] F. Loumagne, F. Langlais, R. Naslain, Experimental Kinetic Study of The Chemical Vapor Deposition of SiC-Based Ceramics from CH_3SiCl_3/H_2 Gas Precursor. Journal of Crystal Growth, **155(3-4):** 198-204(1995).
[16] O. Ferona, F. Langlais, R. Naslain, J. Thebaultb. On Kinetic and Microstructural Transitions in the CVD of Pyrocarbon from Propane, Carbon, **37(9):**1343–1353 (1999).
[17] T. M. Besmann, B. W. Sheldon, T. M. MossIII and M. D. Kaster, J.Am. Ceram. Soc., 75:2899-2903(1992)
[18] J. N. Burgess, T. J. Lewis, Kinetics of the reduction of methyltrichlorosilane by hydrogen. Chem. Ind. (London), 9:76-77(1974)

II. Nanomaterials and Nanotechnologies

SYNTHESIS OF HEMATITE-ZIRCON-SILICA NANO COMPOSITE AS A NON TOXIC CERAMIC PIGMENT BY SOL-GEL METHOD

Maryam Hosseini Zori[1]*

1- Assistant Professor of the Department of Inorganic Pigments and Glazes, Institute for Colorants, Paint and Coatings, P. O. Box 16765-654, Tehran, Iran; mhosseini@ icrc.ac.ir

ABSTRACT

Hematite as a natural and non toxic red ceramic pigment has been known since prehistoric times but the colour shade of hematite becomes unstable with temperature and needs to be protected with a suitable matrix. The best red shades are obtained by the inclusion of hematite in silica casings. Zircon has the best thermal and chemical stability but it is more rare and expensive than silica, so in this research a fraction of zircon is substituted with silica crystals. A Sol-Gel method has been applied in order to improve the inclusion efficiency of hematite into silica and zircon crystals; Iron sulfate was used as Fe precursor and matrix agents were zirconium chloride and colloidal silica. Continuous changes in color were measured by comparing L*-a*-b* values of the heated samples. TEM analysis on calcined powders shows hematite single crystals with spherical morphology and diameter of 5-10nm that were occluded with silica-zircon crystals successfully. Due to its chemical and thermal stability, the pigment of hematite-silica-zircon system may be considered as a suitable red pigment for ceramic manufacturing by fast firing cycles.

Keywords: Ceramic, Inclusion pigment, Non toxic red pigment, Hematite-Silica-zircon, Nano Composite

Contacting Author: Prof. Maryam Hosseini Zori

Department of Inorganic Pigments and Glazes, Institute for Colorants, Paint and Coatings, P. O. Box 16765-654, Tehran, Iran;

TEL: +98-9125784171

FAX: +98-2122947537

E-mail address: mhosseini@ icrc.ac.ir

Introduction

In ceramic applications including glazes, ceramic bodies and porcelain enamels, pigments are dispersed in the media and most do not dissolve. In conclusion, powders used for coloring ceramics must show thermal and chemical stability at high temperature and must be inert to the action of molten glass (frits or sintering aids).[1] These characteristics limit ceramic pigments to a very small number of refractory systems which are fully reacted and relatively inert to the matrix in which they are dispersed.[2-3] This need for great chemical and thermal stability has dominated research and development in recent years especially towards new red or pink pigments. In particular the interest is directed to the development of inclusion pigments which make utilizable colouring substances suffering the industrial thermal and chemical conditions by occluding them in a stable glassy or crystalline matrix (heteromorphic pigments). The inclusion or encapsulation of a reactive, colored or toxic crystal into a stable crystalline matrix, gives a protection effect to the crystal guest by the host crystal. The guest crystals are inactivated into the matrix.

Silica may be considered to have a relatively low price giving it a potential to be used in occluded pigments as a matrix, due to its thermal and chemical stability towards glassy phases.

The aim of this work was to study the optimization of synthesizing red inorganic pigments for ceramic applications. In order to improve the inclusion efficiency of hematite into silica and zircon matrixes, the aqua sol-gel route has been applied as chemical processes which improve microstructural characteristics and control particles morphology [4-5].

Experimental

Samples of the $SiO_2 - Fe_2O_3 - ZrSiO_4$ were prepared using the Sol-Gel method. A concentrated aqua solution was prepared by adding iron sulphate ($FeSO_4.7H_2O$, Merck) in the deionized water, refluxing at 70°C for 30 minute. Then, the required colloidal silica and zirconium chloride (Merck) was added to the aqueous solutions by Drops of concentrated solution. The system was continuously stirred and kept at 70°C until the pH stabilized equal to 5. The resulting light yellow gel was dried at 110°C and then fired. In order to determine the effects of firing temperature, the powders were fired at temperatures ranging from 900 to 1100°C in an electrical furnace with a soaking time of 3 h. The fired samples were micronised, wet milled in water and finally dried at 110°C.

To identify the crystalline phases that were present in the raw and fired samples, X-ray diffraction patterns were collected using a conventional powder technique in a Siemens Diffractometer (D500 mod) employing Cu Ka Ni-filtered radiation. To define the color developed about the samples, a UV-Vis spectrophotometer with analytical software for color measurements (PERKIN ELMER Spectrometer Lambda 19, UV/VIS/NIR, Standard Observer: 10°) has been used. L*,a*,b* color parameters have been measured following the CIE (Commission International de l'Eclairage) colorimetric method. In this method, L* is the lightness axis (black 0)→white (100)), a* is the green (−) →red (+) axis, and b* is the blue (−) →yellow (+) axis.

Powders microstructure characterization and morphology of the occluded hematite has been studied by transmission electron microscopy (Jeol JEM 2010).

Results and Discussion

Morphology of the hematite particles can be detected just by TEM analysis because they are very fine and occluded by the matrix. The spherical nano hematite crystals have been successfully occluded in silica and zircon particles after firing. Figure 1 report the TEM images of a not calcined sample. In this case, the sample is constituted of very fine and spherical particles. Location of Iron chloride crystals were not detected. It seems that they are occluded by the amorphous phase because the X ray pattern of the dried powder in Figure 1c did not show any light dot related to a crystalline plate. Figure 2 is related to the STA analysis and shows all of the reactions have been take placed before 800°C and total decrease of weight percent is 38.4%. The main reaction in mentioned aqua Sol-Gel method was:

$$2FeSO_4 + ZrCl_4 \rightarrow Zr(SO_4)_2 + 2FeCl_2$$

Due to heat treatment obtained zirconium sulfate and iron chloride have been decomposed and oxidized to very fine particles of zirconium oxide and hematite respectively. Therefore the real precursor of red colored agent of this synthesized pigment (hematite) is Iron chloride that it will affect on hematite morphology [5]. It seems that very fine and aggressive zirconium oxide particle in situ react with nano silica particle while the crystal growth and diffusion have been taking place. High surface area energy of the fine particles has been caused to reducing of reaction temperatures.

CIELab values of dried gel and fired powder samples are reported in Table 1. Base on the D65 standard of colorimeter results in Table 1, the red factor equals to 19.069 and it is very near to yellow factor (19.08). These data report that the obtained pigment after calcinations has red brown shade.

According to XRD results, it can be seen in Figure 3; interested three phases of hematite, cristobalite and zircon have been crystallized after calcination in 1000°C in the samples and before of this temperature just hematite can be detected (not shown).

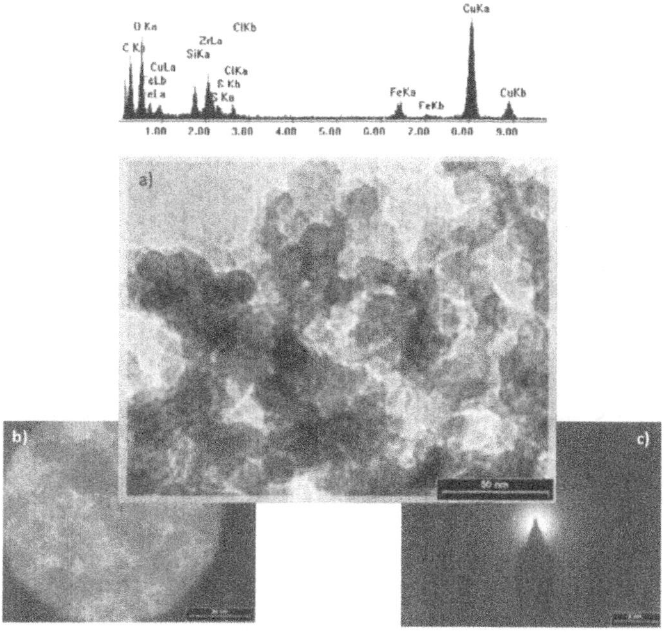

Figure 1: TEM Micrographs of the dried powders and its EDX a) image, b) selected area diffraction zone and c) X ray pattern of it that indicates amorphous powder

Table 1: CIELab values of powder samples a) dried gel and b) calcination temperature was 1000°C

	Standard	L*	a*	b*	C*	h°
a	D65	82.304	-2.448	60.57	60.622	92.314
	A	84.925	6.291	59.44	59.772	83.958
	CWF_2	84.523	-2.472	68.33	68.379	92.072
b	D65	57.797	19.069	19.08	26.974	45.014
	A	61.133	22.142	24.22	32.819	47.572
	CWF_2	59.22	14.084	21.97	26.1	57.342

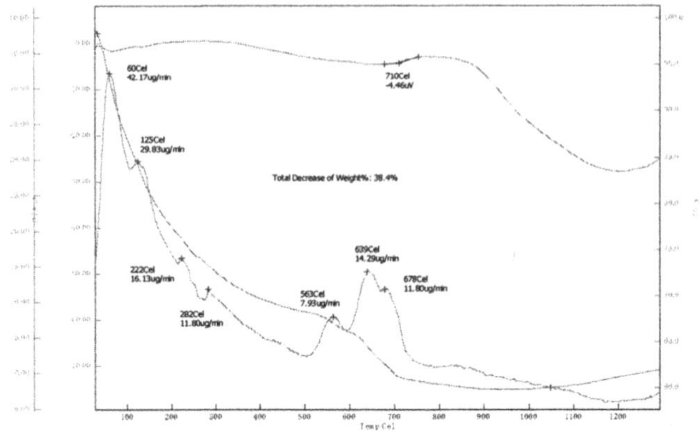

Figure 2: Simultaneous thermal analysis (TG and DTA) of the dried powder

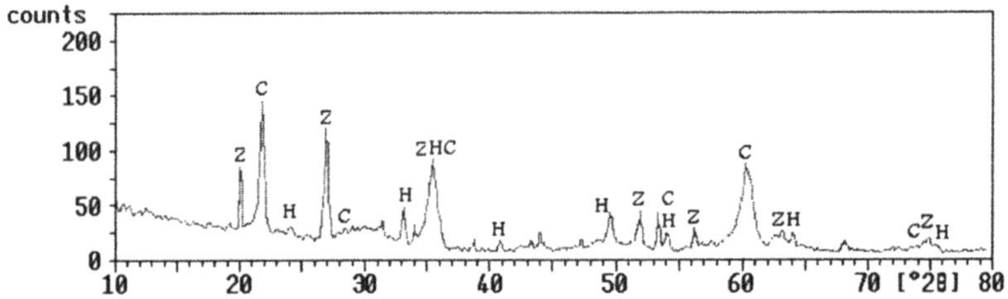

Figure 3: XRD patterns of a) dried powder b) fired sample after calcination in 1000°C,
H: Hematite, C: Cristobalite, Z: Zircon

Figure 4 and 5 are TEM Micrographs of the fired powders at 1100°C/3h and at 1000°C/3h respectively. Those present the size, morphology and location of the hematite particles that have been occluded in the silica-zircon matrix. Even, it can be seen the planes of a hematite crystal that are regular like single crystals. TEM Micrographs shows black spherical circles with 5-10 nm diameters in the sintered and uniformed of grey matrix. EDX analysis of the black spherical circles has detected elements of iron, zirconium and silicate, therefore figure 5a) indicate that black spherical circles are hematite particles with spherical shapes and presence of Zr and Si is due to encapsulation of hematite by the matrix, as can be seen similar to egg (red sign). Figure 5b) is related to EDX analysis of grey background of the same powder (far from black spherical circles) and shows the matrix contain just Zr and Si without any dissolved Fe ions. Increasing of sintering temperatures at 1100°C/3h did not show any effect on hematite morphology but it might be important about red shade because of oxygen reduction [3].

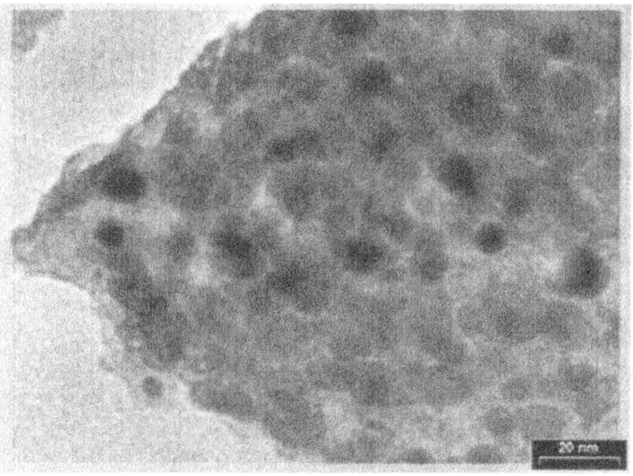

Figure 4: TEM Micrographs of the fired powders at 1100°C/3h

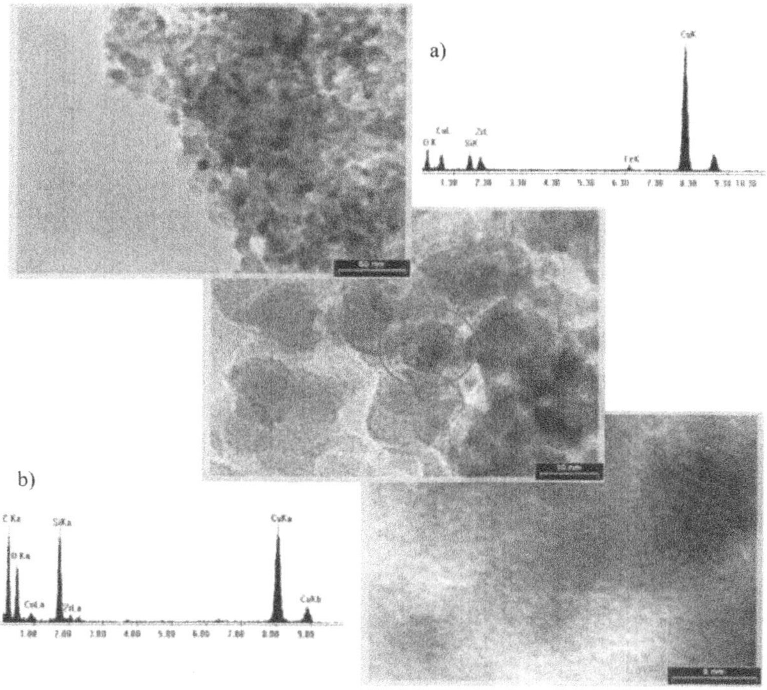

Figure 5: TEM Micrographs of the fired powder at 1000°C/3h with different magnifications and EDX analysis of the a) black spherical circles and b) grey background of this powder

Conclusion

In order to prepare a hematite–cristobalite-zircon inclusion red ceramic pigment, Sol-Gel process with the colloidal silica, Zirconium cloride and iron (II) sulfate were been synthesized.

Nano-sized and homogeneous hematite particles were obtained into sintered cristobalite-zircon matrix after heat treatment at 1000°C/3h and 1100°C/3h. Occluded hematite particles have been spherical shapes with 5-10 nm diameters. Changes of sintering temperatures did not show any effect on hematite morphology but it was important about red shade, inclusion efficiency and thermal-chemical stability of the pigments.

Due to its high inclusion efficiency, this heteromorphic pigment may be considered to be a suitable red pigment for ceramic applications.

References

[1] Bondioli F, Ferrari AM, Leonelli C, Manfredini T., Mater Res Bull,33(5), 723-9,1998.

[2] Bondioli F, Manfredini T, Siligardi C, Ferrari AM., J Am Ceram Soc, 88(4), 1070-1, 2005.

[3] Hosseini Zori M, Taheri E, and Mirhabibi AR., Ceram Int, 34, 491–496, 2008.

[4] Bondioli F, Manfredini T., Am Ceram Soc Bull, 79(2), 68-70, 2000.

[5] Hosseini Zori M., F. Bondioli, T. Manfredini, E. Taheri-Nassaj, Dyes and Pigments, 77, 53-58, 2008.

FORMATION OF NANOCRYSTALLINE α-ALUMINAS IN DIFFERENT MORPHOLOGY FROM GEL POWDER AND BOEHMITE POWDER: A COMPARATIVE STUDY

Xiaoxue Zhang [a*], Yanling Ge [b], Simo-Pekka Hannula [b], Erkki Levänen [a], Tapio Mäntylä [a]
[a] Department of Materials Science, Tampere University of Technology, P.O. Box 589, FI-33101 Tampere, Finland
[b] Department of Materials Science and Engineering, Helsinki University of Technology, P.O. Box 6200, FI-02015 TKK, Finland

ABSTRACT

Nanocrystalline α-alumina powders in different morphology obtained from gel powder and boehmite (AlOOH) powder are compared in this study. The boehmite powder in flaky morphology was prepared by reacting the gel powder with water. Spherical α-alumina was obtained by calcining the gel powder at 1000 °C for 6 h, while rod-shaped α-alumina was synthesized by calcining the boehmite powder at 1000 °C for 40 h. Phase transformation and morphology are characterized by X-ray diffraction (XRD), thermo-gravimetric method (TG-DSC) and transmission electron microscopy (TEM). The formation scheme of the spherical and rod-shaped morphology is comparatively illustrated and furthermore the phase transformation to α-alumina is discussed.

INTRODUCTION

α-Alumina, as one of the most important ceramic materials, has been widely used in many applications as a structural material and a functional material based on its mechanical, electrical and optical properties. [1-2] Recently more attention has been drawn to synthesize nanocrystalline α-alumina to further improve its mechanical properties. Meanwhile, it is believed that the morphology of alumina particles can also affect the mechanical properties, [3] for example nanosized ball-shaped and plate-shaped α-aluminas have improved fracture strength and toughness. [4] Therefore, nanocrystalline α-alumina in different morphology is required for different advanced engineering and structural applications. The ball-shaped and plate-shaped morphology of α-alumina have been commonly reported. Also the influence of precursors and additives such as fluorides to the morphology of α-alumina has been widely studied. [5]

To synthesize α-alumina, boehmite powder and gel powder are two common starting materials. In calcining, boehmite powder and gel powder pass different transformation series and finally form α-alumina at about 1200 °C. [6] Many efforts have been done to lower the phase transformation temperature of α-alumina and addition of seeding such as α-Fe_2O_3, [7-8] AlF_3 [5] and alumina sol [9] into the starting material is a typical way to provide heterogeneous nucleation sites for α-alumina to lower the phase formation temperature.

We have reported a novel formation of nanosized boehmite powder in flaky morphology by a modified sol-gel route. [10] The sol-gel method is a well-known chemical synthesis route with high purity, high chemical homogeneity, lower calcination temperatures and good control of particle size. [11] It is a versatile method not only to synthesize nanocrystalline powder but also nanostructured functional thin films on various substrates. [10, 12-14] By calcining the as-synthesized nanosized boehmite powder, transition alumina nanorods and nanoparticles were obtained. [10] Meanwhile nanocrystalline α-alumina with novel morphology was obtained by calcining such boehmite powder at a relatively low temperature of 1000 °C without any seeding material. [15] The α-alumina nanocrystallites in size of 5 nm were observed for the first time, and more interestingly the 5

* Corresponding author: Xiaoxue Zhang, Tel: +358 40 849 0197, Fax: +358 3 3115 2330, Email: xiaoxue.zhang@tut.fi

nm-nanocrystallites organize themselves into nanorods with the widths of about 15 nm and the lengths of about 50-250 nm. The formation of the novel morphology and the detailed process parameters have been reported and discussed elsewhere. [16] In this paper, the goal is to compare the nanocrystalline α-alumina in spherical morphology obtained from the gel powder to that in rod-shaped morphology prepared from the boehmite powder. Though the boehmite powder was obtained from the gel powder, the boehmite powder and gel powder resulted in nanocrystalline α-alumina in different morphology. The formation scheme of the spherical and rod-shaped morphology is comparatively illustrated and furthermore the phase transformation to α-alumina is discussed.

EXPERIMENTAL

The raw materials are aluminium tri-sec-butoxide (denoted as $Al(O\text{-}sec\text{-}Bu)_3$, $C_{12}H_{27}AlO_3 > 97$ %, VWR), isopropyl alcohol (denoted as i-PrOH, $C_3H_7OH > 99$ %, VWR) and ethyl acetoacetate (denoted as EAcAc, $C_6H_{10}O_3 > 98$ %, VWR). The flow chart in figure 1 describes the experimental procedures in this study. Firstly, 3 g $Al(O\text{-}sec\text{-}Bu)_3$ and 30 ml i-PrOH were mixed and stirred at room temperature for 1 h. Then 2 ml EAcAc was introduced, and the solution was stirred for 1 h. Finally the mixture of distilled water and i-PrOH was added for mild hydrolysis. After stirring for 2 h, the precursor was dried at 120 °C into fine powder, which is named as gel powder. The gel powder was then reacted with boiling water for 10 min to form boehmite (AlOOH). The boehmite-containing suspension was then dried at 120 °C to obtain fine boehmite powder. The gel powder and the boehmite powder were two different starting materials of the following calcinations in a tube furnace at 1000 °C for varying hours in air atmosphere with a heating rate of 13 °C/min and a cooling in furnace.

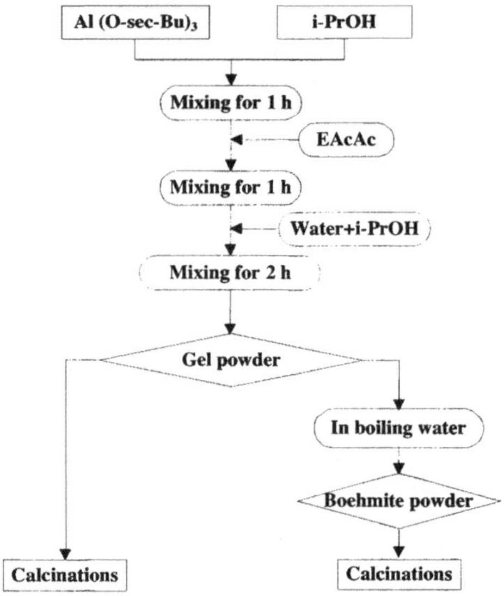

Figure 1. Flow chart of the experimental procedures.

Crystalline phase was determined by X-ray diffraction (XRD, Kristalloflex D-500, Simens) using Cu Kα radiation from two-theta 20° to 80° with a step size of 0.01° and a count time of 1 s per step. Phase transformation was studied by thermo-gravimetric testing (TG-DSC, STA 409, Netzsch) from 30 to 1200 °C with a heating rate of 10 °C /min. The sample weight was 20.1 mg. Morphology was studied by a transmission electron microscope (TEM, Jeol JEM 2010) and a high resolution transmission electron microscope (HRTEM, Tecnai F20 S-Twin). The specific surface area was calculated by the Brunauer-Emmett-Teller (BET) equation using the data in a P/P_0 range of 0.05-0.4 from the N_2 adsorption/desorption isotherm obtained by a gas sorption experiment (Coulter Omnisorp 100 CX, Beckman Coulter Inc.).

RESULTS AND DISCUSSION

The XRD patterns of the samples obtained by calcining the gel powder and the boehmite powder at 1000 °C for different hours are shown in figure 2. The gel powder is possibly amorphous indicated by its XRD pattern. The solid squares signify the peaks of the α-alumina according to JCPDS data (card No. 10-173). When the gel powder was calcined for 0.5 h, α-alumina coexisted with θ-alumina. When the dwell time was increased to 6 h, θ-alumina was transformed to α-alumina completely. On the other hand, when the gel powder was reacted with water, boehmite was formed as revealed by the XRD pattern indexed according to the JCPDS card 74-1895. When such boehmite powder was calcined at 1000 °C, no peaks of α-alumina were observed for 1 h. α-Alumina started to form after 5 h and pure α-alumina was obtained in 40 h. That is, the boehmite powder needs much longer dwell time than the gel powder to transform completely to α-alumina at 1000 °C.

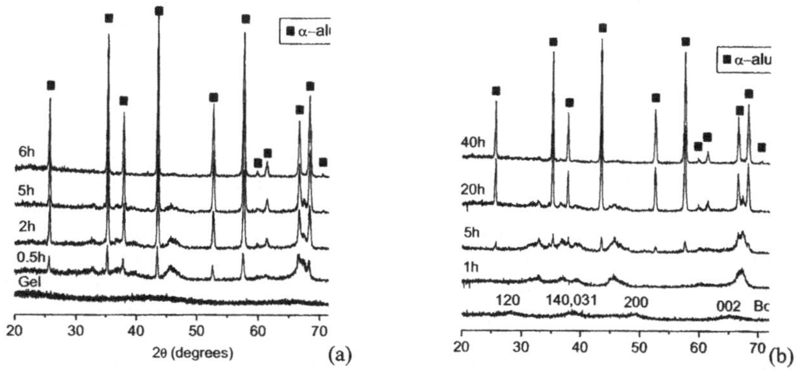

Figure 2. XRD patterns of the samples obtained by calcining the gel powder at 1000 °C from 0.5 h to 6 h (a) and of the samples prepared by calcining the boehmite powder at 1000 °C from 1 h to 40 h (b). The XRD patterns of the gel powder and the boehmite powder are also included.

The phase transformation series of the gel powder and the boehmite powder under calcining were revealed by TG-DSC curves in figure 3. The gel powder undergoes three stages of decomposition. The first stage under 140 °C is the liberation of the adsorbed water from the gel pores. The second stage at 140-260 °C corresponds to the removal of the structural water in the gel network and organic residues, while the third stage at 260-500 °C is due to dehydroxylation. [17] The total weight loss is about 60%. As known, gel powder transforms to η-alumina, then θ-alumina, and finally α-alumina. [6] Therefore the exothermic peak around 421 °C is due to the formation of η-alumina, the peak around 816 °C is due to the formation of θ-alumina, and finally the peak around 1100 °C is due to the formation of α-alumina.

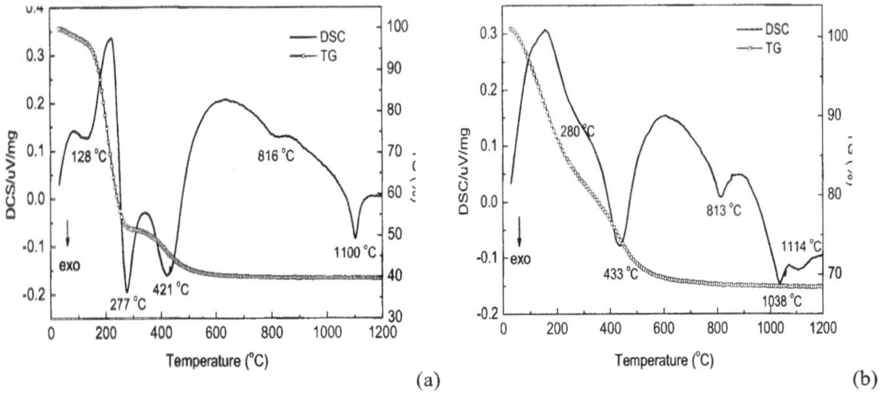

Figure 3. TG-DSC curves of the gel powder (a) and of the boehmite powder (b).

The boehmite powder also undergoes three stages of decomposition. The total weight loss is about 32%. The boehmite powder has different phase transformation series as compared to that of the gel powder. Boehmite forms first γ-alumina, then δ-alumina, θ-alumina and finally α-alumina. [6] Therefore the exothermic peak around 433 °C is due to the dehydroxylation of boehmite into γ-alumina. The peak around 813 °C is from the formation of δ-alumina and the peak around 1038 °C due to the formation of θ-alumina, and finally the peak around 1114 °C from the formation of α-alumina. Although the formation temperature of α-alumina is around 1100 °C as shown in the TG-DSC study, with longer calcining times, pure α-alumina can be obtained already at 1000 °C.

When the gel powder was the starting material, the obtained alumina has spherical morphology, which is shown in figure 4. At 1000 °C for 0.5 h, α-alumina coexisted with θ-alumina, with a specific surface area of 57 m^2/g. Strong agglomeration can be seen from figure 4 (a). However, crystallites in size of 5 nm can be clearly observed. When the gel powder was calcined at 1000 °C for 6 h, θ-alumina transformed completely to α-alumina with a specific surface area of 5.8 m^2/g. The loss in the specific surface area is due to the increased agglomeration/aggregation as well as the transformation from θ-alumina to α-alumina and its grain growth. Particles of α-alumina in size of about 100 nm are shown in figure 4 (c). However, agglomerated small crystallites of α-alumina can be observed, see figure 4 (d). On the other hand, when the boehmite powder was the starting material, the formed alumina has rod-shaped morphology, shown in figure 5. When the boehmite powder was calcined at 1000 °C for 5 h, δ-alumina and θ-alumina coexisted with α-alumina, with a specific surface area of 66 m^2/g. The corresponding TEM image is given in figure 5 (a), where the nanorods are in widths of about 10 nm and lengths of about 50 nm. When the boehmite powder was calcined at 1000 °C for 40 h, pure α-alumina was formed with a specific surface area of 8.2 m^2/g. The nanorods grow wider and longer as shown in figure 5 (b). Moreover, small crystallites in size of about 5 nm can be seen on the surface of the nanorods, see figure 5 (c).

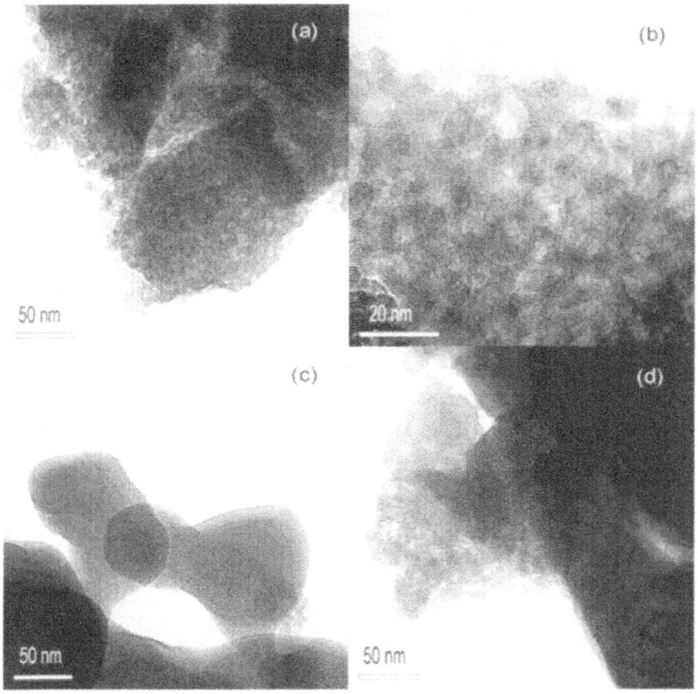

Figure 4. TEM images of the samples obtained by calcining the gel powder for 0.5 h (a-b) and 6 h (c-d) at 1000 °C.

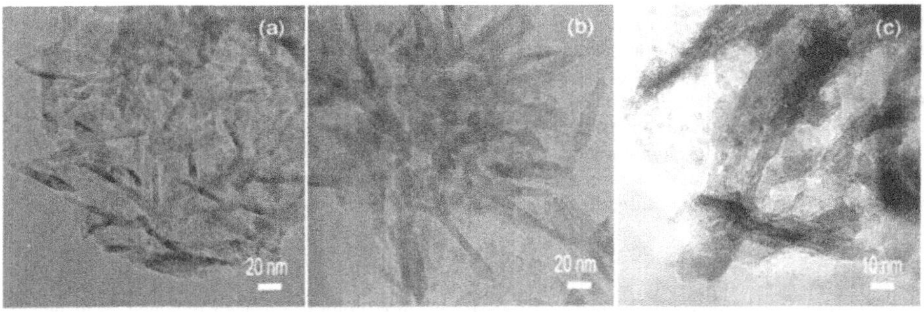

Figure 5. TEM images of the samples prepared by calcining the boehmite powder for 5 h (a) and 40 h (b-c) at 1000 °C.

The schematic illustration of the formation of nanocrystalline α-alumina in different morphology is shown in figure 6. The morphology of the starting material plays a very important role in the morphology of the resulting nanocrystalline α-alumina. The gel powder obtained by drying the sol-gel precursor has a specific surface area of 576 m^2/g. The large surface area suggests that the gel powder consists of tiny particles. When the gel powder was calcined, η-alumina crystallites formed, and then transformed into θ-alumina which agglomerated into larger particles. Finally α-alumina crystallites were formed at 1000 °C for 6 h, which aggregated into large particles, as shown in figure 4 (c). On the other hand, when the gel powder reacted with water, boehmite flakes were formed with a specific surface area of 303 m^2/g. The formation of the flakes results from the preferential growth due to the presence of weak hydrogen bonds and interaction between the solvent molecules and the surface OH^- groups via hydrogen bonds. [18] By calcining the boehmite powder, α-alumina nanorods were formed at 1000 °C for 40 h. The morphology of the starting boehmite powder plays an important role. The stacking and orientation of the original boehmite flake layers are preserved, which is responsible for the final morphology of the α-alumina nanorods. A more detailed study of such nanocrystalline α-alumina has been reported elsewhere. [15-16]

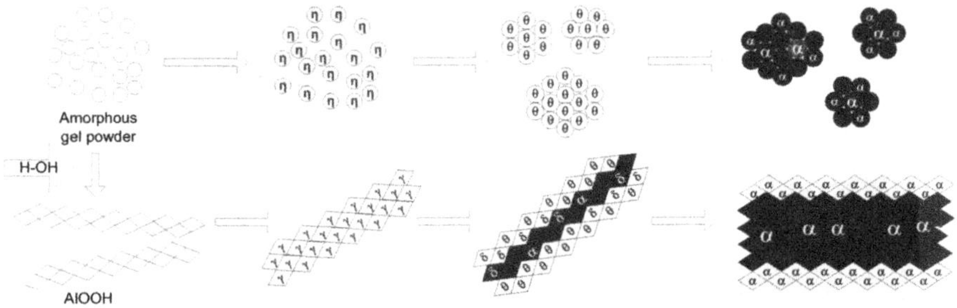

Figure 6. Schematic illustration of the formation of the nanocrystalline α-alumina in different morphology from the gel powder and the boehmite powder.

Meanwhile this study reveals that a much shorter dwell time was needed to form α-alumina from the gel powder than the boehmite powder at 1000 °C. This is possibly due to the difference in the crystallinity and the morphology of the gel powder and the boehmite powder. The gel powder is amorphous, while the boehmite is nanocrystalline with several broad peaks in the XRD pattern. Amorphous state is less stable than the crystalline phase, therefore the activation energy required for the transformation to α-alumina will be lower for the amorphous gel powder. [9] Meanwhile, the flaky morphology of the boehmite makes less available contact points which might help to retard the possible diffusion controlled reaction leading to grain growth followed by phase transformation.[7]

CONCLUSIONS

Nanocrystalline α-alumina in spherical and rod-shaped morphology has been obtained from gel powder and boehmite powder at 1000 °C without any seeding material. The morphology of the gel powder and the boehmite powder plays very important role in the morphology of the resulting nanocrystalline α-alumina. The large surface area of the gel powder suggests that the gel powder is formed of tiny particles which results in the α-alumina with spherical morphology. However, the boehmite powder has flaky morphology. The stacking and orientation of the original boehmite flake layers are preserved during the calcination to form the α-alumina in rod-shaped morphology. Meanwhile the boehmite powder needs longer dwell time than the gel powder to transform to α-alumina completely, which results possibly from the flaky morphology and the better crystallinity of the boehmite powder than the gel powder.

ACKNOWLEDGMENTS

The present study was supported by Finnish National Graduate School on New Materials and Processes. The authors also thank Elina Huttunen-Saarivirta for the TG-DSC measurements and Mari Honkanen for the TEM testing.

REFERENCES

[1]J. Li, Y. Wu, Y. Pan, W. Liu, Y. Zhu and J. Guo, Agglomeration of α-Al_2O_3 Powders Prepared by Gel Combustion, *Ceram. Int.* **34**, 1539-42 (2008).

[2]Y. Wu, Y. Zhang, G. Pezzotti and J. Guo, Influence of AlF_3 and ZnF_2 on the Phase Transformation of Gamma to Alpha Alumina, *Mater. Lett.* **52**, 366-69 (2002).

[3]H. Lu, H. Sun, A. Mao, H. Yang, H. Wang and X. Hu, Preparation of Plate-like Nano α-Al_2O_3 Using Nano-Aluminum Seeds by Wet Chemical Methods, *Mater. Sci. Eng. A* **406**, 19-23 (2005).

[4]Y. Q. Wu, Y. F. Zhang, X. X. Huang and J. K. Guo, Preparation of Platelike Nano Alpha Alumina Particles, *Ceram. Int.* **27**, 265-68 (2001).

[5]H. J. Kim, T. G. Kim, J. J. Kim, S. S. Park, S. S. Hong and G. D. Lee, Influence of Precursor and Additive on the Morphology of Nanocrystalline α-Alumina, *J. Phys. Chem. Solids* **69**, 1521-24 (2008).

[6]L. D. Hart, Alumina Chemicals: Science and Technology Handbook, The American Ceramic Society Inc, Westerville, Ohio, 1990, p19.

[7]M. Kumagai and G. L. Messing, Enhanced Densification of Boehmite Sol-Gels by α-Alumina Seeding, *J. Am. Ceram. Soc.* **67**, c230-31(1984).

[8]J. L. McArdle and G. L. Messing, Transformation, Microstructure Development and Densification in α-Fe_2O_3-Seeded Boehmite-Derived Alumina, *J. Am. Ceram. Soc.* **76**, 214-22 (1993).

[9]H. J. Youn and K. S. Hong, Low Temperature Formation of α-Alumina by Doping of An Alumina Sol, *J. Col. Inter.Sci.*, **211**, 110-13(1999).

[10]X. Zhang, M. Honkanen, E. Levänen and T. Mäntylä, Transition Alumina Nanoparticles and Nanorods from Boehmite Nanoflakes, *J. Cryst. Growth* **310**, 3674-79 (2008).

[11]H. K. Farag and F. Endres, Studies on the Synthesis of Nano-Alumina in Air and Water Stable Ionic Liquids, *J. Mater. Chem.* **18**, 442-49 (2008).

[12]X. Zhang, M. Honkanen, V. Pore, E. Levänen and T. Mäntylä, Effect of Heat Treating Gel Films on the Formation of Superhydrophobic Boehmite Flaky Structures on Austenitic Stainless Steel, *Ceram. Int.*, (2008) in press.

[13]X. Zhang, M. Honkanen, M. Järn, J. Peltonen, V. Pore, E. Levänen and T. Mäntylä, Thermal Stability of the Structural Features in the Superhydrophobic Boehmite Films on Austenitic Stainless Steels, *Appl. Surf. Sci.* **254**, 5129-33 (2008).

[14]X. Zhang, M. Järn, J. Peltonen, V. Pore, T. Vuorinen, E. Levänen and T. Mäntylä, Analysis of Roughness Parameters to Specify Superhydrophobic Antireflective Boehmite Films Made by the Sol-Gel Process, *J. Eur. Ceram. Soc.* **28**, 2177-81 (2008).

[15]X. Zhang, Y. Ge, S. Hannula, E. Levänen and T. Mäntylä, Nanocrystalline α-Alumina with Novel Morphology at 1000 °C, *J. Mater. Chem.* **18**, 2423-25 (2008).

[16]X. Zhang, Y. Ge, S. Hannula, E. Levänen and T. Mäntylä, Process Study on the Formation of Nanocrystalline α-Alumina with Novel Morphology at 1000 °C, submitted to *J. Mater. Chem.*

[17]P. Padmaja, P. K. Pillai, K. G. K. Warrier and M. Padmanabhan, Adsorption Isotherm and Pore Characteristic of Nano Alumina Derived from Sol-Gel Boehmite, *J. Porous Mat.* **11**, 147-55 (2004).
[18]S. C. Kuiry, E. Megen, S. D. Patil, S. A. Deshpande, S. Seal, Solution-Based Chemical Synthesis of Boehmite Nanofibers and Alumina Nanorods, *J. Phys. Chem. B* **109**, 3868-72 (2005).

SYNTHESIS AND IN VITRO RELEASE OF GENTAMICIN FROM CaMCM-41/PLLA COMPOSITE MICROSPHERES

Yufang Zhu [1,2]*, Stefan Kaskel [2]
[1] ICYS-Sengen, National Institute for Materials Science (NIMS), 1-2-1 Sengen, Tsukuba, Ibaraki, 305-0047, Japan. [2] Institut für Anorganische Chemie, Technische Universität Dresden, Mommsen strasse 6, Dresden, 01069, Germany
* Email: ZHU.Yufang@nims.go.jp

ABSTRACT

Composite microspheres based on Ca-doped mesoporous silica and poly(L-lactic acid) (CaMCM-41/PLLA) were prepared by a solid-in-oil-in-water emulsion/solvent evaporation technique. Using gentamicin as a model drug, gentamicin was loaded in the microspheres by the adsorption method, and the in vitro release of gentamicin from the microspheres was evaluated in phosphorus buffer saline (PBS) solution at 37 °C. The results showed that the release rates of gentamicin from CaMCM-41/PLLA composite microspheres were much lower than that from pure PLLA microspheres. On the other hand, CaMCM-41 incorporation into PLLA had the ability to induce hydroxyapatite formation and deposited on the surface of CaMCM-41/PLLA composite microspheres during gentamicin releasing in PBS. Therefore, CaMCM-41/PLLA composite microspheres could potentially be used as a local drug release system for bone filling.

INTRODUCTION

The study of biomaterials for bone filling is one of the most interesting fields in orthopedic surgery[1]. Some biomaterials have been considered for this application, such as hydroxyapatite, bioactive glasses, bioceramics and biopolymeric cements, etc[2]. These materials can fill the defect of dead spaces caused by surgical intervention over traumatized or damaged bone, due to their excellent biocompatibility and integration with the osseous tissue. However, a serious trouble associated with bone filling is the osteomyelitis incidence, which is the inflammation of bone caused by a pyogenic organism[3].

There are many techniques for treating this trouble, such as systemic antibiotic administration, surgical debridement, wound drainage, and implant removal. All these treatments give the patients extra sufferings. Interestingly, the use of local antibiotics release in the implanted site offers a favorable strategy to treat the infection, which can maintain the antibiotics in the desired therapeutic range with a single dose and reduce the need for follow-up care. To date, the introduction of an appropriate drug release system into the bone implant site has been widely investigated. One strategy to realize the local drug release in the implanted site focuses on biopolymers[4-5], but they may not be suitable for bone repair as filling materials since most of the biopolymers are not able to chemically bond to living bone. Another strategy is using bioactive ceramics or bioactive glasses[2], but they did not allow for a sustained release over more than a few days. Therefore, many researchers focused their interest on developing implantable materials with local drug release combined with biopolymers and bioactive inorganic materials, which would be able to enhance bone growth and also could release antibiotics at the most critical inflammation-infection step[6-7]. For example, Li et al. reported the preparation of PHBV/wollastonite composite microspheres, and 90% of the total amount of gentamicin can be released from these composite microspheres after soaking in PBS or SBF for 22 days[6].

Recently, there has been increased interest in mesoporous silica materials for utilization as drug carrier in the field of controlled drug release, to meet the need for prolonged and better control of drug administration. Amorphous mesoporous silica materials have been investigated as drug supports due to their non-toxic nature, adjustable pore diameter, and very high specific surface area with abundant Si-OH bonds on pore surface. Several research groups have investigated mesoporous silica materials, such as MCM-41 and SBA-15, used as drug delivery systems and showed a controlled release property[8-9]. However, pure mesoporous silica may not be suitable for bone repair as filling materials since it lacks bioactivity[10]. Recently, there are a few reports on the synthesis of calcium doped mesoporous silica[11-12], and these materials show biocompatibility, biodegradation and bioactivity. For example, Diaz et al. reported that mesoporous HA-Silica composite biomaterials can be prepared from Ca-doped mesoporous silica (SBA-15), and the composite biomaterial was bioactive[11].

In this paper, we have successfully synthesized Ca-doped mesoporous materials CaMCM-41. Subsequently, the composite microspheres based on CaMCM-41 and poly(L-lactic acid) (CaMCM-41/PLLA) were also prepared by a solid-in-oil-in-water (s/o/w) emulsion/solvent evaporation technique. Gentamicin was absorbed into the microspheres to obtain a drug delivery system. In vitro release behaviors of gentamicin were evaluated in PBS solution at 37 °C. In addition, the degradation of CaMCM-41incorporated in the composite microspheres was also investigated.

EXPERIMENTAL

Preparation of CaMCM-41/PLLA composite microspheres: Ca-doped mesoporous silica CaMCM-41 was synthesized following a previous reported method.[12] CaMCM-41/PLLA composite microspheres were prepared by a solid-in-oil-in-water (s/o/w) emulsion/solvent evaporation technique as previously reported[6]. Briefly, 0.5 g of PLLA (Fluka) powder was dissolved in 20 ml of chloroform to form a solution with a concentration of 2.5% (w/v). A certain amount of CaMCM-41 was mixed with the PLLA solution and the mixture was stirred for 2 h to form a homogenous solution. The CaMCM-41/PLLA mixture was then added dropwise into 200 ml 1% (w/v) polyvinyl alcohol (PVA) (MW=86000, Fluka) solution. The mixture was vigorously stirred until the solvent evaporated completely. The resulting microspheres were washed three times with deionized water and then collected by filtering. Subsequently, these microspheres were dried in vacuum. The pure PLLA microspheres were prepared using the same method without addition of CaMCM-41. The compositions and sample names are listed in Table 1.

Loading and in vitro release of gentamicin: Loading of gentamicin into the pure PLLA and CaMCM-41/PLLA microspheres was carried out in PBS solution (pH 7.4) at room temperature for 24 h, as described by Sivakumar[13]. 0.3 g of microspheres was immersed in 10 ml of PBS containing 100 mg of gentamicin (10 mg/ml). After 24 h, the microspheres were separated by filtering and dried at 25 °C for 48 h. The estimation of gentamicin loaded in the microspheres was carried out through an indirect method, by determining the difference in gentamicin concentration before and after loading.

In vitro release of gentamicin from PLLA and CaMCM-41/PLLA composite microspheres (50 mg) was carried out at 37 °C in 10 ml of PBS solution. The concentration was determined by UV-visible spectroscopy. The analysis was carried out by measuring the absorbance values at the absorbance of gentamicin at the wavelength of 256 nm[14]. Before determination, a calibration curve was recorded. The release medium was withdrawn at the predetermined time intervals, and replaced with fresh soaking medium each time.

Characterization method: SEM was carried out on a Zeiss DMS 982 Gemini field emission scanning electron microscope at 4.0 kV. TEM was performed using a JEOL 2100F electron microscope operated at 200 kV. The UV/Vis absorption spectra were measured using a Shimadzu UV-1650PC spectrophotometer. N_2 adsorption-desorption isotherms were obtained on Nova 2000 pore analyzer at 77 K under continuous adsorption condition. The ionic concentration of the solution was analyzed by inductively coupled plasma atomic emission spectroscopy (ICP–AES; Varian Co., USA).

RESULTS AND DISCUSSION

Figure 1 shows the representive SEM and TEM images of CaMCM-41. It can be observed that the morphology of CaMCM-41 is ellipsoidal and spherical, they are well dispersed and the average size is around 100-200 nm. Furthermore, CaMCM-41 has worm-like mesopores. N_2 adsorption-desorption isotherm and corresponding pore size distribution are shown in Figure 2. The type IV isotherm curve with a well-defined step between 0.2 and 0.4 of P/P_0 indicates that CaMCM-41 has a mesoporous structure. Correspondingly, the pore size distribution (inset of Figure 3) of CaMCM-41 shows a narrow pore distribution with the peak pore size of 2.7 nm. The sample with a specific surface area of 548 m^2/g and the single point adsorption total volume of 0.375 cm^3/g at $P/P_0 = 0.699$ was obtained using the Brunauer-Emmett-Teller (BET) and Barrett-Joyner-Halenda (BJH) methods, respectively.

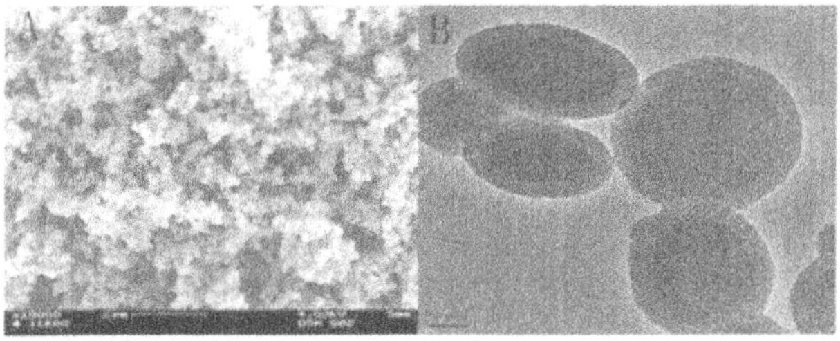

Figure 1 SEM (A) and TEM (B) images of CaMCM-41

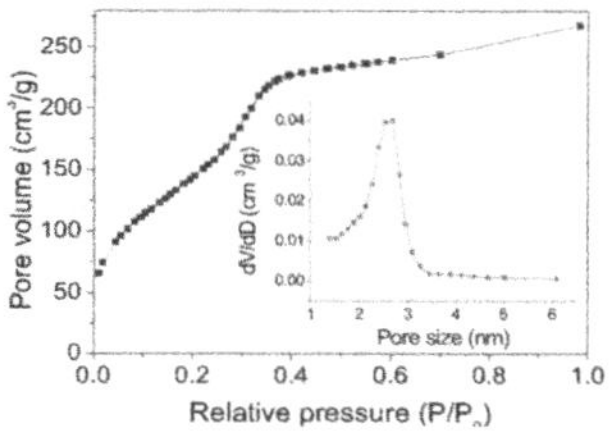

Figure 2 N_2 adsorption-desorption isotherm and pore size distribution

Table 1 The compositions and parameters of different samples

Samples	Composition ($m_{CaMCM-41}$:m_{PLLA})	S_{BET} (m^2/g)	Gentamicin loading (mg/g)	Release rate (days)[a]
PLLA	0	29	11.4	9
CaMCM-41/PLLA (1:9)	1:9	58	36.5	16
CaMCM-41/PLLA (3:7)	3:7	138	73.5	16

[a] the time for 90% of the total amount of gentamicin in the sample.

Figure 3 shows the representive SEM images of composite microspheres with different amounts of CaMCM-41. It can be found that pure PLLA and CaMCM-41/PLLA composite microspheres are spherical. For pure PLLA microspheres, the size is mainly in range of 50-70 μm and the surface of microspheres is smooth. After incorporation of CaMCM-41 into PLLA, the CaMCM-41/PLLA composite microspheres exhibit the porous surface and some particles are observed on the surface. It can be explained that some CaMCM-41 particles had a tendency to migrate towards the aqueous phase during the formation of composite microspheres due to the hydrophilicity of CaMCM-41. With increasing the amount of CaMCM-41 incorporated in PLLA, the pores increased and the surface became much rougher. Futhermore, the mesoporous structure of CaMCM-41 was not destroyed in the composite microspheres after the preparation process. The surface areas of pure PLLA microspheres is 29 m^2/g, while CaMCM-41/PLLA(1:9) and CaMCM-41/PLLA(3:7) can reach 58 m^2/g and 138 m^2/g, respectively (Table 1). The increased surface areas are attributed to the incorporation of CaMCM-41, and also facilitate a higher drug loading in composite microspheres.
After immersing the microspheres in 10 ml of gentamicin solution at the concentration of 10 mg/ml

for 24 h, the loading amounts of gentamicin were calculated to be 11.4 mg/g, 36.5 mg/g and 73.5 mg/g for pure PLLA microspheres (Table 1), CaMCM-41/PLLA(1:9) and CaMCM-41/PLLA(3:7) composite microspheres, respectively. Therefore, CaMCM-41 played an important role on higher drug loading capacity.

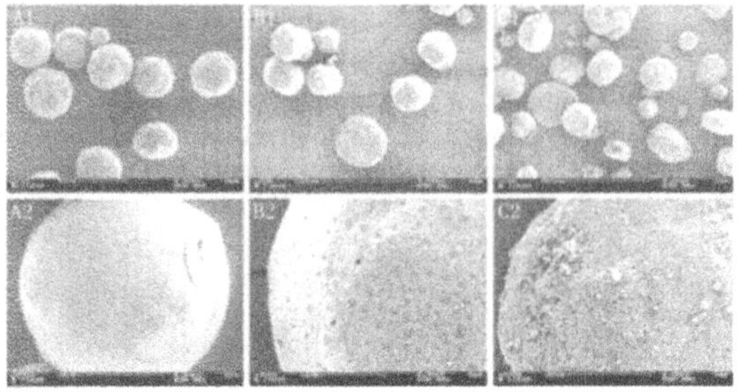

Figure 3 SEM images of pure PLLA and CaMCM-41/PLLA composite microspheres: (A1-2) PLLA; (B1-2) CaMCM-41/PLLA (1:9) and (C1-2) CaMCM-41/PLLA (3:7)

Figure 4 shows the release profiles of gentamicin from the microsperes in PBS solution for 22 days. It can be observed that gentamicin in CaMCM-41/PLLA coposite microspheres exhibited a fast release during the first 12 h, which is similar to that from pure PLLA microspheres. This could be related to the release of gentamicin from the microsphere surface. After 12 h, the release rates of gentamicin from pure PLLA microsphers and CaMCM-41/PLLA composite microspheres decreased. However, gentamicin released more slowly from the composite microspheres than from the pure microspheres. For pure PLLA microspheres, 90% of the total amount of gentamicin was released after 9 days, but CaMCM-41/PLLA composite microspheres can last over 16 days. For CaMCM-41/PLLA composite microspheres system, besides some gentamicn molecules were entrapped in PLLA , many gentamicin molecules were loaded in the mesopores of CaMCM-41 through physical adsorption and hydrogen bonding. These gentamicin molecules must diffuse longer pathways from the mesopores to PBS solution. Furthermore, the release rate of gentamicin loaded through hydrogen bonding was controlled by the equilibrium between bonding with CaMCM-41 and dissolution in aqueous medium, which also results in the slower release rate. In addition, the release rates of gentamicin from CaMCM-41/PLLA(1:9) and CaMCM-41/PLLA(3:7) composite microspheres were close to each other.

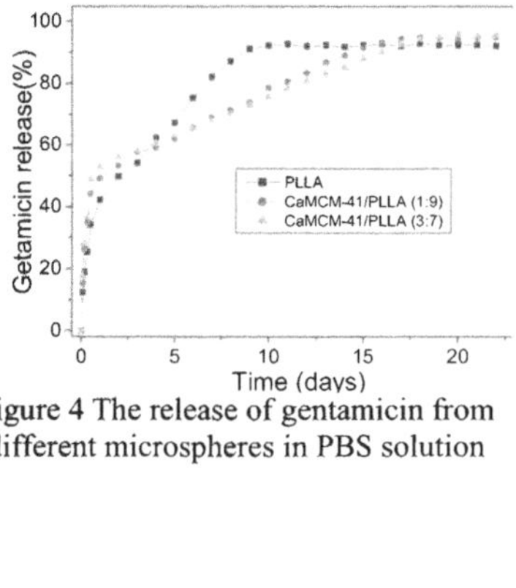

Figure 4 The release of gentamicin from different microspheres in PBS solution

Figure 5 shows SEM images of pure PLLA and CaMCM-41/PLLA composite microspheres after gentamicin releasing in PBS solution for 22 days. It can be observed that a layer composed of particles coverd the surface. EDX analysis confirmed that the chemical composition can be assigned to a hydroxylcarbonate apatite phase (Ca/P ratio ≈1.66). In contrast, there was no hydroxyapatite particles on the surface of pure PLLA microspheres. The results indicated that CaMCM-41 incorporation into PLLA had the ability to induce hydroxyapatite formation. On the other hand, there more pores appeared on the surface of the microspheres as compared to that of the microspheres before gentamicin releasing, which is attributed to the degradation of PLLA and CaMCM-41.

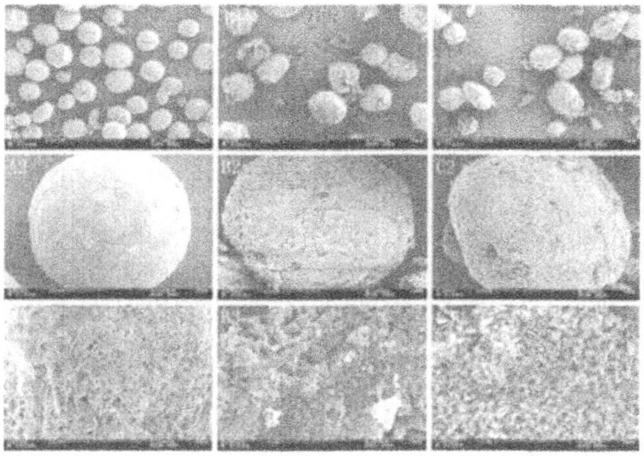

Figure 5 SEM images of (A1-3) PLLA, (B1-3) CaMCM-41/PLLA(1:9) and (C1-3) CaMCM-41/PLLA (3:7) after gentamicin releasing in PBS.

Figure 6 shows the changes in ion concentrations of Ca, P and Si of PBS solutions after CaMCM-41/PLLA composite microspheres soaking. It was obvious that Si ion concentrations increased rapidly within the first 3 days of soaking, and then continued to increase at a slower rate up to 21 days. Ca ion concentrations increased slowly within the soaking period. Furthermore, CaMCM-41/PLLA(3:7) showed a more intensive release of both Si and Ca ions as compared with CaMCM-41/PLLA(1:9). In contrast to the increase of Si and Ca ion concentration, P ion concentration of PBS solution decreased gradually through the whole soaking period, and the decrease rate for CaMCM-41/PLLA(3:7) was faster than for CaMCM-41/PLLA(1:9). The results indicated that CaMCM-41 degraded and released Si and Ca ion during soaking in PBS soluion, and then formed hydroxyapatite in the presence of PO_4^{3-}.

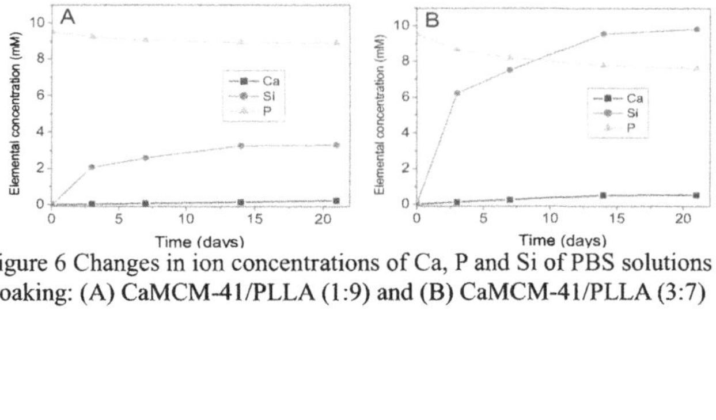

Figure 6 Changes in ion concentrations of Ca, P and Si of PBS solutions soaking: (A) CaMCM-41/PLLA (1:9) and (B) CaMCM-41/PLLA (3:7)

CONCLUSION

In this study, CaMCM-41/PLLA composite microspheres were prepared and the release behavior of gentamicin from the microspheres was investigated. The results showed that gentamicin release was controlled from CaMCM-41/PLLA composite microspheres at a relatively lower release rate as compared to that from pure PLLA microspheres in PBS solution, and the sustained release can last three weeks. The analysis of the microspheres after gentamicin release demonstrated that hydroxyapatite particles can be deposited on the surface of CaMCM-41/PLLA composite microspheres. Therefore, these results suggested that CaMCM-41/PLLA composite microspheres might be useful as a local drug release system for bone filling.

REFERENCES

[1] C. A. Shapoff, D. C. Alexander, A. E. Clark: Clinial use of bioactive glass particulate in the treatment of human osseus defects. *Compend Contin Educ. Dent.* **18**, 352 (1997).

[2] A. Krajwski, A. Ravaglioi, E. Roncari, P. Pinasco: Porous ceramic bodies for drug delivery. *J. Mater. Sci., Mater. Med.* **12**, 763 (2000).

[3] J. Ciampolini, K. G. Harding: Pathophysiology of chronic bacterial osteomyelitis. Why do antibiotics fail so often? *Postgrad. Med.* **76**, 479 (2000).

[4] D. Sendil, I. Guresel, D. L. Wise, V. Hasirci: Antibiotic release from biodegradable PHBV microparticles. *J. Control. Release,* **59**, 207 (1999).

[5] D. G. Wallace, J. Rosenblatt: Collagen gel systems foe sustained delivery and tissue engineering, *Adv. Drug Deliv. Rev.* **55**, 1631 (2003).

[6] H. Li, J. Chang: Preparation, characterization and in vitro release of gentamicin from PHBV/wollastonite composite microspheres. *J. Controlled Release,* **107**, 463 (2005).

[7] J. M. Xue, M. Shi: PLGA/mesoporous silica hybrid structure for controlled drug release. *J. Controlled Release,* **98**, 209 (2004).

[8] M. Vallet-Regí, A. Rámila, R. P. Del Real, J. Pérez-Pariente: A new property of MCM-41: drug delivery system. *Chem. Mater.* **13**, 308 (2001).

[9] Y. Zhu, J. Shi, W. Shen, X. Dong, J. Feng, M. Ruan, Y. Li: Stimuli-responsive controlled drug Release from a hollow mesoporous silica sphere/polyelectrolyte multilayers core-shell structure. *Angew. Chem. Int. Ed.* **44**, 5083 (2005).

[10] P. Horcajada, A. Ramila, K. Boulahya, J.G. Callet, M. Vallet-Regi: Bioactivity in ordered mesoporous materials. *Solid State Sci.* **6**, 1295 (2004).

[11] A. Diaz, T. Lopez, J. Manjarrez, E. Basaldella, J. M. Martinez-Blanes, J. A. Odriozola: Growth of hydroxyapatite in a biocompatible mesoporous ordered silica. *Acta Biomaterialia,* **2**, 173 (2006).

[12] X. Li, L. Zhang, X. Dong, J. Liang, J. Shi: Preparation of mesoporous calcium doped silica spheres with narrow size dispersion and their drug loading and degradation behavior. *Micropor. Mesopor. Mater.* **102**, 151 (2007).

[13] M. Sivakumar, K. Panduranga Rao: Preparation, characterization and in vitro release of gentamicin from coralline hydroxyapatite–gelatin composite microspheres. *Biomaterials* **23**, 3175 (2002).

[14] Y. Zhang, M. Zhang: Calcium phosphate/chitosan composite scaffolds for controlled in vitro antibiotic drug release. *J. Biomed. Mater. Res.* **62**, 378 (2002).

HIGHLY ORDERED CUBIC MESOPOROUS COBALT OXIDE BY AN ACCURATELY CONTROLLED INCIPIENT WETNESS TECHNIQUE

Limin Guo, Xiangzhi Cui and Jianlin Shi*
State Key Laboratory of High Performance Ceramics and Superfine Microstructure,
Shanghai Institute of Ceramics, Chinese Academy of Sciences
Shanghai, 200050, People's Republic of China

ABSTRACT

Highly cubic ordered cobalt oxides have been successfully synthesized from KIT-6 by an accurately controlled incipient wetness approach. The adding volume of cobalt nitrate and absolute ethanol were determined by the pore volume of KIT-6, and the synthesis procedure is effective and economical. Furthermore, the obtained mesoporous cobalt oxides have better mesostructure comparing with those of former reports. The X-ray diffraction, N_2 sorption isotherms, transmission electron microscopy (TEM), energy-dispersive spectroscopy (EDS) and Fourier-transform infrared spectroscopy were used to characterize the as-synthesized mesoporous cobalt oxides.

INTRODUCTION

Mesoporous materials with a transition-metal oxide framework have great potential for applications in catalysis, photocatalysis, sensors and electrodes because of their characteristic catalytic, optical and electronic properties. [1] The synthesis of ordered transition-metal oxides has stimulated extensive researches over the past few years. The most widely used approach is the direct synthesis *via* soft templates. [2] However, direct synthesis of these kinds of mesoporous materials using surfactants is not always a success. One of the difficulties is the easy crystallization of most of these oxides, which is accompanied by structural collapse, during mesostructure formation and removal of the organic templates. [3] Correspondingly, ordered mesoporous materials by the direct synthesis usually have amorphous or semi-crystalline framework, which limits their applications. Recently, J. Lee and co-workers[4] have reported a strategy of "combined assembly by soft and hard chemistries" to directly synthesize thermally stable and highly crystalline mesoporous transition-metal (group-Ⅳ and group-Ⅴ) oxides. The mesoporous materials are 2D ordered with the pore size about 24nm, but the pore structure and size were difficult to control. More attractively, the ordered and uniform pore structure, large surface areas and large pore volumes of mesoporous silica materials make them perfect candidates to serve as "hard templates".[5] So scientists have utilized mesoporous silica as "hard-template" to synthesize mesoporous transition-metal oxides, and developed many methodologies to facilitate the respective metal precursors into the silica pores, such as "microwave-digested silica templating", [5] vinyl-functionalized, [6] amino-functionalized, [7] bi-solvents, [8] repeated impregnation techniques[9] , and so on. Due to the harsh conditions in removing silica templates either by hot NaOH (2 mol/L) or HF (10%), ordered mesoporous carbons (CMK-type) were also successfully used as templates for some metal oxides. [10-14] Although various soft- and hard-templating approaches have been developed and many kinds of mesoporous metal oxides were synthesized in the past few years, they usually suffer from the drawbacks of multiple, tedious steps and poorly crystallized walls or low pore structure symmetry.[4] In addition, the present hard-templating route usually employs large extra amount of precursor solution and long time impregnation to let the precursor enter the pores, therefore, only a very small portion of the precursors and solvents was finally impregnated into the pore channels and becomes frameworks. An incipient wetness technique has been reported to synthesize mesoporous oxide using mesoporous silica as template, however, the obtained oxide, e.g., ceria, as we could found up to date, showed deteriorated ordering of mesoporous structure.[15] Herein, highly ordered mesoporous transition-metal oxides with crystalline walls has been synthesized by an accurately controlled incipient wetness technique. This method, with the accurate control on the amount of added precursor solution and the adding rate, make the full use of the hydrophilic nature of the silica surface and the large capillary force of the nano-pore channels, is facile and highly efficient.

Spinel-type cobalt oxide (Co_3O_4) was selected as a model material because it is potentially useful for applications in catalysis, sensors, magnetic materials, and energy storage as electrodes in lithium-ion batteries. [6,16]

EXPERIMENTAL

Synthesis of mesoporous cobalt oxide

Mesoporous silica KIT-6 was obtained following the procedure described previously. [17] Pore volume of the as-synthesized KIT-6 is 1.2 cm^3/g.

In a typical synthesis of cubic ordered mesoporous Co_3O_4, 1.8g cobalt nitrate ($Co(NO_3)\bullet 6H_2O$) was dissolved in 1.5mL ethanol. The resultant cobalt nitrate solution was dropwise added into 1g KIT-6 powder, and this process took several minutes. The wetted powder was then dried at 333K. After that the powder was heated slowly to 573K and calcined at that temperature for 3h. The resulting powder was treated twice with a 2M hot NaOH solution in water to remove the silica template, followed by washing with distilled water several times, and drying at 373K. This sample with once wetting is named as Co_3O_4-1. The same procedure could be repeated, but the precursor solution is 1.6g cobalt nitrate in 1.3ml ethanol, and the powder was then calcined at 773K for 3h. The mesoporous Co_3O_4 obtained by the twice wettings was designated as Co_3O_4-2.

Characterization techniques

Low-angle powder XRD patterns were recorded with a Rigaku D/Max-2550V diffractometer using Cu Kα radiation (40kV, 40mA) with a step width 0.002° and a scanning rate 0.6°/min. The wide-angle powder XRD patterns were recorded with a step width 0.02° and a scanning rate 4°/min.N_2 adsorption-desorption isotherms were measured on a Micrometitics Tristar 3000 system at 77K. Prior to the measurements, all samples were degassed at a temperature of 403K for 12h.

TEM images were obtained with a JEM-2010 electron microscope at 200kV. EDX spectra were collected from an attached Oxford Link ISIS energy-dispersive spectrometer fixed on the JEm-2010 electron microscope. FTIR spectra were obtained in the range of 400-4000cm^{-1} using a Nicolet 7000-C with a resolution of 8 cm^{-1}.

RESULTS AND DISCUSSION

Following the synthesis procedure given in the experimental section, two samples of cubic ordered mesoporous Co_3O_4-1 (once wetting) and Co_3O_4-2 (twice wetting) were obtained. Figure 1 shows the low-angle X-ray diffraction patterns of the template KIT-6, Co_3O_4-1 and Co_3O_4-2. All the XRD patterns have a very sharp diffraction peak (211) and two or more weak peaks, which are characteristic of a 3D hexagonal (Ia3d) structure.[17] Although the (211) peak intensities of Co_3O_4-1 and Co_3O_4-2 in Figure1 are a little lower than that of the template, the other two weak but clearly distinguishable peaks at $2\theta = 1.2^\circ$ (220 plane) and 1.7-1.8° indicate a long-range periodically ordered pore structure with cubic symmetry in the resultant mesoporous cobalt oxide. Such a long range ordering of Co_3O_4-2 is better than that of Co_3O_4-1's as judged from the low-angle XRD patterns, and the fact was further confirmed by the TEM images (see below). Unit-cell parameters of 22.6 nm for Co_3O_4-1 and 22.8 nm for Co_3O_4-2 were calculated from the position of the (211) reflections, which are a little smaller than that of the template (23.0nm). This reduction in the unit-cell size can be attributed to the structure contraction in the calcination procedure during the preparation of mesoporous cobalt oxide. The lattice contraction of Co_3O_4-1 is slightly higher than Co_3O_4-2 due to a little higher precursor loading amount in template for the latter. However, the overall contractions of the templated Co_3O_4 are very limited with the present approach, indicating the very high efficiency of the precursor loading even with only once wetting. The facts can be further illuminated by TEM images and pore size distribution curve (see below).

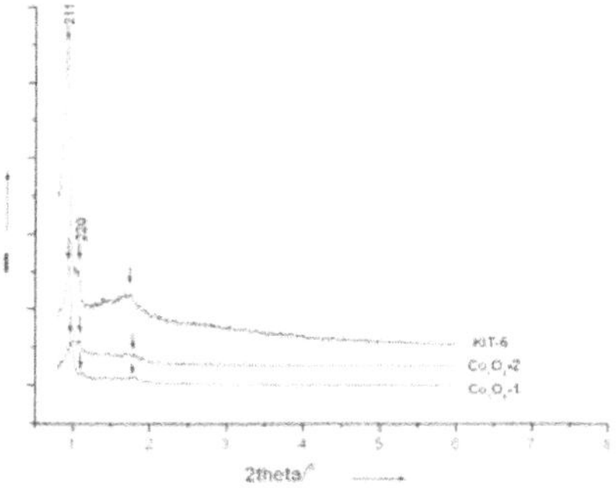

Figure 1. Low-angle XRD patterns of KIT-6, Co_3O_4-1 and Co_3O_4-2

The wide-angle XRD patterns of the mesoporous cobalt oxide (Figure 2) shows broadened peaks, suggesting that this cobalt oxide is crystallized within the confined mesoporous channels. All the diffraction peaks of the mesoporous cobalt oxides match well to the spinel lattice structure known to exist for the bulk cobalt oxide (JCPDS NOS. 42-1467). The average crystallite sizes calculated from the peak broadening of 311 reflections by applying the Scherrer equation are 13.0 nm for Co_3O_4-1 and 14.4 nm for Co_3O_4-2, respectively, which is larger than the pore diameter of the template (9.4nm). The average crystallite sizes also calculated from the peak broadening of 220 reflections are 12.0 nm for Co_3O_4-1 and 12.4 nm for Co_3O_4-2, respectively. This can be understood considering that the simple analysis with the Scherrer equation assumes spherical particles, and therefore calculates the average of the particle size in all three dimensions.

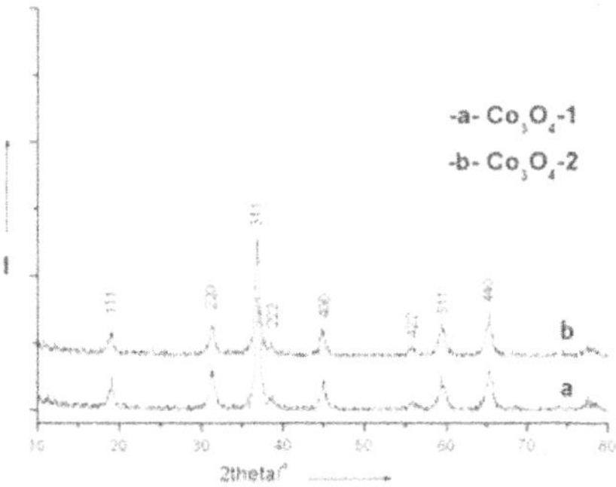

Figure 2. Wide-angle XRD patterns of the mesoporous Co_3O_4

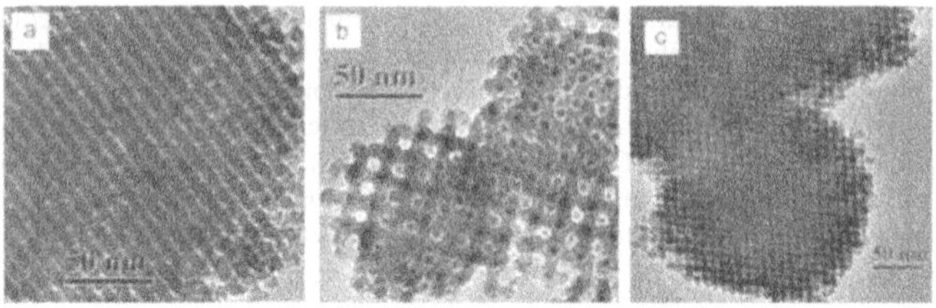

Figure 3. TEM images of the mesoporous Co_3O_4-1 recorded along the a) [100], b) [111] and c) [531] directions

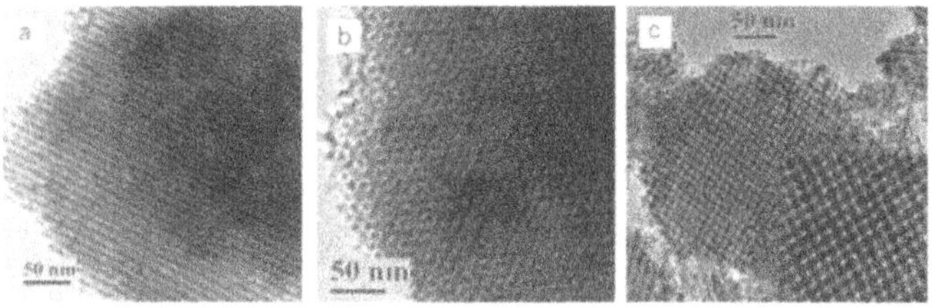

Figure 4. TEM images of the mesoporous Co_3O_4-2 recorded along the a) [100], b) [111] and c) [531] directions

TEM images of Co_3O_4-1 and Co_3O_4-2 were recorded along the [100], [531] and [111] directions (Figure 3 and 4), which confirms a cubic structure over a large area. Examinations in a wide range of particles demonstrate that they all have the ordered mesoporous structure. These images support the conclusion that the as-synthesized cobalt oxides have well-ordered mesoporous structured as indicated by the one sharp and several weak peaks in the low-angle XRD patterns. From TEM images along different directions of Co_3O_4-1 and Co_3O_4-2, it can be clearly found that the mesoporous structure of Co_3O_4-2 possesses a more highly ordered mesoporous structure. This is in accordance with the results of low-angle XRD diffraction (Figure 1). EDX analysis (Figure 5) and FTIR spectra (Figure 6) were used to confirm the absence of silica template after the removal. No Si signals can be detected in the EDX spectrum and the typical FTIR adsorption peaks of Si-O-Si ($780cm^{-1}$ and $1000\text{-}1250\ cm^{-1}$) and Si-OH ($960cm^{-1}$) have disappeared in Co_3O_4-1 and Co_3O_4-2.
Figure7 show N_2 adsorption-desorption isotherms and the corresponding pore size distribution curves for as-synthesized mesoporous cobalt oxides. The isotherms have the characteristic type-Ⅳ shape with a marked leap in the adsorption branch at relative pressures p/p_0 between 0.3 and 0.6. The mesoporous cobalt oxides exhibits a narrow pore size distribution (calculated by the adsorption branches using BJH method), and the pore sizes at the most probabilities are 3.3nm for Co_3O_4-1 and 2.8nm for Co_3O_4-2, respectively. Here, Co_3O_4-2 has smaller pore as more cobalt precursor was loaded into the pore channels. The surface areas (calculated by BET method) are $97.7\ m^2/g$ and $79.2\ m^2/g$ for Co_3O_4-1 and Co_3O_4-2, respectively.
The all above results show that highly ordered cubic mesoporous cobalt oxides with crystalline walls can be successfully synthesized with the incipient wetness approach by carefully controlling the amount of added precursor solution and the adding rate. First, this approach results in gradual wetting/loading of precursors into the pore channels and the air within the pore channels can be

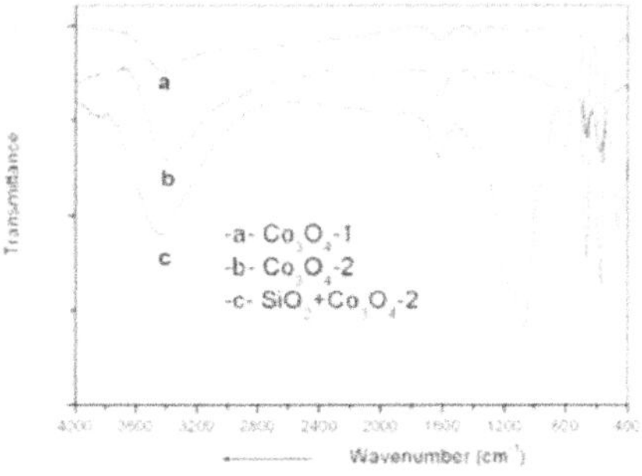

Figure 5. FTIR spectra of Co_3O_4-1 (a), Co_3O_4-2(b) and Co_3O_4-2 without removing the silica template(c)

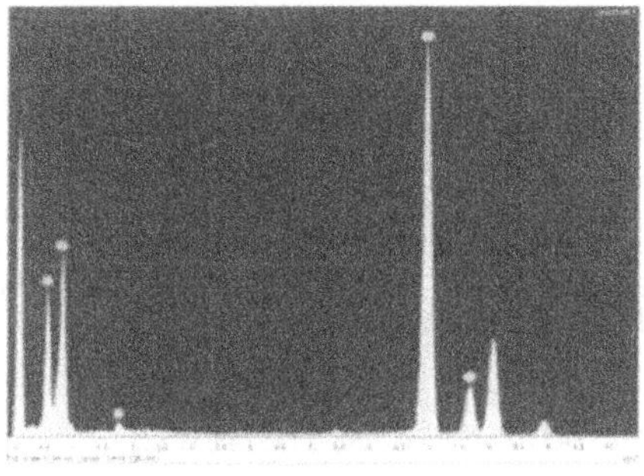

Figure 6. EDX spectrum of Co_3O_4-2

effectively driven out during the wetting process, therefore high loading of cobalt precursors into pore channels can be ensured for even once wetting with little precursor materials deposited on the outer surface of the matrix; second, the adding amount of cobalt nitrate can be calculated and therefore precisely controlled by considering the pore volume of the mesoporous silica template, i.e., the volume of the cobalt nitrate used should be equal or a little higher than the pore volume of the template; third, the concentration of the precursors solution should be reasonably high to obtain high loading, but should be a little lower that the saturation level, as the solvent, ethanol, is easy to evaporate during the dropwise wetting process, therefore slightly larger volume of the solvent in the solution of cobalt nitrate (with lower concentration than the saturation) will be helpful in preventing cobalt nitrate from depositing and blocking the entrance of mesoporous channels during wetting process on one aspect; on the other aspect the solution volume cannot be too large, i.e., concentration cannot be too low, or else the extra amount of solution may trap the air within the pore channels resulting in limited loading amount of cobalt nitrate and precursor depositing on the outer surface of the matrix. Therefore, both the precursor amount and the nitrate solution concentration should be carefully controlled to ensure the high efficient loading of the precursor into pore channels without depositing on the outer surface. The incipient wetness process usually takes several minutes for the preparation of small amount of samples (about 1g

Co_3O_4-2 could be obtained with 1g KIT-6 template) in laboratory, and long time stirring/refluxing is not needed. Even for the large amount samples preparation, similarly a multi-drop wise wetting technique (e.g., by solution spraying) can also be adapted.

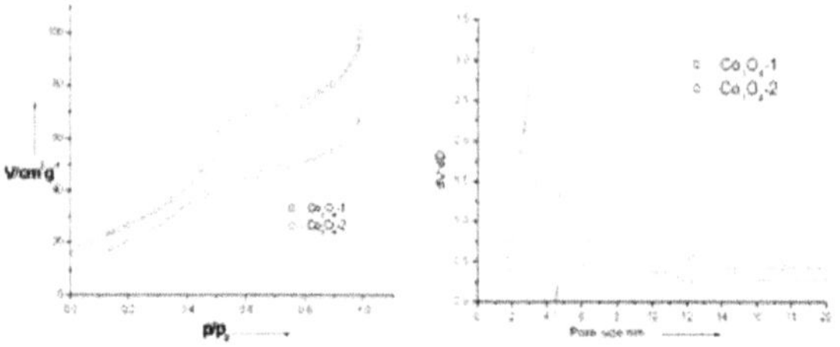

Figure 7. a) N_2 adsorption-desorption isotherms for mesoporous Co_3O_4-1 and Co_3O_4-2, and b) corresponding pore size distribution curves.

CONCLUSION

In summary, highly cubic ordered cobalt oxides with crystalline walls have been synthesized by an accurately controlled incipient wetness approach using mesoporous silica as hard template and cobalt nitrate as the precursors. Compared with the previously reported solution impregnation processes, this technique is facile, time-saving and economical. This technique can be used to synthesize ordered mesoporous metal oxides, and we expect that with this approach various kinds of mesoporous oxides or other materials can be synthesized, which may benefits their applications in many areas such as in catalysis, sensors, clean energy and electronics.

FOOTNOTE

*Corresponding author Tel.: +86-21-52412714; Fax: +86-21-52413122.
E-mail address: jlshi@sunm.shcnc.ac.cn

ACKNOWLEDGEMENTS

The authors gratefully acknowledge the support of this research by National Natural Science Foundation of China (Grant No. 20633090) and Chinese Academy of Sciences (Grant No.KJCX2.YW.M02) .

REFERENCES

[1] X. He and D. Antonelli, Recent Advances in Synthesis and Applications of Transition Metal Containing Mesoporous Molecular Sieves, Angew. Chem. Int. Ed., **41**, 214-29(2002).

[2] D. M. Antonelli and J. Y. Yin, Synthesis of Hexagonally Packed Mesoporous TiO_2 by a Modified Sol-Gel Method, Angew. Chem. Int. Ed., **34**, 2014-7(1995).

[3] F. Schüth, Non-siliceous Mesostructured and Mesoporous Materials, Chem. Mater., **13**, 3184-95(2001).

[4] J. W. Lee, M. C. Orilall, S. C. Warren, M. Kamperman, F. J. Disalvo and U. Wiesner, Direct access to thermally stable and highly crystalline mesoporous transition-metal oxides with uniform pores, Nat. Mater., **7**, 222-8(2008).

[5] B. Z. Tian, X. Y. Liu, H. F. Yang, S. H. Xie, C. Z. Yu, B. Tu and D. Y. Zhao, General synthesis of ordered crystallized metal oxide nanoarrays replicated by microwave-digested mesoporous silica, Adv. Mater., **15**, 1370-4(2003).

[6] a) Y. Q. Wang, C. M. Yang, W. Schmidt, B. Spliethoff, E. Bill and F. Schüth, Weakly ferromagnetic

ordered mesoporous Co_3O_4 synthesized by nanocasting from vinyl-functionalized cubic Ia3d mesoporous silica, Adv. Mater., **17**, 53-6(2005); b) A. Rumplecker, F. Kleitz, E.-L. Salabas and F. Schüth, Hard Templating Pathways for the Synthesis of Nanostructured Porous Co_3O_4, Chem. Mater., **19**, 485-96(2007).

[7] a) K. Zhu, B. Yue, W. Zhou and H. He, Preparation of three-dimensional chromium oxide porous single crystals templated by SBA-15, Chem. Commun., **1**, 98-99 (2003); b) C. Dickinson, W. Zhou, R. P. Hodykins, Y. Shi, D. Zhao and H. He, Formation Mechanism of Porous Single-Crystal Cr_2O_3 and Co_3O_4 Templated by Mesoporous Silica, Chem. Mater., **18**, 3088-96(2006); c) J. Parmentier, L. A. Solovyov, F. Ehrburger-Dolle, J. Werckmann, O. Ersen, F. Bley and J. Patarin, Structural Peculiarities of Mesostructured Carbons Obtained by Nanocasting Ordered Mesoporous Templates via Carbon Chemical Vapor or Liquid Phase Infiltration Routes, Chem. Mater., **18**, 6316-23(2006).

[8] F. Jiao, A. Harrison, A. H. Hill and P. G. Bruce, Mesoporous Mn_2O_3 and Mn_3O_4 with Crystalline Walls, Adv. Mater., **19**, 4063-6(2007).

[9] F. Jiao, K. M. Shaju and P. G. Bruce, Synthesis of Nanowire and Mesoporous Low-Temperature $LiCoO_2$ by a Post-Templating Reaction, Angew. Chem. Int. Ed., **44**, 6550-3(2005).

[10] a) A. H. Lu, W. Schmidt, A. Taguchi, B. Spliethoff, B. Tesche and F. Schüth, Taking Nanocasting One Step Further: Replicating CMK-3 as a Silica Material, Angew. Chem. Int. Ed., **41**, 3489-92(2002); b) A. H. Lu, W. Schmidt, B. Spliethoff and F. Schüth, Synthesis and characterization of nanocast silica NCS-1 with CMK-3 as a template, Chem. -Eur. J., **10**, 6085-92(2004).

[11] M. Kang, S. H. Yi, H. I. Lee, J. E. Yie and J. M. Kim, Reversible replication between ordered mesoporous silica and mesoporous carbon, Chem. Commun., **17**,1944-5(2002).

[12]a) J. Roggenbuck and M. Tiemann, Ordered Mesoporous Magnesium Oxide with High Thermal Stability Synthesized by Exotemplating Using CMK-3 Carbon, J. Am. Chem. Soc., **127**, 1096-7(2005); b) J. Roggenbuck, G. Koch and M. Tiemann, Synthesis of Mesoporous Magnesium Oxide by CMK-3 Carbon Structure Replication, Chem. Mater. **18**, 4151-6(2006).

[13] T. Waitz, M. Tiemann, P. J. Klar, J. Sann, J. Stehr and B. K. Meyer, Crystalline ZnO with an enhanced surface area obtained by nanocasting, Appl. Phys. Lett., **90,** 1231081-3(2007).

[14] X. Y. Lai, X. T. Li, W. C. Geng, J. C. Tu, J. X. Li and S. L. Qiu, Ordered Mesoporous Copper Oxide with Crystalline Walls, Angew. Chem. Int. Ed., **46**, 738-41(2007).

[15] S. C. Laha and R. Ryoo, Synthesis of thermally stable mesoporous cerium oxide with nanocrystalline frameworks using mesoporous silica templates, Chem. Commun., **17**, 2138-9 (2003).

[16] P. Poizot, S. Laruelle, S. Grugeon, L. Dupont and J. M. Tarascon, Nano-sized transition-metaloxides as negative-electrode materials for lithium-ion batteries, Nature, **407**, 496-9(2000).

[17] F. Klertz, S. H. Choi, R. Ryoo, Cubic Ia3d large mesoporous silica: synthesis and replication to platinum nanowires, carbon nanorods and carbon nanotubes, Chem. Commun., **17**, 2136-7 (2003).

PREPARATION OF Fe_3O_4 NANOPARTICLES BY TWO DIFFERENT METHODS

Mingxin Geng, Futian Liu, Zengbao Zhao
School of Materials Science and Engineering, University of Jinan.
Jinan, Shandong, 250022, China

ABSTRACT

Fe_3O_4 magnetic nanoparticles were prepared by both hydrothermal and microemulsion methods. The advantages and disadvantages of the two methods were compared through characterization and analysis. The synthesized products were characterized by X-ray diffraction (XRD), transmission electron microscopy (TEM). The grain sizes were determined by laser grain size analysis meter and the B-H curve was measured by alternating current gradient magnetor. The results show that the products prepared by both methods are pure Fe_3O_4 phase with narrow particle size distribution and have good superparamagnetism. The particles made by hydrothermal method showed better magnetism and their grain size was narrower.

KEY WORDS

Fe_3O_4; nanoparticles; hydrothermal method; microemulsion method; magnetism

1 INTRODUCTION

Magnetic nanoparticles are an important category of nanomaterials. They not only share some common characteristics of nanometer materials, but also have other exclusive characteristics such as super-smooth magnetism, apparent magnetism. A variety of methods to synthesize magnetic nanoparticles have been put forward. Fe_3O_4 magnetic nanoparticles are a spinel ferrite oxide. Its advantages include simple preparation, low price, non-toxicity and so on. Fe_3O_4 magnetic nanoparticles are humidity sensitive and magnetic-responsive. They can be used in the fields of high-density, nuclear magnetic resonance, drugs controlled release.[1-4]

There are many methods for synthesizing magnetic Fe_3O_4 nanoparticles. Among them, hydrothermal method and microemulsion method are two new methods which have been developed in recent years.

The hydrothermal method[5-6] uses water as the reaction medium in a tailored airtight reaction vessel (high-pressured cauldron), creating a high temperature and pressure environment through heating the reaction vessel, which enables the dissolution and recrystallization of the less dissolvable and nondissolvable reactants. Compared with other preparation methods, hydro-thermal method applies a low reaction temperature (<250°C) and is a simple procedure. The composition, size and the morphology of the crystals can be easily controlled by changing the reaction conditions, such as the reaction temperature, pH, the reactant concentration, the amount of the surfactant, etc and in the meantime, the recrystallization process results in a product with a high purity.

Microemulsion method[7] is an effective way developed in recent years to prepare superfine particles. Microemulsion is a thermodynamic stable system which is prepared by surfactant, oil phase, water phase and lacquer solvent with proper ratios. And it has some characteristics such as diaphanous, low viscosity and isotropy. Since the disperse drops are very tiny and homogeneous, further agglomeration of the formed fine particles can be effectively avoided. The final nanoparticles are therefore exhibit a narrow particle size distribution, good crystal shape and good dispersion . A wide range of particle sizes can be achieved by controlling the relative rate of nucleation and crystal growth through the adjustment of the water volume of micro-emulsion dripping and the consistency of various reactants.

This paper prepared magnetic Fe_3O_4 nanoparticles with the both methods separately. The obtained results are compared.

2 EXPERIMENTAL

HYDROTHERMAL METHOD

$FeCl_2 \cdot 4H_2O$ and $FeCl_3 \cdot 6H_2O$ solution were mixed with a given mol ratio in a beaker and stirred in a water-bath at a constant temperature. After adding excessive NaOH solution, a proper amount of neopelex was added in the mixture. After being stirred for 30min, Fe_3O_4 monomer precursor was prepared. The obtained homogeneous yellow solution was transferred to a Teflon-lined stainless-steel autoclave and sealed. Then it was heated at 160°C in an oven. After a reaction of 8h, the autoclave was cooled to room temperature. The product was collected, washed with deionized water and ethanol, and then vacuum-dried.

MICROEMULSION METHOD

$FeSO_4 \cdot 7H_2O$ and $FeCl_3 \cdot 6H_2O$ were dissolved in 10ml deionized water and then were transferred to a three-necked flask under the protection of nitrogen atmosphere. The flask was put in a water-bath at a constant temperature. 100ml methylbenzene was added in the flask and stirred for 20 min after adding a certain amount of neopelex. NaOH solution which was used as precipitant was added dropwise to the emulsion. Then the mixture was stirred for 1 h, followed by the addition of 10ml ethanol under continuous stirring. The reaction was stopped after 0.5 h. Then the flask was cooled to room temperature. Finally The obtained magnetite particles were washed with deionized water and ethanol, and then dried in vacuum.

3 RESULTS AND DISCUSSION

The XRD patterns of products synthesized by the hydrothermal and microemulsion methods are shown in Figure 1, respectively.

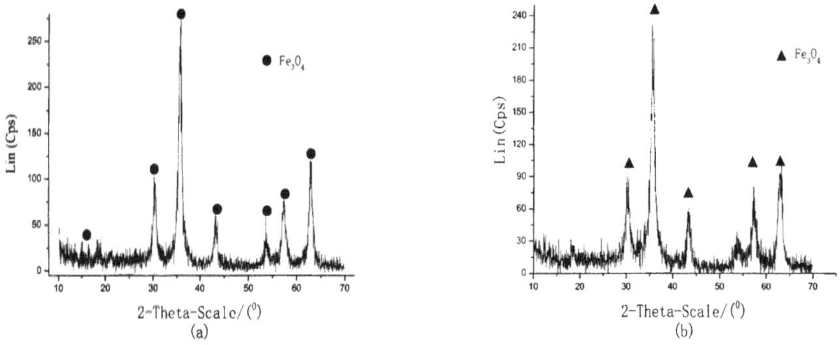

Figure 1. a. TheXRD pattern of the the Fe_3O_4 prepared by hydrothermal method ; b.The XRD pattern of the Fe_3O_4 prepared by microemulsion method

Both patterns show that the obtained products are well-crystallized. The diffraction peaks all can be indexed to Fe_3O_4 crystals. No impurity phase is identified. According to the Scherrer formula:

$$D = \frac{K\lambda}{\beta\cos\theta} \quad (1)$$

(K =0.89; λ =1.54056; D is the average size of crystal particle; β is a half width of the highest

peak; θ is half of the diffraction angle)
The calculated crystal sizes of the samples are 10 nm and 15nm in case of using the hydrothermal and microemulsion methods, respectively.

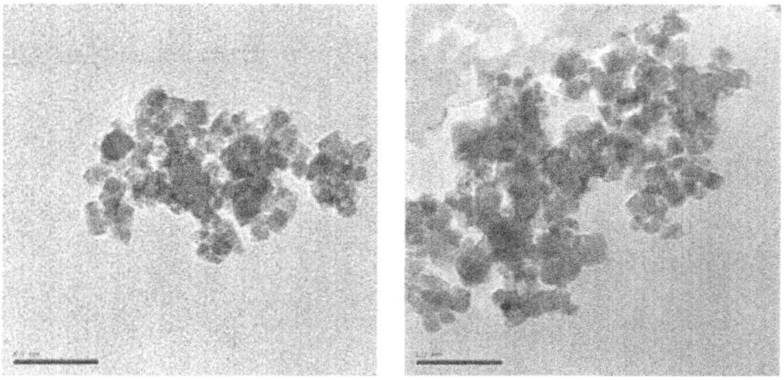

(a) (b)

Figure 2. a.TEM image of Fe_3O_4 prepared by hydrothermal method ; b. TEM image of Fe_3O_4 prepared by microemulsion method

Figure 2 depicts the TEM images of the Fe_3O_4 nanoparticles. From these Figures, it can be seen that the Fe_3O_4 nanoparticles synthesized have uniform particle sizes. Figure 2.a shows that the average grain diameter of the product prepared by hydrothermal method is about 10nm. Figure 2.b shows that the average grain diameter of the product prepared by microemulsion method is about 15nm but the particles have a certain degree of soft agglomeration. The particles themselves are magnetic and nanoparticles have a large surface area so that the mutual attraction between the particles is large. A comparison of the two images indicates that product prepared by the microemulsion method is better dispersed.

The particle size of the products measured by the laser grain size analyzer are shown in Figure 3.a and Figure 3.b.
Figure 3.a shows most particles which are prepared by hydrothermal method are in the range of 40-100nm. Figure 3.b shows most particles which are prepared by microemulsion method distribute in the range of 40-150nm. The grain diameter determined by laser grain size analysis meter is large because the particles have a certain degree of agglomeration and the grain sizes determined should be grain sizes of agglomerated big particles.
It also can be seen that the nanocrystals prepared by hydrothermal method have a narrower particle size distribution than those prepared by microemulsion method. The degree of agglomeration of the product obtained by hydrothermal method is smaller.
The magnetization curves of the prepared samples are shown in Figure 4.
Figure 4 show that the saturation intensity of product prepared by hydrothermal method is $62.7Am^2/kg$ and the saturation intensity of product prepared by microgalactic method is 60.9 Am^2/kg. The coercive force and residual magnetization of two samples are relatively small, which can be considered to have superparamagnetism. The reason that samples have some residual magnetism and the coercive force may be: there are some large particles which may be greater than the super-paramagnetic critical size. However, the proportion of large particles in samples is very

small, so residual magnetism and the value of coercive force of samples are not large and samples still have superparamagnetism. It is confirmed that the magnetism of Fe_3O_4 nanoparticles made by two methods were not significantly different. Comparatively, product prepared by hydrothermal method is better.

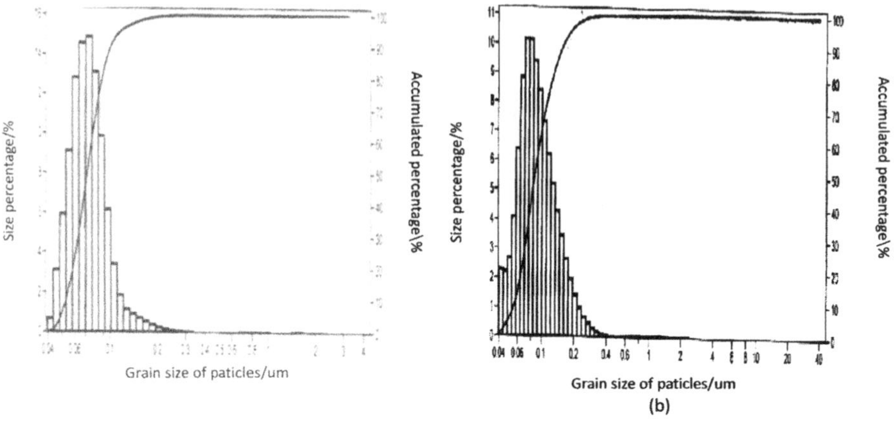

Figure 3. a.Grain size pattern of Fe_3O_4 prepared by hydrothermal method b.Grain size pattern of Fe_3O_4 prepared by microemulsion method

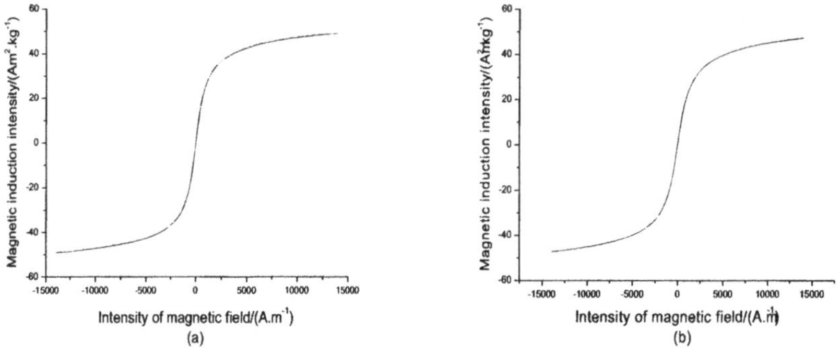

Figure 4 a. B-H curve of Fe_3O_4 prepared by hydrothermal method; b. B-H curve of Fe_3O_4 prepared by microgalactic method

4 CONCLUSION

Fe_3O_4 magnetic nanoparticles were prepared by hydrothermal method and microemulsion method. It is confirmed the products prepared by both methods were pure Fe_3O_4 magnetic nanoparticles and exhibit good superparamagnetism.

(1) The average grain diameter of the product prepared by hydrothermal method is about 10nm and the average grain diameter of the product prepared by microemulsion method is about 15nm. The grain size of particles synthesized by the hydrothermal method were narrower.

(2) The saturation intensity of product prepared by hydrothermal method is 62.7Am^2/kg and the saturation intensity of product prepared by microgalactic method is 60.9 Am^2/kg. Product synthesized by hydrothermal method has better magnetism.

In conclusion, Fe_3O_4 magnetic nanoparticles which were prepared by hydrothermal method show some advantages. Using hydrothermal method can be conducive to the improvement of magnetic properties. Moreover hydrothermal preparation is carried out in a closed container resulting in relatively high-pressure [(0.3 ~ 4)] MPa to avoid the components volating and improve the purity of the product.

REFERENCES

[1]Z. H. Zhao, S. W.Yao, W. G. Zhang. Preparation and Current Status of Fe_3O_4 Magnetic Nanoparticles, *Chemical Industry and Engineering Progress*, **24**, 865-868(2005).

[2]M. Bai, J. Zhou. Nanomagnetic Materials and its Development, *Information Recording Materials*, **13**, 38-39(2002).

[3]R. Massart. Preparation of Magnetite Nanoparticles, *IEEE Trans Magn*, **17**, 1247-1250(1981).

[4] J. C. Dubois, P. Exbrayat, M. L. Couble. Effect of New Machinable Ceramic on Behavior of Rat Bone Cells Culture Dinvitro, *J. Biomed Mater Res*, **43**, 215-225(1998).

[5]Q. R. Geng, R. L. Jiang, G. F. Liu, Y. C. Liu, G. C.Zhang, X. Guo. Preparation of Fe_3O_4 Magnetic Fluid by Hydrothermal Method and Its Characterization, *J. of University of Science and Technology of Suzhou(Engineering and Technology)*, **19**, 51-53(2006).

[6]Q. X. He, W. Z. Li, Q. Q. Chen, Preparation of Nano-Fe_3O_4 Powder by Hydrothermal Method, *J. of Guangxi University(Engineering and Technology)*, **29**, 170-174(2004).

[7]Q. X. He, H. Yang, Q. Q. Chen, Study on Preparative Condition of Magnetic Fe_3O_4 Nanoparticle Synthesized From Method of Microemulsification, *Magn Mater Devices*, **2**, 9-11(2003).

NANO-ZIRCONIA/MULLITE COMPOSITE CERAMICS PREPARED BY IN-SITU CONTROLLED CRYSTALLIZATION FROM THE Si-Al-Zr-O AMORPHOUS BULK

Liang Shuquan *, Zhong Jie, Zhang Guowei, Tan Xiaoping
School of Material Science and Engineering, Central South University, Changsha 410083
P.R.China

Abstract

Zirconia-mullite nano-composite ceramics were fabricated by in-situ controlled crystallizing from the Si-Al-Zr-O amorphous bulk, which were first heat treated at 900~1000℃ for nucleation, then treated at higher temperature for crystallization to obtain ultra-fine zirconia-mullite composite ceramics. The effects of treating temperature and ZrO_2 addition on mechanical properties and microstructure were analyzed. A unique structure in which there are a lot of near equiaxed t-ZrO_2 grains banding to mullite by partially coherent inter phase boundary and fine micro-cracks had been developed for the samples with 15wt% zirconia addition treated at 1150℃. This specific microstructure was much more effective in toughening and strengthening ceramics matrix and resulted in the best mechanical properties with 520MPa flexural strength and 5.13 $MPa{\cdot}m^{1/2}$fracture toughness respectively. Either higher zirconia addition or higher crystallization temperature would lead to a larger size rod-like ZrO_2 and mullite grains to be developed, which were of negative effect on mechanical properties of this new composite ceramics.

Keywords: Si-Al-Zr-O amorphous bulk; crystallization, Zirconia-mullite composite; structure, mechanical property

1. Introduction

Mullite and its composite ceramics have achieved outstanding importance as potential candidates for high temperature structured applications due to their favorable thermal and mechanical properties [1-4]. However, wider applications would be obtained if only their low flexural strength (150MPa) and low fracture toughness (1.8 $MPa{\cdot}m^{1/2}$) could be improved. Many strategies have been developed to improve the mechanical properties of mullite ceramics such as adding ZrO_2 component [5-7] and dispersing SiC particles [8], carbon nanotubes [9] and other micro- or nanoparticles in the mullite ceramic matrix as reinforcing phases.

Dispersing metastable tetragonal zirconia (t-ZrO_2) particles in a mullite matrix is a well-known and relatively cheap route to reinforce mullite [6, 7]. Particularly, by adding some stabilizing agents, transformation of tetragonal zirconia(t-ZrO_2) to monocline zirconia(m-ZrO_2) in the cooling process could be prohibited and a better toughening effect could be obtained[6-8, 10, 11].

Recently, some other non- conventional ways have been used to prepare zirconia-mullite ceramics [12-17]. In our previous work, nano-zirconia/mullite composite ceramics were prepared by in-situ controlled crystallization from the Si-Al-Zr-O amorphous bulk, the crystallization kinetics and the general view on the performance were discussed for the first time. In this article, the detail crystallization behavior, structure changes and their effect on mechanical properties for this new nano composite ceramics with high performance have been evaluated.

2. Experimental procedure

2.1. Samples preparation

The batch powder contents were 30~45wt%SiO_2、30~40wt%Al_2O_3、10~25wt%ZrO_2 and a small amount of MgO,CaO as additives, which were mixed and homogenized by ball milling with zirconia balls for 10 hours. Then 20g of mixed powders were put into Al_2O_3 crucible which was heated in air to temperatures in the range of 1 620～1700℃ for 2~4 hours in an electric furnace. The homogeneous flux was thereafter poured into a stainless mold and deeply cooled with liquid nitrogen to produce amorphous bulks. Afterward, the as received amorphous bulks were first treated at 900~1000℃ for nucleation, then treated at higher temperature for crystallization to obtain nano-zirconia / mullite composite ceramics. Samples were numbered as Z15, Z18 and Z20 indicated zirconia weight percentage of 15wt%, 18wt% and 20wt% respectively.

2.2. Characterization

Phase compositions were measured by X-ray diffraction using Japanese D/MAX 2500VB instrument in a step-scanning mode with Ni-filtered Cu-Kα as the radiation source and the radiation is over a range of 10～80°. The volume fractions of tetragonal zirconia (V_t) are calculated by the following equations [18]:

$$V_t = 1 - V_m \qquad (1)$$

V_m is the volume fraction of m-ZrO_2, which could be calculated by

$$V_m = \frac{PX_m}{1+(P-1)X_m} \qquad (2)$$

Where X_m is the integrated intensity ratio, and $P = 1.340$

$$X_m = \frac{I_{m(\bar{1}11)} + I_{m(111)}}{I_{m(\bar{1}11)} + I_{m(111)} + I_{t(101)}} \qquad (3)$$

Here I_m and I_t are the peak heights of m-ZrO_2 and t-ZrO_2.

After crystallization, the bulk density of samples was measured using the Archimedes' technique. The micro structure was examined by scanning electronic microscope (SEM) using a Siri-on200 microscope. The bulk samples for SEM observation were etched by 1vol% hydrofluoric acid-water solution after polished and washed 3 times with deionized water. The crystallization status of t-ZrO_2 was examined by selected-area electron diffraction to powder samples using a Tecnai G2 20 S-TWIN transmission electron microscopy (TEM). Super fine structures were observed by high resolution electron microscope (HREM). The flexural strength of the samples was determined from a three-point bending test. The fracture toughness (K_{IC}) was measured by using an indentation micro-crack method with a load of 10 Kg and a holding time of 15 s.

3. Results and discussion

3.1. Phases state and structure analysis

Fig.1 is the X-ray diffraction patterns of Z15, Z18 and Z20 samples treated at different temperatures. The nucleation test results at 900℃, 1000 ℃ for Z20 sample are also given; the precipitated phases at different crystallization temperatures are listed in Table 1, respectively.

Z20 sample was still amorphous of pre-heated at 900℃. When increase the temperature up to 1000℃, the small amount of t-ZrO_2 was precipitated firstly. When the temperature increases up to 1100℃, the mullite, cristobalite, m-ZrO_2 were precipitated. When the heating temperature was higher than 1150, for all samples there are no more new phases generated. The main phases were t-ZrO_2, mullite, cristobalite, m-ZrO_2 and a trace of cordierite.

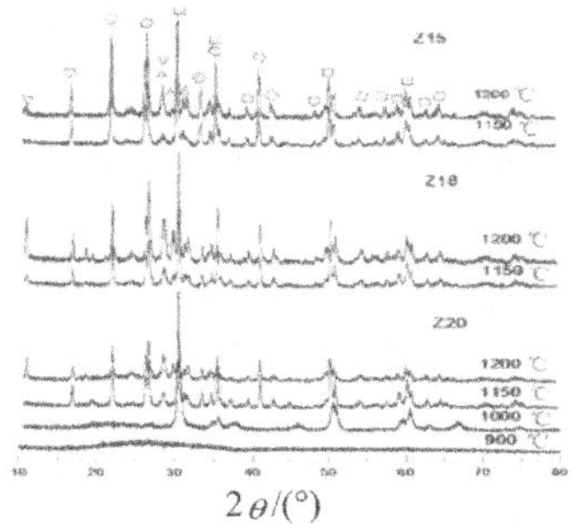

Fig.1 XRD patterns of Z15, Z18 and Z20 samples treated at different temperature.
● – mullite; ▲—m- ZrO_2; ▼ – cordierite;
■ – t-ZrO_2; ◆ – SiO_2

Table 1 Phase components of Z20 samples heat-treated at different temperature

Temp. /℃	Phases
900	amorphous
1000	t-ZrO_2
1100	t-ZrO_2, mullite, cristobalite, m-ZrO_2
1150	t-ZrO_2, mullite, cristobalite, m-ZrO_2, cordierite

Fig.2 shows the SEM micrograph of Z20 heat-treated at 900℃. It indicates the phase segregation occurred in the amorphous bulk, for the preparation for new phase crystallization. Fig.3 is the TEM dark field micrograph of Z20 sample heat treated at 1000℃. The particles with bright colour are

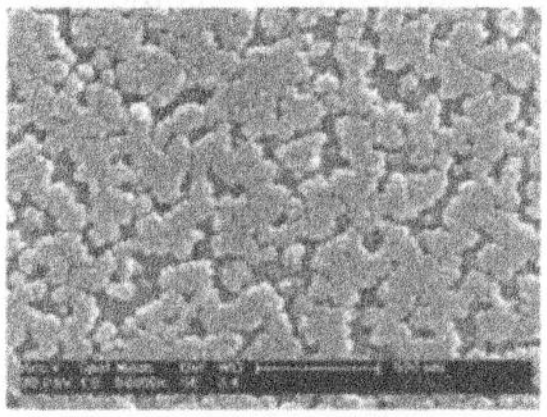

Fig.2 SEM micrograph of Z20 pre-treated at 900℃

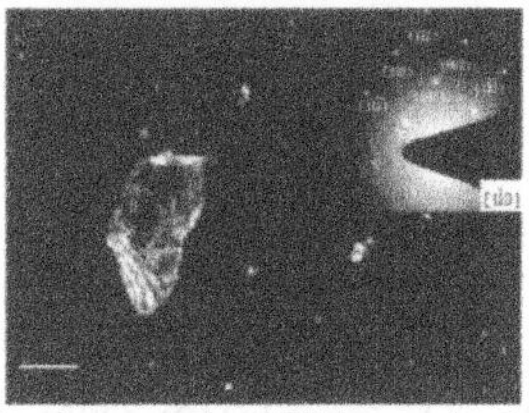

Fig.3 TEM dark field micrograph of Z20 powder heat-treated at 1000℃

ZrO_2, smaller than 50nm. The selected area diffraction spots for bright colour particles is inserting at the right corner, and the analysis results to the diffraction spots showed that ZrO_2 is of the tetragonal crystal structure(t-ZrO_2). This was in agreement with XRD results, as shown in Fig.1.

In order to promote the crystallization process for mullite and other new phases, it is better to increase the heat-treating temperature, such as up to 1150℃ or 1200℃. The experimental results show that the overall best materials are achieved at 1150℃ heat treatment. Although a higher temperature, such as 1200℃, is helpful for speeding up the new phase forming, but the higher temperature is much harmful to the mechanical properties of this new materials due to the specific structure change, which will be discussed later on. Fig.4 is TEM micrograph of Z20 sample treated at 1150℃, from which it could be found that the structure is very fine and uniform. Therefore, this method can be used to fabricate nano-zirconia/mullite composite ceramics and has a lot of advantages, such as unnecessary use of nano size starting powder.

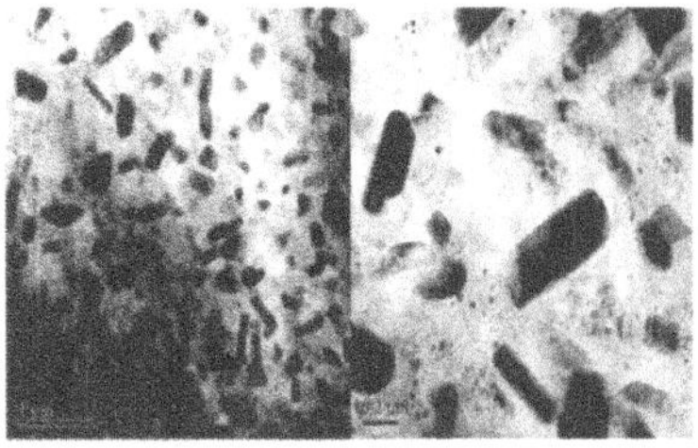

Fig.4 TEM micrograph of Z20 sample treated at 1150℃

In order to understand the structural change well, a further super fine structure change analysis has been made. In terms of the related equilibrium phase diagram, Mg^{2+} and Ca^{2+} could be dissolved into t-ZrO_2 lattice to form the t-ZrO_2 solid solution t-ZrO_2 (ss), and Al^{3+} could not be dissolved into t-ZrO_2. But in recent years, G. Kimmel found that Al^{3+} could be dissolved into t-ZrO_2 lattice to form a special t-ZrO_2 solid solution. Here it was indicated by t-ZrO_2 (k-ss).Compared two similar groups of crystal planes (112), (200) and (103), (211)of the normal t-ZrO_2 respectively, as shown in Fig.5, it is obvious that the 2θ values of (112), (200) were consistent with the normal t-ZrO_2 well, but the 2θ values of (112), (103) were not. In addition, from Fig.5, it also could be found that the 2θ value differences increased with the treated temperature increase. This means the lattice of t-ZrO_2 (ss) dilated due to the Al^{3+} solution.

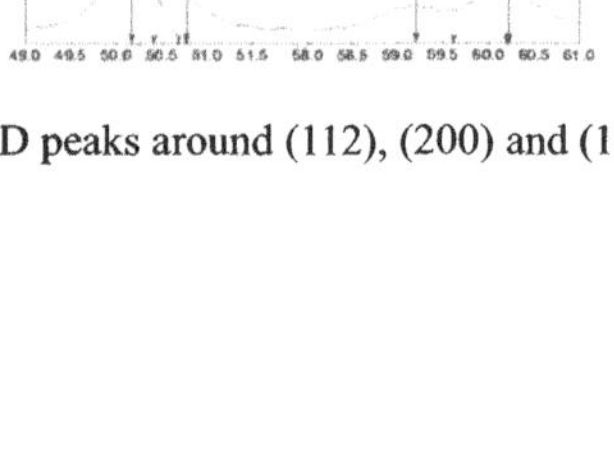

Fig.5. Partially details of XRD peaks around (112), (200) and (103), (211) for t-ZrO_2 (ss).

Fig.6 is the SEM with energy spectrum micrograph of Z20 heat-treated at 1300℃. Some Al^{3+},$Mg^{2+}$and Ca^{2+} were detected in the lattice of t-ZrO_2. Some point defects could be observed in high-resolution TEM micrograph of ZrO_2 particles, seeing Fig.7

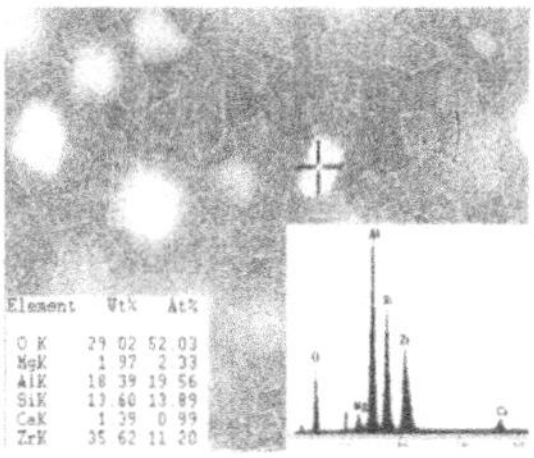

Fig.6 SEM and energy spectrum micrograph of Z20 heat-treated at 1300℃

Fig.7 High-resolution TEM micrograph of ZrO_2 particles in Z20

Fig.8-a, b, c, d show XRD intensity changes for the main planes of t- ZrO_2 (k-ss), m- ZrO_2, mullite and cordierite respectively.

Fig.8-a shows that the crystallization of the t-ZrO_2(k-ss) reached the peak at 1150℃, then decreased significantly with the temperature increased to 1200℃ and 1250℃. From Fig.8-b, it could be seen that the intensity of m-ZrO_2 was linearly increased with the heat-treated temperature increase. Hence, the crystallization of ZrO_2 might be finished at 1150℃. Fig.8-c demonstrates the crystallization of mullite which began at 1100℃, and reached to the maximum at 1150℃, then

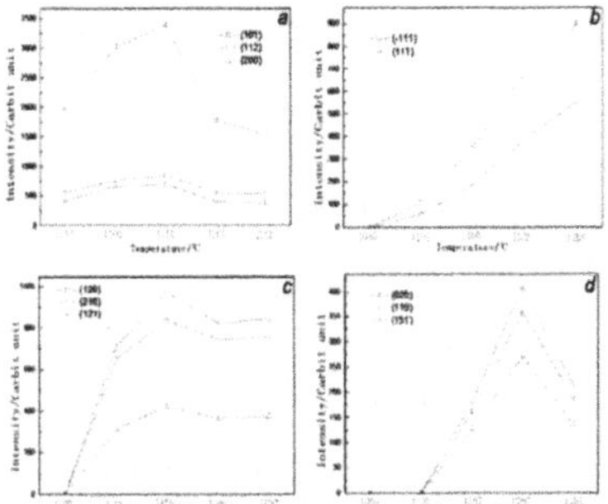

Fig.8. X-ray diffraction intensity of the major phases of Z20 sample.
a: t-ZrO_2 (k-ss) b: m-ZrO_2 c: mullite d: cordierite

crystallization volume changed a little with temperature increased. This meant the structure of mullite was stable. Fig.8-d shows the crystallization of cordierite, which began at 1150℃, and reached the peak at 1200 ℃, and decreased significantly at 1250℃.Since the formation of cordierite needed some Al^{3+}, Mg^{2+} from the t-ZrO_2(k-ss) solid solution, it affected the structure of t-ZrO_2 (k-ss) to a great extent. Amended by Rietveld Diffraction Peak Shape Fit Method, the calculated values of the volume of t-ZrO_2(k-ss) cells and c/a ratio of t-ZrO_2 are showed in Fig.9 and Fig.10. From Fig.9 and 10, it could be found that there was a wave fluctuation between 1150℃ and 1200℃.

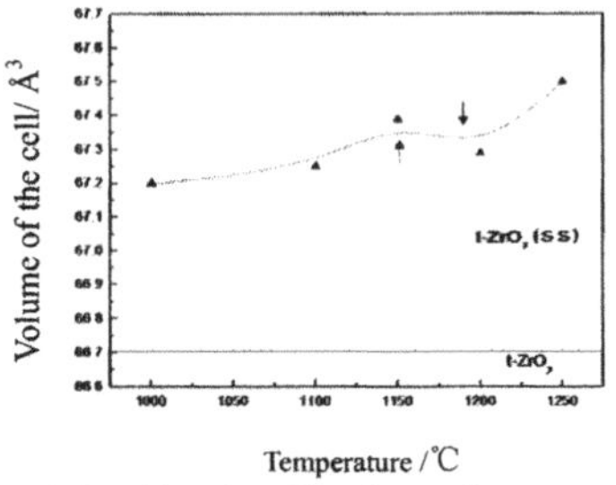

Fig.9 .Volume of t-ZrO_2 (k-ss) cells of samples treated at different temperature.

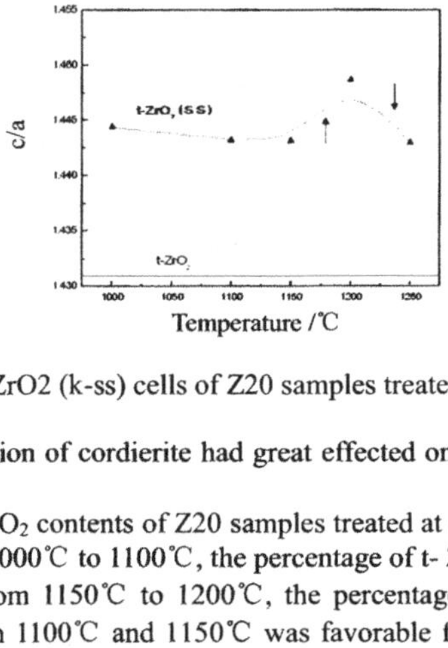

Fig.10 c/a ratio of t-ZrO2 (k-ss) cells of Z20 samples treated at different temperature.

These indicated that the formation of cordierite had great effected on the structure and phase state of t-ZrO_2 (k-ss) .

Table 2 shows the t-ZrO_2, m-ZrO_2 contents of Z20 samples treated at different temperature. Raising heat-treated temperature from 1000℃ to 1100℃, the percentage of t- ZrO_2 increased by 62.4%, and with the temperature raised from 1150℃ to 1200℃, the percentage decreased about 29%. This means the temperature between 1100℃ and 1150℃ was favorable for preserving high content of

Table 2 The t-ZrO_2、m -ZrO_2 content of treated at different temperature Z20 samples (vol.%)

Temp/℃	1000	1100	1150	1200	1250
t-ZrO_2	8.5%	13.8%	12.4%	8.8%	6.5%
m-ZrO_2	-	0.7%	3.1%	6.7%	9.0%

t- ZrO_2. Compared to the maximum content, the content of t- ZrO_2 decreased about 36.2% and 55.8% after treated at 1200℃ and 1250℃ respectively. Lower content of t- ZrO_2 was unfavorable for high performance material preparation, so the optimum treatment temperature was in the temperature range of 1100 ℃ to 1150℃.

3.2. Mechanical properties

Fig.11 shows the flexural strength and fracture toughness of samples Z15, Z18 and Z20 treated at 1150℃ and 1200℃ for 1 hour crystallization, respectively. From the results, it could be found that the mechanical properties decreased as the heat treatment temperature increasing. Samples heat-treated at 1150℃ for an hour have better mechanical properties than that treated at 1200℃, but with an exception of the fracture toughness of Z18. The Z15 samples heat-treated at 1150℃ have the conditional optimized properties with 520MPa flexural strength and 5.13MPa·m$^{1/2}$ fracture toughness, respectively, as shown Fig.12.

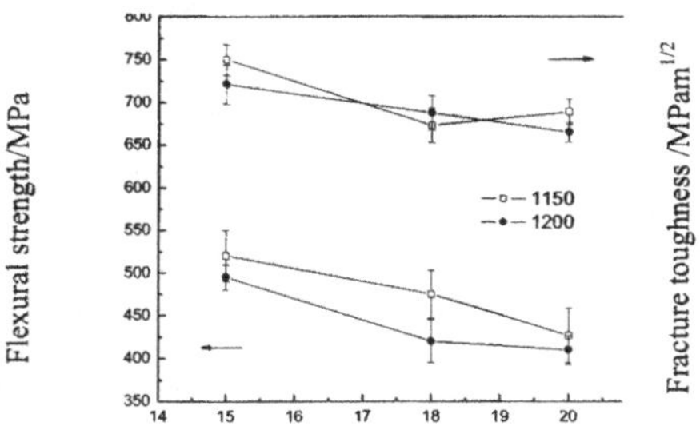

Fig.11. Mechanical properties of Z15, Z18 and Z20 samples treated at different temperature

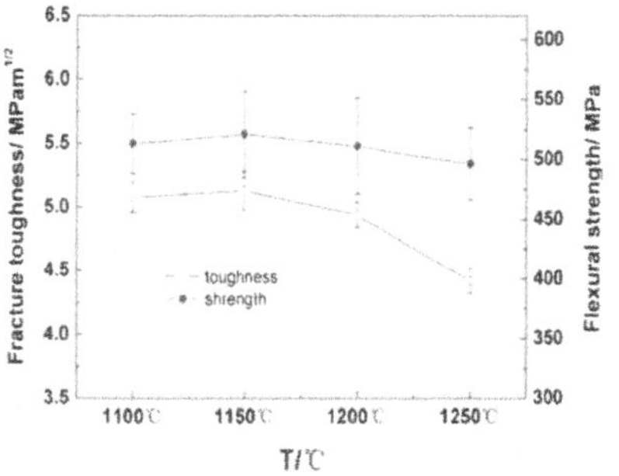

Fig.12. Effects of crystallization temperature on mechanical properties of sample Z15

The flexural strength is 40% higher than that of the zirconia-toughened mullite ceramics which were fabricated by conventional methods [10]. This result demonstrated that in order to obtain better mechanical properties, it was better to control the heat-treatment temperature to not be over 1150℃. From the Fig.11, it also could be found that the mechanical properties decreased with the content increasing of ZrO_2 in starting materials after the content of ZrO_2 over 15wt%. This means that the further increase of zirconia after 15wt% ZrO_2 addition has little help for the mechanical property improvement. That is, the optimum content of ZrO_2 in starting materials should not be over 15wt%.

3.3. Microstructure

The average densities of Z15, Z18 and Z20 samples were given in Table 3. The densities increase with both the ZrO_2 content increasing and increasing of heat-treating temperature. Since the conditional best mechanical properties is obtained by Z15 sample treated at 1150℃, there should be other structural factors to play important role in the determination of the sample mechanical properties, except for the density and the volume fractions of t-ZrO_2.

Table 3 The average densities(g/cm^3) of samples heat-treated at 1150℃ and 1200℃

Samples	1150℃	1200℃
Z15	2.66	2.67
Z18	2.97	3.01
Z20	3.15	3.19

Fig.13 shows the back-scattered scanning electron micrographs of Z15 and Z18 samples, heat-treated at1150℃ and 1200℃ for 1 hour respectively. As signed in pictures, the particles with bright colour are ZrO_2 grains, and the dark or gray area is other phases. It could be found that the Z15 samples heat-treated at 1150℃ contained a lot of very small ZrO_2 grains, only about 60nm. This is a very important for the better mechanical property obtainment. The other important obvious characteristics in Fig.13a are that there are a lot of micro-cracks in this new ceramic composite. These micro-cracks will give their contribution to the improvements of the mechanical properties by micro-cracking toughening mechanism. With the increase of the heat treatment temperature, such as to 1200℃, the grains grew significantly. Particularly, the ZrO_2 grains grew very rapidly. Most of them had a size of 200 nm, but less than 300nm, as shown in Fig.13 b. There was 1.4vol% of t-ZrO_2 to be transformed into m-ZrO_2 according to the calculation. The other obviously structural change is the decrease of micro-cracks in the matrix, compared with the Z15 sample heat treated at 1150℃. These two important structural changes resulted in the mechanical properties difference for Z15 samples treated at 1150 ℃.and 1200℃.

Fig.13 c and d are SEM micrographs of Z18 samples heated at 1150℃ and 1200℃, respectively. With the increase of the heat treatment temperature, the average grain size of ZrO_2 particles increased from 200nm in Fig.13 c to 500nm in Fig.13 d. This resulted in 2.3vol% t-ZrO_2 phase to be transformed into m-ZrO_2 according to the calculation. On the other hand, it could be found that there were much less micro-cracks in the matrix for the Z15 sample heat treated at 1200℃ (Fig.13 b). Therefore, only poor mechanical properties could be obtained for Z18 samples, as shown in Fig.11. The flexural strength drops about 10%, and fracture toughness drops about 20%.

Comparing Fig.13 a and Fig.13 c, it could be found that the zirconia addition in starting materials had significant influence on microstructure for this new composite ceramics. In Z18 sample, both the ZrO_2 and mullite grains are of much larger grain size than in Z15 samples. The largest grain for both ZrO_2 and mullite approaches 300 nm approximately. In addition, the zirconia addition increase obviously reduces the micro-cracks in the matrix. These structural changes resulted in the mechanical properties difference for Z15 and Z18 samples treated at 1150 ℃, which was indicated in Fig.11.The third important aspect is the effect of the zirconia addition in starting materials on the ZrO_2 and mullite grain morphology. The ZrO_2 grains in the samples contained less than 15%wt zirconia are near equiaxed, see Fig 13 a, but in higher zirconia content samples, like Z18, both ZrO_2 and mullite grains are of rod-like grain morphology. The t-ZrO_2 volume fraction decrease markedly when the samples contained more rod-like grains according to the calculations. This is because the rod-like t-ZrO_2 grains are easily to be grown over the critical size, and result in tetragonal-monocline transformation. This process will be accelerated by both the higher heat treating temperature and higher zirconia addition in starting materials. The mechanical properties decrease linearly with the decrease of t-ZrO_2 volume fraction. Therefore, equiaxed ZrO_2 grains are more effective in toughening mullite martrix. The conditional optimum content of zirconia should be about 15wt%.

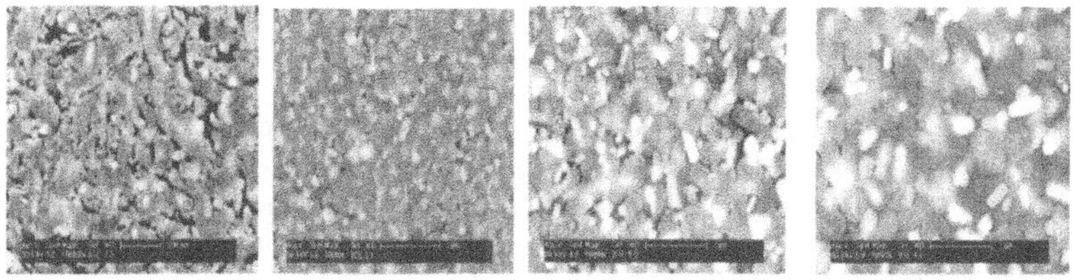

Fig.13 SEM micrographs of Z15, Z18 and Z20 samples treated at different temperature.
a: Z15,1150℃/1h ,b: Z15,1200℃/1h ,c: Z18,1150℃/1h , d: Z18,1200℃/1h

Fig.14 High-resolution TEM micrograph of t- ZrO_2 particles in Z15

In order to understand the structure relationship between t-ZrO_2 and mullite, the high resolution TEM observation was made to the Z15 tested sample, as shown in Fig.14, from which it could be seen that the partially coherent inter phase boundary was developed. This indicated the strength of the boundary between t-ZrO_2 and mullite is strong enough. This structural characteristic is of great importance for obtaining the high performance composite ceramics.

4. Conclusions

The Si-Al-Zr-O amorphous bulks as-received by deep cooling could be used to prepare a new homogenous nano zirconia/mullite composite ceramics by in-situ controlled crystallization. The main phases were zirconia and mullite. Zirconia starts to be precipitated at around 1000℃ in t-ZrO_2 phase state, and mullite at about 1100℃. The Z15 sample which was treated at 1150℃ for crystallization will develop a unique structure with a lot of near equiaxed t-ZrO_2 grains and micro-cracks, and have the best mechanical properties. The flexural strength and fracture toughness were 520MPa and 5.13 $MPa{\cdot}m^{1/2}$ respectively. Either higher zirconia addition or higher crystallization temperature would lead a larger size rod-like ZrO_2 and mullite grains to be developed, which are of negative effect on mechanical properties. The formation of cordierite had great effect on the structure and phase state of t-ZrO_2 (k-ss) and produced harmful results for high performance materials preparation. The partially coherent inter phase boundaries developed between t-ZrO_2 and mullite was helpful for the improvement of the materials performance.

References

[1] Eugene Medvedovski. Alumina–mullite ceramics for structural applications. *Ceramics International,* 2006, 32: 369~375.

[2] Feng-tao Lan, Ke-zhi Li , He-jun Li, Qian-gang Fu. A cordierite-mullite anti-oxidation coating forcarbon/carbon composites. *Carbon,* 2007,45: 2692~2716.

[3] Zhang H.Y., Maljkovic N., Mitchell B.S., Structure and interfacial properties of nanocrystalline alumina/mullite composites. *Mater. Sci. Eng,* 2002, A326: 317~323.

[4] Ananthakumar S., Jayasankar M., Warrier K.G.K.. Microstructural, mechanical and thermal characterisation of sol-gel derived aluminium titanate-mullite ceramic composites. *Acta Materialia,* 2006,54: 2965~2973

[5] Khor K A , Yu L G, Li Y. Spark plasma reaction sintering of ZrO_2—mullite composites from plasma sphe-roidized zircon/ alumina powers. *Mater Sci Eng A,* 2003, 339 (12) : 286~296.

[6] Jin Xihai, Gao Lian, Kan Yanmei, Chen Yuru,Yuan Qimin. Influence of Nb_2O_5 on the mechanical performances and toughening mechanism of ZrO_2 in ZTM-Al_2O_3. *Journal of Inorganic Materials,* 2000,15 (6) : 1009~1014.

[7] Ebadzadeh T, Ghasemi E. Effect of TiO_2 addition on the stability of t-ZrO_2 in mullite-ZrO_2 composites prepared from various starting materials. *Ceramics International,* 2002, 28(4): 447~450

[8] Hong J.S., Huang X.X., Guo J.K., Li B.S., Gui L.H.. Strengthen and toughening of mullite ceramics by SiC particles and Y-TZP. *J. Inorg. Mater.* 1990,5 (4): 340~345.

[9] Wang Jing, Kou Hua-min, Liu Xue-jian,Pan Yu-bai, Guo Jing-kun. Reinforcement of mullite matrix with multi-walled carbon nanotubes. *Ceramics International,* 2007, 33:719~722.

[10] Huang Yangfeng, Xie Gensheng, Xiao Hanning . The influence of CeO_2 in ZTM ceramics prepared by in-situ sintering. *Ceramics,* 2006 6 : 9~11

[11] Garrido L B , Aglietti E F. Reaction-sintered mullite-zirconia composites by colloidal processing of alumina-zircon-CeO_2 mixtures. *Mater Sci Eng A,* 2004, 369 (12) : 250~257.

[12] Garrido L.B., Aglietti E.F., Martorello L. , Camerucci M.A., Cavalieri A.L.. Hardness and fracture toughness of mullite–zirconia composites obtained by slip casting. *Materials Science and Engineering A,* 2006, 419: 290~296

[13] Zhao S.K., Huang Y., Wang C.An., Huang X.X., Guo J.K.. Sinterability of $ZrSiO_4$/a-Al_2O_3 mixed powders. *Ceram. Int,* 2003, 29: 49~53.

[14] Maitra S., RahamaA. n, SarkaA. r, Tarafdar A.. Zirconia–mullite materials prepared from semi-colloidal route derived precursors. *Ceramics International,* 2006,32: 201~206

[15] Liang Shu-quan , Li Shao-qiang , Tan Xiao-ping , Tang Yan , Zhang Yong. Crystallization behavior of Si-Al-Zr-O amorphous bulk. *The Chinese Journal of Nonferrous Metals.* 2005,15(8):1189~1193

[16] TAN Xiao-ping , LIANG Shu-quan , LI Shao-qiang , TANG Yan. Preparation of ZrO2 —mullite nano composite ceramics. *J. Cent. South Univ.* 2005,36(5):790~794.

[17] Monica Popa, Jose M, Calderon Moreno. Crystallization of gel-derived and quenched glasses in the ternary oxide Al_2O_3-ZrO_2-SiO_2 system. *J Non-cryst Solids.* 2002, 297: 290~300.

[18] Toraya H , Yoshimura M , Somiya S. Calibration curve for quantitative analysis of the monoclinic—tert ragonal ZrO_2 system by X—ray diff raction . *J Am Ceram Soc* , 1984 , 6 (2) : 112~11

PREPARATION AND CHARACTERIZATION OF Er:Gd_2O_3 POWDERS

ZHANG Rong, QIN Lian-Jie,* WANG Bo, FENG Zhi-Qiang, GE Ru
School of Environmental and Material Engineering, Yantai University, 32 Qingquan Road, Yantai 264005, Shandong Province, China

ABSTRACT
The co-precipitation method and sol-gel combustion method were used to prepare Er:Gd_2O_3 powders. The process of heat disassembly, phase composition, morphology and purity of the powders were investigated by TG-DTA, XRD, SEM and FTIR, respectively. The powders having cubic phase were obtained by sol-gel combustion method when citric acid and EDTA were used as combination fuel. They are fluffy, porous and agglomerated and its grains are irregular in morphology. Besides oxalate being generated, $Gd_2O_2CO_3$ is generated in the process of heat treatment by oxalic acid co-precipitation method. The decomposition temperatures of oxalate and $Gd_2O_2CO_3$ are about 401°C and 642°C. The grains prepared by acid co-precipitation method are flaky, having uniform size and clear crystal boundary. After 900°C heat treatment, the average length of the flaky grains is about 1μm, and the average thickness is about 20nm. The powders prepared by these two methods are a cubic phase, and easily absorb CO_2 to form CO_3^{2-} on the surface of the grains.

INTRODUCTION
Rare earth ions have been paid much attention to for their particular optical properties and magnetic properties. Rare earth compound have been widely used as laser materials, luminescent materials, coloring agents of ceramics or glass and so on[1]. The optical materials doped erbium ion have been widely investigated[2-5] for this ion has abundant energy levels. The energy levels ($^2H_{11/2}\rightarrow{}^4I_{15/2}$ and $^4S_{3/2}\rightarrow{}^4I_{15/2}$) of erbium ion have a large emission section and are easy to achieve upconversion, so erbium ion is a good active ion as upconversion phosphors materials. The energy level transition($^4I_{13/2}\rightarrow{}^4I_{15/2}$) can emit out 1.5 μm eye-safe laser radiation and the ion is also a well active ion for the eye-safe laser material. Gd_2O_3 powder with a cubic structure is a good host material for its excellent photics and thermal properties. It not only can be used as upconversion phosphors material but also can be used to prepare transparent ceramics as a host material.

Co-precipitation method, sol-gel method, freeze drying method, spray pyrolysis method and combustion method are usually used to prepare powders. Powders prepared by the combustion method have small size, high purity and good chemical stability. Urea, glycine, carbohydrazide, citric acid and so on were used as the fuel in this paper. According to the previous report, Gd_2O_3 powders have a monoclinic structure[6] when citric acid was used as the fuel. While we used the citric acid and EDTA as combination fuel, the Gd_2O_3 powders prepared had the cubic structure in this paper. As comparison, oxalic acid co-precipitation was also used to prepare the powders.

EXPERIMENTAL

Powders synthesis

Oxalic acid co-precipitation method

Er_2O_3 and Gd_2O_3 (Er_2O_3 doped concentration 5%) with purity of 99.999% were used as starting materials. The starting materials were dissolved into the nitric acid solution. Then, a suitable amount of oxalic acid was dissolved into distilled water. The rare earth solution was added into the oxalic acid solution at suitable speed. The pH value of the mixture was adjusted to 3.5~4.0 using

**Corresponding author.* E-mail address: lianjieqin@126.com

ammonia, and let stand for 24h. The precipitate was washed with distilled water and ethanol, and then dried at 80°C for 20 hours in an oven to obtain the precursor. The precursor was heat-treated with different temperatures for 2h in a muffle.

Sol-gel combustion method

Er_2O_3 and Gd_2O_3 (Er_2O_3 doped concentration 5%) with purity of 99.999% were used as starting materials. The starting materials were dissolved into the nitric acid solution. Then, according to the mol number of rare ions used in our experiments, a certain of citric acid and EDTA was dissolved into distilled water. The rare earth solution was added into the mixture of citric acid and EDTA. After that, the solution was stirred below 100°C until the sol became gel. The wet gel was dried to drain away the superabundant water, thereby, transforming it to dry gel in an oven at 80°C. Then, the temperature of the oven was adjusted to 200°C. The black porous sponge-puffy stuff as the precursor was obtained from the dry gel combustion in an oven at 200°C. The precursor was then milled to fine power and heat-treated with different temperatures for 2h in a muffle.

Experimental Measurement

The process of heat disassembly of the powders was analyzed by thermal analyzer (ZRY-2P, Shanghai, 10°C/min). The crystal structure and phase type of the powder were determined by X-ray diffractometer (XRD-7000, Japanese, 40Kv, 30mA, CuKa). The morphology of the powder was examined by scanning electron microscopy (SEM, JSM-5600LV, Japanese, 15kV). The purity of the powder was analyzed by Fourier transform infrared spectrometer (IRPrestige-21, Japanese, $4600cm^{-1}$~ $400cm^{-1}$).

RESULTS AND DISCUSSION

Thermal behavior analysis

Figure 1 shows the TG-DTA curves of the obtained precursor with oxalic acid co-precipitation method heated at a rate of $10°C·min^{-1}$ in air. It is noted that there are mainly three stages of weight loss. The first weight loss step (4.1%) is in the range of 209°C to 313°C accompanied by a small endothermic peak near 227°C in the DTA curve owing to the loss of the internal combined water and NH_4^+ in the precursor. The second weight loss step (37.4%) is noticed between 323°C and 500°C accompanied by a small endothermic peak near 401°C in the DTA curve due to the decomposition of the oxalate. The third weight loss step (3.6%) is from 618°C to 690°C, and a small endothermic peak appears near 642°C in the DTA curve because of the decomposition of the $Gd_2O_2CO_3$ into Gd_2O_3, and there is an exothermic peak near 610°C in the DTA curve caused by powder crystallizing. After 690°C, the TG and DTA curves are all stable.

Figure 2 shows the TG-DTA curves of the dry gel and the precursor obtained with sol-gel combustion method heated at a rate of $10°C·min^{-1}$ in air. Both dry gel and precursor only have a big weight loss step in their TG curves. This weight loss (57.9%) is in the range of 226°C to 282°C accompanied by a strong exothermic peak near 281°C in the DTA curve owing to oxidation and reduction reaction between carbonaceous groups and NO_3^-. The weight loss of precursor (73.0%) is from 220°C to 241°C, and an exothermic peak appears near 240°C in the DTA curve because of the oxidation and reduction reaction between remaining unburned carbonaceous groups and NO_3^-.

XRD analysis

With the heat-treatment of different temperatures, powders prepared by oxalic acid co-precipitation method were characterized by XRD, and the results are shown in Fig.3. It can be seen that the precursor is amorphous, and the precursor with 600°C heat-treatment is also amorphous. The

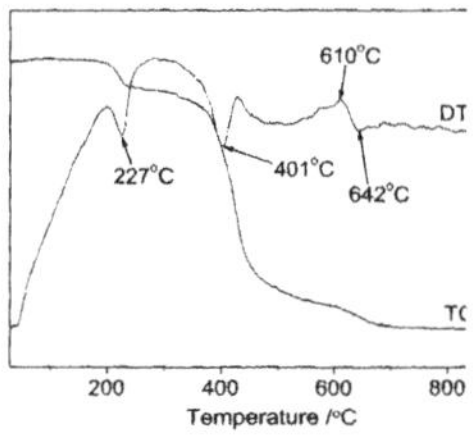

Figure 1. TG-DTA curves of precursor

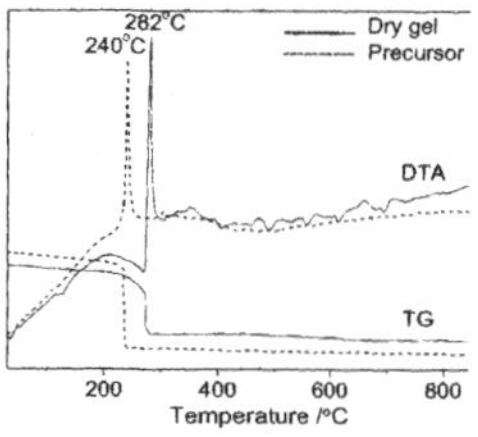

Figure 2. TG-DTA curves of dry gel and precursor

diffraction peaks of the precursor with 800°C heat-treatment are all in good agreement with those of the JCPDS card, which indicates that the pure cubic phase Gd_2O_3 is formed. Figure 4 shows the XRD curves of the powder prepared by sol-gel combustion method. As is shown in this figure, the precursor is amorphous, but the diffraction peaks of the precursor after 600°C heat-treatment perfectly coincide with the JCPDS of cubic phase Gd_2O_3. When pure citric acid was used as fuel, Gd_2O_3 powder prepared by Chen Si-shun et al was a monoclinic phase[7]. However citric acid and EDTA are used as fuel combination instead of pure citric acid, powders prepared by sol-gel combustion method are a cubic phase. With the increasing of heat treatment temperature, diffraction peaks of the powders prepared by these two methods both become more and more sharp.

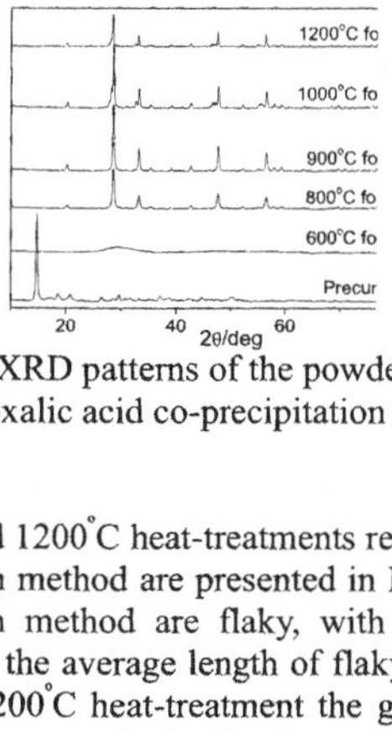

Figure 3. XRD patterns of the powders prepared by oxalic acid co-precipitation method

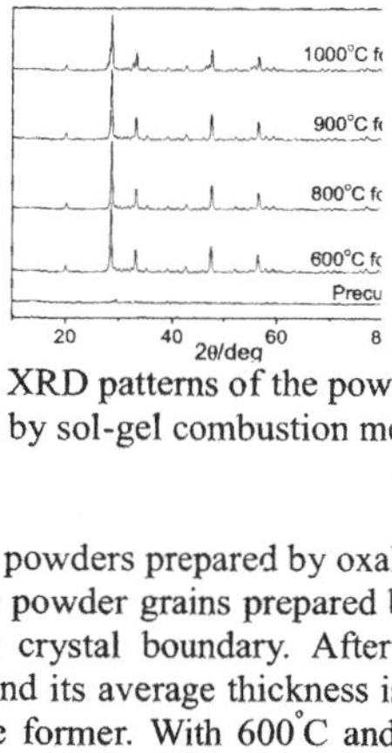

Figure 4. XRD patterns of the powders prepared by sol-gel combustion method

SEM analysis

With 900°C and 1200°C heat-treatments respectively, the images of powders prepared by oxalic acid co-precipitation method are presented in Fig.5. It is found that the powder grains prepared by acid co-precipitation method are flaky, with uniform size and clear crystal boundary. After 900°C heat-treatment, the average length of flaky grains is about 1 μm, and its average thickness is about 20nm. After 1200°C heat-treatment the grains are bigger than the former. With 600°C and 900°C heat-treatments respectively, the images of powders prepared by sol-gel combustion method are presented in Fig.6. After 600°C and 900°C heat-treatments powders are fluffy, porous and agglomerated, whose grains are irregular in morphology.

FTIR analysis

The FTIR patterns of the precursor with 600°C and 1000°C heat treatments prepared by oxalic acid co-precipitation method are shown in Fig.7. In the pattern of the precursor, the peak near $3445cm^{-1}$ is assigned to the vibration of the hydroxyl group. The peaks near $3188cm^{-1}$, $3076cm^{-1}$, $1451cm^{-1}$

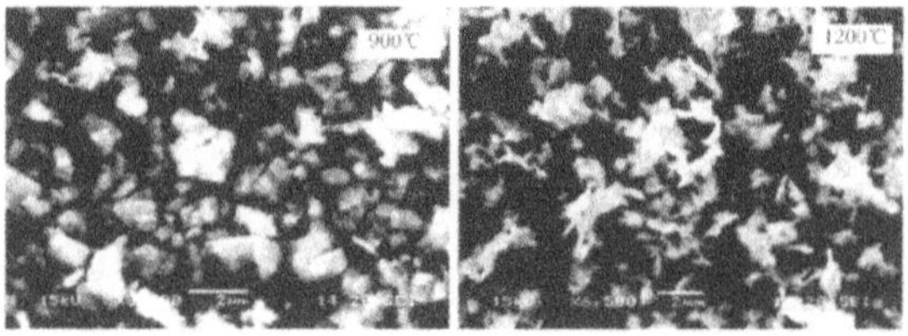

Figure 5. SEM images of the powders prepared by oxalic acid co-precipitation method

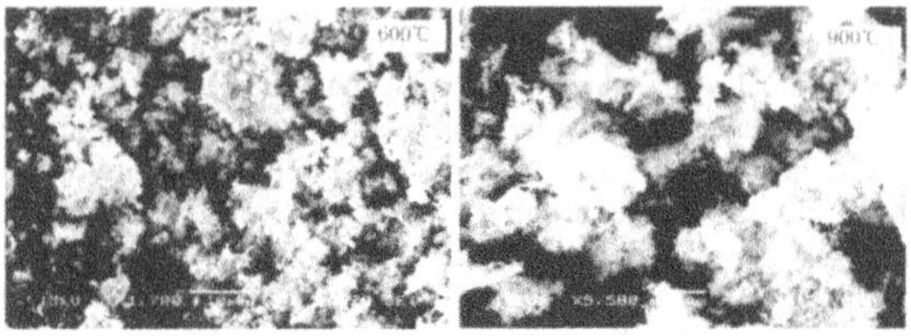

Figure 6. SEM images of the powders prepared by sol-gel combustion method

are corresponding to the vibration of NH_4^+. The peaks near 2862cm^{-1}, 1362 cm^{-1}, 1323cm^{-1}, 807cm^{-1} are associated with the vibration of the NO_3^-. The peak near 1635cm^{-1} is assigned to the vibration of COO^-. The peak near 1747cm^{-1} is due to the vibration of C=O. The peak near 490cm^{-1} is attributed to the vibration of Gd(Er)-O bond. In the pattern of the precursor with 600°C heat-treatment, the peak near 3445cm^{-1} is due to the O-H bond vibration of H_2O absorbed by powders. The peak near 400cm^{-1} is assigned to the vibration of Gd(Er)-O bond. The peaks near 1498cm^{-1}, 1388 cm^{-1}, 850cm^{-1} are attributed to the vibration of CO_3^{2-}, which indicates that $Gd_2O_2CO_3$ has been formed after 600°C heat- treatment. In the pattern of the precursor with 1000°C heat-treatment, the absorption peaks appearing near 540cm^{-1} and 440cm^{-1} are attributed to the vibration of metal Gd(Er)-O bond, which indicates that $Gd(Er)_2O_2CO_3$ has been resolved into $Gd(Er)_2O_3$, and there are still three absorption peaks of CO_3^{2-} near 1478cm^{-1}, but these three peaks get weak obviously, which is probably due to the absorption of CO_2 from ambient atmosphere.

Figure 8 shows the FTIR patterns of the precursor with 600°C and 1000°C heat treatments prepared by sol-gel combustion method. In the pattern of the precursor, a strong band at 3660cm^{-1}~2430cm^{-1} is assigned to the vibration of the hydroxyl group. The peak near 1602cm^{-1} is attributed to the

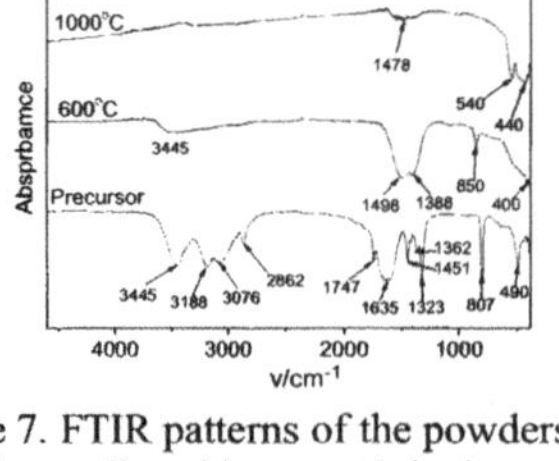

Figure 7. FTIR patterns of the powders prepared by oxalic acid co-precipitation method

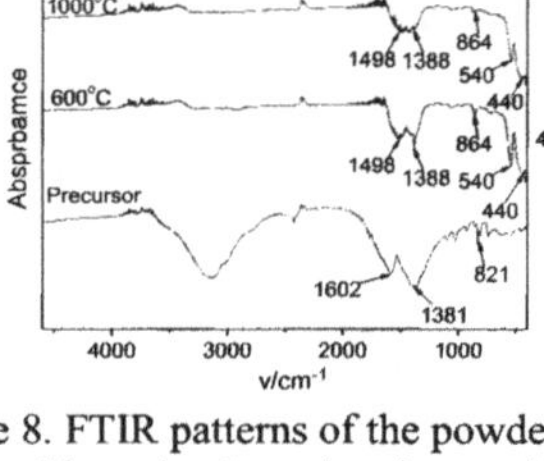

Figure 8. FTIR patterns of the powders prepared by sol-gel combustion method

vibration of -COO^{2-}. The peaks near 1381cm^{-1}, 821cm^{-1} are due to the vibration of NO_3^-. It is concluded that -COO^{2-} and NO_3^- can't react completely in an oven at 200°C, they need a higher reaction temperature. In the spectrum of the powder with 600°C heat- treatment, the peaks near 1498cm^{-1}, 1388cm^{-1}, 864 cm^{-1} are assigned to the vibration of CO_3^{2-}, which is probably due to the absorption of CO_2 from the combustion of carbonaceous groups rather than ambient atmosphere. These three peaks are still after 1000°C heat- treatments. the absorption peaks appearing near 540cm^{-1} and 440cm^{-1} are attributed to the vibration of metal Gd(Er)-O bond.

CONCLUSION

Powders prepared by sol-gel combustion method are fluffy, porous and agglomerated, whose grains are irregular in morphology. Powders prepared by acid co-precipitation method are flaky, having uniform size and clear crystal boundary. When citric acid was used as the fuel, Gd_2O_3 powders have a monoclinic structure, but we used the citric acid and EDTA as combination fuel, Gd_2O_3 powders own a cubic structure, which improved the performance of powders in some respect. Er:Gd_2O_3 powders prepared by these two methods are both easy to absorb CO_2 to form CO_3^{2-} on the surface of the grains. By general analysis, the oxalic acid precipitation method is better than the sol-gel combustion method to prepare Er:Gd_2O_3 powders.

ACKNOWLEDGEMENTS

This work is supported by the *natural science foundation of Shandong* province (Grant No. Y2007F37).

REFERENCES

[1]L. H. Yang, and H. W. Song, Synthesis and Luninescence Properties of Nanowfires and Nanoslices of Gd_2O_3:Er^{3+}/Yb^{3+}, *Chinese Journal of Lumnescence*, **27**, 987-990(2006).

[2]D. L. Zhang, P. R. Hua, E. Y. B. Pun, L. Sun, and Y. H. Xu, Emission Characteristics of Er^{3+} in Vapor-transport-equilibrated Er/Zn-codoped $LiNbO_3$ Crystals, *J. Lumin.*, **128**, 1709-1718(2008).

[3]Q. Wu, L. W. Yang, Y. X. Liu, C. F. Xu, Z. G. Shang, Y. Zhang, and Q. B. Yang, Frequency Up-Conversion Properties of Er^{3+}/Yb^{3+} Co-Doped Zinc Oxide, *Spectroscopy and Spectral Analysis*, **28**, 1473-1478(2008).

[4]J.A. Capobianco, F. Vetrone, T.D. Alesio, G. Tessari, A. Speghini and M. Bettinelli, Optical spectroscopy of nanocrystalline cubic Y_2O_3:Er obtained by combustion synthesis, *Chem. Phys.*, **2**, 3203-3207(2000).

[5]R. Kapoor, C.S. Friend, A. Biswas, and P.N. Prasad. Highly Efficient Infrared-to-Visible Energy Upconversion in Er^{3+}:Y_2O_3, *Opt. Lett.*, **25**, 338-340(2000,.

[6]S. S. Chen, X. H. Chen, M. J. Ding, and X. S. Niu, Preparation and Photoluminescence Gd_2O_3:Eu Phosphor Powder, *Journal of The Chinese Ceramic Society*, **35**, 178-181(2007).

[7]C. M. Jiao,W. Q. Lu, Q. Cheng, and P. F. Wang, Thermal Decomposition of Nanosized Oxalate Prepared by Microemulsion, *Chinese Journal of Inorganic Chemistry*, **22**, 166-170(2006).

III. Ceramics in Energy Conversion Systems

CMC MATERIALS AND BIOMORPHIC SiSiC FOR ENERGY APPLICATIONS

B. Heidenreich, J. Schmidt, Sandrine Denis, Nicole Lützenburger [(1)]
J. Göring, P. Mechnich, M. Schmücker [(2)]

[(1)] DLR – German Aerospace Center
Institute of Structures and Design
Pfaffenwaldring 38-40
D-70569 Stuttgart, Germany

[(2)]DLR – German Aerospace Center
Institute of Materials Research
Linder Höhe
51147 Köln, Germany

ABSTRACT

Ceramic materials offer a high thermal and chemical stability and are therefore potential candidates for high temperature applications in severe environments, where metals can not longer be used. In future energy applications, high process temperatures > 1200 °C are required to increase the efficiency, to lower the fuel consumption, and to decrease the emissions. In order to achieve these goals, novel ceramic materials and manufacturing processes for complex structures are under development.

1. INTRODUCTION – CERAMIC MATERIALS FOR ENERGY APPLICATIONS

At DLR, three different classes of ceramic materials have been developed for the use in energy applications: Oxide/oxide and non oxide Ceramic Matrix Composites (CMC) as well as monolithic SiSiC materials.

The use of CMCs as liner material for gas turbines is a key concept to increase efficiency and reduce emissions. Replacement of metallic components by thermally stable ceramics allows reducing the amount of cooling air significantly. This will not only increase overall efficiency but allows lean combustion concepts. Moreover, ceramic components are required for future turbine technologies based on hydrogen combustion.

High temperature ceramic heat exchangers (HX) either in tube-in-tube or plate type design are promising candidates for the use in harsh corrosive and combustion environments. In contrast to tubes, plate-type HX with integrated flow channels can contribute to an increasing efficiency of the heat transfer. Those HX can be used for heat recovery processes, thermo chemical splitting reactions or within externally fired combined cycles.

For heating and drying processes porous burners can be used due to their high radiation output. The typically high brittleness and low thermal shock resistance of monolithic ceramics could be overcome by the build up of thin walled, highly porous SiSiC structures, offering structural integrity even at high temperature gradients, caused by extreme heating and/or cooling rates or by locally inhomogeneous temperature distributions inside the structure of such burners.

The main aspect of design and dimensioning of CMC combustion liners is the integration of CMC components into metallic structures. Therefore the different thermal expansion of the CMC components and the metallic support structure has to be taken into account carefully. Based on an anisotropic material model and a failure criterion suitable for the CMC material, Finite Element Analysis supports the design of the liner.

2. CMC HOT GAS LINERS FOR GAS TURBINES

Combustor materials require sufficient resistance to high temperature corrosion and thermal stability for operating times > 10,000 h at maximum surface temperatures of 1300 °C. A high resistance to cyclic fatigue and creep along with non-brittle, damage tolerant fracture behavior is mandatory. All-oxide ceramic matrix composites can meet these requirements and hence are promising materials for combustion liners in aircraft and stationary gas turbines with reduced cooling air consumption.

In institute-spanning, interdisciplinary projects oxide CMCs are being developed and tested for aircraft combustor liners. The main aim of these projects is the cooling air reduction in gas turbines, along with the development of specific cooling concepts for the ceramic composites having very low thermal conductivity. Performance was tested in model combustion chambers by the DLR Institutes of Propulsion Technology and Combustion Technology. The attachment concepts for the hot ceramic tiles onto the cold metallic support structure were developed at the DLR Institute of Structures and Design..

The CMC liner is composed of single curved shingles screwed together at radially oriented brackets (Fig. 1). The resulting CMC structure is positioned by the brackets in the metallic structure. However, both structures, the CMC liner and the metallic housing, are able to expand radially freely at any temperature during service.

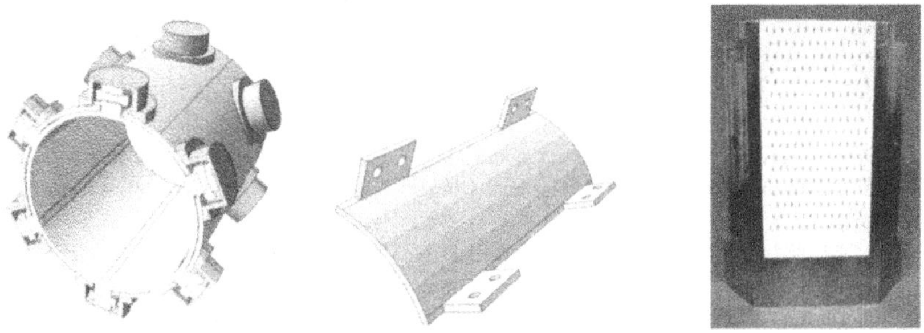

Figure 1. Schematic view of a design concept for CMC combustion chambers, based on joined segments (left and center). WHIPOX combustion chamber shingle with laser drilled cooling channels for rig testing (right).

At the DLR Institute of Materials Research an all-oxide CMC (WHIPOX = wound highly porous oxide) consisting of alumina fibres Nextel 610 or mullite based fibre (Nextel 720, both 3M) and an alumina or mullite matrix, respectively, has been developed in recent years [1, 2].

Mullite-based CMCs typically offer higher creep stability than alumina-based composites but display lower thermal conductivity. Lower thermal stability of alumina-based materials,

however, can be accepted, since service temperature is significantly lower for these materials as a result of the better cooling efficiency. Therefore material development was focused on alumina

CMCs, i.e. Nextel 610 fibres and virtually pure alumina matrices.
The chemical stability of mullite and alumina is a serious issue for long-term application of oxide/oxide CMCs in combustion environments due to the presence of water-vapour rich (exhaust) gases. Under highly dynamic flow conditions of powerful industrial burners and combustors, mullite and alumina are prone to decomposition and volatilization. The application of chemically resistant environmental barrier coatings (EBCs) is considered a solution for the corrosion problem. Due to its thermodynamic compatibility and low recession rate up to high temperatures (>1400°C), yttria stabilized zirconia (Y-ZrO_2, YSZ) is another attractive EBC material for alumina- and mullite-based CMCs. Low thermal conductivity of ZrO_2 coatings additionally provides thermal protection. At DLR, different types of ZrO_2-based coatings were developed for WHIPOX-type oxide/oxide CMCs [2].
Mechanical tests using bending, tensile and compressive load conditions including the determination of elastic constants of the orthotropic material were carried out under room and high temperature conditions. For creep tests in tension four testing devices were established and creep tests longer than 6,000 h were carried out with different CMC qualities. The WHIPOX CMCs show much better creep resistance compared to state-of-the-art metallic combustor materials.
Calculations and tests in a high pressure cooling rig shortly will demonstrate the reduction of cooling air using the all-oxide CMC WHIPOX as thermal protection system in combustion chambers.
At the DLR Institute of Structures and Design the activities are focused on the development of non-oxide CMC materials and structures for hot gas liners in gas turbines. Within the "Engine 3E" project, which was financed through the German Aviation Research Programme (Luftfahrtforschungsprogramm), the first investigations on the development of SiC long fibre reinforced ceramic tiles for use in the combustion chamber of an aero engine started in 1995. Thereby highly efficient gas turbines with staged combustion were in the focus, leading to demanding operation conditions: The material used must have suitable properties which withstand long operating times of up to 20,000 hours, at high temperatures (1300-1600 °C) and in high corrosive or oxidative stress environment.
The application conditions require the use of fibre reinforced materials which, in addition to a high thermal and oxidative stability, also have sufficient processability through the availability of textile products. Carbon fibres are not suitable due to their low oxidation resistance, because a long-term oxidation protection under the given transient operating conditions with high temperature gradients (thermal shock) is not possible. Therefore research studies were conducted with commercially available SiC fibres (e.g. Nicalon NL 207, Tyranno Lox M). These fibres demonstrate a good thermal and chemical resistance and are available as drapeable 2D fabrics. Using the Liquid Silicon Infiltration process (LSI), composites were produced, whose matrices were largely free of unreacted carbon. The precursors used were chosen so that their remaining shares of carbon after pyrolysis could be completely converted to silicon carbide during the siliconization step, or later removed via oxidation. Due to their limited thermal resistance and their tendency to recrystallize at elevated temperatures, the implementation of the SiC fibres under normal conditions of the LSI process, would have lead to complete fibre degradation. Accordingly, the conditions of siliconization (temperature, holding time) as well as the porosity and its distribution within the matrix had to be adjusted. By varying the temperature of pyrolysis or fibre pre-treatment, microstructures were

realized which enabled a successful conversion to so-called SiC/C-SiC materials.

The prequalification of these materials was carried out by static and cyclic oxidation tests at 1200-1300 °C. First test components (80 x 30 x 3 mm) were produced (Fig.2) and successfully tested on the high pressure sector test rig at DLR. The test conditions chosen - 20 cycles at 20 bar, approximately 1200 °C, in air for a total duration of 4 hours - are representative for operating conditions.

Figure 2. SiC/C-SiC combustion chamber shingles for rig tests, manufactured via LSI.

Due to the fact, that SiC fibres without coating were used, the fibres were partially attacked by the highly reactive Si during siliconization. The resulting SiC/C-SiC material was characterized by low fracture toughness and damage tolerance as well as by a high brittleness. Currently, new SiC/SiC materials based on high temperature resistant SiC fibres, like Tyranno SA, are in development. To protect the fibres during the LSI process and to obtain a weak fibre matrix interphase the fibres are PyC coated via rapid CVI.

3. HIGH TEMPERATURE HEAT EXCHANGERS

In the scope of the European project "Prediction of the Lifetime Behaviour for C/C-SiC Tubes as High and Ultrahigh Temperature Heat Exchangers" (HITHEX, CEC contract No. G5RD-CT-2000-00218) ceramic tubular components have been manufactured and tested. These tubes shall be used in bayonet type heat exchangers (HX) (Fig. 3), e.g. in the Externally Fired Combined Cycle (EFCC) processes. For the long term use, the hot gas turbine must be isolated from the combustion gases by integrating an HX system. The ceramic HX should be creep resistant, gas-tight, thermo-shock resistant and stable against hot gas corrosion and oxidation at temperatures of about 1200-1400 °C. Ceramic HX tubes made of C/C-SiC were already tested in coal combustion chambers. These tests have shown that the attack of water vapour and coal ashes at high temperatures limits the lifetime of the HX components. The corrosion of the uncoated CMC was mainly due to the presence of metals like iron or alkali metals like sodium. When liquid coal slag comes in contact with the surface, silicides or silicates are formed and especially attack the SiC

matrix. On the other hand samples with an applied multilayer BoraSiC®-cordierite outer surface coating are much more resistant. The corrosion of the ceramic components by water vapour can be reduced by such improved environmental barrier coatings (EBCs) [3].
Silicon melt infiltrated and gas-tight C/SiSiC is one favourite material for the development of an inexpensive compact HX in plate design (Fig. 3) for the thermo chemical hydrogen production. The compact offset fin plate heat exchanger concept has been developed to meet the functional and cost goals, which will serve as the intermediate heat-exchanger (IHX) to transfer high temperature heat from a helium-cooled high temperature nuclear reactor to a liquid salt intermediate loop, which couples to hydrogen production loops. The IHX uses offset fin (OSF) structures with fin widths and heights in the mm scale. The detailed local and global thermal mechanical stress analyses show that the OSF design can tolerate large pressure and temperature difference from two fluid sides. Leak-tight pyrolytic carbon coatings have been successfully applied on C/SiSiC specimens and excellent helium hermeticity was obtained [4].

Figure 3. Bayonet type heat exchanger assembly (left). Prototypical ceramic HX stack in plate design (right)

4. RADIATION HEATERS BASED ON HIGHLY POROUS SiC BURNERS

For heating and drying purposes, e.g. in paper industry, ceramic porous burners are currently under development. The porous burner technology is based on the stabilization of combustion reactions within an inert open cell porous ceramic structure (Fig. 4). The materials should be stable against thermal cycling (thermal gradient > 100 K/s) and active oxidation. In comparison to conventional free flame burners the combustion in porous structures offers exceptional advantages, e.g. low emissions, high power modulation range, small scale sizes and high radiation output. Within the German funded BMWi-project CERPOR (Optimization of ceramic components for the porous burner technology, FKZ 16INO182) degradation mechanism were investigated and the results were used to create novel ceramic structures with an improved durability.
The most promising material for the combustion area (zone C) is Si-infiltrated SiC, which should

have a pore size distribution of about 8-10 ppi (pore diameter ~3 mm) so that lateral flame propagation is possible and the combustion is stabilized. For the fabrication of such porous SiSiC ceramics DLR carried out a new technology based on C/C sheets and lamellae. These basic materials can be combined to lightweight (porosity ~80 Vol.-%, density ~0.6 g/cm^3) 3D stacks. Through the variation of the amplitude and number of lamellae per inch, the porosity and orientation of the pore channels could be tailored in a wide range. Best results from durability tests were obtained with structures, which are composed of oriented pore channels. Suitable structures should have angles ( ) of about = 50 ±10°. The results from burner rig tests (Fig. 4) with improved components are very promising, since no significant oxidation or degradation could be observed after 1.939 h and 10.800 start-ups [5]. From the industrial point of view a lifetime of about 3 years and some thousands cycles are required and probably can be fulfilled by these structures.

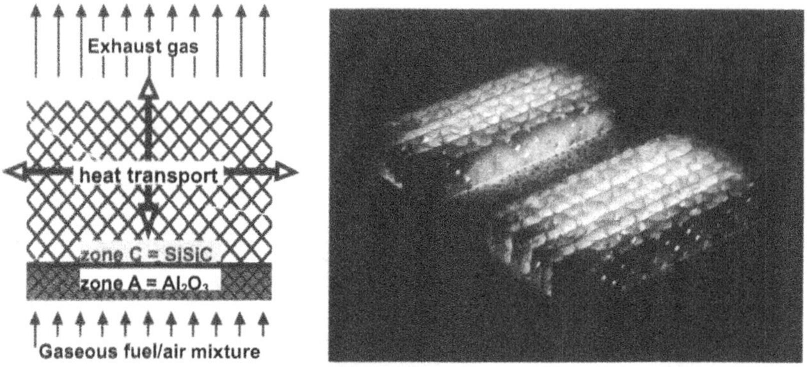

Figure 4. Schematic setup of a porous burner for drying and heating purposes (left). 3D cardboard like ceramic SiSiC structure on the test bench at 1400 °C (right)

5. SUMMARY AND OUTLOOK

At DLR, oxide and non oxide CMC as well as not fibre reinforced SiSiC materials and structures have been developed successfully for energy applications. In first rig tests.WHIPOX materials and also nonoxide SiC/C-SiC materials showed a high potential for the use as combustion chamber shingles in gas turbines, due to typical CMC properties like high temperature and thermal shock resistance.

C/C-SiC tubes for high temperature heat exchangers in coal combustion chambers could withstand highly aggressive environments including water vapour and liquid coal slags, especially coated with multilayer EBC based on B_4C, SiC and Cordierite.

Thin walled structures made of C/SiSiC and SiC materials based on biocarbon and carbon/carbon preforms showed excellent long term stability in porous burner systems and are in development for

high temperature heat exchangers in plate design.
For combustion chamber shingles, future work will be focused on environmental and thermal barrier coatings for oxide and nonoxide CMC, to obtain long term stability and to increase service temperatures and overall efficiency. Additionally, CMC materials based on newly developed nonoxide fibres, like SiBNC are a main topic. Intensive testing at realistic conditions will be necessary for further development and a future integration of CMC materials in gas turbines. At DLR an interdisciplinary team of scientists in different institutes are working together in internal programmes as well as in close cooperation with potential industrial users. Thereby the whole spectrum including material research, structural design, component manufacturing and rig testing as well as quality assurance and non destructive testing are available for a goal oriented development.

References:

1. J. Göring, B. Kanka, M. Schmücker, H. Schneider, A Potential Oxide/oxide Ceramic Matrix Composite for Gas Turbine Application, Proc. ASME / IGTI Turbo Expo; 2003

2. M. Schmücker, P. Mechnich, All-Oxide Ceramic Matrix Composites with Porous Matrices, in W. Krenkel (ed.) "Ceramic Matrix Composites, Fibre-Reinforced Ceramics and their application" Wiley-VCH, Weinheim, 2008, 205-229

3. J. Schmidt; J. Schulte-Fischedick; E. Cordano, C. Mao, V. Liedtke, R. Fordham: CMC tubes based on C/C-SiC with high oxidation and corrosion resistance, Proceedings of the 5th international conference on high temperature ceramic matrix composites, ed. by M. Singh, R. Kerans, E. Lara-Curzio, R. Naslain, published by the American Ceramic Society, p. 531-536, 2004

4. P. Peterson; H. Zhao; F. Niu; W. Wang; J. Schmidt; J. Schulte-Fischedick: Development of C-SiC ceramic compact plate heat exchangers for high temperature heat transfer applications, Proc. of AIChE Annual Meeting,12.-17. November, San Francisco, 2006

5. J. Schmidt; M. Scheiffele: Fabrication and Testing of corrugated 3D SiSiC ceramics for porous burner applications, Industrial Ceramics, Vol. 27, No. 2, S. 127-130, 2007

CRYSTALLIZATION, MICROSTRUCTURE AND PHYSICAL PROPERTY OF NEW TYPES OF BOROSILICATE GLASS-CERAMICS

Shufeng Song, Zhaoyin Wen, Liu Yu , Qunxi Zhang, Jingchao Zhang, Xiangwei Wu
Shanghai Institute of Ceramics, Chinese Academy of Sciences, Shanghai 200050, P. R. China
Corresponding author. Tel.: +86-21-5241-1704 E-mail: zywen@mail.sic.ac.cn

ABSTRACT:

In this work, we reported new types of glass-ceramics by adding Li_2CO_3 to a borosiliate glass. The effects of the mole ratio of SiO_2/Li_2O and the thermal treatment conditions on the crystallization, microstructure and physical propertoes of the new glass-ceramics were investigated. Based on the XRD analysis, it was assumed that the mole ratio of SiO_2/Li_2O dominates the mechanism of the crystallization of the glass-ceramics. Furthermore, the crystallization mechanism of the new glass-ceramics was systematically concluded. The new glass-ceramics possessed a high TEC of about $10.4\sim12.8\times10^{-6}/^{o}C$ at 0-400 °C. A densified structure was obtained for the glass with SiO_2/Li_2O mole ratio of 1.5:1. The glass with SiO_2/Li_2O mole ratio as 2:1 displayed high flexural strength as high as 102.1MPa.

INTRODUCTION:

Borosilicate glasses are widely used in various applications, like optical communication, glass to metal seals, sealing glass in sodium-sulfur battery, ion exchange materials, nuclear waste immobilization, etc. [1-6]. Comprehensive studies have focused on the structural aspects of borosilicate glasses using the techniques such as FTIR, Raman and 29Si,11B MAS AMR[7-9]. However, the mechanical strength of the glasses still restricts their applications. Additionally, the thermal expansion coefficient (TEC) plays a key role in determining the applications for different types of glasses and glass-ceramics materials [10].

Most of the borosilicate glasses are phase separable. Such glass generally consists of a chemically durable silica-rich and a less durable boratio of rich phase [11]. Generally the cristobalite precipitates due to the phase separation. It was thought that the cristobalite is an unfavorable transformation product in terms of the thermal expansion behavior. Many studies have been reported on preventing the cristobalite crystallization from the borosilicate glasses. Five kinds of dopants $MgCO_3$,$CaCO_3$, $SrCO_3$,$BaCO_3$ and $MnCO_3$ were found to effectively induce amorphous silica acid being transformed into quartz [12]. Bailey reported that amorphous silica with the added CaO, $CaSiO_3$ or $CaCO_3$ was converted to quartz by heating at 1070~1100 °C [13]. Takeuchi and coworkers explored the mechanism of the quartz formation in the silica gel and silica glass by mixing with various substances as intentional additives [14]. They pronounced that Li_2CO_3 produced the largest amount of quartz(94±4%) among all the additives and had the lowest onset temperature of quartz formation. However, most authors did not study the properties of the glasses with the quartz precipitates. It is required to combine the crystalline behavior with the properties of glasses to well develop their applications.

In this study, we obtained new type of glass-ceramics with high-strength and high thermal expansion coefficient by adding Li_2CO_3 to a borosilicate glass. The differences in the crystallization, microstructure and physical properties of the new glass-ceramics were systematically studied by considering the change of the mole ratio of SiO_2/Li_2O. Moreover, the mechanism of the crystalization behavior of the new glass-ceramics was assumed.

EXPERIMENTAL

Table Ⅰ. Composition of the studied glasses

Glass	Composition of the glasses(wt.%)					
	SiO_2	Al_2O_3	B_2O_3	Na_2O	K_2O	Li_2O
G1 (hypo,molSiO_2:Li_2O=1.5:1)	52.1	3.2	18.0	4.5	5.0	17.3
G2 (stoich,molSiO_2:Li_2O=2:1)	55.6	3.2	18.0	4.5	5.0	17.3
G3(hyper,molSiO_2:Li_2O=2.33:1)	57.2	3.2	18.0	4.5	5.0	17.3

The chemical compositions of the glasses were given in Table 1. The bulk glasses were prepared by conventional melting-quenching method.

Differential scanning calorimeter (DSC, Netzsch 409PC) was employed to record the crystallization temperature of the samples. Measurements were carried out in the temperature range of 50—1000 °C at a heating ratio of 10 °C /min.

The crystal phases in the glass-ceramics were determined by XRD analysis. All instruments were precisely and identically set to ensure a high precision to obtain the integral peak area. The microstructure of the fresh fractured cross section of the glass-ceramics was observed by SEM. The thermal expansion coefficient (TEC) was calculated from room temperature to 500 °C at a heating rate of 5 °C/min in the dilatometry analyser (NETZSCH, DIL402PC). The flexural strength was determined in a 3-point bend test at a constant strain ratio of 0.5mm/min.

RESULTS AND DISCUSSION

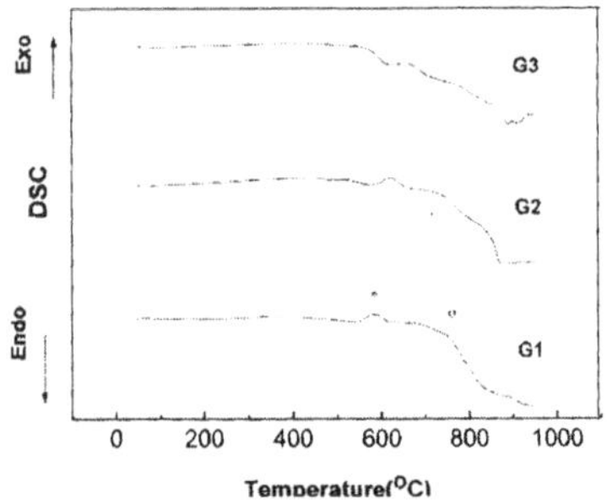

Figure 1. DSC curves for the three glasses

Fig. 1 shows the DSC curves of the three glasses. An obvious exothermic peak at approximately 600 °C and a broad exothermic peak at 700-850 °C were observed for each glass. The first crystallization temperature was thus fixed at around 600 °C. Since the second exothermic peak was broad and unobvious, the second crystallization temperature was chosen at 700 °C、750 °C、800 °C and 850 °C to observe the crystalline behavior. For a better comparison, the conditions of the heat treatment were set identically for all three glass-ceramics.

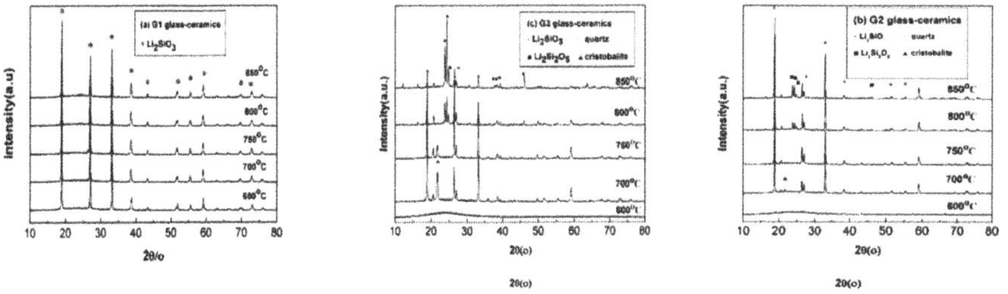

Figure 2. XRD patterns of G1,G2 and G3 glass-ceramics sintered at different temperature.

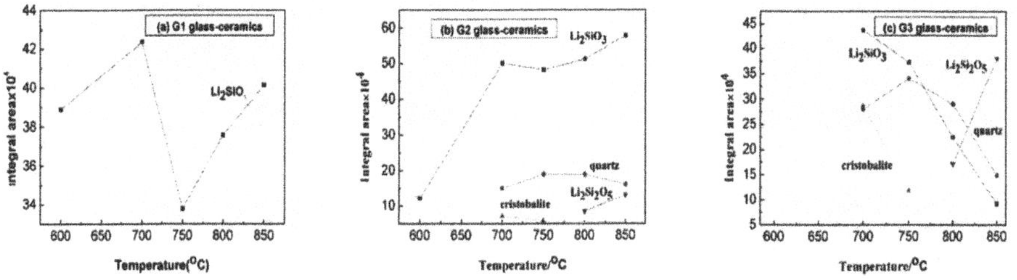

Figure 3 . Curves of the peak maxmum integral area of G1,G2 and G3 glass-ceramics as a function of mole ratio of SiO2/Li2O and heat treatment conditions.

The maximum integral area calculated from Fig.2 was shown in Fig. 3. It was seen in Fig. 2(a) that the G1 glass devitrified mainly Li_2SiO_3 crystals at 600~850 °C. It is interesting that the Li_2SiO_3 amount is the maximum at 700 °C, as shown in Fig.3(a). As shown in Fig.2(b), the G2 glass devitrfied Li_2SiO_3 at 600 °C, quartz and cristobalite at 700 °C, $Li_2Si_2O_5$ at 800 °C, respectively. Fig.3(b) shows that the cristobalite is vanishing while the quartz is maximum at 750 °C in G2 glass-ceramics sintered at 800 °C for 2h, and the Li_2SiO_3 crystals amount increased at 600~850 °C and the $Li_2Si_2O_5$ crystals amount increased at 800~850 °C. Fig. 2(c) revealed that the G3 glass devitrfied Li_2SiO_3 at 600~850 °C, quartz at 700 °C, cristobalite and $Li_2Si_2O_5$ crystals at 800 °C, respectively. The crystallization behavior of the G3 glass-ceramics was shown in Fig.3(c).

The crystallization behavior of Li_2SiO_3 of the G1 glass was similar to that of G2 glass. However, G1 precipitated mainly Li_2SiO_3 crystal at 600~850 °C, while G2 precipitated not only Li_2SiO_3 crystal but also $Li_2Si_2O_5$, quartz and cristobalite crystals. The variation trends of the quartz, cristobalite and $Li_2Si_2O_5$ crystals of G2 was similar to that of G3, while the crystallization behavior of Li_2SiO_3 crystal was different.

Based on the XRD analysis, it was assumed that the mole ratio of SiO_2/Li_2O dominated the mechanism of the crystallization behavior of the new glass-ceramics. For SiO_2/Li_2O =2.33:1, the $Li_2Si_2O_5$ crystal precipitation in the glass-ceramics might follow the following two reactions:

$$Li_2SiO_3(crystal)+SiO_2(l)= Li_2Si_2O_5(crystal) \quad (1)$$
$$SiO_2(quartz)+Li_2O(l) = Li_2Si_2O_5(crystal) \quad (2)$$

For SiO_2/Li_2O =2:1,

$$SiO_2(quartz)+Li_2O(l) = Li_2Si_2O_5(crystal) \quad (3)$$

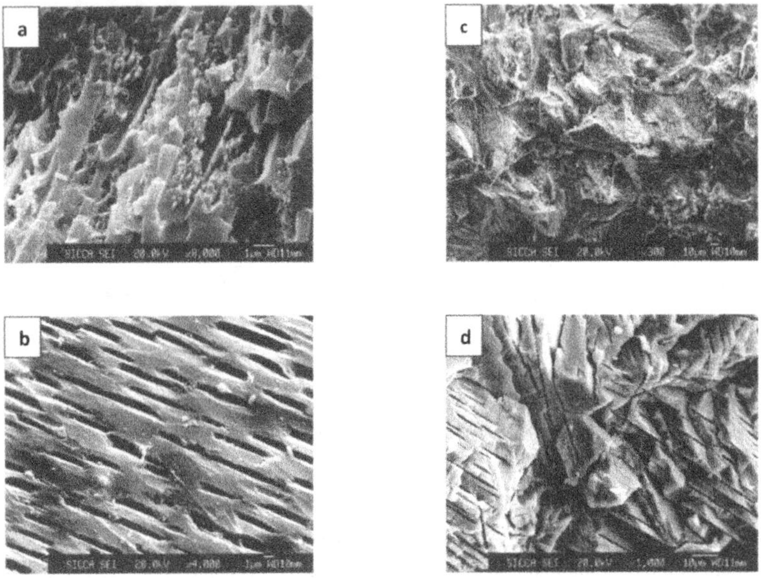

Figure 4. SEM photographs of the fresh cross section of the new glass-ceramics (a) G1 glass-ceramic sintered at 850 °C for 2h, (b) G2 glass-ceramic sintered at 850 °C for 2h, (c) G3 glass-ceramic sintered at 750 °C for 2h, (d) G3 glass-ceramic sintered at 850 °C for 2h.

The SEM micrographs in Fig.4 showed that the morphology of the new glass-ceramics were closely related to the mole ratio of SiO_2/Li_2O and the heat treatment conditions. The morphology of the G1 glass-ceramic sintered at 850 °C for 2h was showed in Fig.4(a). Li_2SiO_3 crystals formed as a rod-like dispersion in a glass matrix and a densified structure was observed. Fig. 4(b) revealed the formation of microcrack in the G2 glass-ceramic sintered at 850 °C for 2h. It is interesting to find that the microcrack is very regular and separately distributed. As shown in Fig.4(c), the G3 glass-ceramic sintered at 750 °C for 2h showed a densified structure. However,

random microcracks were observed in the G3 glass-ceramic sintered at 850 °C for 2h.
To some extent, the formation of microcracks was probably related to the quantity and the category of the precipitated crystals. As known, the TEC and internal stress strongly depended on the category of the crystals.

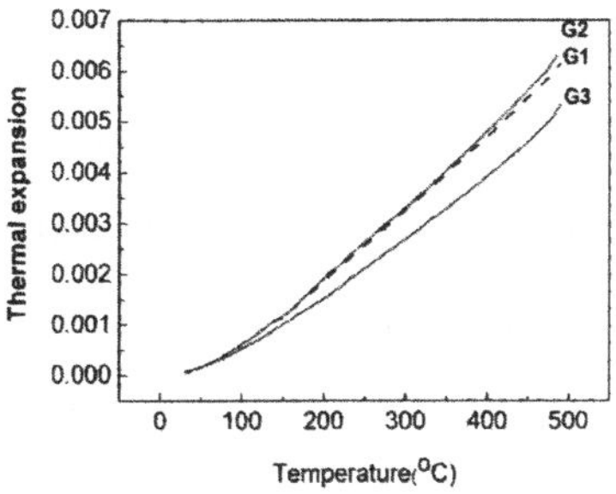

Figure 5. Thermal expansion curves of G1, G2 and G3 glass-ceramics sintered at 850 °C for 2h.

Fig. 5 shows the thermal expansion behavior of the glass-ceramics sintered at 850 °C for 2 h. As seen, the thermal expansion curves of the G1 and G2 glass-ceramics were almost overlapped from room temperature to 400 °C. The slope of the expansion curve of the G3 glass-ceramic was smaller than that of the G1 and G2 glass-ceramics.

The TEC of the glass-ceramics was determined by the crystalline phases and the residual glass phase. As supposed above, the mole ratio of SiO_2/Li_2O determined the mechanism of the crystallization of the glass-ceramics. After sintered at 850 °C for 2h, the major crystal phase and its content of the G1 glass-ceramic were similar to those of G2. It resulted in a similar TEC for the G1 and G2 glass-ceramics. However, $Li_2Si_2O_5$ was the major crystal phase instead of Li_2SiO_3 for the G3 glass-ceramic sintered at 850 °C for 2h. Since the TEC of $Li_2Si_2O_5$ was smaller than that of Li_2SiO_3 and the crystal content of G3 glass-ceramic was comparable with that of the G1 or G2, G3 showed the minimum TEC.

The mechanical strength of the G2 and G3 glass-ceramics sintered at 850 °C for 2h were tested. The mean flexural strength was 102.1MPa.The flexural strength of G3 was 82.0 MPa. The irregular microcrack of the G3 glass-ceramics can easily join together due to the crack extension during compression. It resulted in lower flexural strength.

CONCLUSIONS

In this work, we developed new types of glass-ceramics by adding Li_2CO_3 to a borosilicate glass. The crystallization mechanism of the new glass-ceramics was assumed based on the experimental analysis. As found, the mole ratio of SiO_2/Li_2O dominated the mechanism of the crystallization behavior. The mole ratio of SiO_2/Li_2O and the heat treatment conditions had significant effects on the microstructure of the new glass-ceramics. Densified structure was obtained for the glass of SiO_2/Li_2O =1.5:1. The new glass-ceramics have high TEC about $10.4\sim12.8\times10^{-6}$/°C from room temperature to 400 °C. Additionally, the glass-ceramics (SiO_2/Li_2O =2:1) possessed high flexural strength as high as 102.1MPa.

ACKNOWLEDGEMENT

The authors would like to thank Prof. S.R. Wang for the thermal expansion coefficient measurement. This work was financially supported by NSFC Project No. 50672114, Research Project of Chinese Science and Technology Ministry No. 2007BAA07B01 and 973 Project of China No. 2007CB209700.

REFERENCES

[1]J. Lee, T. Yano, S. Shibata, M. Yamane, Structural evolution during Cu+/Na+ ion-exchange in the system Na2O–Al2O3–SiO2, *J. Non-Cryst. Solids,* 246, 83 (1999).

[2]Walter George Budgen and Peter Raymond Smith, Glass seal for sodium-sulfphur cells. GB 2207545,1989

[3]Dong-Sil Park, Cliften Park, Louis Navias, N. Y. Schenectady, Sodium resistant sealing glasses. *US4268313,*1981.

[4]K. Matusita, J.D. Mackenzie, Low expansion copper aluminosilicate glasses, *J. Non-Cryst. Solids*, 30 ,285 (1979).

[5]R.K. Mishra, V. Sudarsan, A.K. Tyagi, C.P. Kaushik, K. Raj, S.K.Kulshreshtha, Structural studies of ThO2 containing barium borosilicate glasses , *J. Non-Cryst. Solids,* 352, 2952 (2006).

[6]O. Pietl, E.D. Zanotto, Thermal shock properties of chemically toughened borosilicate glass, *J. Non-Cryst. Solids,* 247, 39 (1999).

[7]K.El-Egili, Infrared studies of Na_2O-B_2O_3-Si_2O and Al_2O_3- Na_2O-B_2O3-Si_2O glasses. *physica B* 325, 340-348 (2003)

[8]Jayshree Ramkumar,etc., Structural studies on boroaluminosilicate glasses, *J. Non-Cryst. Solids*, 354, 15-16, 1591-1597(2008).

[9]K. Takahashi, A. Osaka and R. Furuno, Network structure of sodium and potassium borosilicate glass systems. *J. Non-Cryst. Solids*, 55, 1, 15-26(1983).

[10]G.H. Beall, K. Chyung, J.E. Pierson, Negative CTE β-eucryptite glass-ceramics for fiber bragg grating, in: *Proceedings of the XVIII International Congress on Glass (CD-ROM)*, 5–10(1998).

[11]M.Arbab, etc. The effect of RO oxides on mirostructure and chemical durability of borosilicate glasses opcified by P_2O_5. *Ceramics Internationa,l* 33, 943-950 (2007).

[12]L.S.Birks and J.H.Schulman. *Am.Mineral.* 35, 1035 (1990).

[13] D.A.Bailey. *Am.Mineral.* 34, 601 (1949).

[14]N.Takeuchi, S.Yamane, S.Ishida* and H.Nanri. Conversion of silica gel and silica glass mixed with various metal oxides into quartz , *J. Non-Cryst. Solids,* 203, 369—374 (1996).

A STUDY OF Al_2O_3 AND YSZ CERAMIC SUPPORTS FOR PALLADIUM MEMBRANE

M. Kitiwan and D. Atong*
National Metal and Materials Technology Center
114 Thailand Science Park, Paholyothin Rd., Klong1, Klong Luang, Pathumthani, 12120 Thailand
*E-mail: duangdua@mtec.or.th

ABSTRACT

The palladium membrane is well-known for using in hydrogen separation process. However, the free-standing Pd is expensive and gives low hydrogen flux. The ceramic substrate can be used as a membrane support because it provides many advantages for the palladium membrane, for example, increase mechanical strength and thermal stability, reduce membrane thickness and especially achieve higher hydrogen permeation. In this study, the porous tubular Al_2O_3 and YSZ substrate were successfully fabricated from extrusion method. The evolution of density, porosity, flexural strength and microstructure were investigated after sintering at the temperature ranging from 1200 to 1450°C. The effect of sintering temperature was significant when the tubular support was fabricated from the small particle size of YSZ ceramic.

INTRODUCTION

Nowadays, the demand for hydrogen energy is growing extensively. Although hydrogen production cost is high relative to conventional fuel; hydrogen energy is the key for dealing with global concerning such as climate change. Hydrogen offers significant benefit as a clean fuel when utilized through fuel cell – the efficient energy conversion system. Besides, hydrogen is a potential solution for energy crisis when hydrogen-rich gas is derived from domestic renewable source via biomass gasification. The recovery of high purity hydrogen can be accomplished by employing the membrane separation technology. Many advantages gain from integrating hydrogen separation membrane into gasification process. For example, the syn-gas produced from gasifier at elevated temperature enhances the catalytic activity of membrane and thus reduces overall energy consumption of process. Also, combining those extracting reaction and separation step in compact unit can supply hydrogen fuel to the energy system conveniently.

Palladium and its alloy represent as one of the most viable membrane to purify hydrogen. Dense palladium-based membranes have been used in many applications where ultra-high purity hydrogen is required[1-2]. The important feature of membrane is not only selectivity for hydrogen but also permeability, namely that the hydrogen flux permeate through palladium membrane should be considerably high. The membrane with much thinner layer improves the rate of hydrogen permeation effectively. In order to minimize the membrane thickness along with maintaining mechanical strength, the palladium was always proposed as a composite membrane which involves thin palladium deposited on the outer surface of porous substrate.

The palladium membranes had been deposited onto various kinds of porous materials such as Vycor glass, stainless steel, alumina (Al_2O_3) and zirconia (ZrO_2). As an outstanding candidate of substrate, porous ceramics tube was chosen in this study because they are significantly durable in higher operating temperature. Moreover, tubular shape provides larger membrane area and have withstood the higher pressure compared to others configurations. The physical appearances of support cannot be neglected. Since the resistance of the porous support influences to hydrogen permeation behaviors through palladium composite membranes, appropriate pore size and thickness of substrate would increase the hydrogen flux. It was reported by I.J. Iwuchukwu et al.[3] that the suitable palladium film deposited on alumina support of about 2mm thickness and porosity of about 50% provided the hydrogen fluxes in excess.

The whole study of this research has been divided into two parts: preparation of porous substrate and deposition of thin palladium membrane. This paper reveals only the first one, i.e. the fabrication of porous ceramic tubes by extrusion method. Early ceramic supports for palladium membrane were made of Al_2O_3 [4-5]. In recent work, we attempted to examine the properties of two kinds of ceramic materials, Al_2O_3 and YSZ (yittria stabilized zirconia). The extrusion of those ceramic materials was carried out by mixing with additives in various portions. After that, they were sintered at temperature between 1200 – 1450 °C in order to investigate the effect of sintering temperature on pore size and porosity of porous support. The mechanical strength was also inspected to clarify the most appropriate sintering temperature for each ceramics support.

EXPERIMENTAL

The porous tubes were prepared from commercial alumina ceramic powders (Al_2O_3, average particle size 1.7 μm, AL-45, Showa Denko) and zirconia stabilized with 8 mol% of yittia (YSZ; average particle size 90 nm, TZ-8YS, Tosoh). Powders were mixed with 11-13% organic binder and 18-22% distilled water to observe the workability of ceramic dough. The procedure used to prepare ceramic tubes involves three steps. First, all precursor materials were blended together in a high speed mixer (MHS-100, Miyazaki Iron Work). Second, the mixture was placed in a three-roll mill machine (80S/1585, EXAKT Apparatebau) to form ceramic dough. Then mixture was kept aging at low temperature overnight. Last and most important, the ceramic dough was formed into a cylindrical shape by extruder (FM-30-1, Miyazaki Iron Work). The green samples were then dried and sintered at various temperatures from 1200 to 1450°C with a dwelling time of 2 hours.

The particle sizes of starting materials were analyzed by a laser light scattering technique (Mastersizer-S version 2.19, Marvern). The density of sintered specimen was determined by Archimedes method. Porosity and pore size distribution were analyzed by mercury porosimetry method (Quantachome Instrument). The microstructures of the samples were observed with a scanning electron microscope (SEM; JEOL, JSME5410). The mechanical strength of sintered specimen was measured by three-point bending method using Universal Testing Machine (55R4502, Instron). The test was conducted at the cross-head speed of 1 mm sec^{-1}. The following equation was applied to calculate the three-point bending of ceramic tubes[6]:

$$\sigma = \frac{8 d_2 L}{\left(d_2^4 - d_1^4\right)\pi} P_{\max} \tag{1}$$

,where σ is the flexural strength (MPa); P_{max} is the maximum load at break point (N); L is the distance between fulcrum (m) which was fixed at 4 mm, d_1 and d_2 are the inner and outer diameter of tube (m) respectively.

RESULTS AND DISCUSSION

Powder characterization

The particle size of starting ceramic powder is one of the most important parameter which has a profound impact on extrusion process. For particles with smaller size, they have higher surface areas, which require large amount of additives to produce extrudable ceramic. The particle size analysis by a laser light scattering technique showed that Al_2O_3 powder had a narrow size distribution at 1.86 μm, whereas YSZ had a bimodal-size characteristic, in which those two peaks appeared at 0.47 μm and 4.42 μm. The microstructures of as-received ceramic powder are showed in Fig. 1(a) and (b). The morphology of Al_2O_3 powders are in irregular shape, while YSZ ones are in the form of granules which are consisted of several nano-size particles.

Figure 1. SEM micrographs of as-received powder of (a) Al_2O_3 and (b) YSZ

Tubular ceramic manufacture

The aspect for the extrusion is frequently concerned with plasticity of the ceramic dough. This is because ceramics are non plastic material when mixed with water; therefore, it is necessary to add some additives to improve plasticity. High plasticity could enhance the workability of the mixture. However, the excess quantity of additive could obstruct the high sinter density of final products; thus led to an attempt to reduce specific surface area of the ceramic powder by calcination[7]. In this work, the porous ceramic is desirable because of its use as a substrate for palladium; therefore large amount of additives were become benefit to the tubular support. The various amounts of additive and water used in the producing of tubular ceramic dough were investigated to succeed the extrusion.

Table I illustrates that the sufficient binder and right quantity of water adding to ceramic played an important role in manufacturing of tubular ceramic. The fabrication of Al_2O_3 dough was obtained according to our previous successful effort[8]. The amount of water and binder required for preparation of alumina dough was not quite suitable for YSZ case; thus resulted in dry and dusty dough. The reason might be because of the different nature of the two powders, YSZ powder is very dry and has less humidity than Al_2O_3 powder. The bimodal size of YSZ also led to a closely packed particle which trapped water among particle packing. Therefore, more water is needed in order to form suitable dough. The YSZ-2 composition which contained the 20 wt% of water yielded in an appropriate level of fluidity and stickiness of dough. If the water portion increased to

22 wt% (YSZ-3), the YSZ dough was too wet and soft which might cause the deformation of extruded tube. With such an increasing amount of binder for YSZ-4, the YSZ dough was hard which might contribute to the crack of product surface and eventually blocked the extruder. Hence, the composition of YSZ-2 was found to be suitable for a subsequent extrusion because it exhibited proper dough with uniformity and had enough strength for handling as well as no crack and deformation during drying. It was also noted that the extrusion of YSZ-2 could be performed continuously without trouble.

Table I. The various compositions of ceramic dough and the appearances of resulting mixture

Formulas	Compositions (%wt)			Mixture appearances
	Ceramic mass	Binder	DI-water	
Al_2O_3	100	12	18	Good dough, success extrusion
YSZ-1	100	12	18	Dry and dusty dough
YSZ-2	100	12	20	Good dough, success extrusion
YSZ-3	100	12	22	Too soft dough
YSZ-4	100	13	20	Hard dough

When combined right amount of binder with the sufficient water to the ceramic powder by the high speed mixer, all ingredients were combined homogeneously within a short time of 5 minutes. Afterwards, three-roll mill machine was used to introduce shear into the mixture body. The extrusion machine used in this experiment comprises of vacuum and kneader parts. During extrusion, the vacuum helped removing air in the ceramic dough, while the kneader gave excellent homogeneous mixing. The progressive twin screws enabled unit to produce high pressure feeding of viscous dough. The extrusion of Al_2O_3 and YSZ green tubes were successfully performed through the die with outer of 6 mm and inner diameter of 4 mm. The extruded tubes from both two type of ceramic dough gave smooth surface and consistent cross-section. The pressures during extrusion for Al_2O_3 and YSZ dough were 4.5 MPa and 5.2 MPa, respectively. However, the effect of the extrusion pressure on the properties of final product did not quite significant since dough had enough green strength which could retain its shape. The extruded tubes were dried in hot-air oven at 50°C for 3 hours and then cut into the length of 8 mm prepared for further sintering investigation. The degree of powder packing of green tubes shown in Fig. 2 exhibited obviously dissimilar

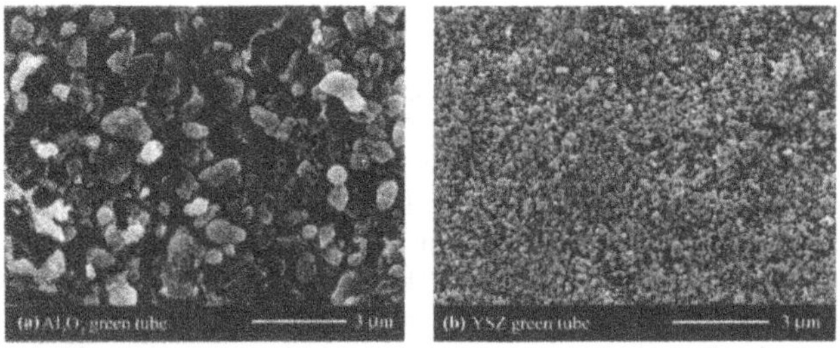

Figure 2. SEM micrographs of green body of extruded tubes of (a) Al_2O_3 and (b) YSZ

morphology. It was found that larger particle size of Al_2O_3 led to inhomogeneous powder packing. On the contrary, large agglomeration of YSZ powder had become broken down into smaller particles and rearranged in a dense closely packed structure.

Effects of Sintering Temperature

The evaluation of tube density was monitored after sintering at temperature ranging from 1200 to 1450°C as demonstrated in Fig. 3(a). The density of Al_2O_3 tube did not changed significantly after sintering at high temperature. The low level of densification might be due to loose powder compaction. This result suggested that the density of support was restricted by particle size. The YSZ tube fabricated from smaller particles which packed densely as a result it was sintered to high density in a shorter time. This is inferred that the densification would rise easily with the homogeneous morphology of primary particle. The porous supports are generally intrinsic brittle material, thus the mechanical property have to be strong enough to provide the strength to the top membrane. Figure 3(b) shows how the strength of porous tube varies with the sintering condition. The mechanical strength developed with the increasing of temperature. In comparison, the flexural strength of sintered YSZ tubes was considerably superior to that of Al_2O_3. The flexural strength of both Al_2O_3 and YSZ were found to be dependent on their densities.

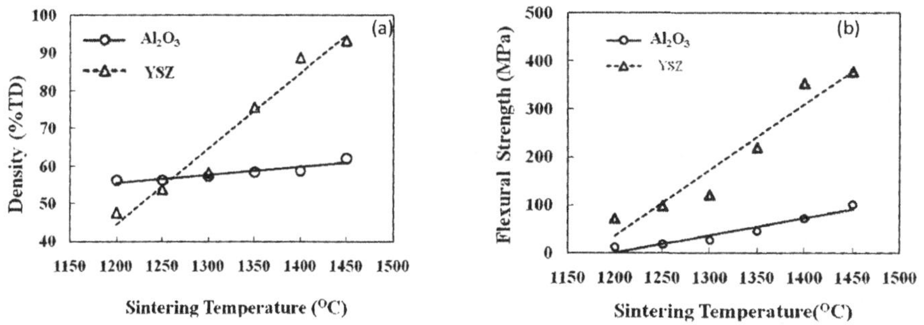

Figure 3. Effect of sintering temperture on (a) density and (b) fluxural strenght of tubular ceramic.

After the sintering process, the binder was eliminated then left a homogeneous pore distributed in ceramic tube body. The effect of sintering temperature on porosity and pore diameter was further investigated as illustrated in Fig.4. The porosity of Al_2O_3 was gradually decreased with the increasing of sintering temperature from 60% to 40%. As for YSZ tube, at sintering temperature below 1300°C, the porosity of YSZ was comparable to the Al_2O_3 ones. However, the obviously change in porosity took place when the sintering was raised up from 1300 to 1450°C. It was though to be the small particle contributed to high densification rate. The maximum pore size distributions of Al_2O_3 were in the same range between 0.25 to 0.32 μm, but those for YSZ tube were continuously decreased with rising temperature from 0.24 to 0.04 μm. The reducing of pore size of YSZ sample was due to shrinkage of the pore occurred during densification. According to solid state sintering, the driving force during sintering is the reduction of surface area. The particles with smaller size promote fast densification; the pores are consequently smaller and finally disappear[7]. These results were corresponding with the density, the YSZ tubes had remarkably high densification rate thus resulted to rapid decreasing in porosity and pore size. While the density of Al_2O_3 tubes were gradually changed; therefore, their porosities were slightly decreased and pore size were almost the same.

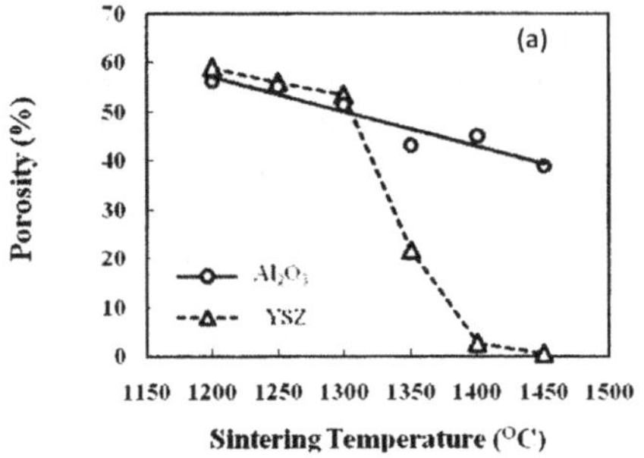

Figure 4. Influence of sintering temperture on (a) porosity and (b) pore diameter of sintered ceramic.

Many essential aspects are required from membrane support, including high strength over pressure difference and lower resistance to gas flow. High density support provides high mechanical strength but it might obstruct the permeability of the permeated gas. With increasing sintering temperature, density and mechanical strength of ceramic support were increased, while theirs porosity and pore size were decreased. Therefore, the optimization of these parameters should be further investigation by gas permeability test.

CONCLUSIONS

The extrusion of Al_2O_3 and YSZ had been successfully carried out with optimize addition of binder and water. The plasticity of ceramic dough played an important role in the continuous extrusion of ceramic tube with constant cross section and smooth surface. The starting particle size had enormous influence to density after sintering and also affected to other properties including mechanical strength, porosity and pore size. The YSZ had finer particle size, with increasing sintering temperature, the density and mechanical strength improved, while porosity and pore size rapidly decreased. In contrast, Al_2O_3 had rather coarse particle size, those physical and mechanical properties were not much variable with sintering temperature.

ACKNOWLEDGMENTS

The authors would like to thank to National metal and materials technology center, Thailand, for financial support (MT-B-51-END-07-057-I).

REFERENCES

[1] Y. Swesi, D. Ronzeb, I. Pitaulta, R. Dittmeyerd and F. Heurtauxe, Purification Process for Chemical Storage of Hydrogen for Fuel Cell Vehicles Applications, *Int. J. Hydrogen Energy*, **32**, 5059 – 5066 (2007).

[2] B.K.R. Nair and M.P. Harold, Hydrogen Generation in a Pd Membrane Fuel Processor: Productivity Effects during Methanol Steam Reforming, *Chem. Eng. Sci.*, **61**, 6616 – 6636 (2006).

[3] I.J. Iwuchukwu, A. Sheth, Mathematical Modeling of High Temperature and High-Pressure Dense Membrane Separation of Hydrogen from Gasification, *Chem. Eng. Pro.*, **47**, 1292-1304 (2008).

[4] D.A.P.Tanaka, M.A.L. Tanco, S. Niwa, Y. Wakui, F. Mizukami, T. Namba and T.M. Suzuki, Preparation of Palladium and Silver Alloy Membrane on a Porous α-Alumina Tube via Simultaneous Electroless Plating, *J. Membr. Sci.*, **247**, 21–27 (2005).

[5] S. Abate, G. Centi, S. Perathoner and F. Frusteri, Enhanced stability of catalytic membranes based on a porous thin Pd film on a ceramic support by forming a Pd–Ag interlayer, *Catalysis Today*,**118**, 189–197 (2006).

[6] C. Zhang, Z. Xu, X. Chang, Z. Zhang and W. Jin, Preparation and characterization of mixed-conducting thin tubular membrane, *J. Membr Sci*, 299, 261–267 (2007).

[7] Y. Du, N.M. Sammes and G.A. Tompsett, Optimisation parameters for the extrusion of thin YSZ tubes for SOFC electrolytes, *J. Eur. Cer. Soc.*, **20**, 959-965 (2000).

[8] P. Aungkavattana, K. Hemra, D.Atong, S. Wongkasenjit, N. Kuanchertchoo, S. Kulprathipanja; Alumina ceramic formulation for extrusion process; Thai Patent no. 0601005855 (2006).

[9] H. Bissett, J. Zah and H.M. Krieg, Manufacture and Optimization of Tubular Ceramic Membrane Supports, *Powder Technology*, *181*, 57–66 (2008).

SYNTHESIS OF OLIVINE ($LiFePO_4$) and Ni/OLIVINE ($LiFePO_4$) CATALYSTS FOR UPGRADING SYN-GAS PRODUCTION

D. Atong[1], C. Pechyen[2], D. Aht-Ong [2,3], and V. Sricharoenchaikul [4]*

[1]National Metal and Materials Technology Center, 114 Thailand Science Park, Thailand

[2]Department of Materials Science, Faculty of Science, Chulalongkorn University, Thailand

[3]National Center of Excellence for Petroleum, Petrochemicals, and Advanced Materials, Chulalongkorn University, Thailand

[4]Department of Environmental Engineering, Faculty of Engineering, Chulalongkorn University, Thailand

*E-mail: viboon.sr@chula.ac.th

Metallic nickel as active phase doped on olivine compounds can be used as catalysts for upgrading syngas production via pyrolysis or gasification. Phospho-olivine ($LiFePO_4$) was chosen as catalyst support because of its favorable activity in pyrolysis and tar cracking, along with its high attrition resistance. $LiFePO_4$ was synthesized by co-precipitation synthesis using Lithium phosphate, phosphoric acid, and ferric citrate n-hydrate as starting materials. The wet powder obtained was then heated at 140°C for 12 h, ground fired under Ar up to 500-1000°C for 24 hours, and air quenched to obtain crystallized $LiFePO_4$. The synthesis of Ni/Olivine was carried out by wet impregnation of synthesized olivine supports with $Ni(NO_3)_3$ solutions for 6 hours. After drying, the catalyst sample was calcined in air at 800°C for 2 h and then reduced at 900°C under H_2 atmosphere. The best condition of synthesis Ni/olivine catalyst was where olivine was calcined at 700°C in which a single-phase and well-crystallized olivine is indicated. Lower calcination temperature did not yield crystallized olivine; while higher calcination temperature led to smooth support surface with lower surface area. SEM micrograph showed presence of Ni particles formed on the surface of $LiFePO_4$ support and the average particle size was around 1-5 μm.

INTRODUCTION

Gasification of biomass is a potential renewable energy option to produce useful fuel gases such as syngas or pure hydrogen. One of the major issues in biomass gasification is how to deal with the tar formed during the process. Tars are complex mixture of condensable hydrocarbons which include single ring to five ring aromatic compounds along with other oxygen containing hydrocarbons[1]. Tar can be eliminated by thermal cracking or by the use of catalysts. The catalytic gasification process is an attractive technological alternative to deal with tar and to produce high yield of syngas. Steam is one of the most commonly used gasification agents because high percentage of hydrogen can be obtained during the process. Many researchers have proved the usefulness and effectiveness of calcined olivine and nickel based steam reforming catalysts on decreasing tar yield [2-3]. The catalyst can increase the reaction rate of the steam as well as participate in secondary tar decomposition reactions. Therefore, the catalyst improves the quality of the gas product and reduces tar content in the process. Besides adding active bed materials also prevents agglomeration tendencies and subsequent coking of the bed. Nickel and olivine catalysts have been proven to be very active in terms of tar reduction and it shows excellent catalytic activity, resistance of coking and sulfur poisoning [4].

The aim of this paper is to show that olivine can be an interesting support for nickel, giving a system with high attrition resistance and strong linking with nickel. The present work showed the synthesis of olivine support and its evolution with calcinations temperature. The nickel/olivine catalyst was then synthesized via impregnation method.

EXPERIMENTAL PROCEDURE

Olivine as catalyst support

$LiFePO_4$ was prepared by a co-precipitation method using Li_3PO_4, phosphoric acid (0.85 H_3PO_4.0.15 H_2O) and ferric citrate n-hydrate ($FeC_6H_8O_7.nH_2O$) as starting materials. Lithium phosphate (1M) and phosphoric acid (1M) were dissolved in 200 ml of deionized water. Ferric citrated n-hydrate (1M) was dissolved in 500 ml of deionized water (boiling water), and the two solutions were combined and concentrated on a hot plate until a wet powder with high viscosity was formed. The wet powders were placed in an oven and heated at 140 °C for 12 hours. The dried powders were grounded before firing at a heating rate of 10°C/min under Ar up to 500-1000 °C, held for 24 hours, and the samples were then air quenched to obtain crystallized $LiFePO_4$. The synthesis olivine catalyst was crushed and sieved to particle size between 20-30 mesh.

Catalyst preparation

The Ni/olivine catalysts were prepared by wet impregnation of Phosphor olivine ($LiFePO_4$) with an excess of nickel salt solution. The samples were calcined under air at 800°C for 2 hours and then reduced at 900°C. These different ways of preparation a series of Ni/olivine catalysts were described in Table I.

Table I. Formulation of olivine and Ni/olivine catalysts

Formulation code	Calcination temperature (1st)	Impregnation with $Ni(NO_3)_2$	Calcination temperature (2nd)	Reduction temperature
[1] $LiFePO_4$	-	-	-	-
[2] $LiFePO_4$-500	500	-	-	-
[3] $LiFePO_4$-700	700	-	-	-
[4] $LiFePO_4$-900	900	-	-	-
[5] $LiFePO_4$-1000	1000	-	-	-
[6] $LiFePO_4$-500-Ni	500	Ni	-	-
[7] $LiFePO_4$-700-Ni	700	Ni	-	-
[8] $LiFePO_4$-900-Ni	900	Ni	-	-
[9] $LiFePO_4$-500-Ni-800	500	Ni	800	-
[10] $LiFePO_4$-700-Ni-800	700	Ni	800	-
[11] $LiFePO_4$-900-Ni-800	900	Ni	800	-
[12] $LiFePO_4$-500-Ni-800-900	500	Ni	800	900
[13] $LiFePO_4$-700-Ni-800-900	700	Ni	800	900
[14] $LiFePO_4$-900-Ni-800-900	900	Ni	800	900

Characterization of olivine and Ni/olivine catalysts

The thermogravimetry (TG,DTG-100, Mettler Telledo) was performed to investigate the thermal stability of the hydrothermally synthesized $LiFePO_4$. Particle-size of olivine and Ni/olivine were characterized with laser scattering (Malvern Instrument 2000). The BET-surface area was determined by means of N_2 chemisorptions on Micromertics ASAP-2000 equipment. Phase and microstructure were characterized by X-ray diffraction (XRD) on Bruker AXS diffractometer using Cu Kα radiation and by scanning electron microscopy (SEM) on a JEOL JSM-6480LV microscope apparatus coupled to energy dispersive X-ray spectroscopy (EDXS), respectively.

RESULTS AND DISCUSSION

$LiFePO_4$ olivine compounds synthesized by co-precipitations appeared visually in yellowish color (Fig 1). The TGA/DTA results of the $LiFePO_4$ precursor performed under flowing nitrogen was shown in Fig.2. While the broaden endothermic peaks attributed to water evaporation was found at temperatures between 60°C and 160°C, the small endothermic peak at 250°C caused by organic compound decomposition are also observed. Whereas the peak exhibited at 500°C is due to the crystallization of $LiFePO_4$[5]. This result corresponded well with the continuous weight loss since ambient temperature to ~ 550°C. The observed initial weight loss before 160°C, the step weight loss between 170°C and 270°C, and the final weight loss from 450-550°C corresponded to the elimination of absorbed water, decomposition of organic compounds, and crystallization of phosphate, respectively. Therefore, it suggests that the precursor should be calcined above 550°C to obtained $LiFePO_4$ crystallized phase.

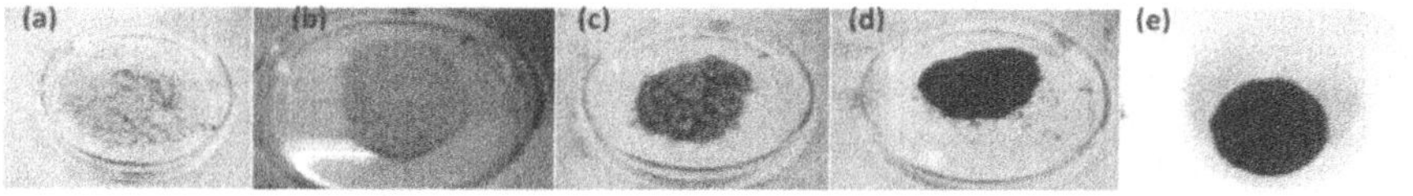

Figure1. The photograph of the (a) fresh $LiFePO_4$, (b) $LiFePO_4$ calcined at 500°C, (c) $LiFePO_4$ calcined at 700°C, and (d) $LiFePO_4$ calcined at 900°C

The olivine calcined at various temperatures appeared in different color (Fig. 1). The yellowish color of olivine calcined at 500°C was quite similar to the fresh olivine. At higher calcination temperatures, the powders turned into gray, purple, and finally black at calcination temperature of 1000°C where the powder sample also melted. This indicated phase transformation with heat treatment which was agreed well with the results from the XRD analysis showed in Fig. 3. It was found that olivine phase did not formed in the sample heat-treated at 500°C as expected, which is in consistent with TGA/DTG results showed in Fig. 2. As the calcination temperature increased, the diffraction peaks corresponding to the olivine structure emerged which indexed on an orthorhombic olivine structure type (space group: Pmnb). No second phase is found. For sample calcined at 900°C, all diffraction lines attributed to the olivine type phase $LiFePO_4$become prominent, an enhanced degree of crystallinity have been realized, which are evident from the sharp diffractograms of increased intensity.

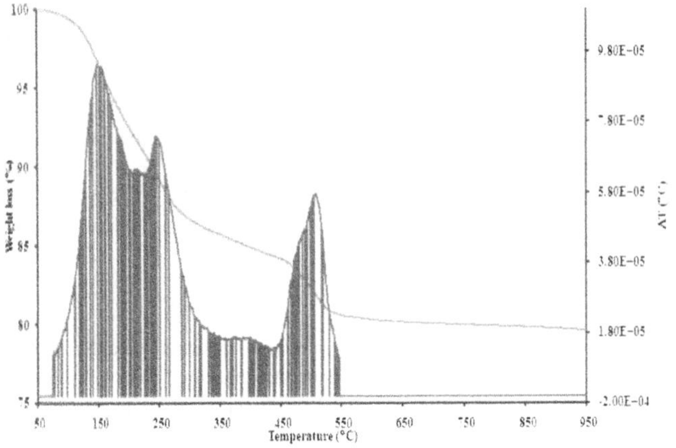

Figure 2. The TGA/DTG curves for the $LiFePO_4$ olivine recorded over the temperature range from ambient to 950°C at heating rate of 10C min^{-1} in nitrogen gas at 20 ml min^{-1} flow rate

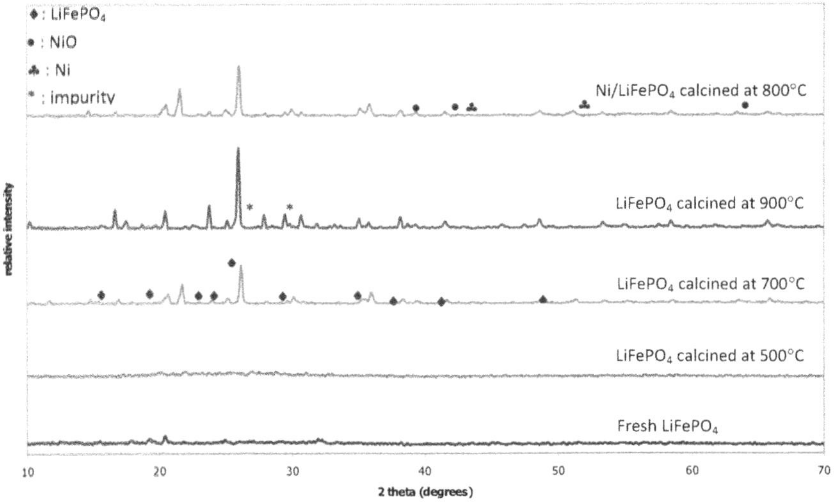

Figure 3. X-ray diffraction patterns of (a) fresh $LiFePO_4$ olivine, (b) $LiFePO_4$ calcined at 500°C, (c) $LiFePO_4$ calcined at 700°C, (d) $LiFePO_4$ calcined at 900°C, and (e) Ni/$LiFePO_4$ calcined at 800°C, with $LiFePO_4$ calcined at 700°C

However, a minor impurity phases were also detected which might be because a higher sintering temperature sped the crystal phase transformation process, thus miscellaneous reaction might occur. XRD of the Ni/ $LiFePO_4$ catalyst calcined at 800°C, with a support calcined at 700°C, shows that the olivine phase is maintained along with trace amount of phases related to Ni (Ni, NiO, Ni $(OH)_2$, $Ni(NO_3)_2 6H_2O$). This was thought that the nickel containing particles decreased in size or more probably their insertion into the olivine structure[6].
The SEM of the fresh olivine shows porosity of this support (Fig. 4). The particle size was around 96μm with BET surface area of 1.94 m^2 g^{-1} (Table II). The particle size of calcined $LiFePO_4$ is bimodal where the popular peak around smaller particle size is about 56.16, 20.37, and 58.22μm for samples calcined at 500, 700 and 900°C, respectively. The BET surface of the $LiFePO_4$ calcined at 500, 700 and 900°C, were 3.11, 5.28 and 2.09 m^2g^{-1}. The enhancement of the BET surface may attribute to the decrease of particle size after calcinations, and the obvious improvement of the total surface area of $LiFePO_4$ calcined at 700°C could be ascribed to the formation of porous structure in $LiFePO_4$ material itself. After calcination at 900°C, the porosity observed on fresh olivine disappears, more compact grains are formed as evidenced by a larger particle size (58.22 μm) and lower surface area (2.09 m^2g^{-1}).

Figure 4. SEM photographs of (a) Fresh $LiFePO_4$olivine, (b) $LiFePO_4$ calcined at 500°C, (c) $LiFePO_4$ calcined at 700°C and (d) $LiFePO_4$calcined at 900°C, (e) Ni/$LiFePO_4$ with $LiFePO_4$ calcined at 500°C, (f) Ni/$LiFePO_4$ with $LiFePO_4$ calcined at 700°C, (g) Ni/$LiFePO_4$ with $LiFePO_4$ calcined at 900°C

On the scanning electron micrographs of the Ni/olivine catalyst calcined at 800°C (Fig. 4 e-g), a deposit of individual and cluster of grains (probably NiO) with size between 1-5 μm can be observed. Ni/olivine catalyst with support calcined at 500°C (Fig. 4e) showed grain growth of $LiFePO_4$ compared to before impregnation with nickel salt, which was due to the second heat treatment after impregnation at 800°C. Some areas of catalyst with support calcined at 700°C (Fig. 4f) are similar to what is obtained after calcination at 500°C with deposited particles of same size but more linked to the support. Other areas show a nickel oxide with bigger deposited grains. Calcination at 800°C of Ni/olivine catalyst with support calcined at 900°C (Fig. 4g) led to the formation of a very smooth surface of the support, similar to that obtained with calcined olivine before impregnation, but with a less deposition and insertion of Ni crystallites into the olivine structure. This confirmed by EDX analysis. The elements of catalysts with olivine support calcined at 700°C contained 8.12% Ni, 41.98% P, and 49.12%Fe. Whereas, the one with support calcined at 900°C had lower amount of nickel, 5.39%. Smooth surface with a smaller surface area structure might cause a lower degree of Ni deposition. The reduction of Ni/Olivine powders at 900°C was not successful because powders were melted down to the crucible. The necessity of lower reduction temperature would be investigated further, along with the catalytic activity of Ni/Olivine catalysts.

Table II. Particle size, BET surface area, and element of olivine and Ni/olivine catalysts

Formulation code	BET	Particle size distribution			EDX analysis		
		D10	D50	D90	(% Ni)	(% P)	(% Fe)
[1] $LiFePO_4$	1.94	10.21	96.26	455.59	-	44.34	53.51
[2] $LiFePO_4$-500	3.11	7.85	56.16	217.94	-	40.36	55.78
[3] $LiFePO_4$-700	5.28	6.39	20.37	64.31	-	39.23	58.08
[4] $LiFePO_4$-900	2.09	8.66	58.22	236.18	-	38.88	58.56
[5] $LiFePO_4$-1000	-	-	-	-	-	-	-
[6] $LiFePO_4$-500-Ni	2.82	9.75	105.56	464.88	14.68	35.35	48.64
[7] $LiFePO_4$-700-Ni	5.16	2.89	55.31	352.85	16.74	32.01	50.11
[8] $LiFePO_4$-900-Ni	1.79	8.02	117.80	603.30	12.9	31.43	55.28
[9] $LiFePO_4$-500-Ni-800	2.28	8.11	159.26	753.77	6.35	44.45	48.33
[10] $LiFePO_4$-700-Ni-800	4.06	11.08	84.58	340.44	8.12	41.98	49.12
[11] $LiFePO_4$-900-Ni-800	1.13	27.69	298.55	803.99	5.39	39.87	51.21
[12] $LiFePO_4$-500-Ni-800-900	N/A	N/A	N/A	N/A	N/A	N/A	N/A
[13] $LiFePO_4$-700-Ni-800-900	N/A	N/A	N/A	N/A	N/A	N/A	N/A
[14] $LiFePO_4$-900-Ni-800-900	N/A	N/A	N/A	N/A	N/A	N/A	N/A

N/A : powders were melt

CONCLUSION

The Ni/$LiFePO_4$ powders were successfully synthesized by a co-precipitation and wet impregnation methods. $LiFePO_4$ olivine was obtained via co-precipitation of Li_3PO_4, phosphoric acid and ferric citrate n-hydrate. Calcination of olivine with Ar for 24 hours at 700°C resulted in a single-phase and well-crystallized olivine. Lower calcination temperature of 500°C did not yield crystallized olivine; while higher calcination temperature of 900°C led to smooth support surface with lower surface area. It was observed that the porous structure remained for the calcined olivine; higher calcinations temperature tended to decrease particle size of olivine while increase the BET surface area. After calcinations at 900°C, surface area and close pores decreased due to melting of olivine surface. SEM reveals the homogenity of the pre-calcined and calcined samples of olivine. However, after impregnation, nickel compounds were formed on the surface but not quite uniform. In some positions nickel concentrations as high as 8 wt% is detected for olivine samples calcined at 700°C. Smooth surface with a smaller surface area for olivine samples calcined at 900°C led a lower degree of Ni deposition. Therefore the optimize condition for synthesis Ni/$LiFePO_4$ was where olivine support was calcined at 700°C. In addition, this preparation method uses inexpensive starting materials and operates in mild synthetic conditions, and therefore, it may provide a feasible way for industrial production of $Ni/LiFePO_4$ catalyst for tar removal in the future.

ACKNOWLEDGEMENT

The authors sincerely acknowledge Thailand Graduate Institute of Science and Technology, TGIST, (TG-33-09-49-030D) and National Metal and Materials Technology Center (Project No. MT-B-49-END-07-007-I) for their financial support. Special thanks to the Department of Materials Science, Department of Environmental Engineering and National Center of Excellence for Petroleum, Petrochemicals, and Advanced Materials, Chulalongkorn University, Thailand for helping out with different characterization techniques.

REFERENCES

[1] R.Coll, J. Salvado, and D. Montane, Steam reforming model compounds of biomass gasification tars: conversion at different operating conditions and tendency towards coke formation, *Fuel Process. Technol.*, 74,19-31, (2001).

[2] L. Devi, and K.J. Ptasinski, A review of primary measures for tar elimination in biomass gasification processes, *Biomass Bioenergy*, 24, 125-140, (2003).

[3] D. Sutton, and B. Kelleher, Review of literature on catalysts for biomass gasification, *Fuel Process. Technol.*, 73, 155-173, (2001).

[4] T.Wang, and J. Chang, Novel catalyst for cracking of biomass tar, *Energy Fuels*, 19, 22-27., (2005).

[5] R.Dominko, M. Bele, M. Gaberscek, , M. Remskar, D. Hanzel, J.M. Goupil, S. Pejovnik, and J. Jamnik, Porous olivine composites systhesized by sol-gel technique. *Journal of Power Sources*, 153, 274-280, (2006).

[6] C.Courson, E. Makaga, C. Petit, and A. Kiennemann, Development of Ni catalysts for gas production from biomass gasification. Reactivity in steam- and dry-reforming., *Catalysis Today*, 63, 427–437, (2000).

FABRICATION AND CHARACTERIZATION OF CERMET MEMBRANE FOR HYDROGEN SEPARATION

S. Vichaphund and D. Atong*
National Metal and Materials Technology Center,
114 Thailand Science Park Paholyothin Rd., Klong 1, Klong Luang, Pathumthani 12120 Thailand
*E-mail: duangdua@mtec.or.th

ABSTRACT

In this work, a cermet (ceramic-metal) membrane was fabricated to separate hydrogen from syngas generation process. An Al_2O_3 disk support (13 mm diameter) was prepared by uniaxially pressing Al_2O_3 powder to form monolithic shape at 7 MPa and sintered at 1000-1200 °C for 2h. The pore size of 150-160 nm and 50-60 % porosity were determined by Hg porosimetry. The Al_2O_3 support was then soaked into 10wt% Nickle (II) nitrate hexahydrate solution. Ni solution was applied onto both front and back sides of membrane each for 10 min. After soaking process, the coated Ni-Al_2O_3 membrane was dried at 25°C for 5 h, then at 100°C for 24 h and calcined at 900°C for 2h. The soaking-drying-firing sequence was repeated ten times to confirm a sufficient amount of Ni was deposited. Finally, calcined Ni-Al_2O_3 membrane was reduced under H_2 atmosphere at 910°C for 2h. The crystalline phases of membrane were investigated by XRD. The microstructure and elemental distribution of membranes were also characterized by scanning electron microscopy (SEM) and energy dispersive X-ray spectrometer. The pore size and porosity were determined by Hg porosimetry.

Key words: composite membrane, hydrogen separation, Nickle, Alumina

INTRODUCTION

Hydrogen, is an important gas for many applications including fuel cell technologies for transportation[1-8]. There have been many efforts to develop membranes for hydrogen separation using materials such as palladium, silica, alumina and ceramic-metals. Pure Pd and Pd-based membranes are very attractive because of their high selectivity of hydrogen from gas mixtures due to the high solubility and mobility of hydrogen in the Pd lattice [2, 3, 7, 9-11]. Unfortunately, there are several disadvantages of Pd-based membranes. Pd is very expensive and scarce [4, 5], thus the use of Pd-based membranes has been limited to small scale applications. Another disadvantage is that Pd-based membranes may become brittle at low temperature in a hydrogen atmosphere due to a phase transition between the α-phase (Face centered cubic; FCC) and β phase (Body centered cubic; BCC) in the Pd-H system [1, 2, 12-14]. Moreover, another major technical disadvantage of palladium membranes in most applications is their high sensitivity to chemicals such as sulphur, chlorine and CO [15, 16]. For these reasons, it has become necessary to develop non-palladium-based or low-cost hydrogen permeation membranes. Ceramic-metal membranes (cermet membranes) seem to be a suitable alternative because it requires low-cost materials and simple fabrication techniques. Furthermore, the ceramic phase in cermet membranes can improve mechanical rigidity and thermal stability. There are various researches of ceramic-metal membranes such as Ni/ceramic with different preparation methods including soaking-rolling, electroless plating, and impregnation.[6, 14, 17] The reason for using nickel in cermet membranes is nickel's good hydrogen adsorption capacity, and lower material cost than palladium. However, there have been limited studies dealing with the use of nickel as an effective hydrogen separator [6, 14].

In this work, we report on the preliminary results from the fabrication and characterization of Ni-Al_2O_3 membranes. The effect of sintering temperatures on membrane support was investigated. The fabricated membranes were characterized by X-ray diffractometry (XRD), scanning electron microscopy (SEM), energy dispersive X-ray spectrometer including X-ray mapping (EDS). In addition, the pore size and porosity were determined by Hg porosimetry.

EXPERIMENTAL PROCEDURE

The first step consists in the preparation of an Al_2O_3 disk support. The Al_2O_3 powder was mixed with 5wt% PVA (polyvinyl alcohol, 98-99% with MW 85,000-146,000) by conventional ball milling with distilled water using alumina media. After milling, the slurry was dried at 100°C for 24h and then sieved through 100 mesh. The sieved powder was uniaxially pressed to form a disk shape with a diameter of 13 mm at 7 MPa. Then, the Al_2O_3 disk was sintered for 2 hours at 1000°C or at 1200 °C. The second step consists in the preparation of Ni-Al_2O_3 membrane by soaking the Al_2O_3 support into 10wt% Ni solution by a rotary vacuum pump. Ni solution was applied onto both the front and back sides of the Al2O3 disk support each for 10 min. After the soaking process, the coated Ni-Al_2O_3 membrane was dried at 25°C for 5 h, then at 100°C for 24 h and calcined at 900°C for 2h. The soaking-drying-firing sequence was repeated ten times. Finally, the membranes were further treated at 910°C for 2 h under a reducing gas (H_2, 99.99%).

After fabrication, the density of membrane was measured by dimension calculation. The crystalline phases of cermet membrane were investigated by using X-ray diffraction (XRD; model JEOL, JDX-3530). Scanning electron microscope (SEM; JEOL, JSM-6301F) was used to investigate the microstructure of Ni/Al_2O_3 membrane. For analysis of nickel dispersion of Ni/Al_2O_3 membrane, energy dispersive X-ray spectroscope (EDS; Oxford Inca 300 and 350) with X-ray dot mapping was used. In addition, the pore size and porosity were determined by mercury porosimetry.

RESULTS AND DISCUSSION

Physical properties of Ni-Al_2O_3 composite membrane

Table I lists the physical properties of Ni-Al_2O_3 membranes with Al_2O_3 support disks sintered at either 1000°C or 1200°C. For the same calcination condition, 900°C, it was noticed that the densification of Al_2O_3 support played an important role to the densification of the membrane. As the sintering temperature increased, the density of Al_2O_3 support increased which led to higher density of Ni-Al_2O_3 membranes. The shrinkage of Ni-Al_2O_3 membranes also increased with sintering temperature. The Ni-Al_2O_3 membrane where Al_2O_3 support sintered at 1200°C had the highest density of 2.168 g/cm^3.

Phase analysis

XRD patterns of Ni -Al_2O_3 membranes calcined at 900 are shown in Fig.1 (a). The composite membrane composed of alumina (alpha-phase, corundum (Al_2O_3, JCPDS:10-0173)) and nickel aluminum oxide ($NiAl_2O_4$, JCPDS: 10-0339) as an intermediate phase, which results from the interaction between Ni salt and alumina support disk. However, after reduction in H_2, the peaks associated with the $NiAl_2O_4$ phase disappeared. The only phases present were alumina and nickel

(Ni, JCPDS:4-0850) (Fig. 1 (b)). From these results, it can be concluded that a reduction temperature of 910°C was high enough to reduce the intermediate phase.

Table I Physical properties of Ni-Al_2O_3 composite membrane calcined at 900 °C

Ni-Al_2O_3 membrane	Density (g/cm^3)	Diameter Shrinkage (%)	Thickness shrinkage(%)
Al_2O_3-1000°C	1.828	0.399	0.842
Al_2O_3-1200°C	2.168	5.987	6.854

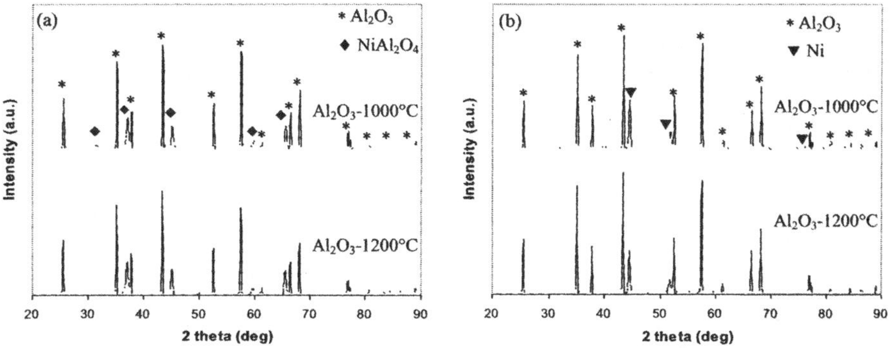

Figure 1. XRD patterns of Ni -Al_2O_3 membranes preparing from (a) calcination at 900°C and (b) calcinations followed by reduction at 910 °C

Pore size and porosity

The pore size and % porosity of Ni-Al_2O_3 composite membranes measured by mercury porosimeter are listed in Table II. It can be seen that the porosity of Al_2O_3 support decreased significantly from 60 to 50% with increasing sintering temperature, while the pore size hardly changed. It is assumed that the alumina support experienced intermediate stage of sintering. After preparation of Ni-Al_2O_3 membrane, the Ni impregnated Al_2O_3 samples were calcined at 900°C. This re-heating step resulted in a decrease in the porosity of membrane as expected. The porosity of Ni-Al_2O_3 membrane with Al_2O_3 support sintered at 1000°C decreased from 60 to 53%. The same trend was observed in the composite membrane with Al_2O_3 support sintered at 1200°C; but with less significance, from 50 to 48%. The calcination process did not improve the densification of Al_2O_3 support sintered at 1200°C as much as the one sintered at 1000°C. Furthermore, it was noticed that a narrow pore size distribution was obtained after coating with the Ni solution. The pore sizes for Ni-Al_2O_3 membranes were 0.145 and 0.140 micron for Al_2O_3 support sintered at 1000 and 1200°C, respectively. It was found that the decrease in the porosity and pore size was due to Ni deposition on/into the Al_2O_3 support successfully. Overall, the Ni-Al_2O_3 composite membrane prepared with an alumina substrate sintered at 1200°C, resulted in higher densification, smaller pore size, and less porosity than that prepared using a support sintered at 1000°C. However, the difference in porosity and pore size for

these two membranes was not significant.
The mechanical strength of the membranes and their gas permeability will be determined in the future to determine the optimum sintering temperature of the alumina support.

Table II. The pore size and porosity of Ni-Al_2O_3 composite membranes measured by mercury porosimeter

Materials	Temperature (°C)	Pore size (μm)	Total Porosity (%)
Al_2O_3-Support			
Al_2O_3 support disk	1000	0.156	59.9
Al_2O_3 support disk	1200	0.151	49.9
Ni-Al_2O_3 membrane with			
Al_2O_3 support disk-1000°C	900	0.145	52.9
Al_2O_3 support disk-1200°C	900	0.140	48.5

The microstructure of Ni -Al_2O_3 membranes

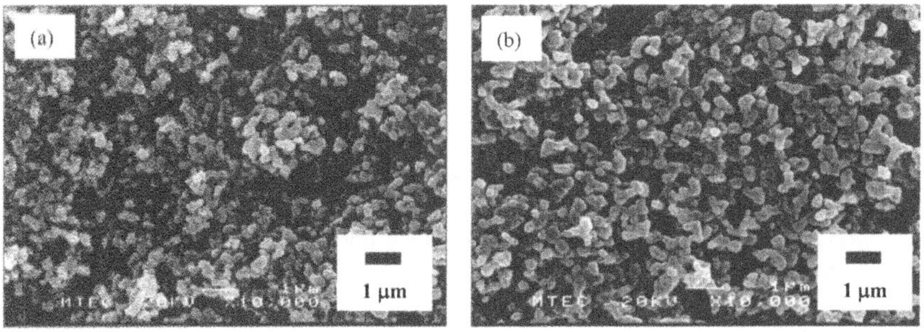

Figure 2. SEM images of Ni -Al_2O_3 membranes calcined at 900 °C: (a) Surface of membranes with Al_2O_3 support sintered at 1000°C; (b) Surface of membranes with Al_2O_3 support sintered at 1200°C, respectively

Fig. 2 demonstrated the microstructures of Ni -Al_2O_3 membranes. It can be seen that the microstructure of both membrane surfaces composed of interconnected submicron particle size with less than 1 micron in size. The interconnected porosity where pores are connected to the surfaces of the membrane still appeared on both cases. However, it was noticed that a denser structure is obtained on the surface of membranes with alumina support sintered at 1200°C. Larger particle sizes along with a greater connection between particles were observed.
The elemental analysis of Ni-Al_2O_3 membrane estimated by EDS is presented in Table III. It can be seen that the elements of the composite membrane with Al_2O_3 support sintered at 1000°C contained 17.3 wt% nickel (Ni), 46.4 wt% aluminium (Al), and 36.3 wt% oxygen (O). Whereas the membrane with Al_2O_3 support sintered at 1200°C had lower amount of nickel, 13.2 wt%. The lower porosity and smaller pore size might be a reason of a less degree of impregnation. However, it was confirmed by X-ray mapping as shown in Fig. 3 that the distribution between Ni and Al_2O_3 was quite uniformly dispersed on both membrane surfaces; one with support sintered at 1000°C and

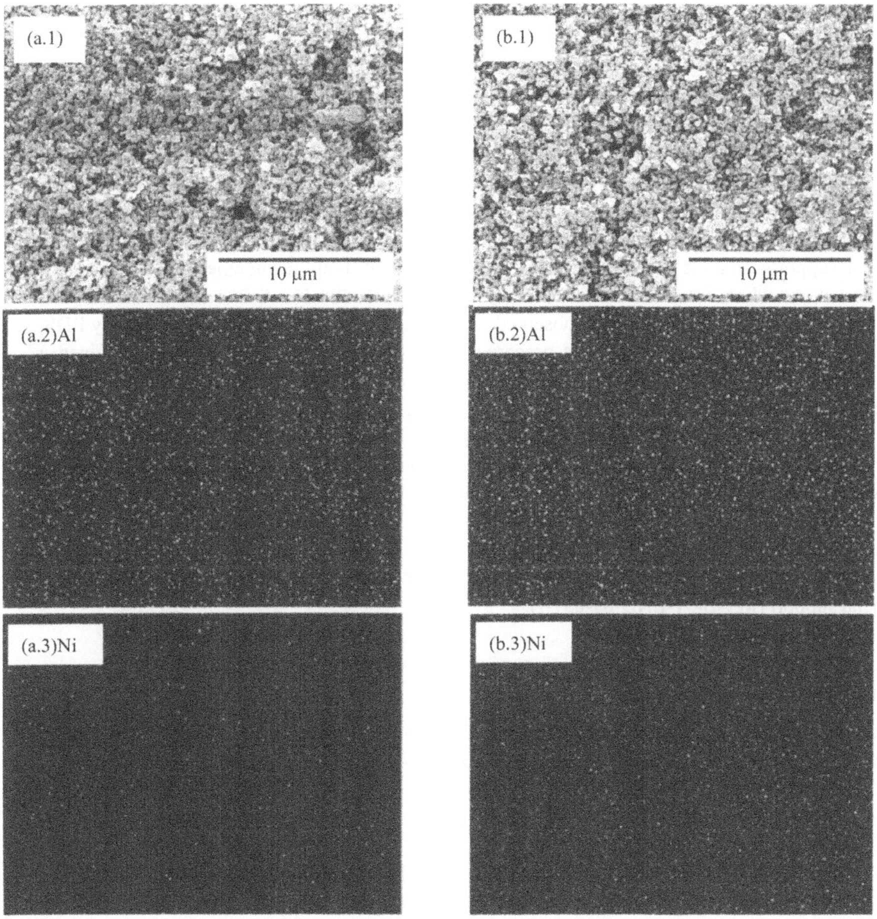

Figure 3. X-ray mapping of elemental constituents within the microstructure of Ni -Al_2O_3 membrane at 900 °C at: (a) Surface of membranes with Al_2O_3 support sintered at 1000°C; (b) Surface of membranes with Al_2O_3 support sintered at 1200 °C, respectively

Table III The element analysis of Ni-Al_2O_3 membrane estimated by EDS

Ni-Al_2O_3 membrane	Elements (wt%)		
	Ni	Al	O
Al_2O_3 support-1000°C	17.3	46.4	36.3
Al_2O_3 support-1200°C	13.2	48.4	38.4

another at 1200°C.

CONCLUSIONS

The fabrication of Ni-Alumina membranes for hydrogen separation applications via soaking-drying-firing technique was studied. XRD analyses showed that the reduction process at 910°C could completely transform the intermediate phases of nickel aluminum oxide ($NiAl_2O_4$) to the desirable single Ni and Al_2O_3 phases. Comparing the membranes fabricated from support sintered at different sintering temperatures, it was found that the 1200°C –alumina supported membrane resulted in higher densification, smaller pore size, and less porosity than membrane with alumina supports sintered at 1000°C. SEM micrographs of both membranes revealed similar microstructures where interconnected submicron particle size with less than 1 micron in size and interconnected porosity structure were observed. From the elementary analysis determined by EDS, it was confirmed that both membranes prepared from soaking-drying-firing method contained a concentration of Ni between 13 and 17 wt%. The distribution between Ni and Al_2O_3 particles was quite uniformly dispersed as confirmed by X-ray mapping. From these preliminary results, it can be concluded that the soaking-drying-firing technique is a feasible route to fabricate Ni-Al_2O_3 composite membranes. In future work, the need to sinter the alumina support at 1200°C will be assessed through mechanical strength and permeability testing.

ACKNOWLEDGEMENT

The authors would like to thank to National Metal and Materials Technology Center, Thailand for financial support (MT-B-51-END-07-057-I).

REFERENCES

[1] S. Haag, M. Burgard and B. Ernst, Pure Nickel Coating on a Mesoporous Alumina Membrane: Preparation by Electroless Plating and Characterization, *Surf. Coat. Technol.*, **201**, 2166-73 (2006).

[2] D.-W. Kim, Y. J. Park, J.-W. Moon, S.-K. Ryi and J.-S. Park, The effect of Cu Reflow on the Pd-Cu-Ni Ternary Alloy Membrane Fabrication for Infinite Hydrogen Separation, *Thin Solid Films*, **516**, 3036-44 (2008).

[3] S.-K. Ryi, J.-S. Park, S.-H. Choi, S.-H. Cho and S.-H. Kim, Fabrication and Characterization of Metal Porous Membrane made of Ni Powder for Hydrogen Separation, *Sep. Purif. Technol.*, **47**, 148-55 (2006).

[4] G.F. Tereschenko, M.M. Ermilova, V.P. Mordovin, N.V. Orekhova, V.M. Gryaznov, A. Iulianelli, F. Gallucci and A. Basile, New Ti-Ni Dense Membranes with Low Palladium Content, *Int. J. Hydrogen Energy*, **32**, 4016-22 (2007).

[5] A. Basile, F. Gallucci, A. Iulianelli, G.F. Tereschenko, M.M. Ermilova and N. V. Orekhova, Ti-Ni-Pd Dense Membranes-The Effect of the Gas Mixtures on the Hydrogen Permeation, *J. Membr. Sci.*, **310**, 44-50 (2008).

[6] C.-Y. Yu, B.-K. Sea, D.-W. Lee, S.-J. Park, K.-Y. Lee and K.-H. Lee, Effect of Nickel Deposition on Hydrogen Permeation Behavior of Mesoporous γ-Alumina Composite Membranes, *J. Colloid Interface Sci.*, **319**, 470-76 (2008).

[7] M. Kanezashi and M. Asaeda, Hydrogen Permeation Characteristics and Stability of Ni-doped Silica Membranes in Steam at High Temperature, *J. Membr. Sci.*, **271**, 86-93 (2006).
[8] J. S. Hardy, E. C. Thomsen, N.L. Canfiled, J. V. Crum, K. S. Weil and L. R. Pederson, Development of Passive Hydrogen Separation Membranes made from Co-synthesized Nanoscale Cermet Powders, *Int. J. Hydrogen Energy*, **32**, 3631-39 (2007).
[9] S. Haag, M. Burgard and B. Ernst, Beneficial effects of the Use of a Nickel Membrane Reactor for the Dry Reforming of Methane: Comparison with Thermodynamic Predictions, *J. Catal.*,**252**, 190-204 (2007).
[10] A. Li, W. Liang and R. Hughes, Fabrication of Dense Palladium Composite Membranes for Hydrogen Separation, *Catal. Today*, **56**, 45-51 (2000).
[11] S. Tosti, Supported and Laminated Pd-based Metallic Membranes, *Int. J. Hydrogen Energy*, **28**, 1445-54 (2003).
[12] Y. Swesi, D. Ronze, I. Pitault, R. Dittmeyer and F. Heurtaux, Purification Process fpr Chemical Storage of Hydrogen for Fuel cell Vehicles applications, *Int. J. Hydrogen Energy*, **32**, 5059-66 (2007).
[13] J. Gabitto and C. Tsouris, Hydrogen Transport in Composite Inorganic Membranes, *J. Membr. Sci.*, **312**, 132-42 (2008).
[14] B. Ernst, S. Haag and M. Burgard, Permselectivity of a Nickle/Ceramic Composite Membrane at Elevated Temperatures: A New Prospect in Hydrogen Separation?, *J. Membr. Sci.*, **288**, 208-17 (2007).
[15] S. Gopalakrishnan and J.C.D. da Costa, Hydrogen Gas Mixture Separation by CVD Silica Membrane, *J. Membr. Sci.*, **323**, 144-47 (2008).
[16] V. Sebastian, Z. Lin, J. Rocha, C. Tellez, J. Santamaria and J. Coronas, Improved Ti-silicate Umbite Membranes from the Separation of H_2, *J. Membr. Sci.*, **323**, 207-12 (2008).
[17] C. Courson, E. Makaga, C. Petit and A. Kiennemann, Development of Ni Catalysts for Gas Production from Biomass Gasification. Reactivity in Steam- and Dry-reforming, *Catal. Today*, **63**, 427-37 (2000).

POROUS CERAMICS FOR HOT GAS CLEANING; DEGRADATION MECHANISMS OF SiC-BASED FILTERS CAUSED BY LONG TERM WATER VAPOUR EXPOSURE

Pirjo Laurila and Tapio Mantyla
Department of Materials Science, Tampere University of Technology
Korkeakoulunkatu 6, FI-33720 Tampere, Finland

ABSTRACT

Hot gas filters are one of the key components for the advanced coal and biomass based power generation like pressurised fluidised-bed combustion or integrated gasification combine cycle processes. Insufficient reliability of hot gas filter materials is among the main obstacles of these more efficient combined steam and gas turbine power processes to replace the traditional ones. Knowledge about the effects of water vapour and thermal transients on microstructure and its connection to the mechanical properties of the filter materials is important for the understanding of the material degradation process at complex operation environments, for lifetime prediction and for further development of the materials. In this study, the effect of high temperature water vapour on the microstructure of two advanced commercially available SiC-based silicate-bonded hot gas filters was characterised with scanning electron microscope, a quantitative X-ray diffraction (XRD) method, Archimedes' method and by chemical analysis. Crystallization of an amorphous binder and oxidation of SiC were found but the rate of oxidation was low. There was a clear difference in the resistance to crystallization and oxidation of the two materials and they showed different trends of apparent density as function of time and the amount of water vapour.

INTRODUCTION

Hot gas filters are one of the key components for the advanced coal and biomass based power generation like pressurised fluidised-bed combustion (PFBC) or integrated gasification combine cycle (IGCC) processes that utilize both steam cycle and gas turbine.[1,2] The issue of reliability and the availability factor are the main obstacles for the applications of hot gas cleaning systems. There are several causes for the failure of such systems, such as the design of filtering unit, the type of filter material, candle design, thermal transients and residual ash deposits[3]. Although considerable advancement has been obtained in avoiding these, there still exist gaps in the understanding of the detailed failure mechanism. The complex chemistry of the environments combined with the changing thermal loads makes it difficult to determine the degradation mechanisms in pilot tests simulating the real environments, especially when the testing history includes several parameter changes during the lifetime of the component. In this study we have tried to determine the role of water vapour at high temperature and thermal transients in possible changes of microstructure of two commercial silicate bonded SiC materials in order to understand their degradation mechanisms in more complex operation environments, for lifetime prediction and for further materials development. Water vapour is present both in combustion and gasification processes.

MATERIALS AND METHODS

The SiC-based silicate-bonded filters were supplied by Schumacher Umwelt- und Trenn Technik

GmbH, Germany and they are denoted as A (DSL N 10-20) and B (DSL N 10-20). The filters were membrane covered tube elements, but in this study the focus is on the support structure. The strength of type A was known *a priori* to be better than that of type B. The support structure is a composite of SiC grains and heterogeneous binder containing an amorphous aluminium silicate phase, mullite and cristobalite. The major differences in the phase morphology of as-received materials are that material A has a coarser grain size of mullite than material B and that cracks are found frequently from the binder of material B but in material A cracks are usually found only when the two SiC grains are closer than about 10 m. The elemental compositions of the amorphous phase in the binder of materials A and B as-received is nearly identical containing about 82.5 wt.-% SiO_2, 11.5 wt.-% Al_2O_3 and total of 6 wt.-% Na_2O and K_2O. The amount of Na_2O in material B, 1.4 wt.-%, is nearly double of that found in material A. The aluminium to alkaline ratio is 1.69 for material A and 1.54 for material B. More detailed characterization of materials is described in ref. 4.

Three bulk (candle pieces) exposures at 850°C with and without water vapour were completed to cause microstructural changes in order to find out their possible role in changes in mechanical properties of materials. The coding and conditions of the bulk exposures are described in Table 1. The water vapour pressure in the exposures corresponded to the atmospheric pressure. Thermal cycling between 150 and 400 °C was done without water vapour in order to see the effect of thermal shock caused by back pulse cleaning and the phase transformation of silica.

In addition small-scale exposures at 870°C were used to study further the role of water vapour with 20 vol.-% and 30 vol.-% water vapour in air in an alumina tube furnace. Phase morphology was studied with scanning electron microscope (SEM). To distinguish between the glassy and crystalline phases some samples were etched in 2-% HF-solution. Phase composition was determined by a quantitative X-ray diffraction (XRD) analysis based on the internal standard, CaF_2, method[4]. The chemical composition of the amorphous phase in the binder was determined locally with EDS analysis in SEM. A bulk chemical analysis of the amorphous phases in binder was performed by Induction Coupled Plasma Mass-Spectrometry (ICP-MS). The samples were crushed powders of as-received and 500 h at water vapour bulk exposed (w) materials. The glassy phase was dissolved in HF-HNO_3 solution. Open porosity, apparent and bulk densities of each specimen were determined by Archimedes' method before and after the tests.

Table 1. The exposure environments and coding of the bulk exposures. In each exposure some samples were removed at early stage of the exposure, their time and/or number of thermal cycles are given in parenthesis.

Exposure	Time at max. T [h]	Temperature [°C]	No. of thermal cycles	Water feed into chamber
Thermal cycling (c)		150-400	(6) / 31 or 30*	No
Water vapour only (w)	(100) / 500	850	-	Yes (2-5 ml/min)
Water vapour and cycling (w+c)	(98) / 455	<150-850	(4) / 8	Yes (2 ml/min)

RESULTS

Phase morphology: Exposures to high temperature water vapour had clearly different effects on the materials. In as-received condition typical features of the microstructure of both materials weresmooth, amorphous areas and needle shaped mullite grains on the pore side surface of the binder. On polished cross-sections the pore shape was smooth and spherical, and in many areas the side surface of pore was amorphous. Another common feature was the cristobalite layer on the SiC surfaces next to the amorphous binder. The binder itself contained several typical areas of microstructure in both materials[4] but the completely amorphous areas and mullite in silica rich matrix were common for both.

Microstructural changes were not found from either material after thermal cycling (c). This was expected since the peak temperature of 400°C is low and the hold time at peak temperature was short. In exposures with high temperature water vapour the nucleation of small crystallites occurred on the surface of material A. It was the clear after the bulk exposure (w) which had the longest continuous hold at the peak temperature of 850°C. The EDS analyses suggested that the small crystals on pore side surface are of pure SiO_2. Otherwise there was no clear difference between the water vapour exposed (w) and the as-received or thermally cycled (c) material A.

In material B extensive crystallization of the binder surfaces on the pore sides occurred in both bulk exposures with water vapour (w and w+c). Crystallized pore side surfaces of material B were also found after 500 h small-scale exposures at 20-% and 30-% water vapour but not as clearly as after the bulk exposures with the higher level of water vapour (w). A feature found only in material B after the bulk water vapour exposures (w and w+c) was the precipitates grown outward from the originally smooth convex of the pore surface. Strong crystallization of the binder surface on pore sides was evident. Some evidence of crystallization and grain growth within the binder were also found; the finger-like features next to the SiC grains in the Fig. 1a. suggest crystal growth from the amorphous binder in the water vapour exposure. This wavy frontline of the non-etched areas was present in the as-received and thermally cycled materials B, but in the water vapour exposed materials the finger-like pattern was much more pronounced and with smaller scale of waviness on it.

No clear oxidation layers were found from the as-received samples. A gap of ~50 nm was present, but it was not possible to determine if it was because of the etching or because of a loose SiC-binder boundary or just an edge effect of secondary electron imaging. No significant oxidation layers were found from material A after the 500 hours bulk exposure (w). Possible oxidation layers of the thickness of 100 nm or less were detected in a few locations when a thin layer, 10 m or less, of the binder covered SiC, Fig. 1b. Clear oxidation layers of the thickness up to 500 nm were found from material B from similar locations, Fig. 1c. These layers in material B were also detectable at lower magnifications. However, when there was a thick layer of the binder on SiC, oxidation layers were absent or were significantly thinner supporting the view that the binder is protective.

Phase composition: The XRD results showed an increase in the amount of cristobalite due to high temperature water vapour exposure (w). This is in accordance with the results of SEM investigation. The XRD results showed also a slight increase in the amount of mullite in the exposed samples of material A, it is questionable due to low exposure temperatures, low intensity and scatter of the results.

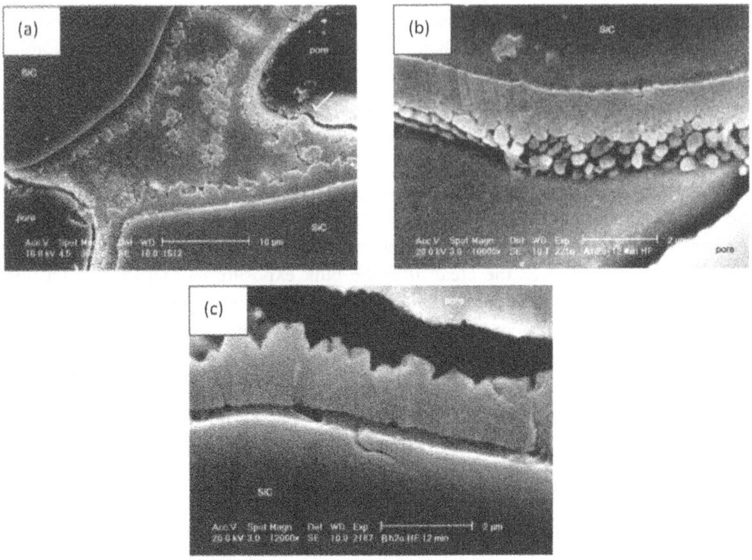

Figure 1. Examples of oxidation layers on SiC, (a) Cross-section of B after 500h exposure at 850°C in atmospheric pressure of water vapour (w). (b) Oxidation layer about 100 nm thick in A after 500 h exposure at 850°C. (c) Oxidation layer about 500 nm thick in B after 500 h exposure at 850°C atmospheric pressure of water vapour (w).

Chemical analysis: Some changes in the composition of the amorphous phase of the binder were found after the water vapour exposures. Table 2 presents the compositions of the amorphous phase between the SiC grains according to EDS analyses and the ratio of aluminium Al to alkalis (Na+K) in the samples. The amount of SiO_2 in material A after 500 h water vapour exposure (w) was

Table 2. Compositions of the amorphous phase in the as-received (ref) and 500 h water vapour exposed (w) materials determined with EDS in oxide form. Weight fractions are given with their standard deviations. The aluminium to alkalis ratio was calculated from the mole fraction of the oxides.

Oxide/Sample	A ref	A w	B ref	B w
Na_2O	0.8±0.10	0.9±0.16	1.4±0.22	1.6±0.19
Al_2O_3	11.5±0.65	12.5+-0.95	11.4±0.68	11.3±0.79
SiO_2	82.6±0.95	80.6±1.17	82.5±0.86	81.6±0.85
K_2O	5.1±0.44	6.0±-0.26	4.6±0.25	5.3±0.39
Al/(Na+K)	1.69	1.56	1.54	1.34
No. of analyses	25	18	21	30

decreased when compared to the composition of the as-received material. The difference was 2 wt.-% or 1.5 mol-%. The glass was then enriched in alumina (Al_2O_3) and potassium oxide (K_2O). There was some decrease in the SiO_2 content of the glass in material B but not as clearly as in material A. The Al/(Na+K) ratio decreased in both materials but more clearly in material B.

Porosity and density: The results of material A suggest a small increase in apparent density and a modest increase in open porosity due to the water vapour exposures, but for material B the changes were negligible. The change in apparent density of material A was small but increased with both time and the amount of water vapour. Further, the data from bulk exposures (w) fit the trend well; the shorter total time at 850°C in vapour with (w+c) exposure can explain the somewhat smaller change in density. If no weight changes have occurred in material A, as the small-scale exposure data suggests, the increase in porosity corresponds the increase in density. However, currently we have no explanation for the mechanism that would be responsible for the increase in porosity or apparent density.

The small decrease or practically no change in density observed for material B in all exposures and for material A in exposures without water is as expected: the density of cristobalite is 2.33 g/cm^3 whereas the estimated density of the amorphous phase is 2.45 g/cm^3 thus crystallization of SiO_2 should lead to a decrease in density. The most extensive crystallization was observed in material B after the bulk exposures in water vapour. Also the oxidation leads to the decrease in density, density of SiC is 3.17 g/cm^3 and of amorphous SiO_2 2.2 g/cm^3. A thin oxidation layer formed on cut and plain SiC grain surfaces in all exposures. Despite that the oxidation was minor in material A and again the most effective in material B after the bulk exposures with water vapour.

DISCUSSION

Devitrification of binder at the pore surfaces has been frequently reported for similar and first generation clay bonded hot gas filters after the use in real combustion conditions[5-8]. In the current study similar devitrification was qualitatively found to increase as function of time and the amount of water vapour in water vapour environment. Further, oxidation of SiC occurred in both materials after 500 h exposure at 850°C atmospheric water vapour. The results of the current study are in accordance with the results of similar filters operated up to 1500 h in pilot plant in varying combustion environments[8]. Our earlier evaluations of the filter material performance have indicated severe oxidation of SiC and this combined with the crystallization of the oxidation layer and cracking due to thermal stresses and mismatch due the phase transformation of SiO_2 has been pointed as a mechanism to degrade the strength and operation lifetime of SiC-based clay bonded hot gas filters.[13]

After the most severe oxidation exposure of the current study (w), the oxidation layers were rare and only about 100 nm thick for A and 500 nm for B. For material B an estimate for the parabolic oxidation rate constant[10] less than 0.0005 m^2/h was calculated from the oxidation scale thickness of 500 nm (Fig.1c). The value is low when compared to parabolic rate constants for SiC in pure atmospheric oxygen at 900°C. However, direct comparison is misleading since in hot gas filters SiC is covered by the binder. This means that water needs time for diffusion through the binder before the oxidation occurs. In addition an initial oxidation layer of unknown thickness may be present.

The presence of small amounts of aluminium or alkalis in the oxidation layer has been found to increase the oxidation rate of SiC significantly[10-13], both alkalis and aluminium are present in the binder phase. It is also well known that oxidation of SiC is more severe in water vapour than in oxygen due to the higher solubility of water than oxygen in silica[10,14]. Against this background the oxidation rate found in this study is less than expected and the binder surrounding the SiC seems to be protective.

CONCLUSIONS

Crystallization of amorphous binder and oxidation of SiC were found but the rate of oxidation was less than expected. Furthermore there was a clear difference in the resistance to crystallization and oxidation of the two materials although there was only small difference in the composition of the amorphous phase of the materials as-received. The two materials also showed different trends of apparent density as function of time and amount of water vapour.

The amount of crystallization of the amorphous binder to cristobalite was qualitatively found to increase as function of time and the amount of water vapour present. Crystallization was minor in material A and clear in material B. Also the quantitative XRD results support this.

Anyway, the found changes in microstructure do not explain directly why there was a decrease in the strength of materials A, but not in material B, in the same water vapour exposures at high temperatures.

ACKNOWLEDGEMENTS

This research has received funding of the Academy of Finland (decision nr. 54316). Mrs. Merja Ritola is acknowledged for sample preparation. Mr. Terho Kaasalainen is acknowledged for constructing equipments, sample preparation and assistance in completing the exposures.

REFERENCES

[1]A. Lehtovaara and W. Mojtahedi, *Bioresource Technology*, **46**, 113-118 (1993).

[2]M. A. Alvin, T. E. Lippert, E. E. Smeltzer and G. J. Bruck, in Proc. 'Advanced Coal-Based Power & Environmental Systems '98 Conference', July 21-23, 1998, Morgantown, West Virginia, USA.

[3]S.D.Sharma, M. Dolan, D. Park, L. Morpeth, A. Ilyushechkin, L. McLennan, D.J. Harris and K.V. Thambimuthu, *Power Technology*, **180**, 115-121 (2008)

[4]P. Pastila, V. Helanti, A.-P. Nikkilä and T. Mäntylä: *Advances in Applied Ceramics*, **104**, 65-72 (2005)

[5]M. A. Alvin, Materials at High Temperatures, **14**, 355-364 (1997)

[6]J. E. Oakey, T. Lowe, R. Morrel, W. P. Byrne, R. Brown and J. Stringer: Materials at High Temperatures, **14**, 371-381 (1997)

[7]M. A. Alvin: in 'High Temperature Gas Cleaning', (ed. A. Dittler et al.), Vol. II, 455-467; 1999, Karlsruhe, Germany, Institut für Mechanische Verfahrenstechnik und Mechanik der Universität Karlsruhe.

[8]Commission of the European Communities Directorate-General for Energy: 'Demonstration of Hot-Gas Cleaning with Hanging and Upright Ceramic Filter Candles', Final technical report, Phase II, SF/3001/93, Berlin, Germany, 2002.

[9]V. Helanti, A.-P. Nikkilä, P. Pastila and T. Mäntylä: in 'High Temperature Gas Cleaning', (ed. A. Dittler et al.), Vol. II, 405-413; 1999, Karlsruhe, Germany, Institut für Mechanische Verfahrenstechnik

und Mechanik der Universität Karlsruhe
[10] N. S. Jacobson: J. Am. Ceram. Soc., **76**, 3-28 (1993)
[11] V. Pareek and D. A. Shores, J. Am. Ceram. Soc., **74**, 556-563 (1991)
[12] S. C. Singhal and F. F. Lange: J. Am. Ceram. Soc., **58**, 433-435 (1975)
[13] Z. Zheng,R. E. Tressler and K. E. Spear: Corrosion Science, **33**, 545-556 (1992)
[14] E. J.Opila: J. Am. Ceram. Soc., **77**, 730-736 (1994)

IV. Solid Oxide Fuel Cells (SOFCs): Materials and Technologies

DEVELOPMENT OF NANO-STRUCTURED YSZ ELECTROLYTE LAYERS FOR SOFC APPLICATIONS VIA SOL-GEL ROUTE

Feng Han, Tim Van Gestel, Robert Mücke, Hans-Peter Buchkremer
Institute for Energy Research 1 (IEF-1), Forschungszentrum Jülich GmbH
Leo-Brandt-Strasse, D-52425 Jülich, Germany
Email: f.han@fz-juelich.de

ABSTRACT

Nano-structured YSZ thin films are promising candidates to enhance the electrolyte performance of solid oxide fuel cells operating at the intermediate temperature. In this work, the sol-gel synthesis was utilized for fabrication of very fine YSZ nanomaterials with a narrow particle size distribution and an average particle size of 7 nm. On top of the macroporous anode substrates, colloidal sols with graded particle size distribution were coated separately to form mesoporous intermediate layers, which served as the support for fine nano-sol coatings[1]. Through controlling the experimental parameters, the graded colloidal sols were prepared with an average particle size in the range of 40-120 nm. The microporous YSZ layer prepared from a fine nano-sol showed a thickness of 50-200 nm and improved the gas-tightness of the electrolyte. The overall thickness of the multi-layered YSZ electrolyte was less than 3 μm after the thermal treatment. Electrical performance was measured after applying a LSCF cathode. The measured open circuit voltage (OCV) was comparable to that of sintered cells, indicating that a gas-tight electrolyte was fabricated. At 700 mV cell voltage, the cells showed a power density of 588 mW/cm^2 at 1073 K.

INTRODUCTION

Sol-gel process is proven to be an attractive fabrication method of multi-component oxide ceramics[2]. In addition to the achieved homogeneity and purity of the products, the sol-gel method also enables a lower phase-formation and sintering temperature in comparison to the conventional sintering of powder[3]. With good size scaling possibility, colloidal sol-gel materials are suitable for depositing layers on macroporous substrates to serve as support of polymeric sol-gel derived layers preventing infiltration of the sol.

The objectives pursued in fabrication of electrolyte layers for SOFC application via sol-gel route are the reduction of the thickness, the minimization of crack/pin-holes and the decrease of the sintering temperature. The reduction of the thickness would allow decreasing SOFC operating temperature, whereas a lower sintering temperature would reduce manufacturing cost significantly and impede the reaction between the YSZ electrolyte layer and CGO diffusion barrier layer when using LSCF cathode. Additionally, the composition of the sols and their properties should be tailored in order to achieve optimal YSZ layer formation properties and manufacturing conditions[4,5].

In this work a combination of polymeric and colloidal sol techniques has been investigated as an alternative method to fabricate electrolyte layers on warm-pressed anode substrates. Sol-gel thin film deposition is applied on dense or very fine porous substrates. Therefore, multilayer structure electrolyte layers are manufactured. Gas-tightness and the electrochemical performance were tested.

EXPERIMENTAL

Anode substrates (Ni/8YSZ) with an average thickness of about 1 mm and a size of 7×7 cm^2 were produced by warm pressing using a so-called Coat-Mix® material[6]. The anode supports was calcined at 1000 °C[7,8].

Stable YSZ colloidal sol was prepared by mixing 8YSZ nano-powder (Tosoh Cooperation, Tokio, Japan and Sigma-Aldrich GmbH, Steinheim, Germany) with diluted nitric acid solution. Polymeric

sol with median particle size distribution (d_{50}) smaller than 10 nm was prepared at room temperature. Diethonolamine (DEA) was first diluted in n-propanol and then added into Zirconium (IV)-propoxide and Yttrium (II)-butoxide mixture with the Y^{3+}:Zr^{4+} ratio of 16:92 (Sigma-Aldrich GmbH, Steinheim, Germany). After one hour stirring, n-propanol solution of HNO_3 was added into the DEA modified precursor solution at a dropping speed of 1.5ml/min, meanwhile co-hydrolyzation took place and a homogeneous YSZ sol formed.
The sols with graded particle size was spin coated on macroporous warm press anode support using a Delta 80T2 spin coater (Suss Micro Tek AG, Garching, Germany). The coated samples were calcined at 600 °C for 10 minutes in a repaid thermal process oven (Xerion Advanced Heating Ofentechnik GmbH, Freiberg, Germany) with a heating rate of 1K/s. The process of coating and calcination was repeated to obtain the desired configuration of graded multilayer electrolyte structure. Finally the samples were sintered at 1400 °C. Then, the polymeric sol was spin coated on colloidal-sol-derived layer and sintered at 600 °C.
The cathode material, $La_{0.58}Sr_{0.40}Fe_{0.80}Co_{0.20}O_{3-\delta}$ was synthesized by spray drying method. After calcination and ball-milling, the cathode powder was screen printed on a 5x5 cm^2 half cell with a electrolyte prepared by sol-gel method, using an ethyl cellulose binder and a terpineol-based solvent[9]. The area of the cathode layers was 4×4 cm^2 with a thickness of about 50 μm. The cathode was dried and no sinter process was carried out before the electrochemical cell test.
Particle size analysis was carried out on a Horiba dynamic laser scattering particle size analyser. Powder X-ray diffraction measurements were performed on a Siemens D500 diffratometer with a monochromated Cu Kα radiation source. SEM analysis was done using a Zeiss Ultra55 Electron Microscope. The gas tightness of non reduced half cells (substrate, anode, and electrolyte, dimensions:
50×50 mm) was measured by a helium leak test apparatus (HTL 260, Pfeiffer Vacuum GmbH, Asslar, Germany). Electrochemical measurements of single cells were performed in an alumina test housing placed inside a furnace. All electrochemical data were obtained by direct current methods using a current-control power supply type Gossen 62N-SSP500-40 (Gossen-Metrawatt GmbH, Nürnberg, Germany), and a computer-controlled data acquisition system including a data logger type NetDAQ 2640A (Fluke, Eindhoven, The Netherlands). H_2 (3% H_2O) with a flow rate of 1000 ml/min was used as test fuel gas.

RESULT AND DISCUSSION

The XRD pattern of the calcined 8YSZ powder obtained by polymeric sol-gel route is shown in Figure 1, indicating the powder with a pure cubic structure[10]. Average domain (τ) of the powder was estimated by Scherrer equation[11]:

$$\tau = K\,\lambda / \beta_\tau \cos\theta \qquad (1)$$

β_τ is the line broadening due to the effect of small crystallite. K is the so-called shape factor, which usually takes a value of about 0.9 assuming the particle shape is spherical. λ is the wavelength of the radiation, λ_{Cu} = 0.154056 nm. From the equation, the calculated domain size of the three samples were 3.1 nm (450 °C), 3.9 nm (600 °C) and 5.1 nm (900 °C), approximately, which reveals the growth tendency of the crystallite during the thermal treatment.
The particle size distributions (PSD) of representative sols are shown in Figure 2. Via wet chemistry routes and physical treatments, such as centrifuge separation, the d_{50} of colloidal sols with graded particle size varied in the range of 40 nm to 120 nm were obtained (Figure1a, b). Using this colloidal sol to form intermediate support layers, infiltration of fine polymeric sol (d_{50}=7 nm) can be significantly reduced or avoided. By using a strong complexing ligand diethonolamine (DEA) as

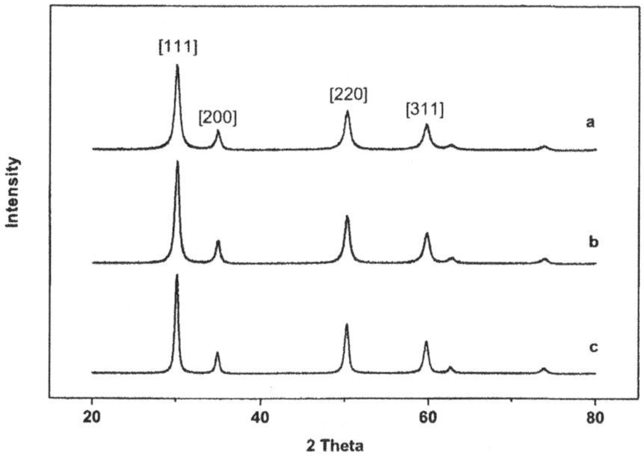

Figure 1. XRD pattern of the calcined 8YSZ powder obtained from sol-gel process, firing at: 450°C, (b) 600°C, (c) 900°C

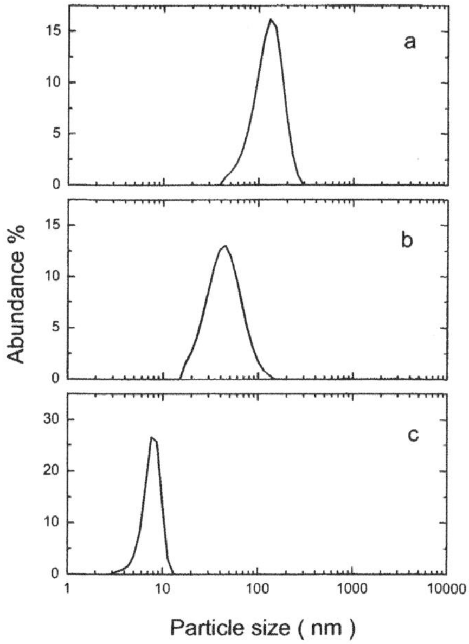

Figure 2. Various YSZ sols with graded particle size distributions: (a) colloidal sol with d_{50} of 120 nm, (b) colloidal sol with d_{50} of 40 nm, (c) polymeric sol with d_{50} of 7 nm.

precursor modifier, rapid hydrolysis reaction of the alkoxide and the precipitation of inhomogeneous hydroxide particles were avoided in the polymeric sol-gel preparation. DEA reacted with the alkoxides to form mixed complexes which were more difficult to hydrolyse than alkoxy groups[12]. The median particle size (d_{50}) of the as-prepared YSZ polymeric sol was 7 nm and the size distribution was much narrower than that of colloidal sols (Figure 2c).

The substrate morphology has a major influence on the microstructure of the coatings. Aiming to obtain polymeric sol-gel layers on macroporous anode supports, a graded layer configuration was proposed. The roughness and pore size of the substrates was adjusted through coating of intermediate support layers, which should be dense or with very fine mesoporous structure. To achieve this objective, a mesoporous and crack-free layer with thickness of around 2-3 µm was firstly deposited by spin coating of colloidal sol with graded particle size on anode substrates (Figure 3a, b). In between each coating, a mild calcination temperature (600 °C) was applied to maintain the mesoporous structure of the sample because the capillary force created by the uniform mesopores was favourable to the formation of homogeneous layers. A rapid thermal treatment program sped up the sample preparation progress and every single coating cycle was finished within 1 hour. As the percolation of Ni in the substrate is of crucial importance for the conductivity of the anode, a thermal treatment temperature as high as 1400 °C was applied for the samples[13]. Moreover, the shrinkage of the anode substrates also facilitated the densification process of the electrolyte.

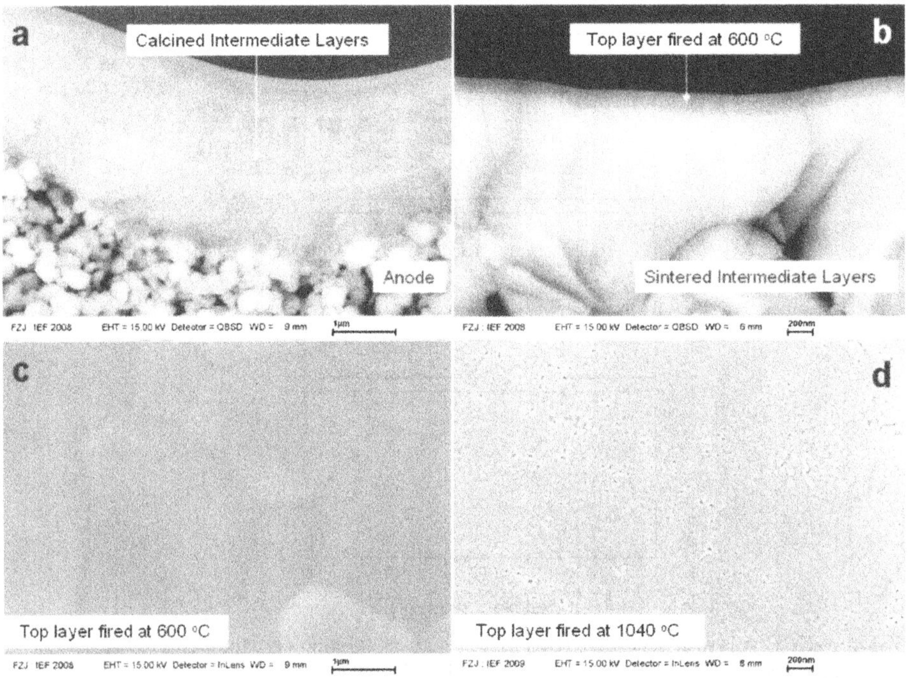

Figure 3. Micrographs of sol-gel YSZ layers: (a) fracture surface of intermediate sol-gel layer (b) fracture surface of top sol-gel layer on sintered intermediate electrolyte layers, (c) surface of top sol-gel layer, fired at 600 °C (d) surface of top sol-gel layer with large pores, fired at 1040 °C,

The polymeric sol showed very good layer forming behaviour on the sintered electrolyte intermediate support layers and a transparent layer of 150 nm in thickness was fabricated on intermediate layer after the thermal treatment at 600 °C for 2 hours (Figure 3b, c). However, the top layers did not sustain a thermal treatment higher than 700 °C and large pores were formed on the surface of the electrolyte (Figure 3d).

The helium leakage value was tested on the half cells in the oxidized state before the coating of cathode. Helium is suitable to be used as the test gas because its molecular is smaller than all the gases under the operation condition of SOFC cells[14]. The helium flowed through the half cell at a pressure difference of 1000 hPa. The shown values in this test were normalised to the measurement area and to a pressure difference of 100 hPa, which was typical for an SOFC stack. One large squared area of 16cm² was measured per cell. The leak rate value of samples without polymeric sol-gel layers was around $5\cdot10^{-4}$ (hPa*dm³/(s*cm³). By applying a polymeric sol-gel layer on top of the colloidal-sol-derived intermediate layers, the leak rate value of samples was improved to $1.5\cdot10^{-5}$ (hPa*dm³/(s*cm²).

The current density data shown in Figure 4 were calculated by dividing the current by the cathode area. At 700 mV cell voltage, the best cell showed the current density of 840 mA/cm² generating the power density of 588 mW/cm² at 1073 K. The area specific resistances of the cell were between 400 to 600 mΩcm². The measured open circuit voltage (OCV) was approximately 1100 mV, which was identical to that of conventional cells, indicating the as-prepared electrolyte layers were gas-tight, also seen in Figure 5a.

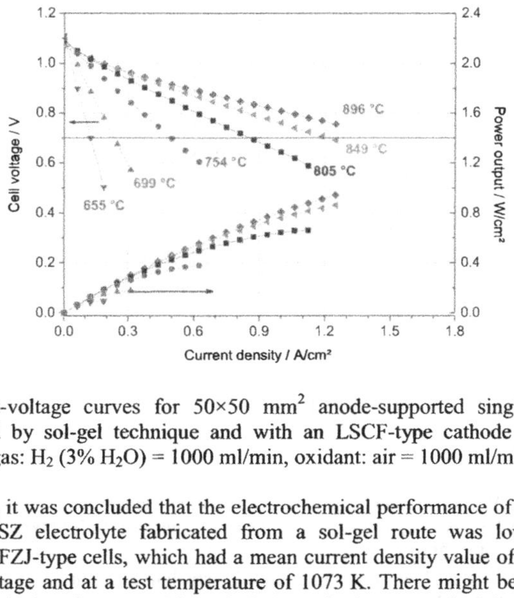

Figure 4. Current-voltage curves for 50×50 mm² anode-supported single cells with a YSZ electrolyte applied by sol-gel technique and with an LSCF-type cathode as a function of the temperature (fuel gas: H_2 (3% H_2O) = 1000 ml/min, oxidant: air = 1000 ml/min).

From these results, it was concluded that the electrochemical performance of the as-prepared single cells with an 8YSZ electrolyte fabricated from a sol-gel route was lower than that of the "State-of-the-Art" FZJ-type cells, which had a mean current density value of around 1800 mA/cm² at 700 mV cell voltage and at a test temperature of 1073 K. There might be three reasons for the limited electrochemical performance. First of all, the cathode was partly delaminated during the cell test. Secondly, the pores in the top electrolyte layers resulted in incomplete contacts with the intermediate electrolyte layers (Figure 5b). The last and most important reason was the diffusion of Strontium (Sr) from the LSCF cathode into the electrolyte during the operation of electrochemical cell test, which formed isolating layer of Sr-zirconate[9] (Figure 5b), such as $SrZrO_3$.

In order to realize the sol-gel method as a competitive technique in manufacturing of electrolyte layers for SOFC applications, further optimizations of fabrication procedures are required in the future work. For example, surface treatment of the substrates and proper thermal treatment of all layers are proposed to avoid delamination of the cathode. Deposition of a $Ce_{0.8}Gd_{0.2}O_{1.90}$ barrier layer will protect the electrolyte from the diffusion of Strontium and efficiently avoid formation of isolating zirconate layers.

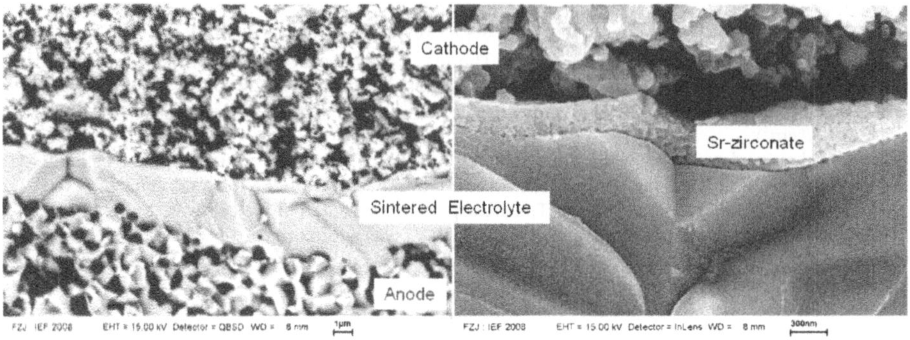

Figure 5. Fracture surface SEM images of single cell: (a) multi-layer sturcture, (b) zirconate layer on top of the electrolyte.

CONCLUSION

Dense electrolytes with a thickness of 3 μm have been deposited on warm-pressed anode substrates by a graded multi-layer sol-gel coating method. By applying a LSCF cathode, current density of about 840 mA/cm^2 and power density of 588 mW/cm^2 has been realized. It could be demonstrated that it is possible to produce highly dense YSZ electrolyte for SOFC application via sol-gel route. Enhanced electrochemical performances of single cells are expected in the future work through optimizations of fabrication procedures.

ACKNOWLEDGEMENTS

We thank Dr. Werner Fischer for XRD investigation, Dr. Doris Seblod for SEM characterization, and Dr. Vincent Haanappel for the electrochemical measurements. Thanks are also due to colleagues for supporting in the sample preparation and Dr. Norbert Menzler for the helpful discussions.

REFERENCES

[1]T. Van Gestel, D. Sebold, W. A. Meulenberg, and H. P. Buchkremer, Development of Thin-Film Nano-Structured Electrolyte Layers for Application in Anode-Supported Solid Oxide Fuel Cells, Solid State Ionics, **179 (11-12)**, 428-437(2008).

[2]K. Mehta, R. Xu, and A. V. Virkar, Two-layer Fuel Cell Electrolyte Structure by Sol-Gel Processing, J. Sol-gel Sci. and Tech., **11**, 203-207(1998).

[3]B. G. Prevo, D. M. Kuncicky, and O. D. Velev, Engineered Deposition of Coatings from Nano- and Micro-Particles: A Brief Review of Convective Assembly at High Volume Fraction, Colloidal and surface A, **311**, 2-10 (2007).

[4]Q.X. Fu, C. R. Xia, S. W. Zha, J. G. Cheng, R. R. Peng, D. K. Peng, G. Y. Meng, The Key Materials Used for Reduced Temperature Ceramic Fuel Cells, Key Engineering Materials, **224-2**, 159-162 (2002).
[5]J.M. Serra, S. Uhlenbruck, W. A. Meulenberg, H. P. Buchkremer, D. Stöver, Nano-Structuring of Solid Oxide Fuel Cells Cathodes, Topics in Catalysis, **40**, 123-131 (2006).
[6]D. Simwonis, A. Naoumidis, F.J. Dias, J. Linke, A. Moropoulou, Material Characterization in Support of the Development of an Anode Substrate for Solid Oxide Fuel Cells, J. Mater. Res., **12**, 1508-1518 (1997).
[7]D. Stöver, H.P. Buchkremer, J.P.P. Huijsmans, in: W. Vielstich, A. Lamm, H.A. Gasteiger (Eds.), Handbook of Fuel Cells, vol. 4, John Wiley and Sons Inc., Chichester, 1015 (2003).
[8]D. Stöver, H.P. Buchkremer, F. Tietz, N.H. Menzler, in: J. Huijsmans (Ed.), European Fuel Cell Forum 2002, Proceedings of the 5th European Solid Oxide Fuel Cell Forum, Lucerne, Switzerland, July 1-5, 1 (2002).
[9]S. Uhlenbruck, N. Jordan, D. Sebold, H.P. Buchkremer, V.A.C. Haanappel and D. Stöver, Thin Film Coating Technologies of $(Ce,Gd)O_{2-\delta}$ Interlayers for Application in Ceramic High-Temperature Fuel Cells, Thin Solid Films, **515**, 4053-4060(2007).
[10]M. Morinaga, J.B. Cohen, X-Ray Diffraction Study of $Zr(Ca,Y)O_{2-\chi}$ I. The Average Structure, Acta Cryst., **A35**, 789-795 (1979).
[11]R. Jenkins and R. Snyder, Introduction to X-ray Powder Diffractometry, John Wiley & Sons Inc., New York, 90, (1996).
[12]L. Cot, A. Ayral, J. Durand, C. Guizard, N. Hovnanian, A. Julbe and A. Larbot, Inorganic Membranes and Solid State Sciences, Solid State Sci., **2**, 313-334 (2000).
[13]Elke Wanzenberg, Herstellung und Charakterisierung von dünnen Elektrolytschichten auf mikrostrukturell modifizierten Anodensubstraten für die Hochtemperatur-Brennstoffzelle, Ph. D. Thesis, Berichte des Forschungszentrums Jülich, **JUEL-4027**, 67 (2003).
[14]R. Mücke, Sinterung von Zirkoniumdioxid-Elektrolyten im Mehrkagenverbund der Oxidkeramischen Brennstoffzelle (SOFC), Ph. D. Thesis, Schriften des Forschungszentrum Jülich, Reihe Energie & Umwelt, **Band 9**, 59 (2008).

DEVELOPMENT OF SINGLE-CHAMBER SOLID OXIDE FUEL CELLS: PERFORMANCE OPTIMIZATION AND MICRO-STACK DESIGNS *

Bo Wei[1,2]**, Zhe Lü[1], Xiqiang Huang[1], Mingliang Liu[1], Dechang Jia[2] and Wenhui Su[1]

[1] Center for Condensed Matter Science and Technology, Harbin Institute of Technology, Yikuang Street 2#, Harbin, 150080, China

[2] Institute of advanced ceramics, School of materials science and engineering, Harbin Institute of Technology, Yikuang Street 2#, Harbin, 150080, China

ABSTRACT:

Single-chamber solid oxide fuel cells (SC-SOFCs) have received increasing attention for portable applications. In this paper, we report our recent progress on button cell performance optimization and specially, the design and evaluation of micro-stacks. Two linear stacks and a novel star-shaped stack were proposed, and the latter one generated a maximum power output of 430 mW, which stably powered an USB fan successfully, indicating this design is very promising for micro power generations.

1 INTRODUCTION

Recently, micro fuel cell systems with high power density have received increasing attention around the world, due to their potential application in portable power generation [1-3]. Single-chamber solid oxide fuel cells (SC-SOFCs), which is based on the selectivity of electrodes toward fuel-oxygen mixture, are considered as a promising alternative to conventional batteries [2]. Sealeant-free operation of SC-SOFCs provides several advantages over dual-chamber SOFC, such as compact stack design, simple gas management, quick start-up and improved shock/mechanical resistances.

Using readily available hydrocarbon fuels, high power densities have been achieved for SC-SOFCs [4, 5], which are even comparable with dual-chamber performance. For micro-stacks, Suzuki et. al. have fabricated a $Ce_{0.8}Sm_{0.2}O_{1.9}$ (SDC) electrolyte supported module with 1.5 V open-circuit voltage (OCV) and about 17 mW maximum power at 550°C [6]. Anode supported cells can significantly reduce the ohmic resistance and thereby reaching higher output. With two anode-supported cells, Shao and Haile have developed a thermally self-sustaining micro-stack [2], which generated a power output of 350 mW using propane as fuel and successfully powered an MP3 player. In this paper, our recent advances on SC-SOFCs are presented, including single cell performance optimization and the design and evaluation of micro-stacks.

2 EXPERIMNTIAL

In these studies, yttria-stabilized zirconia (YSZ) thin films, NiO+YSZ cermet, and modified $La_{0.7}Sr_{0.3}MnO_3$ (LSM) perovskite were used as electrolyte, anode and cathode, respectively. The details of the single cell preparation can be found elsewhere [7]. Specially, $Ce_{0.8}Sm_{0.2}O_{2-\delta}$ nano-grains were introduced to porous LSM backbone via a wet infiltration method to improve the cell performance in methane-oxygen mixture [7].

For linear stack assembly, the pre-reduced (Ni+YSZ anodes) single cells were connceted one by one using pored Ag sheets. A 3-cell stack and a 7-cell stack were arranged. While for the novel star-shaped design, 4 single cells were carefully attached to an alumina ceramic holder one by one using silver paste, and 3 adjacent cells were connected through silver wires, finally forming the star-shape stack.

The single cell/micro-stacks were tested in a flow-through quartz tube by four-terminal method, and a schematic view of testing setup was given in Figure 1. Electrochemical properties were evaluated by a Solartron SI 1287 electrochemical interface and a Solartron SI 1260 frequency response analyzer. Under open circuit condition, impedance spectra were collected in the frequency range of

91 kHz to 0.1 Hz. Mass flow controllers were used to control the gas composition containing nitrogen, methane and oxygen. Two *K*-type thermocouples were used to monitor the differences between furnace temperature (T_f) and the cell temperature (T_c). The microstructure of cathodes was revealed by a SEM (Hitachi, S-4700).

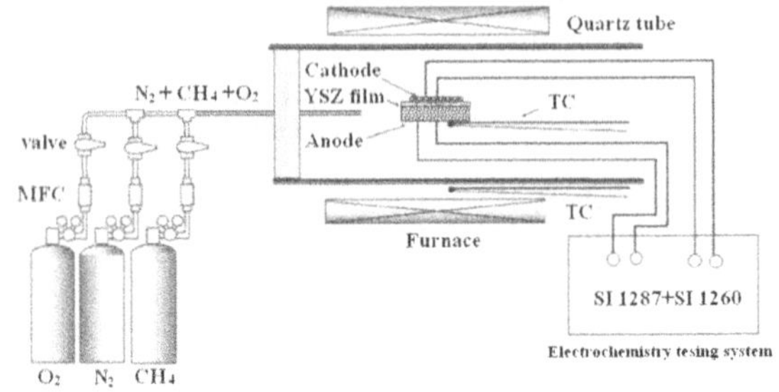

Figure 1. Schematic diagram of the testing system

3 RESULTS AND DISCUSSIONS

Figure 2. SEM microstructures of (a) pure LSM and (b) impregnated cathodes.

Figure 2 compared the microstructures of the LSM based cathodes before and after solution impregnation treatment. For pure cathode (Figure 2a), the LSM grains were well contacted after sintering with homogenous particle size of 0.2-0.3μm. After inflitration, many nanosized $Ce_{0.8}Sm_{0.2}O_{2-\delta}$ particles (about 30nm) were introduced to porous LSM framework. They tended to aggregate at or near the interfaces of LSM grains, and some of them are located on LSM surfaces. These nanoparticles were very active torwards oxygen reduction, which accordingly improved the cathode and cell performances as proved below.

Figure 3 compared the discharge performance of the cells without and with SDC-impregnation at CH_4:O_2 =2:1(active area: 0.5 cm^2). For the cell with pure LSM cathode, maximum power densities (MPD) achieved 79 mW cm^{-2} and 100 mW cm^{-2} at the T_f of 650 °C and 750 °C, respectively. When using SDC-modified cathode, the cell performance was significantly improved with corresponding values of 240 mW cm^{-2}, 404 mW cm^{-2}, which are about 3 times and 4 times higher than the previous cell. This enhancement can be mainly attributed to the introduced ionic conducting SDC nanoparticles, which obviously extended the electrochemical reaction zone. Accordingly, the impregnated cathodes were used in following stacks. Note that the actual cell temperatures were 35-50 °C higher than furnace temperatures due to exothermic partial oxidation reactions of methane to syngas (H_2 and CO).

Then, we fabricated a linear stack with 3 cells connected in series by pored Ag sheets (total active area: 3×0.5 cm^2), the arrangement of the stack was sketched like the insert in Figure 4a. The discharge performance and impedance spectra of the stack are shown in Figure 4a and 4b,

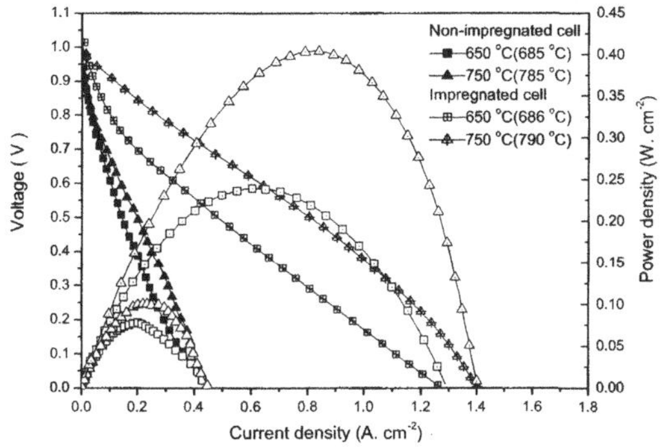

Figure 3. The performance comparison of the single cells without and with SDC impregnated LSM cathode.

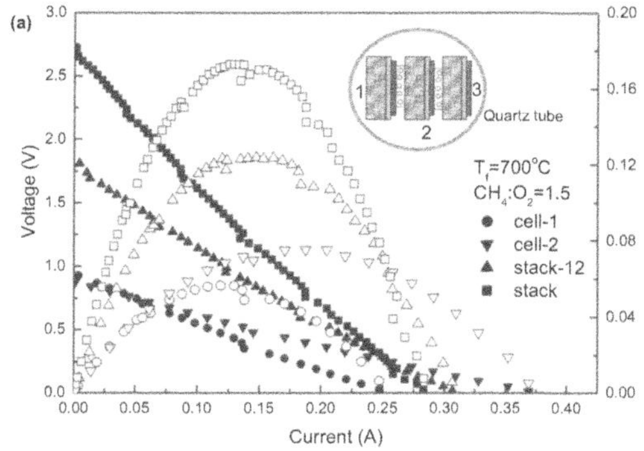

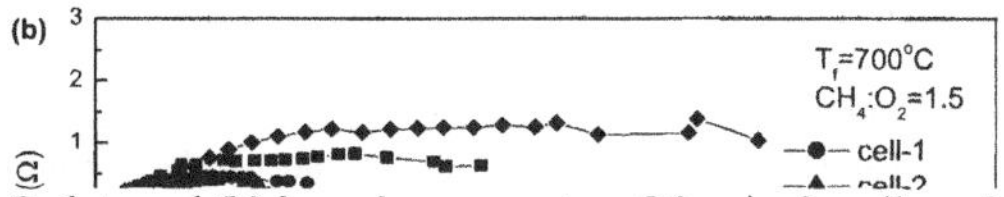

Figure 4. (a) I-V and I-P plots and (b) impedance spectra of the single cells and stacks for the 3-cell linear stack, (insert: cross-section view of the stack).

respectively. The OCV of the 3-cell stack was about 3 times higher than the single cell, and the maximum output was about 170mW. It can be found that the performance of two single cells was not identical; the maximum power and short circuit current of cell-2 were higher than that of cell-1. The impedance spectra also indicated a smaller total resistance for cell-2. It also can be found that, for both single cells and micro stacks, the ohmic resistances only accounted for less than 10% of the entire resistance, indicating that the performances of both single cells and stacks were mainly limited by the electrode polarization resistances, which still need further improvement.

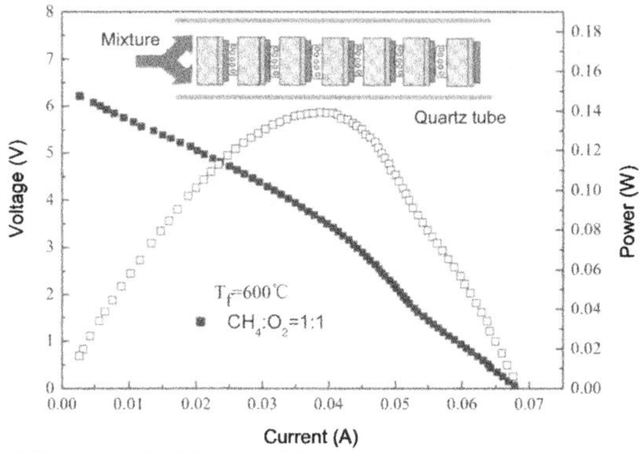

Figure 5. I-V and I-P plots and of the 7-cell linear stack, (insert: cross-section view of the stack).

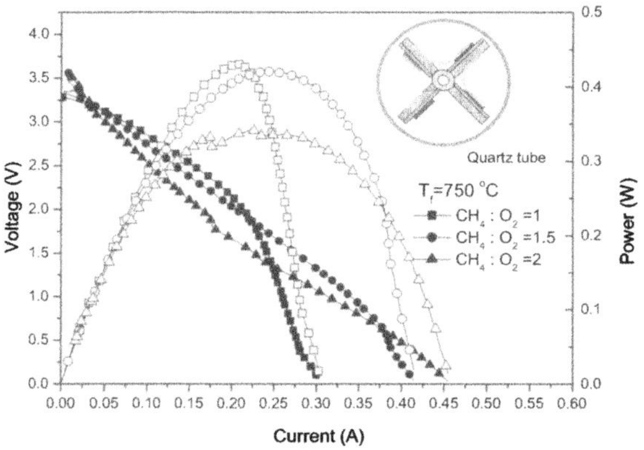

Figure 6. I-V and I-P curves of the star-shaped stack. (insert) schematic view of the stack.

Further, 7 single cells were connected into a linear stack (total active area: 7×0.5 cm^2). But its length was much longer than the diameter of the tube, we had to place it along the tube like the insert figure, and the anode of first cell faced the mixture directly. Figure 5 showed the discharge profiles. The OCV of the stack was relative good (6.3V), but the stack output was quite low with the value of 140mW at 600°C. Correspondingly, the average value was estimated as only 40mW cm^{-2}, much smaller than the value of a single cell. It was considered that the performance was primarily limited by the uneven gas distribution and the uneven performance of single cells, demonstrating the importance of the gas distribution management for SC-SOFC micro-stacks.

To solve this problem, a novel star-shaped design was proposed (see insert for the schematic figure). Good symmetry of this design can ensure the electrodes of each cell paralleling to the gas flow direction, which was favorable for high output. Figure 6 showed the discharge profiles of the star-shaped stack (total active area: 4×0.25 cm^2). At the furnace temperatures of 750°C, a high output of 430mW was obtained at CH_4:O_2=1. When increasing the methane content, its output tended to decrease. The average peak power densities of the single cell were estimated about 300-430 mW cm^{-2}, which was similar to single cell performance. The open circuit voltages were considerably high with the values of 3.4-3.5V. Clearly, high output and stack voltage have been obtained, which successfully powered an USB fan stably during the testing.

4 CONCLUSION

With nano-SDC infiltrated LSM cathode, the single cell performance was obviously improved. Our micro-stack results proved that the gas flow geometry critically affects the stack performance. The performance of linear micro-stacks decreased with increasing cell number, probably due to the uneven gas/temperature distribution. The symmetric star-shaped design exhibited attractive output, making it very promising for portable power sources.

FOOTNOTES

*This research was supported by the Ministry of Science and Technology of China under contract of No. 2007AA05Z139.

* *Contacting Author: Dr. Bo Wei,
E-mail: bowei@hit.edu.cn, TEL: +86-451-86418420, FAX: +86-451-86418420.

REFERENCES

[1] C. K. Dyer, Fuel Cells for Portable Applications. *J. Power Sources*, **106**, 31-34 (2002).

[2] Z. P. Shao, S. M. Haile, J. Ahn, P. D. Ronney, Z. L. Zhan, S. A. Barnett, A Thermally Self-Sustained Micro Solid-Oxide Fuel-Cell Stack with High Power Density. *Nature*, **435**, 795-98 (2005).

[3] A. Bieberle-Hutter, D. Beckel, A. Infortuna, U. P. Muecke, J. L. M. Rupp, L. J. Gauckler, S. Rey-Mermet, P. Muralt, N. R. Bieri, N. Hotz, M. J. Stutz, D. Poulikakos, P. Heeb, P. Muller, A. Bernard, R. Gmur, T. Hocker, A Micro-Solid Oxide Fuel Cell System as Battery Replacement. *J. Power Sources*, **177**, 123-30 (2008).

[4] Z. P. Shao, J. Mederos, W. C. Chueh, S. M. Haile, High Power-Density Single-Chamber Fuel Cells Operated on Methane. *J. Power Sources*, **162**, 589-96 (2006).

[5] T. Hibino, A. Hashimoto, T. Inoue, J. Tokuno, S. Yoshida, M. Sano, A Low-Operating-Temperature Solid Oxide Fuel Cell in Hydrocarbon-Air Mixtures. *Science*, **288**, 203-33 (2000).

[6] T. Suzuki, P. Jasinski, H. U. Anderson, F. Dogan, Single Chamber Electrolyte Supported Sofc Module. *Electrochem. Solid State Lett.*, **7**, A39-A93 (2004).

[7] B. Wei, Z. Lü, X. Huang, M. Liu, K. Chen, W. Su, Enhanced Performance of a Single-Chamber Solid Oxide Fuel Cell with an SDC-Impregnated Cathode. *J. Power Sources*, **167**, 58-63 (2007).

DEVELOPMENT OF BUNDLE/STACK FABRICATION TECHONOLOGY FOR MICRO SOFCS

Toshio Suzuki, Toshiaki Yamaguchi, Yoshinobu Fujishiro, and Masanobu Awano
National Institute of Advanced Industrial Science and Technology (AIST)
2266-98 Anagahora, Shimo-shidami, Moriyama-ku,
Nagoya, 463-8560, Japan

Yoshihiro Funahashi
Fine Ceramics Research Association (FCRA)
2266-99 Anagahora, Shimo-shidami, Moriyama-ku,
Nagoya, 463-8561, Japan

ABSTRACT

Micro tubular solid oxide fuel cells (SOFCs) were shown to have high thermal durability under quick start-up/ shut-down operation and to be operable at lower temperatures, and thus, it is expected to realize cost effective, compact and high performance power sources using them. In this study, two types (type A and B) of micro SOFC stacks were proposed and demonstrated using micro SOFC bundles, which consist of micro tubular SOFCs under 2 mm diameter and porous cathode matrices made of (La, Sr)(Co, Fe)O_3. The type A was constructed using four bundles, vertically connected in series, and fuel and air were applied using ceramic manifolds. The type B consists of three bundles, horizontally connected in series, and fuel was applied using ceramic manifolds. The type A stack (volume 0.8 cm^3) showed over 2 W and 3.65 V OCV at 490 °C, while the performance of the tyep B stack whose volume is 1 cm^3 was shown to be 2.8 V OCV and maximum power output of 1.5 W at 500 °C, applying air only by natural convection.

INTRODUCTION

Commercialization of power sources using solid oxide fuel cells is relied on the development of high performance SOFCs and cost effective module fabrication technology. Currently, decrease of operating temperature under 700 °C becomes one of main research targets because it can decrease material degradation, prolong stack life time, and reduce cost by utilizing metal materials for stack fabrication. Therefore, numbers of studies related to reduced temperature SOFCs have been reported. The approaches to reduce operating temperature have widely been reported: for examples, (1) using a new electrolyte, cathode and anode materials [1-9], (2) reducing the thickness of the electrolyte using traditional electrolyte materials such as Y doped ZrO_2 [10, 11], (3) introducing new structure for electrolytes [12- 14].

Cell design is also an important factor to improve the performance of SOFC stack/module [15-18]. Use of small diameter SOFC may also give opportunity to reduce operating temperature by increasing the volumetric power density [19]. Thus, they are expected to accelerate the progress of SOFC systems which can be applied to portable devices and auxiliary power units for automobile.

Our study aims to develop fabrication technology for new micro tubular SOFCs under 2 mm and their bundles and stacks. Here we proposed new stack deigns using micro bundles consistign of micro tubular cells and porous cathode matrices as a support of the cells. In this study, two types (type A and B) of micro SOFC stacks were fabricated and demonstrated. The type A was constructed using four micro bundles, vertically connected in series, and fuel and air were applied using ceramic manifolds. The type B consists of three bundles, horizontally connected in series, and fuel was applied using ceramic manifolds, while the air was applying only by natural convection. Both stacks were tested at/below 500 °C to seek the possibility of new application such as portable power sources.

FABRICATION OF MICRO TUBULAR SOFC BUNDLES/STACKS

Figure 1 shows the processing procedure of micro tubular SOFC bundles. Anode tubes were made from NiO powder (Seimi Chemical co., ltd.), $Gd_{0.2}Ce_{0.8}O_{2-x}$ (GDC) (Shin-Etsu Chemical co., ltd.), poly methyl methacrylate beads (PMMA) (Sekisui Plastics co., ltd.), and cellulose (Yuken Kogyo co., ltd.). After adding the proper amount of water, these powders were mixed using a mixer 5DMV-rr (Dalton co., ltd.) in a vacuumed chamber. The tubes were extruded from the clay using a piston cylinder type extruder (Ishikawa-Toki Tekko-sho co., ltd.). An electrolyte was prepared on the surface of the anode tube by dip-coating a slurry which consists of the GDC powder used in the anode tube preparation, solvents (methyl ethyl ketone and ethanol), binder (poly vinyl butyral), dispersant (polymer of an amine system) and plasticizer (dioctyl phthalate), and co-sintered at 1400 °C for 1 h in air. The anode tubes with electrolyte were, again, dip-coated using a cathode slurry, which was prepared in the same manner using $La_{0.6}Sr_{0.4}Co_{0.2}Fe_{0.8}O_{3-y}$ (LSCF) powder (Seimi Chemical, co., ltd.), the GDC powder, and organic ingredients. After dip-coated, the tubes were dried and sintered at 1000°C for 1 h in air to complete a cell.

A cathode porous support (matrix) to bundle the tubular SOFCs was prepared using LSCF powder (Daiichi Kigenso Kagaku Kogyo co., ltd.), poly methyl methacrylate beads (PMMA) and cellulose (Yuken Kogyo co., ltd.). These powders were mixed with a proper of water, and extruded from a metal mold using a screw cylinder type extruder (Miyazaki Tekko co., ltd.). The microstructure of the cathode matrices was controlled by changing the amount and diameter of the pore-former, the grain size of the starting LSCF powder and sintering temperatures. Detailed are described elsewhere [20].

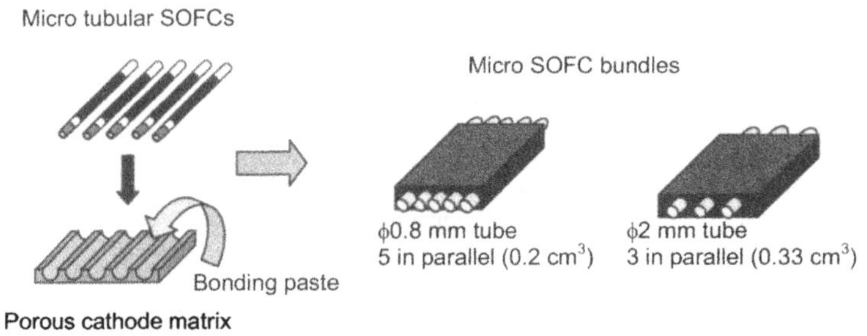

Fig. 1 (a) Fabrication procedure of the micro tubular SOFC bundles

For placing tubular SOFCs in the cathode matrix, a bonding paste was used, prepared by mixing the LSCF powder, the binder (cellulose), the dispersant (polymer of an amine system), and the solvent (diethylene glycol monobutyl ether). The paste was painted on the surface of the cathode matrices, followed by the placement of tubular cells and sintered at 1000 °C for 1 h in air. Two types of bundles were prepared; a bundle with five ϕ0.8 mm tubes and a bundle with three ϕ2 mm tubes whose volumes are 0.2 and 0.33 cm^3, respectively.

The SOFC bundles were completed by applying sealing layer and interconnects. The sealing layer was formed by applying a glass paste (prepared from SiO_2-B_2O_3-RO(R: alkali metal) base glass powder, AGC Co., Ltd.). The glass treatment temperature was 700 °C for 1 h. Ag wire and sheet were used as interconnects which were fixed by using Ag paste. The current collectors for the anode (attached on the sealing layer at the side of the bundle) and for the cathode (the other side of the bundle) was placed for each bundle, which allows easy assembling of the stack electrically connected in series.

Two types (type A and B) of micro SOFC stacks were prepared as shown in Fig. 2 using those micro tubular SOFC bundles. The type A was constructed using four micro bundles with ϕ0.8 mm tubes, vertically connected in series after applying sealing layer and interconnects to each bundle. Fuel and air were applied using ceramic manifolds as shown in Fig. 2 (a). The type B consists of three bundles, sealing layers and interconnects, and fuel manifolds. The size of the stack A without fuel manifold is 1 x 1 x 0.8 cm (~ 0.8 cm^3) and the size of the stack B without fuel manifold is 1 x 3 cm with the thickness of 3.3 mm (~ 1 cm^3).
Note that the micro SOFC stacks can be flexibly designed depending upon application use using the fabrication technology developed in this study.

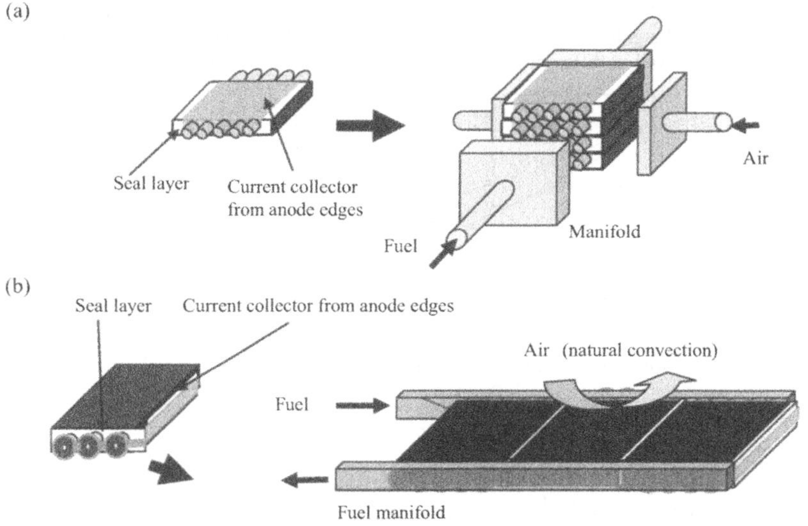

Fig. 2: Conceptual images of (a) type A (vertical stack) and (b) type B stack (horizontal stack)

Experimental set up of the type A stack is shown in Fig. 2 (a). Gas manifolds for fuel and air for inlet and out-let were fixed to the stack with the thermocouples to monitor stack temperature and gas out let temperatures. The discharge characterization was investigated by using a Parstat 2273 (Princeton Applied Research) in DC 4 point probe measurement. The Ag wire was used for collecting current from anode and cathode sides, which were both fixed by Ag paste. Hydrogen (humidified by bubbling water at room temperature) was flowed at the rate of 100 mL min^{-1} and the air was flowed at the rate of 500 mL min^{-1} at the cathode side.
The performance of the type B stack was examined using wet hydrogen (bubbling in H_2O at room temperature) with the flow rate of 50 mL min^{-1}. The module was placed in the round-shaped furnace with open holes at the top and bottom of the furnace, which allows air to be supplied to the module by natural convection. The temperature of the module was monitored using a thermocouple placed at the center of the stack, which was also used for controlling furnace temperature. Four wires were attached to the stack for the measurement. The discharge characterization was investigated by using a Solartron 1260 frequency response analyzer with a 1296 Interface. No

backpressure of the fuel was applied for this measurement.

PERFORMANCE OF MICRO TUBULAR SOFC STACKS

Figure 3 shows the performace of the type A (four bundles in series connection) stack with the volume of 0.8 cc. The open circuit voltage obtained at 448 and 490 °C stack temperature was 3.6 V (0.9 V per bundle), which was similar to that of single micro tubular cell. Note that outlet gas temperaures were always higher due to reacitons inside the stack. The output powers of 0.44, 2.0 W were obtained for 448, 490 °C stack tempertures, respectively corresponding to 1.6, 2.5 W cc^{-1}. Total electrode area of the tubular SOFCs is 5 cm^2 and thus, the power density of 0.4 W cm^{-2} was obtained at 490 °C stack temperature. Compared to the single cell performance (0.32 W cm^{-2} @0.7 V at 500 °C), the bundle performance of the stack turned out to be 0.25 W cm^{-2} @0.7 V at 490 °C, which is lower than that of signle cell. This difference could be resulted from uneven temperature distribution in the stack as well as different fuel flow rate.

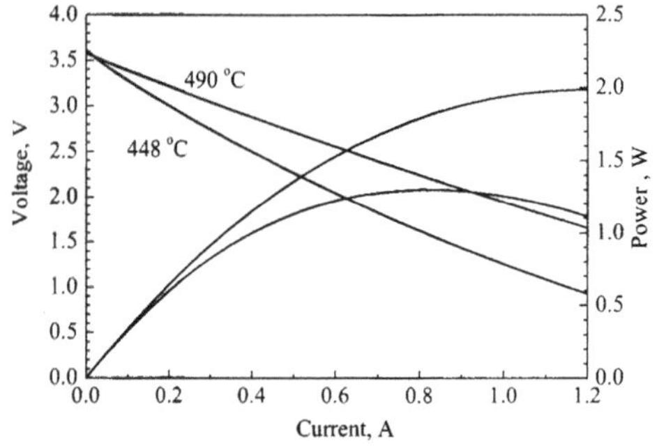

Fig. 3: The performance of the type A stack.

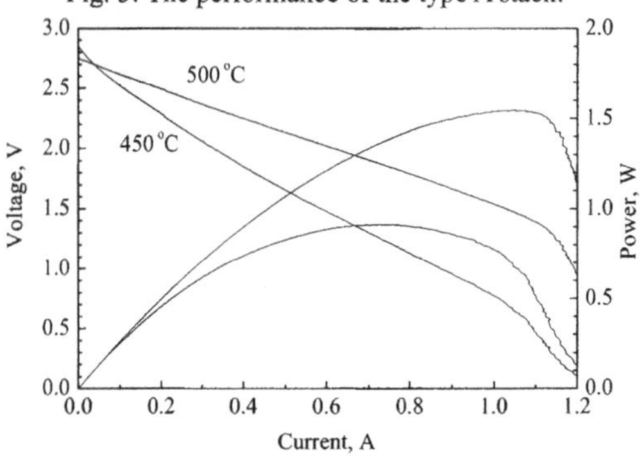

Fig. 4: The performance of the type B stack.

Figure 4 shows the result of the discharge characterization of the type B stack for various operating temperatures at the fuel flow rate of 50 mL min^{-1}. The performance of the stack was shown to be 0.91 and 1.54 W of maximum power outputs obtained at 450 and 500 °C operating temperatures, respectively. As can be seen, sharp drops in the range of $I > 1.1$ A were observed which corresponded to around 40-50 % fuel utilization range. This can be correlated to the depletion of oxygen at the cathode side due to the method of supplying air. Total electrode area of the SOFCs was 5.65 cm^2 and therefore, maximum power densities of 0.16 and 0.27 W cm^{-2} were obtained at 450 and 500 °C operating temperatures, respectively. The fuel utilization of the type B stack at peak power density appeared to be 34.9 and 49.4 % at the operating temperature of 450 and 500 °C, respectively. Further improvement of the fuel utilization could rely on cell/stack arrangement which is under investigation.

CONCLUSION

Fabrication and characterization of tubular SOFC stacks designed for low temperature operation were shown. Two types of stacks were proposed and demonstrated using the micro tubular SOFC bundles; the type A was a vertically assembled stack, four-story cube-type stack with the volume of about 0.8 cm^3. The performance of the type A stack was shown to be 3.6 V OCV and 2 W maximum output power under 500 °C operating temperature. Another type of the stack was also fabricated, which consists of three bundles and fuel manifolds, where the air was applied only by natural convection. Open circuit voltages of the type B stack were shown to be 2.85 and 2.73 V at 450 and 500 °C operating temperatures, with the maximum power outputs of 0.91 and 1.54 W. Overall, both types of stacks showed reasonably sufficient performance for application use.
The performances of both stacks were shown to be sufficient for application use. In addition, these bundle designs allow easy fabrication of stacks with any output power, and voltage, and therefore, use of the micro SOFC bundles for stack fabriation could be ideal, especially for portable SOFC systems. The stacks are still under development and the performance is expected to be improved by optimizing the stack components as well as operating conditions. Temperature distribution and gas flow inside the stacks are also under investigation using simulation to optimaize stack design as well.

ACKNOWLEDGEMENT

This work had been supported by NEDO, as part of the Advanced Ceramic Reactor Project.

REFERENCES

[1] Steele BCH, *Mat. Sci. and Eng. B* (1992) 13 p.79.
[2] Ishihara T, Matsuda H, Takita Y. *Solid State Ionics.* (1995) 79 p.147
[3] Bohn HG, Schober T. *J. Am. Cer. Soc.* (2000) 83 p.768.
[4] Shao Z, Haile SM. *Nature* (2004) 431 170-173.
[5] Steele BCH. *Solid State Ionics* (2000) 129 p.95.
[6] Kuroda K, Hashimoto I, Adachi K, Akikusa J, Tamou Y, Komado N, Ishihara T, Takita Y. *Solid State Ionics* (2000) 132 p.199.
[7] Yoon SP, Han J, Nam SW, Lim TH, Oh IH, Hong SA, Yoo YS, Lim HC. *J. Power Sources* (2002) 106 p.160.
[8] Steele BCH, Heinzel A. *Nature* (2001) 414 345-352.
[9] Simner SP, Bonnett JF, Canfield NF, Meinhardt KD, Sprenkle VL, Stevenson JW. *Electrocehm. Solid State Letters* (2002) 5 p.A173.
[10] de Souza S, Visco SJ, DeJohnge LC. *J. Electrochem. Soc.* (1997) 144 p.L35.
[11] Huang H, Nakamura M, Su PC, Fasching R, Saito Y, Prinz FB, *J. Electrochem. Soc.* (2007) *154* (1) B20-B24
[12] Eguchi K, Setoguchi T, Inoue T, Arai H. *Solid State Ionics* (1992) 52 p.165.
[13] Hibino T, Hashimoto A, Asano K, Yano M, Suzuki M, Sano M. *Electro. Solid State Letters* (2002) 5 p.A242.
[14] Yan J, Matsumoto H, Enoki M, Ishihara T. *Electrochem. Solid-Sate Lett.* (2005) 8 (8) A389-A391.

[15] Sammes NM, Du Y, *Int. J. Appl. Ceram. Technol.* (2007) 4 [2] p.89-102
[16] Sammes NM, Du Y, Bove R, *J. Power Sources* (2005) 145 p.428–434.
[17] Kendall K, Palin M. *J. Power Sources* (1998) 71 p.268-270.
[18]Yashiro K, Yamada N, Kawada T, Hong J, Kaimai A, Nigara Y, Mizusaki J, *Electrochemistry* (2002) 70 No.12 p.958-960.
[19] Sarkar P, Yamarte L, Rho H, Johanson L, *Int. J. Appl. Ceram. Technol.* (2007) 4 [2] p.103-108
[20] Funahashi Y, Shimamori T, Suzuki T, Fujishiro Y, Awano M, *ECS Transactions* (2007) 7 (1) p.643-649

AN OVERVIEW OF SCANDIA STABILIZED ZIRCONIA ELECTROLYTE DEVELOPMENT FOR SOFC APPLICATION

K. Ukai[1]*, M. Yokoyama[1], J. Shimano[1], Y. Mizutani[1] and O. Yamamoto[2]
[1]*SOFC Group, Fundamental Research Dept., Toho Gas Co., Ltd. 507-2*
[2]*Faculty of Engineering, Mie University Japan*

ABSTRACT

Scandia Stabilized Zirconia (ScSZ) shows good potential for SOFC application because of its high electrical conductivity. But, the fundamental study such as suitable composition, mechanical properties and chemical stability are insufficient to compare with Yttria Stabilized Zirconia (YSZ) which is promising material for SOFC application.

We have investigated the electrical and mechanical properties of various composition of ScSZ and power generation performance, long-term stability and reliability of SOFC. The electrolyte-supported cell (ESC) using 10Sc1CeSZ showed good power density of $0.6W/cm^2$ at 1023K and long term-stability. Furthermore, to improve the mechanical reliability of ESC, we have developed the Piston on Ring (POR) method to measure the strength of actual shape cell and found that the strength of cell is affected by the thickness of electrodes and interlayer.

Then high performance and high reliability SOFC single cell using 10Sc1CeSZ electrolyte were achieved.

INTRODUCTION

YSZ is recognized as the most promising material for SOFC electrolyte. However, ionic conductivity of YSZ is not good enough to improve the cell performance. Therefore, many works were reported to improve the ionic conductivity of YSZ, or to investigate new electrolyte material. For an example of new material investigation, Lanthanum gallate was noticed on its high oxide ion conductivity, but mechanical strength is poor. Also, ceria compounds showed good electrical conductivity, but it is difficult to prevent electrical leakage because of its mixed conducting of oxide ion and electron.

Considering the status of new material development, we have tried to improve the oxide conductivity in the zirconia compounds [1, 2]. ScSZ is well known that it has highest electrical conductivity, but the electrical, chemical and mechanical properties of ScSZ as the electrolyte of SOFC were not thoroughly investigated, when we decided to start SOFC development.

Therefore, we have studied the properties of ScSZ for SOFC application. This paper briefly overview our ScSZ development.

ELECTLICAL PEOPERTY OF ScSZ ELECTROLYTE

Suitable composition for cubic phase

At the beginning stage of this study, we investigated the optima composition of ScSZ. It is well known that cubic phase zirconia has highest ion conductivity. Therefore, we tried to make several ScSZ samples, which have different amounts of scandium additives, and confirmed that cubic phase was obtained by addition of over 8-mol% scandium into zirconia. The electrical conductivity of 8 to 12 mol% ScSZ ranges from 0.25 to 0.38 S/cm at 1273K, which is twice or third higher than YSZ[3].

Crystal phase stability during thermal cycle

The temperature dependence of conductivity of several compositions of ScSZ is shown in Figure 1 (a). The conductivity of 10 to 15mol% ScSZ were significantly changed around 560K. It was caused by the crystal phase transition between cubic and rhombohedral. The phase transition accompanied volume change, and may lead to the cell broken. To prevent the phase transition, we tried small amount of the other additives in ScSZ as following compositions.

($11mol\%Sc_2O_3$-$89mol\%ZrO_2$)$99wt\%$-$Al_2O_3 1wt\%$ (11ScSZ1A)

$10mol\%Sc_2O_3$-$1mol\%Y_2O_3$-$89mol\%ZrO_2$ (10Sc1YSZ)

$10mol\%Sc_2O_3$-$1mol\%CeO_2$-$89mol\%ZrO_2$ (10Sc1CeSZ)

Figure 1 (b) shows the temperature dependence of the conductivity of above compositions of ScSZ and 8YSZ. The conductivity of 11ScSZ1A, 10Sc1YSZ and 10Sc1CeSZ was not changed. It means that the phase transition was prevented by small amount of the additives. Besides, the activation energy of 10Sc1CeSZ was the lowest value, and SOFC using 10Sc1CeSZ as an electrolyte is expected to operate at a low temperature range.

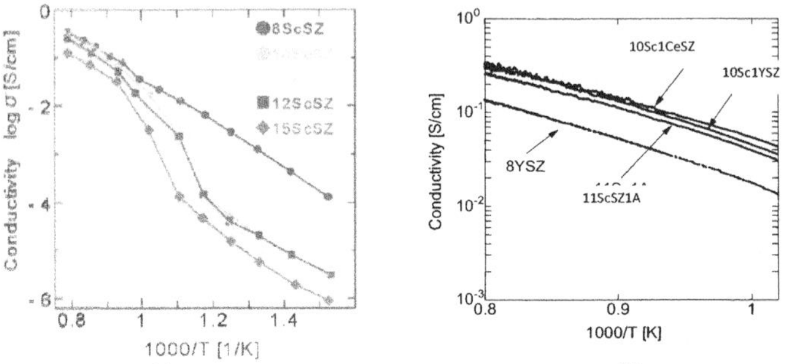

(a) (b)

Fig.1 Temperature dependence of conductivity

Long-term stability of ScSZ electrolyte

The electrolyte of SOFC is exposed to high temperature for long period. It is well known that the electrical conductivity of zirconia electrolyte is significantly degraded by long-term exposure of high temperature atmosphere.

Figure 2 shows the electrical conductivity change of ScSZ and YSZ against time at high temperature. In the initial period of several hundred hr, 8ScSZ and 8YSZ showed a significant electrical conductivity decrease when exposed at 1273K. On the other hand, 11ScSZ1A and 10Sc1CeSZ

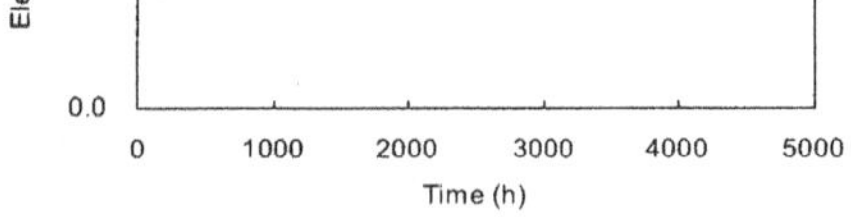

Fig.2 The change of electrical conductivity against time at 1273K

showed no significant decrease of electrical conductivity with the exposure at 1273K. They seemed to be a suitable material for SOFC application.

It is well known that stabilized zirconia has an aging effect in electrical conductivity, and many aging mechanisms, such as phase stability, ordering of the vacancies in crystal lattice, and the segregation of the impurities at the grain boundaries have been proposed. Figure 3 shows the changes of the Raman spectra of 8ScSZ and 11ScSZ with exposure period, respectively. The Raman spectrum of 8ScSZ corresponds to those of a mixture of cubic and tetragonal zirconia.

The shifts of the Raman spectra at 250cm^{-1} increases with annealing period suggest that the formation of the tetragonal phase of zirconia. In addition, this shift disappeared with re-sintering. However, no significant changes of the Raman spectra are observed in 11ScSZ. These results indicate that the electrical conductivity decreases with exposure at high temperature, and is caused by cubic to tetragonal phase transformation and 11ScSZ has long term phase stability at 1273K[4].

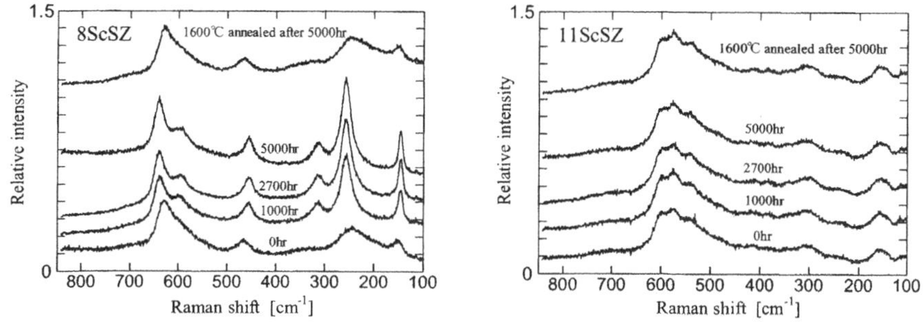

Fig.3 Raman spectra of 8ScSZ and 11ScSZ annealed at 1273 K for various periods

MECHANICAL PEOPERTY OF ScSZ ELECTROLYTE

Mechanical properties such as bending strength and fractural toughness of zirconia compound mainly depend on the crystal phase. In the case of cubic phase ScSZ material, the bending strength at room temperature is around 250 MPa, which was measured according to JIS R1601 method. The small amount of additives is also effective to strengthen the electrolyte, the bending strength of 10Sc1CeSZ showed 340MPa [5].

PERFORMANCE OF ELECTROLYTE SUPPORTED CELL

Above results, 10Sc1CeSZ is expected to be the good material for SOFC application. The electrolyte supported Cell (ESC) using 10Sc1CeSZ for intermediate temperature SOFC was developed in this study. We have studied various electrode materials to match the 10Sc1CeSZ electrolyte and find that the specifications in Table 1 show good cell performance.

Table1 Cell Specifications

	Materials	Thickness	Specifications
Anode	Ni-10Sc1CeSZ cermet	20μm	NiO:ScSZ=45:55Wt%
Electrolyte	10Sc1CeSZ	250μm	
Interlayer	GDC	10μm	
Cathode	LSCF-SDC	20μm	LSCF:SDC=80:20wt%

The power generation characteristics of ESC at 1023K with alumina manifold and Pt mesh current collector was examined and the maximum power density of $0.6W/cm^2$ was obtained. This initial performance is considered to be good enough for system application. But the long-term stability was not so good. The degradation of ESC was mainly caused by the agglomeration of Ni in anode and the unsatisfactory protection effect of GDC interlayer. After improvement of these points, the cell shows good long-term stability as shown in Figure 4.

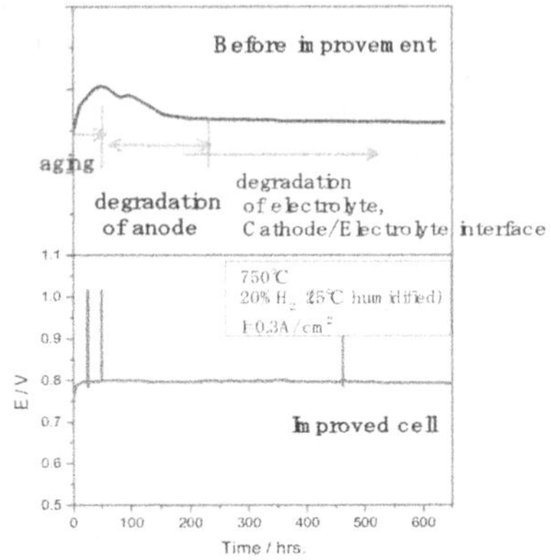

Fig.4 Voltage change of SOFC cell during operation

EXAMINATION BY PISTON ON RING METHOD

Fracture behavior of electrolyte

Usually, mechanical properties of ceramic material were measured with a simple shape such as rectangular bar. But the strength of the actual cell depends on the shape of itself. Therefore, we tried to measure the strength of actual shape cell by piston on ring (POR) method. Thin disk-shaped zirconia specimen was fabricated by tape casting method and sintered at 1623 K. Figure 5 shows the schematic representation of POR testing apparatus. A diameter of upper loading piston was 10 mm. A circle of the support balls were arrayed at diameter of 30 mm under the thin disk specimen (φ 40x0.25 mm). Cross head speed was 1.0 mm/min. The fracture test was performed at R.T., 573, 873 and 1073 K. Strength value was calculated using the finite element method, because the calculation

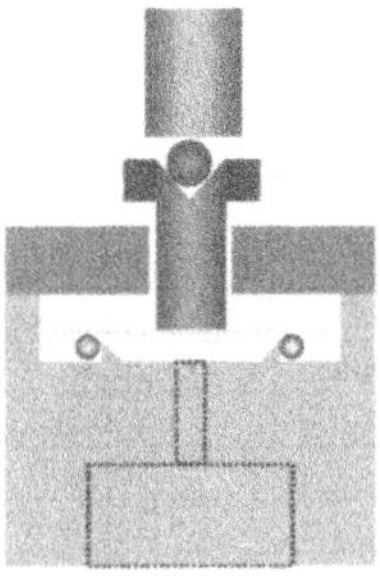

Fig.5 POR testing apparatus

formula in ASTM could not be applied by the large deflection condition. At first, we examined the relationship between POR and 4-point bending method, and confirmed that the POR test showed a good agreement with that of bulk sample examined by the 4-point bending method [6]. This result suggested that the POR test was applicable to the estimation of fracture strength with actual size cell. Although the fracture strength was almost same from room temperature until 573 K, it rapidly decreased over 873 K, and the fracture strength at 1073 K was almost half of that at room temperature. In order to confirm the thermal degradation, the fracture strength of specimen which was performed the heat treatment at 1023K was measured. The fracture strengths of heated-specimens were almost the same as that of as-sintered specimens, as shown in Figure 6. Thermal degradation mechanism was not acted in the experimental conditions of this study. In order to confirm the phase transformation of ScSZ with temperature, the high temperature XRD analysis was performed. Figure 7 showed the XRD data measured at the specified temperatures of fracture testing conditions.

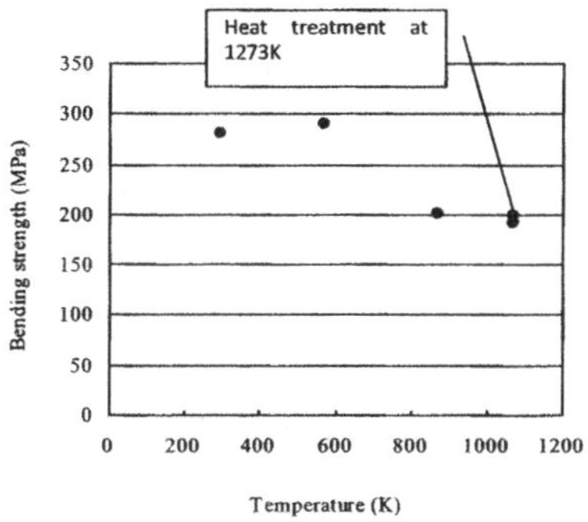

Fig.6 The effect of heat treatment at 1073K on the fracture strength of ScSZ electrolyte.

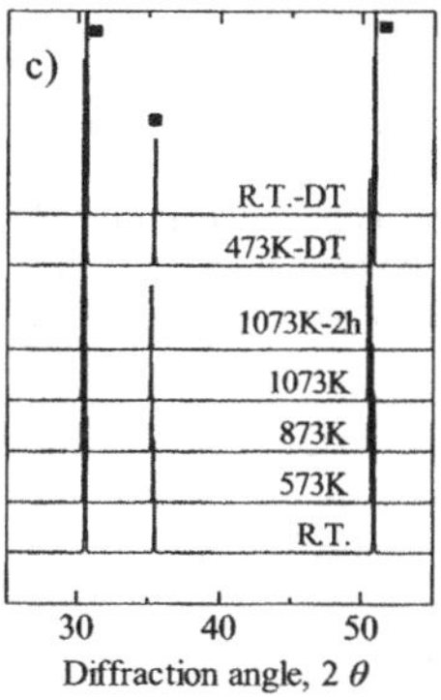

Fig.7 Evaluation of phase changes of ScSZ electrolyte by high temperature XRD analysis at elevating and decreasing temperature.

Fracture behavior of cell specimen

Fig. 8 shows a change of the fracture load of the cell specimen at the room temperature by POR test. The thickness of electrolyte specimen (0.2 mm) is difference from that of cell specimen (0.25 mm). The fracture load of the cell specimen decreased in compared with electrolyte. The fracture load was measured in the condition that the cathode side exerted the tensile stress side (lower surface) was smaller than anode side. In order to confirm the thermal degradation during sintering process, heat treatment was performed to the electrolyte at sintering temperature of 1598 K. Because of the fracture load of heated specimen was almost the same as the as-sintered specimen, the thermal degradation could be ignored. The fracture load of the cell specimen after removed electrodes using the grinder was almost identical to that of electrolyte. Therefore, the decrease of the fracture load of those cell specimens was possibly due to the residual stress caused by the sintering process of the electrodes.

Cell strength is important for practical use. It is necessary to examine exactly relations of cell processing conditions and its fracture strength.

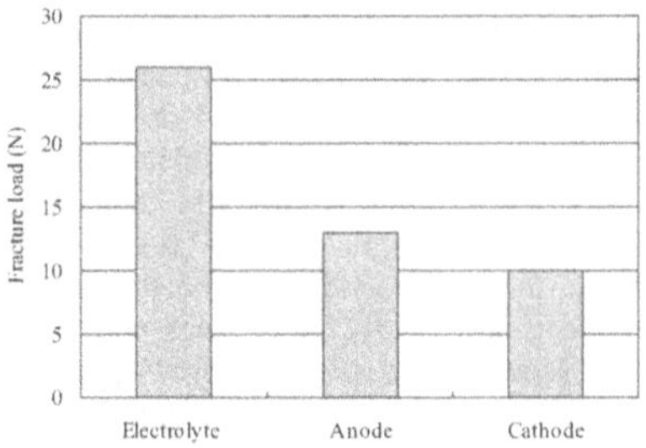

Fig.8 The effect of sintering electrodes on the fracture strength of ScSZ electrolyte measured by POR test.

CONCLUSION

We have investigated the properties of ScSZ for SOFC application and found that 10Sc1CeSZ showed acceptable electrical and mechanical properties. The ESC using 10Sc1CeSZ as electrolyte showed the good power generation characteristics and long-term stability. We also developed the Piston on Ring (POR) method to measure the strength of actual shape cell, and found that the electrodes affected the strength of the made cell.

References

[1] Stricker, D.W. and Carlson W.G. , J. Am. Ceram. Soc. 48:286- (1965)
[2] Etsel, T.H. and Flengas, S.N., Chem. Rev. 70:339- (1970)
[3] Y.Mizutani, M.Tamura, M.Kawai and O.Yamamoto, Solid State Ionics, 72, 271- (1994).
[4] K.Nomura, Y.Mizutania, M.Kawai, Y.Nakamura, and O. Yamamoto, Solid State Ionics, 132 235- (2000)
[5] K. Ukai, Y. Mizutani, Y. Kume and O. Yamamoto, Proc. 7th symp. Solid Oxide Fuel Cells, 375- (2001).
[6] S. Honda, S. Mizuno, S. Hashimoto, Y. Iwamoto, H. Awaji, J. Shimano, K. Ukai and Y. Mizutani, Proc. of the 24th Japan-Korea Int. Seminar on Ceramics, 195- (2007)

FABRICATION OF Ni-GDC ANODE SUBSTRATE BY TAPE CASTING PROCESS

A. Fu Chang Jing [1], B. Chan Siew Hwa [1]*, C. Liu Qing Lin [1], D. Ge Xiao Ming [1]
[1] *Division of Thermal & Fluids Engineering, School of Mechanical & Aerospace Engineering, Nanyang Technological University*, Singapore 639798

ABSTRACT

Lowering the operational temperature of Solid Oxide Fuel Cells (SOFCs) is important for their practical applicability. Ceria based materials, especially gadolinium doped ceria (GDC), are widely considered as good candidates for electrolytes of intermediate temperature SOFCs (IT-SOFCs). In the past few years, we have studied in detail the electrochemical performance of GDC-based composite electrodes, such as Ni-GDC anode and LSCF-GDC cathode. In this paper, the Ni-GDC anode-substrate for IT-SOFCs has been successfully manufactured by the tape-casting technique. The characteristics of the slurry and the green tapes were investigated in order to optimize the slurry compositions for tape casting. Results showed that the anode-substrate prepared with the optimized slurry composition presented excellent microstructure after being sintered in air at 1400℃ for 5 h and reduced in hydrogen at 800℃ for 0.5 h.

INTRODUCTION

Solid oxide fuel cell (SOFC) is a promising power generation system due to its high energy conversion efficiency, low environmental pollution and fuel flexibility. However, the high operation temperatures of conventional SOFC can lead to complex materials problems [1]. Lowering the SOFC operating temperature from 1000℃ to 600~800℃ is of great significance for the use of lower cost materials, such as interconnector and the other peripheral parts, and the improvement of the long-term stability as a result of reduced thermal degradation and thermal cycling stress. In order to lower the SOFC operating temperature, two significant parameters should be addressed, one is increasing the catalytic activity of the electrodes; the other is minimizing the resistance of the electrolyte. Reducing the electrolyte thickness to the range of 10~15 μm and developing the electrolyte with higher conductivity than yttria-stabilized zirconia (YSZ) are needed. This has led to the development of lanthanum gallate and doped ceria [2-5], as well as the development of electrode-supported SOFC [6]. Anode-supported SOFC becomes the focus of investigation because it provides not only superior electric conductivity and good mechanical strength, but also minimal chemical interaction with the electrolyte during the firing process. Gadolinium doped ceria (GDC) is considered to be one of the best ceria-based solid electrolyte material [7]. Therefore, preparing Ni-GDC anode supported SOFC can successfully solve the problems caused by the reduced SOFC operating temperature. Many techniques are adopted in preparing such anode substrates. Tape casting technique is a well-known colloidal shaping technique for large-area, thin, flat ceramic sheet or membranes [8]. Compared with other preparation techniques, tape-casting has the advantages of low operating costs, short manufacture period, and steady performances. In recent years, this technique is also used in SOFC studies and shows its outstanding advantages in technique and economy.

This paper used traditional organic solvent-based tape casting method to prepare Ni-GDC anode substrate for fabrication anode-supported planar IT-SOFCs. The variation of the tape casting parameters is controlled to the extent which can guarantee to get flat large area anode green tapes (over 10x10 cm^2). Firstly, the slurry composition was optimized to adjust the green tapes characteristics, Secondly, we focused on the binder burnout and sintering, the microstructure and porosity analysis of the anode were also studied.

EXPERIMENTAL

Preparation of the Ni-GDC anode substrate

The powder of NiO and GDC were purchased from fuel cell materials Inc. The anode powders consisted of NiO and GDC in a mass ratio of 65 to 35, were rotary ball-milled adequately. The slurry compositions are listed in table 1. The homogenized powders added with solvents and dispersants were ball-milled for 24 h. After the plasticizer and the binder were added, another 24 h ball-milling was followed. The anode slip prepared by the previous procedures was filtrated, and then de-aired by a vacuum air pump. Tape casting was performed using a tape caster for thin film (Richard E. Mistler Inc.), in which slurry was coated onto Mylar carrier film at a rate of 3 $cm \cdot s^{-1}$. The dried green tape was taken off and cut into the desired shapes. The green tape specimens were pre-sintered to remove all the organic additives at 180~700℃ with a very low temperature elevating speed of 0.1~0.5℃$\cdot min^{-1}$, and then further sintered to 1000~1500℃ at an elevating speed of 3~5℃$\cdot min^{-1}$. The holding time of the sintered green tape was 2~5 hours at the highest temperature to obtain the strong NiO-GDC anode substrate. The descending speed of the sintered tape is controlled at 2~5℃$\cdot min^{-1}$ from the highest temperature to the room temperature.

Table 1 the slurry composition of the green tapes

Component	Function	Quantity (wt.%)
NiO/GDC	Ceramic powder	40~55
Ethanol/ Methyl ethyl ketone (MEK)	Solvent	30~40
Glycerol trioleate	Dispersant	1~3
Triethanolamine (TEA)		
Phthalic acid diethyl ester (PHT)	Plasticizer	3~6
Polyethylene glycol (PEG400)		
Polyvinyl butyral (PVB)	Binder	3~8

Characterization of the Ni-GDC cermet

Microstructures and morphology of the anode cermet before and after reduction were investigated by the S-4700 scanning electronic microscope. Thermogravimetric analysis (TGA) was used to confirm the green tape sintering parameters by TAS-100 TGA. Porosities of the Ni-GDC anode were examined by Archimedes method.

RESULTS AND DISCUSSION

Optimization of the slurry composition

A higher dispersant content was required for the cermet powder of finer particle size. The binder and plasticizer contents are optimized according to the slip viscosity and the green tapes properties. Figure 1 shows the viscosity of the slurry as a function of the dispersant compositions. The first stage in the tape casting process involves dispersing the ceramic powders in a solvent with the addition of binders and plasticizers to yield the proper slip rheology. The dispersant nature and content are optimized by rheology measurements using a Brookfield III viscosimeter. Two dispersants were chosen for the dispersion of the Ni-GDC anode slip: Glycerol trioleate and Triethanolamine (TEA). The powder, the solvent and the dispersant were ball milled for 24 h before viscosity control. The evolution of the electrolyte slip viscosity at 30 s^{-1} versus the dispersant content (in wt.% of the total amount of mineral powder) is shown on Fig. 1. The lowest viscosity was observed for 2.2 wt.% of TEA dispersant. Stabilization of the dispersion for this dispersant is brought about by the mechanism of electrostatic repulsion, whereas the steric hindrance mechanism is reported for Glycerol trioleate.

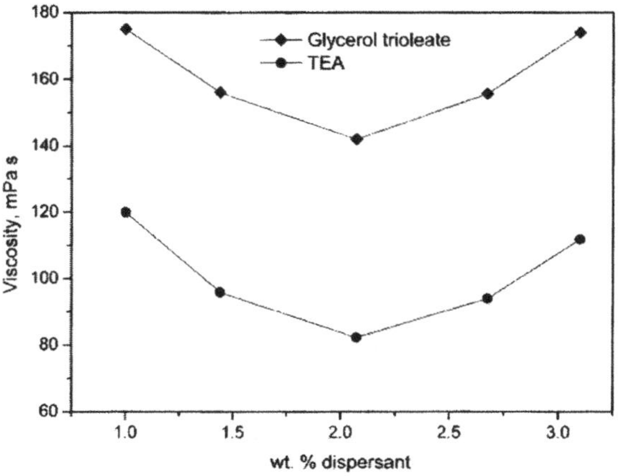

Fig.1 Optimization of the dispersant content for the Ni-GDC anode slip

The next step was the addition of Polyvinyl butyral (PVB) binder, as well as Phthalic acid diethyl ester (PHT) and Polyethylene glycol (PEG400) plasticizer. Anode slips were produced with a binder content between 3 and 8 wt.% and the plasticizer total content was between 3 and 6 wt.%. The observations made on the green tapes are presented in Table 2. Formulations D and E led to tapes with optimum properties: good removal from carrier surface, no cracking, high green density, good microstructural homogeneity as shown in Fig.2. For the cermet powders, a binder content of 7 wt.% and a plasticizer content of 6 wt.% were optimized in the same way.

Table 2 Properties of electrolyte green tapes

Sample	Binder (wt.%)	Plasticizer (wt.%)	Thickness (μm)	Observation
A	3	3	256	Cracks
B	4	4	258	Difficult to removal
C	5	5	242	Difficult to removal
D	6	6	267	No defects
E	7	6	235	No defects
F	8	6	249	Difficult to removal

TGA analysis and the green tapes sintering

Figure 3 shows the loss ratio of the Ni-GDC green tape with the rising of the sintering temperature in air from 25℃ to 1000℃. According to this figure, we can see that the green tape weight losses reach maximum (18 wt.%), when the sintering temperature is up to 310℃. Then the weight of the tape remains almost constant, indicating that the organic additives are completely removed. Thus, we confirmed the Ni-GDC green tapes sintering formulation that the green tapes were heated under air in two stages: a binder burnout stage (0.5℃$\cdot$min^{-1} up to 400℃) to allow the organics combustion, according to a thermogravimetric analysis, and a sintering stage (3~5℃$\cdot$min^{-1} up to 1350~1500℃).

Fig.2 Photo of the Ni-GDC green tape

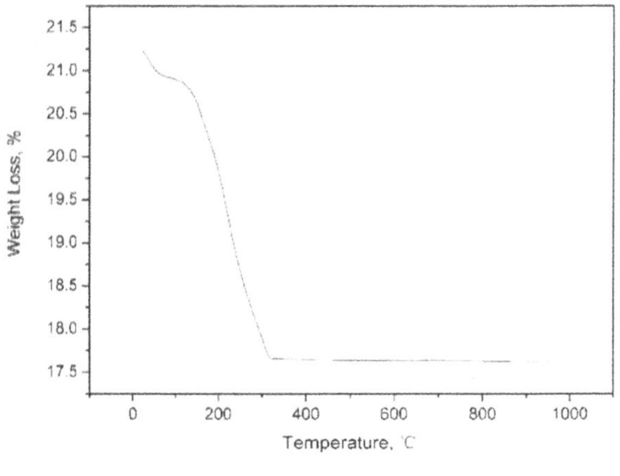

Fig.3 TGA measurement result of Ni-GDC green tapes in air

Microstructure analysis of the Ni-GDC cermet

Tape casting is actually a method for producing thin, dense ceramic tapes. It has been possible to demonstrate, however, that porous substrates permeable to gases can also be produced by this technique using slips with suitable additives. Figure 4 shows the microstructure of the Ni-GDC cermet sintered at 1400℃ in air. After sintering and reduction at 800℃ for 0.5h in hydrogen, the relatively coarse pore channels provide the Ni-GDC cermets with high gas permeability. On the other hand, the sintered agglomerates form a stable GDC skeleton, where the Ni-phase is evenly distributed and well-connected. This GDC matrix acts as a barrier against Ni-agglomeration. Clearly, the parameters such as the electrode porosity, the pore size and pore morphology are expected to influence the transport of gas species and thus the accompanying concentration polarization. Both samples show the homogeneous distribution of Ni. The porosity of the samples is between 45% and 48%, indicating a finer porous structure.

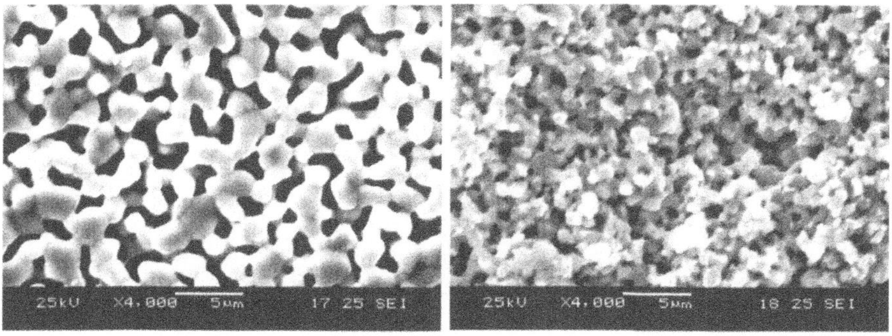

(a) Surface (b) Cross-section

Fig.4 SEM images of the Ni-GDC cermet after reduction at 800℃ for 0.5h

CONCLUSIONS

Tape casting of Ni-GDC anode substrate was successfully performed in this paper. The slurry composition and the sintering parameters of the green tapes were optimized to obtain dense samples with a suitable thickness and a good flatness. The porosity levels remain sufficient for the anode with the finest ceramic powders. Further investigations will be performed to confirm and improve these results by selecting feasible porosity former, by adjusting the green densities of the tapes or by testing a coarser NiO powder and a finer GDC electrolyte powder in order to construct perfect anode microstructure.

FOOTPRINTS

* Contacting Author: Prof. Chan Siew Hwa
E-mail address: SHCHAN@ntu.edu.sg

REFERENCES

[1] C.S. Song, Catalysis Today 77 (2002)17.
[2] M. Feng, J.B. Goodenough, Eur, J. Solid State Inorg. Chem., 31,663-670 (1994).
[3] T. Ishihara, H. Matsuda, Y.Takita, J.Am. Ceram. Soc., 116, 3801-3810 (1994).
[4] B.C. H. Steele, J. Power Sources, 49, 1-7 (1994).
[5] K.M. Myles, R. Doshi, R.Kumar, M. Krumpelt, in Proceedings of the 1st European SOFC Forum, European Fuel Cell Group, Ltd., Lucerne, Switzerland, Oct 3-7, 983-990 (1994).
[6] H. Ohrui, K. Watanabe, M. Arakawa, J. Power Sources, 112, 90-96 (2002).
[7] A. Tsoga, A. Naoumidis, D. Stover, Solid State Ionics, 135, 403-410 (2000).
[8] D. Simwonis, H. Thulen, F.J. Dias, A. Naoumidis, D. Stover, 'Properties of Ni/YSZ porous cermets for SOFC anode substrates prepared by tape casting and coat-mix process', Journal of Materials Processing Technology, 92-93, 107-111 (1999).

V. Ceramics in Environmental Applications

INFLUENCE OF LATTICE STRAIN ON THE $Ce_{0.5}Zr_{0.5}O_2$ AND Al_2O_3 DOPED $Ce_{0.5}Zr_{0.5}O_2$ CATALYTIC POWDERS

Chia-Che Chuang, Hsing-I Hsiang, and Fu-Su Yen
Particulate Materials Research Center, Department of Resources Engineering, National Cheng Kung University, No.1 University Road, Tainan 70101, Taiwan

ABSTRACT

$Ce_{0.5}Zr_{0.5}O_2$ has been widely used in three-way catalyst (TWC). However, the thermal resistance is deteriorated at high temperature due to the phase separation of $Ce_{0.5}Zr_{0.5}O_2$. The phase separation and crystallite growth of $Ce_{0.5}Zr_{0.5}O_2$ could be inhibited by adding Al ions. In this study, both the $Ce_{0.5}Zr_{0.5}O_2$ and Al_2O_3 doped $Ce_{0.5}Zr_{0.5}O_2$ catalytic powders were synthesized via chemical co-precipitation method. The phase separation of $Ce_{0.5}Zr_{0.5}O_2$ samples was observed as the calcination temperature was raised above 1100 °C. However, the Al_2O_3 doped $Ce_{0.5}Zr_{0.5}O_2$ samples could maintain single phase till 1200°C. The Williamson-Hall method was employed on the X-ray peak profile of these powders to determine the variation of lattice strain. The phase separation of $Ce_{0.5}Zr_{0.5}O_2$ samples was observed due to the lattice strain induced by the occurrence of oxygen vacancies as the calcination temperature was raised to 1100 °C. The Al_2O_3 doped $Ce_{0.5}Zr_{0.5}O_2$ samples had similar phenomenon but the phase separation temperature was delayed to 1200 °C due to the diffusion barrier effect of Al ions.

INTRODUCTION

CeO_2 has been widely applied in many fields, including catalysis, fuel cell and glass polishing technologies. In current three-way catalyst (TWC) for automotive pollution control, the unique reducing and oxidizing properties of CeO_2 allow the catalyst to enlarge the operating air/fuel (A/F) ratio window [1, 2]. The redox property of CeO_2 is greatly enhanced by incorporation of zirconium ions (Zr^{+4}) into the lattice to form a solid solution [2-5]. But, the redox property of $Ce_xZr_{1-x}O_2$ solid solution is aggravated at high temperature due to the phase separation of $Ce_xZr_{1-x}O_2(0<x<1)$. Thus, the promotion of thermal resistance and specific surface area has been developed by doping Al_2O_3 into CeO_2-ZrO_2 system [2, 6]. Many researches have investigated on the characteristics of phase, composition and structure which are correlated to the redox properties of CeO_2-ZrO_2 system [1, 3, 7, 8]. However, there are few literatures to discuss the phase separation mechanism of $Ce_{0.5}Zr_{0.5}O_2$ [4, 9]. In this study, the crystallite size and the lattice strain effects on the phase separation of $Ce_{0.5}Zr_{0.5}O_2$ were investigated. In addition, the effects of doping Al ions into $Ce_{0.5}Zr_{0.5}O_2$ on the crystallite growth and phase separation were also examined.

EXPERIMENTAL PROCEDURES

The ACZ sample (Al_2O_3 doped $Ce_{0.5}Zr_{0.5}O_2$, the molar ratio of Al:Ce:Zr = 4:1:1) was synthesized by the coprecipitation method [10]. In a typical synthetic condition, the stoichiometric amount of ammonium cerium(IV) nitrate ($(NH_4)_2Ce(NO_3)_6$, J.T. Baker, >99%) , zirconium(IV) dinitrate oxide ($ZrO(NO_3)_2$, Sigma-Aldrich, >99%) and aluminum nitrate($Al(NO_3)_3 \cdot 9H_2O$, Showa, >98%) were dissolved in distilled water. These solutions were mixed and slowly dripped into an aqueous NH_4OH solution. The pH value of the co-precipitating solution was maintained at 9.0 by adding

10 % NH_4OH solution. The obtained precipitate was filtered and thoroughly washed using distilled water and isopropanol to remove the anion impurities, and finally dried at 80 °C for 48 h. The obtained cakes were ground into powder, and then the powder was calcined at different temperatures (500-1200 °C) for 2 h (heating rate 5 °C/min) in air atmosphere. The same synthesis route was employed for the preparation of the CZ (molar ratio of Ce:Zr = 1:1) samples.
Powder X-ray diffraction (XRD) data was collected using Siemens D5000 (CuKα_1 radiation, 40 KV, 40 mA). Crystallite size was estimated from XRD data using the Scherrer equation [2]. The lattice strain was calculated from Williamson-Hall method [11].

$$\sqrt{\left(B_{exp}{}^2 - B_{inst}{}^2\right)}\cos\theta = \left(\frac{K\lambda}{D}\right) + (2\varepsilon\sin\theta) \qquad (1)$$

Here, B_{exp} and B_{inst} are the FWHM of experiment and the FWHM of instrument, respectively. The Scherrer constant K equals 0.94, λ is the X-ray wavelength, θ is the diffraction angle, D is the average crystallite size and ε is the lattice strain. The BET specific surface area measurement was performed using a standard nitrogen adsorption-desorption technique (Micromeritics ASAP2020). The TEM images of ACZ and CZ samples were observed by Hitachi HF-2000(FE TEM).

RESULTS AND DISCUSSIONS

Figure 1 shows the XRD patterns of the CZ samples calcined within the temperature range of 500-1200 °C. A single phase of $Ce_{0.5}Zr_{0.5}O_2$ solid solution was observed when the calcination temperature was below 1000 °C. As the calcination temperature was raised above 1100 °C, the single phase separated into Ce-rich and Zr-rich phases [4].

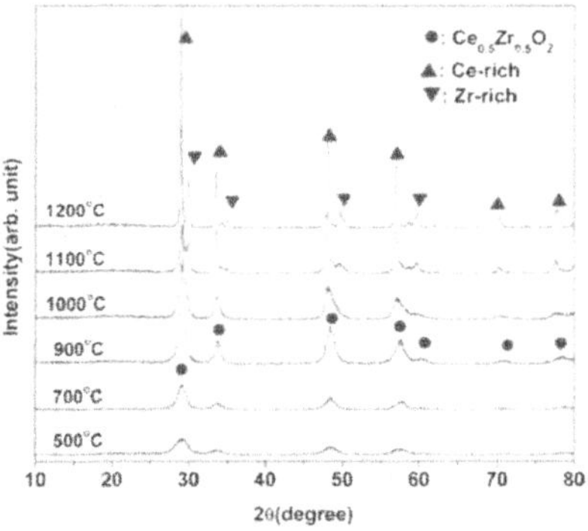

Figure 1. XRD patterns of CZ samples calcined at different temperatures.

XRD patterns of the ACZ samples calcined within the temperature range of 500-1200 °C are shown in Figure 2. Interestingly, a single phase of $Ce_{0.5}Zr_{0.5}O_2$ accompanied with a θ-Al_2O_3 phase was observed in ACZ samples calcined at below 1100 °C. However, a tiny peak adjacent to $Ce_{0.5}Zr_{0.5}O_2$ (111) (2θ= 29 °) belonging to the Zr-rich phase was observed at 1200 °C, indicating that the phase separation for ACZ samples occurred as the calcination temperature was raised above 1200°C.

Conventionally, the phase transformation temperature of θ to α-Al_2O_3 occurs at about 1100 °C. However, the θ to α-Al_2O_3 transformation temperature and the phase separation temperature of $Ce_{0.5}Zr_{0.5}O_2$ for ACZ samples were both delayed to higher than 1200 °C [10].

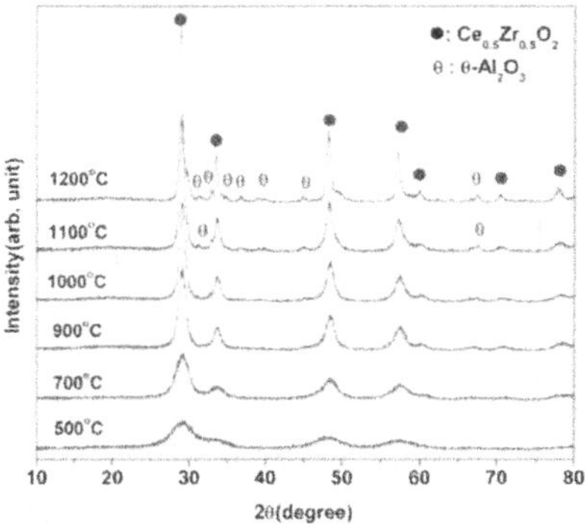

Figure 2. XRD patterns of ACZ samples calcined at different temperatures.

The crystallite sizes of $Ce_{0.5}Zr_{0.5}O_2$ as a function of calcination temperature for CZ and ACZ samples are shown in Figure 3. The crystallite sizes of $Ce_{0.5}Zr_{0.5}O_2$ for CZ samples are larger than that for ACZ samples. Besides, the crystallite sizes of CZ increased rapidly with increasing temperatures above 1100 °C and accompanied with the occurrence of the phase separation of CZ based on the XRD results. Kenevey et al. [4] have found that in the case of $Ce_{0.5}Zr_{0.5}O_2$, there exists a critical size around 15 nm, beyond which phase separation occurs due to the surface energy effect. The crystallite size beyond 15 nm and the occurrence of phase separation are observed after calcining at 1100 °C for CZ samples, and 1200 °C for ACZ samples, which is consistence with the observation of Kenevey.

It is noted that for CZ samples, the crystallite sizes decreased from ~12.5 to ~10.9 nm as the calcination temperature was increased from 900 to 1000 °C. This may be due to the occurrence of the phase separation for CZ samples during calcinations within the temperature range of 900-1000 °C, which led to the broadening of XRD diffraction peak [4]. The crystallite growth rate of ACZ sample was slower than that of CZ sample (Figure 3). This reveals that the doped Al ions via coprecipitation route could suppress the crystallite growth of $Ce_{0.5}Zr_{0.5}O_2$ effectively.

The lattice strains of CZ and ACZ samples calcined at various temperatures are shown in Figure 4. It shows that the lattice strain increased rapidly (from 0.74 to 2.76 %) as the temperature was raised from 700 to 1000 °C for CZ samples. As the temperature was increased above 1100 °C, the lattice strain decreased drastically to 0.52 % and the phase separation of CZ samples occurred simultaneously. We suggest that the lattice strain of CZ is increased due to the occurrence of oxygen vacancies during heating [8]. At temperatures above 1100 °C, the lattice strain was beyond the limit of tolerance, which resulted in the phase separation of $Ce_{0.5}Zr_{0.5}O_2$ to release the excessive lattice strain. For the ACZ samples, the lattice strain was 3.3 % at 500 °C due to the tiny nano-particle effect. The lattice strain (1.78 %) was larger than twice of that of CZ samples at 700 °C due to the larger amount of oxygen vacancies created by doping Al ions compared to the CZ samples. The lattice strain for the ACZ samples was maintained at 1.5~2 % from 700 to 1100 °C, then dropped gradually to 0.71 % at 1200 °C, due to the appearance of phase separation of $Ce_{0.5}Zr_{0.5}O_2$.

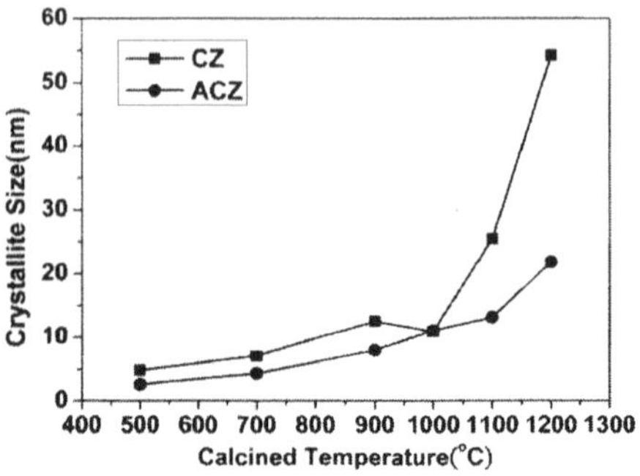

Figure 3. Crystallite sizes of $Ce_{0.5}Zr_{0.5}O_2$ for CZ and ACZ samples calcined at various temperatures.

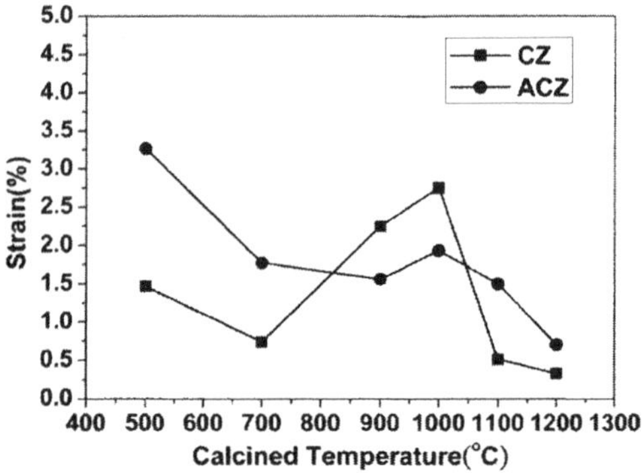

Figure 4. Lattice strains of CZ and ACZ samples calcined at various temperatures.

The BET specific surface area values of CZ and ACZ samples calcined at various temperatures are shown in Figure 5. The BET specific surface area values of ACZ samples are all much higher than those of CZ samples, which are consistent with the results of the crystallite size calculated from XRD (Figure 3). The higher specific surface area values of ACZ samples can be attributed to the suppression of crystallite growth resulted from the doping Al ions.

The TEM micrographs of CZ and ACZ samples are shown in Figure 6 and 7, respectively. The particle sizes of CZ and ACZ are nanometer scale and the average particle sizes of ACZ are smaller than that of CZ, which are consist with the previous XRD results. The estimated average particle size of CZ sample are 10~15 nm at 700 °C (Figure 6(a)) and 25~30 nm at 1000 °C(Figure 6(b)),

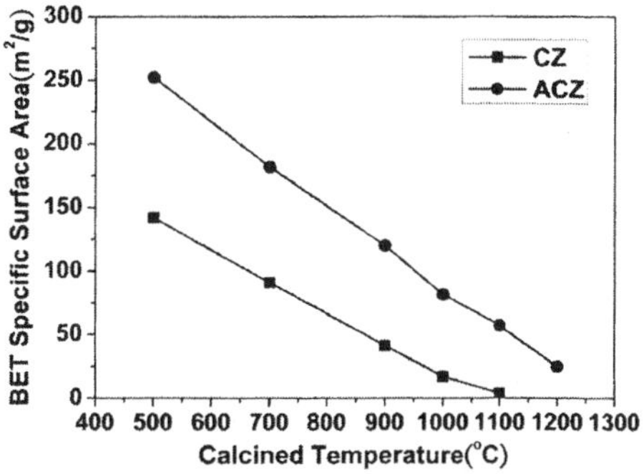

Figure 5. Specific surface area values of CZ and ACZ samples calcined at various temperatures.

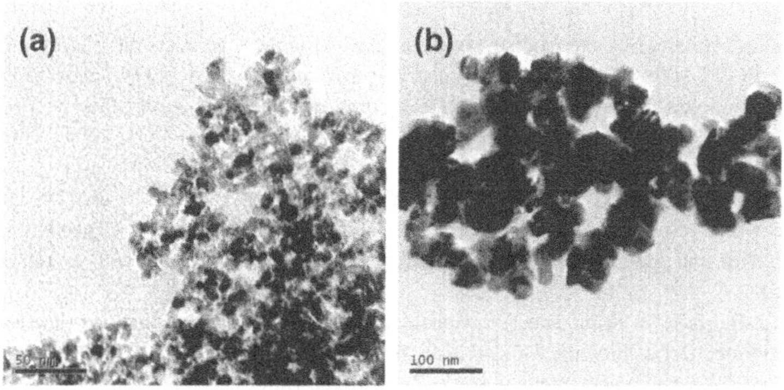

Figure 6. TEM micrographs of CZ samples calcined at (a) 700 °C/2h; (b) 1000 °C/2h.

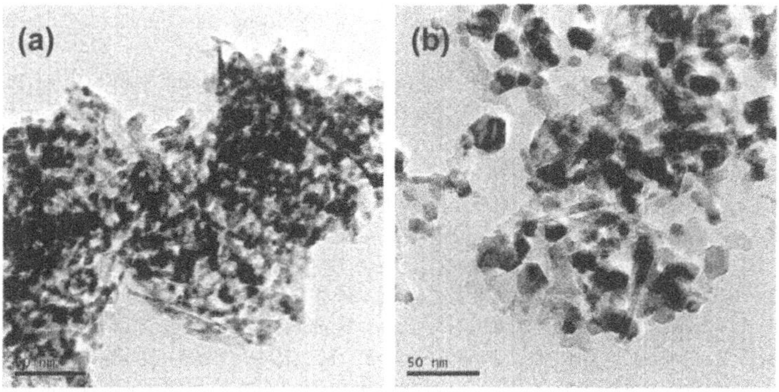

Figure 7. TEM micrographs of ACZ samples calcined at (a) 700 °C/2h; (b) 1000 °C/2h.

respectively. However, the average particle size calculated from TEM observations is larger than twice of that from Scherrer equation of XRD (11.1 nm) for CZ sample calcined at 1000 °C. The difference was due to the broadening XRD peak of the phase separation of $Ce_{0.5}Zr_{0.5}O_2$ at 1000 °C. For ACZ samples, the estimated average particle size are 5~10 nm at 700 °C(Figure 7(a)) and 15~20 nm at 1000 °C(Figure 7(b)), respectively. From the TEM images, the occurrence of rod-like particles is observed in ACZ samples, which will be further investigated in the future.

CONCLUSION

In this study, we have concluded that both the crystallite growth and the phase separation of $Ce_{0.5}Zr_{0.5}O_2$ have been inhibited by doping Al ions. The lattice strain is another key factor to induce the phase separation of $Ce_{0.5}Zr_{0.5}O_2$ beside the particle size effect. The phase separation phenomena of $Ce_{0.5}Zr_{0.5}O_2$ will be further systematically studied based on the crystalline structure, defect chemistry and thermodynamics to clarify this issue.

Acknowledgment

This work was supported by the Ministry of Economic Affairs of the Republic of China through contract (92-EC-17-A-08-S1-023) and the Ministry of Education, Taiwan, R.O.C. under the NCKU Project of Promoting Academic Excellence & Developing World Class Research Centers.

REFERENCES

[1]P. Fornasiero, R. Di Monte, G. R. Rao, J. Kaspar, S. Meriani, A. Trovarelli, and M. Graziani, Rh-Loaded CeO_2-ZrO_2 Solid-Solutions as Highly Efficient Oxygen Exchangers: Dependence of the Reduction Behavior and the Oxygen Storage Capacity on the Structural-Properties, *J. Catal.*, **151**, 168-177, (1995).

[2]M. H. Yao, R. J. Baird, F. W. Kunz, and T. E. Hoost, An XRD and TEM Investigation of the Structure of Alumina-Supported Ceria-Zirconia, *J. Catal.*, **166**, 67-74, (1997).

[3]R. Di Monte and J. Kaspar, Nanostructured CeO_2-ZrO_2 Mixed Oxides, *J. Mater. Chem.*, **15**, 633-648, (2005).

[4]K. Kenevey, F. Valdivieso, M. Soustelle, and M. Pijolat, Thermal Stability of Pd or Pt Loaded $Ce_{0.68}Zr_{0.32}O_2$ and $Ce_{0.50}Zr_{0.50}O_2$ Catalyst Materials under Oxidising Conditions, *Appl. Catal. B: Environ.*, **29**, 93-101, (2001).

[5]F. Zhang, C.-H. Chen, J. C. Hanson, R. D. Robinson, I. P. Herman, and S.-W. Chan, Phase in Ceria-Zirconia Binary Oxide (1-x)CeO_2-xZrO_2 Nanoparticles: The Effect of Particle Size, *J. Am. Ceram. Soc.*, **89**, 1028-1036, (2006).

[6]R. Di Monte, P. Fornasiero, S. Desinan, J. Kaspar, J. M. Gatica, J. J. Calvino, and E. Fonda, Thermal Stabilization of $Ce_xZr_{1-x}O_2$ Oxygen Storage Promoters by Addition of Al_2O_3: Effect of Thermal Aging on Textural, Structural, and Morphological Properties, *Chem. Mater.*, **16**, 4273-4285, (2004).

[7]P. Fornasiero, G. Balducci, R. Di Monte, and J. Kaspar, "Modification of the Redox Behaviour of CeO_2 Induced by Structural Doping with ZrO_2, *J. Catal.*, **164**, 173-183, 1996.

[8]R. G. Wang, P. A. Crozier, R. Sharma, and J. B. Adams, Nanoscale Heterogeneity in Ceria Zirconia with Low-temperature Redox Properties, *J. Phys. Chem. B*, **110**, 18278-18285, (2006).

[9]H.-Y. Zhu, T. Hirata, and Y. Muramatsu, Phase Separation in 12 mol% Ceria-Doped Zirconia Induced by Heat Treatment in H_2 and Ar, *J. Am. Ceram Soc.*, **75**, 2843-48, (1992).

[10]C.-C. Chuang, H.-I. Hsiang, J. S. Hwang, and T. S. Wang, Synthesis and Characterization of Al_2O_3-$Ce_{0.5}Zr_{0.5}O_2$ Powders Prepared by Chemical Coprecipitation Method, *J. Alloys Compd*, In Pressed, (2008).
[11]T. Wang, X. D. Fang, W. W. Dong, R. H. Tao, Z. H. Deng, D. Li, Y. P. Zhao, G. Meng, S. Zhou, and X. B. Zhu, Mechanochemical Effects on Microstructure and Transport Properties of Nanocrystalline $La_{0.8}Na_{0.2}MnO_3$ Ceramics, *J. Alloys Compd*, **458**, 248-252, (2008).

MICROSTRUCTURE AND PROPERTIES OF CORDIERITE-BONDED POROUS SiC CERAMICS PREPARED BY IN SITU REACTION BONDING

Shifeng Liu [a, b], Yu-Ping Zeng [a], Dongliang Jiang [a]
[a] Shanghai Institute of Ceramics, Chinese Academy of Science, Shanghai, 200050, China
[b] Graduate School of the Chinese Academy of Science, Beijing, 100039, China
* yuping-zeng@mail.sic.ac.cn

ABSTRACT

An in situ reaction technique was adopted for the preparation of cordierite-bonded porous SiC ceramics in air from α-SiC, α-Al_2O_3 and MgO, using graphite to adjust the porosity of porous SiC ceramics. During sintering, the surface of SiC was oxidized to SiO_2 and then the oxidation-derived SiO_2 reacted with α-Al_2O_3 and MgO to form cordierite, resulting in the bonding of SiC particles. Microstructure of cordierite-bonded porous SiC ceramics was investigated as a function of graphite content. In addition, properties such as thermal shock resistance, permeability, high temperature oxidation resistance as well as acid and alkaline endurance of cordierite-bonded porous SiC ceramics were studied.

INTRODUCTION

Porous SiC ceramics attract increasing attention in the application for hot gas filtration in recent years because of their superior properties, such as low thermal expansion coefficient, good thermal shock resistance, excellent mechanical and chemical stability at elevated temperatures.[1,2] However, a high sintering temperature is required for the preparation of SiC ceramics owing to the strong covalent nature of Si-C bond, which limits the practical application of porous SiC ceramics.[3] In order to realize the low-temperature fabrication of porous SiC ceramics, secondary phase-bonded porous SiC ceramics have been developed. She et al.[4] developed a unique oxidation-bonding technique for the fabrication of porous SiC ceramics at low temperature and the oxidation bonded porous SiC ceramics exhibit high strength as well as good oxidation and thermal shock resistance. Furthermore, Ding et al.[5] fabricated mullite-bonded porous SiC ceramics by an in situ reaction bonding technique. Compared with the oxidation bonded porous SiC ceramics, the mullite-bonded porous SiC ceramics possess better high temperature stability and oxidation resistance. However, a high temperature of 1500 °C is still necessary for extensive mullitization.

Recently, we have successfully prepared cordierite($2MgO{\cdot}2Al_2O_3{\cdot}5SiO_2$)-bonded porous SiC ceramics from SiC, Al_2O_3, MgO and graphite in air by an in situ reaction bonding process at a relatively low temperature of 1350 °C.[6] The as-fabricated cordierite-bonded porous SiC ceramics exhibit high mechanical strength. In the present work, we further investigate the microstructure, thermal shock resistance, permeability, high temperature oxidation resistance as well as acid and alkaline endurance of the cordierite-bonded porous SiC ceramics.

EXPERIMENTAL PROCEDURE

Commercially available α-SiC powder (99.4% purity, 10.0 μm, Weifang Kaihua Silicon Carbide Micro-powder Co. Ltd., Weifang, China) α-Al_2O_3 (99.9% purity, 0.6 μm, Wusong, Chemical Fertilizer Factory, Shanghai, China) and MgO (>98% purity, Shanghai Tongya Chemical

Technology Co. Ltd., Shanghai, China) were used as the starting materials. Graphite powder (99.4% purity, Qingdao Huatai Lubricant Sealing Science and Technology Co. Ltd., Qingdao, China) with an average particle size of 10.0 μm were employed as the pore-forming agent.

The powder mixtures of SiC, Al_2O_3, MgO and C as listed in Table I were ball-milled in ethanol for 24 h to obtain homogeneous slurries, where the weight ratio of Al_2O_3 to MgO was fixed to the value of 2.53, which is equal to the ratio of Al_2O_3 to MgO in the stoichiometric composition of cordierite. Adding proper quantities PVB (Polyvinyl Butyral) as binder. After being dried in a dry oven at 80 °C and sieved through a 75-mesh screen, the mixed powders were bidirectionally pressed into the rectangular specimens with dimensions of ~5.0×10.0×50.0 mm under a 56 MPa pressure using a steel die. The specimens were sintered in air at 1200-1350 °C for 2 h with a heating and cooling rate of 5 °C/min.

Table I. Compositions of the Powder Mixtures for Porous SiC Ceramics Synthesis.

Mixture	SiC : (Al_2O_3+MgO) : C (wt.)	Content of C (vol.%)
1	8 : 2 : 0	0
2	8 : 2 : 2.26	25
3	8 : 2 : 4.52	40
4	8 : 2 : 6.78	50
5	8 : 2 : 10.18	60

Open porosity was determined by the Archimedes method with distilled water as the liquid medium. Pore size distribution was characterized by the mercury porosimetry (Model Pore-Sizer 9320, Micromeritics, USA). Phase analysis was conducted by X-ray diffraction (XRD) (Model RAX-10, Rigaku, Japan) with Cu Kα radiation. Microstructures were observed by scanning electron microscopy (SEM) (Model KYKY-EM3200, KYKY, China). Flexural strength was measured via the three-point bending test (Model AUTO-GRAPH AG-1, Shimadzu, Japan) with a support distance of 30 mm and a cross-head speed of 0.5 mm/min, four specimens were tested to obtain the average strength and standard deviation. Thermal shock test was conducted by the conventional water-quenching technique. Briefly, the specimens were heated to the preset temperature at a rate of 5 °C /min and held for 0.5 h. Then, the specimens were quenched into a water bath, where the temperature was maintained at 20 °C, and the residual strength of the specimens subjected to the water-quenching was determined by the three-point bending test. Gas permeability was evaluated by a home-made apparatus with nitrogen gas as a permeation medium.

RESULTS AND DISCUSSION

Reaction bonding behavior and microstructure

It is known SiC starts to oxidize at ~750 °C. The initial product is amorphous silica, which begins to crystallize into cristobalite at ~1100 °C.[7] Fig. 1 shows the XRD patterns of porous SiC ceramics sintered at different temperatures for 2 h. At 1200 °C, obvious cristobalite peaks were found. In addition, a substantial spinel forms as the result of the diffusive reaction between Al_2O_3 and MgO in solid state. α-cordierite peaks appear at 1300 °C and their intensity is found to increase notably as the sintering temperature is further increased to 1350 °C. On the other hand, the peak intensity of spinel does not have an obvious change as temperature increases. These results can be explained by the mechanism of cordierite formation. Based on the work by Shi et al.,[8] the dissolution of MgO and Al_2O_3 into SiO_2 results at the initial stage where α-cordierite forms, and the solid solution starts to turn into α-cordierite at ~1250 °C. At a high temperature of ~1290 °C, solid-state reaction between spinel and cristobalite takes place and forms α-cordierite.[6] As the temperature is further increased, Mg-Al-Si-O glass forms at ~1320 °C and this glass is further crystallized to α-cordierite at ~1350 °C.[8] Therefore, a large amount of cordierite has formed after sintering at 1350 °C for 2 h and the SiC particles were bonded each other by the reaction-derived cordierite.

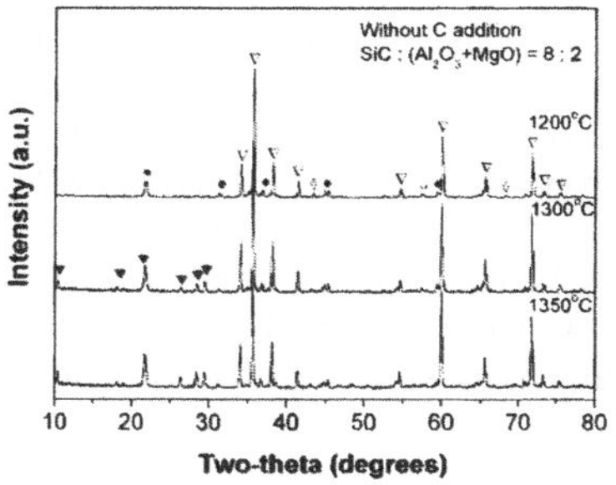

Figure 1. XRD patterns of the specimens sintered at indicated temperature, (∇, SiC; ◊,α-Al_2O_3; •, Cristobalite; ♦, Spinel, ▼,α-cordierite).

Figure 2. SEM micrograph of fracture surfaces of cordierite-bonded porous SiC ceramics (a) without C addition and (b) with 40 vol.% C, sintered at 1350 °C for 2 h, where the weight ratio of SiC to Al_2O_3+MgO was 8 : 2.

Fig. 2a shows the typical microstructure of porous SiC ceramics without C addition sintered at 1350 °C for 2 h. The specimen exhibits a structure with connected pores and the well-developed necks are formed between SiC particles. Clearly, the pores are derived from stacking SiC particles. Typical microstructure of porous SiC ceramics with 40 vol.% C addition is shown in Fig. 2b. Compared with the specimen without C addition, the specimen with 40 vol.% C addition exhibits larger porosity and average pore size. Since 10.0 μm SiC and 10.0 μm graphite are used, the pores originated from the burnout of C are much larger that those originated from the stacking of SiC particles. Therefore, the average pore size of the specimen with 40 vol.% C addition is much larger than that of the specimen without C addition. It is anticipated that using graphite as the pore-forming agent can adjust the porosity and pore size of porous SiC ceramics effectively.

Pore size distribution

Fig. 3 shows the pore size distribution of porous SiC ceramics without and with 40 vol.% C. Both specimens take on a narrow pore size distribution. In addition, it can be seen from Fig. 4 that the average pore size in the specimen with 40 vol.% C addition is much larger than that of the specimen without C addition. This is in agreement with the above SEM results. As expected, the pore size distribution of porous SiC ceramics without C addition exhibits a unimodal distribution. It has known that the pores of porous SiC ceramics with 40 vol.% C addition are derived from burning out C and stacking SiC particles. However, it is interesting to find that the pore size distribution of porous SiC ceramics with 40 vol.% C addition also exhibits a unimodal distribution. This should be attributed to the fact that most of the voids among SiC particles are contacted with C particles in the green body when C content is relatively high. As a result, the majority of voids are merged into the pores formed by burning out C after sintering.

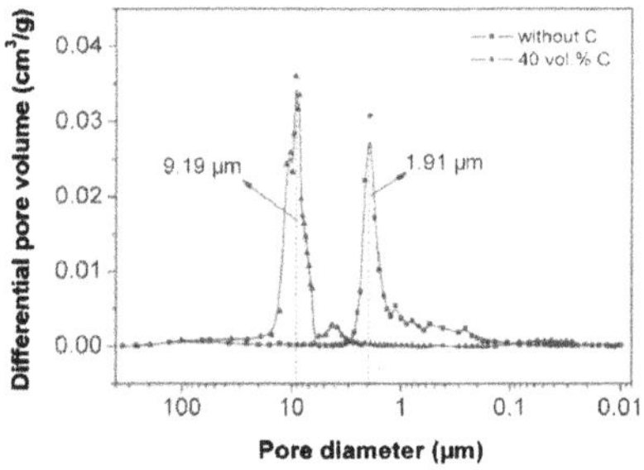

Figure 3. Pore size distribution of porous SiC ceramics without and with 40 vol.%C addition, sintered at 1350 °C for 2 h.

Permeability and thermal shock resistance

Fig. 4 shows the plots of N_2 gas flux versus pressure drop for porous SiC ceramics with different C contents. The N_2 gas flux nearly increases linearly with the pressure drop. In addition, it can be seen that the N_2 gas flux increases sharply with C content. This is attributed to the increases of open porosity and average pore size.

Porous SiC ceramics with 25 vol.% C addition sintered at 1350 °C for 2 h (open porosity and flexural strength were 44.5 % and 26.0 MPa, respectively) was adopted for thermal shock tests. Fig. 5 shows the residual strength of the quenched specimens as a function of quenching severity. Obviously, the specimen exhibits a gradual strength reduction. This can be attributed to the crack-pore interactions. Furthermore, due to the low thermal expansion coefficients of cordierite and SiC, good thermal shock resistance was achieved for the specimen.

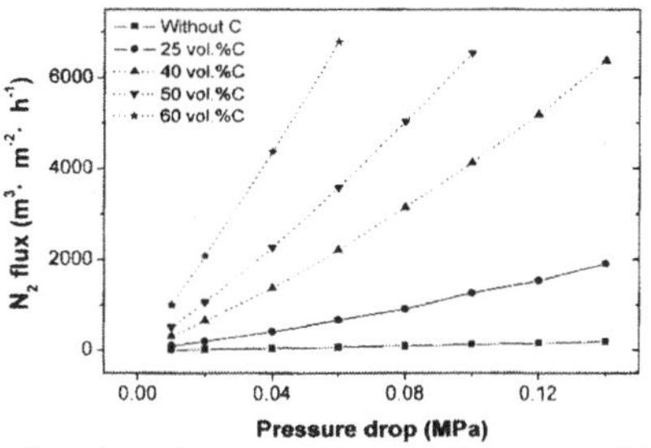

Figure 4. N_2 gas flux as a function of pressure drop for the porous SiC ceramics with different C contents, sintered at 1350 °C for 2 h.

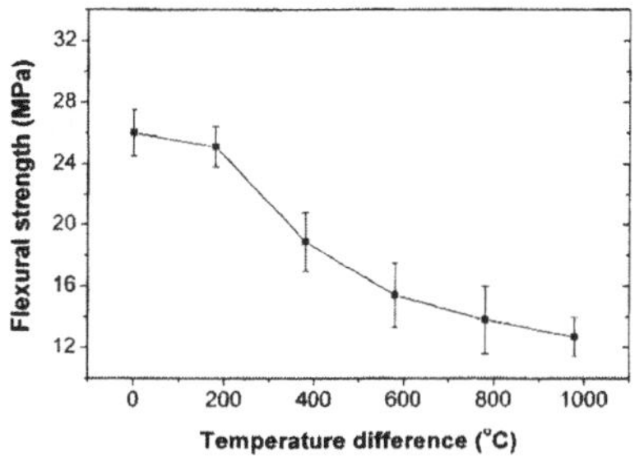

Figure 5. Residual flexural strength as a function of quenching temperature difference for porous SiC ceramics.

High temperature oxidation resistance as well as acid and alkaline endurance

25 vol.% C-added specimen sintered at 1350 °C for 2 h was selected to investigate the high temperature oxidation resistance as well as acid and alkaline endurance of cordierite-bonded porous SiC ceramics. The cordierite-bonded porous SiC ceramics possess good oxidation resistance. After oxidizing in air at 1000 °C for 80 h, the specimen only exhibits a 0.77 % increase in weight and a 19.1 % decrease in flexural strength. This should be attributed to the fact that the SiC particles ware covered by cordierite and SiO_2 in the cordierite-bonded porous SiC ceramics. Acid and alkaline endurance of porous SiC ceramics were evaluated by corroded the specimen under a boiling state for 1 h in the thermal H_2SO_4 solution (20 wt.%) and NaOH solution (2 wt.%), respectively. After acid corroding, the decreases in weight and flexural strength were 2.6 % and 27.7 %, respectively. Correspondingly, 6.1 % and 56.2 % decreases in weight and flexural strength were exhibited after alkaline corroding. These results indicate that the specimen had good acid endurance but relatively bad alkaline endurance.

CONCLUSIONS

Cordierite-bonded porous SiC ceramics were fabricated in air by an in situ reaction technique from α-SiC, α-Al_2O_3 and MgO, using C as pore-forming agent. During sintering, the surface of SiC was oxidized to SiO_2 and then the oxidation-derived SiO_2 reacted with α-Al_2O_3 and MgO to form cordierite, resulting in the bonding of SiC particles. The cordierite-bonded porous SiC ceramics have a microstructure with connected pores and well-developed necks. The porosity and pore size of porous SiC ceramics can be effectively adjusted by adding C. Due to the low thermal expansion coefficients of cordierite, the cordierite-bonded porous SiC ceramics possess good thermal shock resistance. In addition, as-fabricated cordierite-bonded porous SiC ceramics exhibit good high temperature oxidation resistance and acid endurance, but relatively bad alkaline endurance.

ACKNOWLEDGMENT

This work was supported by Science and Technology Commission of Shanghai Municipality under Contracts No. 07JP14093 and No. 08JC1420300.

REFERENCES

[1]P. Pastila, V. Helanti, A. P. Nikkila, and T. Mantyla, Environmental Effects on Microstructure and Strength of SiC-Based Hot Gas Filters, *J. Eur. Ceram. Soc.*, 21, 1261-8 (2001).

[2]S. Heidenreich, and B. Scheibner, Hot Gas Filtration with Ceramic Filters: Experiences and New Developments, *Filtr. and Separat.*, 39, 22-5 (2002).

[3]R. Riedel, G. Passing, H. Schonfelder, and R. J. Brook, Synthesis of Dense Silicon-Based Ceramics at Low Temperatures, *Nature*, 355, 714-7 (1992).

[4]J. H. She, J. F. Yang, N. Kondo, T. Ohji, S. Kanzaki, and Z. Y. Deng, High-Strength Porous Silicon Carbide Ceramics by an Oxidation-Bonding Technique, *J. Am. Ceram. Soc.*, 85, 2852-4 (2002).

[5]S. Q. Ding, S. M. Zhu, Y. P. Zeng, and D. L. Jiang, Fabrication of Mullite-Bonded Porous Silicon Carbide Ceramics by *in situ* Reaction Bonding, *J. Eur. Ceram. Soc.*, 27, 2095-102 (2007).

[6]S. F. Liu, Y. P. Zeng, and D. L. Jiang, Fabrication and characterization of cordierite-bonded porous SiC ceramics, *Ceram. Int.*, 35, 597-602 (2009).

[7]S. Q. Ding, S. M. Zhu, Y. P. Zeng, and D. L. Jiang, Effect of Y_2O_3 Addition on the Properties of Reaction-Bonded Porous SiC Ceramics, *Ceram. Int.*, 32, 461-6 (2006).

[8]Z. M. Shi, K. M. Liang, and S. R. Gu, Effects of CeO_2 on Phase Transformation towards Cordierite in MgO-Al_2O_3-SiO_2 System, *Mater. Lett.*, 51, 68-72 (2001).

FABRICATION OF LIGHTWEIGHT CLAY BRICKS FROM RECYCLED GLASS WASTES

Vorrada Loryuenyong[1, 2,*], Thanapan Panyachai[1], Kanyarat Kaewsimork[1], and Chatnarong Siritai[1]
[1]Department of Materials Science and Engineering, Faculty of Engineering and Industrial Technology, Silpakorn University, Nakhon Pathom, Thailand
[2]National Center of Excellence for Petroleum, Petrochemicals and Advanced Materials, Bangkok, Thailand
(*Corresponding author. Email address: vorrada@su.ac.th)

ABSTRACT

In this study, glass wastes were used as a raw material in clay brick manufacturing. Percentages of substitution with glass wastes in clay mixture were considered, and the effects of glass content up to 45 wt.% were investigated. The results indicated that glass wastes can be efficiently utilized environmentally-friendly as a mixture in clay bricks. The amount of glass wastes in the form of powder in the clay mixture was a very important factor determining the properties and the microstructure of the bricks. With proper amount of glass wastes, clay bricks with considerable physical, mechanical and thermal properties could be obtained. The compressive strength as high as 26-45 MPa, water absorption less than 3% and thermal conductivity of 0.9 W/mK at 40°C were achieved for bricks containing 15-30 wt.% glass content, fired at 1100°C. Addition of rice husks in the clays containing the glass wastes would lower the thermal conductivity but reduce the mechanical strength of the bricks.

INTRODUCTION

In general, ceramic products require high firing temperatures. The manufacture of clay-based bricks, for example, is normally fired at temperatures between 1000°C and 1200°C, depending on the clay types. Advanced ceramics require even higher operating temperatures to affect the sintering process or to achieve full densification. The firing process, therefore, causes the production of ceramics to be energy and cost intensive. Accordingly, a new technology that has the potential to minimize the energy consumption and to lower the energy costs is desirable.

Nowadays, there is a great concern regarding to the increasing amount of industrial wastes. The disposal of these wastes is one of the issues that have received a lot of attention and a high demand for the safety of the environment. One technique used to reduce such wastes is by recycling, and different kinds of wastes have already been recycled in ceramic industry. By substituting raw materials with recycling glass wastes as an alternative ceramic raw material or as a fluxing agent, for example, firing temperatures as well as the manufacturing cost can be reduced. Previous researches have been reported their uses for a variety of applications including stoneware, tiles, bricks and concrete [1]-[8]. Glass mixture is believed to induce the vitrification in clay bodies, resulting in higher density, lower %water absorption and lower drying shrinkage [4,8].

Tucci et. al [6] has shown the benefits of adding soda-lime scrap glass powders into a porcelain stoneware tile mix. The study showed that soda-lime glasses could act as a good fluxing agent, and the replacement of glasses by 10 wt.% resulted in better mechanical characteristics. Shayan and Xu [7] used fine glass powders in order to prepare the concrete with good strength and appropriate drying shrinkage. Topcu and Canbaz [8], however, have observed that the use of glass wastes in

concrete caused a decrease in concrete density as well as the compressive strength and tensile strength with increasing glass contents. As a consequence, the process optimization and control of glass recycling for a particular application or particular interest is needed to be understood.
In the present work, the effects of glass addition, ranging from 0 to 45 wt.%, on the physical, mechanical and thermal characteristics of the clay bricks were studied. The preparation of lightweight clay bricks by using a combination of glass wastes and rice husks were also preliminary investigated.

EXPERIMENTAL PROCEDURE

Glass wastes and ball clays used in this experiment were obtained from local sources. The chemical analysis of the raw materials was determined by X-ray fluorescence analysis, as shown in Table I. Glass wastes were first ball milled and screened to 18 mesh (1 mm-opening size), generating a powder material. The particle-size distribution test was carried out for recycled glass powders, using sieve size analysis (Table II). Additions of 0, 15, 30 and 45 wt.% of glass powders to the clay body (0G, 15G, 30G, and 45G) were carried out in batches, and at least 6 samples were prepared for each batch. Clay bricks containing 5 wt.% rice husks (5R) were used as a reference. Bricks that contain both glass wastes and rice husks were also prepared. The addition of rice husks would reduce the weight of the bricks but might reduce the mechanical strength.

Table I. Chemical composition of the raw materials.

Composition	Mass (%) Ball Clay	Mass (%) Glasses
MgO	0.33	1.56
Al_2O_3	31.4	0.35
SiO_2	50.8	70.49
K_2O	1.66	0.21
CaO	0.13	24.40
TiO_2	0.68	0.15
Fe_2O_3	2.01	2.80
Na_2O	<0.05	-
P_2O_5	0.07	-
CuO	-	0.03
SrO	-	0.01
MnO	0.01	-
Cr_2O_3	<0.01	-
V_2O_5	0.01	-
LOI	12.6	-

The test samples were prepared by mixing the raw materials in different proportions. Water was added, in the amount of 30% by weight based on the total weight, to combine clay mixture. The clay mixture was soaked overnight to get a homogeneous mixture. The soaked mixture was then mechanically stirred and then kneaded by hands on the gypsum board. After the kneading, the clay was formed into lightly-lubricated wooden molds that gave the bricks their shape. The bricks were removed from the molds approximately one day after casting and were allowed to dry in air until constant weight was achieved. The dried bricks were then fired in an electric laboratory furnace at maximum temperatures of 1100°C, with the ramping rate of 3°C/min and 60 min holding time. The samples were naturally cooled down to room temperature in the furnace.

Archimedes' method based on ASTM C67 [9] was used to determine the bulk density, apparent density and %water absorption. The weight loss of bricks was obtained by weighing the bricks before and after firing. The compressive strength and modulus of rupture were determined, using Shimadzu-UH100A universal testing machine, on the test samples with half-brick (6.5 × 7.0 × 4.0 cm^3) and full-brick (6.5 × 14.0 × 4.0 cm^3) sizes, respectively. The phases occurred in the fired clay bodies were determined with X-ray diffraction (XRD), conducted on a Bruker C8 Advance diffractometer. The thermal conductivity was measured by the steady-state method. The scanning electron microscope (SEM) micrographs were used for surface morphology analysis.

RESULTS AND DISCUSSION

Table II shows the particle size distribution of glass powders. The predominant particle size was in the range of 0.1-0.6 mm. The physical and thermal property data of clay bricks with various glass proportion (15G, 30G, and 45G), compared with 5R and 30G5R (30 wt.% glass wastes + 5 wt.% rice husks) bricks are shown in Table III. Without a shrinkage barrier such as rice husks and glass wastes, pure clay bricks or 0G samples (not shown in the table) had very large drying shrinkage lying between 8-9%, and even severer during firing. Large amount of linear shrinkage could increase the risk of appearance of cracks and dimensional defects in bricks. It has been observed that no 0G bricks were survived without cracking, and as a consequence, 5R bricks were used as a reference for this study. The addition of 5 wt.% rice husk lowered both the drying and the firing shrinkages of the bricks. The degree of firing shrinkage was lower than 15G, 30G, and 45G bricks.

Among bricks containing glass wastes, the percentage of linear shrinkage decreased as the amount of glass wastes added in the mixture increased. Normally, linear shrinkage is an important factor to determine the degree of densification during firing. The results, however, indicated that the bricks became less dense (lower bulk density) with the addition of glass wastes. Similarly, %water absorption increased as glass wastes were added in the brick composition, and once the increment surpassed 30 wt.%, the values increased drastically. The clay bricks with glass content of 45 wt.% have water absorption as high as 13%. One interesting observation was that glass particles were likely to be forced out toward the surfaces of all bricks. When the glass content was increased to 45 wt.%, a large number of glass particles were forced out and evidently melted and formed a thin layer on the brick surfaces. A preliminary experiment showed that with smaller particle size of glasses, this problem can be avoided.

A reduction in bulk density and an increase in %water absorption could be attributed to the increasing amount of opened pores, which may be introduced into the bricks due to several reasons such as glass particles oozing out onto the brick surface. An increase in apparent density with increasing glass content could be a result of the vitrification in clay bodies, causing fired clay to be fused together. Nevertheless, the bricks containing glass wastes were found to have higher bulk density and less %water absorption than the reference bricks containing risk husks (5R). From the results, bricks containing 30 wt.% glass wastes yield the optimum physical characteristics. The addition of rice husks to the clay mixture reduced the bulk density and increased %water absorption of these bricks.

Table II. Sieve size analysis for recycled glass powders.

Sieve size (μm)	Cumulative passing (%)
75	0.06
106	0.30
150	0.71
300	20.75
425	63.30
600	94.02
1000	99.80

Table III. Physical and thermal properties of clay bricks glass wastes/rice husks and fired at 1100°C.

Bricks	%Drying shrinkage	%Firing shrinkage	Bulk density (g/cm^3)	Apparent density (g/cm^3)	%Water absorption	Thermal conductivity (W/mK) at 40°C
15G	4.7±1.3	9.0±1.3	2.0±0.1	2.1±0.1	2±0.7	0.86
30G	4.6±1.4	8.7±1.1	2.1±0.1	2.2±0.1	2.6±1.0	0.89
45G	4.3±1.1	5.1±1.9	1.8±0.1	2.4±0.2	13.0±1.0	0.71
5R	6.0±1.8	6.3±1.7	1.8±0.0	2.5±0.1	15.6±1.5	0.45
30G5R	4.1±1.3	7.8±2.2	1.2±0.0	2.1±0.0	18.0±0.7	0.55

Figure 1 shows the amount of firing weight loss from each brick at various compositions. The results indicated that the substitution of raw materials with glass wastes could reduce the weight of the bricks. Firing weight loss increased as the amount of glass wastes added into the mixture increased. The maximum weight loss occurred for the 30G5R samples, which contain both glass wastes and rice husks. Therefore, it was possible to produce lightweight bricks with considerable properties by adding proper amount of glass wastes and rice husks.

In relation to mechanical properties (Figure 2), the bar graphs indicated that the strength of the bricks depended greatly on the amount of glass wastes. Both the compressive strength and the modulus of rupture of the samples decreased with increasing glass content. Nevertheless, the mechanical strength achieved at all proportion was much larger than 5R bricks, especially for the compressive strength measurement. In addition, the strength values were acceptable and eventually met the minimum requirement for some load-bearing applications such as a paving brick subjected to light traffic in less severe environment [10]. Based on above results, the optimum combination of properties with the compressive strength as high as 26-45 MPa and water absorption less than 3% was obtained for the bricks containing 15-30 wt.% glasses, fired at 1100°C. Further adding rice husks to 30G clays causes a significant reduction of the mechanical strength.

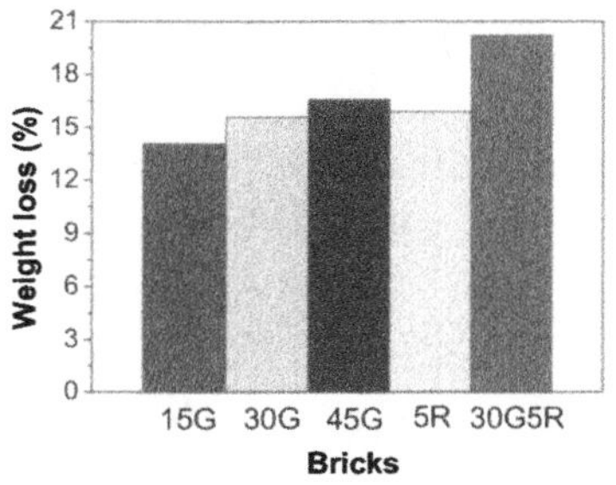

Figure 1. Firing weight loss of clay bricks glass wastes/rice husks and fired at 1100°C.

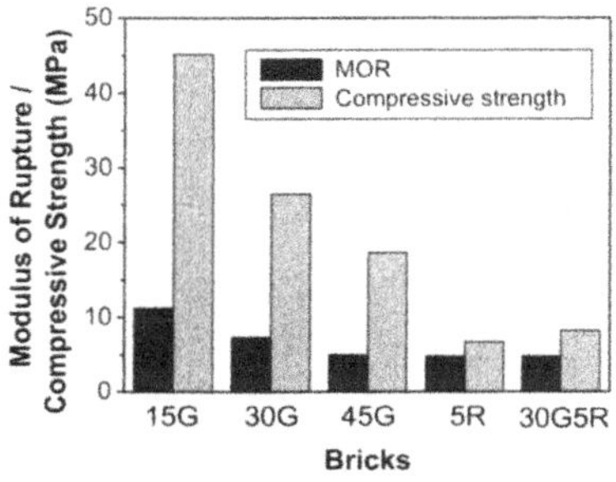

Figure 2. Mechanical properties of clay bricks containing glass wastes/rice husks and fired at 1100°C.

Thermal conductivities of clay bricks were showed in Table III. All bricks containing glass wastes (15G, 30G, 45G and 30G5R) had higher conductivity than that of reference clay bricks (5R). Thermal conductivity is an important factor to determine thermal insulating characteristics of materials. One method of reducing the thermal conductivity is to introduce porosity in clay bodies. Higher porosity due to the presence of rice husks (i.e. 30G5R) could lower thermal conductivity of bricks containing glass wastes (i.e. 30G).

Figure 3 shows the SEM micrographs of the fired clay bricks with different glass contents. From the figure, glass phases were distributed uniformly in the clay bodies. Increasing glass content caused an increase in the glass phases in the bricks. The crystalline phases identified in the bricks containing 0-30 wt.% glasses fired at 1100°C are shown in Figure 4. The patterns indicated the presence of new phases, the amount of which increased with increasing percentage of glasses. These new phases were determined to be cristobalite and albite high. The smaller intensities of quartz and mullite phases in 15 wt.% and 30 wt.% bricks, compared with the fired pure clay brick is due to the lower alumina content, which prevents the formation of mullite.

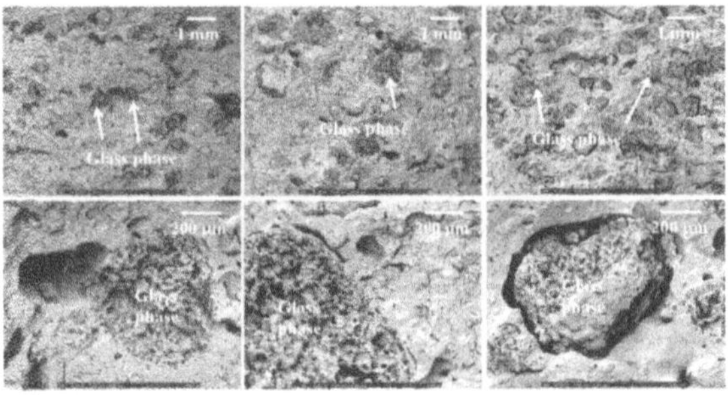

Figure 3. SEM micrographs of clay bricks fired at 1100°C.

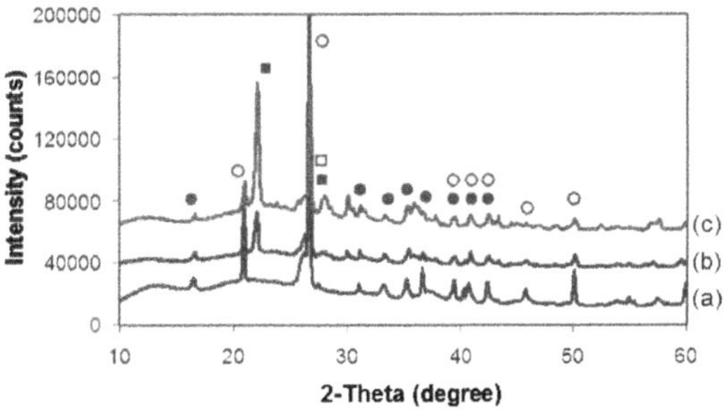

Figure 4. XRD patterns of the bricks fired at 1100°C: (a) 0G, (b) 15G and (c) 30G. ○ - quartz, ● - mullite, ■ - cristobalite, □ - albite high.

CONCLUSION

The objectives of this study were to investigate the physical, mechanical and thermal properties of the clay bricks containing wasted glasses. The results indicated that glass wastes can be mixed with clay in different proportions to produce good quality bricks. The properties of the clay bricks depend on the amount of added glass wastes. The compressive strength and the modulus of rupture of the bricks obtained in this study ranges from 18-45 MPa and 5-11 MPa, respectively, which is suitable in a wide range of applications and even in some load-bearing structures. The optimum combination of properties was obtained for the bricks containing 15-30 wt.% glasses. The thermal properties of the clay bricks containing glass wastes could be improved by the addition of rice husks.

REFERENCES

[1]I.W. Brown and K.J.D. Mackenzie, Process design for the production of a ceramic-like body from recycled waste glass. Part 1. The effect of fabrication variables on green strength, *J. Mater. Sci.*, **17**,

2164–2170 (1982).
[2]R.V. Manukyan and N. S. Davydova, Use of waste in the ceramic industry, *Glass Ceram.*, **53** (7–8), 247–248 (1996).
[3]Y. Lingart, Imitation of natural material tiling using waste glass, *Glass Technol.*, **39** (2), 42–43 (1998).
[4]E. Rambaldi, W.M. Carty, A. Tucci, and L. Esposito, Using waste glass as a partial flux substitution and pyroplastic deformation of a porcelain stoneware tile body, *Ceram. Inter.*, **33**, 727-733 (2007).
[5]E. Rambaldi, A. Tucci and L. Esposito, Use of recycled materials in the traditional ceramic industry, *Ceram. Inter.*, **1–2**, 13–23 (2004).
[6]A. Tucci, L. Esposito, E. Rastelli, C. Palmonari, E. Rambaldi, M.E. Tyrell and A.H. Goode, Use of soda-lime scrap-glass as a fluxing agent in a porcelain stoneware tile mix, *J. Eur. Ceram. Soc.*, **24**, 83-92 (2004).
[7]A. Shayan and A. Xu, Value-added utilization of waste glass in concrete, *Cement Concrete Res.*, **34**, 81-89 (2004).
[8]I.B. Topcu and M. Canbaz, Properties of Concrete Containing Waste Glass, *Cement Concrete Res.*, **34** (2), 267-74 (2004).
[9]ASTM C-67 (1992) Standard Test Method of Sampling and Testing Brick and Structural Clay Tile.
[10]ASTM 902-92 (1992) Standard Specification for Pedestrian and Light Traffic Paving Brick.

ACKNOWLEDGEMENT

This work is supported by Industrial and Research Projects for Undergraduate Students (IRPUS No. I350A05004). The authors also wish to thank Department of Materials Science and Engineering, Faculty of Engineering and Industrial Technology, Silpakorn University, National Center of Excellence for Petroleum, Petrochemicals and Advanced Materials and Clean Glass Ltd. for supporting and encouraging this investigation.

THE PERFORMANCE OF GEOPOLYMER BASED ON RECYCLED CONCRETE SLUDGE

Z.X.Yang, N.R.Ha, M.S.Jang, K.H.Hwang, B.S.Jun*, and J.K.Lee**
(Eng. Res. Inst., GyeongSang Nat'l Univ., Jinju, 660-701, S. Korea,
*Kyungnam University, Korea, **Chosun University, Korea)
(Corresponding to khhwang@gsnu.ac.kr)

ABSTRACT

Geopolymer has a very low rate of green house gas release when compared to Ordinary Portland cement. In this study, the component of geopolymer is recycled concrete sludge, metakaolin and water glass. NaOH was used as alkaline activator. To improve the mechanical properties, silica fume was added as a bonding matrix from 0%~10% to replace part of concrete sludge, and the specimens were cured in the air, then their mechanical properties like compressive strength and bending strength were measured and the microstructures were investigated.

INTRODUCTION

Concrete usage around the globe is second only to water. An important ingredient in the conventional concrete is the Portland cement. The production of one ton of cement emits approximately one ton of carbon dioxide to the atmosphere. Moreover, cement production is not only highly energy-intensive, next to steel and aluminium, but also consumes significant amount of natural resources. As a relatively new material, geopolymer concrete offers the benefits as a construction material for sustainable development. It utilized waste materials such as recycled concrete sludge, fly ash and etc. Nowadays, more and more researches are focused on environment-friendly construction materials. To reduce CO_2 release, geopolymer concrete is expected to replace the traditional Portland cement based concrete. It's reported that geopolymer based concrete releases only 1/6 CO_2 compared to those of based on Ordinary Portland Cement (OPC). But because of the relative high cost of metakaolin, which is the main component of geopolymer, the application of geopolymer construction material is limited. On the other hand, already huge volumes of concrete sludge are generated around the world, most of that is not effectively used, and a large part of it is disposed in landfills, which also caused construction waste pollution. So the solution of that kind of problem should be regarded.

Geopolymer bonded mortar with full metakaolin as raw material for construction applications can reach a 3-day compressive strength above 800Kgf/cm^2 by our previous research. From that point we realized geopolymer can be developed as an excellent binder to replace cement if the cost could be reduced.

This research describes the attempt of using recycled concrete sludge as the raw material to fabricate green geopolymer mortar. NaOH was used as alkaline activator. To improve the mechanical properties, silica fume was added as a bonding matrix from 0%~10% to replace part of concrete sludge, and the specimens were cured in the air, then their mechanical properties like compressive strength and bending strength were measured and microstructures were investigated.

EXPERIMENTAL PROCEDURE

In this study, geopolymer was prepared with metakaolin, recycled concrete sludge powder, silica fume and alkaline activators. Metakaolin was commercially obtained and concrete sludge was supported free of charge by company. The alkaline activators, sodium hydroxide and soluble glass, were used to activate alumino-silicate oxide.

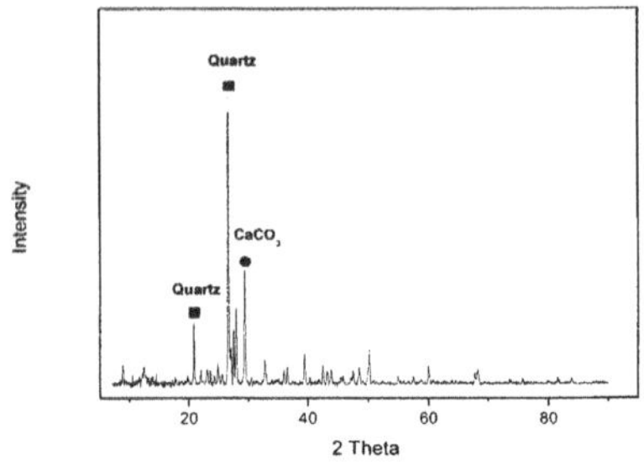

Fig.1. XRD pattern of recycled concrete sludge

Test 1, to test the effect of metakaolin on the properties of sludge based geopolymer, the concrete sludge powder and silica sand were well mixed with different content of metakaolin addition from 10%~40% (all the percentage here is wt% of powder matrix) by the aid of vibrating machine. The powder to sand ratio was 30:70 by weight. Sodium hydroxide was dissolved into water to get a solution with concentration about 10M. Then NaOH solution and water glass were mixed together and cooled down to about 20℃. After that, the mixture was poured into the vibrating bowl to fabricate castable geopolymer mortar. The mortar was filled into steel mould with dimension of 150x15x15mm and 50x50x50mm for bending and compressive strength test respectively. After cured at room temperature for 24hr, demoulded specimens were taken physical properties test and microstructures were also observed by Scanning Electron Microscope (SEM).

Test 2, to investigate the use of silica fume on sludge based geopolymer, the concrete sludge

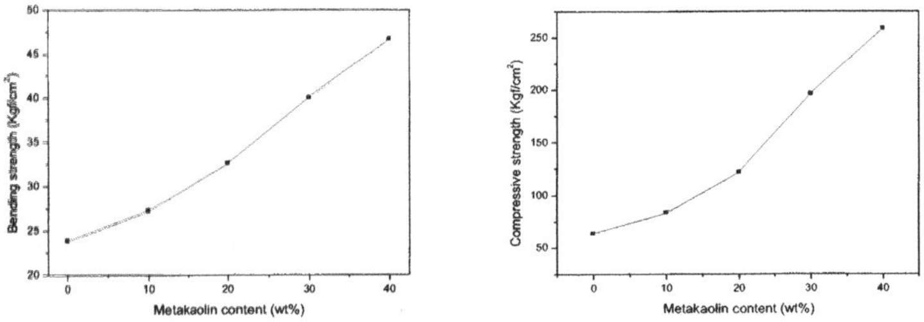

(a) (b)

Fig.2 Variety of a) bending strength and b) compressive strength of sludge-based geopolymer with metakaolin addition

powder and silica sand were well mixed with different content of silica fume addition from 2%~10%. The other processes were the same as test 1.

RESULT AND DISCUSSION

The XRD pattern of recycled concrete sludge was shown as Fig.1. The main phase of concrete is $CaCO_3$ and quartz, which partially provides SiO_2 for geo-polymerization reaction. The microstructure of sample that contained sludge only was shown in Fig.4 (a). Without metakaolin addition, there's no active alumino-silicate existed, which is the main component for geopolymer formation. And the SiO_2 contained in sludge powder was not active enough to fabricate strong bond. So an incompact structure with many pores and cracks was found, due to weak bonding between particles.

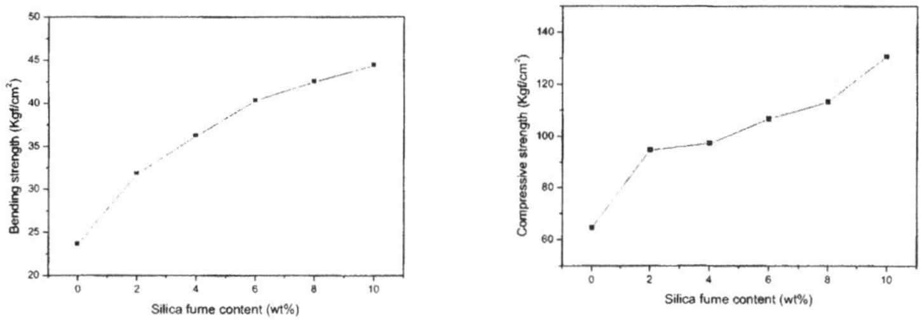

Fig.3. Variety of a) bending strength and b) compressive strength of sludge-based geopolymer with silica fume addition

The specimens' bodies became compacter when metakaolin was added. The geopolymer synthesization was much enhanced with the increase of metakaolin content. Sialate bridges (Si-O-Al-O-) formed by the introduction of alumino-silicate. The in-situ formed geopolymer bonded sludge particles and silica sands very well, leading to a higher bending and compressive strength which was shown as Fig.2. The density was also increased with metakaolin content and the apparent porosity decreased. The microstructures of metakaolin contained sample showed a smoother surface and well bonded boundary between sludge particles and sands with less cracks and pores from Fig 4.(b). There's only few cracks existed and a homogeneous structure was observed compared to the coarse grains with no additives as shown in Fig.4 (a).

When the fine silica fume added, it offered the SiO_2 with high activity. That SiO_2 played a very important role to form the siloxo bridges (-Si-O-Si-O-) in geo-polymerization processing. These bridges chains bonded particles firmly; consequently, both bending strength and compressive strength were enhanced by silica fume addition that could be seen as Fig.3. Strength about more than 130Kgf/cm^2, which was a satisfiable value for mortars, could be reached by 10% addition. And Fig.4.(c) showed a much denser and compacter matrix structure compared with those samples without silica fume addition.

Fig.4. Microstructures of a) concrete sludge only, b) 20% metakaolin addition and c) 10% silica fume addition

CONCLUSION

To fabricate construction appliable geopolymer in a cost-effective manner, recycled concrete sludge was used as raw material, metakaolin powderand silica fume were prepared to optimize the properties.

1. Only concrete sludge contained sample showed poor mechanical properties because of the weak bonding. The microstructure with big cracks and large pores among matrix and sands was found.
2. When adding metakaolin powder, by introducing alumino-silicate, the in-situ formed geopolymer acted as binder to the raw materials, polymerization bonded particles together well, lead to a better performance. Both bending strength and compressive strength of mortars could meet common construction uses. Geopolymer stabilized the waste in a cost-effective and environment-friendly way.
3. The silica fume offered active SiO_2 which was profitable for geopolymer production. So the addition of silica fume also improved the mechanical properties of mortars and made structure much denser. But the amount of that should be considered because of its high cost.
4. Though the raw material was cheap recycled concreted sludge waste, the other components like NaOH and water glass still had a relatively high cost. The substitute of those materials should be found and the properties of this kind of construction material could be further enhanced.

ACKNOWLEDGEMENT

This work was supported by Brain Korea 21 Project.

REFERENCES

[1]. Davidovits, J.: "Geopolymers: Inorganic polymeric new materials," J. Thermal Analysis, 37 (1991) 1633-1656

[2]. Davidovits, J.: "Mineral polymers and methods of marking them", US Patent 4,349,386 (1982)

[3]. Davidovits, J.: "Chemistry of Geopolymeric Systems, Terminology", in Geopolymer '99 International Conference. 1999. France

[4]. J.T. Gourley and G. B. Johnson. "Developments in Geopolymer Precast Concrete", Proceedings of the World Congress Geopolymer 2005, 139-143

[5]. Ke-Liang LI, Guo-Hong HUANG, et al. "Early Mechanical Property and Durability of Geopolymer", Proceedings of the World Congress Geopolymer 2005, 117-120

STRUCTURE AND MICROWAVE DIELECTRIC PROPERTIES OF THE $2.02Li_2O$-$1Nb_2O_5$-$1TiO_2$ CERAMICS

Qun Zeng*[1], Wei Li[2], Jing-kun Guo[3]
[1*] School of Information Photoelectric Science and Technology, South China Normal University, Guangzhou, 510006, China;
[2] School of Materials Science and Engineering, East China University of Science and Technology, Shanghai, 200237, China;
[3] Shanghai Institute of Ceramics, Chinese Academy of Sciences, Shanghai, 200050, China

ABSTRACT

A new low-firing dielectric ceramic with Li_2O: Nb_2O_5: TiO_2 mole ratio of 2.02: 1: 1 has been found and investigated. Two phases, "M-phase" and an unknown phase constitute a special structure in the ceramic. The unknown phase has a size of about 0.5~1 μm is oriented in the "M-phase" phases. This new ceramic has a low sintering temperature (~1100°C) and good microwave dielectric properties of high permittivity (ε_r~58), high $Q*f$ value up to 7500GHz and low temperature coefficient (τ_f ~24ppm/°C), which indicate that the new microwave dielectric ceramic could be a suitable candidate for low-firing materials.

INTRODUCTION

Recently, much attention has been paid to the development of the low temperature co-fired ceramics (LTCC), which enable miniaturization of the dimensions of the multilayer devices and a reduction in cost in the wireless communication industry[1-3]. For LTCC materials, in addition to the appropriate dielectric constant (ε_r), low dielectric loss ($tg\delta = 1/Q$), and near zero temperature coefficient of resonant frequency (τ_f), the low sintering temperature (T_s) is a critical requirement due to the low melting temperature of the highly conductive and inexpensive internal electrode metals, such as silver (961 °C), copper (1064 °C) and their alloys[4]. However, most of the known commercial microwave dielectric materials could not be used as LTCC materials for their high processing temperatures of about 1300 °C or higher[5]. So, there is considerable interest in the development of materials with low sintering temperature. Up to now, in order to lower the sintering temperature, three kinds of methods have been commonly explored: (1) addition of low melting point compounds such as V_2O_5, B_2O_3 and glass[6-9], (2) chemical processing for smaller particle sizes of starting materials, such as the sol-gel and co-precipitation methods[10-11], and (3) search for new material systems with low intrinsic sintering temperatures (normally $\not>$1100°C). The first method is widely used to lower the sintering temperature of developed dielectric materials. However, in many cases, they may induce a significant deterioration in dielectric properties due to the large amounts of liquid-phase-forming additives. The chemical techniques require a complicated procedure, which means that higher cost and longer processing time would be inevitable. So, with the rapid development and increasing requirements, there is always much attention paid to searching for new materials with low intrinsic sintering temperatures (normally $\not>$1100°C).

The Li_2O-Nb_2O_5-TiO_2 system materials have aroused much attention in recent several years due to their low intrinsic sintering temperatures (~1100°C) [12-16]. In this paper, we present the synthesis, characterization, and dielectric properties of a new sort of composition based on the Li_2O-Nb_2O_5-TiO_2 system.

EXPERIMENTAL PROCEDURE

The ceramics were prepared using the conventional mixed oxide route. Starting materials were high-purity oxide powders (>99.5%) of Li_2CO_3, Nb_2O_5 and TiO_2. Stoichiometric proportions of the above raw materials (2.02:1:1 by mole) were mixed in ethanol using zirconia balls as milling media for 24 h. The wet mixture was dried and calcined at 850 °C for 6h. The powder was re-milled in ethanol with zirconia balls in a plastic jar and then uniaxially pressed under the pressure of 150 MPa into disks measuring 16 mm in diameter and 6-8 mm in thickness. The disks covered with crucibles were sintered at 1100 °C.

The XRD patterns of the sintered ceramic were examined by the X-ray diffraction (XRD) analysis with Rigaku RINT2000 (Cu Kα radiation generated at 40kV and 40mA). The densities of the ceramics were measured by the Archimedes method. The polished and thermally etched surface morphologies were observed with electron probe X-ray microanalyser (EPMA) (JXA-8100). The EPMA samples were polished and thermally etched at about 70 °C below their sintering temperatures for about 30 minutes to reveal their grain structures.

The dielectric constant (ε_r) and the quality values Q at microwave frequency were measured using the Hakki-Coleman's dielectric resonator method[17], and modified and improved by Courtney[18]. A vector network analyzer (E8363, Agilent Technologies, Loveland, CO, USA) was employed in the measurement. The temperature coefficient of resonant frequency (τ_f) was measured in the temperature range from -25 to +85 °C. The τ_f value was defined as follows:

$$\tau_f = \frac{(f_{85} - f_{-25})}{110 f_{25}} * 10^6 \qquad (1)$$

where f_{85}, f_{-25}, and f_{25} were the resonant frequencies at 85,-25 and 25 °C, respectively.

RESULTS AND DISCUSSION

The X-ray diffraction (XRD) pattern of the specimen sintered at 1100 °C is presented in Figure 1. It is obvious that the ceramic is composed of two phases, the "M-phase" solid solution phase[12,19] and an unknown phase. Obviously, the main phase is "M-phase". The few unknown phases couldn't be identified clearly yet according to the actual Powder Diffraction File (PDF) database, and might be clarified in the future.

Figure 2 shows the secondary electron image of a polished and thermally etched surface of the $2.02Li_2O$-$1Nb_2O_5$-$1TiO_2$ (LNT) ceramic composite sintered at 1100 °C. The bulk density of the ceramic sintered without use of any additives at 1100 °C is about 4.02 g/cm^3. From Figure 2, it could be found that besides some pores, there exist two kinds of grains in the composite: the long-plate shaped grains and the spot-like grains. The two kinds of grains constitute a very special structure in the ceramic. The spot-like grains are in the long-plate shaped grains, and they are arrayed with the orientation of long-plate shaped grains. Combined with the XRD results, it could be concluded that the long-bar shaped grains are in "M-phase" structure, and the spot-like grains are the unknown phase. The grain sizes of the unknown phase are about 0.5~1μm. Due to the electron beam size of energy dispersive spectra (EDS) would be larger than the grain size of the unknown phase, the exact composition of the unknown phase could not be identified. Although the detailed forming mechanism of this special structure is unclear, it could be reasonably deduced that the forming of the spot-like grains oriented would have effects on the microwave properties of the LNT

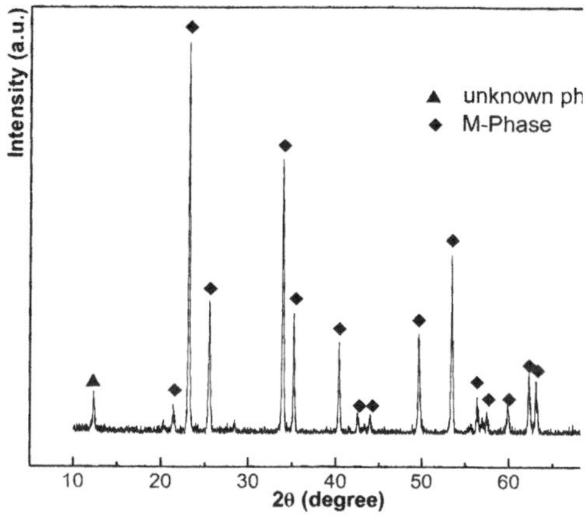

Figure 1 X-ray diffraction pattern of the ceramic pellet sintered at 1100 °C. ▲- unknown phase, ◆-"M-Phase".

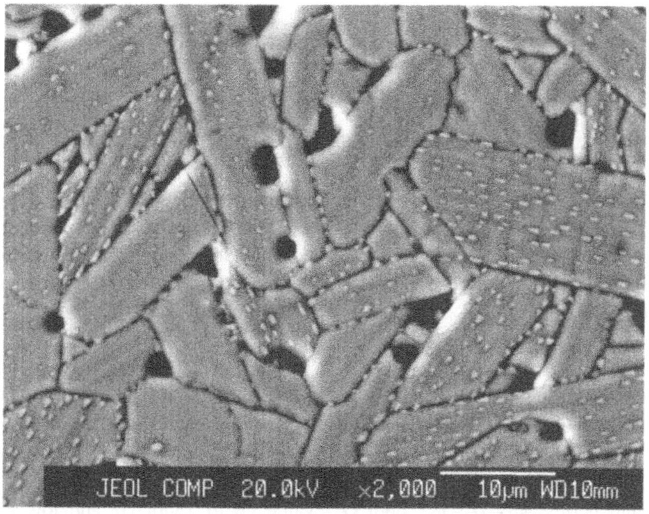

Figure 2 The EPMA image of polished and thermally etched surface of the LNT ceramic sintered at 1100 °C.

ceramics.

The microwave dielectric properties were measured under TE_{011} mode. The ceramic sintered at 1100 °C shows relatively good microwave dielectric properties of high permittivity ($\varepsilon_r \sim 58$) and high $Q*f$ value up to 7500 GHz. Moreover, Figure 3 presents the resonant frequency of the ceramic as a function of the temperature. The τ_f value of the ceramic sintered at 1100 °C calculated by Equation (1) is about 24 ppm/ °C, which is relatively low.

In addition, it could also be found that the ε_r value of the LNT ceramic is close to that of the "M-phase" $Li_{0.95}Nb_{0.65}Ti_{0.45}O_3$ (ε_r=58.4), nevertheless, the τ_f value of this LNT ceramic is positive (τ_f = 24 ppm/°C) and its absolute value is lower than that of $Li_{0.95}Nb_{0.65}Ti_{0.45}O_3$ ceramic (τ_f = -31ppm/°C)[12]. Herein, a sort of dielectric ceramics with $\varepsilon_r \sim$ 58 and near-zero τ_f value might be obtained easily by compositing this new ceramic with the $Li_{0.95}Nb_{0.65}Ti_{0.45}O_3$ ceramic. Therefore, this new ceramic could be a very promising microwave dielectric ceramic.

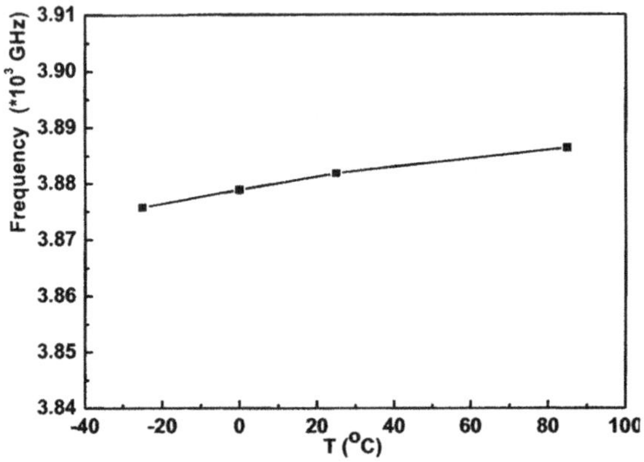

Figure 3 Variation of the resonant frequency of the ceramic as a function of the temperature.

Due to the LNT ceramic has low intrinsic sintering temperature (~1100 °C), it would be relatively easy to further decrease its sintering temperature to about 900 °C for LTCC applications. Currently, we are exploring ways to lower the sintering temperature by adding small amounts of low melting point oxides. In this regard, the introduction of B_2O_3 has been found to be very effective and will be reported in a forthcoming publication[20].

CONCLUSION

In summary, a new low-firing dielectric ceramic with Li_2O: Nb_2O_5: TiO_2 mole ratio of 2.02:1:1 has been investigated in this study. The ceramic is composed of two phases, the "M-phase" and an unknown phase. The main phase in the ceramic is "M-phase". The two phases constitute a very special structure in the new ceramic. This new ceramic shows good microwave dielectric properties of a relatively high permittivity (ε_r~58), high $Q*f$ value up to 7500GHz and low temperature coefficient (τ_f ~24ppm/°C), which was obtained via sintering at 1100°C. Obviously, the new dielectric ceramic could be a promising candidate of low-firing microwave dielectric ceramic materials.

REFERENCES

[1]H. Jantunen, T. Kangasvieri, J. Vähäkangas, S. Leppävuori, Design Aspects of Microwave Components with LTCC Technique, *J. Eur. Ceram. Soc.*, **23,** 2541-48 (2003).
[2]A. Chaouchi, S. d'Astorg, S. Marinel, M. Alioua, $ZnTiO_3$ Ceramic Sintered at Low Temperature with Glass Phase Addition for LTCC Applications, *Mater. Chem. Phys.*, **103**, 106-111 (2007).
[3]S. Solomon, H. P. Kumar, L. Jacob, J.K. Thomas, M. R. Varma, $Ln(Zr_{1/3}Ti_{2/3})TaO_6$ (Ln = Ce, Pr, Nd and Eu): A Novel Group of Microwave Ceramics, *J. Alloy. Compd.*, **461**, 675-7 (2008).
[4]T. Takada, S.F. Wang, S.Yoshikawa, et al., Effects of Glass Additions on $(Zr,Sn)TiO_4$ for Microwave Applications, *J. Am. Ceram. Soc.*, **77,** 2485-88 (1994).
[5]C. C. Chen, T. E. Hsieh, I. N. Lin, Effects of Composition on Low Temperature Sinterable Ba–Nd–Sm–Ti–O Microwave Dielectric Materials, *J. Eur. Ceram. Soc.*, **23,** 2553-58 (2003).
[6]P. Liu, E. S. Kim, S. G. Kang, H. S. Jang, Microwave Dielectric Properties of $Ca[(Li_{1/3}Nb_{2/3})_{1-x}Ti_{3x}]O_{3-\delta}$ Ceramics with B_2O_3, *Mater. Chem. Phys.*, **79**, 270-2 (2003).
[7]C. L. Lo, J. G. Duh, D. S. Chiou, W. H. Lee, Low-Temperature Sintering and Microwave Dielectric Properties of Anorthite-Based Glass-Ceramics, *J. Am. Ceram. Soc.*, **85,** 2230-35 (2002).
[8]X. M. Chen, Y. H. Sun, X. H. Zheng, High Permittivity and Low Loss Dielectric Ceramics in the BaO-La_2O_3-TiO_2-Ta_2O_5 System, *J. Eur. Ceram. Soc.*, **23,** 1571-74 (2003).
[9]M. Z. Jhou, J. H. Jean, Low-Fire Processing of Microwave $BaTi_4O_9$ Dielectric with BaO–ZnO–B_2O_3 Glass, *J. Am. Ceram. Soc.*, **89,** 786-91 (2005) .
[10]H. Kagata, T. Inoue, J. Kato, I. Kameyama, Low-Fire Bismuth-Based Dielectric Ceramics for Microwave Use, Jpn. J. Appl. Phys., **31,** 3152-55 (1992).
[11]Y. Xu, G. Huang, Y. He, *Ceram. Int.*, **31** (2005) 21-25.
[12]A. Y. Borisevich, P. K. Davies, Microwave Dielectric Properties of $Li_{1+x-y}M_{1-x-3y}Ti_{x+4y}O_3$ ($M=Nb^{5+}$, Ta^{5+}) Solid Solutions, *J. Eur. Ceram. Soc.*, **21,** 1719-22 (2001).
[13]A. Y. Borisevich, P. K. Davies, Effect of V_2O_5 Doping on the Sintering and Dielectric Properties of *M*-Phase $Li_{1+x-y}Nb_{1-x-3y}Ti_{x+4y}O_3$ Ceramics, *J. Am. Ceram. Soc.*, **87,** 1047-52 (2004).
[14]D. H. Kang, K. C. Nam, H. J. Cha, Effect of Li_2O–V_2O_5 on the Low Temperature Sintering and Microwave Dielectric Properties of $Li_{1.0}Nb_{0.6}Ti_{0.5}O_3$ Ceramics, *J. Eur. Ceram. Soc.*, **26,** 2117-21 (2006).
[15]Q. Zeng, W. Li, J. L. Shi, J. K. Guo, Fabrication and Microwave Dielectric Properties of a New LTCC Ceramic Composite Based on Li_2O-Nb_2O_5-TiO_2 System, *Mater. Lett.*, **11,** 3203-6 (2006).
[16]Q. Zeng, W. Li, J. L. Shi, J. K. Guo, A New Microwave Dielectric Ceramic for LTCC Applications, *J. Am. Ceram. Soc.*, **89,** 1733-35 (2006).
[17]B.W. Hakki, P.D. Coleman, A Dielectric Resonator Method of Measuring Inductive Capacities in the Millimeter Range, *IRE Trans. MTT.*, **8,** 402-410 (1960).
[18]W.E. Courtney, Analysis and Evaluation of a Method of Measuring the Complex Permittivity and Permeability Microwave Insulators, *IEEE Trans. MTT.*, **18,** 476-485 (1970).
[19]M. E. Villafuerte-Castrejón, A. Aragón-Piña, R. Valenzuela, A. R. West, Compound and Solid-solution Formation in the System Li_2O-Nb_2O_5-TiO_2, *J. Solid State Chem.*, **71,** 103-108 (1987).
[20]Q. Zeng, W. Li, J. L. Shi, J. K. Guo, in preparation.

PHOTOLUMINESCENCE PROPERTIES AND X-RAY PHOTOELECTRON SPECTROSCOPY OF ZnO MICROTUBES SYNTHESIZED BY AN AQUEOUS SOLUTION METHOD

Liwei Lin, Masayoshi Fuji*, Hideo Watanabe and Minoru Takahashi
Ceramics Research Laboratory, Nagoya Institute of Technology
10-6-29, Asahigaoka, Tajimi, 507-0071, Japan
*Email: Fuji@nitech.ac.jp

ABSTRACT

The effects of annealing treatment on the photoluminescence (PL) characteristics of ZnO microtibes, which have been synthesized by introducing ammonia water into zinc chloride aqueous solution, were investigated using a spectrofluorometer. X-ray photoelectron spectroscopy (XPS) is useful for investigating the bonding state of zinc and oxygen by analyzing the energy of photoelectrons. The results in this investigation revealed that the luminescence mechanism was associated primarily with oxygen vacancies. The PL results demonstrated that the central intensity of light emission shifted from orange (582 nm) to red (612 nm). XPS results showed that the concentration of oxygen vacancies decreased after the annealing treatment at 600 °C for 2 h under an ambient atmosphere.

INTRODUCTION

Zinc oxide, as a wurtzite-type semiconductor, has a wide band-gap of 3.37 eV at room temperature and a large exaction binding energy of 60 eV, and thus would have a good number of applications in various fields such as solar cells, photodetectors and luminescence devices.[1-3] The green and blue emissions have been frequently reported, which is caused by defects such as oxygen vacancies, zinc vacancies, interstitial zinc and antisite oxygen.[4-6] There are few literatures about the orange-red emission of ZnO microtubes.

We have developed a novel method for synthesis of hexagonal ZnO microtubes.[7-9] In this paper, the luminescence characteristics of the ZnO microtubes were investigated based on the analyses of XPS. The surface features, such as oxygen vacancies, surface bonding and surface adsorption, will have significant effects on the luminescence properties. The orange emission and red emissions of the ZnO microtubes are found, although the intensity of emission peaks is weak.

EXPERIMENTAL SECTIONS

The preparation of ZnO hexagonal microtubes was synthesized by introducing ammonia water into the conical glass flask with the aqueous solution (400ml) of zinc chloride (purity 98 % from Wako, Japan), of which zinc ionic concentration was 0.5 M. The temperature was controlled by the oil bath at 90 °C under stirring. When the pH arrived 7.5, the white precipitates in the reaction solution were filtrated and used for further examination after drying at 90 °C. The annealing treatment of ZnO tubes were carried out using an Infrared Image Furnace (MS-E44-AN, ULVAC SINKU-RIKO, Yokohama, Japan) for 2 h in air. The increasing and decreasing rates of the temperature were set at 10 °C/min, which was controlled by a temperature program controller (TPC-1000, ULVAC, Japan). Morphology of the products was examined by field microscopy scanning electron microscope (SEM, JSM-7000F JEOL, Japan). The photoluminescence characteristics of ZnO tubes were investigated by photoluminescence measurement using a spectrofluorometer (JASCO FP-6500) with an excitation wavelength of 340 nm. The phase composition of the ZnO particles was investigated by X-ray photoelectron spectroscopy (XPS) on a SSX-100 (Surface Science Instrument) instrument. A monochromatic Al $K\alpha$ x-ray radiation was used as an excitation source in the XPS measurement. The energy resolution of the instrument at 20 eV of the excitation energy is 0.05 eV as estimated from the full width at half maximum of the XPS Ag $3d_{5/2}$ of a pure silver target. The emitted photoelectrons were detected using a hemispherical analyzer at pass energy of 20 eV for the C 1s high resolution XPS peaks.

RESULTS AND DISCUSSION

Figure 1 is photoluminescence spectrum of ZnO microtubes without annealing. The PL spectrum possesses a visible emission band. The visible emission centered at about 582 nm ranging from 400 nm to 660 nm is associated with oxygen vacancies. The broadening of the visible emission can be ascribed to abundant surface defects. A large quantity of defects and impurities on the surface of ZnO microtubes can provide new states which can contribute to visible luminescence centers and broaden the visible emission band.[10]

Figure 2 shows PL spectrum of ZnO microtubes annealed for 2 h at 600 °C in air. It is discernible that the annealed ZnO microtubes have a unique redshift in the emission centered at 612 nm. When ZnO microtubes were annealed in air, the number of oxygen vacancies in ZnO microtubes would decrease. From XPS results shown in Table 1, zinc composition of ZnO microtubes decreases from 72.755 % to 41.613 % after annealing treatment, while oxygen composition of ZnO microtubes increases from 27.246 % to 58.388 %. This means that the ZnO microtubes turn into oxygen-rich particles from zinc-rich particles. As the result, the oxygen vacancy-related visible emission intensity decrease, leading to the visible emission center position shifting from 582 nm to 612 nm. The mechanism of the redshift in the emission of ZnO microtubes is attributed to the oxygen vacancies and zinc vacancies. On the other hand, in Figure 3, the SEM images show that the annealed ZnO microtubes have smooth surface, but almost all of ZnO tubes are broken.

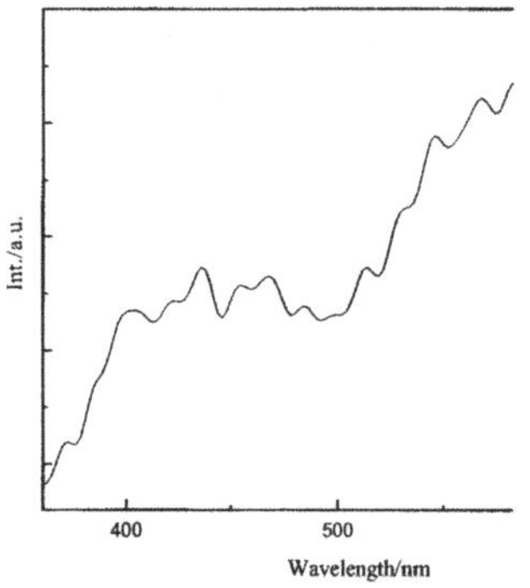

Figure 1. room temperature PL spectrum of ZnO microtubes without annealing

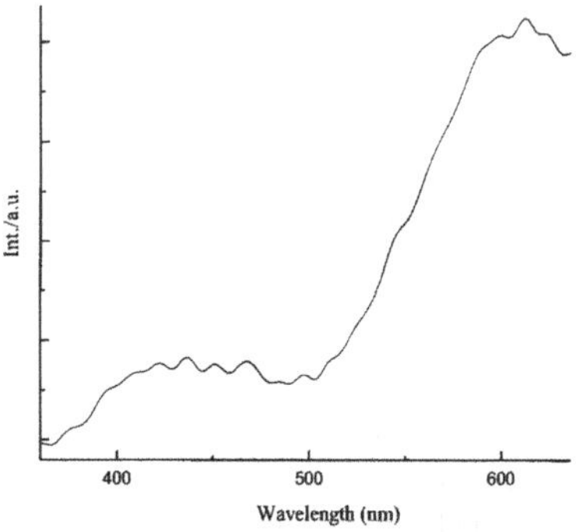

Figure 2. room temperature PL spectrum of ZnO microtubes annealed for 2 h at 600 °C in air

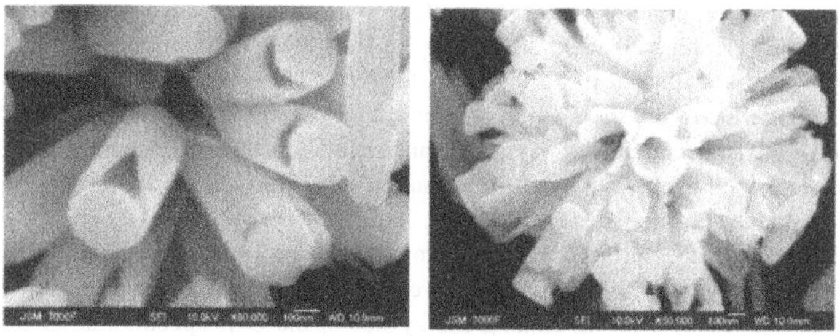

Figure 3. SEM images of ZnO tubes with annealing treatment

Table 1 Composition of ZnO microtubes from XPS analysis before and after annealing treatment

	Without annealing	Annealed at 600 °C for 2 h
Zn atom percentage (%)	72.755	41.613
O atom percentage (%)	27.246	58.388
Zn/O	2.67	0.71

Figure 4 shows a typical wide-scan spectrum of as-synthesized ZnO microtubes. The photoelectron peaks of the main elements, Zn and O, are obtained. No other impurities are detectable in the XPS spectrum. From the statistical results of XPS, the atom ratio of zinc to oxygen is about 2.67 for ZnO microtubes, which shows that ZnO microtubes are zinc-rich, that is to say, there are many oxygen vacancies localizing in the surface of ZnO microtubes.

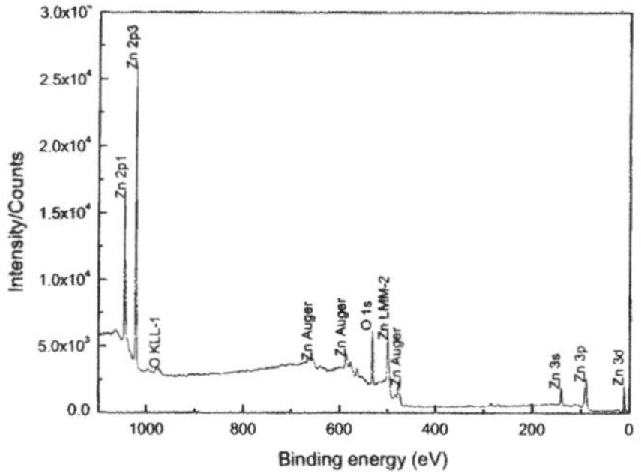

Figure 4. The survey scan for XPS spectrum of ZnO microtubes

The XPS spectra of ZnO microtubes exhibit a main core level binding energy of Zn-$2p_{3/2}$ located at 1022.066 eV, as shown in Figure 5. And it agrees well with NIST XPS database.[11] It is of interest to note that there is a binding energy peak at 1021.859 eV. The origin for the appearance of this peak remains unclear. It may be related to the oxygen vacancies that tend to form on the surface layer initially to compromise the surface shrinkage. Therefore some of surface zinc ions would acquire an extra terminated bond, creating a new binding energy of 1021.859 eV. Another consideration is the sampling depth which is estimated to be 0.5-5 nm for XPS. This implies that considerable XPS detection is, in fact, derived from the interior zinc ions, which are no longer fully coordinated with oxygen ions due to the an occurrence of oxygen vacancies in a few layers beneath the surface. As a result, these zinc ions with the broken bonds as a result of oxygen vacancies are responsible for the binding energy peak to be positioned at 1021.859 eV.

The typical O 1s peaks in the surface of ZnO microtubes can be consistently fitted by Gaussian, which contains a strong peak at 530.827 eV and a weak peak at 532.081 eV, as shown in Figure 6. The component on the low and strong binding energy side of the O 1s spectrum at 530.827 eV is attributed to O^{2-} ions on the wurtzite structure of hexagonal Zn^{2+} ion array in the Zn-O bonds. The peak related to the high binding energy around 532.081 eV is attributed to the presence of chemisorbed or dissociated oxygen and OH species on the surface of ZnO microtubes.

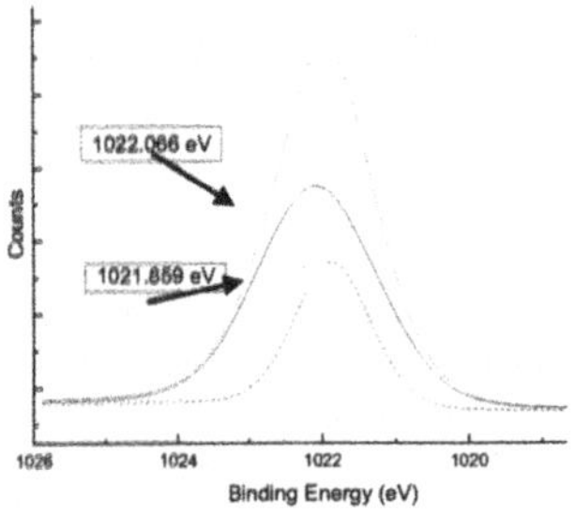

Figure 5. Zinc 2p3 spectra of ZnO microtubes

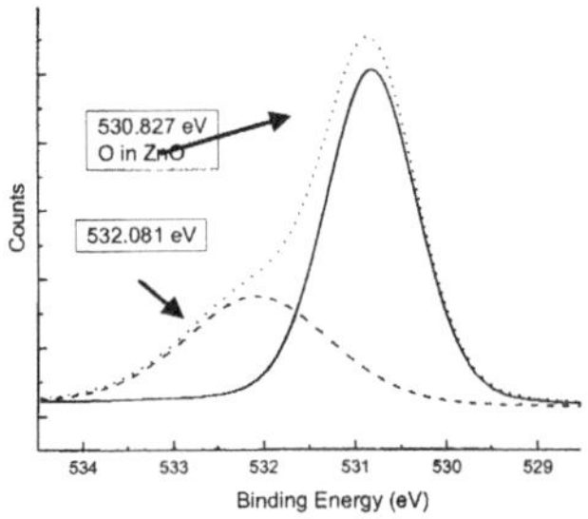

Figure 6. Oxygen 1s spectra of ZnO microtubes

CONCLUSIONS

The chemical composition of ZnO microtubes, based on XPS results, reveals that the concentration of oxygen increase after annealing treatment, that is to say, the oxygen vacancies decrease. The oxygen vacancy-related visible emission intensity decreases, while the surface defect-related visible emission intensity increases, leading to the visible emission center position shifting from 582 nm to 612 nm. From XPS analysis, the new zinc binding energy of 1021.859 eV is associated with the broken bonds per zinc, which generate the oxygen vacancies.

REFERENCES

[1]J. Tornow and K. Schwarzburg, Transient electrical response of dye-sensitized ZnO nanorod solar cells, *J. Phys. Chem. C*, 111 [24], 8692-8698 (2007).

[2]S.J. Young, L.W. Ji, S.J. Chang, S.H. Liang, K.T. Lam, T.H. Fang, K.J. Chen, X.L. Du and Q.K. Xue, ZnO-based MIS photodetectors, *Sensors Actuat. A-Phys.*, 141 [1], 225-229 (2008).

[3]H. Lim, D. Lee and Y. Oh, Gas sensing properties of ZnO thin films prepared by microcontact printing, *Sensors Actuat. A-Phys.*, 125[2], 405-410 (2006).

[4]D H Zhang, Z Y Xue and Q P Wang, The mechanisms of blue emission from ZnO films deposited on glass substrate by r.f. magnetron sputtering, *J. Phys. D: Appl. Phys.*, 35 [21], 2837-2840 (2002).

[5]S.M. Abrarov, Sh.U. Yuldashev, T.W. Kim, S.B. Lee, H.Y. Kwon and T.W. Kang, Dominant ultraviolet-blue photoluminescence of ZnO embedded into synthetic opal, *J.Lumin.*, 114 [2], 118-124 (2005).

[6]Q. P. Wang, D. H. Zhang, Z. Y. Xue and X. J. Zhang, Mechanisms of green emission from ZnO films prepared by rf magnetron sputtering, *Opt. Mater.*, 26 [1], 23-26 (2004).

[7]L.LIN, Y. HAN, M. FUJI, T. ENDO, X. Wang and M. TAKAHASHI, Synthesis of Hexagonal ZnO Microtubes by a Simple Soft Aqueous Solution Method, *J. Ceram. Soc. JPN*, 116[2], 198-200 (2008).

[8]Y. Han, L. Lin, M. Fuji, and M. Takahashi, A Novel One-step Solution Approach to Synthesize Tubular ZnO Nanostructures, *Chem. Lett.*, 36, 8, 1002-1003 (2007).

[9]L. Lin, H. Watanabe, M. Fuji, T. Endo, S. Yamashita and M. Takahashi, Synthesis of ZnO Microtubes by a Facile Aqueous Solution Process, *J. Am. Ceram. Soc.* (In press)

[10]Y. Tong, Y. Liu, C. Shao, Y. Liu, C. Xu, J. Zhang,Y. Lu, D. Shen and X. Fan, Growth and Optical Properties of Faceted Hexagonal ZnO Nanotubes, *J. Phys. Chem. B*, 110, 14714-14718 (2006).

[11]Tery L. Barr, Mengping Yin and Shikha Varma, Detailed x-ray photoelectron spectroscopy valence band and core level studied of select metals oxidations, *J. Vac. Sci. Technol. A*, 10, 2383-2390 (1992).

[illegible]

THE DYNAMICS OF WATER MOLECULES ON YVO_4 PHOTO-CATALYST SURFACE

Mitsutake Oshikiri, Akiyuki Matsushita, and Jinhua Ye
National Institute for Materials Science
Tsukuba, Ibaraki, 305-0041, Japan

Mauro Boero
University of Tsukuba, Japan.
Tsukuba, Ibaraki, 305-8577, Japan

ABSTRACT

The dynamics of water molecules and the adsorption properties on the surface of YVO_4, a photo-catalyst, were investigated using the first principles molecular dynamics simulation. This system has shown excellent performance in the production of both hydrogen and oxygen in the ultraviolet region when using a co-catalyst, NiO_x; however, its water molecule adsorption properties are poorly understood. Here we show that imperfectly oxygen coordinated V sites (i.e., not four-fold oxygen coordinated vanadium but three-fold oxygen coordinated vanadium) exposed on the catalyst's surface play a central role in the dissociation of water molecules. By simulating the H_2O adsorption process, we were able to infer that the dissociation of water at these imperfectly oxygen coordinated V sites promotes proton reduction, triggering H_2 generation. On the other hand, the imperfectly oxygen coordinated Y sites cannot dissociate H_2O molecules: they simply cause stable non-dissociative H_2O adsorption, as in the case of the imperfectly oxygen coordinated Bi sites in the $BiVO_4$ photo-catalyst system.

INTRODUCTION

To date, TiO_2-based materials have been the most extensively studied for use as photo-catalysts. However, the efficiency of oxygen and hydrogen production from water by sunlight irradiation is still lower than that of electrolysis using solar cells. The photocatalytic properties of numerous metal oxides other than TiO_2, including material groups based on vanadium oxide, have been explored in attempts to overcome this difficulty. $BiVO_4$, $InVO_4$ and YVO_4 are typical examples. It is known that $BiVO_4$ (monoclinic structure) can produce oxygen up to approximately 520 nm by using a sacrificial reagent, $AgNO_3$, but no hydrogen generation has yet been reported [1]. $InVO_4$ shows hydrogen evolution in the visible wavelength range (from ultraviolet (UV) to ~ 600 nm), even from pure water [2]; however, regrettably, oxygen cannot be produced in any way. On the other hand, YVO_4 has shown surprisingly high efficiencies in both O_2 and H_2 production with a co-catalyst, NiO_x, although the activity is unfortunately limited to just the UV region [3].

These interesting contrastive catalytic properties prompted our previous investigation of the details of water molecule adsorption behaviors on the surface of $BiVO_4$, using a first principles molecular dynamics simulation [1]. It is demonstrated there that the Bi site is not active in water molecule dissociation. Aiming to gain a comprehensive understanding of the water molecule adsorption process in those materials, we have carried out in this study a continued investigation of the role of the V and Y sites of YVO_4 in the adsorption process.

STRUCTURAL PROPERTIES OF YVO_4 CRYSTAL

A schematic crystal structure of YVO_4 using polyhedral representation is shown in Figure 1. The crystal structure is the zircon-type, in the space group $I4_1/amd$, and the conventional cell parameters a, b and c are 7.12 Å, 7.12 Å and 6.29 Å, respectively. The more detailed crystal parameters of YVO_4 can be found in reference [4]. The YVO_4 crystal includes two kinds of polyhedra: a VO_4 tetrahedron and a YO_8 polyhedron. Each V site is surrounded by four oxygen atoms (four oxygen-coordinated vanadium: 4c-V) with an atomic distance of 1.71 Å between the V and the O. Each Y is surrounded by eight oxygen atoms (8c-Y) with a Y-O distance of either 2.29 Å (for four of the eight Y-O bonds) or 2.44 Å (for the other Y-O bonds). The shortest V-O distance is 1.7 Å and that of Y-O is about 2.3 Å. The shortest V-V, Y-Y, O-O, and V-Y distances are about 3.9 Å, 3.9 Å, 2.6 Å and 3.1 Å, respectively. This YVO_4 crystal structure is also characterized by the feature that every VO_4 tetrahedron is isolated by a YO_8 polyhedron, whereas every YO_8 polyhedron is linked to a neighboring YO_8 polyhedron by a shared edge.

COMPUTATIONAL DETAILS

First principles dynamic simulations were performed within the Car-Parrinello scheme (CPMD) [5] using a Becke-Lee-Yang-Parr (BLYP) gradient-corrected approach [6]. The valence-core interaction was taken into account via norm-conserving Troullier-Martins pseudopotentials [7] for the V, Y, and O atoms. For H, a Car-von Barth BLYP pseudopotential was used. In the cases of V and Y, the use of semi-core states was needed to obtain a good description of both the geometry and the energetics. The electrons of V 3s, 3p, 3d, 4s; Y 4s, 4p, 4d, 5s; O 2s, 2p; and H 1s were included in the valence electrons. Valence wave functions were expanded as plane waves with an energy cut-off of 80 Ry. A fictitious electronic mass of 1200 a.u. and an integration step of 5.0 a.u. ensured good control of the conserved quantities. The surface was represented by a slab using a super cell, whose bottom layers were kept fixed to the bulk crystal, while the rest of the structure was fully relaxed. An empty space with thickness greater than 10 Å was prepared above the relaxed crystal surface. Stoichiometric, non-charged systems were employed. The temperature of ionic dynamics was controlled by a velocity-rescaling algorithm. In the present study, we focused on the (010) surface of YVO_4.

The super cell size is 2a × (a + empty space) × 2c = 14.23 Å × (7.12 + 11.39) Å × 12.57 Å (= 3.31 nm^3) and the size of the simulated surface, obtained by cleaving at the broken line in Figure 1, is 2a × 2c = 178.9 $Å^2$, roughly identical to the size of one mono-molecular H_2O layer including 18.6 H_2O molecules at 1 atom at 300 K. In this study, a slab also including one 6c-Y site and one 3c-V site, in addition to the 7c-Y and 4c-V sites, was prepared on the surface. This model was created by shifting one O atom on the cleavage surface to the opposite side of the slab by applying crystallographic translation along the (0, b, 0) vector, typical of a conventional unit cell of the bulk crystal. The super cell had sixteen Y, sixteen V, and sixty-four O atoms plus 35 H_2O molecules.

RESULTS AND DISCUSSION

During the thermal equilibrium state at 300 K, a few H_2O molecules approached the 7c-Y (Y_1, 6, 7), 6c_Y (Y_5), and 3c-V (V_25) sites (see Figure 2. (a) and (b)). As far as the 7c-Y and 6c-Y sites are concerned, we could not observe any approach of water molecules to distances shorter than about 2.17 Å. However, toward the final stage of our simulation, two water molecules, whose O atoms are labeled as O_106 and O_124 in Figure 2 (a), approached the 7c-Y catalytic sites labeled

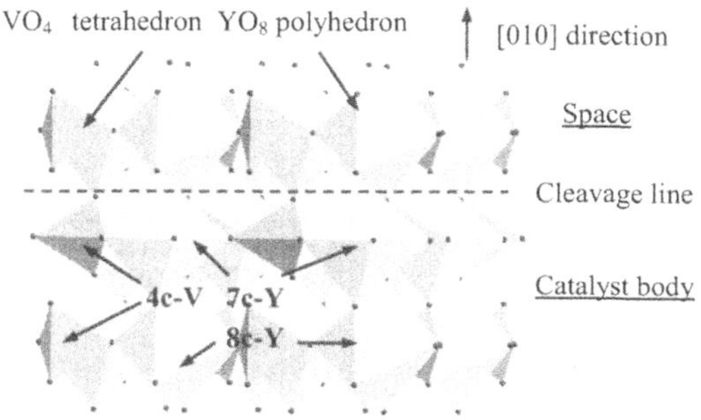

Figure 1. Crystal structure of YVO_4 and the model surface for the simulation.

as Y_1 and Y_6 at distances of about 2.5 Å; these structures are likely to be stable and the adsorption energy was roughly estimated to be about 0.6 eV / molecule, although the H_2O molecules are undissociated.

On our picosecond time scale simulation, a water molecule turned out to be attracted to the 3c-V (V_25) site and underwent dissociation. The snapshot of Figure 2 (b) shows the beginning of the dissociation process with the water molecule already bound to the catalytic site V_25. The time evolutions of the interatomic distances O_116 – V_25 and O_116 – Y_5 are reported in Figure 3 (a) and the red broken line indicates the instant (0.52 ps) in which the dissociation occurred. After dissociation, the average O_116 – V_25 bond length is 1.78 Å and displays small and regular fluctuations typical of a stretching mode. The dissociative adsorption energy was about 2 eV /

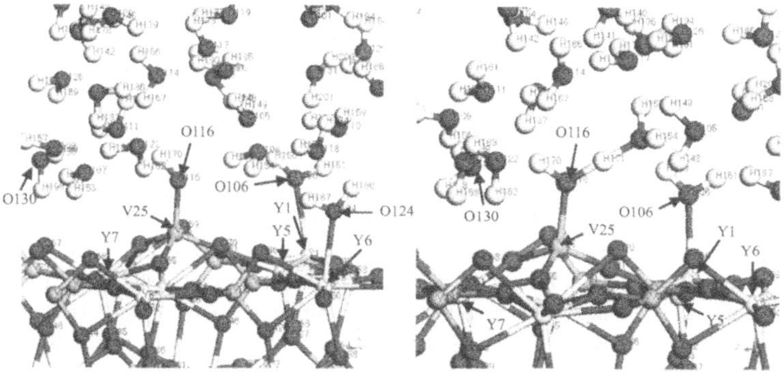

(a) (b)

Figure 2. (a) Snapshot of the catalyst surface with some adsorbed water molecules after one ps molecular dynamics at 300 K. One water molecule is dissociated and the hydroxyl group H_170-O_116- originating from the H_2O molecule is adsorbed onto the V_25 site. (b) Snapshot of the catalyst surface during the water molecule dissociation at the 3c-V site.

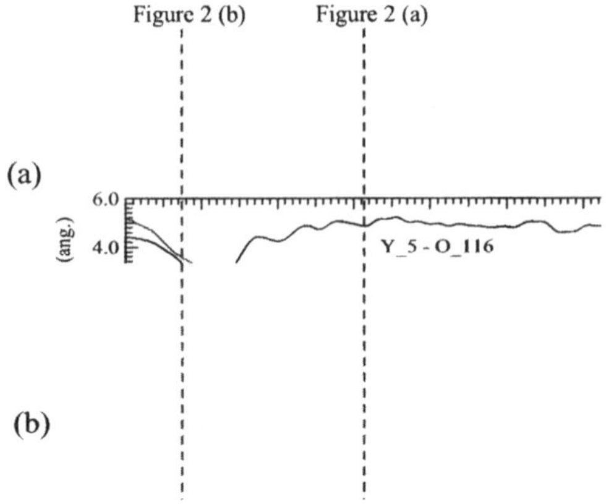

Figure 3. Time evolutions of the main atomic distances during the dynamics at 300 K for the system shown in Figure 2 (a) and (b). (a) Distances between the O atom belonging to the water molecule H_170-O_116-H_171 and the V_25 or Y_5 sites. The red broken line indicates the instant at which the water molecule dissociation occurs. (b) Distances of the O atoms of water molecules approaching the exposed Y sites.

molecule. As expected, the O atoms of the H_2O molecules never come close to V sites if these were coordinated by four O atoms, but water molecules can point H atoms towards the O atoms of a VO_4 tetrahedron, if these O atoms are facing the surface, and form hydrogen bonds of $O_{sub}\cdots H_{water}$.

From the point of view of the electronic structure, the wave functions of the unoccupied states corresponding to the configuration of Figure 2 (b) are noteworthy. When the water molecule starts the dissociation process, a wave function projection shows that the weights of the H_1s orbital components of hydrogens belonging to the H_2O molecule undergoing the dissociation (i.e., H_170 and H_171) are approximately 1.1 % and 0.7 %, respectively, in the wave function of the lowest unoccupied state. These percentages, although small, are much larger than the H_1s components of any other water molecule, which never exceed $10^{-2} \sim 10^{-4}$ %. The other major components of the lowest unoccupied state are V_3d 64 %, O_2p 25 %, and Y_4d 5.5 %.

By looking at the various contributions to the unoccupied and occupied electronic states, the transition moment $< \Phi_{\text{unoccupied state i}}(\mathbf{r}) \mid \boldsymbol{r} \mid \Phi_{\text{occupied state j}}(\mathbf{r}) >$ in the system, including the YVO_4 slab and H_2O molecules, can be roughly split into the various contributions as $< \Sigma\, \alpha_i \phi_{H\_1s}(\mathbf{r}-\mathbf{R}_{H\_i}) + \Sigma\, \beta_i \phi_{V\_3d}(\mathbf{r}-\mathbf{R}_{V\_i}) + \Sigma\, \gamma_i \phi_{Y\_4d}(\mathbf{r}-\mathbf{R}_{Y\_i}) + \Sigma\, \lambda_i \phi_{O\_2p}(\mathbf{r}-\mathbf{R}_{O\_i}) \mid \boldsymbol{r} \mid \Sigma\, \eta_i \phi_{O\_2p}(\mathbf{r}-\mathbf{R}_{O\_i}) >$. Here, $\mathbf{R}_{X\_i}$ means the position of the atom number i belonging to the chemical species X, while $\alpha_i, \beta_i, \gamma_i$ ... are the coefficients of linear expansion. Considering the component $< \phi_{H\_1s}(\mathbf{r}-\mathbf{R}_{H\_i})$ of $\Phi_{\text{unoccupied state i}}(\mathbf{r}) \mid \boldsymbol{r} \mid \Phi_{\text{occupied state j}}(\mathbf{r}) >$, the configuration corresponding to the dissociation process is crucial to understanding whether or not the proton is able to catch the photo-excited electron; in fact, the H_1s component in the unoccupied states increases and the atomic distance between H and O_{slab} becomes shorter when a water molecule is dissociated. As a consequence, the term $< \phi_{H\_1s}(\mathbf{r}-\mathbf{R}_{H\_i})$

of $\Phi_{\text{unoccupied state } i}(\mathbf{r}) \mid \boldsymbol{r} \mid \Phi_{\text{occupied state } j}(\mathbf{r}) >$ is expected to increase.

However, since most of the photo-excited electrons are most likely transferred to V sites, they should be transferred to the H sites, minimizing any possible source of loss, particularly during the proto-reduction process. This might be realized, for instance, by a co-catalyst. However, this goes beyond the scope of the present study and will be taken on in a forthcoming research project.

CONCLUSION

Our studies here have revealed that two requirements must be satisfied when developing a high-performance catalyst. Firstly, the system needs to be designed to bring water molecules close to the catalyst; and secondly, imperfectly oxygen-coordinated cation sites (which do not result in a net oxygen deficit in the system) need to be securely located on the surface of the catalyst.

REFERENCES

[1]A. Kudo, K. Ueda, H. Kato, I. Mikami, Catalysis Lett. **53**, 229 (1998); M. Oshikiri, M. Boero, J. Phys. Chem. B **110**, 1988 (2006)

[2]J. Ye, Z. Zou, M. Oshikiri, A. Matsusita, M. Shimoda, M. Imai, and T. Shishido, Chem. Phys. Lett. **356,** 221 (2002); M. Oshikiri, M. Boero, J. Ye, Z. Zou, and G. Kido, J. Chem. Phys. **117,** 7313 (2002).

[3]J. Ye, Z. Zou, M. Oshikiri, T. Shishido, Materials Science Forum, **423-425**, 825 (2003)

[4]J. A. Baglio, G.. Gashurov, Acta Cryst. **B24,** 292 (1968)

[5]R. Car, and M. Parrinello, Phys. Rev. Lett., **55**, 2471 (1985); CPMD, Copyright IBM Corp. 1990-2001, Copyright MPI für FKF, Stuttgart, 1997-2004.

[6]A. D. Becke, Phys. Rev. A **38**, 3098 (1988); C. Lee, W. Yang and R. G. Parr, Phys. Rev. B **37**, 785 (1988)

[7]N. Troullier and J. L. Martins, Phys. Rev. B **43**, 1993 (1982)

PREPARATION OF SILICON CARBIDE HOLLOW SPHERES BY A TEMPLATE METHOD

Lei ZHANG[1], Jiu-jun YANG[1,2], Xue-ping WANG[1], Feng-chun WEI[1]
[1] School of Materials Science and Engineering, Zhengzhou University. [2] Department of Materials Engineering, Tianjin Institute of Urban Construction.
[1]Zhengzhou, He Nan, 450052, China. [2]Tianjin, 300384, China.

ABSTRACT

Hollow silicon carbide (SiC) spheres have been synthesized by a microwave heating and carbothermal reduction method with carbon spheres as template and fly ash (a solid waste from coal-fired power plant) as silica source. X-ray diffraction and scanning electron microscope were employed to characterize the morphology, structure of the products. The results show that hollow spheres prepared at 1300 ℃ under argon atmosphere have a hollow core and SiC shell structure. The shell of a hollow SiC sphere is composed of a lot of irregular SiC nanowires with 5-20 μm in length and 50-500 nm in diameter which belongs to the β-SiC. Moreover, the formation mechanism of the hollow SiC spheres is also discussed.

1. INTRODUCTION

Materials with hollow sphere structures have attracted much attention due to their excellent physical and chemical properties, and they are technologically important for a variety of application such as optics, electrics, magnetics, catalysis, acoustics, catalyst support, etc. [1-4].

Silicon carbide (SiC) is an important carbide for its high mechanic strength, good thermal conductivity and chemical stability [5]. Hollow SiC spheres with high surface area used as catalyst support can enhance its catalytic performance and maybe an alternative traditional material applied in harsh environments. Therefore synthesizing hollow SiC spheres is an important step for further application. Chen [6] reported a solid–gas reaction approach to fabricate SiC hollow spheres and the sphere size can be controlled from the microscale to nanoscale. Qian [7] reported the synthesis of hollow SiC nanospheres by the reaction between $SiCl_4$, C_6Cl_6 and sodium. Kim [8] obtained ordered assemblies of hollow SiC nanospheres and filled SiCN nanospheres by infiltrating three-dimensional ordered macroporous (3DOM) carbon as a sacrificial template with low molecular weight pre-ceramic polymers. Shi [9] synthesized the SiC hollow spheres with a high surface area by carbothermal reduction technique using SiO_2@ PPy as a template. Guo [10] employed strongly acidic ion exchange resin spheres exchanged by Fe^{3+} cation as a template and then obtained a core-shell structured α-Fe_2O_3/SiC composite sphere by the carbothermal reduction of the resin spheres and commercial silica. Though the synthesis of hollow SiC spheres was reported in many literatures, using fly ash as raw materials to prepare hollow SiC sphere has not been reported up until now.

Fly ash is a by-product of coal combustion in power stations all over the world, which are mainly composed of SiO_2 and Al_2O_3. The stockpiling of fly ash will bring about serious environmental pollution and economic problems associated with its disposal. Results from investigations [11-13] show that fly ash has good potential and promising applications in the ceramics area, such as synthesis of SiAlON, preparation of SiC powder and glass–ceramics.

In the present study, we successfully prepared hollow SiC spheres by carbothermal reduction reaction between fly ash and carbon black, the morphology and structure of hollow SiC spheres were characterized. Moreover, the formation mechanism of the hollow SiC spheres is also discussed.

2. EXPERIMENTAL PROCEDURE

2.1 Starting materials

Fly ash used in this paper were bought from The Second Long-Gang power plants (Yuzhou, He Nan), containing as main chemical components SiO_2 (61.70 wt %), Al_2O_3 (25.64 wt %), Fe_2O_3 (6.13 wt %) and CaO (2.84 wt %). SEM image (Fig.1) shows that fly ash is mainly microspheres with size ranges from 1 μm to 40 μm. The XRD pattern of fly ash is shown in Fig. 2. All sharp peaks can be identified to either quartz (SiO_2) or mullite ($3Al_2O_3 \cdot 2SiO_2$). It also shows an obvious broad hump in the background between approximately 2θ =18° and 30°, which results from the presence of amorphous materials.

Figure 1. SEM image of fly ash

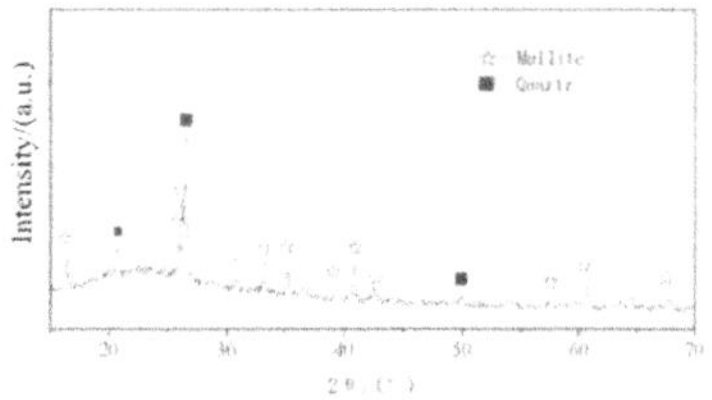

Figure 2. XRD pattern of fly ash

2.2 Preparation of hollow SiC spheres

The procedure of synthesis of macroscopic SiC hollow spheres is as following: (1) Carbon black is passed over 0.2mm sieves to obtain carbon spheres with a diameter of more than 0.2 mm, which formed during the screening through physical adsorption effect. Carbon spheres used not only as macroscopic templates but also as microwave absorption materials to accelerate the heating rate. (2) Fly ash were mixed with carbon spheres (n(C): n (SiO_2) = 4.2), then the obtained mixture was placed in a MW–L0316V microwave furnace to perform carbothermal reduction reaction under argon atmosphere. It was then heated to 1300℃-1400℃ for 0.5 h with the heating rate of 10 ℃/min. (3)The products were calcined at 600 ℃ for 4 h in the air to eliminate the residual carbon .

2.3 Characterization

X-ray diffraction (XRD) patterns were measured by a PHILIPS X'Pert Pro. diffractometer using Cu Ka radiation (λ = 0.154 nm). The morphologies of all mentioned samples were observed by a field emission scanning electron microscope (SEM, JSM- 6700F).

3. RESULTS AND DISCUSSION

3.1 Structure of hollow SiC spheres

Under 1300~1400 ℃, all prepared samples are a mixture of spheres and fine powder. The fine powder can be easily separated from spheres by sieving. The XRD patterns of spheres and fine powder obtained at 1300 ℃, namely (a) and (b) are given in Fig. 3. All of the four peaks with *d* value of 2.516, 2.179, 2.154 and 1.314 in Fig. 3(a) are indexed to the crystalline facets of β-SiC [(111), (200), (220) and (311), respectively], indicating that spheres are almost single SiC phase. Fig. 3(b) shows that the main phases of powder are SiC, Al_2O_3 and mullite. SiC and Al_2O_3 are products of carbothermal reduction reaction, while mullite phase results from the unreacted fly ash. After eliminating the residual carbon black, the SEM image of hollow SiC spheres obtained by microwave heating (1300 ℃, 0.5 h) is shown in Fig 4. It can be seen that the spheres grow well and the diameter ranges from 200 to 1000 μm. Some broken spheres are shown in Fig.4 (a) and (b) reveal their hollow structure. Fig.4 (b) also shows that the hollow sphere has a shell thickness of

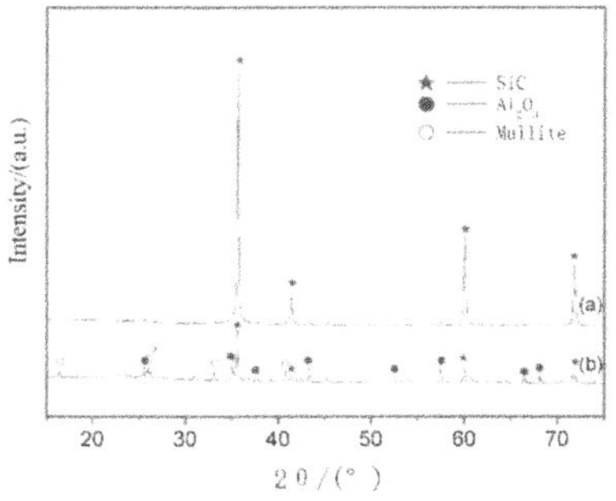

Figure3. XRD patterns of products; (a) spheres, (b) powder.

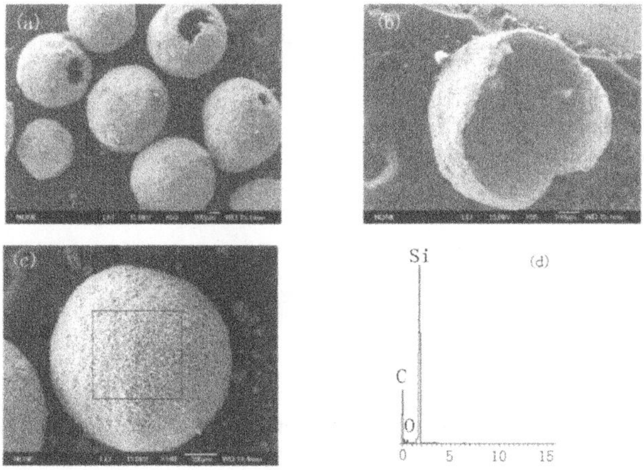

Figure 4. SEM images of hollow SiC spheres; (a) hollow SiC spheres, (b) a broken sphere with obvious hollow structure, (c) a complete SiC sphere, (d) EDS spectrum on selected section of sphere.

about 20 μm, almost 1/40 of its diameter. From the EDS spectrum (Fig.4(d)), the hollow sphere is mainly composed of the silicon, carbon and oxygen, which is the SiC composition and the existence of oxygen may result from a few of SiC oxidized during the calcined process at 600 ℃ in the air.

The microstructure of the hollow SiC spheres is shown in Fig. 5. From Fig. 5(a) and (b), it can be seen that the outside surface and inner side of a hollow SiC sphere is rough and consists of a lot of irregular SiC nanowires with length of 5~20 μm and diameter of 50-500 nm. Most of these nanowires are entangled with each other to form the shell of hollow SiC spheres. The microstructure of the shell is shown in Fig. 5(c) and (d), the selected section shows that the shell also consists of twisted SiC nanowires just like the outside surface and inner side of hollow SiC spheres.

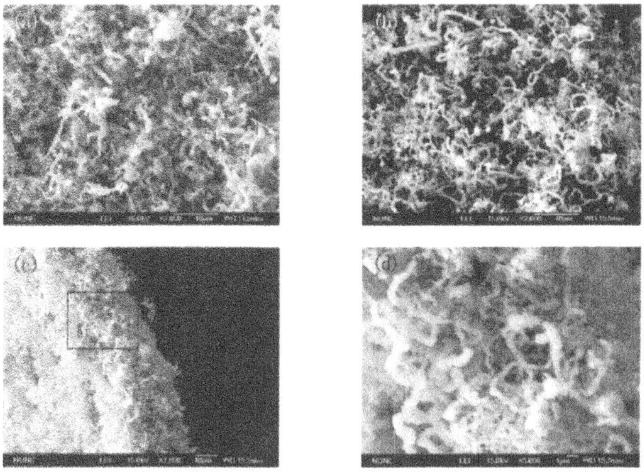

Figure 5. SEM images of the microstructure of hollow SiC sphere; (a) outside surface, (b) inner side, (c) shell and (d) selected section of shell.

3.2 Formation mechanism of hollow SiC spheres

A probable mechanism for the formation of such hollow SiC spheres is presented in Scheme 1

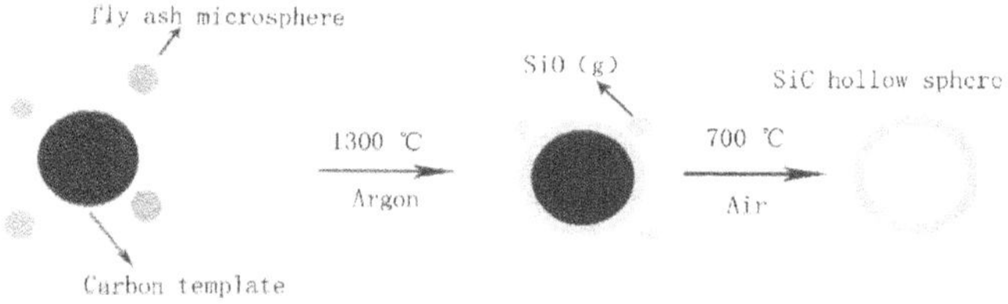

Scheme 1 Schematic mechanism for the formation of hollow SiC spheres.

It is generally agreed that SiC is produced by the reaction [14, 15]:

$$SiO_2 + 3C \rightarrow SiC + 2CO \qquad (1)$$

This is an overall reaction, and it usually includes following reactions:

$$SiO_2 + C \rightarrow SiO + CO \quad (2)$$

$$SiO + 3CO \rightarrow SiC + 2CO_2 \quad (3)$$

$$SiO + 2C \rightarrow SiC + CO \quad (4)$$

Generally, the reaction (3) produces SiC whiskers and nanowires[15,16], while the reaction (4) gives SiC with morphology of that of the carbon, namely shape-memory synthesis [17]. Due to the formation of a large quantity of SiC nanowires in the shell of the hollow SiC sphere, we believe that the SiC is formed through the reaction (3). The reason for SiC nanowires not growing along the <111> direction but forming twisted wires may be ascribed to the fluctuation of supersaturation, which would disturb the balance of the growth. As the reaction progressed, there is a constant consummation of SiO_2, and then mullite is involved to the reaction (reaction (5)). Thus, the sphericity of fly ash microsphere is destroyed.

$$3Al_2O_3 \cdot 2SiO_2\,(s) + 2C \rightarrow 3Al_2O_3\,(s) + 2SiO\,(s) + 2CO\,(g) \quad (5)$$

In fact, the formation of the SiC hollow sphere is far more complicated than we described, and it should be studied further.

4. CONCLUSION

In summary, we have presented a simple approach to synthesize hollow SiC spheres. The size of the hollow spheres depending on the size of the template carbon spheres. The shell of the spheres consists of a lot of twisted SiC nanowires with length of 5~20 μm and diameter of 50-500 nm, which is formed through a gas phase reaction. This technique presents a convenient method to synthesize hollow SiC spheres and an effective way to utilize fly ash.

ACKNOWLEDGMENT

This work was financially supported by National Natural Science Found of China under Grant No. 50472030 and 50772071.

REFERENCES

[1]S. Kidambi, J.h. Dai, Jin Li, and M.L. Bruening, Selective Hydrogenation by Pd Nanoparticles Embedded in Polyelectrolyte Multilayers, *J. Am. Chem. Soc.,* **126**, 2658 -2659(2004).

[2]J. Kim, S. Yoon, and J. Yu, Fabrication of nanocapsules with Au particles trapped inside carbon and silica nanoporous shells, *Chem. Commun.*, 790-791(2003).

[3]G Ibarz, L Daehne, E Donath, and H Moehwald, Smart micro-and nanocontainers for storage, transport, and release, *Adv. Mater.*, **13**, 1324-1327(2001).

[4]A Bourlinos, N Boukos, and D Petridis, Exchange resins in shape fabrication of hollow inorganic and carbonaceous inorganic composite spheres, *Adv. Mater.*, **14**, 21-24(2002)

[5]H.H.Ye, N. Titchenal, Y. Gogotsi, and F. Ko, SiC nanowires synthesized from electrospun nanofiber templates, *Adv. Mater.*, **17**, 1531–1535(2005).

[6]Yong Zhang, Er-Wei Shi, Zhi-Zhan Chen, Xiang-Biao Li, and Bing Xiao, Large-scale fabrication of silicon carbide hollow spheres, *J. Mater. Chem.*, **16**, 4141–4145(2006).

[7]G.Z. Shen, D. Chen, K.B. Tang, Y.T. Qian, and S.Y. Zhang, Silicon carbide hollow nanospheres, nanowires and coaxial nanowires, *Chem. Phys. Lett.,* **375**,177–184(2003).

[8]H.Wang, J.S. Yu, X.D. Li, and D.P. Kim, Inorganic polymer-derived hollow SiC and filled SiCN sphere assemblies from a 3DOM carbon template, *Chem. Commun.*, 2352–2353(2004).

[9]ZHUYu-Fang, FANG Ying, LIUYa-Yun and SHI Jian-Lin, Preparation of SiC Hollow Spheres from SiO2@PPy Core/Shell Structure, *Chem. J. Chinese. U.*, **27**, 23-25(2006).

[10]Xiangyang Wu, Guoqiang Jin, Lianxiu Guan, Hu Cao, and Xiang-Yun Guo, Preparation and characterization of core-shell structuredα-Fe2O3/SiC spheres, *Mat. Sci. Eng. A-Struct.*, **433**, 190-194(2006).

[11] Qi Qiu, Vladimir Hlavacek, and Svante Prochazka. Carbonitridation of Fly Ash. I. Synthesis of SiAlON-Based Materials. *Ind. Eng. Chem. Res.*, **44**, 2469-2476(2005).

[12] Wang hongjie, Wang yonglan, and Jin zhihao, SiC powders prepared from fly ash, *J. Mater. Process. Tech.*, **117**,52-55(2001).

[13] Soon-Do Yoon, and Yeon-Hum Yun, An advanced technique for recycling fly ash and waste glass, *J. Mater. Process. Tech.*,**168**, 56–61(2005).

[14] HP Martin, R Ecke, and E Muller, Synthesis of nanocrystalline silicon carbide powder by carbothermal reduction, *J. Eur. Ceram. Soc.*,**18**,1737-1742(1998).

[15] X. K. Li, L. Liu, Y. X. Zhang, Sh. D. Shen, Sh. Ge and L. Ch. Ling, Synthesis of nanometre silicon carbide whiskers from binary carbonaceous silica aerogels, *Carbon*, **39**,159–165(2001).

[16] G.W. Meng, Z. Cui, L.D. Zhang, and F. Phillipp, Growth and characterization of nanostructured β-SiC via carbothermal reduction of SiO_2 xerogels containing carbon nanoparticles, *J. Cryst. Growth.*, **209**, 801-806 (2000).

[17] M.J. Ledoux, S.H. Hantzer, C.P. Huu, J. Guille, and M.P. Desaneaux, New synthesis and uses of high-specific-surface SiC as a catalytic support that is chemically inert and has high thermal resistance, *J. Catal.*,**114**,176–185(1988).

NONDESTRUCTIVE TESTING OF DEFECT IN A C/SiC COMPOSITE

Hui Mei*, Xiaodong Deng, and Laifei Cheng
National Key Laboratory of Thermostructure Composite Materials, School of Materials Science, Northwestern Polytechnical University, Xi'an, 710072, China

ABSTRACT

In this paper, thermography, X-ray radiography, and industrial computed tomography (CT) were used to detect the blind holes drilled in the back side of a C/SiC composite panel. Diameters and depths of the hole defect were measured by nondestructive testing methods tentatively. Results show that it is very easy for thermography to measure the diameter (D) and depth (d) of each hole defect through a series of thermal images with time, and the errors in diameter measurement decrease with the increase of D/d. X-ray radiography can also detect the hole defects of the C/SiC composites in different brightness, whereas CT is applicable to further determine the exact localization of the defects of interest in all three-dimensional spatial directions.

KEYWORDS: Ceramic matrix composites; Nondestructive testing; Defects

INTRODUCTION

C/SiC is a kind of new promising thermal structural material in many aeronautic and astronautic applications. The quality control and assurance of the C/SiC composites, however, has always been a challenge due to their anisotropy and inhomogeneity.[1] It is essential to use Nondestructive Testing (NDT) technology to assure the quality and reliability of the C/SiC composites during their manufacture and service, deepen the understanding of their internal defects, and guide the material design and application.

At NASA in the USA advanced digital radiography, high resolution computed tomography (CT), thermography, ultrasound, acoustic emission and eddy current systems have demonstrated the maturity and success for application to the shuttle wings, airline rudders and tails, thruster chamber assemblies, combustion liners and other composite components.[2-4] At Astrium of Germany (formerly DASA), diverse standard procedures for the non-destructive testing of C/SiC components, such as thermography, X-ray technology and ultrasonic technology are in use. With help of the NDE methods, possible production defects such as delaminations, pores, and cavities, etc. as well as component conditions before and after testing are to be detected.[5]

This paper highlighted the recent efforts to apply three NDT methods, i.e., the infrared thermography, X-ray radiography, and industrial computed tomography (CT), to detect the blind holes in a C/SiC composite panel, and then to evaluate the abilities of these used NDT techniques for detecting the artificial defects.

* Corresponding author. Tel.:+86-29-88494616; fax: +86-29-88494620.
E-mail address: phdhuimei@yahoo.com (Dr. H. Mei)

EXPERIMENTAL DETAILS

Preparation of the defect laminates with blind holes

The materials used in this investigation are always C/SiC composites, which were processed by the classical chemical vapor infiltration (CVI).[6] CVI processing is used to deposit the PyC interphase on the carbon fibers and infiltrate the SiC matrix into the perform pores.

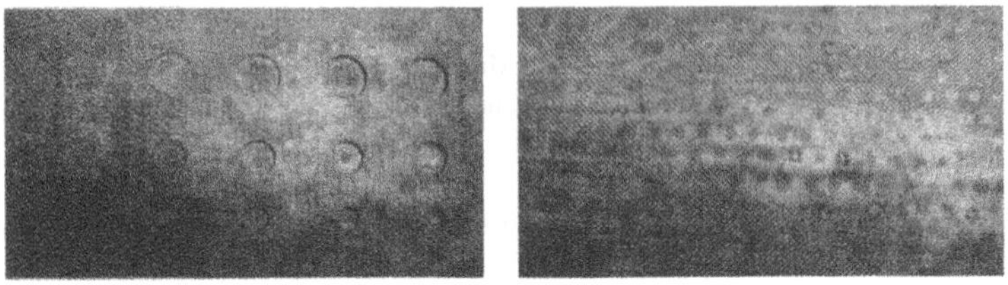

a. back side b. detect side

Figure 1. Photograph showing the standard C/SiC sample with drilled blind holes on the other side.

To simulate air voids or delamination defects in the composites, a 2D C/SiC composite panel was drilled with a series of blind holes (flat bottom) in different diameters and at different depths. Figure 1 shows a photograph of the standard 2D C/SiC composite sample with the blind hole array on the back side. On the detect side, these holes filled with air are invisible, which are considered to simulate delamination or air voids. The drilled blind holes from bottom to top were arranged in four rows according to the different diameters: $\Phi_A = 5$ mm, $\Phi_B = 10$ mm, $\Phi_C = 15$ mm, and $\Phi_D = 20$ mm. The distance between the hole bottom and the detect surface are (right to left): 1.5 mm, 2.0 mm, 2.5 mm, 3.0 mm, and 3.5 mm for the holes in diameter $\Phi = 5$, 10, and 15 mm; and 0.5 mm、 3.0 mm、 3.5 mm、 4.0 mm, and 4.5 mm for the holes in diameter $\Phi = 20$ mm.

Nondestructive testing methods

Thermography

Thermal-wave imaging can obtain surface and sub-surface defect information of a sample as a novel NDE tool. For application to CMC materials, a commercial infrared thermography system EchoTherm® (Thermal Wave Imaging, Inc) is currently used. The IR imager is a commercial radiometer with a cooled 240H × 320V-element GaAs focal plane array detector. The radiometer produces images at both 30 frames per second output (video frame rate, in an RS170, format compatible with standard video equipment) and 60 frames per second output in a 14-bit, RS422 digital format. External optics, consisting of a wide-angle lens, using germanium optical elements, were used to increase the system field-of-view by a factor of approximately two. The expanded field-of-view of this lens is 20° horizontally and 15° vertically. Heat application is achieved by directing the output of two 2400 Joule xenon flash lamps contained within a hood assembly that helps to focus the energy onto the inspection surface.

The C/SiC test panel was irradiated with the flash of light momentarily. After the flash, the surface radiates energy much like reflection of light. The irradiated surface also conducts heat into the sub-surface. With time, we first get the IR image of the surface and subsequently the images of the internal sample at deeper thickness as the heat is conducted.

X-ray radiography

Digital X-ray radiography was used in this investigation. The sample under test was placed between an X-ray source and a digital detector, and the source was enabled for a set time and energy to expose the detector. X-rays are absorbed increasingly by atoms with increasing atomic number, so the resulting images on detector show changes in material density as changes in intensity. For more information on radiographic NDT, see Bossi et al. [7]

Computed tomography

Conventional X-ray radiography and real-time radiography (RTR) suffer from the loss of three-dimensional (3D) information because of structural superposition. X-ray CT provides two-dimensional (2D) density images of cross sections through an object. "Stacking" contiguous CT images (i.e., slices) provides accurate 3-D information. Basic principles of the CT inspection methods were described in detail in the work by Green et al.[8] In the present work, the computed tomography of the C/SiC composite panel was conducted on a high energy CT system with the latest technologies and faster algorithms. The system allowed a spatial resolution of up to 1 lp/mm for detecting changes in density as well as defects.

RESULTS AND DISCUSSION

Thermographic imaging

For the composite panel with the drilled blind holes on the back side, a time sequence of thermal images was obtained by thermography. Figure 2 presents a series of thermographic results from t = 0.017s to 2.002s. These thermal images with time sequence exhibit a clear evolution of the hidden holes from invisibility to visibility at different depths.

The blind hole can be measured in diameter and depth by thermography when the thermal intensity of each hole reached a maximum. The size measurement of each hole in the thermal image corresponds to a separate maximum time t_p because the thermal intensity of each hole reached the maximum value at different time t_p, depending on the diameter and depth as [8]

$$t_p = \sqrt{2\pi}\frac{Dd}{\alpha} \qquad (1)$$

where D and d are diameter and depth of the blind hole from the detect side, and α, thermal diffusivity. Therefore, it is very easy to measure the diameter and depth of each hole from these thermal images. Compared to the designed size of the blind holes, thermographic measurements give relative errors of each measurement result as

$$e_r = \frac{|S_m - S_d|}{S_d}\% \qquad (2)$$

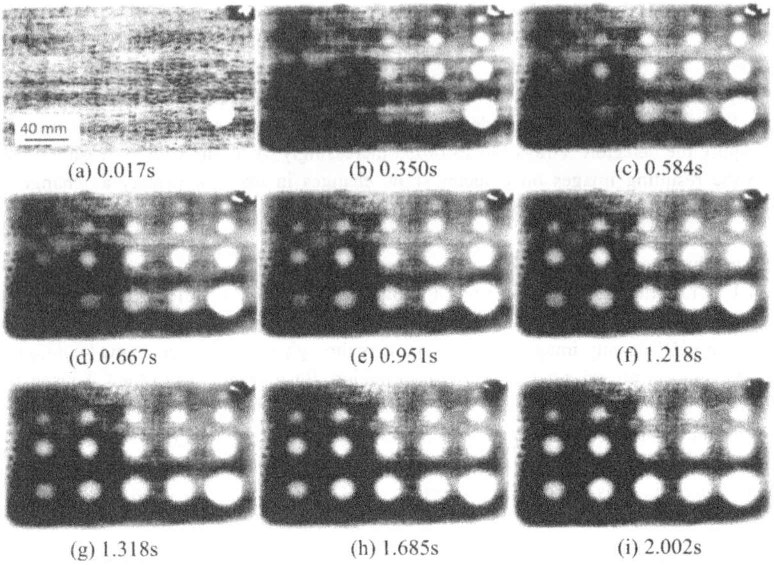

Figure 2. Thermal images of the blind holes in C/SiC sample with different times.

where, S_m and S_d signify the measured and designed sizes of the artificial defects, respectively. As an example, the relative errors in diameter measurement were plotted in Figure 3 as a function of the ratio of diameter to depth D/d. From Figure 4, it can be concluded with regard to usefulness of thermography as below,

a) D/d < 2, the blind holes are invisible for thermography;
b) D/d ≥2, the blind holes are visible for thermography and the mean errors in diameter measurement are below 20% or even lower with higher D/d.

Therefore, the condition D/d = 2 is a critical value, above which the thermography is applicable to detect the delamination or air voids in the C/SiC composite panel. More importantly, the errors in diameter measurement decrease with the increase of D/d; and within the detectable range of D/d ≥2, the errors in depth measurement seem to decrease with depth increase.

X-ray radiographic imaging

Figure 4 shows the X-ray radiographic image of the blind holes in the C/SiC sample panel. The defects can be detected by radiography with clear boundaries, but in different brightness. Structural superposition of the lower density air defects with the higher density C/SiC baseline composites in X-ray radiographic image led to color change of the detected air holes. The closer the distance between the blind hole bottom and the detect side of the composite panel, the less the absorption of the X-ray and the brighter the radiographic image of the defect. As can be seen from Figure 4, the air hole defects in the right are always brighter than those in the left. This is because the depths of the left air holes from the detect side are larger than those of the right ones. In each row of the air hole defects, the brightness of defect image became weaker and weaker with an increase of the hole depth.

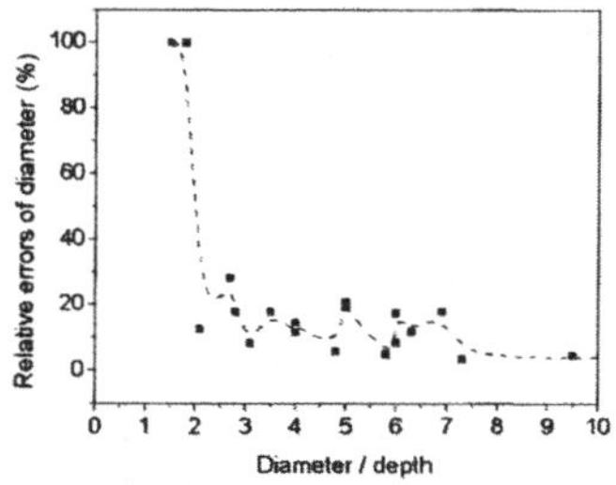

Figure 3. Relationship between the relative errors of the diameter measurements and ratio of diameter to depth.

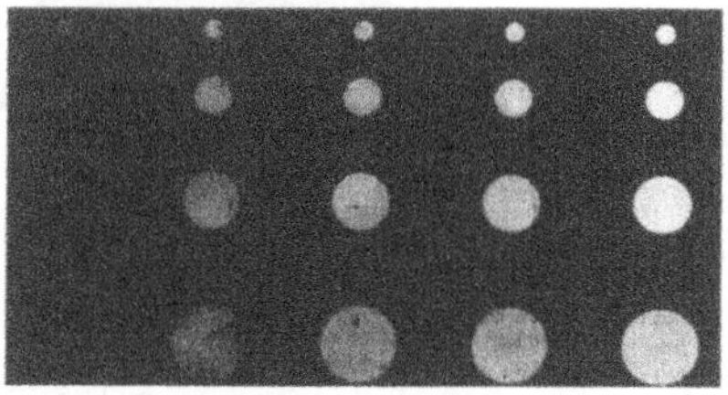

Figure 4. X-ray radiographic image illustrating the hidden hole defects in the other side of the composite panel.

Computed tomographic imaging

Figure 5 shows an example of CT scan results taken from the C/SiC composite panel with blind holes. The differing geometries, dimensions, and depths of the artificial hole defects are clear and unambiguous to be seen in the top slice and the first (D row), third row (B row) slice views. Compared to the above thermography and X-ray radiography, CT measurements can accurately estimate the actual diameter for each predefined hole defect with the clear boundary, independent of the time and depth according to the top slice in Figure 5a. All the hole diameters in Figure 5a were measured by CT and then listed in Table I. It can be seen from Table I that the measured hole diameters by CT coincide extremely well with designed ones. It must be noted that due to machining errors, here the designed diameters have some deviation from the predefined diameters described above in section of the 'defect panel preparation': i.e., $\Phi_A = 5$ mm, $\Phi_B = 10$ mm, $\Phi_C = 15$ mm, and $\Phi_D = 20$ mm.

Table I. CT measurement results of the diameter of the blind holes in the C/SiC composite panel

Row No.	A			B			C			D			E		
	S_m	S_d	e_r	S_m	S_d	e_r	S_m	S_d	e_r	S_m	S_d	e_r	S_m	S_d	e_r
D	21.5	21.24	1.22	21.5	21.24	1.22	21.5	21.40	0.5	21.5	21.44	0.3	20.8	20.78	0.1
C	15.9	15.24	4.3	15.5	15.08	2.8	15.5	15.18	2.1	15.9	15.46	2.8	15.5	15.30	1.3
B	10.2	10.06	1.4	10.2	10.08	1.2	10.2	10.08	1.2	10.6	10.38	2.1	10.6	10.32	2.7
A	5.7	5.56	2.5	5.3	5.26	0.8	5.3	5.34	0.7	5.3	5.42	2.2	4.9	5.26	6.8

S_m — the measured diameter of the hole defects; S_d — the designed diameter of the hole defects; e_r — the relative error

Generally, a decisive advantage of the CT measurement lies in the exact localization, in particular the visualization of the depth and thickness of the thin defects, as well as in the simple estimation of the defect size in all three spatial directions by means of the reconstructed imaging analysis.[5] Subsequently, the depth of each hole defect can be determined by the cross-sectional slices such as the first row hole slice in Figure 5b (D row) and the third row hole slice in Figure 5c (B row). For example, the depths of the third (B) row 10 mm diameter holes measured by CT are: 1.60 mm, 2.00 mm, 2.58 mm, 3.20 mm, and 3.64 mm, which nearly approach the designed depths of 1.5 mm, 2.0 mm, 2.5 mm, 3 mm, and 3.5 mm.

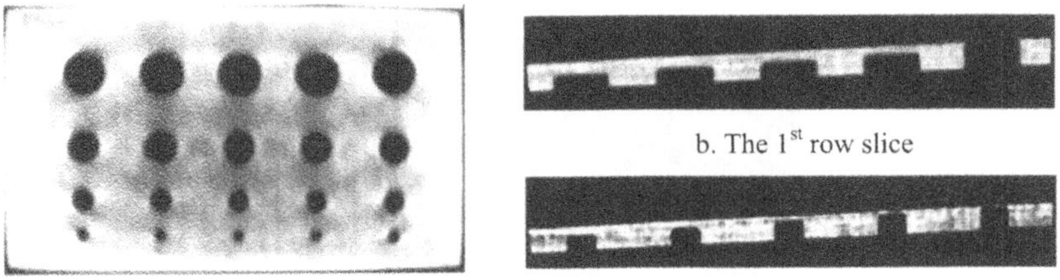

b. The 1st row slice

a. Top slice

c. The 3rd row slice

Figure 5. Typical Computed Tomography images show geometry, diameter, and depth of the hole defects in the composite panel (all dimensions in mm).

SUMMARY AND REMARKS

In this paper, simulated defects of the blind holes drilled in the other side of a C/SiC composite panel were successfully detected by three NDT methods: i.e., thermography, X-ray radiography and computed tomography. Thermography provides the defects in C/SiC composite with relatively clear two-dimensional images with time. Radiography can present the two-dimensional images of the blind hole defects, but image definition depends on the defect depth: the deeper the blind hole bottom from the detect side of the composite panel, the more the absorption of the X-ray and the darker the defect radiographic image. Finally CT can further determine the exact localization of the defects of interest in all three-dimensional spatial directions.

ACKNOWLEDGEMENTS

This work has been financially supported by NPU Foundation for Fundamental Research (NPU-FFR-JC200806), NPU Foundation for Flying Star, and Natural Science Foundation of China (Contract No. 50820145202). The authors also gratefully acknowledge the Program for Changjiang Scholars and Innovative Research Team in university (PCSIRT).

REFERENCES

[1] W. Krenkel, J. M. Hausherr, T. Reimer and M. Frieb, 28th international conference on advanced ceramic and composites B, John Wiley & Sons, Inc., Ohio, 2004, pp.49-58.

[2] J. G. Sun, M. J. Verrilli, R. Stephan, T. R. Barnett and G. Ojard, Nondestructive evaluation of ceramic matrix composite combustor components, NASA/TM-2003-212014, April 2003.

[3] E. I. Madaras, W. P. Winfree, W. H. Prosser, R. A. Wincheski, K. E.Cramer, Nondestructive evaluation for the space shuttle's wing leading edge, AIAA 2005-3630, July 2005.

[4] W. P. Winfree, E. I. Madaras, K. E. Cramer, P. A. Howell, K. L. Hodges, J. P. Seebo and J. L. Grainger, NASA langley inspection of rudder and composite tail of American airlines flight 587, AIAA 2005-2253, April 2005.

[5] S. Schmidt, S. Beyer, H. Immich, H. Knabe, R. Meistring and A. Gessler, Ceramic matrix composites: a challenge in space-propulsion technology applications, *Int. J. Appl. Ceram. Technol.*, 2(2),85–96(2005).

[6] H. Mei, Y. D. Xu, L. F. Cheng et al, Nondestructive evaluation and mechanical characterization of a defect-embedded ceramic matrix composite laminate, *Int. J. Appl. Ceram. Technol.*, 4 (4),378–386(2007).

[7] R. H. Bossi, F. A. Iddings, G. C.Wheeler, in 'Nondestructive testing handbook', (ed. P. O. Moore), Radiographic testing, vol. 4, 1991, Columbus, OH, American Society for Nondestructive Testing.

[8] W. H. Green, J. M. Wells, Nondestructive characterization of impact damage in metallic/nonmetallic composites using X-ray computed tomography imaging, NASA ARL-TR-2399, February 2001.

ACKNOWLEDGEMENTS

[illegible]

REFERENCES

[illegible]

VI. Advanced Structural Ceramics

FABRICATION OF BARIUM ALUMINOSILICATE-SILICON NITRIDE-CARBON NANOTUBE COMPOSITES BY PRESSURELESS SINTERING

Bo Wang[1], Jian-Feng Yang[1] *, Ji-Qiang Gao[1], and Koiichi Niihara[2]
[1] Sch. Mater. Sci Eng., Xi'an Jiaotong Univ., China.
[2] Extreme Energy-Density Res. Inst., Nagaoka Uni. Tech., Japan
*E-mail: yang155@mail.xjtu.edu.cn

ABSTRACT

Barium aluminosilicate (BAS)–silicon nitride composites reinforced with different amounts (1,3 and 5wt%) and different types (1-2 um and 5-15um in length, both 20nm in diameter) of multiwall carbon nanotubes (MWNTs) have been fabricated by pressureless sintering. The effect of CNTs on the microstructure, compositional investigations, as well as mechanical characterization of these composites was investigated. Near fully compacted composites with good properties have been obtained by pressureless sintering. Carbon nanotubes have been preserved in the inter-granular places and had good adherence to matrix grains after the high-temperature processes. Moreover, large content of CNTs may inhibit the densification and the $\alpha\rightarrow\beta$-Si_3N_4 transformation of the composites. BAS glass served as an effective liquid phase sintering aid to preserve the carbon nanotubes and promote the densification of the CNT reinforced composites.

Keywords:

Silicon nitride; Carbon nanotubes; Barium aluminosilicate; Microstructure; Pressureless Sintering;

Introduction

The extraordinary mechanical, thermal and electrical properties of carbon nanotubes (CNT) have prompted intense research into a wide range of applications in structural materials, electronics, and chemical processing.[1-3] Attempts have been made to develop advanced engineering materials with improved or novel properties through the incorporation of carbon nanotubes in selected matrices (polymers, metals and ceramics).[4-6] But the use of carbon nanotubes to reinforce ceramic composites has not been very successful. So far, only modest improvements of properties were reported in CNTs reinforced silicon carbide[4] and silicon nitride matrix composites,[7] while a noticeable increase of the fracture toughness and of electrical conductivity has been achieved in CNTs reinforced alumina matrix composites.[8]

In order to obtain dense CNT-dispersed Si_3N_4 ceramics by fully using the benefits of CNTs, it should be concentrated on the following tasks: (i) fast densification of CNT-dispersed Si_3N_4 composites, (ii) optimization of processing parameters to avoid the damage of carbon nanotubes at high temperatures, (iii) interfacial engineering optimization of the interfacial bonding between CNTs and matrix. Therefore, a good additive system should form a liquid phase at a low liquid eutectic temperature, and the liquid phase will be crystallized later completely into a compound with a high melting point. Barium aluminumsilicate ($BaO{\cdot}Al_2O_3{\cdot}2SiO_2$, BAS) meets these requirements.[9,10] In this composite, the BAS glass-ceramic serves not only as a liquid phase sintering aid for the silicon nitride phase transformation and fast densification, but also remains as a structural matrix. Besides, the CNTs were not destroyed during conventional high temperature hot-pressing of CNT-reinforced BAS composites and had good binding with the BAS matrix grains without any obvious interfacial reaction or amorphous layer.[10]

Although there have been considerable achievements in terms of crystalline ceramic composites reinforced with CNTs, no study has been carried out for glass-ceramics composites and no pressureless sintering technology was used either to address the complicated shape. The objective of this study was to fabricate dense barium aluminosilicate-silicon nitride-carbon nanotube composites by pressureless sintering. The compositional, microstructure, mechanical properties and toughening mechanisms of the resulting composites were investigated.

Experimental Procedure

$BaCO_3$ (purity>99%) was mixed with 32wt%SiO_2 (purity>99%) and 27.1wt% Al_2O_3 (purity>99%) powder by wet milling in anhydrous alcohol for 24 h in a plastic bottle. After milling, the slurry was dried, and the obtained powder mixture was sieved, and sintered at 1300°C for 2h to obtain BAS glass powder. The BAS powders were subsequently pulverized and screened through 150 μm screen.

The starting materials were α-Si_3N_4 (over 95% α-phase content and 1.5 mass% oxygen content, Shanghai, China; mean particle size: 0.5μm), 30 wt% BAS powder, and MWNTs (supplied by Shenzhen NANO tech. Port Co., Ltd., China), had dimensions of 20nm in diameter 1-2 um and 5-15um in length. The amount of CNT was changed from 0 to 10 wt% of the total amount of the other raw powders. To disperse the nanotubes homogeneously, 3wt% of dispersant was added to ethanol, whose ratio was 65 ml to 1 g of CNT, to prepare a slurry using an ultrasonic vibrator for 1h. The slurry was mixed with the 30wt% BAS/Si_3N_4 slurry with 2wt% of the dispersant and then the final mixtures were ball-milled in anhydrous alcohol using high-purity Si_3N_4 balls for 24h in a plastic bottle. After milling, the slurry was dried, sieved, and uniaxially pressed to form rectangular bars measuring 30mm • 30mm • 5mm.. The green bodies were sintered in a furnace (High multi-5000, Fijidempa Co. Ltd., Osaka, Japan) at 1750°C for 2h under argon-gas pressures of 0.6 MPa. Heating rates from 10°–20°C/min were used. The samples were covered with Si_3N_4–BAS–BN powder mixture to protect the samples from decomposition and deformation.

The bulk density of the sintered products was measured by the Archimedes displacement method. The theoretical density of the specimens was calculated according to the rule of mixtures. Crystalline phases of the resultant samples were identified by XRD (D/MAX-2400X, Rigaku Co., Tokyo, Japan) analysis. The specimens were machined into test bars for flexural strength measurement. The flexure strength was measured by three-point bending method with a 20 mm span at a cross-head speed of 0.5 mm/min at room temperature. The fracture toughness was determined by single-edge-noteched-beam (SENB) method at room temperature with 20mm span at a cross-head speed of 0.05mm/min. Each final value was averaged over five measurements. Microstructure observation was carried out using high-resolution by SEM (JSM-7000F).

Results and Discussion

X-ray diffraction patterns for different amount of the CNTs reinforced 30wt.%BAS/Si_3N_4 composites after sintering at 1750°C are shown in Figure 1. Phase identification consisted primarily of β-Si_3N_4 and hexacelsian $BaAl_2Si_2O_8$ in composite with 0, 1wt% CNTs (Fig. 1a), whereas, a trace of residual α-Si_3N_4 phase was detected in composite sintered with 5 wt% CNTs (Fig. 1b). It indicated that the α- to β-phase transformation was slightly restrained with the addition of the CNTs. The heterogeneous nucleation is the dominant mechanism in the α- to β-phase transformation during the liquid-phase sintering of Si_3N_4.[11] No information on the CNTs was obtainable by this technique, which indicated that XRD is not effective in revealing the presence of CNTs. It also shows no signs of silicon carbide peaks, which were seen by Tatami et al[12] for samples with CNTs content exceeding 2 wt%. Only the hexacelsian $BaAl_2Si_2O_8$ phase was revealed in the sintered samples without any other crystalline phases or non-crystalline phases, indicating the excellent crystallization capability of BAS glass. No monoclinic BAS was detected due to the sluggish transformation of hexacelsian to celsian phase.[13]

The density of CNT reinforced BAS/Si_3N_4 composites prepared in an argon-gas furnace as a fraction of CNT content is shown in Figure 2. For both additive levels, relative densities in excess of ~79 % of theoretical density are obtained when sintering at 1750°C or above, for a period of two hours. The reference sample without carbon nanotube addition possesses the highest degree of densification, and the relative density decreases with the increasing CNT content. In the case of

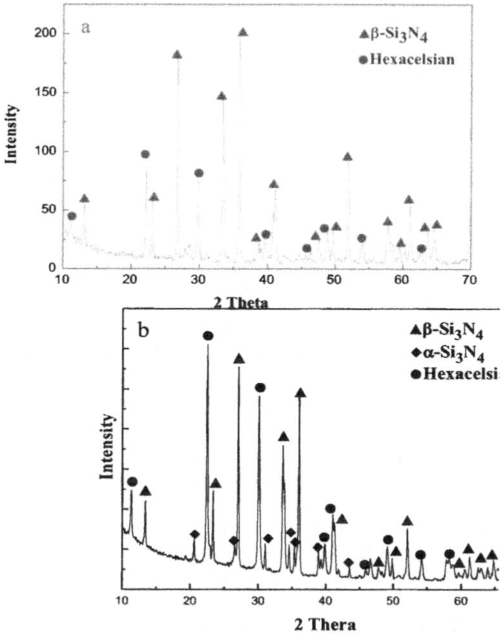

Fig. 1. XRD patterns of CNT reinforced 30%BAS/Si_3N_4 composites sintered at 1750°C/1h: (a) sample with 0,1wt.% CNTs (b) with 5 wt.% CNTs.

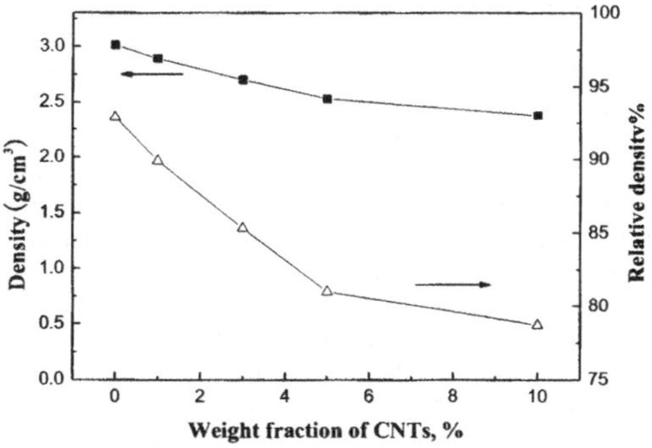

Fig. 2. The density of BAS-Si_3N_4-CNT composites sintered at 1750°C/1h as a fraction of CNT content

1.wt% CNT addition for BAS/Si_3N_4 composite, ~90 % of theoretical density, a higher densification level of 2.89 g·cm^{-3}, can be achieved by pressureless sintering. It indicated that the BAS glass served as an effective liquid phase sintering aid for the carbon nanotubes reinforced Si_3N_4 composite to attain nearly full densification. BAS has been proven to be an effective sintering aid to the fast densification of ceramic composites.[14] In general, it is very difficult to fabricate dense ceramic composites with high CNT contents via a conventional powder process, because CNTs greatly inhibit the grain growth, which is detrimental to the material densification.[8, 11,15] On the other hand, larger CNTs content does not benefit the distribution, which can also reduce the density. In this case, the dense carbon nanotubes reinforced composites were believed to possess superior mechanical properties.

Representative microstructures of the NaOH-etched surface of the BAS/Si_3N_4 and CNT reinforced BAS/Si_3N_4 composites by pressureless sintering are shown in Figue 3a and 3b. A grain growing process accompanied by a phase transformation can be observed in the microstructure. Each of these microstructures shows bimodal distribution with some β-Si_3N_4 grains of high aspect ratio embedded in a smaller grained matrix developed during the sintering process. The CNT-BAS/Si_3N_4 composite incorporated with 1% CNTs (Fig. 3b) shows a measurable increase in the numbers of larger sized β-Si_3N_4 grains in comparison with the BAS/Si_3N_4 composite (Fig. 3a). It also shows a measurable increase in the numbers of grains of average diameter 0.4 microns or less in comparison with the BAS/Si_3N_4 material. According to these results, it was suggested that a high densification (large amount of liquid phase formed by BAS) can in principle promote the evolution of a fine Si_3N_4 grain microstructure. At first, under the assumption of a diffusion-controlled grain-growth mechanism, it was thought that, in the presence of large amount of low viscous glass, both diffusion and grain growth would be favored. Thus, rapid densification and a fast coarsening of the microstructure were expected to occur. In addition to the influence of the amount of liquid, the actual dissolution rate of α-Si_3N_4 within the liquid can affect the probability of β-Si_3N_4-nuclei formation. For the BAS/Si_3N_4 material it was, therefore, assumed that the large amount of BAS glass resulted in a fast dissolution and rapid saturation of Si_3N_4 in the liquid. This in turn favors a higher number of β-Si_3N_4-nuclei formed upon sintering, since oversaturation was locally achieved faster compared to α-Si_3N_4 which contains fewer glass. All of the newly formed nuclei would simultaneously grow by diffusion-controlled Ostwald ripening.

The CNT additions appear to have caused a degree of refinement in the grain structure. Their presence has increased the fraction of small acicular grains and has also caused an increase in the number of fine (low average-diameter) grains. The formation of bimodal distribution microstructures coupled with increased acicular grains in a silicon nitride with CNT additions can be explained on the basis of the CNTs providing nucleation sites for β-grains during the complex liquid phase sintering process.[16] If the dispersion of CNTs is sufficiently uniform, the increased number of nucleation sites they provide should produce a finer grain structure with some small acicular grains as seen in CNT-BAS/Si_3N_4. The lower sintering temperature (1750°C) used in this study also promotes the formation of low average-diameter grains. The presence of higher aspect ratio β-Si_3N_4 grains in CNT-BAS/Si_3N_4 was due to the larger amount of small acicular β-grains grown from nucleation sites provided by CNTs supplied adequate space for grain growing.

Moreover, large size β-Si_3N_4 grains involved in the presence of individual CNTs were presented in the micrograph (Fig. 3c), which indicated a good dispersion of CNTs in the ceramic/glass matrix. It can be seen that the CNTs have a good bond with the Si_3N_4 grains and BAS matrix (Fig. 3d) without any obvious interfacial reaction, suggesting that the nanotubes were not damaged during the pressureless sintering. The carbon nanotubes can connect many β-Si_3N_4 grains to form web-like microstructure which potentially contribute to high fracture toughness and toughing, giving rise to a self-reinforced microstructure.

Figure 4 shows the microstructure of the fracture surfaces of CNT-BAS/Si_3N_4 composites. It can be seen that the carbon nanotubes are dispersed in the microstructure. The CNTs are located mainly in the inter-granular places and they are presenting good adherence to BAS grains. It was suggested that high densification could in principle hinder carbon nanotube reacting with Si_3N_4 and the large amount of liquid phase could promote the dispersion of carbon nanotubes in the grain-boundary phase. It could be concluded that the final bimodal distribution Si_3N_4 microstructure with homogeneously dispersed CNTs, when keeping processing conditions constant, was dominated by the liquid that forms during high-temperature sintering. The toughening mechanisms should be the CNTs pullouts, crack bridging, crack deflection. It has been reported that the presence of an ideal CNT–BAS interfacial structure suitable for crack deflection and the pullout mechanism.[8] Since the elastic modulus of the CNTs is much higher than that of the BAS matrix, the modulus-load-transfer also increases toughness by transferring stresses at a crack tip to regions remote from the crack tip,

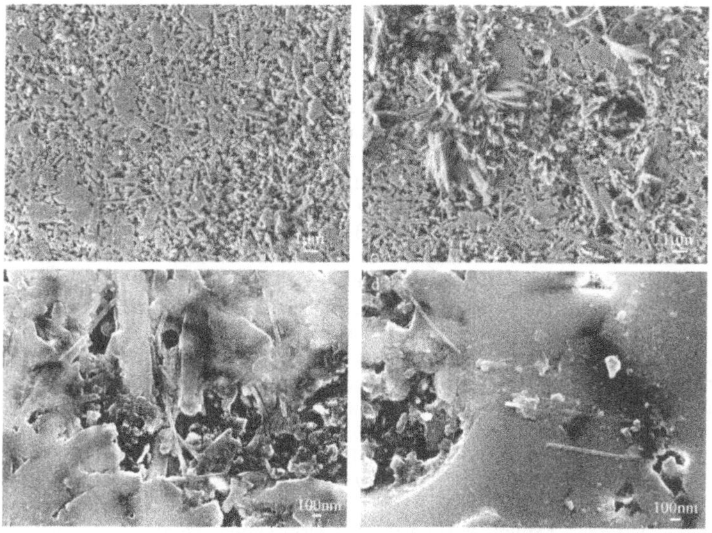

Fig. 3. SEM micrograph of the NaOH-etched surface of 1wt% CNTs-BAS/Si_3N_4 composite sintered at 1750°C/1h. (a) reference sample without CNTs (b) (c) (d) with 1wt.% CNTs.

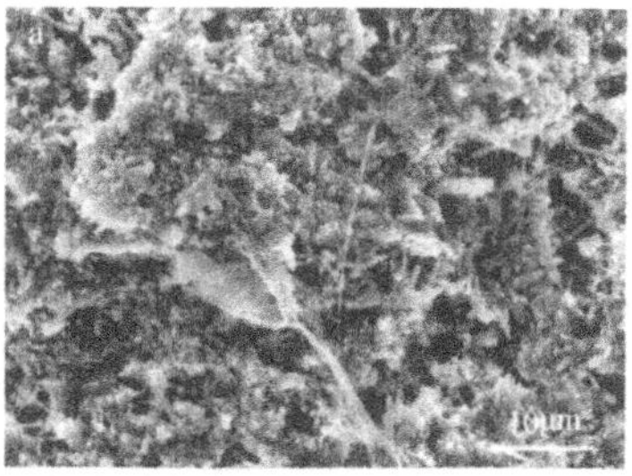

Fig. 4. SEM micrograph of the typical fracture surfaces of 1wt.% CNTs reinforced 30%BAS/Si_3N_4 composite sintered at 1750°C/1h.

hence decreasing the stress intensity at the crack tip.[8] High room temperature strength of 556±40 MPa and fracture toughness of 6.5MPa·m$^{1/2}$ were obtained for BAS/Si_3N_4 composites with a 1wt% addition of CNTs, and it is considered to be due to the finer microstructure and toughening mechanisms.

CONCLUSION

Dense BAS-Si_3N_4-BAS composites were successfully fabricated by pressureless sintering. Carbon nanotubes showed good contact to the BAS and Si_3N_4 grains, and served as crystallization sites and seeds for Si_3N_4 grains, which contribute to the formation of bimodal distribution microstructures coupled with increased acicular grains in BAS/Si_3N_4 composites. The excessive CNTs content may inhibit the $\alpha \rightarrow \beta$-Si_3N_4 phase transformation. BAS glass served as an effective liquid phase sintering which supplied sufficient liquid phase aid to promote the densification of the CNT composites and the dispersion of carbon nanotubes in the grain-boundary phase.

REFERENCE

1. S. Ijima, Helical microtubules of graphitic carbon, Nature., 354, 56–8 (1991).
2. S. Rochie, Carbon nanotubes: exceptional mechanical and electrical properties, Ann. Chim. Sci. Mater. 25, 529–32 (2000).
3. A. Peigney, Tougher ceramics with nanotubes, Nat. Mater., 2, 15–6 (2003).
4. R.Z. Ma, J. Wu, B.Q. Wei, J. Liang, and D.H. Wu, Processing and properties of carbon nanotubes-nano-SiC ceramic, J. Mater. Sci., 33, 5243–6 (1998).
5. X. Wang, N. P. Padture, and H. Tanaka, Contact-demage-resistant ceramic/single-wall carbon nanotubes and ceramic/graphite composites, Nat. Mater., 3, 539–44 (2004).
6. J.P. Tu, Y.Z. Yang, L.Y. Wang, X.C. Ma, and X.B. Zhang, Tribological properties of carbon-nanotube-reinforced copper composites, Tribol. Lett., 10, 225–8 (2001).
7. Cs. Balazsi, Z. Konya, F. Weber, L.P. Biro, and P. Arato, Preparation and characterization of carbon nanotube reinforced silicon nitride composites, Mater. Sci. Eng. C., 23, 1133–7 (2003).
8. G.D. Zhan, J.D. Kuntz, J. Wan, and A.K. Mukherjee, Single-wall carbon nanotubes as attractive toughening agents in alumina-based nanocomposites, Nat. Mater., 2, 38–42(2003).
9. K.K. Richardson, D.W. Freitag, and D. Hunn, Barium Aluminosilicate Reinforced In Situ with Silicon Nitride, J. Am. Ceram. Soc., 78, 2662–8 (1995).
10. F. Ye, L. Liu, Y. Wang, Y. Zhou, B. Peng, and Q. Meng, Preparation and mechanical properties of carbon nanotube reinforced barium aluminosilicate glass-ceramic composites, Script. Mater., 55, 911–4 (2006).
11. P.F. Becher, Microstructural Design of Toughened Ceramics, J. Am. Ceram. Soc., 74, 255–69 (1991).
12. J. Tatami, T. Katashima, K. Komeya, T. Meguro, and T. Wakihara. Electrically conductive CNT-dispersed silicon nitride ceramics, J. Am. Ceram. Soc., 88-10, 2889–93 (2005).
13. F. Ye, J.M. Yang, L.T. Zhang, W.C. Zhou, Y. Zhou, and T.C. Lei. Fracture Behavior of SiC-Whisker-Reinforced Barium Aluminosilicate Glass-Ceramic Matrix Composites, J. Am. Ceram. Soc., 84, 881–3 (2001).
14. F. Ye, S. Chen, and M. Iwasa. Synthesis and properties of barium aluminosilicate glass–ceramic composites reinforced with in situ grown Si_3N_4 whiskers, Script. Mater., 48, 1433–8 (2003).
15. A. Peigney, C.H. Laurent, E. Flahaut, and A. Rousset, Carbon nanotubes in novel ceramic matrix nanocomposites, Ceram. Int., 26, 677–83 (2000).
16. Cs. Balazsi, F. Weber, Zs. Kover, Z. Shen, Z. Konya, Zs. Kasztovszky, Z. Vertesy, L. P. Biro, I. Kiricsi, and P. Arato, Application of carbon nanotubes to silicon nitride matrix reinforcements, Current. App. Phy., 6, 124–30 (2006).

NONLINEAR FINITE ELEMENT ANALYSIS OF CONVECTIVE HEAT TRANSFER STEADY THERMAL STRESSES IN A ZrO_2/FGM/Ti-6Al-4V COMPOSITE EFBF PLATE WITH TEMPERATURE-DEPENDENT MATERIAL PROPERTIES

Yangjian Xu [1]*, Daihui Tu [2], Chunping Xiao [1]
[1] Dept.Engrg Mech., Hebei Engrg Univ., China. [2] Dept. Appl Chem., Hebei Engrg Univ., China.
Handan City, Hebei Province, 056038, China

ABSTRACT

In order to study the steady thermal stresses in a ZrO_2/FGM/Ti-6Al-4V three-layered composite EFBF plate with temperature-dependent material properties under convective heat transfer boundary, the analytical model for the steady thermal stresses in the composite plate is established. Starting from heat conduction law, based on thermoelasticity theory, we derive the finite element basic equation of the one-dimensional heat conduction of the composite plate using variational principle. We present a Sinpson method for the solution of steady thermal stress formulas of the composite plate. From FORTRAN language we design the calculation software to obtain numerical results. When T_a=400K and T_b=1 700K, the steady thermal stress distributions of the composite plate are obtained, and compared with those of the composite graded three-layered plate with constant material properties and with nongraded two-layered plate with temperature-dependent material properties. The results show the various degrees of the effects of thickness and composition and porosity in FGM layer and the convective heat transfer coefficients on the surfaces on thermal stresses of the composite EFBF plate. It was also found that the temperature dependency of the material properties is one of the most important factors in the accurate evaluation of the thermal stress and the thermal stress in graded three-layered plate is gentler and the maximum tensile stress reduces by 51.8%. The results provide the foundations of theory calculation for the design and application of the composite plate.

INTRODUCTION

Functionally graded material (FGM) is a new type of nonhomogeneous composite material with special characteristics duo to arbitrarily distributed and continuously varied material properties. Therefore, FGM has received considerable attention in the field of structural design subjected to extremely high thermal loading [1-2]. Because it is used widely in high temperature working environment such as aviation and nuclear reactors, and so on, it is important to analyze the thermal stress filed of the body made of the material. Particularly noteworthy is to consider the problem with temperature-dependent material properties. Obata [3] and Tanigawa [4] researched thermal stress of pure FGM plate using perturbation and laminated analytical method, respectively. Huang [5] analyzed the thermal elastic limitation of four-layered composite plate with an interlayer of FGM. But these methods are too complex so as to lead to a complicated equation system, and are not convenient for engineering application. Therefore, Xu [6] studied the problem of transient thermal stress of pure FGM plate under convective heat transfer boundary using NFEM. Based on the above-mentioned research work, starting from the heat conduction law, this paper will discuss the convective heat transfer steady thermal stress problem of a ZrO_2/FGM/Ti-6Al-4V three-layered composite EFBF plate with temperature-dependent material properties by the NFEM and the Sinpson method, expecting the analytical results obtained to be more close to actual engineering conditions and to obtain some instructive conclusions for the production and application of ceramics / metal composite plate with an interlayer of FGM.

MODEL OF ANALYSIS

As shown in Figure 1, we now consider the steady thermal stress field distributions of a three-layered infinite long composite EFBF plate made of pure metal (Ti-6Al-4V) and pure ceramics (ZrO_2) with an interlayer of FGM. We have the following assumptions. (1) The lower layer of three-layered plate is pure metal; $k_m(T)$, $E_m(T)$, $\alpha_m(T)$ and $\upsilon_m(T)$ denote thermal conductivity rate, Young's modulus, the coefficient of linear thermal expansion and Poisson's ratio of the pure metal layer, respectively, and the layer thickness is h_1. The middle is continuous and arbitrary variant FGM gradient layer; $k(T, y)$, $E(T, y)$, $\alpha(T, y)$ and $\upsilon(T, y)$ denote the above material properties of FGM gradient layer, and the layer thickness is h_2=h_{FGM}. The upper layer is pure ceramics; $k_c(T)$, $E_c(T)$, $\alpha_c(T)$ and $\upsilon_c(T)$ denote the above material properties of pure ceramic layer, and the layer thickness is h_3. (2) Initially, the plate is under the stress-free status; the initial temperature of the plate is T_0; the plate is heated from the lower and upper surfaces by surrounding media with heat transfer coefficients $\zeta_a(T)$ and $\zeta_b(T)$, respectively, and we denote the temperature of the surrounding media by constant T_a and T_b. (3) The periphery of the plate is adiabatic, and there are no heat sources within the plate. Coordinate axis y is chosen as shown in Figure 1, and the interfaces between the layers are perfectly bonded at all times. T is the temperature function. The material's properties for each same ordinate y are homogeneous and isotropic. Subscripts c and m mean ceramics and metal, respectively. The total thickness of the plate is b =h_1+h_2+h_3.

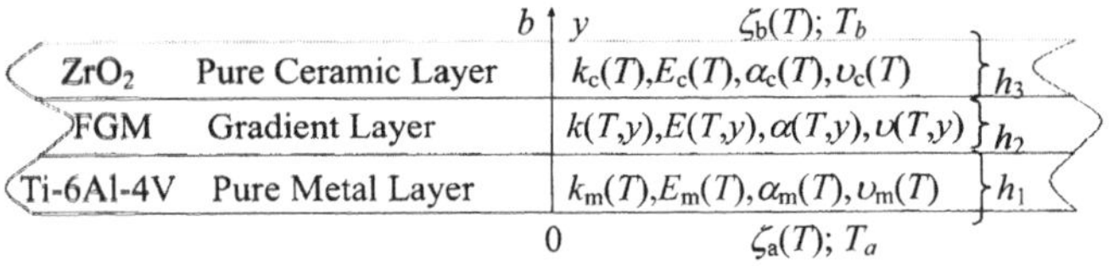

Figure 1. ZrO_2/FGM/Ti-6Al-4V composite plate (considered temperature dependency)

HEAT CONDUCTION ANALYSIS

The steady thermal conduction basic equation of the i th layer of the three-layered composite plate is

$$0 = \frac{d}{dy}\left\{k_i(T_i,y)\frac{dT_i(y)}{dy}\right\}, i = 1,2,3 \tag{1}$$

where $k_i(T,y)$ is the thermal conductivity rate of per layer of the three-layered composite plate (such as i =1, $k_1(T,y)$=$k_m(T)$, the rest on the analogy of this). Thermal conductivity rate of the FGM gradient layer is $k(T,y)$. The convective heat transfer boundary and the conditions of continuity of the temperature in the three-layered composite plate and the heat flux at interfaces are expressed in the following form

$$\left.\begin{aligned} & y = 0, k_m(T_1)\frac{dT_1(y)}{dy} - \zeta_a(T)T_1(y) = -\zeta_a(T)T_a \\ & y = b, k_c(T_3)\frac{dT_3(y)}{dy} + \zeta_b(T)T_3(y) = \zeta_b(T)T_b \end{aligned}\right\}; \quad \left.\begin{aligned} & y_i = b_i,\ T_i(y_i) = T_{i+1}(y_i), b_1 = h_1, b_2 = h_1 + h_2 \\ & k_i(T_i, y_i)\frac{dT_i(y_i)}{dy} = k_{i+1}(T_i, y_i)\frac{dT_{i+1}(y_i)}{dy}, i = 1,2 \end{aligned}\right\} \tag{2}$$

To solve the nonlinear control equation (1) under the condition (2) approximately using FEM, we need to establish relevant functional. The paper adopts one-dimensional nonlinear FEM to solve the above-mentioned one-dimensional heat conduction problem. Under the condition of assumptions in this paper, the element functional [8] (5.14) of one-dimensional steady heat conduction problem under the convective heat transfer boundary condition is

$$\pi_e = \int \frac{k_{n-1}^e}{2}\left(\frac{dT}{dy}\right)^2 dy + \zeta_{n-1}^e\left(\frac{T^2}{2} - \hat{T}_\Gamma T\right) \tag{3}$$

where ζ_{n-1}^e, k_{n-1}^e are convective heat transfer coefficient and the thermal conductivity rate of the element, respectively, and the constant values not the function of y; but these values are different for different element. Order numbers $n = 2,3,\cdots$(e.g.: $n = 2$: y=0, $\zeta_{n-1}^e(0) = \zeta_1^e(0) = \zeta_a(T_0)$; $y = b$, $\zeta_{n-1}^e(b) = \zeta_1^e(b) = \zeta_b(T_0)$, the rest on the analogy of this). $\hat{T}_\Gamma$ is the environmental media temperature, and Γ is the boundary of given convective heat transfer condition. The iteration formats of temperature, thermal conductivity rate and convective heat transfer coefficient are given in the following form

$$T_{n-1} = T_{n-2} + \Delta T_{n-2}; k_{n-1} = k_{n-2} + \Delta k_{n-2}; \zeta_{n-1} = \zeta_{n-2} + \Delta\zeta_{n-2} \tag{4}$$

The criterion of iteration is

$$|(T_{n-1} - T_{n-2})/T_{n-1}| < \varepsilon \tag{5}$$

where ε is the prescribed precision of iteration.

We now consider bar element, and the element length is l^e. Two nodes are denoted by i, j. The trial function of temperature field is linear distribution. Under the convective heat transfer boundary condition, the finite element basic equation of steady heat conduction in the three-layered composite plate is [8]

$$H\ T = Q \tag{6}$$

where H, T and Q denote thermal stiffness matrix, unknown node temperature array and node thermal load array, respectively. The elements h_{rs}^e and q_r^e (r, s =i, j) in matrix H and Q are respectively

$$h_{rs}^e = \frac{k_{n-1}^e}{l^e}(2\delta_{rs} - 1) + \zeta_{n-1}^e\delta_{rj}; \quad q_r^e = \frac{\zeta_{n-1}^e}{2}\hat{T}_\Gamma\delta_{rj} \tag{7}$$

where δ_{rs} is the symbol of Kronecker δ.

THERMAL STRESS ANALYSIS

The strain components ε_{xxi}, ε_{zzi} and stress components σ_{xxi}, σ_{zzi} of the ith layer of the three-layered composite plate are given respectively by the relations [4]

$$\varepsilon_{xxi}(\bar{y}) = \varepsilon_{zzi}(\bar{y}) = \varepsilon_0 + \bar{y}/\bar{r}_0, \sigma_{xxi}(\bar{y}) = \sigma_{zzi}(\bar{y}) = \frac{E_i(T,\bar{y})}{1-\upsilon_i(T,\bar{y})} \times \left[\varepsilon_0 + \frac{\bar{y}}{\bar{r}_0} - \alpha_i(T,\bar{y})T'(\bar{y})\right], i = 1,2,3 \tag{8}$$

where $\bar{y} = y/b$ is dimensionless position coordinate; ε_0 and $1/\bar{r}_0 = b/r$ denote strain component and dimensionless curvature on the $\bar{y} = 0$ plane respectively, and $E_i(T,\bar{y}), \alpha_i(T,\bar{y})$ and $\upsilon_i(T,\bar{y})$ denote Young's modulus, the coefficient of linear thermal expansion and Poisson's ratio of per layer of the three-layered composite plate, respectively (such as i=1, $E_i(T,\bar{y}) = E_m(T)$, $\alpha_i(T,\bar{y}) = \alpha_m(T)$ and $\upsilon_i(T,\bar{y}) = \upsilon_m(T)$, the rest on the analogy of this). $T'(\bar{y})$ is temperature rise. ε_0 and $1/\bar{r}_0$ are unknown constants and they are determined by the mechanical boundary condition. Supposing that the infinite long plate can elongate and bend freely (EFBF), the unknown constants are determined by the following equilibrium equations

$$\sum_i \int_{\bar{y}_{i-1}}^{\bar{y}_i} \sigma_{xxi}(\bar{y})\bar{y}d\bar{y} = 0; \sum_i \int_{\bar{y}_{i-1}}^{\bar{y}_i} \sigma_{xxi}(\bar{y})d\bar{y} = 0 \tag{9}$$

We can obtain by substitution the second of Eq. (8) into Eq. (9)

$$\sigma_{xxi}(\bar{y}) = \frac{E_i(T,\bar{y})}{1-\upsilon_i(T,\bar{y})} \times \left\{ \frac{(B_2D_0 - B_1D_1) + (-B_1D_0 + B_0D_1)\bar{y}}{B_0B_2 - B_1^2} - \alpha_i(T,\bar{y})T'(\bar{y}) \right\} \tag{10}$$

where $B_j = \sum_i \int_{\bar{y}_{i-1}}^{\bar{y}_i} \frac{E_i(T,\bar{y})}{1-\upsilon_i(T,\bar{y})} \bar{y}^j d\bar{y},\ j = 0,1,2;\quad D_j = \sum_i \int_{\bar{y}_{i-1}}^{\bar{y}_i} \frac{\alpha_i(T,\bar{y})E_i(T,\bar{y})T'(\bar{y})}{1-\upsilon_i(T,\bar{y})} \bar{y}^j d\bar{y},\ j = 0,1$ (11)

where B_j, D_j are calculated according to the Simpson numerical integration method. It is necessary to illustrate that E, F, C and B denote elongation, free, clamped and bending, respectively. Such as EFBF denotes that the plate can elongate and bend freely, the rest on the analogy of this.

RESULTS AND DISCUSSION

The volume fractions $V_m(y)$ and $V_c(y)$ of metal (Ti-6Al-4V) and ceramics (ZrO_2) phase, porosity $P(y)$ and material's property in FGM layer and the proof of the validity of the method are shown in the references [6-7]. The total thickness of the plate b is 10 mm, and $h_1 = h_3$ in this paper. The finite element mesh of the ZrO_2/FGM/Ti-6Al-4V composite plate is divided into 640 elements and 641 nodes. There are the nodes at the interfaces of three-layered plate. The smallest side length of the element is 0.015625mm, and T_0=300K. The relative convective heat transfer coefficients on the lower and upper surfaces are denoted by $(\bar{h}_a, \bar{h}_b) = b \times (\zeta_a / k_m,\ \zeta_b / k_c)$.

Effect of FGM Layer Thickness on Thermal Stress

Figure 2 shows the effect of FGM layer thickness on thermal stress. In the metal layer, the thermal stress curve is an almost level line and more gentle, and from h_2=2mm to 6mm, the length of the level line decreases. In the FGM layer, the thermal stress diagram is a rising curve and the tensile stress at the interface between ceramic layer and FGM layer reaches the maximum when h_2=2mm (curve 1). With the increase of the FGM layer thickness, knowing from the curves 1, 2 and 3, thermal stress curves tend to gentle, and the thermal stress gradient of each curve decreases, and the stress distribution in the composite plate is more reasonable, and also the largest tensile stress of the EFBF composite plate reduces by 59.6%. In the ceramic layer, the thermal stress diagram is a falling and slightly bending curve with a steep slope, and the thermal stress gradient is much bigger, and the compressive stress on the ceramic surface reaches maximum when h_2=2mm.

Effect of FGM layer Composition on Thermal Stress

Figure 3 shows the effect of FGM layer composition on thermal stress. The thermal stress curves at interfaces between the layers appear at an obvious turning point, and the thermal stress curves in metal and ceramic layers are almost linear and the slope of each curve is different. In FGM layer, the thermal stress curves change greatly. When M=5 (curve 3), the tensile stress at the interface of between ceramic layer and FGM layer reaches the biggest and appears peak value and sharp point, and the thermal stress curve of FGM layer is concave-down. When M=0.2 (curve 1), the thermal stress curve of FGM layer is concave-up. When M=1(curve 2), the thermal stress curve is relative gentle and smooth, and it is not like M=0.2, 5 (curves 1, 3) that show apparent undulations.

Effect of Temperature-Dependent Material Properties on Thermal Stress

Figures 3, 4 show the effect of temperature-dependent material properties on thermal stress. In metal layer, the thermal stress curves in Figure 3 are more gentle and smooth than those in Figure 4, and the slope of the curves in Figure 3 is smaller than that in Figure 4. In FGM layer, the thermal stress curves in Figure 4 are gentle and smooth, and it does not like curves 1, 2, 3 in Figure 3 that show apparent turning points and undulations. Compared with Figure 4, the maximum compressive stress on the surface of the metal in Figure 3 reduces by 30.1%, and the maximum compressive stress on the surface of the ceramics in Figure 3 increases by 20.0%, and the maximum tensile stress at the interface between the ceramic layer and the FGM layer in Figure 3 increases by 66.6%.

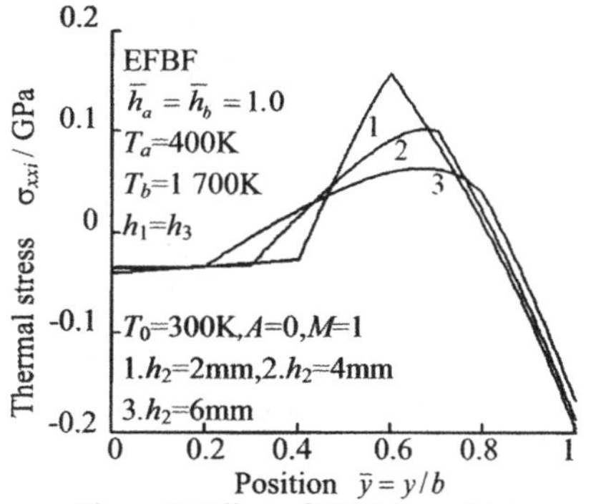

Figure 2. Effect of FGM layer thickness

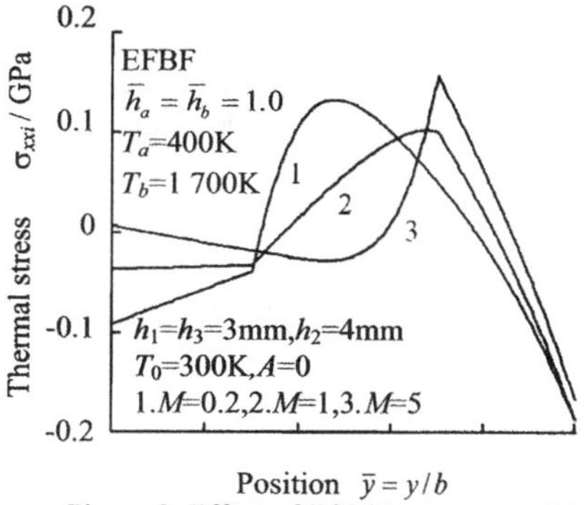

Figure 3. Effect of FGM layer composition

Effect of FGM layer Porosity on Thermal Stress

Figure 5 shows the effect of the FGM layer Porosity on thermal stress. We select the air thermal conductivity rate k_a=0.02757 W/m·K. When A = 0(curve 1), the thermal stress curve is gentle and smooth, and the compressive stress on the surface of ceramics reaches the maximum. With the increase of A, the variations of thermal stress curves become big. When A = 3.99(curve 5), the gradient of thermal stress curve in the pure metal layer and the variations of the thermal stress curve at the bonding interfaces between the three-layered plate becomes big obviously, and the curves appear at a sharp angle, and the maximum tensile stress value of curve 5 at the interface between metal layer and FGM layer is 2.3 times that of curve 1, and also the tensile stress on the surface of ceramics reaches the maximum. Because it is weak in tension, the large tensile stress is unfavorable to the strength of ceramics, and if the tensile stress reaches the limit of ceramics, it will induce damage to the ceramics. During the application, we should pay more attention to this problem.

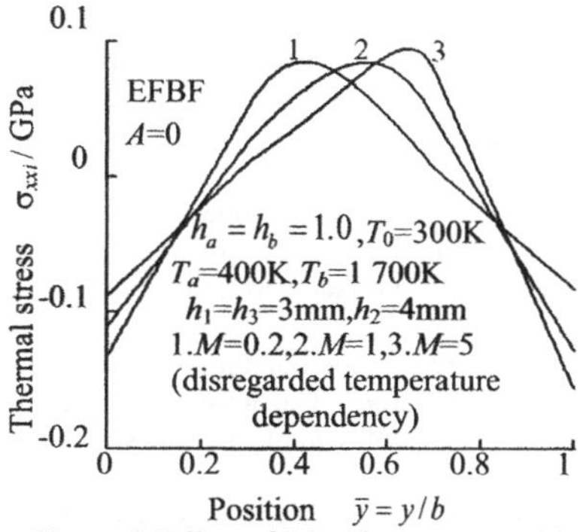

Figure 4. Effect of FGM layer composition

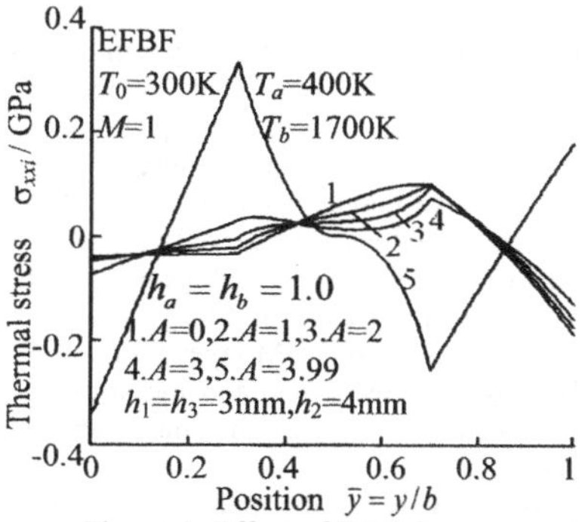

Figure 5. Effect of FGM layer porosity

Effect of Different Composite Plate on Thermal Stress

Figure 6 shows the effect of the different composite plate on thermal stress. The whole variation law of thermal stress curves 1 and 2 is similar in the ceramic and metal layers, but the thermal stress variation at the bonding interface in the nongraded two-layered composite plate becomes large, as shown in curve 2, and the curve appears sharp angle and sharp change, and reaches peak value. Compared with curve 2, the thermal stress curve 1 of the ceramic / metal composite plate with an interlayer of FGM is gentler, and the maximum tensile stress reduces by 51.8%.

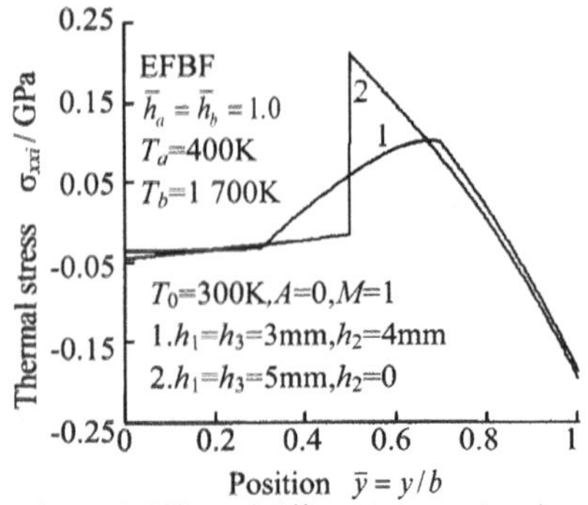

Figure 6. Effect of different composite plate

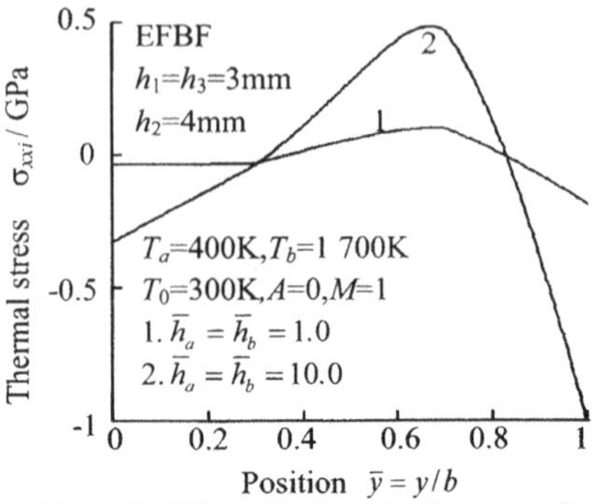

Figure 7. Effect of convective heat transfer

Effect of Convective Heat Transfer Coefficient on Thermal Stress

Figure 7 shows the effect of the convective heat transfer coefficient on thermal stress. With the increase of the convective heat transfer coefficient, the variations of thermal stress curves become big, and the thermal stress curve 1 is more gentle and smooth than curve 2, and the slope of the curve 2 is bigger than that of curve 1. Compared with the curve 1, the maximum compressive stress of the curve 2 on the surface of metal increases 8.95 times, and the maximum tensile stress of the curve 2 at the interface between FGM layer and ceramic layer increases 3.75 times, and the maximum compressive stress on the surface of ceramics increases 5.45 times.

CONCLUSION

(1) With the increase of the FGM layer thickness, the stress distribution in the ZrO_2/FGM/Ti-6Al-4V composite plate is more reasonable, and the maximum tensile stress reduces by 59.6%. When $M=1$, the thermal stress curve is relative gentle and smooth, and it does not like M=0.2, 5 that appear apparent turning points and undulations.

(2) When we take into account the effect of temperature-dependent material properties, compared with the results of constant material properties, the maximum compressive stress on the surface of ceramics increases by 20.0%, and the maximum tensile stress at the interface between ceramic layer and FGM layer increases by 66.6%.

(3) With the increase of FGM layer porosity P, the change of stress at the bonding interface of the three-layered plate increases, and the stress curves appear peak values. The tensile stress on the surface of ceramics reaches the maximum. The tensile stress is unfavorable to the strength of ceramics.

(4) Compared with the thermal stress sharp change at the bonding interface of ceramic-metal two-layered composite plate, the thermal stress of the ZrO_2/FGM/Ti-6Al-4V three-layered composite plate is very gentle, and the largest tensile stress reduces by 51.8%. With the increase of the convective heat transfer coefficients, the variations of thermal stress curves become big, and the maximum tensile stress at interface between FGM layer and ceramic layer increases 3.75 times.

ACKNOWLEDGEMENTS

We would like to thank the Hebei province education department fund (2003136) and Handan city science and technology department fund (0821120081-2) of China for the support of this project.

REFERENCES

[1]Y. Tanigawa, Some basic thermoelastic problems for nonhomogeneous structural materials, *Appl Mech Rev.*, **48**, 287-300 (1995).

[2]Y. Li, Z.M. Zhang, S.Y. Ma, Progress of the study on thermal stress of heat-resisting functionally gradient materials, *Advances in Mech.*, **30**, 571-580 (2000).

[3]Y. Obata, N. Noda, Unsteady thermal stresses in a functionally gradient material plate(Influence of heating and cooling conditions on unsteady thermal stresses), *Trans. JSME.*, Series A, **59**, 1097-1103 (1993).

[4]Y. Tanigawa, T. Akai, R. Kawamura, and N. Oka, Transient heat conduction and thermal stress problems of a nonhomogeneous plate with temperature-dependent material properties, *J. Thermal stresses*, **19**, 77-102(1996).

[5]J. Huang, Y. B. Lü, Thermal elastic limit analysis of layered plates of ceramic/metal functionally graded material, *J. Wuhan Univ. Technol.* (*Trans. Sci. & Engrg.*), **27**, 754-757 (2003).

[6]Y.J. Xu, J.J. Zhang and D. H. Tu, Transient thermal stress analysis of functionally gradient material plate with temperature-dependent material properties under convective heat transfer boundary. *China J. Mech. Engrg.*, **41**, 198-204 (2005).

[7]N. Noda, T. Tsuji, Steady thermal stresses in a functionally gradient material plate with temperature-dependent material properties. *Trans. JSME.*, Series A, **57**, 625-631 (1991).

[8]H. G. Wang, Introduction of thermal elasticity, *Tsinghua Univ. Press*, Beijing (1989).

EFFECT OF MULLITE GRAINS ORIENTATION ON TOUGHNESS OF MULLITE/ZIRCONIA COMPOSITES

Y. K. Tür*, A. E. Sünbül, H. Yılmaz and C. Duran
Dept.Mat. Science and Engineering, GYTE, Turkey.
*E-mail: yktur@gyte.edu.tr

ABSTRACT

The objective of this study was to investigate the effect of crystallographic texture on the fracture toughness in mullite/zirconia ($3Al_2O_3{\cdot}2SiO_2/ZrO_2$) composites. A mixture of Al_2O_3 and $ZrSiO_4$ powders were reactively sintered to obtain mullite/ZrO_2 composites. Tape casting method was used in order to achieve the crystallographic texture through templated grain growth using aluminum borate templates. Four different kinds of mullite/ZrO_2 compositions were prepared by utilizing templates with different lengths. The fracture toughness was determined by the indentation strength method. Presence of template particles increased the fracture toughness of the mullite/zirconia composites. Fracture toughness was 3.5 MPa $m^{1/2}$ for composites with long templates. On the other hand, an R-curve behavior was observed for composites with very short templates and fracture toughness was increased from 3 MPa $m^{1/2}$ to 4.5MPa $m^{1/2}$ with increasing indentation load. The increase in the fracture toughness was attributed to the tortured crack path due to fine but elongated mullite grains.

INTRODUCTION

Fracture toughness of a material is proportional to the energy dissipated during crack propagation and most ceramics have low fracture toughness because they have only one energy dissipation mechanism: namely, surface energy. Toughness can be enhanced if other dissipative mechanisms can be introduced into the microstructure such as grain interlocking[1,2] or phase transformation[3].

Mullite is among the noteworthy structural ceramics and has potential structural applications due to its low thermal expansion coefficient, excellent creep resistance and high temperature strength[4]. However, its engineering applications are limited because of its low fracture toughness (2-3 MPa $m^{1/2}$)[5].

Reaction sintering of alumina with zircon is an affordable and practical way of processing mullite/zirconia composites. It was shown that the addition of acicular aluminum borate templates leads to mullite grain growth around these templates. Texturing of the microstructure of mullite/zirconia composite is achieved by templated grain growth in which templates are aligned by tape casting[6].

In this study the effect of texture on fracture toughness was studied by varying the template's aspect ratio. SEM analyses were done for microstructure characterization. Rocking curve was done to quantify the orientation distribution of mullite grains in the mullite/zirconia composite. Elastic modulus, strength, hardness and toughness of the composites were measured and correlated to the textured microstructure.

EXPERIMENTAL PROCEDURE

Mullite/Zirconia ($3Al_2O_3.2SiO_2$ / ZrO_2) composites were prepared from α-Al_2O_3 (Alcoa, SG3000) and $ZrSiO_4$ (Eczacıbaşı, Doğa) powders. [001] aluminum borate ($9Al_2O_3.2B_2O_3$) whiskers (Shikoku Chemical Co.) were used as templates. 3 wt% TiO_2 (Merck, Rutile type) and 1 wt% MgO (Merck) were added to modify the liquid (or glass) phase.

Appropriate amounts of $ZrSiO_4$, Al_2O_3, TiO_2 and MgO powders were dispersed by ball milling for 24 hrs in an azeotropic mixture of methyl ethyl ketone (MEK) and ethanol (EtOH) (40/60 vol %) using PVB, PEG and BBP. 10 wt % Al-borate templates were added to the slurry in three different ways: i) pulverized templates before ball milling (PL); ii) as received templates before ball milling to break templates (BR); and iii) as received templates stirred in MEK/EtOH mixture and then added to the slurry after ball milling to keep them intact (IN). For comparison purposes, samples with no templates were also prepared (NO). Details of composite sample preparation are given elsewhere[7]. Samples with templates were pressless-sintered at 1500 °C for four hours with a constant heating rate of 7 °C/min in air while samples without templates were also pressless-sintered at two different temperatures: namely 1500 and 1600 °C for four hours. The density of the samples was determined using the Archimedes technique. The tetragonal zirconia content was obtained from X-ray diffraction analysis and the orientation distribution of grains was obtained from rocking-curves using the (002) mullite peak[7].

For mechanical tests 35x 2 x 1.5 mm specimens were prepared, the longitudinal direction of the samples was parallel to the tape casting direction. Elastic modulus of samples was measured by the resonance frequency method according to ASTM standard C1259-94 (Grindo-Sonic MkV, J.W.Lemmens,Belgium). The flexural strength of samples was measured with an electronic universal tester (Model 5569, Instron ltd.) by a three point bending test with a lower span of 25 mm and crosshead speed of 0.25 mm/min, based on ASTM standard C1161-90. Vickers hardness was measured in the range of 10 to 200 N for 10 seconds; eight indentations were made at each load. Fracture toughness was determined by indentation strength in bending method. In this method, after polishing the specimens to a sufficiently fine finish (1 μm for tensile surface, 1200 mesh SiC for all other surfaces), samples were indented on their face centers with Vickers indenter at indentation loads in the range of 10-300 N. Indented samples loaded to failure in three point bending with indent facing tension surface.

RESULTS AND DISCUSSIONS

SEM micrographs from the samples of four different compositions with no (NO), pulverized (PL), broken (BR), and intact (IN) Al-borate templates were shown in Figure 1. In these micrographs mullite grains were darker and it was observed that brighter zirconia grains were distributed throughout and among the mullite grains. Figure 1.a shows that when there was no templates added to the composition, mullite grains were coarse and have a small aspect ratio. Addition of templates resulted in elongated mullite grains. Figure 1.b shows that the microstructure of the sample with pulverized templates contained randomly oriented elongated mullite grains. A fine microstructure was obtained, even though there were some large mullite grains. Figure 1.c and 1.d shows that the microstructure of the samples with broken and intact templates contained mullite grains mostly aligned in the tape casting direction. Also, these grains were much coarser compared to the sample with pulverized templates. Each Al-borate template acted as a nucleation site, which lead to a finer microstructure.

Figure 2 shows orientation distribution of mullite grains in the samples with pulverized, broken and intact templates. The curves indicate that the elongated mullite grain orientations were distributed within 7.4° (i.e., half width at half maximum (HWHM) is 3.7°) in intact templated samples and 12.0° (i.e., HWHM is 6.0°) in broken templated samples. From Mach-Dollase equation[8] the r parameters, which is related to the texture distribution, were found to be 0.19 and 0.26, respectively. On the other hand, rocking curve analysis confirms that there was no grain orientation in pulverized templated samples. X-ray diffraction analysis showed that tetragonal zirconia phase was less than 10% in all samples.

In Table I. the relative density, elastic modulus (E), three point bending strength (σ_f), and Vickers hardness (H) values were reported for the studied composites. Samples with no templates sintered at 1500°C appeared to have a high relative density; however, X-ray diffraction analysis showed that these samples contained $ZrSiO_4$ (i.e., reaction was not completed). At 1600 °C, mullite

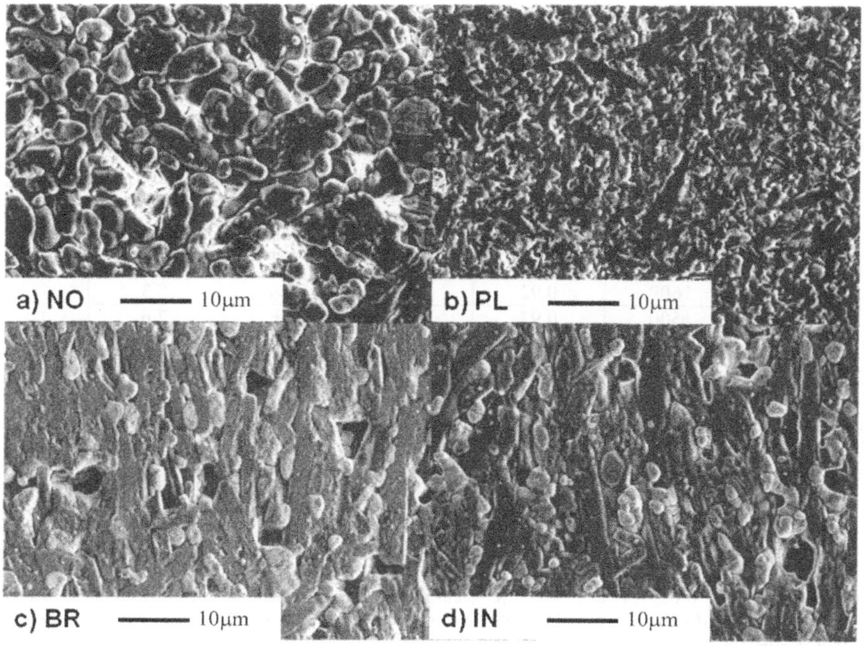

Figure 1. SEM micrographs of samples with a) no b) pulverized, c) broken, and d) intact Al-borate templates sintered at 1500 °C.

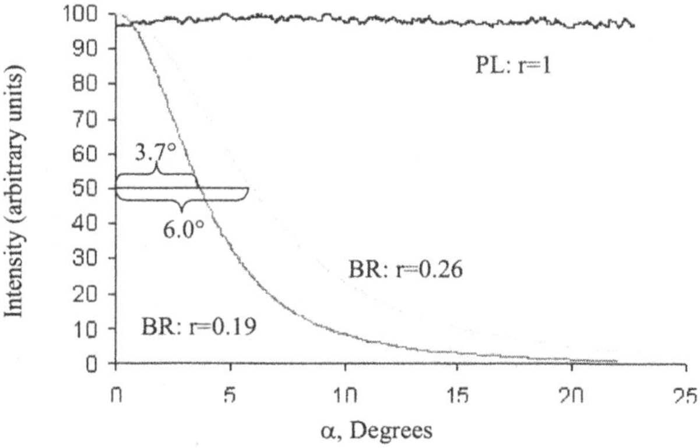

Figure 2. Orientation distribution curves for pulverized, broken and intact templates.

zirconia formation was completed but relative density of the composite was only 92%, resulting in poor mechanical properties. Among template containing samples, the one with broken templates had a relatively lower density because of the entangling of templates which lead to poor packing that cannot be eliminated during sintering. Presence of porosities resulted in degradation of elastic modulus, strength and hardness for broken templated composites. Pulverized template containing samples have the highest strength.

Table I. Strength and hardness of composites.

	Sintering (°C)	Relative Density (%)	E (GPa)	σ_f (MPa)	H (GPa)
NO	1500	0.98	148	175	4.7
NO	1600	0.92	152	128	5.3
PL	1500	0.97	188	306	7.6
BR	1500	0.94	217	279	7.1
IN	1500	0.98	226	298	7.5

Indentation strength in bending was widely used to characterize the crack propagation resistance with increasing crack length[9]. In this method a controlled crack is formed by Vickers indenter and then the specimen is subjected to a bending test. Fracture toughness is calculated by Eq. 1.

$$K_R = \eta\left(\frac{E}{H}\right)^{1/8}\left(\sigma_A P^{1/3}\right)^{3/4} \qquad (1)$$

In this equation, P is the indentation load, σ_A is the stress on the tension surface at failure, and η is fitting coefficient, η =0.62. Since $P \propto c^{3/2}$ (c: crack length), an increase in fracture toughness with indentation load signifies an increased resistance to crack propagation with increasing crack length; so called R-curve behavior.

Figure 3 shows toughness measured by indentation strength method versus indentation load for the four composites with no templated samples sintered at 1600 °C, and pulverized, broken, intact templated samples sintered at 1500 °C. For no templated sample, which had course rounded grains in the microstructure, the fracture toughness was close to 2 MPa $m^{1/2}$ for small indentation loads, and it increased to 3 MPa $m^{1/2}$ as the indentation load was increased. On the other hand, for the pulverized templated sample, which had a fine microstructure with elongated but random mullite grains, the fracture toughness was close to 3 MPa $m^{1/2}$ and increased to 4.5 MPa $m^{1/2}$ with increasing indentation load. For the intact templated sample, which had a course microstructure with oriented mullite grains, the fracture toughness was about 3.5 MPa $m^{1/2}$ and remained constant as indentation load was increased. Highest fracture toughness observed in pulverized templated sample is in agreement with the observed highest bending strength among the studied samples (Table I).

Figure 4 shows the crack propagation in samples with pulverized and intact templates. The fine microstructure obtained by using pulverized templates led to a mixture of intergranular and transgranular crack path. The resulting torturous crack path suggested that grain interlocking was present and the microstructure offered the resistance to the stable crack propagation. Grain interlocking explained the observed R-curve behavior of the toughness values (Fig. 3.b). On the other hand, the coarse microstructure of the intact templated sample led to a mostly transgranular crack path; therefore, no toughness enhancement was observed (Fig. 3.d).

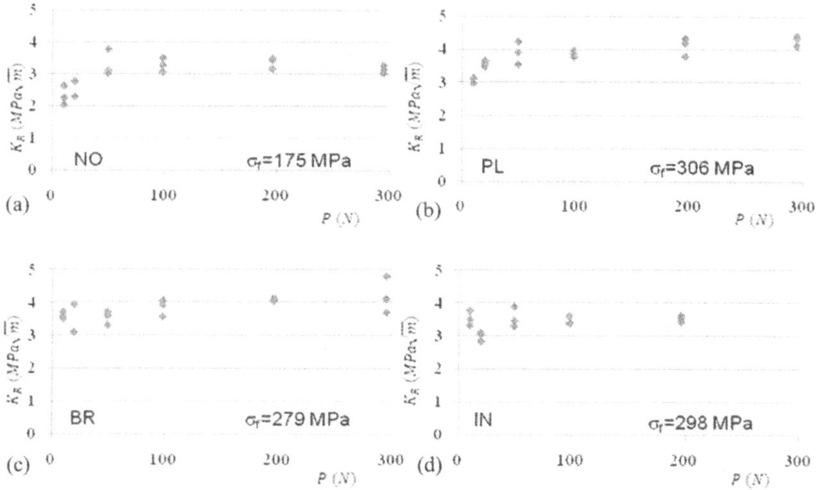

Figure 3. Toughness measured by indentation strength method versus indentation load for the four composites: a) no templated sample sintered at 1600 °C, and b) pulverized, c) broken, d) intact templated samples sintered at 1500 °C.

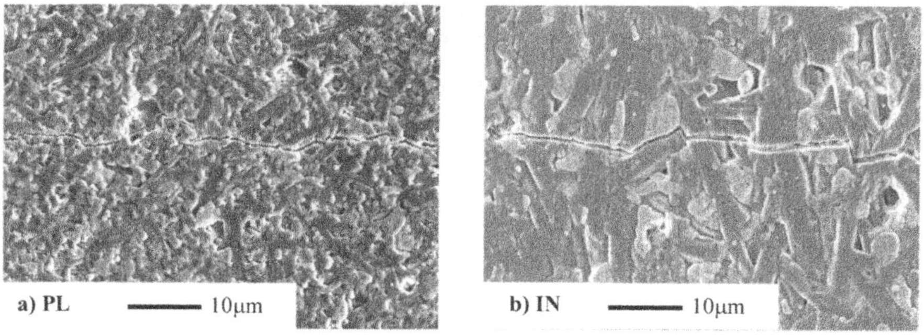

Figure 4. SEM micrographs of crack propagation in samples with a) pulverized and b) intact Al-borate templates sintered at 1500 °C.

CONCLUSIONS

A fine microstructure with randomly oriented elongated mullite grain was obtained when Al-borate templates were added in to the slurry in pulverized form. On the other hand, a course microstructure with elongated mullite grains aligned in the tape casting direction was observed when templates were added in broken or intact forms. Absence of templates resulted in coarse mullite grains with a small aspect ratio. Indentation strength fracture toughness curves showed that the pulverized templated specimen showed an R-curve behavior where toughness increased from 3 to 4.5 MPa $m^{1/2}$. R-curve behavior was attributed to the transgranular and intergranular torturous crack path. Fracture toughness of the intact templated sample was 3.5 MPa $m^{1/2}$ and remained a constant and a transgranular crack path was observed. The highest bending strength was also observed in the pulverized templated sample in accordance with its highest fracture toughness among the studied samples.

ACKNOWLEDGMENT

The financial support of the Scientific and Technical Research Council of Turkey (TUBITAK, Project #:104M229) is greatly acknowledged.

REFERENCES

[1]C.W. Li, D.J. Lee, ve S. C. Lui, R-Curve Behaviour and Strength for In-Situ Reinforced Silicon Nitrides with Different Microstructures, *J. Am. Ceram. Soc.*, **75**, 1777-85 (1992).

[2]A. Khan, H. M. Chan, and M. P. Harmer, Alumina Agglomerate Affects on Toughness-Curve Behavior of Alumina-Mullite Composites, *J. Am. Ceram. Soc.*, **83**, 3089-94 (2000).

[3]R. H. Hannink, P. M. Kelly, and B. C. Muddle, Transformation Toughening in Zirconia-Containing Ceramics, J. Am. Ceram. Soc., **83**, 3, 461-87 (2000).

[4]T.I. Mah, and K. S. Mazdiyasni, Mechanical Properties of Mullite, *J. Am. Ceram. Soc.*, **66**, 699-703 (1983).

[5]T. Huang, M. N. Rahaman, T. Mah, ve T. A. Parthasarathay, Anistropic Grain Growth and Microstructural Evolution of Dense Mullite Above 1550°C, *J. Am. Ceram. Soc.*, **83**, 204-10 (2000).

[6]I.E.Gonenli and G.L.Messing, Texturing of Mullite by Templated Grain Growth with Aluminum Borate Whiskers, *J. Eur. Cer. Soc.*, **21**, 2495-2501 (2001).

[7]C. Duran, and Y.K.Tür, Phase Formation and Texture Development in Mullite/Zirconia Composites Fabricated by Templated Grain Growth, *J. Mater. Sci*, **41**, 3303-13 (2006).

[8]W. A. Dollase, Correction of Intensities for Preferred Orientation in Powder Diffractometry: Application of the March Model, *J. Appl. Crystallogr.*, **19**, 267-272 (1986)

[9]B. Lawn, Fracture of Brittle Solids (2nd ed.) Cambridge University Press (1998).

CONTROLLED CRYSTALLISATION OF GRAIN BOUNDARY-TYPE Y-SIALON GLASS TYPICAL OF THOSE FOUND IN SILICON NITRIDE CERAMICS

Michael J. Pomeroy and Stuart Hampshire
Materials and Surface Science Institute
University of Limerick, Limerick, Ireland

ABSTRACT

Following liquid phase sintering of silicon nitride ceramics with yttria and alumina densification additives, a Y-SiAlON glass remains at the grain boundaries. Crystallisation of such a glass in bulk form has been studied using both differential thermal analysis (DTA) and separate isothermal heat treatments in a tube furnace under nitrogen. The activation energy for the crystallisation process was determined by DTA. The nucleation temperature, Tg + 40°C, which corresponded to the maximum volume fraction of crystalline phases, agreed closely with the optimum nucleation temperature of Tg + 35°C, found from DTA. The optimum crystal growth temperature was observed to be 1210°C and yielded the α- and β-polymorphs of yttrium disilicate. The activation energy for crystallisation was observed to be similar to that for viscous flow of Y-SiAlON glasses. Heat treatments over a range of temperatures resulted in formation of different polymorphs of yttrium disilicate in addition to silicon oxynitride. Careful analysis of DTA crystallisation exothermic peaks were carried out in order to clearly identify polymorphic transformations of yttrium disilicate. Partial substitution of yttrium by lanthanum stabilises the α-polymorph of yttrium disilicate over a wider temperature range.

INTRODUCTION

Oxynitride (M-SiAlON) glasses are found as grain boundary phases in silicon nitride based ceramics as a result of the use of densifying additives that allow liquid phase sintering to occur[1-4]. These additives, such as yttria or one of the rare earth oxides, plus alumina, react with silicon nitride and silica present on the nitride particle surface to form M-Si-Al-O-N (M=Y or rare earth) liquid phases which result in densification of the ceramics and transformation of the α-silicon nitride to the β form[1-4]. The liquid phases cool as intergranular oxynitride glass films[5] and form triple points. The content of the modifying cation (Y or RE) in these intergranular glasses, and the overall volume fractions of glass phase within the silicon nitride ceramic control the mechanical properties[1-3].

The intergranular glass was known to have deleterious effects on high-temperature strength and considerable efforts were made to crystallize the grain boundary glasses in Y doped through controlled post-densification annealing[6-10]. This was, in large measure, very successful, and significant improvements in strength, at room temperature (to >1100 MPa) and high temperature (> 800 MPa at 1300°C) were obtained although there are theoretical limits to the extent of crystallization possible[7].

Bulk oxynitride glasses have been studied extensively and reviewed[11-12] recently. Studies of crystallization[13-14] of bulk Y-Si-Al-O-N glasses show that formation of yttrium disilicate occurs and the morphology changes with temperature. As further growth occurs, yttrium aluminium garnet, YAG, also forms leaving a N-rich glass phase from which silicon oxynitride precipitates. This paper describes the optimisation of crystallization of a similar Y-Si-Al-O-N glass and its implications for grain boundary crystallization of silicon nitride ceramics.

EXPERIMENTAL PROCEDURE

The composition of the Y-Si-Al-O-N glass studied is given in Table I with values of density, microhardness, glass transition temperature (Tg) and crystallisation exotherm temperatures (T_{c1}, T_{c2} and T_{c3}). The value of the first crystallisation exotherm was unaffected by particle size, indicating that crystallisation was a bulk rather than a surface phenomenon.

Table I. Composition of Y-Si-Al-O-N glass and selected glass properties

	Glass composition					Selected property values					
	Y	Si	Al	O	N	density	Micro-hardness	T_g	T_{C1}	T_{C2}	T_{C3}
[eq.%]	28	56	16	83	17	[$g.cm^{-3}$]	[GPa]	[°C]	[°C]	[°C]	[°C]
[at.%]	12.3	18.5	7	54.7	7.5	3.75 ± 0.01	10.2 ± 0.2	985	1136	1185	1270

Specimens of the glass, typically 10 x 10 x 2 mm, were embedded in boron nitride in an alumina crucible and subjected to varying heat treatments, shown in Table II, in a flowing nitrogen environment. In order to determine the optimum nucleation temperature, heat treatments were conducted at intervals of 20°C over a temperature range from T_g-40 to T_g+100 (°C) for 10 hours followed by heating to 1270°C (T_{C3}) for 30 minutes to grow the crystal nuclei for subsequent microstructural and microhardness analysis. Having established the optimum nucleation temperature, the optimum nucleation hold time was established by holding the sample for 2, 4, 10, 16 and 32 h prior to heating to T_{C3} for 30 min to grow the crystals. Having established optimum nucleation temperature and hold time, heat treatments were carried out using these values and then varying the crystal growth hold temperature from 1170 to 1310°C at 20°C intervals. Crystal growth hold times were 30 minutes. Heating and cooling rates for all heat treatments were: ambient to nucleation hold temperature at 20°Cmin^{-1}, nucleation hold temperature to crystal growth temperature at 10°Cmin^{-1} and cooling from the crystal growth temperature at 10°Cmin^{-1}.

Table II Heat treatment conditions to determine optimum nucleation temperature and time and optimum crystal growth temperature

For determination of:	Temperature range [°C]	Time [h]	Temperature range [°C]	Time [h]
Optimum nucleation temperature	T_g-40 to T_g+100	10	T_{C3} = 1270	0.5
Optimum nucleation time	Tg + 40	2 to 32	T_{C3} = 1270	0.5
Optimum crystal growth temperature	Tg + 40	10	1170 to 1310	0.5

Heating rates for all heat treatments up to nucleation hold temperature = 20°C min^{-1}
Heating and cooling rates to crystal growth temperature and cooling to ambient = 10°C min^{-1}

Following heat treatment, density was measured using an Archimedes technique. All samples were subjected to X-ray diffraction (XRD) using Cu K_α radiation. Specimens were also mounted, polished and examined using backscattered electron imaging. These images were analysed using a point counting technique[15] in order to establish the volume fractions of crystalline and residual amorphous phases. The polished samples were also tested to determine microhardness using a diamond indenter operating under a 300g load held for 15s.

Differential thermal analysis, DTA, was also carried out to determine the optimum nucleation temperature according to the method reported by Marotta et al.[16] who concluded that, if samples are held for the same time tn, at each heat-treatment temperature Tn, then ln I (kinetic rate constant for nucleation) is proportional to $\{(1/T_p) - (1/T_p')$, where T_p and T_p' are, respectively the crystallization exotherm temperatures obtained with and without a nucleation hold. Plotting $\{(1/T_p) - (1/T_p')\}$ against nucleation hold temperature gives a bell shaped curve, with the optimum nucleation temperature corresponding to the maximum of this curve. The activation energy (E) is related to heating rate (α), crystallisation temperature T_p and gas constant R such that a plot of $\ln(\alpha^3/T_p^2)$ versus $1/T_p$ yields a straight line of slope -3E/R, when crystallisation is a bulk process[17].

RESULTS

HEAT TREATMENT S: OPTIMISATION OF NUCLEATION TEMPERATURE AND TIME

Fig. 1 shows that the highest fractional crystallisation as found by microstructural analysis occurred at 1025°C corresponding to T_g + 40°C. The microhardness was also a maximum at this temperature and showed an increase over that for the parent glass of 5.3%. Therefore, the optimum nucleation temperature was concluded to be T_g + 40°C. The nucleation hold time giving the most significant level of crystallisation was observed to be 10 hours at T_g + 40°C, and so the optimum nucleation hold time was considered to be 10 hours at 1025°C.

HEAT TREATMENT S: OPTIMISATION OF CRYSTAL GROWTH TEMPERATURE

Fig. 2 shows the effect of crystal growth temperature on the volume fraction of crystal phases and the change in microhardness compared to the Y-Si-Al-O-N glass. It is seen that, following

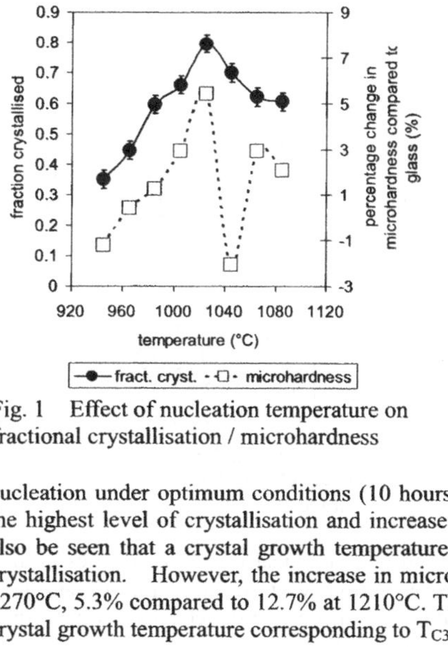

Fig. 1 Effect of nucleation temperature on fractional crystallisation / microhardness

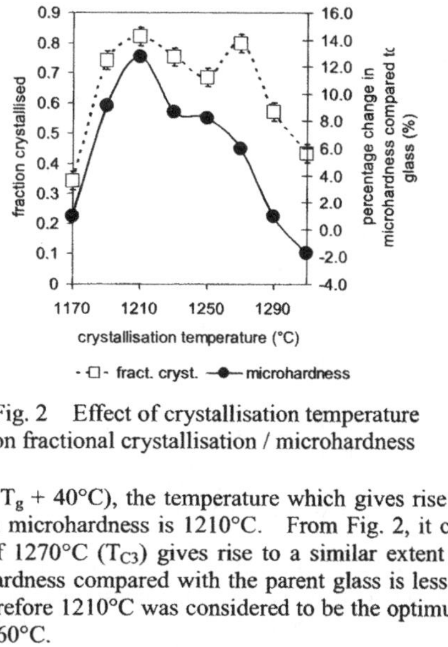

Fig. 2 Effect of crystallisation temperature on fractional crystallisation / microhardness

nucleation under optimum conditions (10 hours / T_g + 40°C), the temperature which gives rise to the highest level of crystallisation and increase in microhardness is 1210°C. From Fig. 2, it can also be seen that a crystal growth temperature of 1270°C (T_{C3}) gives rise to a similar extent of crystallisation. However, the increase in microhardness compared with the parent glass is less at 1270°C, 5.3% compared to 12.7% at 1210°C. Therefore 1210°C was considered to be the optimum crystal growth temperature corresponding to T_{C3} - 60°C.

DTA ANALYSIS OF OPTIMUM NUCLEATION TEMPERATURE AND ACTIVATION ENERGY FOR CRYSTALLISATION

The plot of $\{(1/T_p) - (1/T_p')\}$ against nucleation hold temperature showed that optimum nucleation temperature is 1020°C, i.e. T_g + 35°C, which is in close agreement with the optimum nucleation temperature determined using the heat treatment experiments. A plot of $\ln(\alpha^3/T_p^2)$ versus $1/T_p$ gave good linear correspondence. The activation energy calculated from this slope was 834 kJ.

mol^{-1}. This agrees well with activation energies observed for viscous flow (creep) of Y-Si-Al-O-N glasses.

EFFECT OF NUCLEATION AND GROWTH CONDITIONS ON PHASE FORMATION

Table III shows that the phase assemblage associated with the optimum nucleation temperature (T_g + 40°C) comprises α + β yttrium disilicates, Si_2N_2O and traces of YAG. For nucleation hold temperatures greater than this, the β-disilicate becomes the most prevalent phase. For differing nucleation hold times, changes in phase assemblage involve not only changes in the polymorph of yttrium disilicate but also the formation of the more yttrium rich Y_2SiO_5 at the longer hold times of 16 and 32 hours. Si_2N_2O is only observed in trace amounts except for the optimum hold time of 10 hours. It is also significant that at ≤1170°C, y-yttrium disilicate is the only phase present and therefore this must be the only phase existing after the nucleation hold for 10h at T_g + 40°C (see Table IV).

Accordingly, the phase assemblage corresponding to the optimum crystal growth temperature of 1210°C can only arise if y transforms to α– and β–$Y_2Si_2O_7$. The phase assemblage arising at the second best optimum crystallisation temperature (1270°C) comprises a greater proportion of β-$Y_2Si_2O_7$ (see Table IV) and therefore more of the y or perhaps the α polymorph must have transformed.

The data given in Table IV does not fully correspond with the temperatures of the three exothermic events given in Table II and as a result, the DTA curve was more carefully analysed.

Table III Effect of changing nucleation temperature on phase assemblage

Temp. [°C]	Phases present in order of decreasing X-ray intensity
T_g - 40	α, y
T_g - 20	α, y, β
T_g	α, β, Si_2N_2O
T_g + 20	α, β, Si_2N_2O, YAG*
T_g + 40	β, α, Si_2N_2O, YAG*
T_g + 60	β, α, Si_2N_2O, YAG*
T_g + 80	β, Si_2N_2O, γ, YAG*
T_g + 100	β, Si_2N_2O, γ, YAG*

Table IV Effect of changing nucleation temperature on phase assemblage

Temp. [°C]	Phases present in order of decreasing X-ray intensity
1170	y
1190	y, α, β
1210	α, β
1230	β, α, Si_2N_2O*, YAG*
1250	β, α, Si_2N_2O, YAG*
1270	β, α, Si_2N_2O, YAG*
1290	δ, Si_2N_2O, β, YAG*
1310	δ, Si_2N_2O, β, YAG*

y, α, β, δ, γ – various polymorphs of yttrium disilicate ($Y_2Si_2O_7$),
YAG - yttrium aluminium garnet ($Y_5Al_3O_{12}$), * = trace amounts

Fig. 3 shows a reinterpretation of the original curve and it is clear that the discontinuities in the curve can be associated with the transformations observed from y– to α– to β–$Y_2Si_2O_7$ and the crystallisation of Si_2N_2O. This may be problematic for silicon nitride ceramics where the temperature of operation may differ from that used to crystallise the grain boundary glass.

One approach which may overcome variable polymorphic phase assemblages during crystallisation is to stabilise one single polymorph of the disilicate by doping as was shown for mixed Y-La-Si-Al-O-N glasses[18]. In this case, a glass of the same composition as that studied here, but with 25% of the Y replaced by La, was heated to 1300°C and a single α-yttrium disilicate phase was crystallised. These findings were rationalised by superimposing the average ionic radius for the mixed La-Y glass modifiers on the polymorph stability – ionic radius map given by Maier et al.[19]. This shows that the mixed α–$Y_{1.5}La_{0.5}Si_2O_7$ should be stable up to at least 1350°C before any transformation occurs.

CONCLUSIONS

1) The crystallisation process for a 28Y:56Si:16Al:O:N glass containing 17 equivalent % nitrogen (7.5 at.%) is highly complex due to the fact that all 5 polymorphs of yttrium disilicate can form under varying conditions of heat treatment as well as the additional crystallisation of silicon oxynitride.

2) The optimum nucleation conditions for this Y-Si-Al-O-N glass found from heat treatment experiments and from DTA were in good agreement and correspond to those which offer the best heterogeneous nucleation condition for the transformation of y- to α– and β– yttrium disilicates.

3) In order to obtain a single phase disilicate, partial substitution of Y by a rare earth ion such as La in the Y-Si-Al-O-N glass can be made so that the average ionic radius of the mixed modifier system is compatible with obtaining a single disilicate polymorph.

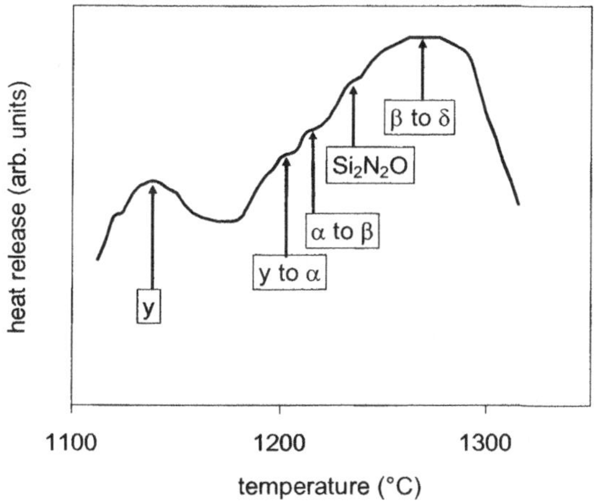

Figure 3 DTA curve showing crystallisation range and superposition of temperatures at which changes in phase assemblage occur (see Table IV)

REFERENCES

[1] F. L. Riley, Silicon Nitride and Related Materials. *J. Am. Ceram. Soc.,* **83** [2], 245–65 (2000).

[2]S. Hampshire, 2009, Silicon Nitride Ceramics, in: *Advances in Ceramic Materials*, Ed. B Ralph, P. Xiao, Trans Tech Publications, Switzerland, *Mater. Sci. Forum,* **606**, 27-41 (2009).

[3]E. Y. Sun, P. F. Becher, K. P. Plucknett, C.-H. Hsueh, K. B. Alexander, S. B. Waters, K. Hirao and M. E. Brito, Microstructural Design of Silicon Nitride with Improved Fracture Toughness: II, Effects of Yttria and Alumina Additives, *J. Am. Ceram. Soc.*, **81** [11], 2831–40 (1998).

[4]A. Tsuge, and K. Nishida, High-Strength Hot-Pressed Si_3N_4 with Concurrent Y_2O_3 and Al_2O_3 Additions, *Am. Ceram. Soc. Bull.*, **57** [4], 424-31 (1978).

[5]P. F. Becher, G. S. Painter, E.Y. Sun, C. H. Hsueh, M. J. Lance, The importance of amorphous intergranular films in self-reinforced Si3N4 ceramics, *Acta Mater.*, **48**, 4493-99 (2000).

[6]A. Tsuge, K. Nishida, and M. Komatsu, Effect of Crystallizing the Grain Boundary Glass Phase on the High-Temperature Strength of Hot-Pressed Si_3N_4 Containing Y_2O_3, *J. Am. Ceram. Soc.*, **58**, 323-26 (1975).
[7]R. Raj, and F. F. Lange, Crystallization of Small Quantities of Glass (or a Liquid) Segregated in Grain-Boundaries, *Acta Metall.*, **29** [12], 1993-2000 (1981).
[8]M. K. Cinibulk, G. Thomas and S. M. Johnson, Grain-Boundary-phase Crystallization and Strength of Silicon-Nitride Sintered with a YSiAlON Glass, *J. Am. Ceram. Soc.*, **73,** 1606-12 (1990).
[9]M. K. Cinibulk and G. Thomas, Fabrication and Secondary-Phase Crystallization of Rare-Earth Disilicate Silicon-Nitride Ceramics, *J. Am. Ceram. Soc.* **75**, 2037-43 (1992).
[10]M. K. Cinibulk, G. Thomas and S. M. Johnson, Strength and Creep-Behavior of Rare-Earth Disilicate Silicon-Nitride Ceramics, *J. Am. Ceram. Soc.* **75**, 2050-55 (1992).
[11]S. Hampshire and M. J. Pomeroy, Oxynitride Glasses, *Int. J. Appl. Ceram. Tech.*, **5** [2], 155-63 (2008).
[12]S. Hampshire, Oxynitride Glasses, *J. Euro. Ceram. Soc.*, **28** [7], 1475-83 (2008).
[13]T. K. Dinger, R. S. Rai. and G. Thomas Crystallization Behavior of a Glass in the Y_2O_3-SiO_2-AlN System, *J . Am. Ceram. Soc.*, **71**, 236-34 (1988).
[14]J.L. Besson, D. Billieres, T. Rouxel, P. Goursat, R. Flynn and S. Hampshire, Crystallization and Properties of a Si-Y-Al-O-N Glass Ceramic, *J. Am. Ceram. Soc.* **76**, 2103-2105 (1993).
[15]R. T. Dehoff and F. N. Rhines, *Quantitative Microscopy*, McGraw–Hill, New York, 1968.
[16]A. Marotta, A. Buri, F. Branda and S. Saiello, Nucleation and crystallisation of $Li_2O.2SiO_2$ Glass—A DTA Study, in *Advances in Ceramics*, Vol. **4**, *Nucleation and Crystallization in Glasses*, American Ceramic Society, Columbus, OH (1982), pp.146–152.
[17]K. Matusita and S. Sakka, "Kinetic-Study on Crystallization of Glass by Differential Thermal-Analysis - Criterion on Application of Kissinger Plot", *J. Non-Cryst. Solids*, **38–39**, 741-46 (1980).
[18]M. J. Pomeroy, E. Nestor, R. Ramesh and S. Hampshire, Properties and crystallisation of rare earth SiAlON glasses containing mixed trivalent modifiers, *J. Am. Ceram. Soc.*, **88** [4], 875-881 (2005).
[19]N. Maier, G. Rixecker and K. G. Nickel, Formation and stability of Gd, Y, Yb and Lu disilicates and their solid solutions, *J. Solid State Chem.* **179**, 1630-35 (2006).

HIGH TEMPERATURE COMPRESSION CREEP BEHAVIOR OF AMORPHOUS SI-B-C-N CERAMICS IN CONTROLLED ATMOSPHERE

Ravi Kumar [1] *, C. Eswarapragada [1], A. Zimmermann [2], and F. Aldinger [3]

[1] Dept. of Metallurgical & Materials Engineering, IIT Madras, Chennai 600036, India
[2] Corporate Research and Development, Robert Bosch GmbH, Germany
[3] Max-Planck-Institute for Metals Research, Stuttgart, Germany.

ABSTRACT

Amorphous Si-B-C-N ceramics obtained from the polymer precursor $(B[C_2H_4\text{-}Si(CH_3)NH]_3)_n$ has been investigated for its high temperature mechanical properties by carrying out compression creep experiments. In order to understand and quantify the deformation behavior more precisely, free of any oxidation effects, the experiments were conducted in argon atmosphere. The results showed only primary creep for all stresses and temperatures tested during the entire creep test time which is typical for these classes of materials. The continuous decrease in creep rates is indicative of the continuous structural changes in these materials. However, there was no quantitative difference between the results obtained in an oxygen-free environment and in atmospheric ambience. It was observed that the shrinkage component of the total strain rate dominates over oxidation and controls the deformation behavior.

INTRODUCTION

Polymer precursor derived amorphous Si-B-C-N ceramics have been investigated to a reasonable extent to understand the high temperature deformation behavior by carrying out extensive compression creep experiments. The studies showed that these materials yield highly creep resistant class of materials. The relatively high creep resistance comes from the fact that these materials lack the presence of any oxidic grain boundary phase, resulting from low melting point sintering additives as in the case of conventionally sintered materials. The deformation in amorphous Si-B-C-N ceramics at high temperatures was analyzed using the free volume model which is generally used to describe the deformation behavior in amorphous metallic glasses [1]. However, this amorphous state of these materials is metastable and these materials transform in to a more stable crystalline state at temperature beyond 1800 °C. The deformation rates exhibited by these transformed nano-crystalline materials indicate an order of magnitude increase in the creep resistance, but their creep resistance is affected by oxidation effects. Extensive compression creep experiments on the nano-crystalline Si-B-C-N ceramics both in atmospheric ambience and controlled atmosphere have shown a strong influence of the deformation behavior and mechanisms on the test atmosphere [2, 3]. However, the investigation of the deformation mechanisms of amorphous Si-B-C-N ceramics has been restricted to only in atmospheric ambience. Hence, the purpose of this study is to study the deformation behavior these materials free of any oxidation effects and to compare the behavior with the results obtained in air.

EXPERIMENTAL

The synthesis of the polymer precursor with the chemical composition $(B[C_2H_4\text{-}Si(CH_3)NH]_3)_n$ and details of the ceramization processes can be found elsewhere [4, 5]. Compression-creep specimens were prepared by cutting and grinding the annealed samples. For carrying out compression-creep experiments in controlled atmosphere, a test frame 'Zwick' with a furnace from the company 'Maytec' was used as shown in Figure 1. Two thermocouples (Pt/Pt 10% Rh) are installed inside the furnace, one for measuring the temperature of the sample and connected to the controller to maintain the creep test temperature, and the other for limiting the temperature to 1530 °C. The heating elements and the radiation shields are made of molybdenum (Figure 1). A rotary-vane vacuum pump in series with a turbomolecular pump which can generate vacuum of the order of 9 x 10^{-6} mbar is used for the evacuation of the chamber. The vacuum generated in the chamber is constantly monitored using a Penning gauge. A 10 kN load cell is installed in the machine, although the frame itself is designed for 100 kN. The load measurement accuracy is of the order of ± 0.2 N. During all the creep tests a continuous flow of argon gas with a flow rate of 1 litre/min is maintained.

Figure 1. Experimental set-up for compression creep showing the sample (**P**) placed between two SiC pads (**S**). Heating and radiation shields are made of molybdenum (**R**).

RESULTS AND DISCUSSION

Time and stress dependence of deformation

The time and stress dependence of deformation is shown in Figure 1 and Figure 2. The results from the compression creep experiments carried out in the stress range of σ = 5 MPa – 200 MPa coupled with a constant temperature of 1400 °C indicate a maximum strain of around 4.5 % after around 200 hours of creep testing time. From Figure 2 it can be clearly observed that the deformation rates at all stresses continuously decrease with time indicating the presence of only primary stage of creep and no asymptotic approach to steady state even after several hours of creep testing time. The deformation rates varied from 10^{-6} s^{-1} at 10^3 s to 10^{-9} s^{-1} to 10^{-8} s^{-1} at 10^6 s indicating nearly two orders of magnitude increase in creep resistance during creep test time. Further, it is seen that the time dependence of the deformation rates is the same at all stresses. However, a distinct stress dependence of deformation rates is observed with the experimental data running parallel for all stresses beyond 2 x 10^4 s which indicates no change in the mechanism of deformation during that time. An experiment carried out at 5 MPa stress is considered as a no-load condition since that is the minimum stress that is required to keep the specimen in position. Also, since these amorphous Si-B-C-N ceramics do experience shrinkage, the strain and strain rate that correspond to this stress is considered purely shrinkage and shrinkage rate respectively.

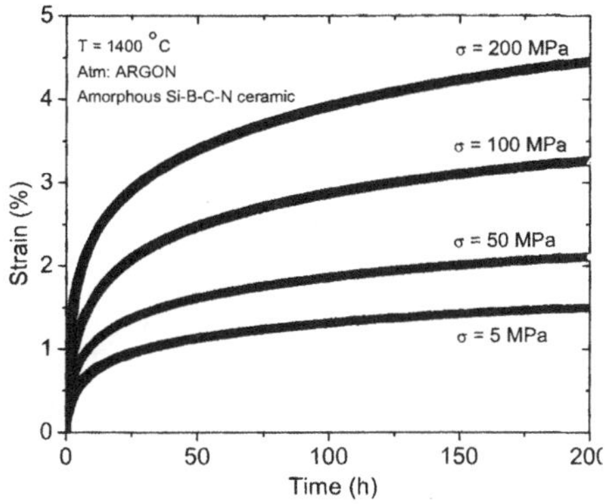

Figure 2. Deformation of amorphous Si-B-C-N ceramics as a function of time for various compressive stresses at a constant test-temperature of 1400 °C.

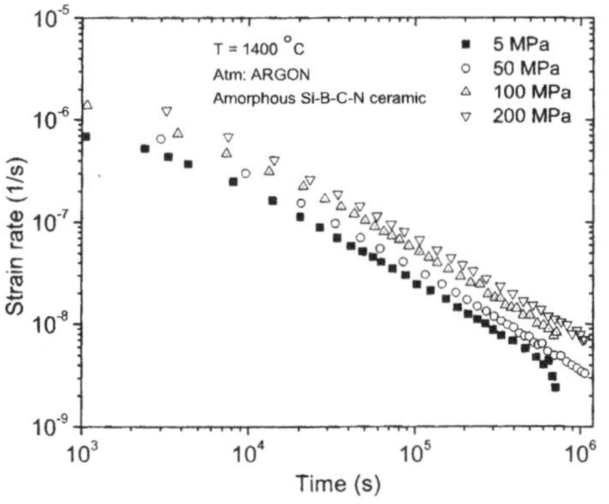

Figure 3. Deformation rates of amorphous Si-B-C-N ceramics at various compressive stresses at a constant temperature of 1400 °C indicating the presence of only primary stage of creep.

Temperature dependence of deformation

The understand the temperature dependence of the deformation rates, compression creep experiments were carried out between 1400 °C to 1500 °C at a constant stress of 50 MPa and the results are shown in Figure 4.

For all the temperatures tested, it is again seen that no approach to steady state is reached during the entire creep test time and increase in creep resistance is more than two orders of magnitude with three orders of magnitude increase in time with strain rates of the order of ~ 4 x 10^{-9} s^{-1}. The experimental data corresponding to creep test temperatures 1400 °C, 1450 °C and 1500 °C (as exemplified in Figure 4) overlap indicating that there is no strong temperature dependence of the strain rates between 1400 °C to 1500 °C.

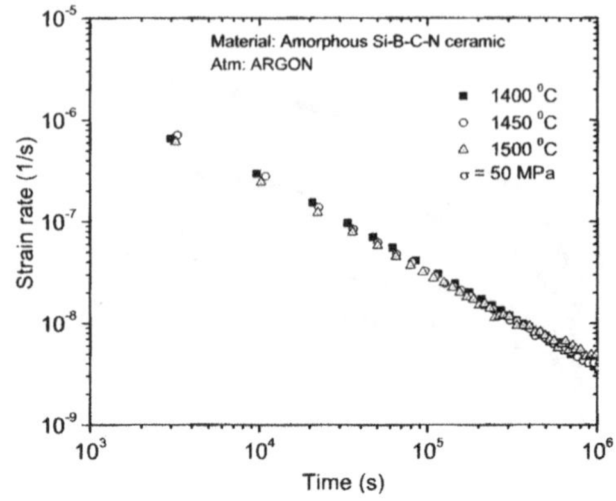

Figure 4. Deformation rates of amorphous Si-B-C-N ceramics at various temperatures coupled with a constant stress of 50 MPa

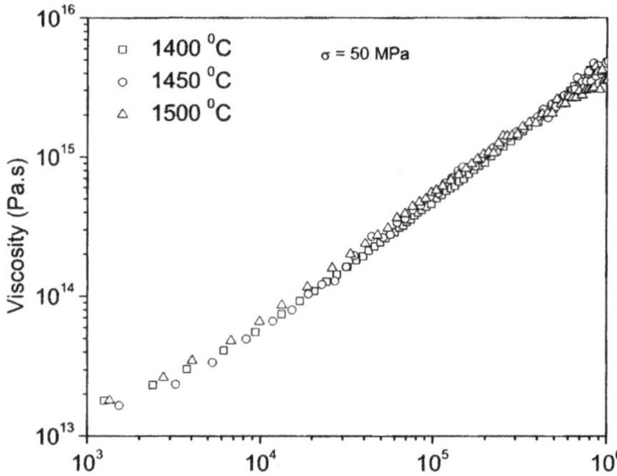

Figure 5. Variation of Newtonian viscosity as a function time for various temperatures at a constant stress of 50 MPa.

The values of viscosity determined assuming Newtonian viscous flow using the expression $\sigma / 3\dot{\varepsilon}$ in the temperature range 1400 °C – 1500 °C at a constant load of 50 MPa is exemplified in Figure 5. A linear increase in the viscosity values with time for the temperatures tested is observed. The viscosities as determined from the creep experiments were as high as ~ 5 x 10^{15} Pa.s at the end of the creep tests and the continued increase in the trend is indicative of the excellent mechanical stability of these materials at elevated temperatures. While a linear increase in viscosity is predicted by the free volume model which was earlier used to understand the deformation mechanisms of these materials, the similar trend observed here justifies the use of such a model to understand the mechanism of deformation.

Influence of atmosphere on deformation rates

Since the purpose of this study was to investigate and analyze the deformation behavior of amorphous Si-B-C-N ceramics free of any oxidation effects, the results obtained in the present study were compared with the results obtained earlier on the same material in atmospheric ambience [1]. Unlike their nano-crystalline counterparts, where the deformation rates were strongly influenced by oxidation effects [2, 3], the results from the amorphous Si-B-C-N ceramics both in air and argon did not show any significant difference. The values of strain rates and viscosities in the stress and the temperature regimes tested in both air and argon were almost the same at all times suggesting the presence of possible competitive mechanisms which suppress the oxidation effects on the deformation behavior of amorphous Si-B-C-N ceramics. Since the total strain rate is composed of an additional shrinkage component in addition to the anelastic and viscous components, the total deformation behavior can be a strong function of the shrinkage component. Hence the effect of oxidation on the shrinkage component determines the overall deformation response of these amorphous materials. A comparison of the shrinkage rate observed in both atmospheric ambience and argon is exemplified in Figure 6.

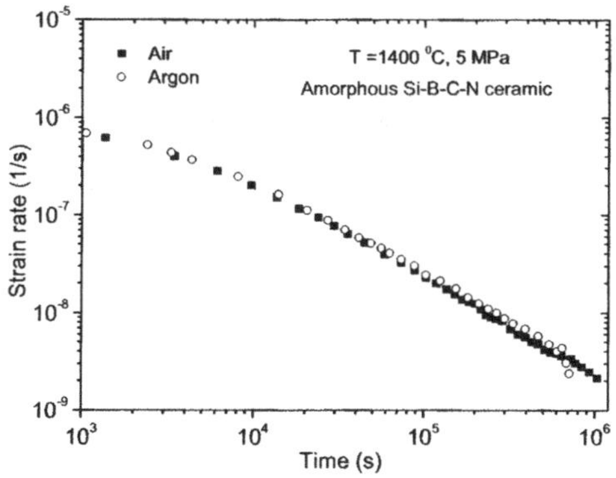

Figure 6. Shrinkage rates of amorphous Si-B-C-N ceramics tested in both atmospheric ambience and controlled atmosphere.

As clearly observed in Figure 6, the shrinkage rates of amorphous materials tested in air and argon overlap hinting the dominant effect of shrinkage over oxidation in the sample tested in air. This clearly shows that shrinkage and oxidation of the amorphous material with time, tested in air are competitive processes and the shrinkage component dominates over the oxidation processes resulting in quantitative similarity in the deformation behavior of amorphous Si-B-C-N ceramics investigated in atmospheric ambience and controlled atmosphere.

CONCLUSION

The high temperature deformation behavior of polymer precursor derived amorphous Si-B-C-N ceramics was investigated in controlled atmosphere to understand the mechanisms of deformation free of oxidation effects. The results showed the existence of only primary stage of creep indicating continuous structural changes in the material, including shrinkage. The strain rates and the viscosities derived from creep experiments showed values in the order of $10^{-8} - 10^{-9}$ s^{-1} and $10^{15} - 10^{16}$ Pa.s respectively. However, comparison of these results with those obtained in atmospheric ambience did not show any substantial variation. No influence of oxidation on the shrinkage component of the deformation rate for sample tested in air indicates the shrinkage behavior to dominate the oxidation effects. This obviously results in quantitative similarity of the deformation behavior of amorphous Si-B-C-N ceramics and the mechanisms per se will not be different as the creep parameters that will be derived in both the cases will be the same.

REFERENCES

[1] M. Christ, G. Thurn, M. Weinmann, J. Bill and F. Aldinger, High-Temperature Mechanical Properties of Si-B-C-N Ceramics and the Applicability of Deformation Models Developed for Metallic Glasses, J. Am. Ceram. Soc., **83**, 3025-32 (2000).

[2] R. Kumar, R. Mager, F. Phillipp, A. Zimmermann, and G. Rixecker, High Temperature Deformation Behavior of Nanocrystalline Precursor Derived Si-B-C-N Ceramics in Controlled Atmosphere, Int. J. Mater. Res., **97,** 626 – 631 (2006).

[3] R. Kumar, F. Phillipp, and F. Aldinger, Oxidation Induced Effects on the Creep Properties of Nano-Crystalline Boron Doped Silicon Carbonitrides, Mater. Sci. and Eng. A, **445 – 446**, 251 – 258 (2007).

[4] R. Riedel, A. Kienzle, W. Dreßler, L. M. Ruswisch, J. Bill, F. Aldinger: Nature (London), **382,** 796 - 798 (1996).

[5] R. Kumar, P. Gerstel, Y. Cai, G. Rixecker, F. Aldinger, Processing, Crystallization and Characterization of Polymer Derived Nano-crystalline Si-B-C-N Ceramics, J. Mater. Sci., **41 [21],** 7088 – 95 (2006).

FOOTNOTES

* Communicating author
Email: nvrk@iitm.ac.in

ACKNOWLEDGEMENTS

Mr. Gerstel is acknowledged for providing the polymer precursor for the experiments. Mr. Mager is acknowledged for the technical support in carrying out the creep experiments. The Max-Planck-Society is gratefully acknowledged for the financial support.

FABRICATION AND PROPERTIES OF Si_3N_4/BN COMPOSITE CERAMICS BY PRESSURELESS SINTERING WITH Yb_2O_3-Al_2O_3-Y_2O_3 AS SINTERING ADDITIVES

Yongfeng Li[1], Ping Liu[1], Guanjun Qiao[1]*, Jianfeng Yang[1], Haiyun Jin[2], Xiangdong Wang[3], and Guojun Zhang[4]
[1] State Key Laboratory for Mechanical Behavior of Materials, Xi'an Jiaotong University, Xi'an, Shaanxi, 710049 China.
[2] School of Electrical Engineering, Xi'an Jiaotong University, Xi'an, Shaanxi, 710049 China.
[3] School of Science, Xi'an Jiaotong University, Xi'an, Shaanxi, 710049 China.
[4] State Key Laboratory of High Performance Ceramics and Superfine Microstructure, Shanghai Institute of Ceramics, Chinese Academy of Sciences, Shanghai, 200050 China.

ABSTRACT

With commercial α-Si_3N_4 and h-BN powders as starting materials, Yb_2O_3, Al_2O_3, Y_2O_3 as sintering additives, Si_3N_4/BN composite ceramics with 25vol% h-BN were fabricated by pressureless sintering. Various amounts of Yb_2O_3 (0, 2, 4, 8, 15wt%) were added, with a constant Y_2O_3/Al_2O_3 weight ratio of 2:1 and total additives (Y_2O_3/Al_2O_3/Yb_2O_3) amount of 15wt%. The densification behaviors, α-β transformation, and room-temperature strength of Si_3N_4/BN composite were investigated. The sintering shrinkage of the samples increased with Yb_2O_3 content increasing, and the highest linear shrinkage was observed for samples containing 4wt% Yb_2O_3. The same trend in flexural strength was also observed. The flexural strength of all the specimens increased with the addition up to 4wt% and changed greatly thereafter. The highest room-temperature flexural strength, 264.3MPa, was obtained when 4wt% Yb_2O_3 was added. Results of XRD patterns revealed that both β-Si_3N_4 and h-BN existed in all the specimens. No α-Si_3N_4 was detected by XRD analysis, implying that the α- to β-Si_3N_4 transformation has been completed during the pressureless sintering. The results show that the Yb_2O_3-Al_2O_3-Y_2O_3 system could act as effective sintering additives for pressureless sintering of Si_3N_4/BN composite.

Key words: silicon nitride; boron nitride; composite ceramics; pressureless sintering; Yb_2O_3-Al_2O_3-Y_2O_3 additives

INTRODUCTION

Silicon nitride (Si_3N_4) is one of the most common ceramic materials for application in many engineering fields because of its high strength, high toughness and good resistance to corrosion, wear and chemicals [1]. However, they have poor machinability due to their brittle and hard nature, that limits them to be manufactured into complex shapes for the required applications. In recent years, attempts have been made to develop Si_3N_4 machinable ceramics by means of introducing hexagonal-BN as weak boundary phase in matrices [2-7]. The cleavage behavior of plate-like h-BN endowed Si_3N_4/h-BN composite with good machinability as BN content was sufficiently high. In addition, such properties as corrosion resistance to molten metal and thermal shock resistance were also greatly improved [4-5].

Si_3N_4/h-BN composites are usually prepared by hot-press sintering. To increase the mechanical properties of Si_3N_4/h-BN composites, a chemical route and subsequent hot-pressing (HP)[2,3,7] were developed. Recently, Li et al [8] also prepared Si_3N_4/BN nanocomposites with high strength and good machinability via a similar chemical route and subsequent spark plasma sintering (SPS). They found that SPS was effective to help obtaining the expected microstructure in Si_3N_4/BN system within only a few minutes and to prevent h-BN from growing as well. To fabricate complex shape components, pressureless sintering offers many technical and economical advantages. But it is still difficult for Si_3N_4/h-BN composites to obtain high density for the poor sinterability of h-BN which has strong covalent nature and plate-like structure. To improve the sintering and densification of Si_3N_4/h-BN composites, it is desirable to reduce the porosity of pressureless sintering products because pores deteriorate mechanical properties of ceramics. However, the pore is a kind of weak boundary phase that can be tailored into materials to enhance the ceramics machinability[9,10]. Kawai and Yamakawa [9] have successfully fabricated porous Si_3N_4 machinable ceramics, which can be machined easily by cemented carbide drills due to the effect of pores. Therefore, both h-BN additive and residual pores can be regarded as weak boundary phases in pressureless sintered Si_3N_4/h-BN composites, which could strongly control the mechanical properties and machinability. Nevertheless, the influence of pores as weak boundary phase on the characteristics of Si_3N_4/h-BN composites fabricated by pressureless sintering has not been reported yet.

As mentioned before, Si_3N_4/h-BN composites are usually prepared with Y_2O_3, Al_2O_3 as sintering aids [3, 4, 7 ,8]. In this paper, to improve the high temperature mechanical properties, Yb_2O_3 was introduced into Si_3N_4/h-BN system as the third composition of sintering aids. With Yb_2O_3, Y_2O_3 and Al_2O_3 as sintering additives, commercial α-Si_3N_4 and h-BN powders as starting materials, Si_3N_4/h-BN composites with 25vol% h-BN were fabricated by pressureless sintering process. The densification behaviors, α-β transformation and mechanical properties of Si_3N_4/BN composite were discussed in detail.

EXPERIMENTS

α-Si_3N_4 powder (99% purity in mass, content of α-Si_3N_4>95%, with an average particle size of 1μm) and h-BN powder (25vol%, i.e., 25vol% h-BN+75vol% (Si_3N_4 + Additives)) (98% purity in mass, with an average particle size of 1μm) were used as starting materials, and various amounts of Yb_2O_3 (0, 2, 4, 8, 15wt%) were added, with a constant Y_2O_3/Al_2O_3 weight ratio of 2:1 and total additives ($Y_2O_3/Al_2O_3/Yb_2O_3$) amount of 15wt%(i.e., 15wt% additives+85wt% Si_3N_4). Wet mixing was performed in anhydrous alcohol for 12h, and the slurry was dried in a rotary evaporator, then sieved to 200 mesh and uniaxially pressed to form rectangular bars. The green compacts were pressureless sintered in a graphite-heater furnace (model High muti-5000, Fuijidenpa Co. Ltd., Osaka, Japan) at 1800°C for 2h in N_2 gas.

The bulk density of the sintered products was measured by the Archimedes immersion technique in distilled water, and theoretical density was estimated by the rule of mixtures. Crystalline phases were identified by X-ray diffraction analysis (XRD, Model D/max-2400X, Rigaku Co., Tokyo, Japan). The bending strength was measured by three-point bending test (sample size 3mm×4mm×20mm, span 16mm, crosshead speed 0.5mm/min). The fracture toughness was determined by the single-edge notched beam (SENB) method (sample size 2mm×4mm×20mm, span 16mm, crosshead 0.05mm/min, gap width≤0.2mm, gap depth 0.5±0.1mm). The Vickers hardness was measured by Vickers indentations method with a load of 49N for 30s. The microstructure was characterized by a field emission scanning electron microscope (FESEM, model JSM7000F, JEOL Co. Ltd., Tokyo, Japan).

RESULTS AND DISCUSSION

Phase composition analysis

Figures 1 shows the XRD patterns of specimens with Yb_2O_3 levels of 2, 4, 8, and 15wt% as sintering additives. According to the results in Fig.1, the main phases in all the specimens are β -Si_3N_4 and h-BN, no α-Si_3N_4 phase was detected, and a complete transformation from α- to β-Si_3N_4 during the pressureless sintering was confirmed by a solution-precipitation process. A small amount of second phase $Y_2Si_3O_3N_4$, formed by the reaction between Y_2O_3, Si_3N_4 and SiO_2 that existed on the surface of Si_3N_4 particles, can be observed in specimens with 2, and 4wt% Yb_2O_3, indicating that Yb_2O_3 and Al_2O_3 were dissolved in the eutectic liquid during the sintering process. When the amount of Yb_2O_3 was increased to 8wt%, an X-ray pattern in Fig.1 shows that $Yb_4Si_2O_7N_2$ was the major secondary phase and $Yb_2Si_2O_7$ is the minor secondary phase. When the amount of Yb_2O_3 was further increased to 15wt%, an increase in the amount of $Yb_4Si_2O_7N_2$ was observed by the XRD pattern in Fig.1. These results are consistent with those of Hyoungjoon Park et al. [11] and X.W. Zhu et al. [12].

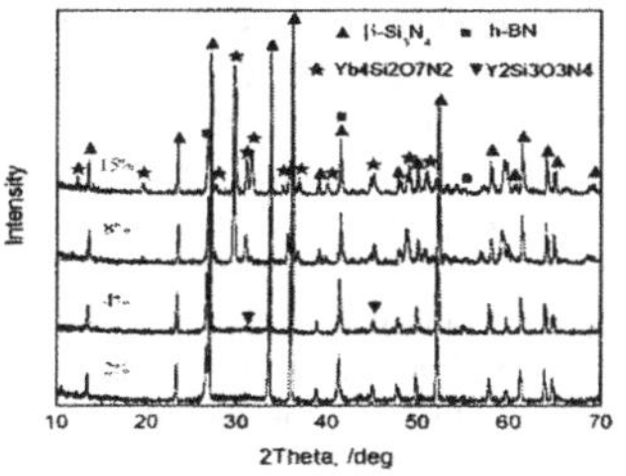

Figure 1. XRD patterns of Si_3N_4/BN composites with various amounts of Yb_2O_3

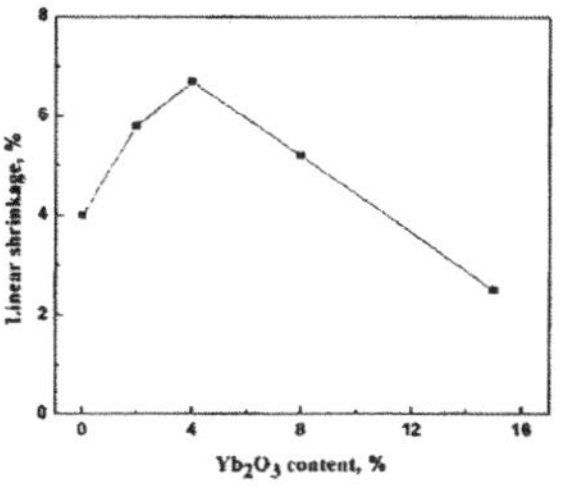

Figure 2. Linear shrinkage of Si_3N_4/BN composites with various amounts of Yb_2O_3

Densification behavior

Figure 2 shows the linear shrinkage of specimens with various amounts of Yb_2O_3. It can be seen that with Yb_2O_3 content increasing, the sintering shrinkage of the samples increased and the highest linear shrinkage of 6.7% was observed for samples containing 4wt% Yb_2O_3, then leaded to a decrease. The same trend in apparent density was also observed. The apparent density of all the specimens increased with the addition up to 4wt% and changed greatly thereafter. The highest apparent density, 2.40g/cm^3, was observed when 4wt% Yb_2O_3 was added. The addition of small amount of Yb_2O_3 in Al_2O_3-Y_2O_3 additives tends to promote densification, while for a high additive amount of 8wt% or 15wt% Yb_2O_3, densification tends to be delayed. With the addition of 15wt% Yb_2O_3, the linear shrinkage is the lowest of 2.5%, while the porosity increases to 32.5%. And specimens containing a smaller amount of Yb_2O_3 clearly exhibit greater density. Densification behavior at 1800°C is considered to be basically dependent on the amount of Yb_2O_3, i.e. the ratios of Re_2O_3/Al_2O_3, which effects the liquidus temperatures and viscosity of liquid phase formed in the sintered body. With the increasing of Yb_2O_3 content, the amounts of Al_2O_3 and Y_2O_3 decrease, thus the corresponding Re_2O_3/Al_2O_3 ratio increase, and the liquidus temperatures of Yb_2O_3-Al_2O_3-Y_2O_3 are higher, thus the liquid phase is considered to be formed at higher temperatures compared with Al_2O_3-Y_2O_3 additives. In addition, the viscosity of Yb_2O_3-Al_2O_3-Y_2O_3 based glass phase, which has a higher liquidus temperature, is greater at the same temperature than that of the Al_2O_3-Y_2O_3 additives. The densification of Si_3N_4/h-BN composites tends to be delayed for a high additive amount of 8wt% or 15wt% Yb_2O_3.

Microstructure and mechanical properties

Figure 3 shows FESEM micrographs of the fracture surfaces of the sintered samples with various amounts of Yb_2O_3. From the FESEM graphs, a typically interlocked microstructure that elongated β-Si_3N_4 grains embedded in the small equiaxed β-Si_3N_4 grains can be observed in all the specimens. And intergranular fractures were also observed. The residual pores were observed near the h-BN platelets (Fig. 3(a-d)). Furthermore, the following microstructure differences could be observed for the samples with various amounts of Yb_2O_3: (1) Differences in the morphological structure of BN particles for various amounts of Yb_2O_3, with the increasing of Yb_2O_3 content, the morphological structures of BN were changed from plate-like (Fig.3(a)) to little spherical or needle-like particles (Fig.3(d)); and (2) Differences in the distribution of BN particles, when 2wt% Yb_2O_3 was added, plate-like BN were distributed among β-Si_3N_4 grains (Fig.3(a)), and with Yb_2O_3 content increased to 4wt%, plate-like BN and Si_3N_4 grains were coated with abundant liquid phase, and the bonding between plate-like BN particle and Si_3N_4 was more closely, but with further increasing Yb_2O_3 content, h-BN particles were agglomerated along the fringe of β-Si_3N_4 grains or in residual pores. Besides the intrinsic defects produced by industrial manufacture process, there are still many defects on BN particles introduced by ball-milling, physical and chemical adsorption, and chemical reaction with O_2 or H_2O in air during the experimental procedure. The higher viscosity of glass means an increase in the extent of bonding between grains and grain boundary phases. With Yb_2O_3 content increasing, the viscosity of Yb_2O_3 based glass increased, which would cause an increase in the extent of the corresponding bonding between BN particles and glass phases. The exposure of BN particle surface to grain boundary would introduce several changes, such as local densification, compressive stress or production of point defects (interstitial atoms and vacancies). With the increasing of viscosity, more defects would form on BN particles, and the extent of layered h-BN particles dissociating at these defects increased, which resulted in the transition from plate-like to little spherical or needle-like structure of BN particles. We will discuss the question in detail elsewhere.

The great difference of microstructure can be explained as follows. With the increasing of Yb_2O_3 content, the viscosity of Yb_2O_3 based glass phase increased. The higher viscosity hinders both the heterogeneous nucleation of grains and the material transport by means of diffusion during solution-reprecipitation, and thus it can promote preferential growth of the pre-existing intrinsic β-Si_3N_4 grains, as shown in Fig.3(c) and (d). However, abnormal larger grain sizes lead to lower mechanical strength of sintered material, because the flaw size relating to fracture becomes larger. On the other hand, from XRD results above, the degrees of crystallization of the liquid phases were improved with the increasing of Yb_2O_3. But the wetting behaviors of liquid phase on BN particles were also changed by the crystallization of liquid phases, which was proven by the corresponding

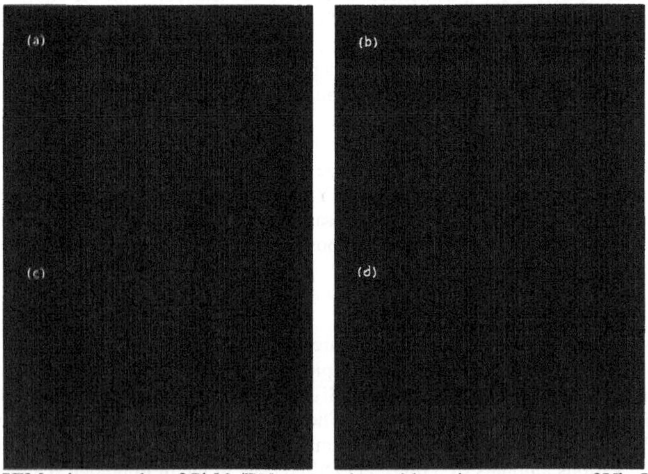

Figure 3. FESEM micrographs of Si_3N_4/BN composites with various amounts of Yb_2O_3: (a) 2wt%, (b) 4wt%, (c) 8wt%, and (d) 15wt%

linear shrinkage behaviors (as indicated in Fig.2) and variety of porosity, though it would improve high temperature strength as well as fracture toughness. Partial crystallization could improve the bonding of plate-like BN and Si_3N_4 grains as can be seen in Fig.3 (b). The mechanical properties of Si_3N_4/BN composites are closely related to their microstructure, especially the residual pores and morphology of BN andβ-Si_3N_4 grains. Thus the room temperature strength of Si_3N_4/h-BN composites increased with the addition up to 4wt% and changed greatly thereafter, as shown in Figure 4. The highest room-temperature flexural strength, 264.3MPa, was obtained when 4wt% Yb_2O_3 was added. The addition of small amount of Yb_2O_3 in Al_2O_3-Y_2O_3 additives could improve the mechanical properties of Si_3N_4/h-BN composites. Residual pores, BN phase and larger β-Si_3N_4 grains become the fracture sources. The main fracture mode was intergranular.

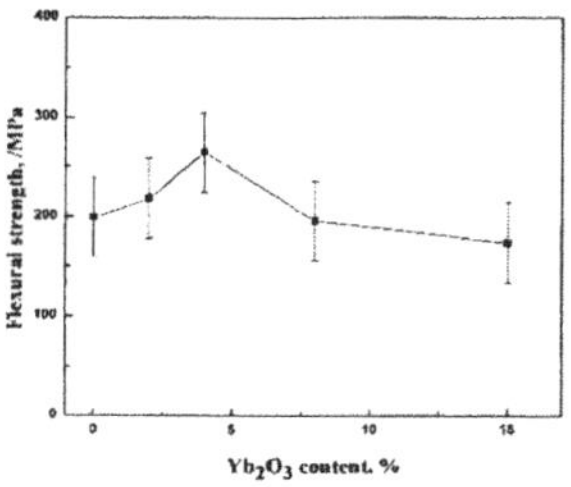

Figure 4. Dependence of fracture strength of Si_3N_4/BN composites on Yb_2O_3 amounts

CONCLUSION

The Si_3N_4/h-BN composites with 25vol% h-BN were fabricated by pressureless sintering with Yb_2O_3-Al_2O_3-Y_2O_3 system as sintering additives. Results of XRD patterns revealed that both β-Si_3N_4 and h-BN were detected in all the specimens, no α-Si_3N_4 was observed, implying that α- to β-Si_3N_4 transformation was completed during the pressureless sintering. The addition of Yb_2O_3 in Al_2O_3-Y_2O_3 additives increased the viscosity of liquid phase, thus changed the wetting behaviors of liquid phase on BN particles and the growth of β-Si_3N_4 grains. The addition of small amount of Yb_2O_3 in Al_2O_3-Y_2O_3 additives could improve the densification and mechanical properties of Si_3N_4/h-BN composites. The results show that the Yb_2O_3-Al_2O_3-Y_2O_3 system could act as effective sintering additives for pressureless sintered Si_3N_4/BN composites.

ACKNOWLEDGEMENT

This work was financially supported by the National Natural Science Foundation of China (Grant No. 50772086) and the High-Tech R & D Program of China (863, Grant No. 2007AA03Z558).

REFERENCE

[1]F. L. Riley, Silicon Nitride and Related Materials, *J. Am. Ceram. Soc.*, **83**, 2094-96(2000).

[2]T. Kusunose, T. Sekino, Y.H. Choa, and K. Niihara, Fabrication and Microstructure of Silicon Nitride/Boron Nitride Nanocomposites, *J. Am. Ceram. Soc.*, **85**, 2678–88(2002).

[3]T.Kusunose, T.Sekino, Y.H. Choa, and K. Niihara, Machinability of Silicon Nitride/Boron Nitride Nanocomposites, *J. Am. Ceram. Soc.*, **85**, 2689-95(2002).

[4]K. S. Mazdiyasni and R. Ruh, High/Low Modulus Si_3N_4–BN Composite for Improved Electrical and Thermal Shock Behavior, *J. Am. Ceram. Soc.*, **64**,415–18(1981).

[5] E. H. Lutz and M. V. Swain, Fracture Toughness and Thermal Shock Behavior of Silicon Nitride Ceramics, *J. Am. Ceram. Soc.*, **75**, 67–70 (1992).

[6]H. Liu and S. M. Hsu, Fracture Behavior of Multilayer Silicon Nitride/Boron Nitride Ceramics, *J. Am. Ceram. Soc.*, **79**, 2452–57 (1996).
[7]L. Gao, X.H. Hai, J.G. Li, Y.G. Li, and J. Sun, BN/Si_3N_4 Nanocomposite with High Strength and Good Machinability, *Mater. Sci. Eng. A,* **415**, 145–48(2006).
[8]Y.L. Li, R.X. Li, J.X. Zhang. Enhanced Mechanical Properties of Machinable Si_3N_4/BN Composites by Spark Plasma Sintering, *Mater. Sci. Eng. A,* **483–484**, 207-10(2008).
[9]C. Kawai, A. Yamakawa, Effect of Porosity and Microstructure on the Strength of Si_3N_4: Designed Microstructure for High Strength, High Thermal Shock Resistance, and Facile Machining, *J. Am. Ceram. Soc.*, **80**, 2705-08(1997).
[10]Z.Q Shi, J.P. Wang, G.J. Qiao, and Z.H. Jin, Effects of Weak Boundary Phases (WBP) on the Microstructure and Mechanical Properties of Pressureless Sintered Al_2O_3/h-BN Machinable Composites, *Mater. Sci. Eng. A,* **492**, 29-34(2008).
[11]H. Park and H. Kim, Microstructure Evolution and Mechanical Properties of Si_3N_4 with Yb_2O_3 as a Sintering Additive, *J. Am. Ceram. Soc.*, **80**, 750-56(1997).
[12]X.W. Zhu, Y. Zhou, and K. Hirao, Effect of Sintering Additive Composition on the Processing and Thermal Conductivity of Sintered Reaction-Bonded Si_3N_4, *J. Am. Ceram. Soc.*, **87**, 1398-400(2004).

EFFECT OF B_4C ADDITIONS ON THE PRESSURELESS SINTERING OF ZrB_2-SiC ULTRA-HIGH TEMPERATURE CERAMICS

Hui Zhang [1, 2], Yongjie Yan [1], Zhengren Huang [1], Xuejian Liu [1], and Dongliang Jiang [1]
[1] Shanghai Institute of Ceramics, Chinese Academy of Sciences, Shanghai, 200050, China.
[2] Graduate School of the Chinese Academy of Sciences, Beijing, 100039, China.

ABSTRACT

Pressureless sintering of ZrB_2-SiC ceramics with different amounts of B_4C additions was investigated in the present work. The relative density, microstructure and mechanical properties of the sintered body were observed to study the role of B_4C additions in the sintering process. The results showed that the excessive B_4C additions may have a negative effect on the density and mechanical properties of the materials. SEM images indicated that the SiC grains were regularly dispersed in the ZrB_2 matrix and there had low residual porosity. In addition, the effect of B_4C additions on the oxidation resistance of ZrB_2-SiC ceramics was studied by observing the mass gain and residual strength of samples after heat treatments.

INTRODUCTION

Ultra-high temperature ceramics (UHTCs), such as zirconium diboride (ZrB_2), HfB_2, ZrC, HfC and so on, are remarkable for their ultra-high melting points (>3000℃) and most of them belong to the early transition metal borides and carbides[1-3]. In this family of materials, ZrB_2-based UHTCs composites have the unique combination of properties such as high strength, high hardness, high electrical and thermal conductivities, good oxidation resistance and chemical attack resistance and the lowest theoretic density[4-7], which makes them attractive for aerospace and other applications such as those associated with atmospheric re-entry, hypersonic flight and rocket propulsion, high-temperature electrodes and crucibles for molten metal contact and so on[8; 9; 10].

Owing to their strong covalent bonding as well as low self-diffusion coefficients, ZrB_2-based ceramics have typically been densified by hot pressing (HP)[5; 6; 11]. Recently, spark plasma sintering (SPS) has been used to densify diborides[12; 13]. Densification of borides has also been achieved by reactive hot processing (RHP)[14; 15]. However, pressureless sintering offers some unique advantages compared with HP, SPS and RHP, including the ability to fabricate components to near-net shape using standard powder processing methods. Above all, the diamond machining of the sintered parts can be obviously reduced and the costs were cut dramatically[16]. On the other hand, it becomes rather difficult to achieve full density of pure ZrB_2-based ceramics by pressureless sintering without any additives due to its intrinsic poor sinterability. Incorporation of additives such as TiB_2[5], Si_3N_4[17], SiC[18;19], Mo[20], Fe[21], Ni[5;11;22], B_4C[5;23], and C[23;24] has been reported to improve the pressureless sintering of ZrB_2-based ceramics. But the addition of metallic or ceramic additives could apparently deteriorate the properties of the ceramic materials at high temperatures. Recently, some progress has occurred in the pressureless sintering of ZrB_2-based UHTCs. Chamberlain et al[7] sintered ZrB_2 to a relative density of ~98% without applied external pressure. However, the introduction of WC (~2vol. %) was thought to from a solid solution with ZrB_2 and promote the pressureless sintering process. Zhang et al[23] densified ZrB_2 at 1850℃ by pressureless sintering to nearly full density. Both WC and B_4C were used as sintering aids to facilitate the sintering process. Zhu et al[25] used B_4C and C as the sintering aids to achieve the nearly full density ZrB_2 by pressureless sintering method. Previous investigations have demonstrated that SiC particulate additions can increase the oxidation resistance of ZrB_2 by promoting the formation of silicate-based glass phases that inhibit oxidation at temperatures between 800℃ and 1700℃[6; 26]. Besides that, the SiC particulates can also improve the sinterability and inhibit the grain growth of the ZrB_2. However, the role of B_4C additions played in the sintering process was not reported till now.

The objective of the present study was to identify the effect of different B_4C additions on the pressureless sintering process to fabricate ZrB_2-SiC ceramics. The relative density, mechanical properties, microstructure and oxidation resistance were investigated as the B_4C contents changed.

EXPERIMENTAL PROCEDURE

Commercially available precursor powders were used as raw materials in this study. The ZrB_2 powders had a purity of >90% and an average particle size of <2.5μm. SiC powders had a purity of >99% and an average particle size of <0.5μm. As the sintering aids, B_4C powders had a purity of >97%, with a particle size <1μm, and the phenolic resin which was also used as adhesives, could produce a 30wt% pyrolyzed carbon.

To produce pellets for the densification study, a certain amount of commercial ZrB_2 powders were mixed with 20vol%SiC and 3wt% phenolic resin and different B_4C additions, and then ball milled for 48h using SiC milling media in ethanol. After milling, ethanol was removed using rotary evaporation. The resulting powders were crushed and then passed through a 100-mesh sieve. The obtained materials contained B_4C in the amount of: 0, 0.2, 0.4, 0.8, 1.0, 2.0, 4.0, 6.0wt% (based on the weight of the ceramic powders). The composite powders were firstly uniaxially dry-pressed and followed by cold isostatic pressing. The green compacts were heated to 900℃ (1hour hold) in vacuum to remove the binder at a rate of 5℃/min. After binder removal, the samples were sintered to 2100℃ (2 hours hold under flowing argon) at 10℃/min. After sintering, the furnace was cooled to room temperature under flowing argon. All the heat-treatment processes were carried out in a high-temperature graphite resistance furnace (High-Multi 10000, Fujidempa kogyo. Ltd., Saitama, Japan).

The crystallite size and morphology of the starting powders were observed by field emission scanning electron microscopy (FE-SEM) (JSM-6700F, JEOL, Tokyo, Japan). The microstructure of the sintered body was observed by electron-probe microanalysis (EPMA) (JXA-8100, JEOL, Tokyo, Japan) with energy-dispersive spectroscopy (EDS). The bulk density was measured using the Archimedes displacement method with water as the immersing medium and the relative density was calculated by dividing the measured bulk density by a calculated theoretical density. The theoretical density was estimated using rule-of-mixtures calculations that assumed the nominal compositions of the batches as specified. Flexural strength was measured by three-point bending method with a span of 30mm and a cross-head speed of 0.5mm/min, using Instron 5566 universal testing machine. And elastic modulus was determined according to the Chinese Standard GB/T 10700-2006 by a bending method. The size of the samples was both nominally 3mm×4mm×36mm. The oxidation resistance was investigated by measuring the mass gain and the residual strength of the specimens after heat treatments at different temperatures.

RESULTS AND DISCUSSION

The particle size and morphology of the ZrB_2 powders were observed by scanning electron microscopy (Fig .1). The reported particle size was ~2.5μm. SEM images revealed that the as-received powders had serious agglomerations, and the particles were primarily irregular in shape. The specific surface area was ~7.7m^2/g.

The density of the sintered body affected the microstructure. The density versus B_4C contents was shown in Fig 2. As the amount of B_4C inclusions increased from 0 to 6wt%, the theoretical density decreased by ~6% for the small density of B_4C. On the other hand, along with the increased B_4C contents from 0.2 to 6.0wt% the relative density of the sintered body decreased from 98.3% to 91.7%. In other words, the relative density was decreased as the B_4C contents increased. The samples without B_4C addition were seriously crooked and obviously not dense, so the further measurements were omitted. As a whole, the relative density was all more than ~90% and the less B_4C inclusion lead to the higher density. The maximum density was 98.3% when 0.2wt% B_4C was added.

Fig 1. Scanning electron microscopy image of the as-received ZrB_2 powders.

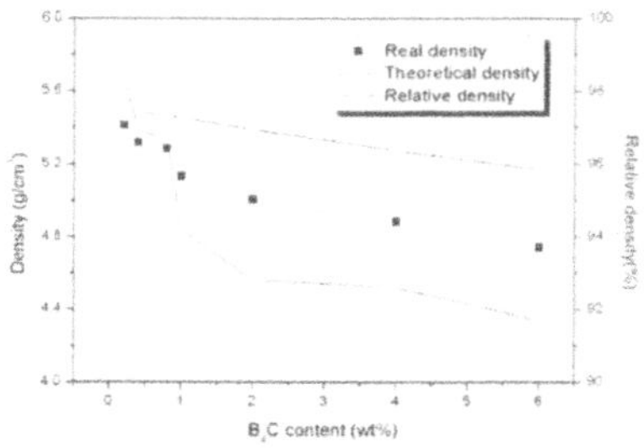

Fig 2. Density of the sintered ZrB_2–SiC ceramics with different B_4C contents.

The microstructure of the sintered ZrB_2–SiC ceramics was characterized by examining the polished cross-sections of the samples. For example, the SEM image of the samples with 0.8wt%B_4C inclusion was shown in Fig 3. It was obviously observed that the SiC grains (dark phase) were regularly dispersed in the ZrB_2 (gray phase) matrix and appeared to be elongated which were consistent with the other reported ZrB_2-SiC ceramics prepared by pressureless sintering[20]. The average size of SiC grains was approximately 8μm, which may attribute to the high temperature and holding time of pressureless sintering process. Because the boron and carbon were all the light elements, it was difficult to examine them by EDS. In addition, from the SEM image there had low residual porosity (the density of the samples was ~96.7% of the theoretical density) and no other phases existed.

Properties of the pressurelessly-sintered ZrB_2-SiC ceramics were shown in Fig 4. It was observed that the flexural strength and elastic modulus almostly had the same trends as the B_4C contents changed except for the first two batches. Along with the B_4C concentration increased, the elastic modulus decreased sharply at the beginning and then as the B_4C inclusion was more than 1wt%, the modulus and strength were all hardly dropped. On the other hand, in the first two batches the strength was low and increased as more B_4C added. The reasons were not clear at present and further research is needed.

The sintered ZrB_2-SiC ceramics with 0.8wt%B_4C additions were taken as an example. They had an average flexural strength of 303±18MPa and elastic modulus of 375±12GPa, which were comparable with the other pressureless sintering results for ZrB_2-SiC ceramics[20]. In addition, crack propagations from Vickers indentation on the polished surface were investigated by back-scattered scanning electron microscopy. On the corner of the indention site, a crack was deflected and split. The cracks tended to propagate through the SiC grains, whereas the cracks tended to propagate between ZrB_2 grains and obvious crack deflections were observed.

Besides that, the oxidation resistances of the sintered ZrB_2-SiC ceramics were also investigated. In this study, the samples with different B_4C additions were oxidized at 1400℃, 1500℃ and 1600℃ with 30min hold respectively. Results showed that in general as the temperature increased the mass gain increased nearly consistently. And it was found that at the relative low temperatures (such as <1500℃), the more B_4C addition, the less mass gain. To the contrary, when the temperature was high (e.g.>1600℃), more B_4C additions were harmful for the oxidation resistance of the materials. On the other hand, the residual flexural strength was a little lower than the room temperature strength before oxidation. As a whole, the proper amount of B_4C additions to some extent improved the oxidation resistance of the materials and whereas the excessive B_4C additions may also deteriorate the oxidation resistance for the evaporation of B_2O_3.

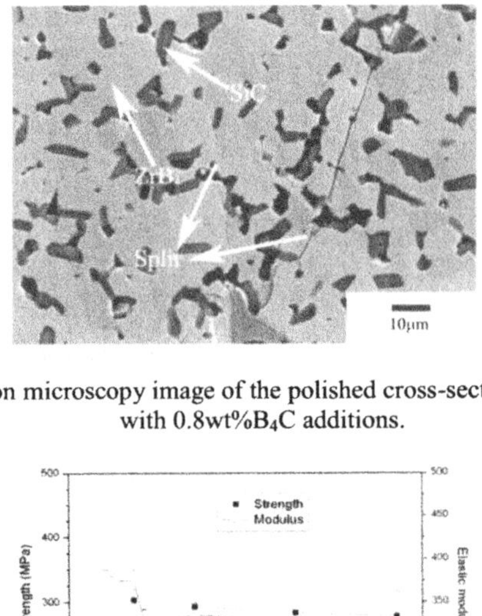

Fig 3. Scanning electron microscopy image of the polished cross-sections of ZrB_2-SiC ceramics with 0.8wt%B_4C additions.

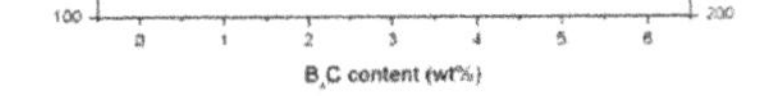

Fig 4. Mechanical properties of the sintered ZrB_2-SiC ceramics with different amount of B_4C additions.

CONCLUSION

ZrB_2-SiC ceramics with different amounts of B_4C additions had been densified by pressureless sintering. It was shown that along with the addition of B_4C increased, the relative density of the sintered bodies were decreased sharply and then hardly. The mechanical properties such as the flexural strength and elastic modulus were nearly in accordance with the trend of the relative density. In addition, the oxidation resistances of the ZrB_2-SiC ceramics were improved by certain amounts of B_4C additions. The mass gain had a few differences at varied temperatures and the residual strength was also nearly 80% of the room temperature strength.

REFERENCES

[1]R.A. Cutler, Engineering Properties of Borides, in "Ceramics and Glasses, Engineered Materials Handbook", ASM International, Materials Park, OH, pp. 787-803 (1992).

[2]C. Mroz, Zirconium Diboride, *Am. Ceram. Soc. Bull.*, **73[6]**, 141-142 (1994).

[3]R. Telle, L.S. Sigl, et al., Boride-Based Hard Materials, in "Handbook of Ceramic Hard Materials", Wiley-VCH, pp. 802-945 (2000).

[4]S.R. Levine, E.J. Opila, et al., Evaluation of Ultra-high Temperature Ceramics for Aeropropulsion Use, *J. European Ceram. Soc.*, **22[14-15]**, 2757-2767 (2002).

[5]F. Monteverde, A. Bellosi, et al., Processing and Properties of Zirconium Diboride-based Composites, *J. European Ceram. Soc.*, **22[3]**, 279-288 (2002).

[6]A.L. Chamberlain, W.G. Fahrenholtz, et al., High-strength Zirconium Diboride-based Ceramics, *J. Am. Ceram. Soc.*, **87[6]**, 1170-1172 (2004).

[7]A.L. Chamberlain, W.G. Fahrenholtz, et al., Pressureless Sintering of Zirconium Diboride, *J. Am. Ceram. Soc.*, **89[2]**, 450-456 (2006).

[8]K. Upadhya, J.M. Yang, et al., Materials for Ultrahigh Temperature Structural Applications, *Am. Ceram. Soc. Bull.*, **76[12]**, 51-56 (1997).

[9]F. Monteverde, S. Guicciardi, et al., Advances in Microstructure and Mechanical Properties of Zirconium Diboride Based Ceramics, *Mater. Sci. Eng. A-Struct. Mater. Prop. Microstruct. Process.*, **346[1-2]**, 310-319 (2003).

[10]S.N. Karlsdottir and J.W. Halloran, Rapid Oxidation Characterization of Ultra-High Temperature Ceramics, *J. Am. Ceram. Soc.*, **90[10]**, 3233-3238 (2007).

[11]A. Bellosi, F. Monteverde, et al., Microstructure and Properties of ZrB_2-based Ceramics, *J. Mater. Process. Manuf. Sci.*, **9[2]**, 156-170 (2000).

[12]T. Tsuchida and S. Yamamoto, MA-SHS and SPS of ZrB_2-ZrC composites, *Solid State Ion.*, **172[1-4]**, 215-216 (2004).

[13]Y. Zhao, L.J. Wang, et al., Preparation and Microstructure of a ZrB_2-SiC Composite Fabricated by the Spark Plasma Sintering-reactive Synthesis (SPS-RS) Method, *J. Am. Ceram. Soc.*, **90[12]**, 4040-4042 (2007).

[14]G.J. Zhang, Z.Y. Deng, et al., Reactive Hot Pressing of ZrB_2-SiC Composites, *J. Am. Ceram. Soc.*, **83[9]**, 2330-2332 (2000).

[15]J.W. Zimmermann, G.E. Hilmas, et al., Fabrication and Properties of Reactively Hot Pressed ZrB_2-SiC Ceramics, *J. European Ceram. Soc.*, **27[7]**, 2729-2736 (2007).

[16]Shi C. Zhang, Greg E. Hilmas, et al., Pressureless Sintering of ZrB_2-SiC Ceramics, *J. Am. Ceram. Soc.*,**91[1]**, 26-32 (2007).

[17]F. Monteverde and A. Bellosi, Effect of the Addition of Silicon Nitride on Sintering Behaviour and Microstructure of Zirconium Diboride, *Scr. Mater.*, **46[3]**, 223-228 (2002).

[18]F. Monteverde and A. Bellosi, Development and Characterization of Metal-diboride-based Composites Toughened with Ultra-fine SiC Particulates, *Solid State Sci.*, **7[5]**, 622-630 (2005).

[19]A. Rezaie, W.G. Fahrenholtz, et al., Effect of Hot Pressing Time and Temperature on the Microstructure and Mechanical Properties of ZrB_2-SiC, *J. Mater. Sci.*, **42[8]**, 2735-2744 (2007).

[20]Y.J. Yan, Z.R. Huang, et al., Pressureless Sintering of High-density ZrB_2-SiC Ceramic Composites, *J. Am. Ceram. Soc.*, **89[11]**, 3589-3592 (2006).
[21]S.K. Mishra, S.K. Das, et al., Effect of Fe and Cr Addition on the Sintering Behavior of ZrB_2 Produced by Self-propagating High-temperature Synthesis, *J. Am. Ceram. Soc.*, **85[11]**, 2846-2848 (2002).
[22]J.J. Melendez-Martinez, A. Dominguez-Rodriguez, et al., Characterisation and High Temperature Mechanical Properties of Zirconium Boride-based Materials, *J. European Ceram. Soc.*, **22[14-15]**, 2543-2549 (2002).
[23]S.C. Zhang, G.E. Hilmas, et al., Pressureless Densification of Zirconium Diboride with Boron Carbide Additions, *J. Am. Ceram. Soc.*, **89[5]**, 1544-1550 (2006).
[24]S.M. Zhu, W.G. Fahrenholtz, et al., Pressureless Sintering of Carbon-coated Zirconium Diboride Powders, *Mater. Sci. Eng. A-Struct. Mater. Prop. Microstruct. Process.*, **459[1-2]**, 167-171 (2007).
[25]S. Zhu, W.G. Fahrenholtz, et al., Pressureless Sintering of Zirconium Diboride Using Boron Carbide and Carbon Additions, *J. Am. Ceram. Soc.*, **90[11]**, 3660-3663 (2007).
[26]W.G. Fahrenholtz, Thermodynamic Analysis of ZrB_2-SiC Oxidation: Formation of a SiC-depleted Region, *J. Am. Ceram. Soc.*, **90[1]**, 143-148 (2007).

TRANSLUCENT AND TOUGHENED Dy-α-SiAlON CERAMICS WITH LiF AS SINTERING ADDITIVE

Qian Liu[1] *, Junming Xue[1,2], and Wei He[1,2]
[1] The State Key Laboratory of High Performance Ceramics and Superfine Microstructure, Shanghai Institute of Ceramics, Shanghai 200050, P. R. China.
[2] Graduate School of the Chinese Academy of Sciences, Beijing 100049, P. R. China

ABSTRACT

SiAlON ceramics show a great potential for application, with excellent thermal shock resistance, good oxidation resistance, good wear resistance, and high hardness. So the present research focusing on its transmittance improvement has been carried out by the application of LiF as a sintering additive and processing control. Also, the interaction relation among sintering, phase evolution, microstructure and optical property has been understood. The samples with a highest transmittance as 73% have been manufactured by hot-press(HP) sintering. A satisfied microstructure, such as a fully developed main phase, uniform morphology, and neat grain boundary, must be a suitable explanation for the transmittance improvement. From the view of microstructure, the duplex morphology of both elongated and equiaxed grains in the SiAlON samples should be responsible for the self-reinforced result, except for the optical transmittance. To a certain extent, the properties of the translucent Dy-α-SiAlON have been reasonably tailored by composition design and process control, with resultant samples possessed a higher transmittance of 65% and also kept a higher fracture toughness of 5.21 $MPa \cdot m^{1/2}$ at the same time.

INTRODUCTION

Since the translucent sintered Al_2O_3 was fabricated in 1962, many new kinds of oxides translucent ceramics have been prepared by different sintering methods, such as MgO, BeO, ZnO, PLZT. Resultantly, some nitrides ceramics with higher translucence, for example Si_3N_4 and its solid solution SiAlON, have also been reported in the past 2-3 decades. It was interesting in that the translucent Si_3N_4 or SiAlON with a hexagonal structure could be obtained by improvement in powder preparation, processing, and use of sintering additives, although the difficulty in sintering[1-5] to extend their application in the related fields.

Combining the dual properties of optical translucence with mechanical toughness, the present investigation was designed to prepare a kind of Dy-α-SiAlON ceramics which is translucent and tough as well. The research was scheduled to find sintering additives and optimize sintering conditions which could produce translucent SiAlON ceramics, and then try to keep a high toughness value. At first, a starting composition was designed based on the phase relationship of RE-Si-Al-O-N systems to obtain a pure α-SiAlON phase, using LiF as a co-additive for lower temperature sintering. Then Dy-SiAlON ceramics disks were found to be translucent in visible and infrared range with the advantage of a higher content of α-SiAlON phase and equiaxed grain microstructure. Considering the brittle property of ceramics, the translucent Dy-α-SiAlON was toughened by an in-situ reinforcement from some elongated α-SiAlON and AlN' grains, undergoing a longer time HP-sintering. It was found that Dy-α-SiAlON disk had a much higher transmittance about 73 % in the infrared range of 3-5 μm. This appears, therefore, attractive and important for Dy-α-SiAlON ceramics to be used as a type of special optical materials. Finally, considerable tests were made on the toughened behavior of Dy-α-SiAlON to figure out the effects of sintering temperature and soaking time on the toughness property, to fit the needs of application for the translucent Dy-SiAlON ceramics as windows or other kinds.

EXPERIMENTAL

According to the formula of $M_{m/v}Si_{12-(m+n)}Al_{m+n}O_nN_{16-n}$, the α-SiAlON ceramic reported here has a nominal composition $Dy_{0.66}Si_{9.0}Al_{3.0}O_{1.0}N_{15.0}$ (M=Dy, m=2n=2.0, v=3). High-purity powders of α-Si_3N_4 (SN-E10, Ube, Japan), AlN (Grade A, Starck, Germany), Dy_2O_3 (Yaolong Chemical Plant, Shanghai, China), and 0.1 wt % LiF (Sinopharm Chemical Reagent Co., Ltd, Beijing, China) were mixed in ethanol, milled using Si_3N_4 balls, dried at 80 °C, sieved, and then a charge of 6.0 g each was loaded directly into a graphite die. The samples were hot pressed at 1600-1650 °C for 60-120 min under a flowing nitrogen atmosphere of 1 atm. The LiF added was used as an important sintering additive to reduce the temperature [6]. The input and output facets of the sintered samples were polished for mechanical and optical measurements.

Bulk density of specimens was measured by the Archimedes principle. The Vickers hardness as well as indentation fracture toughness were determined at room temperature using a Vickers diamond indenter at a 10 kg load for 10 s. Microstructure observation was performed under a transmission electron microscope (TEM; JEOL, Tokyo, Japan; JEM-2010/200CX/2100F). Optical transmissions of the translucent samples in wavelength of 2.5-6.5μm were measured by FTIR spectrometer (EQUINOX55, Bruker, Billerica, MA).

RESULTS AND DISCUSSION

Table I. Density, Transmittance and Mechanical Property of the Sintered Samples

Sample	Temperature [°C]	Soaking Time [min]	Bulk Density [g·cm^{-3}]	Transmittance [%]**	Mechanical Property H_{V10} [GPa]	K_{IC} [MPa·m$^{1/2}$]
16060	1600	60	3.62	63	19.98	3.79
16260	1620	60	3.64	57	19.84	4.65
16560	1650	60	3.66	61	19.74	5.02
16090	1600	90	3.63	68	20.02	3.48
16290	1620	90	3.64	67	19.80	4.84
16590	1650	90	3.65	65	19.71	5.21
160120	1600	120	3.63	68	20.02	4.06
162120	1620	120	3.65	73	20.53	4.14
165120	1650	120	3.64	71	21.01	4.52

Table 1 shows the effects of sintering temperature and soaking time on the densification of the mixed powders and properties of the bulks as well. The bulk density values of the sintered samples are 3.62-3.66 g·cm^{-3}, showing a better densification result. All samples possess excellent mechanical properties, where the hardness values are >19.00 GPa and the fracture toughnesses are about 3.79-5.21 MPa·m$^{1/2}$, with changing of sintering temperature and soaking time. The optical transmittance of most samples is higher than 60%. For the samples sintered at varied temperatures but for same soaking time (60 -120 min), the bulk density value of the all samples was improved when the temperature increased from 1600 to 1650 °C. At the same time, the temperature increase had an obvious influence on mechanical properties.

It is clear that the optical transmittance of the samples sintered for 120 min(T=70-75%) are better than those sintered for 60(T=55-65%) or 90(T=65-70%) min. Therefore, a longer soaking results in a higher optical transmittance. But for the same soaking time, lower sintering temperature leads to a better transmittance, which can be confirmed by the samples sintered at 1600 °C for 60 or 90 min, not including the case of 120 min.

Relationship between the microstructure and optical transmittance

Fig. 1 shows the optical transmittance curves in the range of 2.5-5.5 μm of the samples manufactured by different process. The cutoff wavelength of the samples is located at about 5.0 μm. It is also noted that one absorption peak appears at 2.8 μm. The electron transformation from rear earth ion Dy^{3+} might be a possible explanation for the absorptions, but the mechanism has not been very clear yet.

It is useful to find out the reason for the difference of the transmittance of the samples. So, the observation of morphology and distribution of grains with grain boundaries on the sample 16260, 16290 and 16290 were carried out by TEM, seen

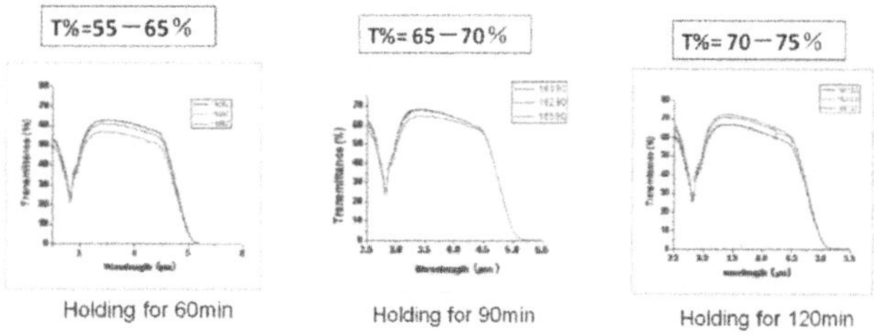

Figure 1. Optical transmittance curves of Dy-SiAlON sintered at different Temperature for varied soaking time, 0.5 mm in thickness, in the range of 2.5-5.5μm.

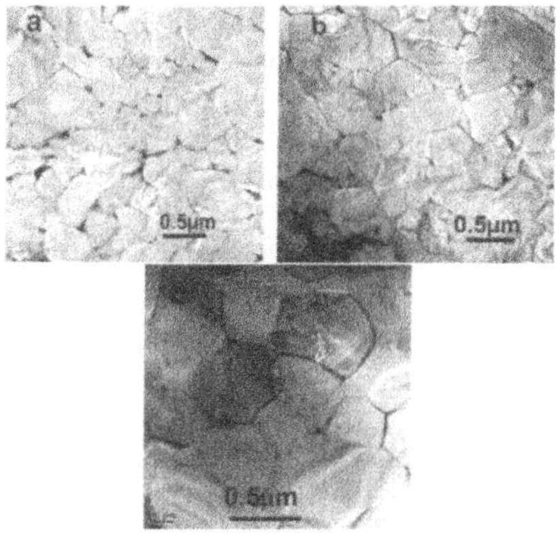

Figure 2. TEM images of samples (a) 16260, (b) 16290, and (c) 162120.

In Fig. 2, where Fig.2(c) uses a higher magnification to show the bigger and uniformer grains. It is clear that Dy-α-SiAlON phase has been formed in the three samples. Most of grain size is less than 1.0 μm. The grain morphology in samples 16290 and 162120 is more uniform and equiaxed than those in sample 16260, showing that a longer soaking time can promote the grain development or growth . It is just the reason why the transmittance of the samples sintered for 90(T=67%) and 120(T=73%) min are better than those sintered for 60 min(T=57%).

Relationship between the microstructure and mechanical toughness

As mentioned above, the sintering temperature and soaking time has an important influence not only on the translucence but also on mechanical properties for Dy-SiAlON. Fig.3 shows the mechanical property variation of Dy-SiAlON sintered at varied temperature for different soaking time, in which the effects of temperature and time on hardness and toughness can be seen obviously. With the increase of temperature, although the K_{IC} values become higher, H_{v10} ones keep a little bit decrease for a same soaking time situation of 60 or 90 min. but show a little bit increase for a soaking of 120 min. Comparatively, when extending the soaking time to 120 min, although the K_{IC} values shows a increasing tendency for the samples sintered from 1600 to 1650 °C, but their relative values are quite lower than those samples soaked for 60 or 90 min, showed in Fig. 3. It should be noticed that the increasing K_{IC} values resulted from the formation of some elongated AlN' in the SiAlON matrix, for all samples HP-pressed under varied conditions, from either 1600 to 1650 °C heating or from 60 to 120 min of soaking. The elongated grains can be seen in Fig.4. It was the AlN' grains, along with elongated α' grains developed during the soaking procedure, that took a role of in-situ reinforcement in the matrix[7-8]. But as a negative-effect, the existence of elongated grains did definitely play a role of decreasing the optical transmittance to a certain extent, owing to the scattering from these rod-like grains.

Table II indicates the transmittance and mechanical properties of those samples self-reinforced, under optimized sintering conditions, at 1650 °C for 60 or 90 min. The self-reinforced samples all possess rather higher K_{IC} values, higher H_{v10}, and higher optical transmittance as well. Fig.5 shows the cross-section images of the rough fracture surface from samples heat-treated at 1600 °C for 90 min, 1620 °C for 90 min, and 1650 °C for 90 min respectively, where the sample 16590 exhibits a self-reinforced mechanism of elongated grains, the so-called mechanism of pull-out of grains. It is the pull-out of grains that exhausted more fracture energy, leading to a increase of fracture toughness. That is why the sample 16590 has a higher K_{IC} values.

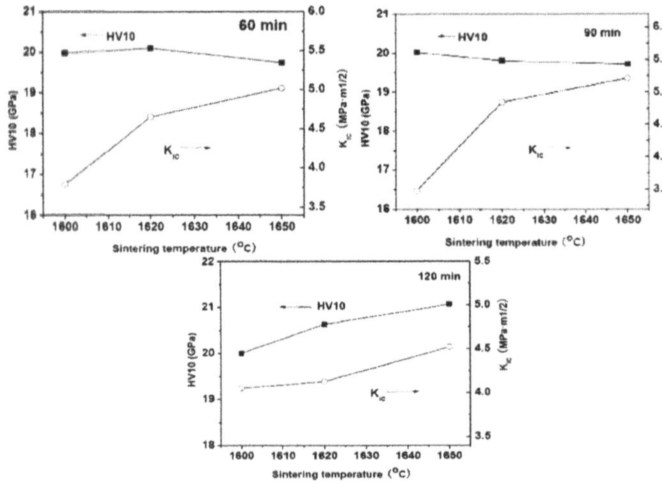

Figure 3. H_{V10} and K_{IC} values of samples sintered at 1600-1650 °C for different soaking time, 60 min, (b) 90 min, and (c) 120 min.

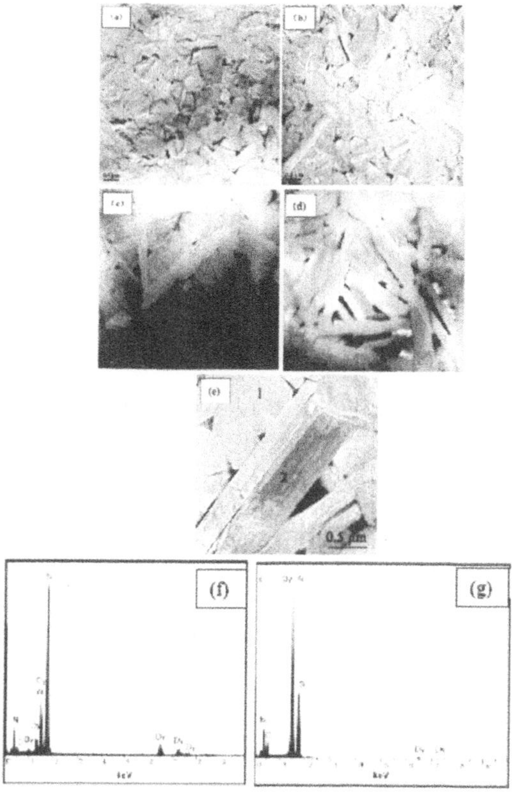

Figure 4. TEM images of samples (a)16090, (b)16290, (c)(d)(e)16590, (f)EDS of α' grain 1 and (g)EDS of AlN grain 2 in Fig.4(e).

Table II. Transmittance and Mechanical Properties of Self-Reinforced Samples

Sample	T_{max}(%)	Fracture Toughness ($MPa{\cdot}m^{1/2}$)	Hardness (GPa)
16560	61	5.02	19.74
16590	65	5.21	19.71

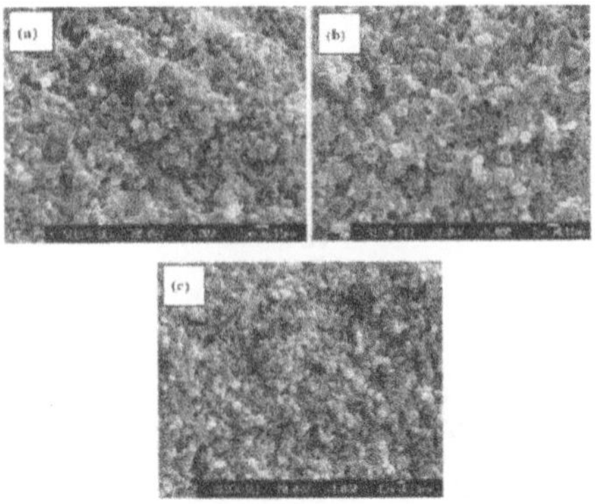

Figure 5. SEM images of fracture surface for (a)16090, (b)16290, and (c)16590.

CONCLUSION

A kind of translucent but tough Dy-SiAlON was successfully prepared by HP-sintered at 1600-1650 °C for 60-90 min using LiF as a co-additive to reduce sintering temperature. The resultant samples possessed a higher transmittance of 65% and also kept a higher fracture toughness of 5.21 $MPa{\cdot}m^{1/2}$ at the same time.

FOOTNOTES

*Contacting Author: Prof. Qian Liu, 1295 Dingxi Road, Shanghai 200050, P. R. China, TEL: +86-21-5241-2612, FAX: +86-21-5241-3122, qianliu@sunm.shcnc.ac.cn

** 0.5mm thickness

REFERENCES

[1] B.S.B. Karunaratne, R.J. Lumby and M. H. Lewis, Rear-Earth-Doped α-SiAlON Ceramics with Novel Optical Properties, *J. Mater. Res.*, **11**, 2790-2794(1996).

[2] X.L. Su, et al., Translucent α-SiAlON Ceramics by Hot Pressing, *J.Am.Ceram.Soc.*, **87**, 730-732(2004).

[3] M.I. Jones, H. Hyuga, K. Hirao, Optical and Mechanical Properties of α/β-SiAlON Composites, J. Am. Ceram. Soc., **86**, 520-5229(2003).

[4] G. Ziegler, J. Heinrich, G. Wotting, Review: Relationship between Processing, Microstructure and Properties of Dense and Reaction Bonded Silicon Nitride, J. Mater. Sci., **22**, 3041-3086(1987).

[5] M. Mimoto, Y. Tajima, Sintering Properties and Applications of SiliconNnitride and SiAlON Ceramics, J. Ceram. Soc. Jap., **99**, 1014-1025(1991).

[6] J.M. Xue, Q. Liu, and L.H. Gui, Lower-Temperature Hot-Pressed Dy-α-SiAlON Ceramics with LiF Additive, J. Am. Ceram. Soc., **90**,1623-1625(2007).

[7] S.V. Okatov, A.L. Ivanovskii, Chemical Bonding and Atomic Ordering Effects in β-SiAlON, International Journal of Inorganic Materials, **3**, 923-930(2001).

[8] V.G. Gilev, IR Spectra and Structure of Si–Al–O–N Phases Prepared by Carbothermal Reduction of Kaolin in Nitriding Atmosphere, Inorg. Mater., **37**, 1224–1229(2001).

PROPERTIES OF SILICON CARBIDE CERAMIC FROM GELCASTING AND PRESSURELESS SINTERING

Jingxian Zhang, Dongliang Jiang, Qingling Lin, Zhongming Chen, Zhengren Huang
State Key Laboratory of High Performance Ceramics and Superfine Microstructures, Shanghai Institute of Ceramics 1295 Dingxi Road, Shanghai 200050, China

Keywords: Silicon carbide; Gelcasting; Mechanical properties

ABSTRACT

In this paper, the properties of green and sintered SiC samples formed by gelcasting were studied based on our previous work. N, N-dimethyl acrylamide (DMAA) and N, N'-methylenebisacrylamide (MBAM) are used as organic monomer and cross-linker respectively. The initiator was 2, 2'-azobis[2-(imidazolin-2-yl) propane]dihydrochloride(AZIP·2HCl). After gelcasting and drying, SiC samples can be densified by pressureless sintering. The properties of green and as-sintered samples were studied and related to the solid content of the gelcasting slurries.

1 INTRODUCTION

Gelcasting is an attractive highly versatile fabrication process to prepare a ceramic green body with high-quality and complex shape[1,2]. In gelcasting, slurry made from ceramic powder and a water-based monomer solution is prepared and poured into a mold, followed by the polymerization *in-situ* to immobilize the particles in a gelled part. Then the samples were removed from the mold when it was still wet, dried and fired. In recent years, gelcasting has been widely studied to produce ceramic materials. Ceramic parts from over a dozen different compositions ranging from alumina-based refractories to high-performance silicon nitride (Al_2O_3, Si_3N_4, SiC,PZT, $BaTiO_3$ etc.) have been produced by gelcasting.

Because the relative densities of green and sintered body in gelcasting process have direct correlation with the solids loading of the slurry, many researchers focus on the preparation of concentrated slurry with low viscosity. Most of them have thought of an optimized circumstance of dispersant, pH, etc., to increase the solids loading of the slurry[3,4].

Silicon carbide (SiC) has a variety of desirable properties, such as high mechanical strength, high chemical stability, good thermal conductivity, low coefficient of expansion, and outstanding erosion resistance. This material has received a great deal of attention as technologically important materials and has been used in various applications including heat engines, gas turbines, and high-temperature energy conversion systems etc.

The objective of this paper was to develop a gelcasting process for SiC ceramics. The properties of SiC slurries and the samples properties at different preparation stage were evaluated. The influence of solid content on the strength of the green and sintered bodies was studied.

2 EXPERIMENTAL PROCEDURES

2.1. Starting materials

Aqueous gelcasting of SiC was carried out using dimethylacrylamide (DMAA) as the monomer, and *N,N*_-methylenebisacrylamide(MBAM,Aldrich) as the crosslinker. The initiator here we used was 2,2'-azobis[2-(2-imidazolin- 2-yl)propane] (AZIP), a kind of water soluble azo-initiator.

Commercially available SiC powder (FCP-15, Saint-Gobain Ceramics Department, Norton, Norway) produced by the Acheson method was used in this study. The average particle size and a specific surface area are 0.58μm and 15.24m^2/g respectively. Boron carbide (Mudanjiang Jingangzuan Boron Carbide Co., Ltd.) with the average particle size as 0.93 μm and a specific surface area as 10.78 m^2/g were used as the sintering additives. Dextrin was used as the carbon source (sintering additives).

2.2. Gelcasting and sintering

In a typical experiment, initially, SiC powder was suspended in a premix solution, which had been prepared by dissolving 12 wt% DMAA (methacrylamide) and MBAM (mithylenebisacrylamide) in a 13:1 ratio in de-ionized water. The slurries were mixed under mechanical agitation while keeping the temperature of the suspension at 0–5 °C with a water bath. To improve the dispersion of SiC powder in the premix solution and the fluidity of the suspension, PEI (at the rate of 0.456 mg/m^2 of SiC) was used as the dispersant[2,4,5]. All the particulate slurries were degassed for 10 min after mechanically mixing by vacuum pumping. After adding the initiator, the suspension was degassed again in order to eliminate the air bubbles trapped inside before casting into moulds. Afterward, the slurries were cast into non-porous molds, which were then allowed to set in water bath at 40 °C for 30 min in order to gel the system. The gelled green bodies were de-molded and dried under controlled humidity conditions to avoid cracking and non-uniform shrinkage due to rapid drying. Rheological measurements were performed on a stress controlled rheometry (SR-5 Rheomeric scientific instrument company, U.S.A.) at 25°C.

Pyrolysis is critical for gelcasting green sample. Sufficient time should be given for the organic materials burnout before the sintering of ceramics to avoid cracks in the microstructure. In our study, TG / DTA was used to determine the temperature schedule. TG curve indicated that the total mass loss in the heat-treated process is about 6.5 wt%, including the water, the dextrin and the gels. The polymers began to pyrolyze at about 200°C and were completely burned out near 600°C. Thus, to allow ample time for the complete burnout of the polymer in gelcasting samples, the heating rate was set for 1°C / min up to 600°C in vacuum with a dwell time of 1 h to take advantage of the carbon produced, similar to that reported in literature for HA[6]. Then the samples were densified by pressureless sintering at 2200 °C for 1 h in an inert atmosphere.

2.3. Characterization of gel-casting bodied and ceramics bodies

The density of green pieces was determined by Hg intrusion porosimetry in a Micromeritics ASAP2010 porosimeter. The relative density and porosity of the sintered samples were determined by Archimedes's method. The microstructure of the composites was observed on the fractured surface by scanning electron microscopy (SEM) (EPMA- 8705Q, HII, Shimadzu, Japan). The flexural strength of green and sintered samples with the size of 3 × 4 × 36 mm^3 were measured by three-point bending, using a span of 30 mm and a crosshead speed 0.5 mm/min. The indentation test was performed in a microhardness tester (IF, AkashIII, Japan), with a Vickers indenter, applying a load of 2.0 kg for 10 s.

3. RESULTS AND DISCUSSIONS

3.1 Properties of green samples

The effect of dispersant, monomers on the rheological properties of SiC slurries as well as the consolidation process has been well studied in our lab[7,8]. Based on the previous work, we kept the monomer content as 12wt% and the DMAA/MBAM ratio as 13:1. In this study, efforts are paid toward the effect of solid content on the properties of green and sintered samples. Fig.1 showed the rheological properties of 50vol% SiC slurries. SiC slurries exhibited a shear thinning behavior at low shear rate (<288 s^{-1}) and mild shear thickening behavior at higher shear rate (>288 s^{-1}), which can be related to the structures formed in slurries. The increase in viscosity is due to a transition from a two- dimensional layered arrangement of particles to a three-dimensional structure. Fig.1 showed that the slurry is well dispersed and meet the requirement for gelcasting: during the casting process, the viscosity is lower under shearing; once deposited, the slurries recover to a high viscosity level to suppress uncontrolled flow and to prevent sedimentation of the ceramic particles during the subsequent drying process.

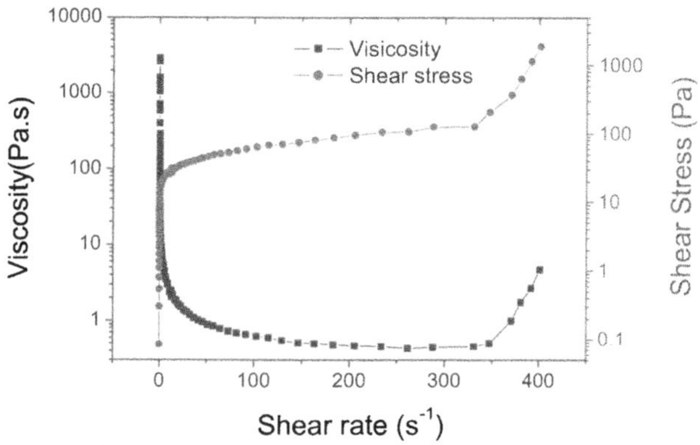

Fig.1 Rheological properties of 50vol% SiC gelcasting slurries

After casting, gelification and drying, well-shaped green bodies were obtained. The optical photographs of the green samples are shown in Fig.2. The variation of the relative density and the linear shrinkage of the green bodies as a function of solid loading are presented in Fig. 3. With the increase in the solid loading, the relative density of the green body increase sharply at the solid loading of 51vol% and then increase gradually thereafter. At high solid loading, this trend becomes

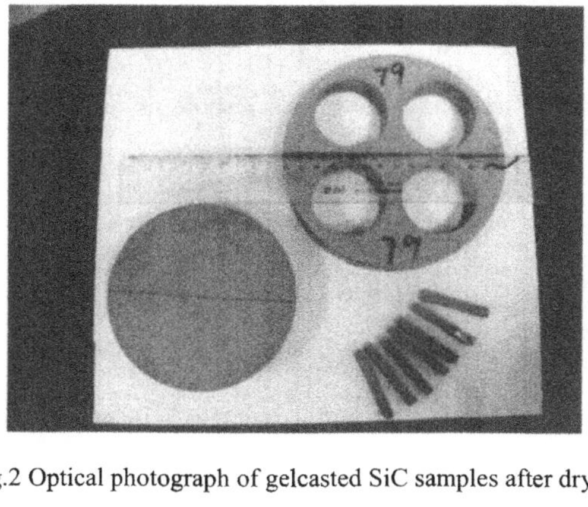

Fig.2 Optical photograph of gelcasted SiC samples after drying

not so clear. This might be due to the difficult in dispersion at high solid content. This trend can be well corresponded to the linear shrinkage curve. The average bending strength as high as 39.9±8.1MPa was obtained for green samples. The high strength comes from the cross-linking gel network and the homogenous packing of particles.

The microstructure of the green body as observed by SEM is shown in Fig. 4. After consolidation, powders in green body compact closely and homogeneously. However, the polymer network was not observed via SEM. The pore diameter distribution, obtained by Hg intrusion porosimetry, showed a monomodal distribution type. The relative density, porosity, and the median pore diameter of green samples were 57.67%, 35.02%, and 10.6nm, respectively.

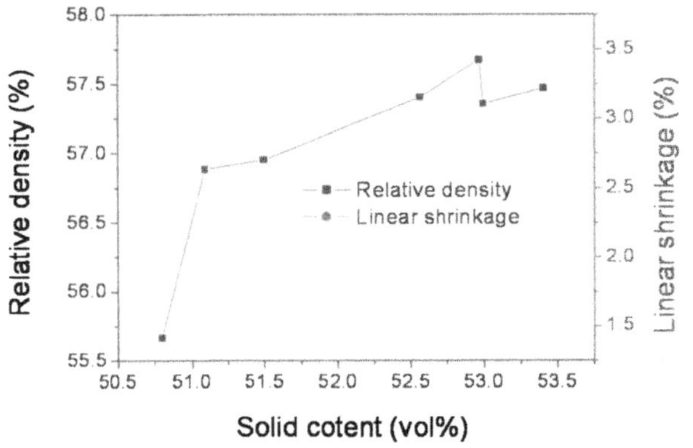

Fig.3 Dependence of the relative density and the linear shrinkage of green gelcasted SiC samples on the concentration of solid in the slurry

Fig.4 Microstructure and pore size distribution of green SiC samples (Solid content 53vol%)

3.2. Sintered samples

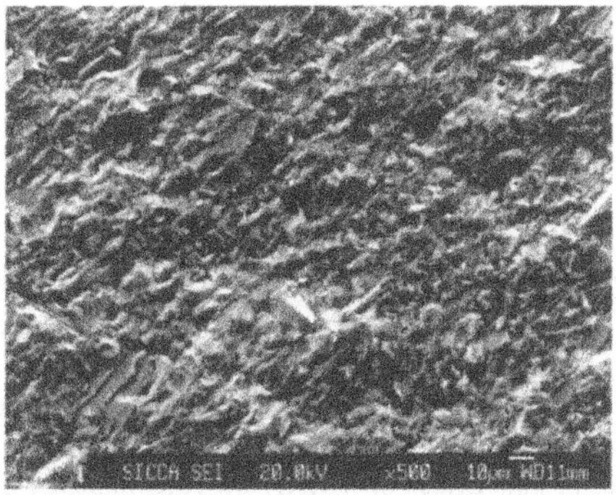

Fig. 5 he fracture surface of SiC samples with the solid content as 53%.

Table 1 Properties of SiC samples after sintering

Solid content	Relative density (%)	Flexural strength (MPa)	Toughness (MPa·s$^{1/2}$)
50.80	97.46±0.18	389±54	3.07±0.12
51.10	97.662±0.15	449±54	3.07±0.21
51.50	97.72±0.07	486±45	3.14±0.22
52.50	97.98±0.06	508±63	3.24±0.27
53.00	97.99±0.17	539±89	3.46±0.31
53.40	97.89±0.69	449±84	3.42±0.34

Table 1 shows the properties of SiC samples via gelcasting forming method and pressureless sintering. Differed to the trend in linear shrinkage with the solid content, there was not much difference in sinter density with respect to the solid content. Sintered densities as high as 98%TD were achieved for all the samples.

Fig. 5 displays the optical photograph and the fracture micrographs (SEM) of SiC samples after sintering. The surface is characteristic of compacted ceramics materials. No cracks and pores are observed. Based on SEM observation, the microstructure is homogeneous, and most of the SiC crystal grains grow too big. This might be due to the high sintering temperature. There are few pores in SiC sintered body. The fracture micrograph showed that the fracture mode is a mixture of intergranular and intragranular type. XRD patterns after sintering is consistent with of SiC 6H phase according to the JCPDS file (72-0018). There were no other phases present.

4. Conclusions

SiC slurries with PEI as the dispersant showed satisfied rheological properties for gelcasting. The green strength of SiC bodies was 39.9±8.1MPa, suitable for green machining. After gel-casting, drying and sintering at 2200°C, SiC samples can be densified to a relative density of 98%. The mechanical properties of the obtained pieces are satisfying, with the flexural strength, toughness and the hardness as 539±89MPa, 3.46±0.31 MPa·s$^{1/2}$ and 26.2±0.97 GPa , respectively at the solid content of 53vol%. The process appears to be highly suitable for manufacturing complex shaped components requiring no post sintering machining.

Acknowledgement

This work was supported by the National Natural Science Foundation of China (No. 50772128), Shanghai Science and Technology Committee (No. 07DJ14001, No.07pj14094), and the State Key Laboratory of High Performance Ceramics and Superfine Microstructures.

Reference

[1] A. C. Young, O. O. Omatete, M. A. Janney, and P. A. Menchhofer, Gelcasting of Alumina,' *J. Am. Ceram. Soc.*, **74**, 612–8 (1991).

[2] O. O. Omatete, M. A. Janney, and R. A. Strehlow, Gelcasting–A New Ceramic Forming Process, *Am. Ceram. Soc. Bull.*, **70**, 1641–9 (1991).

[3] J. X. Zhang, D. L. Jiang, S. H. Tan, L. H. Gui, and M. L. Ruan, Aqueous Processing of SiC Green Sheets I: Dispersant, *J. Mater. Res.*, **17**, 2012–8 (2002).

[4] V. A. Hackley, Colloidal Processing of Silicon Nitride with Poly (Acrylic Acid): I, Adsorption and Electrostatic Interactions, *J. Am. Ceram. Soc.*, **80**, 2315–25 (1997).

[5] W. Li, P. Chen, M. Gu, and Y. Jin, Effect of TMAH on Rheological Behavior of SiC Aqueous Suspensions, J. *Eur. Ceram. Soc.*, **24**, 3679–84 (2004).

[6] B.Q. Chen, Z.Q. Zhang, J.X. Zhang, M.J. Dong and D.L. Jiang, Aqueous gel-casting of hydroxyapatite, *Mater. Sci. Eng. A* **435–436,** 198–203 (2006).

[7] T. Zhang, Z.Q. Zhang, M.J. Dong, J.X. Zhang, Q.L. Lin and D.L. Jiang, The influence of polyethylene imine on the gelcasting of SiC with two different initiators. *J. Am. Ceram. Soc.*, **90**, 3748-3752 (2007).

[8] T. Zhang, Z.Q. Zhang, J.X. Zhang, Q.L. Lin and D.L. Jiang, Preparation of Dense SiC Ceramics by Aqueous Gelcasting, *J. Inorg. Mater.*,**22**, 489-492 (2007).

MICROWAVE DIELECTRIC PROPERTIES OF Nb_2O_3-$Zn_{0.95}Mg_{0.05}TiO_3$+0.25TiO_2 CERAMICS WITH Bi_2O_3 ADDITION

Ying-Chieh Lee[a,*], Hui-Hsiang Huang[b], Wen-Hsi Lee[c], Yen-Lin Huang[a], Shih-Feng Chien[c]
[a] Department of Material Engineering, National Pingtung University of Science & Technology, Pingtung, Taiwan
[b] R&D Department, Chilisin Electronics Corp., Shinchu 303, Taiwan.
[c] Department of Electrical Engineering, National Cheng Kung University, Tainan 701, Taiwan

ABSTRACT

In this study, the effects of Bi_2O_3 addition of up to 10 wt% on sintering characteristics, microwave dielectric properties and microstructures of Nb_2O_3-$Zn_{0.95}Mg_{0.05}TiO_3$ + 0.25TiO_2 (Nb-ZMT′) ceramics, prepared by conventional solid-state routes have been investigated. The sintered ceramic samples were characterized by X-ray diffraction and scanning electron microscopy (SEM). It was found that as the content of Bi_2O_3 increased, the density of the sintered ceramics increased, and the density of Nb-doped samples with 5 wt% Bi_2O_3 was over 98% of theoretical density. The dielectric constant of the Nb-doped samples increased with increasing Bi_2O_3 content. In addition, ε_r of the Nb-doped samples was higher than the undoped samples. The Nb-ZMT′ ceramic with 5 wt% Bi_2O_3 addition sintered at 900 °C exhibited the optimum dielectric properties: Qxf = 12000 GHz, ε_r = 30, and τ_f = -12 ppm / °C.

INTRODUCTION

Recently, the development of commercial wireless technologies has been made rapid progress because of the improved characteristics of dielectric resonators in microwave ranges. This rapid development of the wireless communication implies to design new ceramics sinterable at low temperature, e.g. at around 900 °C and exhibiting good dielectric properties. This low sintering temperature is of primary importance to produce silver co-sintering devices such as silver based multiplayer ceramic capacitors or hybrid circuits [1]. The required specifications in term of dielectrics properties are a high dielectric constant (ε_r > 20), a high quality factor (Q > 10,000) which corresponds to a low dielectric loss (tanδ = $1/Q$) and a temperature coefficient of the permittivity close to zero ppm/°C. However, the sintering temperature of conventional dielectric ceramics used for microwave resonators, filters, and duplexers of portable phones usually ranges from 1200 to 1500°C [2].

There are three compound phases present in the ZnO-TiO_2 phase diagram: Zn_2TiO_4 (cubic), $ZnTiO_3$ (hexagonal), and $Zn_2Ti_3O_8$ (cubic), which is a low-temperature form of $ZnTiO_3$ that exists below 820°C. It has been reported that the zinc titanates ($ZnTiO_3$) can be sintered at 1100°C without sintering aids. However, if a sintering aid is added, it can be fired at temperatures lower then 900°C [3]. This material has a permittivity (ε_r) of 19, a Q value of 3000 at 10 GHz, and a temperature coefficient (τ_f) of −55 ppm/°C. Kim *et al.* [4] studied the $ZnTiO_3$-$x$$TiO_2$ system, where x = 0 – 0.5; the optimum microwave dielectric properties were obtained to be τ_f = -20 ppm/°C, ε_r = 27.5, and Qxf = 14000 GHz for x = 0.25 and sintering temperature of 925 °C. Nevertheless, the pure $ZnTiO_3$ phase is not easily to be obtained because the compound decomposes to Zn_2TiO_4 and TiO_2 at high temperatures [5]. Kim *et al.* [4] also studied zinc titanate with a small amount of magnesium addition, and reported that the $(Zn_{1-x}Mg_x)TiO_3$ phase was stable at a much higher temperature, the microwave dielectric properties of which were τ_f = -90ppm/°C, ε_r = 21, and Qxf =

* Corresponding author, e-mail: YCLee@mail.npust.edu.tw

40000 at x = 0.05 and sintering temperature of 900 °C.

In the previous study [6], the microstructure and dielectric properties of the Bi_2O_3 additive in $Zn_{0.95}Mg_{0.05}TiO_3$+0.25TiO_2 with 1 wt% 3ZnO-B_2O_3 (ZnBO-ZMT′) ceramics have been investigated. It was found that 5% Bi_2O_3 addition can significantly suppress spinel Zn_2TiO_4 phase formation in the ZnBO-ZMT′ ceramics, but the microwave dielectric properties become worse due to low Q value (< 600). Although, the permittivity still was kept at ~24. The main purpose of incorporating the Bi_2O_3 additive in $Zn_{0.95}Mg_{0.05}TiO_3$ + 0.25TiO_2 with 1 wt% 3ZnO-B_2O_3 and 1 wt% Nb_2O_3 (Nb-ZMT′) ceramics is to improve the microwave dielectric properties of sintered samples at lower temperatures, to obtain a homogeneous microstructure, and to suppress spinel Zn_2TiO_4 phase formation in this study.

EXPERIMENTAL PROCEDURES

The starting materials were ZnO, MgO and TiO_2 powders. They were mixed and ground in deionized water with 2 mm zirconia beads for 24 h, until the mean particle size (D50) was approximately 0.35 μm. The optimum composition of ZMT′ determined in previous studies [4,5] is $Zn_{0.95}Mg_{0.05}TiO_2$ + 0.25TiO_2 (ZMT′), which is adopted as the reference composition. The powders were calcined in air at 900 °C for 5 h after ball milling. Nb_2O_3 was added and fixed at 1 wt%. Bi_2O_3 glass was chosen as sintering aids, and it was added in the amounts of 1, 3, 5 and 10 wt%, respectively. Then the calcined powders were milled again for 6 h. After grinding, the mean particle size was measured to be about 0.5 μm. After drying, the powders were pressed uniaxially into pellets in a steel die. Sintering of these pellets was carried out at 880, 900, 920, 940, 960, 980 and 1000 °C for 2 h. The crystalline phases of the sintered ceramics were identified using X-ray diffraction pattern analysis (XRD, Philips X'Pert-MPD). Microstructural observation of the sintered ceramics was performed using scanning electron microscopy (SEM, JEOL. JEL-6400 Japan) equipped with energy-dispersive spectroscopy (EDS). The dielectric characteristics at microwave frequencies (7.25–10GHz) were measured by the Hakki-Coleman dielectric resonator method, where a cylindrically shaped specimen was positioned between two plates. An HP8719C network analyzer was used to measure the microwave property. The dielectric properties were calculated from the frequency of the TE_{011} resonant mode. For convenience, the $Q \times f$ factor was used to evaluate the loss quality, where f is the resonant frequency.

RESULTS AND DISCUSSION

(1) Influence of Bi_2O_3 addition on densification and microstructure characterization of the Nb-ZMT′ ceramics

Figure 1 shows the bulk density of the Nb-ZMT′ ceramics sintered at 900 °C as a function of the amount of Bi_2O_3 additions. The bulk density gradually approached a plateau at above 5 wt%. This was consistent with the previous study in which 1 wt% ZnBO addition in the ZMT′ ceramic was shown to lower the sintering temperature of the ceramic from 1100 ° to 900 °C. Zhang *et al.* [8] reported that during sintering, Bi_2O_3, having a low melting point of 800 °C, exists along the dielectric grains in liquid phase. The glass phase assisted in the densification of dielectrics through liquid-phase sintering. A high sintered density could be obtained at relatively low sintering temperatures; over 98% of the theoretical density, i.e., 4.9 g/cm^3, had been achieved for the specimen with 5 wt% Bi_2O_3 addition sintered at 900 °C for 2 h in air. Further Bi_2O_3 addition up to 10 wt% in the ceramic could also lower the sintering temperature down to 880 °C. The extent of increase in the bulk density of the sintered ceramics depends upon the amount of Bi_2O_3 addition.

Figure 2 shows the XRD results for Nb-ZMT′ with 5 wt% Bi_2O_3 addition sintered at temperatures ranging from 880 ° to 1000 °C. It was interesting to note that no cubic Zn_2TiO_4 phase was present in the Nb-ZMT′ ceramic with 5 wt% Bi_2O_3 addition, even at a temperature as high as 940 °C.

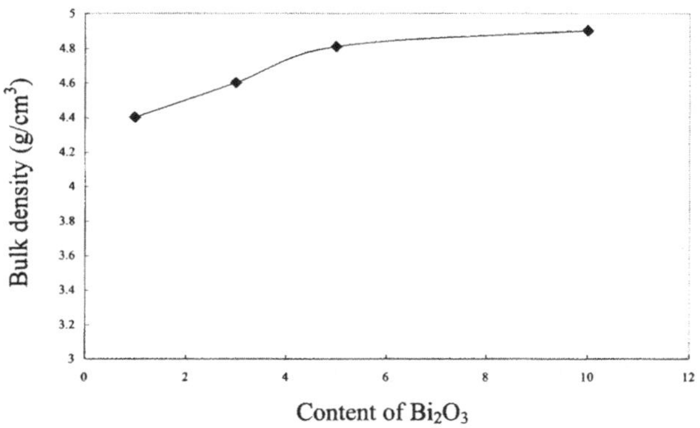

Fig. 1: Bulk density of Nb-ZMT′ ceramics sintered at 900 °C as a function of Bi_2O_3 amount.

From our previous study [6], it was known that the Zn_2TiO_4 phase was formed in ZnBO-ZMT′ ceramics sintered at 920 °C. However, the formation of Zn_2TiO_4 phase in the Nb-ZMT′ ceramics could be restrained by the addition of Bi_2O_3. In addition, significant extra phase of $Bi_2Ti_2O_7$ was found at sintering temperature of 940 °C. Figure 3 shows the XRD results of the sintered dielectrics having the composition of Nb-ZMT′ ceramics with different amounts of Bi_2O_3. When the same heat treatments 900 °C were applied, the microstructure of the ceramics mainly consisted of $ZnTiO_3$ and TiO_2 phases. For the undoped Nb-ZMT′ ceramics sintered at 900 °C, as shown in Fig. 3(a), it can be observed that two well-characterized phases, namely TiO_2 and $ZnTiO_3$ appeared in the ceramic contains [7]. For Nb-ZMT′ ceramics with 3, 5 and 10 wt% Bi_2O_3 additions sintered at

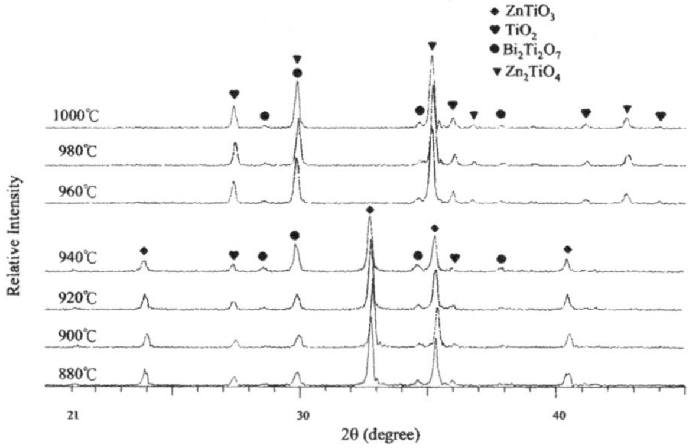

Fig. 2: X-ray diffraction (XRD) spectra of Nb-ZMT′ ceramics with 5 wt% Bi_2O_3 sintered at (a) 880°C, (b) 900°C, (c) 920°C, (d) 940°C, (e) 960°C, (f) 980°C and (g) 1000°C.

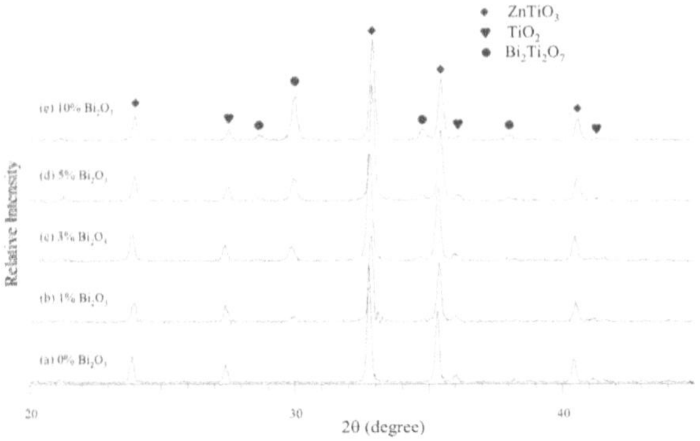

Fig. 3: X-ray diffraction (XRD) spectra of Nb-ZMT′ ceramics with different amount of Bi_2O_3 addition sintered at 900°C for 2 h.

900 °C, as shown in Figs. 3(c), 3(d) and 3(e), respectively, it is found that the major crystalline phases are the same as those of the undoped Nb-ZMT′ ceramics, along with an extra phase identified to be $Bi_2Ti_2O_7$. The new phase must be associated with Bi_2O_3 addition because it clearly appears in specimens with $Bi_2O_3 \geq 3$ wt%.

Figure 4 shows SEM micrographs of the Nb-ZMT′ ceramics sintered at 900 °C with 3 and 10 wt% Bi_2O_3 addition. The microstructures of the specimens looked quite similar. The gray, similar rectangular grains are $ZnTiO_3$ phases, the dark grains are rutile phases (TiO_2) [8], and the white grains are second phases $Bi_2Ti_2O_7$. However, the microstructure of the sintered ceramics showed a lot of changes, that is, larger amount of new second phasewas present, as shown in the Figs. 4(b), for the sample which contains 10 wt% Bi_2O_3. The second phase was further analyzed by energy-dispersive-spectroscopy with a 5 nm beam size, and the results for Nb-ZMT′ ceramics with 10 wt% Bi_2O_3 addition sintered at 900 °C for 2 h are shown in Fig. 5. Fluctuation of titanium content as well as other elements is clearly depicted. Compared with the matrix phase (a), the white part (b) showed that the secondary phase had higher signal intensities in Bi. It was therefore believed that the secondary phase was mainly composed of the $Bi_2Ti_2O_7$ phase. This argument was further supported by the results of X-ray diffraction analysis in the previous study, from which it was concluded that greater Bi_2O_3 addition enhanced the formation of the $Bi_2Ti_2O_7$ phase.

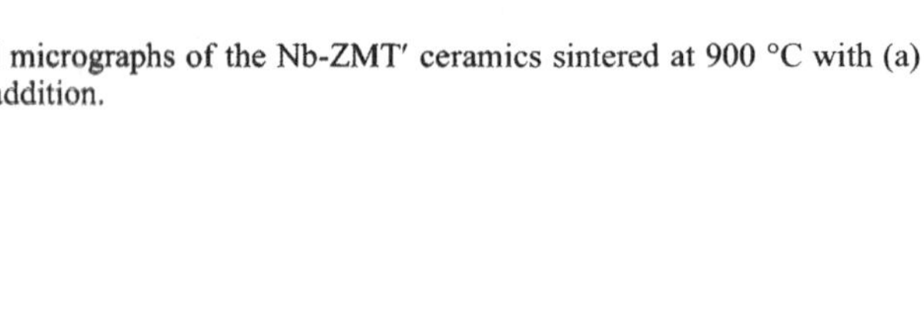

Fig. 4: SEM micrographs of the Nb-ZMT′ ceramics sintered at 900 °C with (a) 3 wt% and (b) 10 wt% Bi_2O_3 addition.

(2) Microwave dielectric properties of the the Nb-ZMT′ ceramics

The dielectric constant of Nb-ZMT′ ceramics was measured at a frequency of 10 GHz, and the results are shown in Fig. 6 (a). The ε_r value of Nb-ZMT′ ceramics was closely related to the Bi_2O_3 addition, e.g., the dielectric constants of the ceramic were 25.1, 28.2, 30.6 and 31.8 for Bi_2O_3 addition at 1, 3, 5 and 10wt%, respectively. It must be pointed out, however, that second phases such as $Bi_2Ti_2O_7$, the grain size and lattice strain of the sintered ceramics (due to Zn^{2+} replaced by Nb^{2+}) may had essential effects on the dielectric constant of the ceramics. However, a trace of a second phase in Fig. 2, identified as $Bi_2Ti_2O_7$ phase, was detected in the sample ≥ 3.0 wt% Bi_2O_3. Because, the addition of Bi_2O_3 effectively led to promote densification of Nb-ZMT′ ceramics as compared with lower Bi_2O_3 content of sample (Fig. 1).

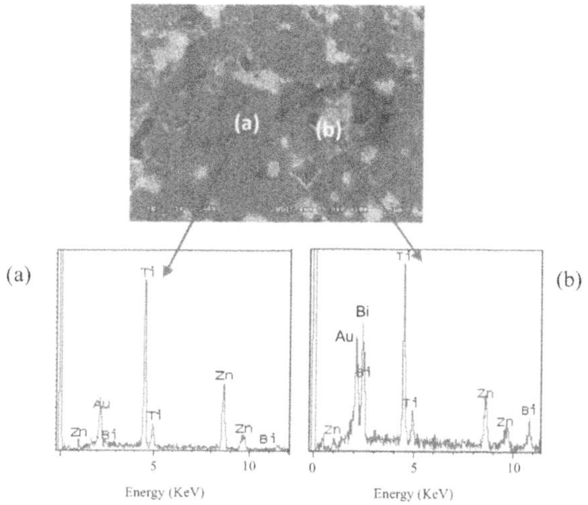

Fig. 5: SEM micrograph and EDS spectra of the Nb-ZMT′ ceramics with 10 wt% Bi_2O_3 addition sintered at 900°C.

The $ZnTiO_3$ ceramic is an interesting material with a negative τ_f of -6.2 ppm/°C. Figure 6 (b) shows the temperature coefficient of the resonant frequency, τ_f, at the maximum Q value as a function of the amount of added Bi_2O_3 for samples sintered at 900°C. The Nb-ZMT′ ceramics without Bi_2O_3 addition had a τ_f around –50 ppm/°C. The Nb-ZMT′ ceramic is known for its temperature-stable characteristic. The τ_f values can be changed to close to zero when Bi_2O_3 was added to the Nb-ZMT′ ceramic. The τ_f value of the Nb-ZMT′ ceramic with 5 wt% Bi_2O_3 addition exhibited the smallest negative value of –12 ppm/°C. Figure 6 (b) also shows the Qxf value of the Nb-ZMT′ ceramics sintered at 900°C as a function of the amount of Bi_2O_3 addition. The Qxf values of the Nb-ZMT′ ceramics without Bi_2O_3 addition were about 4200, which was lower than the samples with Bi_2O_3 addition. When 5 wt% Bi_2O_3 phase was added to the ceramics, the Qxf value increased rapidly from 4200 to 12000. This result was probably due to the high densification of the sintered ceramics. It is known that the microwave dielectric loss is mainly controlled by the second phases formed or the crystal defects and grain boundaries [9,10]. Nevertheless, from the point of view of practical application, ceramics with a low *Qxf* value are not suitable for high-frequency applications.

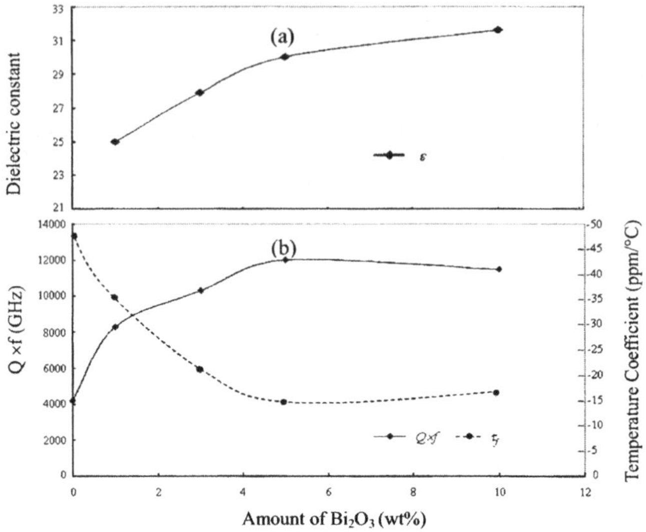

Fig. 6: Dielectric constants, *Qxf* values and temperature coefficients of resonant frequency of the Nb-ZMT′ ceramics as a function of the amount of added Bi_2O_3.

CONCLUSIONS

The sinterability was significantly enhanced and the microwave properties were affected by the addition of Bi_2O_3 for Nb-ZMT′ ceramics. A new secondary phase was found, which was related to Bi_2O_3, since it appeared in specimens containing ≥ 1 wt% Bi_2O_3. The phase composition of the second phase was very similar to $Bi_2Ti_2O_7$. In addition, Zn_2TiO_4 formation in Nb-ZMT′ ceramics can be inhibited by the Bi_2O_3 addition at sintered temperatures of up to 940 °C. In this study, sintering of Nb-ZMT′ ceramics with 5 wt% Bi_2O_3 resulted in over 98% of the theoretical density at 900 °C for 2 h, and the dielectric properties were $Q{\times}f$ = 12000 GHz, ε_r = 30, and τ_f = -12 ppm / °C.

REFERENCES

[1]A. Chaouchi, S. Marinel, M. Aliouat, S. Astorg, J. Eur. Ceram. Soc., 27, 2561-66 (2007).
[2]T. Takada, S. F. Wang, S. Yoshikawa, S. J. Jang, R. E. Newnham, J. Am. Ceram. Soc., 77, 2485 (1994).
[3]F. H. Dulin and D. E. Rase, J. Am. Ceram. Soc., 43, 125 (1960).
[4]H. T. Kim, J. D. Byun and Y. H. Kim, Mater. Res. Bull., 33, 963 (1998).
[5]H. T. Kim, S. Nahm and J.D. Byun, J. Am. Ceram. Soc., 82, 3476 (1999).
[6]Y.C. Lee and W.H. Lee, Jpn J. App. Phys., 44 [4A], 1838-1843 (2005).
[7]K. Haga, T. Ishii, J. I. Mashiyama and T. Ikeda, Jpn. J. Appl. Phys., 31, 3156 (1992).
[8]H. T. Kim, S. H. Kim, S. Nahm and J. D. Byun, J. Am. Ceram. Soc., 82, 3043 (1999).
[9]C. F. Yang, Jpn. J. Appl. Phys., 38, 3576 (1999).
[10]S. I. Hirano, T. Hayashi and A. Hatto, J. Am. Ceram. Soc., 74, 1320 (1991).

FABRICATION OF YTTRIA-STABILIZED ZIRCONIA CERAMICS WITH RETICULATED PORE MICROSTRUCTURE BY FREEZE-DRYING

Yuan Zhang[1,2], Kaihui Zuo[1], Yu-Ping Zeng[1,*]
1.Shanghai Institute of Ceramics, Chinese Academy of Sciences, Shanghai 200050, China
2.Graduate School of the Chinese Academy of Sciences, Beijing 100049, China

ABSTRACT

A freeze-drying process was used to fabricate porous yttria-stabilized zirconia (YSZ) ceramics. Methyl cellulose (MC) was added into the slurry to control the microstructures of the porous YSZ ceramics. The experimental results indicated that the MC concentration and solid loading of YSZ slurry had great influences on the morphologies of pores. After MC addition, the microstructure of porous YSZ developed from the laminated pores to the reticular pores. Comparison with the MC concentration, the solid loading of slurry had an obvious effect on the porosity. The specimens prepared from 5 to 30 vol% YSZ slurries with 2 wt% MC had porosities from 95.60 to 38.64%.

INTRODUCTION

Ceramics with reticulated pore microstructures have wide applications in many fields, including exhaust filters, catalyst supports, gas burners and particle filters due to their three-dimensional network structure.[1] Ceramics with reticulated pores are usually fabricated by replicating struts of a polymer template.[2-5] Polymer replication technology generally needs repeated coating and easily contributes to strut cracks and big shrinkage.[1, 3, 6] However, the freeze-drying process is an efficient way to obtain crack-free struts, because during freeze drying process, solvent is removed by sublimation of solid phase under vacuum, avoiding the drying stresses and shrinkage that may lead to cracks and warp during normal drying process.[7-11]

In the present work, freeze-drying was used to prepare reticulated porous YSZ ceramics and the methyl cellulose (MC) solution was added in the aqueous ceramic slurry to modify the morphologies of pores. The influences of solid loading and MC concentration on microstructure, porosity and bulk density were investigated.

EXPERIMENTAL PROCEDURE

5 mol% Y_2O_3-doped yttria-stabilized zirconia (YSZ, d_{50}=60nm, Farmeiya Advanced Materials Co., Ltd., Jiangxi, China) was used as a raw material. Deionized water was used as a solvent, ammonium polyacrylate (Lopon 885, BK Giulini, Germany) was selected as a dispersant, and 3 wt% aqueous MC (450 g/mol of molecular weight, Sinopharm Chemical Reagent Co.Ltd., Shanghai, China) solution was chosen as a microstructure modifying agent. In a typical procedure, the aqueous YSZ slurry was prepared by mixing the powder with the dispersant in the deionized water. The concentration of dispersant, based on YSZ powder, is 1.5 wt%. The initial slurry with solid loading of 10-30 vol% was ball milled with ZrO_2 balls for 24 h. Then, the MC solution was added into the YSZ slurry and ball milled for another 24 h. The concentration of MC addition based on YSZ powder is from 0 wt% to 4 wt%.

The mixed slurries were poured into columnar molds made of aluminum with the length of 50 mm, width of 10 mm and height of 10 mm, respectively. The wall thickness of the mold was 10 mm. All samples were frozen in a chamber of -18 °C. After complete solidification, the samples were taken off from molds and placed into a vacuum freeze drier (Model TLG-A, Shanghai Zhongke Biotech Co. Ltd., Shanghai, China), followed by drying at 6 Pa and 60 °C for 12 h. Then green bodies were carefully moved into a furnace and heated up to 600 °C at a rate of 2 °C/min holding for 3h to burn out the organic additives. Continually, green bodies were heated up to 1000 °C at 10

°C/min, followed to 1300 °C at a rate of 5 °C/min and maintained for 1h. Finally, the specimens were cooled down to 1000 °C at 10 °C/min, the electric power was then switched off cooling the furnace to the room temperature. Microstructures were observed by scanning electron microscopy (SEM, Model JSM-6700F, JEOL, Akishima, Japan). The porosities and bulk densities were measured by the Archimedes method in distilled water.

RESULTS AND DISCUSSION

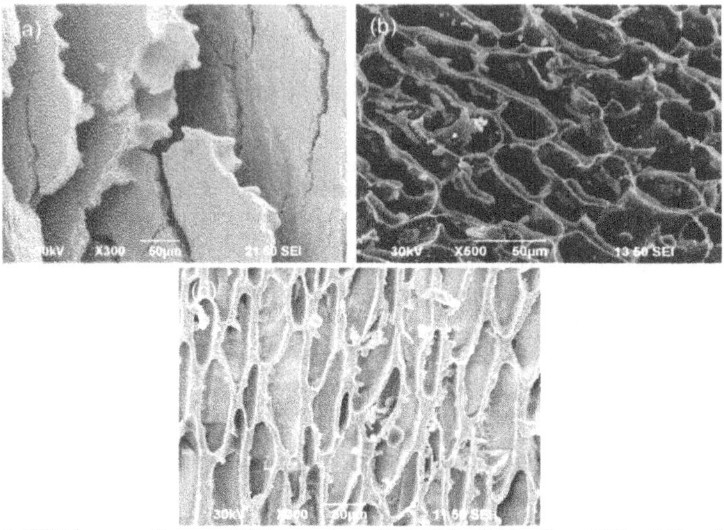

Fig.1. SEM images of porous YSZ ceramics with initial solid loading of 15 vol.% and MC concentration of 0 wt% (a), 2 wt% (b) and 4 wt% (c).

Figure 1 shows the variation of pore morphologies of YSZ ceramics with initial solid loading of 15 vol.% and different MC addition. Without MC addition, the microstructure is composed typically of lamellar pores and ceramic walls (Fig. 1 (a)). In freezing process, ice grows into big crystal and forms laminated pore after the sublimation of ice, and ceramic particles are expelled by the ice crystal to form the walls between laminated pores. Since there are no linkages among the walls, the mechanical properties are weak, resulting in big cracks in the walls. So the YSZ ceramics without MC addition are easily shattered even under very small mechanical force. After adding MC addition, the morphologies of YSZ ceramics are changed, shown in Fig. 1(b, c). The two-dimensional laminated pores transit into the smaller three-dimensional reticular pores due to the linkage among the walls. The results also indicate that there are no cracks and the struts are very dense.

Figure 2 shows the pore morphologies of YSZ ceramics with different initial solid loading and 2 wt% MC addition. When increasing the solid loading from 5 vol.% to 15 vol.%, the obstacles of ceramics particles to the growth of ice crystals increase and the particles are difficult to be repelled, so the thickness of struts raises, the pore sizes decrease and pore shape is shorten obviously. The solid loading only affect the morphologies of pores, however, it doesn't change the reticular structure as the slurries having the MC addition.

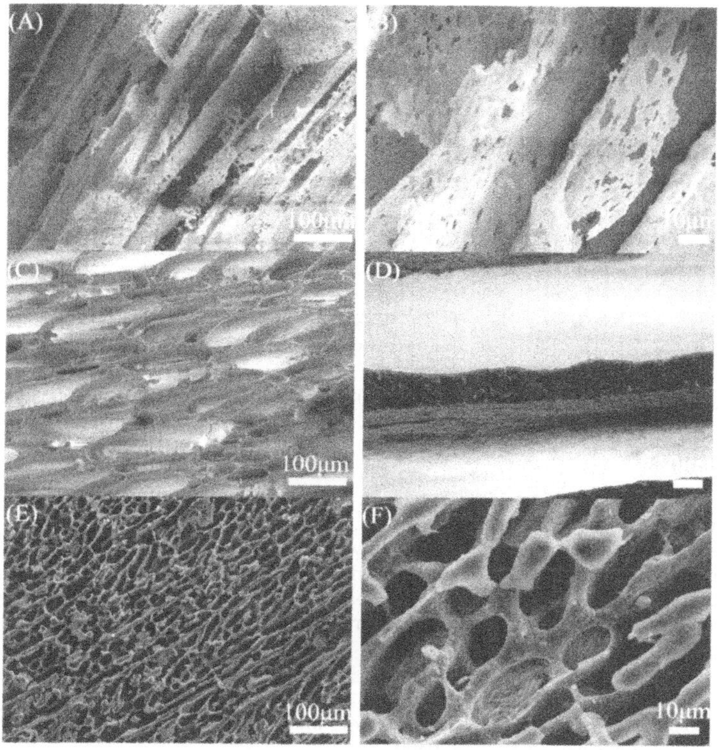

Fig.2. SEM images of porous YSZ ceramics with 2 wt% MC and different initial solid loading of 5 vol% (a, b), 10 vol% (c, d) and 15 vol% (e, f).

The solid loading of slurry and MC concentration not only affects the morphologies of pores, but also affects the porosity. Comparison with samples without MC addition, the porosities of samples with MC addition increase due to the burning of MC. For example, the porosities of samples with initial solid loading of 15 vol.% and MC concentration of 0 and 2 wt% are 69.10 % and 73.12 %, respectively. With more MC addition, the porosities decrease again for more MC can increase the slurry viscosity. The porosity of sample with initial solid loading of 15 vol.% and MC concentration of 4 wt% is 70.09 %, which is lower than that of sample with solid loading of 15 vol.% and MC concentration of 2 wt%. Comparing with the MC addition, the solid loading of slurry obviously affects the porosity. Figure 3 shows the porosity and bulk density of porous YSZ ceramics with 2 wt% MC and different initial solid loading. The results indicate the porosity decreases and bulk density rises as the solid loading increases. When the MC concentration is kept at 2 wt%, YSZ ceramics with initial solid loading of 5 vol.% exhibit a high porosity of 95.60 % and low bulk density of 0.25 g/cm^3.

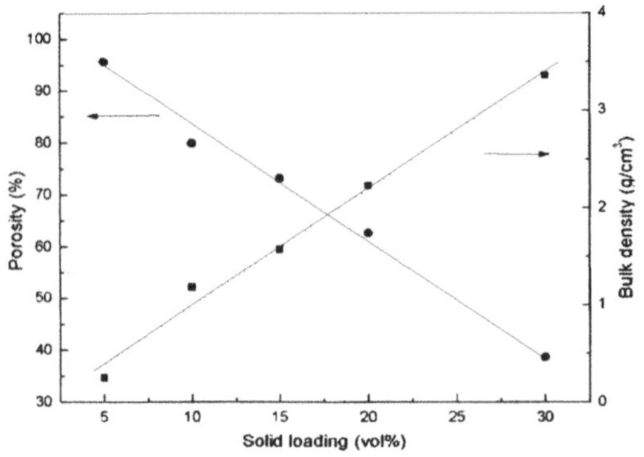

Fig.3. Porosity and bulk density of porous YSZ ceramics with 2 wt% MC and different initial solid loading.

CONCLUSION

Reticulated porous YSZ ceramics with widely porosities range have been fabricated by freeze-drying process. With the addition of MC, the big lamellar pores change into small reticulated and interconnected pores. Comparison with YSZ ceramics without MC addition, the porosities of YSZ ceramics with MC addition increase. With more MC addition, the porosities decrease again. As increasing the initial solid loading of slurry, the pore size and porosity decreases, and pore shape becomes shorted. In a word, both MC concentration and solid loading of slurry both affect the size and morphology of pores.

ACKNOWLEDGMENT

This work was supported by Science and Technology Commission of Shanghai Municipality under Contracts No. 07JP14093 and No. 08JC1420300.

REFERENCES

[1]X. W. Zhu, D. L. Jiang, S. H. Tan, and Z. Q. Zhang, "Improvement in the strut thickness of reticulated porous ceramics," *J. Am. Ceram. Soc.*, 84 [7] 1654-56 (2001)

[2]X. P. Pu, X. J. Liu, F. G. Qiu, and L. P. Huang, "Novel method to optimize the structure of reticulated porous ceramics," *J. Am. Ceram. Soc.*, 87 [7] 1392-94 (2004)

[3]I. K. Jun, Y. H. Koh, J. H. Song, S. H. Lee, H. E. Kim, "Improved compressive strength of reticulated porous zirconia using carbon coated polymeric sponge as novel template," *Mater. Lett.*, 60, 2507-10 (2006)

[4]S. R. Wang, H. R. Geng, L. H. Hui, and Y. Z. Wang, "Reticulated porous mutiphase ceramics with improved compressive strength and fracture toughness," *J. Mater. Eng. Perform.*, 16 [1] 113-8 (2007)

[5]S. Callcut and J. C. Knowles, "Correlation between structure and compressive strength in a reticulated glass-reinforced hydroxyapatite foam," *J. Mater. Sci*, 13, 485-9 (2002)

[6]I. K. Jun, Y. M. Kong, S. H. Lee, and H. E. Kim, "Reinforcement of a Reticulated porous ceramic by a novel infiltration technique," *J. Am. Ceram. Soc.*, 89 [7] 2317-19 (2006)

[7]T. Fukasawa and M. Ando, "Synthesis of porous ceramics with complex pore structure by freeze-drying process," *J. Am. Ceram. Soc.*, 84 [1] 230-2 (2006)

[8]S. Stokols, M. H. Tuszynski, "Freeze-dried agarose scaffolds with uniaxial channel stimulated and

guide linear axonal growth following spinal cord injury," *Biomaterials*, 27, 443-51 (2006)
[9]B. H. Yoon, Y. H. koh, C. S. Park, and H. E. Kim, "Generation of large pore channels for bone tissue engineering using camphene-based freeze casting," *J. Am. Ceram. Soc.*, 90 [6] 1744-52 (2007)
[10]N. Sultana and M. Wang, "Fabrication of HA/PHBV composition scaffolds through the emulsion freezing/feeze-drying process and characterization of the scaffolds," *J. Mater. Sci.*, 19, 2555-61 (2008)
[11]T. Moritz and H. J. Richter, "Ceramic bodies with complex geometries and ceramic shells by freeze casting using ice as mold material," *J. Am. Ceram. Soc.*, 89 [8] 2394-98 (2006)

THE NOTCHED BALL TEST – A NEW STRENGTH TEST FOR CERAMIC SPHERES

Peter Supancic[1,2,*)], Robert Danzer[1)], Zhonghua Wang[1,+)], Stefan Witschnig[1,2)] and Oskar Schöppl[3)]
[1)] Institut für Struktur- und Funktionskeramik, Montanuniversität Leoben, Peter-Tunner Str. 5, A-8700 Leoben, Austria
[2)] Materials Center Leoben, Roseggerstrasse 12, A-8700 Leoben, Austria
[3)] SKF Development Centre Steyr, Seitenstettner Strasse 15, A-4401 Steyr, Austria
[+)] Now at: Kunming University of Science and Technology, Wenchang Road 68, 650093 Kunming, China
[*)] E-mail: phs@unileoben.ac.at

ABSTRACT

In this paper a new strength test for ceramic balls (spheres) is presented. A long thin notch is cut in the equatorial plane of the ball and the ball is then squeezed together perpendicular to the notch. This causes tensile stresses in the surface region of the ball perpendicular to the notch, which are analysed carefully with FE methods. The tensile stress amplitude depends on the bending moment in the notch ligament and on details of the notch geometry. The stress state is almost uniaxial. Therefore the stress amplitude is also (slightly) influenced by the Poisson's ratio.

Strength tests are made on silicon nitride balls with a large (~ 31.8 mm) and a small (~ 4.8 mm) diameter. It is shown that the surface quality of the balls has a significant influence on their strength.

INTRODUCTION

Silicon nitride ceramic balls have been used in highly stressed ball bearings for some years [1]. Compared with steel balls silicon nitride balls have higher hardness, Young's modulus, wear and corrosion resistance, and a lower density. These properties are beneficial for bearing ball materials. For applications in electric power generation the high electrical resistance of silicon nitride is also very beneficial [2]. However, ceramic materials are more brittle than steels, therefore information on the tensile strength of balls is essential, but simple testing methods for balls are still missing.

For large balls (having a diameter of say 15 mm or larger) it is possible to machine bending specimens out of the balls in order to perform three or four point bending tests [3], but this procedure is very costly. The surface of the specimens has to be prepared very carefully to avoid machining damage. By testing these specimens in bending the uniaxial strength of the material is measured in the interior of the balls, which is not particularly useful.

Disc shaped specimens can also be cut out of the balls. Testing can be done by some kind of biaxial disc testing, e.g. using the ring on ring or the ball on three balls test [4, 5]. This is time consuming and costly too, and again the interior of the ball material is tested. The preparation of the tensile loaded surface takes even more care than that of bending specimens, since under a biaxial stress state surface cracks of any direction are possible fracture origins (in the case of uniaxial bending tests, cracks parallel to the stress direction are harmless) [6-8].

Some work has been done to test original balls by squeezing them together. This can be done by squeezing one ball between two plates, to position two balls on top of each other and then squeeze them between plates, or even to position three balls on top of each other and then squeeze them between plates [9]. The first test is expected to be not very reliable: significant tensile stresses only occur in a ring shaped zone around the contact area between plates and the ball (i.e. Hertzian stresses, [10, 11]). The highly stressed zone is very close to the area where the load is transferred from the fixture into the ball. Therefore the actual amplitude of the highest tensile stresses sensibly depends on details of the contact zone, e.g. some plastic deformation of the plates, surface roughness, etc. The loading situation is well defined in the second type of test, where the contact situation between the two balls is symmetrical and therefore free from friction effects. (Remark: by using spherical seats on ends to the pistons the contact situation between the ball and the piston is harmless with respect to failure initiation). Therefore this type of test should be reliable if fracture starts near the middle plane, but this can hardly be observed during testing or later (e.g. by fractography). In the case of three balls, the situation for the ball in the middle is well defined on both contact regions and test results are significant, and reliable if fracture starts in the ring shaped near surface regions around the contact planes of the middle ball. This can be recognised by fractographic analysis of the broken pieces. However, the interpretation of the test results is still a little unclear. Recently it has been claimed that this test does not determine the strength of the balls but some kind of toughness and plasticity of the ball material [12, 13], and a similar conclusion was found recently for the contact loading of tools for metal forming [14].

In this paper a new strength test – the Notched Ball Test - is proposed. A long and narrow notch is cut into the equatorial plane of the ball. The ball is squeezed together diametrally along the axis perpendicular to this plane. High tensile stresses, which are used to determine strength, occur in the surface region of the ball opposite the notch root (in the ligament) [12]. Fracture starts from defects which exist in this region. It is important to note, that in this area the notched ball still has its original surface. Therefore the notched ball test can be used to determine the quality of the ball surfaces.

A similar test – the C-Sphere test – has been proposed in 2007 by an American group [15]. In this case a wide notch is used (the width equals to half of the diameter) and the shape of the notch root is a semicircle. This notch geometry is used to maximise the effective volume and the effective surface of the specimen. But this wide notch is difficult to machine precisely. In our notched ball test a narrower notch is used, having a typical width between 7 to 15 % of the diameter and a typical length of 78 to 85 % of the diameter. The exact geometry of the notch root (e.g. half circle, rectangular, etc.) can be determined after machining and will be used for the determination of the stress field. These notches can be machined using simple commercially available grinding wheels.

THE STRESS FIELD IN THE NOTCHED BALL STRENGHT TEST

The notched ball test has recently been developed at the Institut für Struktur- und Funktionskeramik, University of Leoben, Austria. A deep narrow notch is cut into the equatorial plane of the ball. Then the ball is squeezed together perpendicular to the notch. This causes high tensile stresses in the ball surface opposite the notch. The testing configuration is shown in **Fig. 1**. The highest stressed region is in the ligament of the ball opposite the notch. There the stress field is almost uniaxial and resembles that of a bending specimen (in fact a beam-like ligament opposite the notch is bent). Note that this area is far from the areas in the ball where the load is applied. Therefore the details of the load application (e.g. some plastic deformation of the jig and exact contact conditions) will only have negligible influence on the stress field in the ligament.

A complete analysis of the stress states in notched balls having different notch geometries has been performed using the commercial finite element program package ANSYS®. The testing principle and an example of the results – namely the distribution of the 1st principal stress for a unit load of 1 N - are shown in Fig. 2.a for a ball geometry of $D = 31.75$ mm, which is used in the experimental part of this work (of course the results scale with the diameter). It can be recognized that high tensile stresses occur in a relatively large region at the surface (for example: the black area in Fig. 2.a corresponds with the surface with a first principal stresses above 90 % of the peak value of 0.096 MPa/N at the equatorial plane, position 1). Fig. 2.b shows the course of the first principal stress at the surface of this ball specimen along a path in (path from position 1 to 3), and perpendicular to (path from position 1 to 2), the notch plane.

Fig. 1: Notched Ball test specimen (diameter: D=31.75 mm) during testing. The load is applied by two pistons, which press the specimen diametrally together.

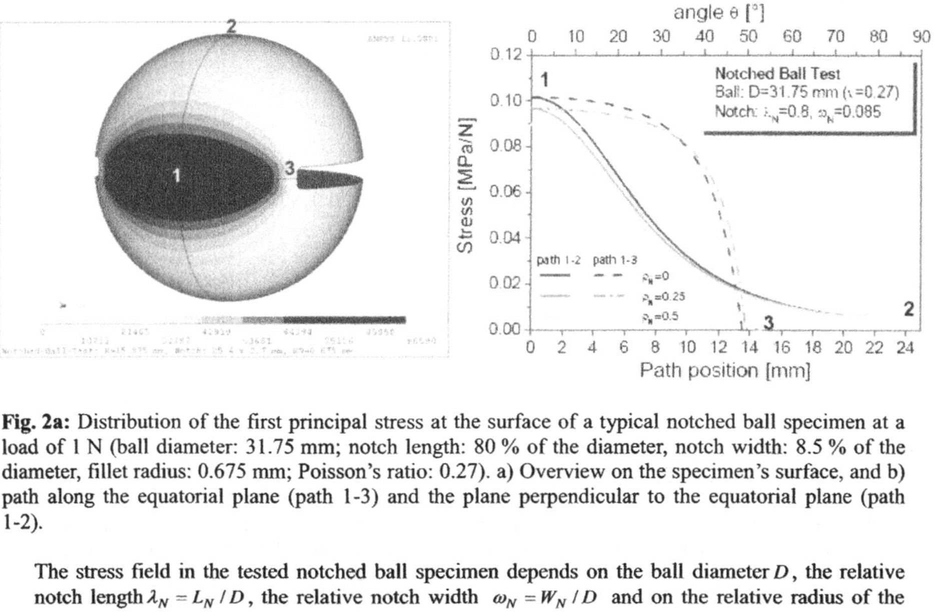

Fig. 2a: Distribution of the first principal stress at the surface of a typical notched ball specimen at a load of 1 N (ball diameter: 31.75 mm; notch length: 80 % of the diameter, notch width: 8.5 % of the diameter, fillet radius: 0.675 mm; Poisson's ratio: 0.27). a) Overview on the specimen's surface, and b) path along the equatorial plane (path 1-3) and the plane perpendicular to the equatorial plane (path 1-2).

The stress field in the tested notched ball specimen depends on the ball diameter D, the relative notch length $\lambda_N = L_N / D$, the relative notch width $\omega_N = W_N / D$ and on the relative radius of the fillet $\rho_N = R_N / W_N$ at the notch base.

The global maximum of the first principal stress (peak stress) is used to determine the strength. Since the stress state at this position is slightly biaxial (the amplitude of the second principal stress is of the order of 10 % of the first principal stress for the specimen geometries analysed here), the strength also depends marginally on the Poisson's ratio, ν.

By analogy with the peak stress in a bending bar, the peak stress in the notched ball test can be represented by the following equation:

$$\sigma = f_N \cdot \frac{6F}{h^2} \quad . \tag{1}$$

where F is the applied force, $h = D - L_N$ is the ligament thickness, and L_N is the length of the notch. The numerical results were used to determine the function f_N which – for notches used in this work – is of the order of one (note: the stress amplitude is not dependent on the ball diameter directly). Recommended values of the ligament thickness are in the range of 15 to 23 % of the ball diameter and for the notch width is 7 to 15 % of the diameter. A power series for the evaluation of $f_N = f_N(\lambda_N, \omega_N, \rho_N, \nu)$ has been fitted to the numerical results of the parametric study, which depends on the relative notch length and width, the relative fillet radius at the notch base and the Poisson's ratio. The fit function describes the data gained by the numerical analysis with an error less than 1 %. The analysis has also been used to check the published results for the C-sphere test. Our results and the results published by [15] match together within 0.5 % error. The results of the analysis will be published elsewhere [16]. In the present study, balls with $D = 31.75$ mm and 4.761 mm respectively have been tested. Notches of width $W_N = 2.7$ mm and 0.62 mm respectively have been machined out of the balls. For these special cases it holds:

$$f_N(\lambda_N, 0.085, \rho_N, 0.27) = \frac{-1.5655\lambda_N^3 + \lambda_N(-1.4589 - 0.6167\rho_N) + \lambda_N^2(2.66 + 0.4222\rho_N) - 0.003691(-60.134 + \rho_N)(1.323 + \rho_N}{(1 - 1.127\lambda_N)\sqrt{\lambda_N(1 - \lambda_N)}} \tag{2.a}$$

for the 31.75 mm balls and

$$f_N(\lambda_N, 0.13, \rho_N, 0.27) = \frac{-1.2295\lambda_N^3 + \lambda_N(-0.8086 - 0.6961\rho_N) + \lambda_N^2(1.862 + 0.4683\rho_N) - 0.00901(-28.361 + \rho_N)(0.4696 + \rho_N)}{(1 - 1.0974\lambda_N)\sqrt{\lambda_N(1 - \lambda_N)}} \tag{2.b}$$

for the 4.761 mm balls, by taking a Poisson's ratio of $\nu = 0.27$. The valid range for the relative notch length is $0.76 \le \lambda_N \le 0.86$ and for the relative radius of fillet at the notch base $0 \le \rho_N \le 0.5$. The dependence of these functions (2.a) and (2.b) on the relative notch length for specimens with different fillet radii at the notch base (i.e. sharp cornered and fully rounded) is shown in **Fig. 3**.

EXPERIMENTAL PROCEDURE AND RESULTS

Commercial silicon nitride bearing balls with a diameter of 31.75 and 4.761 mm respectively were used for this investigation. Of the first kind 60 balls (two sets: A, C) and of the second kind 30 balls (set B) were tested. The balls were weighed and their diameters were measured. The measurement uncertainties of the diameter were ± 1 µm, i.e. they are determined by the precision of the micrometre screw. The density of the ball materials (determined by the ratio of weight/volume) is 3244 ± 2 kg/m^3. These data demonstrate a high quality and reproducibility of manufacture of the balls.

The notches were machined into the balls using a conventional diamond grinding wheel. Special care was taken to position the notch in an equatorial plane of the ball. Since the highest tensile stresses occur at the original surface of the balls the handling of the balls and the machining of the notch was made very carefully to avoid any damage in this area of the surface (but it should be recognised that such damage may still exist in the balls due to grinding damage in production or the handling and transport after production).

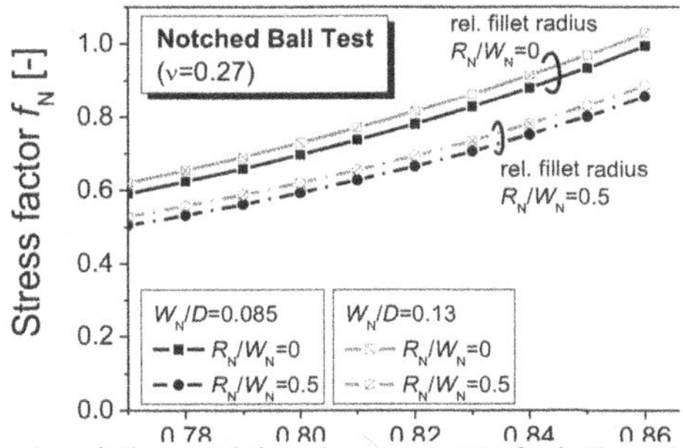

Fig. 3: Factor f_N versus the relative notch length $\lambda_N = L_N / D$ for balls with a relative notch width of 0.085 and 0.13 respectively and for a Poisson's ratio of 0.27. Parameter is the relative fillet radius at the notch base (two extremes are plotted: fully rounded - $\rho_N = 0.5$ - and sharp cornered $\rho_N = 0$). For all relevant specimens geometries the factor is in the order of one (actually between 0.5 and 1).

The smaller balls were machined together as one batch of 30 in one operation, while the larger balls were machined in sub-samples of 6-8. For the data evaluation the length of the notch (i.e. the ligament thickness) was determined for each specimen separately. The fillet radius at the notch base and the width of the notch were determined directly on separate specimens, the dimensions of which correspond directly with the thickness and condition of the grinding wheel. Because the fillet radius increases with progressive wear of the wheel requiring the wheel to be dressed after each cut, it was determined for each subsample separately. The relative fillet radius ρ_N at the notch base was of the order of 0.25 for both types of balls.

The specimens were positioned carefully between the pistons of a universal testing machine using a positioning aid to guarantee that the load is applied perpendicular to the notch plane. Then a preload was applied (about 10 % of the estimated fracture load). Then the aid was removed and the load was increased until fracture. The loading rate was selected in such a way that a test can be done within 10 to 20 seconds. The fracture load was used to determine strength (eq. 1).

Two samples (30 notched ball specimens each) of the large (set A) the small balls (set B) were tested. The results are shown in **Fig. 4** in a Weibull plot. For comparison, strength data determined by the conventional four point bending test are also shown in this diagram. In this case, for the bending test according to EN 843-1, the specimens were cut out of very large bearing balls.

The theory of brittle fracture predicts a size effect of strength [8, 17-21]. It is expected that specimens having a larger (effective) volume have a lower mean strength than specimens with a lower (effective) volume. A strength data extrapolation based on the bending test results is shown in **Fig. 5** (for details of the extrapolation see [19, 22]). It can be recognised that the characteristic strength of the notched ball specimens is significantly lower than the strength predicted by Weibull theory. Probably fracture in bending specimens is caused by a flaw population different from that in the notched ball specimens. Since the notched ball specimens still have the original surface of the balls, surface flaws (produced during machining or handling) may be important. In the case of the bending specimens, the tensile loaded surface was carefully machined according to the procedure in EN 843-1, and the same type of surface flaws cannot exist in these specimens. Therefore their strength is higher than that of the balls.

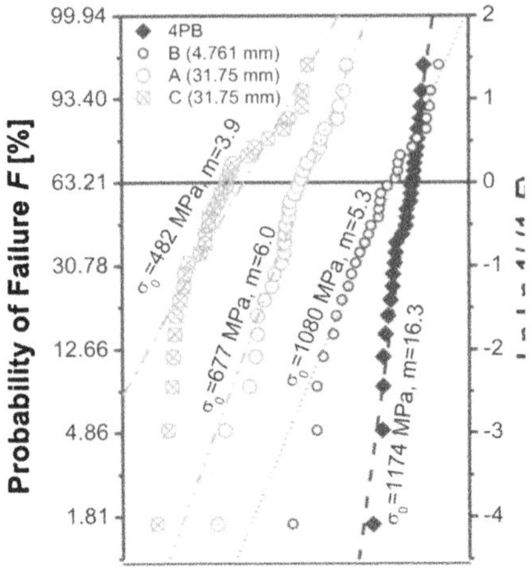

Fig. 4: Strength test results in a Weibull graph: plotted are results of conventional bending tests (4PB) and of the notched ball tests (set A, B and C). It is obvious that the strength depends on the type of the test specimen. (A, B, C and 4PB). The comparison of test results of sets A and C clearly reflects the influence of surface damage on strength.

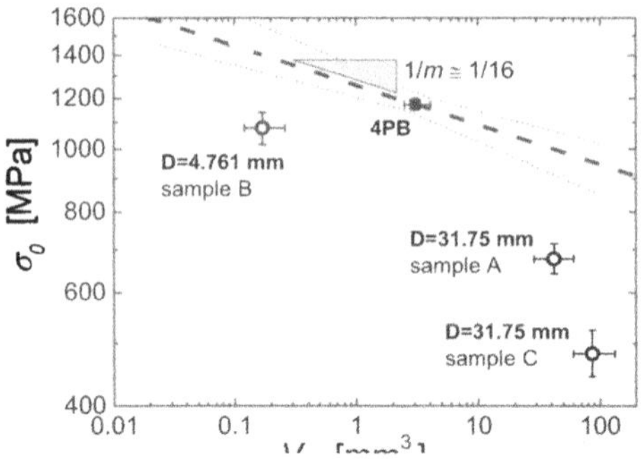

Fig. 5: Characteristic strength versus effective volume in a double logarithmic plot (for details see [23]). The size effect on strength predicted by Weibull theory (based on bending strength data, 4PB) is indicated by the straight line (m=16). It is obvious that the strength data determined by the notched ball test (set A, B and C) do not follow the predicted trend. This is caused by some surface damage, which is different in each of the investigated data sets (A, B and C).

It should be noted that this kind of data evaluation is a very sensitive way to test whether strength data of different samples are consistent with Weibull behaviour or not.

To check this hypothesis a second sample of the large balls (31.75 mm diameter, 30 specimens, set C) was investigated. Surface defects were made visible in a light microscope using a fluorescent dye penetrant. In a majority of the sample of 30 balls large surface flaws could be detected. A typical example is shown in **Fig. 6.a**. The C-shaped crack is similar to cracks produced by Hertzian contact [11]. After inspection, the notch was made in the balls in such a way that a large surface flaw was positioned in the highest loaded area of the specimen. The strength data from these tests are also shown in **Fig. 4** and **Fig. 5**. It can clearly be recognised that the strength of specimens of set C is much lower than that of set A. It should be also noted that the three specimens in which no surface flaws were found during the optical inspection are in fact the specimens with the highest strength in set C.

All fracture surfaces were analysed by fractographic means in order to identify fracture origins. In any case the fracture origins are at or very close to the surface. In the case of set C, the surface flaws identified by the dye penetrant and positioned in the highest loaded area of the specimens were fracture origins in every case. An example is shown in **Fig. 6.b**. The fracture origin is a Hertzian contact crack (a part of a ring crack), which had previously been detected before fracture (**Fig. 6.a**).

In the specimens of set A and B similar fracture origins were found in some, but not in all, cases. In general, the fractographic analysis clearly demonstrates the strong influence of surface damage on strength.

Fig. 6. a) Surface inspection using a dye penetrant makes surface cracks visible. In 27 of 30 specimens of sample C cracks could be detected before strength testing.
b) Fracture surface after strength testing, SEM. These surface cracks are preferential fracture origins and cause a significant reduction of strength. The fracture origin is the same flaw as shown in Fig. a).

CONCLUSIONS

- A new test to determine the tensile strength of spheres has been developed.
- The new test is easy to apply.
- The new test makes the testing of very small balls possible. Even the testing of spheres with a diameter of a few millimetres becomes possible. For such small components alternative strength testing procedures will hardly become possible.
- The testing of large balls is also possible and convenient.
- The strength determined in the notched ball test is strongly influenced by the quality of the ball original surface.
- The notched ball test determines the strength of the ball and not the strength of the material (which is more related to the size and frequency of volume flaws).
- The notched ball test is well situated to determine the surface quality of ceramic balls, as they are used in modern ceramic ball bearings.
- It is proposed to use this test in the quality control of ceramic bearing balls.

ACKNOWLEDGEMENTS

Financial support by the Austrian Federal Government (in particular from the Bundesministerium für Verkehr, Innovation und Technologie and the Bundesministerium für Wirtschaft und Arbeit) and the Styrian Provincial Government, represented by Österreichische Forschungsförderungsgesellschaft mbH and by Steirische Wirtschaftsförderungsgesellschaft mbH, within the research activities of the K2 Competence Centre on "Integrated Research in Materials, Processing and Product Engineering", operated by the Materials Center Leoben Forschung GmbH in the framework of the Austrian COMET Competence Centre Programme, is gratefully acknowledged.

LITERATURE

[1] L. Wang, R. W. Snidle, and L. Gu, "Rolling contact silicon nitride bearing technology: a review of recent research", *Wear,* vol. 246, pp. 159-173, 2000.
[2] H. Köttritsch, "Science Report: Development Centre Steyr", Steyr: SKF Österreich AG, 2007.
[3] EN 843-1, "Advanced Technical Ceramics, Monolithic Ceramics; Mechanical Tests at Room Temperature, Part 1 - Determination of flexual strength", 1995, p. 18.
[4] A. Börger, P. Supancic, and R. Danzer, "The Ball on three Balls Test for Strength Testing of Brittle Discs - Stress Distribution in the Disc", *Journal of the European Ceramic Society,* vol. 22, pp. 1425-1436, 2002.
[5] A. Börger, P. Supancic, and R. Danzer, "The Ball on three Balls Test for Strength Testing of Brittle Discs - Part II: Analysis of Possible Errors in the Strength Determination", *Journal of the European Ceramic Society,* vol. 24, pp. 2917-2928, 2004.
[6] G. D. Quinn and R. Morrell, "Design Data for engineering ceramics: a review of the flexure test", *Journal of the American Ceramic Society,* vol. 74, pp. 2037-2066, 1991.
[7] G. D. Quinn, *Fractography of Ceramics and Glasses* vol. Special Publication 960-16: National Institute of Standards and Technology, 2007.
[8] R. Danzer, T. Lube, P. Supancic, and R. Damani, "Fracture of Ceramics", *Advanced Engineering Materials,* vol. 10, pp. 275-298, 2008.
[9] O. Schöppl, R. Huber, H. Weninger, and H. Köttritsch, "Ceramic Rolling Elements and their Influence on Static Behaviour of Rolling Bearings", in *Science Report: Development Centre Steyr, 2005-2006,* H. Köttritsch, Ed. Steyr: SKF Österreich AG, 2006, pp. 37-44.
[10] H. Hertz, "Über die Berührung fester elastischer Körper", *Journal für die reine und angewandte Mathematik,* vol. 92, pp. 165-171, 1882.
[11] B. R. Lawn, "Indentation of Ceramics with Spheres: A Century after Hertz", *Journal of the American Ceramic Society,* vol. 81, pp. 1977-1994, 1998.
[12] P. Supancic, R. Danzer, W. Harrer, Z. Wang, S. Witschnig, and O. Schöppl, "Strength Tests on Silicon Nitride Balls", *Key Engineering Materials,* p. to be published, 2008.
[13] A. C. Fischer-Cripps and B. R. Lawn, "Stress Analysis of Contact Deformation in Quasi-Plastic Ceramics", *Journal of the American Ceramic Society,* vol. 79, pp. 2609-2618, 1996.
[14] M. Lengauer and R. Danzer, "Silicon Nitride Tools for Hot Rolling of High Alloyed Steel and Superalloy Wires - Crack Growth and Lifetime Prediction", *Journal of the European Ceramic Society,* vol. 28, pp. 2289-2298, 2008.
[15] A. A. Wereszczak, T. P. Kirkland, and O. M. Jadaan, "Strength Measurement of Ceramic Spheres Using a Diametrally Compressed "C-Sphere" Specimen", *Journal of the American Ceramic Society,* vol. 90, pp. 1843-1849, 2007.
[16] P. Supancic and R. Danzer, "A New Strength Measurement of Ceramic Balls using the "Notched Ball Test"", *Journal of the European Ceramic Society,* p. to be published, 2008.
[17] R. Danzer, P. Supancic, and T. Lube, "Failure Statistics Beyond the Weibull Behavior", *Ceramic Engineering and Science Proceedings,* vol. 24, pp. 497-502, 26.-31.1.2003 2003.
[18] R. Danzer, P. Supancic, J. Pascual, and T. Lube, "Fracture Statistics of Ceramics - Weibull Statistics and Deviations from Weibull Statistics", *Engineering Fracture Mechanics,* vol. 74, pp. 2919-2932, 2007.
[19] G. D. Quinn, "Weibull Strength Scaling for Standardized Rectangular Flexure Specimens", *Journal of the American Ceramic Society,* vol. 86, pp. 508-510, 2003.
[20] D. Munz and T. Fett, *Ceramics: Mechanical properties, failure behaviour, materials selection,* Springer series in materials science ed. vol. X. Berlin: Springer Verlag, 1999.
[21] R. Danzer, "Fracture Mechanics of Ceramics - A Short Introduction", *Key Engineering Materials,* vol. 333, pp. 77-86, 2007.

[22] G. D. Quinn, "Weibull Effective Volumes and Surfaces for Cylindrical Rods Loaded in Flexure", *Journal of the American Ceramic Society,* vol. 86, pp. 475-479, 2003.
[23] R. Danzer, T. Lube, and P. Supancic, "Monte-Carlo Simulations of Strength Distributions of Brittle Materials - Type of Distribution, Specimen- and Sample Size", *Zeitschrift für Metallkunde,* vol. 92, pp. 773-783, 2001.

LIQUID PHASE SINTERED α-silicon CARBIDE WITH AlN-Re_2O_3 AS SINTERING ADDITIVE

Yuhong Chen, Laner Wu, Yong Jiang, Youjun Lu, Zhenkun Huang
School of material science and engineering,
North University for Ethnic, Yinchuan, Ningxia, China 750021

Abstract

The densification of α-SiC occurred by liquid-phase sintering mechanism with AlN-R_2O_3(La2O3,Nd_2O_3,Y_2O_3) was studied. The total additive content was fixed at 17.5 wt %. Cold isostatically pressed samples were sintered at 1800-1950℃ under N_2 atmosphere for 1hr. The linear shrinkage and weight loss of the samples were about 17-20% and 2-5% respectively. The mechanical properties and microstructure of sintered samples were investigated. The experimental results showed that the fracture toughness of samples was 6-8MPa.$m^{1/2}$, the hardness was in the range of 18-21GPa and the bending strength was in the range of 400-500 MPa. The SEM of sintered sample showed a fine grained microstructure with equiaxed grains. Fracture mode was intergranular fracture.

Key words: α-silicon carbide, liquid phase, sintering, mechanical properties, microstructure

1. Introduction

Silicon carbide can be pressureless sintered by a solid stated process with the aid of B and C to near full density at temperatures in excess of 2100℃. However, the lower fracture toughness (3 to 4 Mpa·$m^{1/2}$) limit their use in many potential structural applications [1].

It has been known that sintering of SiC can be achieved at relatively lower temperature (1850℃-2000℃)with the addition of oxides (Al_2O_3 and Y_2O_3) *via* liquid phase sintering[2,3]. The resulting material obtained with homogeneous and equiaxed fine-grained microstructure. Oxides like SiO_2 and Al_2O_3, which are normally considered as thermodynamically stable, are prone to react with SiC at temperature of about 2000℃, leading to formation of gaseous products such as CO, SiO and Al_2O.

Al_2O_3+SiC→Al_2O(g)+SiO(g)+CO(g)

In order to suppress these reactions, a powder bed is generally required [4]. Alternatively, the additive system of AlN and rare earth oxides including Y_2O_3, is used where the decomposition of AlN into Al and N_2 can be efficiently controlled by using N_2 atmosphere, leading to lower weight lost [5,8]. The AlN –Y_2O_3 phase diagram indicates that eutectic temperature in this system is about 1850℃ [9]. It might avoid forming a liquid with rather low melting temperature and a coarse surface of ceramic caused by vaporized gases from the reaction of SiO_2 and Al_2O_3-Y_2O_3. Also in this system the intermediate compositions can offer sufficient amount of liquid with melting temperature higher than 1700°C as sintering aid of LPS-SiC. Some studies have been carried out by using rare-earth oxide containing densification aids [10,12]. Our previous study on melting behaviors of SiC and a series of Re_2O_3（1：1mol mixture）has shown that melting temperatures raise with increasing the atomic number of rare earth element (from La to Er and Y)[13]. The aim of this work was to study the sintering behavior of liquid phase sintered SiC with AlN and Re_2O_3 (La_2O_3,Nd_2O_3 , Y_2O_3) additive system and their mechanical properties in detail.

2. Materials and Methods

2.1 Materials

α-SiC powder was manufactured by North university for ethnic. The chemical analysis of the powder was performed by manufacturer and the data was as following: SiC content >97%, free C <

1%, SiO_2 < 1.2wt%; median particle size: D_{50}=0.7μm). AlN powder (D_{50}<0.8μm, purity>98%) was provided by Beijing iron research institute. Y_2O_3, La_2O_3,Nd_2O_3 (purity>99.9% D_{50}= 2-5μm) was provided by Baotou rear earth research institute.

2.2 Methods

2.2.1 Preparation of the green bodies and sintering

SiC powder and 17.5%(mass content) of additives were mixed in an attrition mill for about 1hr in alcohol using Si_3N_4 balls as medium. The compositions of various powder mixtures prepared and the nomenclature used to describe the samples are specified in Table 1.The milled slurry was separated from the milling ball and possible wear debris by screening through 320mesh. The slurry dried in a stirring evaporator and completely dried in a drying oven at 80℃. The dried powder was sieved through 100 mesh. The mixed powder was axial pressed under pressure of 100Mpa and then isostatically pressed under 250MPa.

The rectangular shaped green samples of approximately 10×50×50mm was sintered in a graphite furnace (made by Robert furnace co. China). The samples were put into a graphite crucible using BN powder as separating. A high purity N_2 gas atmosphere was used during sintering. The gas pressure was maintaining at 0.02Mpa during sintering. The samples were sintered at 1800, 1850, 1900, 1950, 2000℃and 2050℃ for 1hr separately.

2.2.2 Measurement of properties

The weight loss and shrinkage of green body and sintered specimen of all samples were measured. Bulk density were measured by Archemede's principle by a water displacement method. The hardness was determined by using a load of 98N in a microhardness test fitted with a Vicker's square indenter (Wolpert U.S.A). The fracture toughness was calculated by the length of the cracks originating from the edges.: $Kic=0.016\ (E/Hv)^{0.5}\times(p/c^{-1.5})$ where Kic is the fracture toughness of the material, Hv is the Vickers hardness, E is the Young's modulus (for LPS-SiC a value of 400 was assumed) c is the crack length(μm) and a is indentation diagonal [14]. The specimens were cut into rectangular beams with dimensions of 3×4×36 mm to test three point bending strength. The tensile edges were beveled to remove stress concentrations and edge flaws caused by sectioning.

Observation of the microstructure has been performed by SEM (ssx-550 Shimadzu Japan) on fracture surfaces and also on finished surface polished by 1μm diamond paste. The phase composition of samples was determined by X-ray diffraction using Cu-K radiation (XRD-6000 Shimadzu Japan) , a step width of 0.2 with an exposure time of 2 degree/min per position.

Tab1 compositions and theory density of mixtures

sample number	AlN /mol%	Y_2O_3 /mol%	Nd_2O_3 /mol%	La_2O_3 /mol%	ρtheoy density /g/cm^3
Sly-1	40	60	0	0	3.40
Sly-2	60	40	0	0	3.38
Sly-3	80	20	0	0	3.34
Sln	60	0	40	0	3.50
Slny	60	20	20	0	3.44
Sla	60	0	0	40	3.47
Slay	60	20	0	20	3.43

3. Result and discussion

3.1 Sintering processing

Heating rates of 20℃/min from ambient temperature to 1600℃ and 10℃/min from 1600℃to final sintering temperature were used.

3.1.1 Sinterability of $SiC-AlN-Y_2O_3$ system
Similar with other works[6,15] , the sintering temperature for completed densification is a function of the additive composition, the best densification behavior does not coincide with the eutectic composition in the $AlN-Y_2O_3$ system, which is about 40mol% AlN as shown in Fig1. It is well known that one important requirement of liquid phase sintering is that there must be good wetting of the solid phase (SiC) by the liquid phase (additive) and there must be a small contact angle θ between the solid SiC and the liquid drops formed by the additive. R.M.Balestra's work showed that at this additive system with 60%mol% AlN had good wettability (θmin≦6°)[16].The viscosity of silicate melts increases with their nitrogen content, in analogy to the glass transition temperatures of oxynitride glasses.

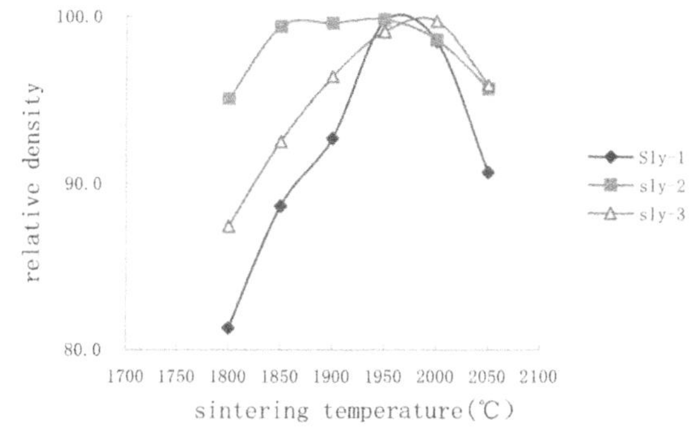

Fig1 density as a function of nitrogen content in the additive

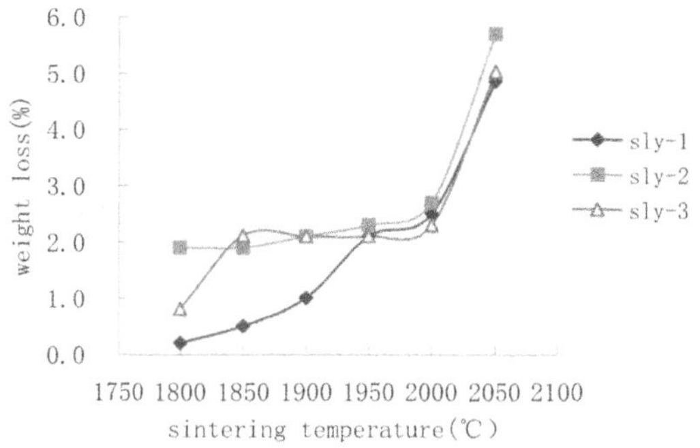

Fig2 Weight loss of samples in sintering

The total weight loss of all full density specimens kept at about 2%, as showed in fig2. When the sintering temperature raised higher than 2000℃, the weight loss of specimens increased to more than 5%, and the diametric shrinkage was less than those in full density temperature. Hence at that temperature, additive decomposition made the density of specimens decreased.
Experimental results showed that SiC-AlN-Y_2O_3 could be fully densified in wide temperature range (1850℃-2000℃), and kept low weight loss around 2% in this range. The surface of specimens remains smooth, indicating that sintering could be done without powder bed.

3.1.2 Sinterability of SiC-AlN-R_2O_3 (R=Nd, La) systems
The best sintered density and weight loss data of specimens of all test used AlN-Re_2O_3 additive system are shown in Table 2. These test results indicated that the specimens wouldn't been fully densified by using AlN-Nd_2O_3 or AlN-La_2O_3 additive system, all these systems showed much higher weight loss than those results reported in gas pressure sintering [17] which indicated much decomposition reaction occurred without N_2 gas protect.

Table 2 sintering density and weight loss of AlN- R_2O_3 system

sample number	Sintering temperature[℃]	Weight loss [%]	ρ_{rel}[%]
Sln	1900	5.9	96.5
Slny	1950	3.1	99.2
Sla	1900	6.9	92.4
Slay	1950	5.1	98.1

Interestingly, AlN-Re_2O_3-Y_2O_3 additive system showed much better sintering behavior than AlN-Re_2O_3 system. Although they also showed more weight loss than AlN- Y_2O_3 system, and for densitification, higher sintering temperature was needed.

3.2 Mechanical properties
Mechanical properties of all densificated specimens are summarized in Table 3. For AlN-Y_2O_3 system specimens, the hardness (Hv) increased with AlN content increasing. AlN-Nd_2O_3-Y_2O_3 additive specimen show higher hardness than that of all other specimens, which has same hardness as SSSiC(21-25GPa)[18]. All specimens have same bending strength in range of 350-500MPa. All specimens have relative higher fracture toughness comparing with SSSiC which is in range of 3-5 MPa·$m^{1/2}$. The SEM picture of crack and the fracture surface are shown in Fig3. That indicated fracture mode was intergranular fracture. Grain refinement and inter-crystal deflection are the main reasons of toughness increasing.

Table 3 mechanical properties of specimens

sample number	Hardness (GPa)	Bending strength(Mpa)	Fracture toughness (MPa·$m^{1/2}$)
sly-1	18.7±0.7	410±4.8	6.8±0.4
sly-2	19.4±0.8	435±42	8±0.7
sly-3	20.8±0.2	481±57	6.1±0.2
slny	22.2±0.2	--	6.9±0.3
slay	18.9±1.1	367±13	6.5±0.3

3.3 Microstructure and phase composition
Typical microstructure of AlN-Y_2O_3 system are shown in Fig4, similar to the microstructure described in previous reports [7,,12,13,19] . The SiC grains are predominantly equiaxed with a

mean grain size of 1-2μm. Relatively little grain growth occurred during densification, indicated that the atomic transport through melt is sluggish. The core-rim structure is found clearly in high AlN content specimen. The XRD pattern of the sample is shown in Fig 5. The major phase is 6H SiC, the minor phases are AlN, Y_2O_3 and $Y_{0.54}$ $(Si_{9.57}Al_{2.43}O_{0.81}N_{15.19})$. The work of Haihui Ye described that sample sintered in 1Mpa N_2 atmosphere, the AlN, $Y_{10}Al_2Si_3O_{18}N_4$, and $Y_2Si_3N_4O_3$ phase were indentified, but in Ar, Y_2O_3, $Y_{10}Al_2Si_3O_{18}N_4$ phase were indentified [8]. Our experimental result is similar to those in Ar, being in lower N_2 atmosphere pressure. For $AlN-Re_2O_3-Y_2O_3$ additive system, the microstructure is similar with $AlN-Y_2O_3$ system, but core-rim structure are hardly found in SEM(Fig6). The XRD pattern of the sample with $AlN-Nd_2O_3-Y_2O_3$ is shown in Fig 7, complex phase of $Y_{0.54}$ $(Si_{9.57}Al_{2.43}O_{0.81}N_{15.19})$ and $Nd_4Si_2O_7N_2$ (YAM') had been found.

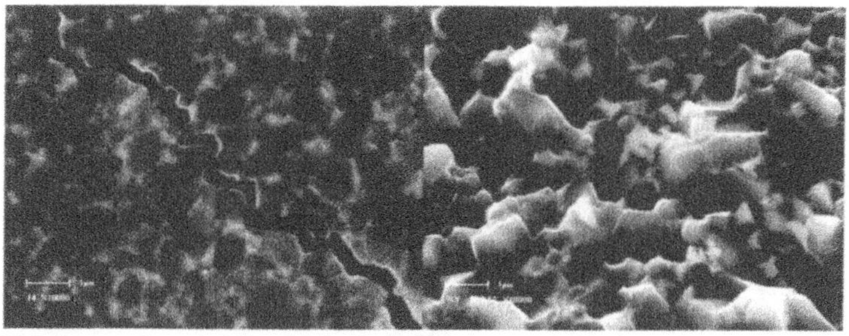

Fig 3 SEM picture of crack deflection and break surface of sly-2 sample (a. crack deflection, b. fracture surface)

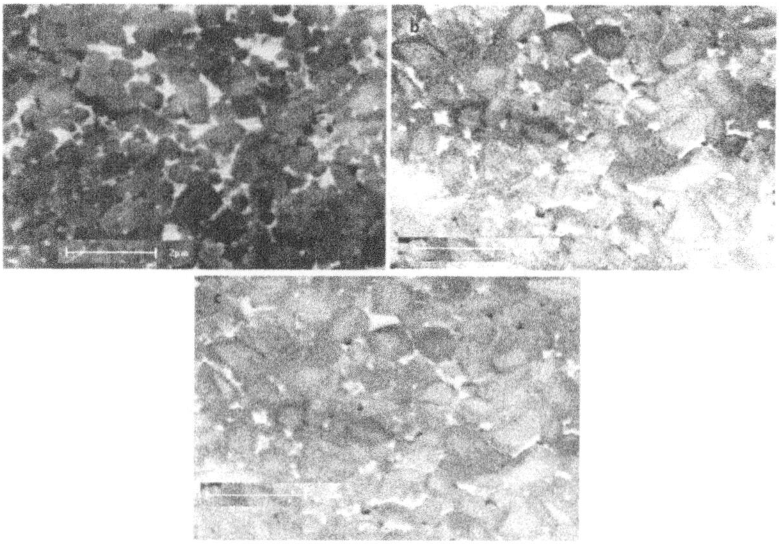

Fig4 Microstructure of sintered sample with $AlN-Y_2O_3$ additive (a sly-1,b sly-2,c sly-3)

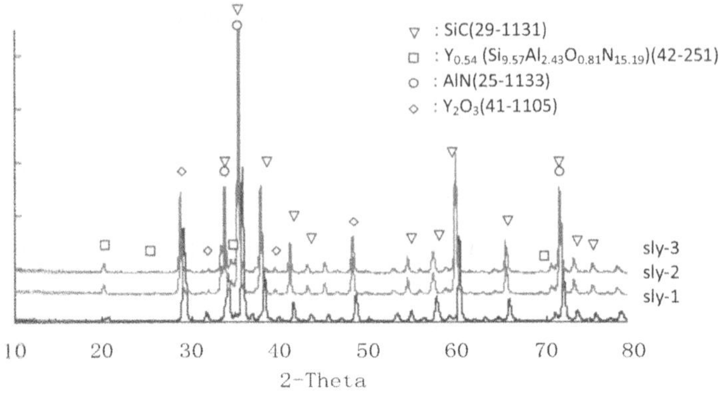

Fig 5 XRD analysis of sintered sample with AlN-Y_2O_3 addtive

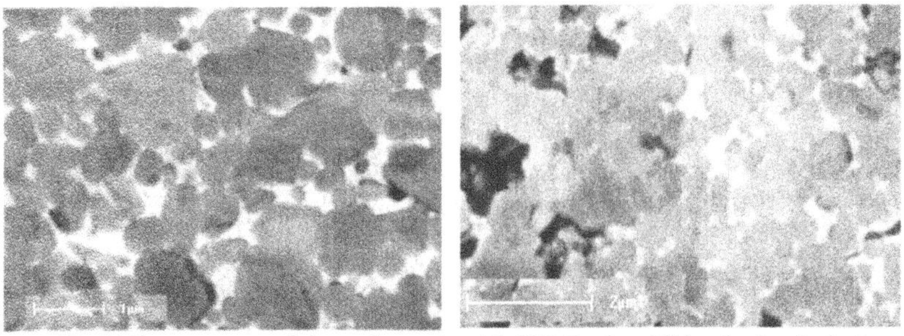

Fig 6 Microstructure of sintered sample with AlN-Re_2O_3 additive (a slny, b slay)

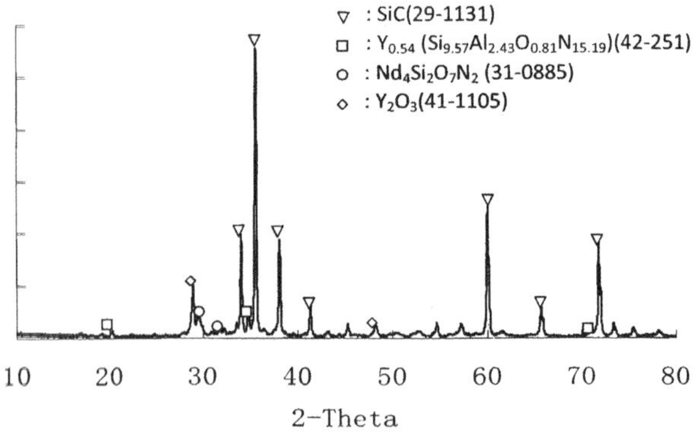

Fig7 XRD analysis of sintered sample with AlN-Nd_2O_3-Y_2O_3addtive

4. Conclusion

4.1 Fully dense SiC ceramics can be obtained by liquid phase sintering with $AlN\text{-}Y_2O_3$ and $AlN\text{-}R_2O_3\text{-}Y_2O_3$ additives. Specimens with 60mol% AlN in $AlN\text{-}Y_2O_3$ additives system show that SiC can be sintered in a wide temperature range (1850℃-2000℃), and keep low weight loss around 2%. The surface of specimens remains smooth, indicating that sintering could be done without a powder bed. $AlN\text{-}R_2O_3\text{-}Y_2O_3$ additives system show higher weight loss around 5%.

4.2 The materials obtained have fine-grained and homogeneous microstructure. The core-rim structure can be found in high AlN content specimen. All specimens have higher fracture toughness in the range of 6-8 $MPa{\cdot}m^{1/2}$. Grain refinement and inter-crystal deflection are main reason of toughness increasing. The XRD analysis indentified that $Y_{0.54}$ $(Si_{9.57}Al_{2.43}O_{0.81}N_{15.19})$ phase had been found in AlN- Y_2O_3 additive system. $AlN\text{-}Nd_2O_3\text{-}Y_2O_3$ additive specimen showed higher hardness, and $Nd_2Si_4O_7N_2$ had been found.

Acknowledgements

This study was supported by National high technical development program (2002AA332110) and Ningxia international cooperation program (NXIC-2006-004). The authors are grateful to Mrs Jiang and Mrs Han for their assistance with XRD analysis.

References

[1] Prochazka S. Sintering of silicon carbide// Proceedings of the Conference on Ceramics for High Performance Applications (Hyannis,MA). Hyamnis: Brook Hill Publishing Co, 1975: 7–13.

[2] Omori M, Takei H. Preparation of pressureless sintering of $SiC\text{–}Y_2O_3\text{–}Al_2O_3$. J Mater Sci,1988, 23(10): 3 744–3 747.

[3] Nitin P. Padture. In situ-toughened silicon carbide. J.Am.Ceram.Soc., 1994,77[2]519-23

[4] TAN Shouhong, CHEN Zhongming, JIANG Dongliang. Liquid phase sintering SiC Ceramics. J. Chin Ceram Soc., (in Chinese), 1998, 26(2): 191–197.

[5]K.Y.Chia,W.D.G.Boecker and R.S.Storm.U.S.Pat., 5,298,470(1994)

[6] Rixecker G, Biswas.K, Wiedmann I and Aldinger.F.. Liquid-phase sintered SiC ceramic with oxynitride additives. J.Ceramic processing research 2000,1:1 12-19

[7]Rixecker G, Wiedmann I, Rosinus A, et al. High-temperature effects in the fracture mechanical behaviour of silicon carbide liquid-phase sintered with $AlN\text{–}Y_2O_3$ additives. J.Euro. Ceram Soc., 2001, 21: 1 013–1 019.

[8]Ye Haihui, Rixecker.G, Siglinde. H, et al. Compositional identification of the intergranular phase in liquid phase sintered SiC. J Euro Ceramic Soc., 2002, 22: 2 379–2 387.

[9] Kouhik Biswas. Liquid phase sintering of SiC ceramics with rare earth sesquioxides [D]. Stuttgart: University of Stuttgart, 2002.

[10] Koushik Biswas Rixecker.G, Aldinger F.. Effect of rare-earth cation additions on the high temperature oxidation behaviour of LPS-SiC. Materials Science and Engineering A 2004,374 56–63

[11] M. Balog, P. Šajgal'ık, M. Hnatko, Z. Len˘céš, F. Monteverde, J. Ke˘ckéš, J.-L. Huangd Nano-versus macro-hardness of liquid phase sintered SiC. J. Euro. Ceramic Soc., 2005,25, 529–534

[12] Koushik Biswas, Georg Rixecker, Fritz Aldinger. Gas pressure sintering of SiC sintered with rare-earth-(III)-oxides and their mechanical properties. Ceramics International 2005,31 703–711

[13] Wu Laner, Chen Yuhong, Jiang Yong, Huang Zhenkun. Liquid sintering of SiC with AlN–Re2O3 Additives. J.Of the Chinese ceramic society 2008, 36， [5] 593-596

[14]G.R.Anstis,P.Chantikul,B.R.Lawn and D.B.Marshall. A critical evaluation of indentation techniques for measureing fracture toughness I Direct crack measurements. J.Am.Ceram.Soc., 1981,64[9]533-543

[15]Magnani,G. and Beaulardi,L.. Properties of liquid phase pressureless sintered SiC-based materials

obtained without powder bed. J.Aus.Ceram.Soc., 2005,41(1),31-36
[16] R.M. Balestra, S. Ribeiro, S.P. Taguchi, F.V. Motta, C. Bormio-Nunes Wetting behaviour of Y_2O_3/AlN additive on SiC ceramics. J. Euro. Ceram. Soc., 2006,26(16) 3881–3886
[17] V.A.Izhevskyi, L.A.Genova,A.H.ABressiani，J.C.Bressiani Liquid phase sintered SiC processing and Transformation controlled microstructure tailoring. Material research, 2000,3[4]131-138
[18] Wu Anhua, Cao Wenbin, Li Jiangtao, Ge Changchun. Solid State Sintered SiC Ceramics. J. of Materials Engineering(China),2001,4,3-5
[19] L.S.Sigl. Thermal conductivity of liquid phase sintered silicon carbide. J. Euro. Ceramic Soc., 2003,23, 1115-1122

PREPARATION OF Si_3N_4 CERAMICS FROM LOW-COST Si_3N_4 POWDER WITH HIGHER β PHASE AND OXYGEN CONTENT *

Yong Jiang**, Laner Wu, Fei Han, Zhenkun Huang
School of MSE, the North University for Ethnics
Yinchuan, 750021, China

ABSTRACT

A low-cost Si_3N_4 powder with the phase composition of $\alpha/\beta = 67/33$ (mass rate) synthesized by self-propagation high temperature synthesis (SHS) method was used to prepare Si_3N_4 ceramics in the present research. The Re(Y, La)$_2O_3$-AlN system was used as a sintering aid. Fully dense Si_3N_4 ceramic with bending strength of 965 MPa was obtained by liquid-phase sintering. More β phase contained in starting powder sites as seeds of β-Si_3N_4 nucleation and promotes the formation of refractory J-phase as medium phase. The effect of the formation of oxygen-richer J-phase on the phase relations of Si_3N_4 with neighboring phases was discussed.

KEYWORDS

low-cost Si_3N_4, phase content, liquid phase sintering, bending strength, ceramics

INTRODUCTION

As well known, fine Si_3N_4 ceramic possesses excellent mechanical properties and is now widely used in many industries as structural materials. Preparing fine ceramics needs fine powders. A high performance Si_3N_4 ceramic can be manufactured from fine and pure α-Si_3N_4 powder with an α-phase content >95% by liquid-phase sintering (LPS). During sintering the transformation of the α-phase to the β-phase promotes densification and the evolution of a network of elongated β-Si_3N_4 grains by means of "solution-reprecipitation" through liquid phase, and hence increases the strength and toughness of the product[1, 2]. However, higher quality powder brings higher cost. Identification of the Si_3N_4 powder that simultaneously possesses a high α-phase content, high chemical quality and a low cost has proved to be challenging.

In recent years, the self-propagation high-temperature synthesis (SHS) technology has been used to produce low-cost Si_3N_4 powder[3-6]. Xinhongxiang Co. (Yinchuan, China) has produced a low-cost SHS-Si_3N_4 powder containing 60%–90% α phase in large amounts, using a short production time and lower cost. Two kinds of the Si_3N_4 powders, named α_6 and α_9 with 60% α + 40% β and 87% α + 13% β content respectively, were used in present work. These powders, with coarse grains (about 2-4 micron), wide granularity distribution, and some impurities, were pretreated[7] before use, in order to get fine, high purity submicron powder. Water was used as grinding-medium instead of alcohol in the pretreatment. The pretreated Si_3N_4 powder contains more oxygen, mass fraction of 3.84% (equivalent 7.2% SiO_2). The (Y, La)$_2O_3$-AlN system[8, 9] was used as a sintering aid for LPS. The relationship between microstructure, mechanical properties of Si_3N_4 ceramic was described. The influence of β phase content in the Si_3N_4 powder upon the phase composition of sintered body and hence on the phase relations of Si_3N_4 with neighboring phases was discussed.

EXPERIMENT

The two kinds of pretreated powders were mixed according to the proportion of $3\alpha_6 : 1\alpha_9$ (labeled as $3\alpha_6\,\alpha_9$). So the contents of α phase of the mixed powder is 67%, β is 33% (mass rate), that is $\alpha/\beta = 67/33$. This powder was mixed for 2 h with ~14 vol.% sintering aid $(Y_{0.8}, La_{0.2})_2O_3$-AlN[8, 9] in alcohol, in a Teflon-Coated jar using Si_3N_4 balls as the milling medium. Both Y_2O_3 and La_2O_3 with 99.9% purity were obtained from the Baotou Institute of Rare Earth, Baotou, China. AlN with 99.9% purity was obtained from the Beijing Institute of Steel and Iron, Beijing, China. The powder mixture was dry-pressed in a 50 mm × 50 mm model under 40 MPa pressure, and further isostatically pressed under 250 MPa as same in our previous work[7]. LPS was carried out in a graphite furnace at 1780℃ for 3 h in N_2 atmosphere. After sintering, the shrinkage, weight loss, and density of the experimental specimens were measured. Density was measured by the Archimedean method. Specimens were cut into 3 mm × 4 mm × 36 mm bars and then polished for the strength test. The bending strength was determined by the three-point measurement method. The broken bars were then subjected to different tests for the measurement of hardness and fracture toughness. The phase compositions of the samples were analyzed by X-ray diffraction (XRD-6000, Shimadzu, Japan). The microstructure was examined on a SSX-550 scanning electron microscope (SSX-550, Shimadzu, Japan), equipped with energy dispersive X-ray (EDX).

RESULTS AND DISCUSSION

Adding additives of 14 vol. %, the pretreated Si_3N_4 powder with phase composition $\alpha/\beta = 67/33$ was liquid-phase sintered to 99% theoretical density (TD). Good densification reflects its excellent mechanical properties which are hardness (14 GPa), fracture toughness (6 $MPa{\cdot}m^{1/2}$) and especially quite high bending strength of 965 MPa (see Table 1).

Table 1. The Properties of Liquid Phase Sintered Si_3N_4 Ceramics.

Specimen code	Density ($g{\cdot}cm^{-3}$)	Relative density (%)	Linearity shrinkage rate Length (%)		Weight loss (%)	Hv (GPa)	K_{Ic} ($MPa{\cdot}m^{0.5}$)	Bending strength (MPa)
			Length	Height				
$3\alpha_6\,\alpha_9$	3.40	99.42	14.83	19.68	2.34	14.14	6.35	965.60★

★ average value of 883.76, 943.54, 984.79, 1107.33, 1056.24, 949.98 and 833.56 MPa.

SEM micrographs of the specimen are shown in Fig. 1. It shows a homogeneous network of elongated grains (gray). The aspect ratio of the rod-shaped grains is about 6. The rod-shaped grains are tightly interlocked with the small equiaxed hexagonal grains (gray) and the liquid grain boundary phase (white, Fig. 1a). It can be expected that more β phase crystal grains in starting powder sites as seeds of β-Si_3N_4 nucleation, promote $\alpha \rightarrow \beta$ phase transformation, and also contribute to develop a structural network of elongated Si_3N_4 grains. Such a microstructure is known to promote strength and fracture toughness of Si_3N_4 ceramics through toughening mechanism of grain pullout (Fig. 1b) and bridge effect in the crack development.

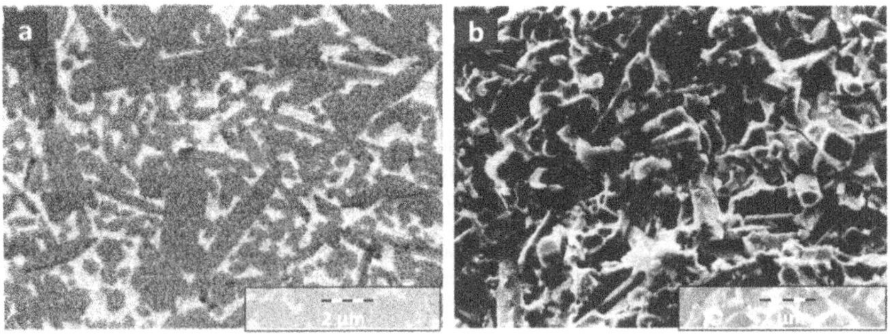

Figure 1. SEM micrographs of sintered specimen, (a) polished surface (b) fracture surface.

The XRD pattern (Fig. 2) of the specimen sintered at 1780℃ under 1 atm nitrogen for 3 h shows β-Si_3N_4 (or β-sialon) as main phase and a refractory J-phase ($2Y_2O_3$•Si_2N_2O) as medium phase, as well as a little M-phase (Y_2O_3•Si_3N_4). The initial α-Si_3N_4 is completely transformed to β-Si_3N_4, but it seems that there still is a trace of Si, which would be a decomposed product of Si_3N_4 or SiO_2 impurity at high temperature. It is obvious that the formation of J-phase in the sintered specimen would be related with more oxygen content (up to 3.84%) in α_6 powder after milling in water. More oxygen content in Si_3N_4 powder would be benefit to form liquid phase at high temperature for LPS, such kind of powders will be more suitable to low-cost industrial applications.

J-phase ($2Y_2O_3$•Si_2N_2O) is oxygen-richer than either M-phase (Y_2O_3•Si_3N_4) or K-phase (Y_2O_3•Si_2N_2O). The formation of J-phase changes the phase relations between Si_3N_4 and its

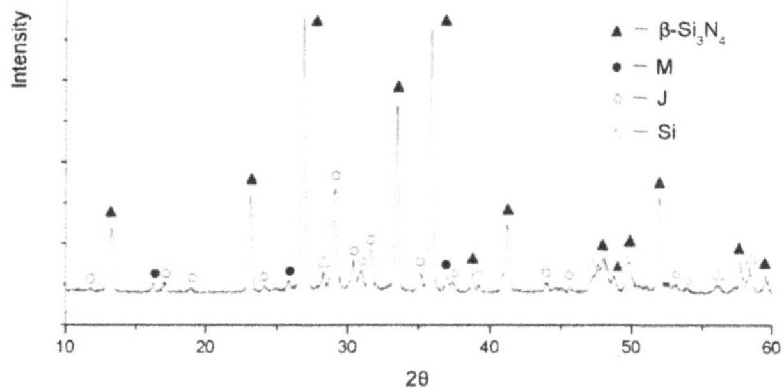

Figure 2. XRD pattern of specimen sintered at 1780℃ for 3 h.

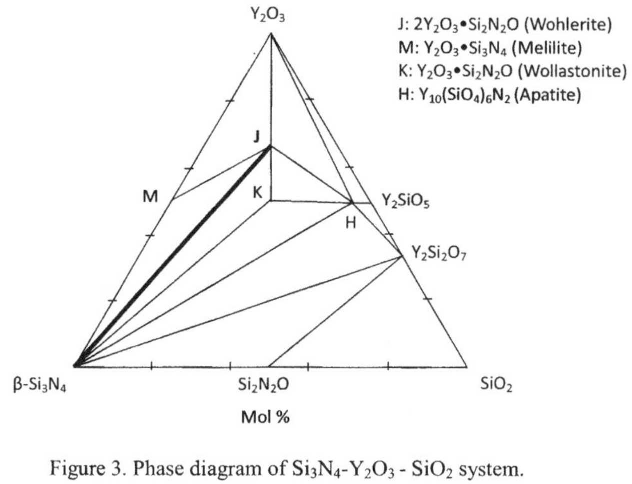

Figure 3. Phase diagram of Si_3N_4-Y_2O_3 - SiO_2 system.

neighboring phases. Now J-phase is compatible with β-Si_3N_4 and M-phase forming a triangle of Si_3N_4-M-J. This equilibrium relationship can be shown in Fig. 3. In which the tie line (bold line) between Si_3N_4 and J (wohlerite) has to instead of the tie line of M (Melilite) – K (Wollastonite), the triangle of Si_3N_4-M-J replaces for one of Si_3N_4-M-K. Latter was reported long time ago[10, 11]. From Fig. 3 it can be suggested that with increasing oxygen content in starting powder system and/or said oxygen partial pressure around sintering environment, the medium phase contained more oxygen composition will be easier to form in the trend of M → J (or K) → H → $Y_2Si_2O_7$. Either of them will be compatible with Si_3N_4 at high temperature.

CONCLUSION

Using starting Si_3N_4 powder with one third β-phase and higher oxygen content, together with a liquid-rich $(Y, La)_2O_3$-AlN additives, fully dense β-Si_3N_4 ceramic with high strength (965 MPa) has been obtained by LPS. Si_3N_4 powder with sufficient amounts of β phase and a little more oxygen has better sinterability and is benefit to form the network of elongated, rot-shaped Si_3N_4, as well as a refractory J phase as medium phase. They contribute to the high strength of sintered bodies. In addition, the compatibility between Si_3N_4 and J phase can be summarized that with increasing oxygen content in starting powder system and/or said oxygen partial pressure around sintering environment, the medium phase contained more oxygen composition will be easier to form in the trend of M → J (or K) → H →$Y_2Si_2O_7$. Either of them will be compatible with Si_3N_4 at high temperature.

FOOTNOTES

* Foundation item: Ningxia Natural Science Fund (NZ0740), China. The Key Lab of Powder Material & Advanced Ceramics (Ningxia, China. 0601)
** Correspondence author: Yong Jiang. Email: jynxyc@126.com

REFERENCES

[1]G. Himsolt, H. Knoch, H. Huebner, and F.W. Kleinlein, Mechanical Properties of Hot-Pressed Silicon Nitride with Different Grain Structures, *J. Am. Ceram. Soc.*, 62, 29-32 (1979).

[2]P.F. Becher, Microstructural Design of Toughened Ceramics, *J. Am. Ceram. Soc.,* 74, 255–269 (1991).

[3]Chen Hong, Mu Bai-chun, Zheng Li-ming, Zheng, Fabrication technique of Si_3N_4 fine powder. Casting, *J. Liaoning Inst. Technol.*, 26, 3, 191-195 (2006).

[4]Zhang Bao-lin, Zhuang Han-rui, X.R. Fu, Self-propagating synthesis of Si_3N_4 with Si powder under high chlorine gas pressure, *J. Chin. Ceram. Soc.*, 20, 3, 241–47 (1992).

[5]S. Yin, Combustion Synthesis, *Metallurgical Industry Publisher, Beijing,* 159–61 (1999).

[6]A.G. Merzhanov, Self-propagating High Temperature Synthesis: Twenty Years of Research and Findings, *VCH Publisher,* New York, 1–4 (1990).

[7]Yong Jiang, Laner. Wu, Peilin Wang, Zhen Kun Huang, Pretreatment and sintering of Si_3N_4 powder synthesized by the high-temperature self-propagation method, *Mat. Res. Bulletin,* 44, 21-24 (2009).

[8]Z.K. Huang, A. Rosenflan, I.W. Chen, Pressureless Sintering of Si_3N_4 Ceramic Using AlN and Rare Earth Oxides, *J. Am. Ceram. Soc.,* 80, 5, 1256-62 (1997).

[9]Z.K. Huang, T.Y. Tien, Solid-liquid Reaction in the Si_3N_4-AlN-Y_2O_3 System, *J. Am. Ceram. Soc.*,79, 6, 1717-1719 (1996).

[10]K.H. Jack, *Mater, Sci. Res.*, 11, 561-578 (1978).

[11]G.Z. Cao, Z.K. Huang, X.R. Fu, D. S, Yang, Phase Relationship of Si_2N_2O-Al_2O_3-Y_2O_3, *J. High Technol. Ceram.*, 1, 2, 119-27 (1985).

MICROSTRUCTURE OF LIQUID PHASE SINTERED SILICON CARBIDE CERAMICS WITH HIGH FRACTURE TOUGHNESS*

Yong Jiang**, Laner Wu, Yuhong Chen, Zhenkun Huang
School of MSE, the North University for Ethnics
Yinchuan, 750021, china

ABSTRACT

By SEM observation and EDX analysis, the relation between the fracture mechanism and the high toughness in the liquid-phase sintered (LPS) SiC ceramic with high toughness was revealed. The effects of two kinds of sintering methods, pressureless (PLS) and hot-pressing (HP), on their microstructure features and the characteristics of fracture mode were investigated. The results show that both the crack-break ways along the grain-boundary and the transgranular fracture were together enable the toughness higher in the PLS specimen, and the crack-break way along the grain-boundary was the key reason for the higher toughness in HP specimen.

KEYWORDS

silicon carbide ceramics, AlN - Re_2O_3 aids, microstructure, fracture toughness

INTRODUCTION

Our previous work reported[1] that PLS and HP dense SiC ceramics were manufactured by using AlN-Re(Y, La)$_2$O$_3$ as sintering additives. They possess excellent mechanical properties, particularly high fracture toughness. Toughness of the PLS sample reached 7 MPa•m$^{1/2}$, while HPed sample up to 10 MPa•m$^{1/2}$. In present work these SiC ceramics with high toughness were made with SEM observation and EDX analysis, in order to reveal the relation between the fracture mechanism and the high toughness. The differences of the fracture modes for two kinds of PLS and HP sintered SiC ceramics were compared.

As common sense, the liquid-phase sintered (LPS) SiC ceramic possesses higher toughness than the solid-phase sintered one. It will concern with liquid phase formation. Liquid phase softens the grain-boundary and promotes the ceramic toughening. This is one of the reasons for high toughness of our SiC ceramic contained sufficient liquid phase. However important effects on the toughening involves many factors, including the microstructure appearance, the characteristic of crystal grains, the behavior of the fracture, the crack-break ways, and so on. These will be described and discussed in this paper.

EXPERIMENT

The dense SiC ceramics were obtained by pressureless sintering (PLS) at 1950°C for 0.5 h protected by N_2 gas and at hot-pressing (HP) of 1850°C for 0.5h protected by N_2 gas. The sintered bodies were grounded on the surface out in 0.5 mm and then machined into the testing bars with dimensions of 3 mm × 4 mm × 36 mm for the three-point bending strength test and of 2.5 mm × 5 mm × 26 mm for the single edge notched beam (SENB) fracture toughness test. Before testing, the bar edges were chamfered and the bars were polished according to the standard of GB/T 6569–2006. For the fracture toughness test, a narrow slit of 0.2 mm × 2.5 mm was made in the middle of bars. The Vickers hardness was tested by using the remains of the samples after the three-point bending strength test[1].

After the hardness and toughness test, the PLS sample with high toughness K_{1c}=7 MPa•m$^{1/2}$ and the HP sample with higher toughness up to K_{1c}=10 MPa•m$^{1/2}$ were used for the research of the microstructure and the fracture mechanism by SEM observation and EDX analysis. As a comparison, the PLS sample with low toughness K_{1c}=3.4 MPa•m$^{1/2}$ was also used in the present work. The SEM photographs were taken on the polished surface and the fracture section of above samples. The trend of crack-break was traced on the SEM map. Samples were mounted with electroconductive glue in an alumina standard。EDX was used for analyzing the components of crystal grain and grain-boundary liquid phase. The instruments of the SSX-550 (Shimadzu) scanning electron microscope were used for SEM observation and EDX analysis.

RESULTS AND DISCUSSION

Microstructure observation

Figure 1 shows the features of polished surface and fracture section of two kinds of liquid-phase sintering of SiC ceramics. As can be seen, the microstructure of PLS sample (Fig. 1a) and HP sample (Fig.1b) is very uniform with small grains and the grains gap was filled with liquid, forming the dense bodies.

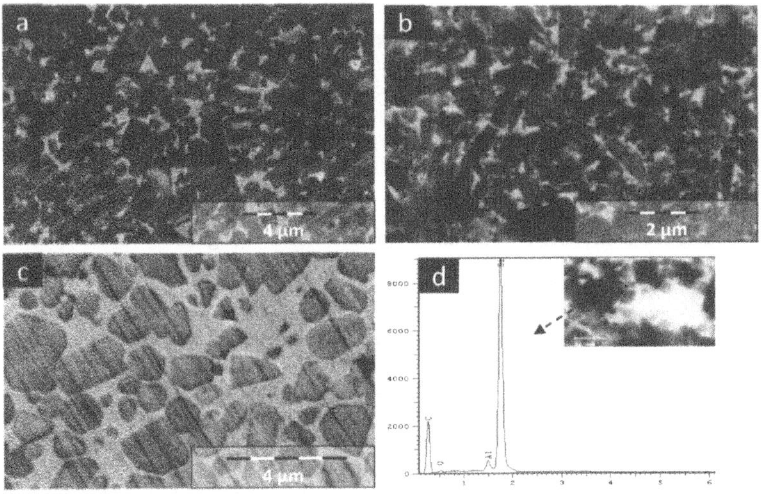

Fig. 1. SEM image and EDX spectrum of the samples, polished surface of PLS (a) and HP (b), (c) the polytypic features of SiC, (d) EDX point analysis.

The PLS samples shows various shapes of grains, like triangle, hexagonal, square, small bar, but close to equiaxed crystals. This crystalline form of diversity reflects the polytypic features of SiC. In the process of sintering, the sufficient amount of liquid promotes the grains well developed through the solution-reprecipitation mechanism. Microstructure of the HP's sample shows more homogeneity with more elongated grains than the PLS's. It indicates that under high pressure the grains grow toward a directed growth trend (Figure 1b). Furthermore the core-rim structures can also be seen on the surface of grains. This phenomenon can be observed more clearly in PLS specimen than in HP specimen (Fig. 1a). Evidently the growth mechanism of the crystal grains is on the base of the solution-reprecipitation process. PLS sample with more core-rim structure was analyzed by EDX. On the SEM map, the element analysis result of the gray grain was shown in Figure 1d. In addition to SiC, there was a little amount of Al and trace O. The trace O might be residual oxygen on the surface of SiC. Besides, a small amount of AlN from additive seemed to enter into SiC, forming SiC-AlN solid solution of $Si_{1-x}Al_xC_{1-x}N_x$. On the liquid phase area (white color) the EDX probed the element components to be Al, rare-earth (La, Y), N, C, as well as Si perhaps from impurity SiO_2 on SiC surface, forming a liquid of nitrogen-contained rare-earth aluminum-silicate as grain-boundary phase. In addition, Both PLS and HP specimens contained fine grains with the size same as or smaller the particles of starting powder, which should be attributed to refinement role of crystal grains in the fusing-recrystallization process. Namely AlN and SiC not only formed the solid solution rim on the SiC grain surface, but also possibly became the new nucleus in the contact surface, through nibbling the original SiC crystal grain to multiply the tiny new crystal grain[2]. Thus it caused original SiC crystal grains minification.

Cross section appearance and crack extension

Figure 2 demonstrates the fracture appearance and the surface crack extension trend of PLS and HP specimens. It can be seen that the appearances of PLS and HP specimens have a very big difference, and the appearances of the sample with different toughness also differ each other. On the fracture surface of PLS specimen (Figure 2 a, b), many pulling out traces of crystal grains (the fracture along the grain-boundary) can be seen clearly. Many cleavage planes (the transgranular fracture) can be seen in the cross section of the specimen with high toughness K_{1c}=7 MPa•m$^{1/2}$ (Figure 2 a), but rare on the low toughness K_{1c}=3.4 MPa•m$^{1/2}$ samples (Figure 2 b), indicating the transgranular fracture plays a certain toughening role in the PLS specimen. The fracture along the grain-boundary consumes the crack extension energy because the way lengthens windingly. The transgranular fracture consumes the main energy to break the strong SiC bonding. When the crack extends along the grain-boundary to meet a crystal grain and is unable to detour forward, crack goes through can be blocked weakens. Therefore, both crack-break ways along the grain boundary and the transgranular fracture together enable the toughness higher in the PLS specimen.

Figure 2b shows many caves on the crossing section, which cause the crystal grains to be loose. This is the characteristic of brittle fracture. It occurs on the sample with low toughness K_{1c} =3.4 MPa•m$^{1/2}$. Moreover it can be observed that the fracture growth stop or detour at the loose plot on the hardness indentation surface of the sample with high toughness (Figure 2 c), explaining that this loose plot (or pores) also has the impediment crack extend. But whether this is the main reason for toughness increment need to be further proven. The transgranular fracture phenomenon is not observed on the HP specimen (Figure 2 d). On the hardness indentation surface, the crack extends along grain-boundary (picture omitted). For the HP specimen, due to the effect of pressure, bigger binding energy exists among the crystal grains compared with PLS one. Therefore the grain boundary is more compact. It will be more difficult for a crack to go through compared with the PLS specimen. The interlaced pillar like crystal grains can be observed on the HP specimen. It might meet a bigger resistance and a longer distance for a crack to go through. Therefore crack breaks along the crystal are the primary cause of higher toughness for the HP sample with high grain boundary energy. Some big crystal grains near 10μm can be found on the cross section of HP specimen (Figure 3a). Perhaps these abnormal crystal grains also play certain nail for the crack to prevent its going through.

The transgranulate break phenomena of crack are not observed in the PLS specimen (Fig. 3b). But it does not mean that there is no transgranular fracture inside the specimen because the superficial crystal grain environment is completely different from inside.

The above analyses are mainly based on the observation of crack break phenomena. Many researches were done about the toughening mechanism concerning with the crack deflect toughening. Sung-Gu Lee, et al[3] and A. De Pablos, et al[4] studied the relationship of the fracture toughness and the crystal grains. The latter carried out the research with the relations between the elongated Si_3N_4 aspect and the related parameters and the fracture toughness. His formula showed that the long crystal grain was the key to enhance the K_{1c}. Further researches need to be carried out to show whether this kind of relations is also available to SiC. At least it can be used to explain qualitatively. HP specimen has longer crystal grains compared with PLS specimen, so its K_{Ic} is much higher.

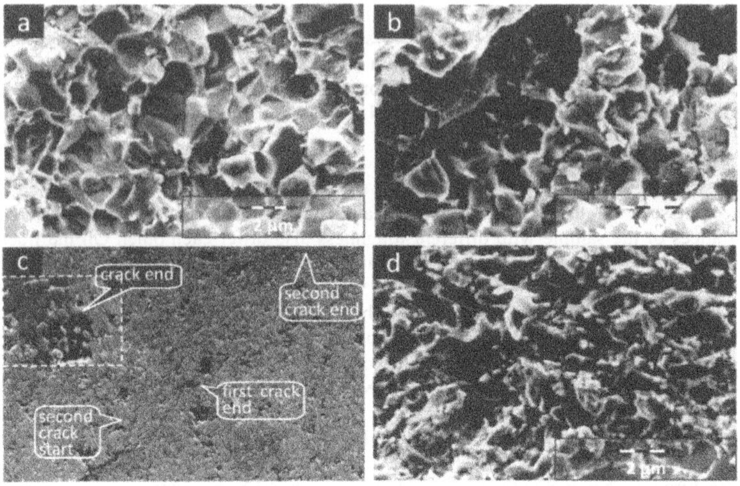

Fig. 2. SEM appearance, PLS specimens fracture toughness: (a) K_{IC}=7 MPa•m$^{1/2}$, (b) K_{IC}=3.4 MPa•m$^{1/2}$, (c) the polished surface crack growth (crack end enlarged view is in dashed line); (d) HP specimen fracture surface: K_{IC}=10 MPa•m$^{1/2}$.

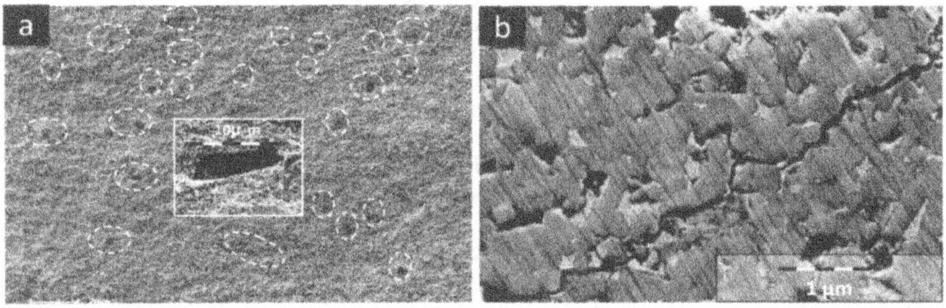

Fig. 3. (a) HP specimen fracture appearance with K_{IC}=10 MPa•m$^{1/2}$ (in dashed line square it is a partially enlarged drawing, showing big crystal grain), (b) Crack extension of PLS specimen.

CONCLUSION

Both the crack-break ways along the grain-boundary and the transgranular fracture together enable the toughness higher in the pressureless sintered specimen. The crack-break way along the elongated interlaced pillar-like crystal grain-boundary is the key reason for higher toughness in the hot-pressing sintered specimen. The compact grains surround by liquid phase with high grain boundary energy is another reason for higher toughening.

FOOTNOTES

* Fundation item: National 863 Plan of China (2002AA33210); The North University for Ethnics (2004Y035)

** Correspondence author: Yong Jiang. Email: jynxyc@126.com

REFERENCES

[1]WU Lan'er, CHEN Yuhong, JIANG Yong, HUANG Zhenkun, Liquid Phase Sintering of SiC With AlN-Re_2O_3 Additives, *JOURNAL OF THE CHINESE CERAMIC SOCIETY,* 36, 5, 593-96 (2008).

[2]Pan Y B, Qm J H, Morita M, et al. The mechanical properties and microstructure of SiC-AlN particulate composite, *J. Mater. Sci.,* 33, 5, 1238(1998).

[3]Sung-Gu Lee and Young-Wook Kim, Relationship between Microstructure and Fracture Toughness of Toughened Silicon Carbide Ceramics, *J. Am. Ceram. Soc.,* 84, 1347-53 (2001).

[4]A. De Pablos, M.I. Osendi, P. Miranzo, Correlation between microstructure and toughness of hot pressed. Si_3N_4 ceramics seeded with β-Si_3N_4 particles, *Ceramics International,* 29, 757-64 (2003).

VII. Advanced Ceramic Coatings

DEVELOPMENT OF ELECTROSPINNING TITANIA WEB FROM SUSPENSION

WD Teng & Nassya M Said
Structural Materials Programme, AMREC, SIRIM, Malaysia
*E-mail: wdteng@sirim.my

ABSTRACT

In view of this, nano titanium dioxide was synthesized and prepared as suspension with PVP solution. The mixture was set to electro-spin into fibre at 18KV. The e-spun fibre was subjected to differential thermal analysis to obtain the sintering profile for the titania fibre. The titania fibre was sintered at 500°C and 1200°C. The sintered phases and microstructural of titania fibre were analysed by using X-ray diffractometer and scanning electron microscope and presented. The cross-linked titania fibre formed a web-like structure maybe reduced the nano risk as compared to the nanoparticles.

KEYWORDS: Nanoparticles, titania nanofibre, electrospinning

1. INTRODUCTION

Titania is one of the most popular photocatalyst materials on the market today [1-4]. It has been widely used in self-cleaning, air purification, water purification, anti-mould, and others. Titania in the form of nanoparticle has been drawing the much researchers and governmental officials' attention on the nano risks. It is a very challenging and expensive task to gauge and study the nano risk. By utilizing electrospinning technique on the possibility of producing a cross-linked nanotitania fibre with web-like structure, this may reduce the nano risk as compared to the nanoparticles or nanofibres. Ceramic nanofibres of alumina-borate, titania, zirconia have been fabricated in previous studies [5-7]. However there is dispute as to the anatase-rutile phase transformation at higher temperature and grain growth [8,9].

Electrospinning is not a new technology for polymer fibre production. It has been known since the 1930's; however, it did not gain significant industrial importance due to the low output of the process, inconsistent and low molecular orientation and poor mechanical properties of the electrospun fibres.

The principle of electrospinning is to use a high voltage electric field to draw a positively charged polymer solution from an orifice to a collector. This creates a cone-jet of solution from the orifice to the grounded collection device. Then it travels to form a stretched jet, whipping motion, and then forms fibres collected on a grounded metal plate. By using electrical forces, the electrospinning process can produce fibres with nanometer diameters. Because of their small diameters, electrospun fibres have drawn great attention for fabrication of ceramic nanofibres.

2. MATERIALS AND METHOD

Materials and Suspensions

Commercially available titanium isopropoxide ($C_{12}H_{28}O_4Ti$ or ($Ti(OiPr)_4$)), ethanol 96%, polyvinylpyrollidone (PVP) 1.3 million Mw, all from Sigma Aldrich were used in the present work.

8ml titanium isopropoxide ($Ti(OiPr)_4$) was mixed with 40ml ethanol and stirred in a clean, conical flask, and then heated to 60°C, then 4ml distilled water was added into warm titanium isopropoxide solution by titration technique with continual stirring by magnetic bar until it is finished. The mixture was covered with foil and continued to be stirred for another hour. While 5.57g PVP and 40ml ethanol were mixed in another clean, lidded glass bottle and stirred for an hour. The quantities were combined to produce stock solution for electrospinning which had in a ratio $Ti(OiPr)_4$ and PVP solutions of 1:1-4 in volume.

Electrospinning

In this experiment, 4ml of the $Ti(OiPr)_4$ solution was mixed with 4ml of the PVP solution to obtain the said stock solution. The solution was then placed inside a 10ml capacity syringe, the excess air was removed, and the syringe was placed in the syringe pump horizontally as shown in Figure 1. The high voltage clamp was then attached at the syringe needle, the pump turned on, then the power supply turned on. The formation of a stable jet was the first attempt with blank PVP solution via an adjustable knob on a variable scale with reading from a multimeter. Therefore the flow rate and voltage were varied to establish a range of electrospinning conditions for $Ti(OiPr)_4$ and PVP mixed solution. The power supply delivers at 18kV in this experiment, at flow rate of 0.5ml/h at a distance of 8cm from the needle tip to the collector.

Heat Treatment and Characterization

The spun fibres were collected at the aluminium foil as the collector. The fibres were analysed using DTA/TG Thermal Analyzer (Rigaku, Japan) from room temperature to 800°C at 5°C/min. From the thermal analysis curves were used to design the firing of fibres profile. The fibres were fired to 500°C at 0.1°C/min, then 5°C/min to 1200°C without soaking in air and cool naturally in the furnace (Superburn, Japan). Phase analysis of the initial spun fibres and fibres fired at 500°C and 1200°C were analysed by X-ray diffraction (XRD) (Geiger-Flex, Rigaku Japan) of the samples was carried out under ambient conditions using Cu-Kα as the radiation source at a scan speed of 0.5° per minute and a step scan of 0.02°. The morphology of the initial spun fibres and fibres fired at 500°C and 1200°C were examined using a scanning electron microscope (SEM, Hitachi Japan).

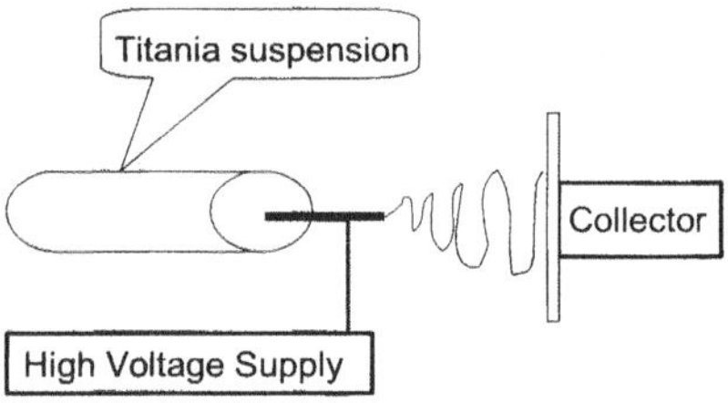

Figure 1: The experiment set up of electrospinning of titania-PVP solution.

3. RESULTS AND DISCUSSION

Fibre and Thermal Analysis

SEM micrograph as shown in Figure 2 revealed the titania web which is the mixture of from 50nm to 3μm fibres. These fibres were fused at the joint forming like a web.

The DTA/TGA graphs showed an initial weight loss of 17% up to 135°C in correspondence to an endothermic reaction. This can be attributed to evaporation of water and ethanol. These fibres had an exothermic peak at 320°C. Between 255°C and 360 °C there is a sharp decrease in weight of 6% in correspondence to a large exothermic reaction with a varying slope. A few reactions are occurring here which can contribute to this; the condensation of titania fibre along with a burnout of PVP polymer chain. The large exothermic hump in the middle can be attributed to this,

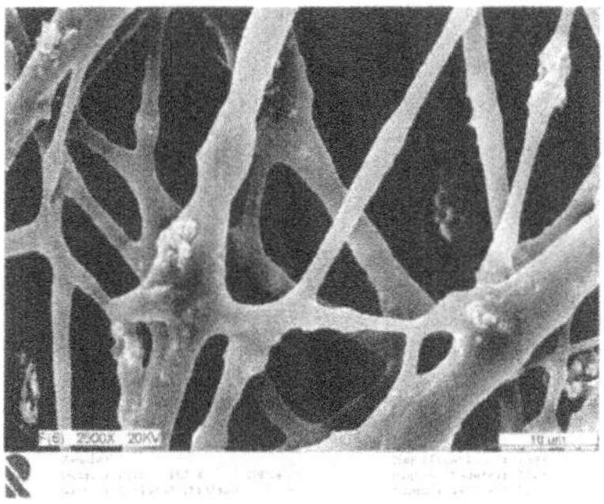

Figure 2: SEM micrograph of titania fibres producing by electrospinning at 18KV and distance of collector of 8cm at flow rate of $0.5mlh^{-1}$.

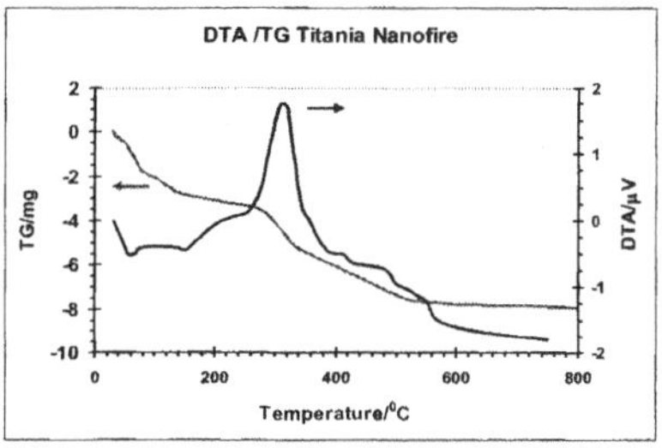

Figure 3: DTA/TG analysis of the initial electrospun titania fibre.

Heat Treatment and Phase Analysis

XRD analysis of the synthesized material revealed the amorphous phase of the initial material. The synthesized titania transformed to anatase phase at 500°C as shown in Figure 4. However, in Figure 5, the anatase transformed into rutile at 1200°C.

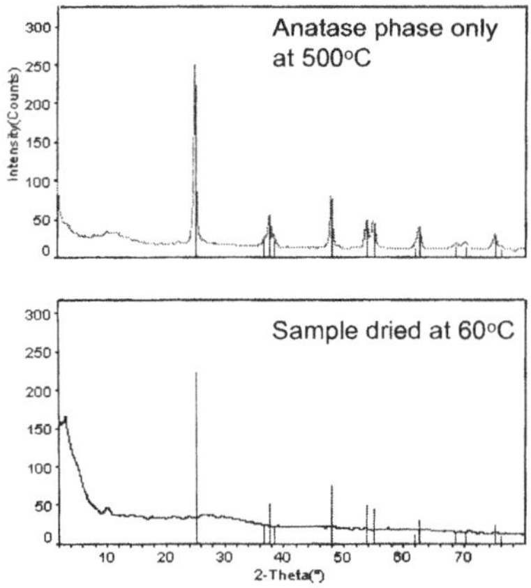

Figure 4: XRD pattern showing amorphous phase of dried fibre (bottom) which transformed into anatase only after firing to 500°C (top).

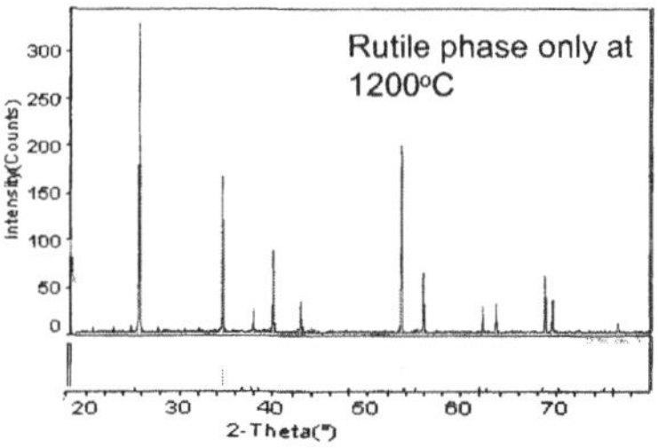

Figure 5: XRD pattern showing the anatase transformed into rutile at 1200°C.

Fibre and Micrograph Analysis

The fibre coagulated and formed bigger fibre with irregular shape as shown in Figure 6 when it was heat-treated at 500°C. The titania nanoparticles joined and formed the fibre that prevent the dispersion of the nanoparticles. As the fibre was heat to 1200°C, the nanoparticles grew into microsizes. However, it formed the continual link fibre like web as shown in Figure 7.

Figure 6: SEM micrograph of titania fibre sintered at 500°C showing the anatase nanoparticles produced by electrospinning.

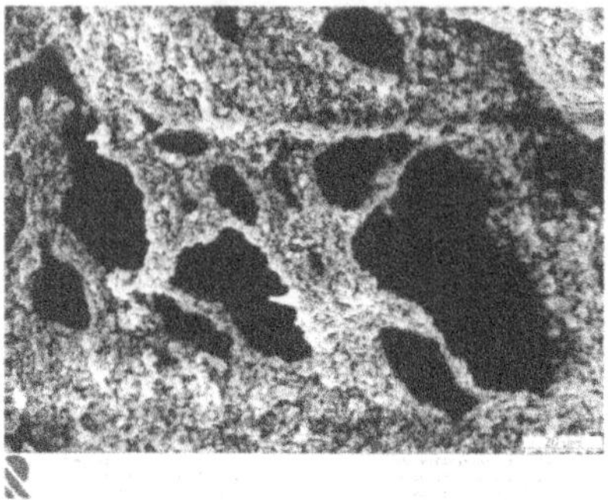

Figure 7: SEM micrograph of titania fibre sintered at 500°C showing the grain growth rutile particles produced by electrospinning.

4. CONCLUSIONS

- The present work indicated that titania fibers were produced successfully by the titania suspension at 18KV at distance of 8cm of the electric field. Although majority of the fibres were in mirco, there are some nanofibre in the midst of the web.
- According to TG measurements, 51wt% of the entire sample is lost due to non TiO2 constituents such as water, ethanol and polymer chains of PVP.
- It is safe to fast firing as assume all of the polymer is burnt out by 500°C, as seemed in TG analysis and XRD analysis.
- The SEM micrograph revealed nanotitania of the broken fibres of anatase sintered at 500°C and grain growth into rutile at 1200°C.
- The cross-linked titania fibre formed a web-like structure maybe reduced the nano risk as compared to the nanoparticles.

ACKNOWLEDGEMENTS

This work was supported under the MOSTI-SF grant No. 03-03-02-SF0026. The authors gratefully acknowledge the technical assistance provided by Azura and Ahmad Sabata.

REFERENCES

1. Klaus P. Kuhn, Iris F. Chaberny, Karl Massholder, Manfred Stickler, Volker W. Benz, Hans, Gunther Sonntag, and Lothar Erdinger. Disinfection of surfaces by photocatalytic oxidation with titanium dioxide and UVA light. *Chemosphere*, 2003,53: p. 71–77
2. D.A. Tryk, A. Fujishima, and K. Honda. Recent topics in photoelectrochemistry: Achievements and future prospects. *Electrochemica Acta*, 2000, 45: p. 2363–2376

3. Michael R. Hoffmann, Scot T. Martin, Choi Wonyong, and Detlef W. Bahnemann. Enviromental applications of semiconductor photocatalysis. *Chemical Review*, 1995. 95(1): p. 69–96
4. Takashi Sato, Yoshiyuki Koizumi, and Masahito Taya. Photocatalytic deactivation of airborne microbial cells on TiO2 loaded plate. *Biochemical Engineering Journal*, 2003, 14: p.149–152,
5. Dan Li, Y.X., Fabrication of titania nanofibers by electrospinning. *Nano Letters*, 2003. 3(4): p. 555-560
6. Sudha Madhugiri, W.Z., John P. Feraris, Kenneth J. ZBalkus Jr., Electrospun mesoporous molecular sieve fibers. *Microporous and Mesoporous Materials*, 2003. 63: p. 75-84.
7. Zhang, H.B. and Edirisinghe, M.J. Electrospinning Zirconia Fiber From a Suspension, *J.Am.Ceram.Soc.*, 89[6] 1870-1875 (2006).
8. Joseph Muscat, Varghese Swamy, Nicholas M. Harrison. First-principles calculations of the phase stability of TiO2. *Physical Review B*, 65, 2002 41.
9. Xing-Zhao Ding, Xiang-huai Liu, Grain growth enhanced by anatase-rutile phase transformation in gel-derived nanocrystalline titania powders. *Journal of Alloys and Compounds*,1997, 248: p. 143-145.

HIGH-SPEED ENGINEERING CERAMIC COATING BY LASER CHEMICAL VAPOR DEPOSITION

Takashi Goto, Teiichi Kimura and Rong Tu
Institute for Materials Research, Tohoku University, Sendai 980-8577, Japan

ABSTRACT

High-speed coating process, such as plasma spray and electron physical vapor deposition (EB-PVD), has been used for thick coating typically thermal barrier coating (TBC). To improve the performance of TBC, a new coating route should be developed. This review briefly introduces conventional high-speed coating processes, and also describes a new laser chemical vapor deposition (LCVD) technique invented by the present authors.

INTRODUCTION

High-efficiency energy usage and high-temperature operation are principal issues for gas turbines of aircraft engine and power generation. A gas-inlet temperature for turbines is increasing year by year and nowadays reaching to almost 1500°C far over melting points of common structural metals. The blade of gas turbine should have high refractoriness, high toughness and creep strength, and then single crystalline Ni-base super-alloys are now commonly utilized. However, the highest temperature to bear in a combustion environment could be less than 1000°C. Therefore, a protective ceramic thermal barrier coating (TBC) combined with an air-cooling system is essential for the higher temperature operation of gas turbines [1].

Ceramic material for TBC should have low thermal conductivity, high thermal-shock resistance and a large thermal expansion coefficient close to metallic substrates. Yttria stabilized zirconia (YSZ) has been widely employed in TBC. However, even a small thermal expansion mismatch between YSZ and metallic substrates would yield a significant thermal stress at the interface, and then cracks might extend during severe heat-cycles resulting in a catastrophic failure of TBC. Although an intermediate band-coat layer, mainly MCrAlY (M: Ni, Co etc.), would relax the thermal stress with improving oxidation resistance, the development of high-performance TBC is a critical issue particularly with controlling microstructure. Atmospheric plasma spray (APS) and electron-beam physical vapor deposition (EB-PVD) have been commonly used for the TBC [2]. This review briefly describes these processes and then introduces a new laser chemical vapor deposition process developed by the present authors.

TBC PROCESS

Since the temperature gradient within TBC would become several 100s°C, the thickness of TBC should be several 100s μm to endure the severe thermal stress. Traditionally, APS and EB-PVD have been available mainly due to their high deposition rates. In APS, YSZ powders are introduced in a plasma torch and melted YSZ are sprayed to substrate. Due to its economic set-up and versatile

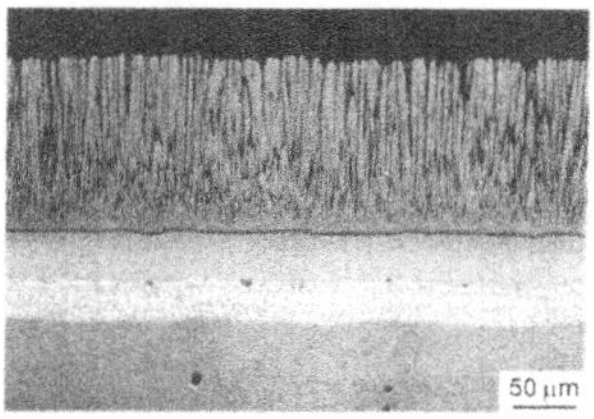

Figure1. Typical cross-sectional microstructure of YSZ coating prepared by EBPVD.

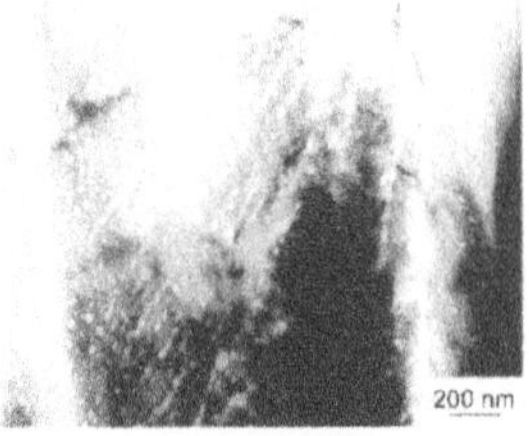

Figure 2. Nano-structure of YSZ coating prepared by EB-PVD.

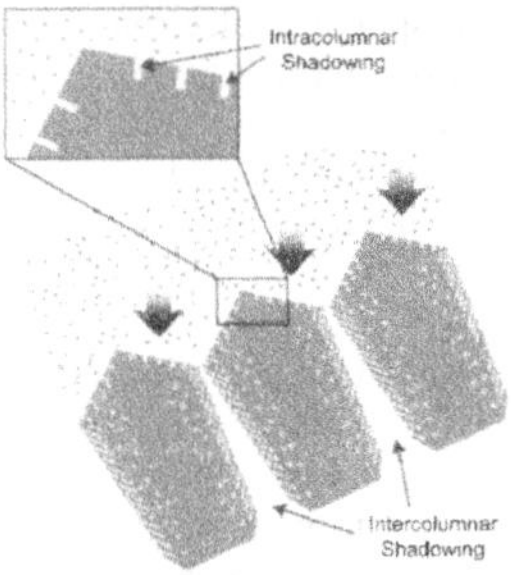

Figure 3. Deposition mechanism (shadowing effect) of EB-PVD.

applications, APS has been widely employed to coat on wide-ranged substrates. However, APS coatings have often contained un-melted powders, voids and cracks, and then its characteristic laminar splat texture would result in delamination. On the other hand, EB-PVD adopts evaporates a high-power electron beam, and YSZ vapor deposits on substrates. Figure 1 demonstrates a typical cross-sectional microstructure of EB-PVD YSZ TBC coating. A well-developed columnar texture would relax a thermal expansion mismatch, preventing delamination due to longitudinal crack extension. It contains a large amount of nano-sized pores in the columnar grains as represented in Fig. 2. These nano-pores significantly reduce the thermal conductivity to less than 1 $Wm^{-1}K^{-1}$, 1/3 to 1/4 of YSZ bulk sintered bodies. Figure 3 demonstrates the formation mechanism of nano-pores in EB-PVD YSZ coating [3]. A shadowing effect may explain the formation of nano-pores and columnar grains.

HIGH-SPEED COATING BY CONVENTIONAL CVD

CVD is advantageous to control the morphology of materials ranging from powder, fine-grained poly-crystal, columnar-grained poly-crystal, dendrite poly-crystal, plate-like single-crystal. Since CVD is a fundamentally atomic or molecular level process of chemical reactions, the nucleation and grain growth at the substrate surface would yield highly-adhered coating with excellent conformal coverage. Although, the preparation of CVD YSZ films has been long studied for applications on mainly solid oxide conductors and buffer layers for thin film oxide super-conductors, the thick and wide-area CVD YSZ coating has been scarcely conducted. Figure 4 summarizes the deposition rates of YSZ films as a function of deposition temperature (T_{dep}) in conventional thermal CVD. The deposition rates have been usually less than several μm/h with the film thickness of below 10 to 20 μm. The source gases were often halides usually $ZrCl_4$ and rCl_3, and metal organic (MO) compounds such as acethylacetonate (acac), methylhepfadionato (thd) and dipivalaylmethanato

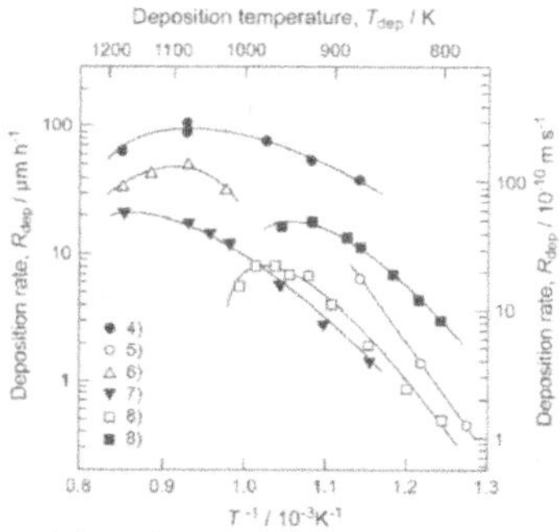

Figure 4. Effect of deposition temperature on the deposition rate of YSZ films in conventional thermal CVD.

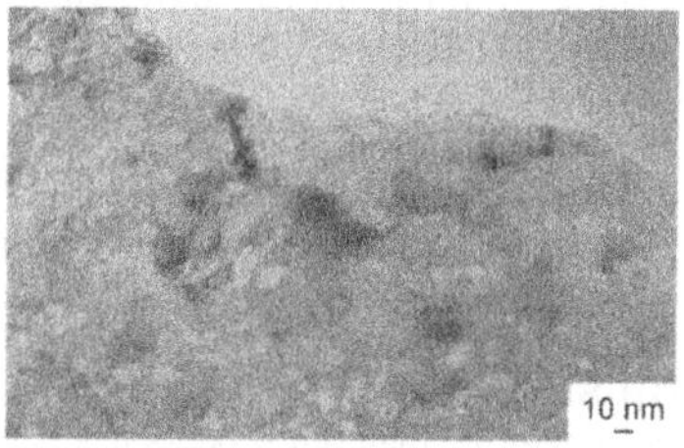

Figure 5. Nano-structure of YSZ film prepared by thermal CVD.

(dpm). Due to low deposition rates of conventional CVD, the increase in deposition rate is essential for the application to TBC. Even with sufficient source gas supply and high T_{dep}, the deposition rate of CVD can not be increased markedly because of homogeneous chemical reactions (powder formation) in a gas phase under a high gas concentration. Therefore, the increase in deposition rate of CVD needs significant efforts by optimizing deposition conditions and modifying equipments. Wahl et al. applied CVD YSZ to TBC, and reported a high deposition rate of 50 μm/h by using $Zr(thd)_4$ and $Y(thd)_3$. We have developed a cold-wall type CVD setup and increased the deposition rate up to 102 μm/h by using $Zr(dpm)_4$ and $Y(dpm)_3$ [4], and thick YSZ films with highly (001) oriented columnar grains containing a large amount nano-pores was obtained as demonstrated in Fig. 5. The nano-pores in EB-PVD YSZ coatings were sphere in shape, whereas those in our thermal CVD was slightly angular and elongated implying some specific crystal planes surrounding the inner-surface of nano-pores. The formation mechanism of nano-pores in thermal CVD can be different from that in EB-PVD because of the different shapes of nano-pores. Figure 6 illustrates a schematic of the nano-pore formation mechanism in thermal CVD [9]. In thermal CVD, the rate-controlling step in a low temperature region can be chemical reaction at the substrate surface. The film would be dense because the nucleation site can be kinks and/or steps located bottom of crystal grains. Since the crystal growth occurs upward along the crystal plane, pores would not form and the film becomes dense. On the other hand, the mass transfer (diffusion) becomes rate-controlling in a high temperature region, where the grain growth takes place as shown in Fig. 6. Due to the mass transfer limited process, chemical reactions are sufficiently high, and the nucleation sites could be top-surface of film where the gas concentration is the highest. The deposition would

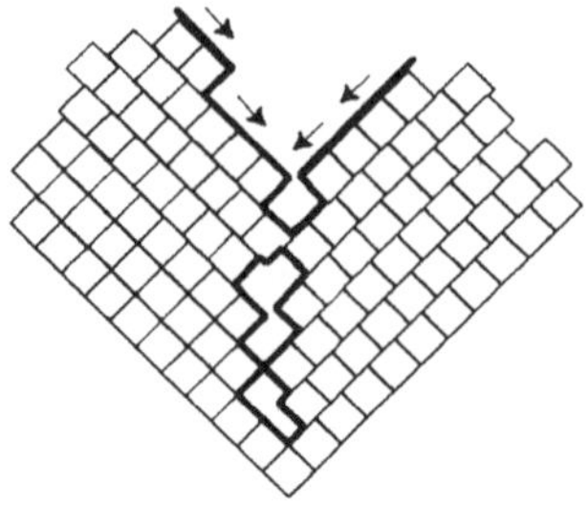

Figure 6. Formation mechanism of nano-pores in thermal CVD.

go downwards and then a large amount of nano-pores can be produced at the boundary among grains. By the effect of significant phonon scattering at the nano-pores, the thermal conductivity reduced to 0.8 $Wm^{-1}K^{-1}$, and CVD YSZ coating on Ni-based super alloy exhibited good heat-cycle performance [10]. However, the main drawback of thermal CVD is slow deposition rates. In order to increase the deposition rate of thermal CVD, plasma-enhanced CVD (PECVD) has been applied to accelerate chemical reactivity of source gases. Prauchat et al. prepared YSZ films by PECVD at a high deposition rate of 100 to 250 μm/h and obtained a total thickness of 65 to 200 μm by using $ZrCl_4$ and $Y(thd)_3$ [11]. The PECVD YSZ coating was quasi-tetragonal (t') phase showing significant (200) orientation and well-developed columnar grain with a low thermal conductivity of 1.6 $Wm^{-1}K^{-1}$.

NEW TECHNOLOGY OF HIGH-SPEED COATING

The application of laser technology to the material process has been sought since 1970, and many processes such as laser welding, laser ablation, laser annealing and laser CVD have been developed [12]. Since the laser has distinct two kinds of energies (heat and light), laser CVD can be generally categorized into two types; pyrolytic laser CVD and photolytic laser CVD [13]. The pyrolytic laser CVD has usually adopted CO_2 laser having a wave length of 10 μm as a heat source. Thermochemical reactions would proceed at a small local region by a focused laser beam. By scanning a laser beam or moving a substrate, small-scale three-dimensional structures or direct patterns of deposit can be fabricated. Nano-dots and nano-fibers have been produced for electronic micro-device applications. TiN and TiB_2 hard-coatings on carbon fiber in several μm thick have been also produced by pyrolytic laser CVD. In photolytic laser CVD, the films can be deposited by photochemical reactions without intensive heating of substrates. However, pyrolytic or photolytic

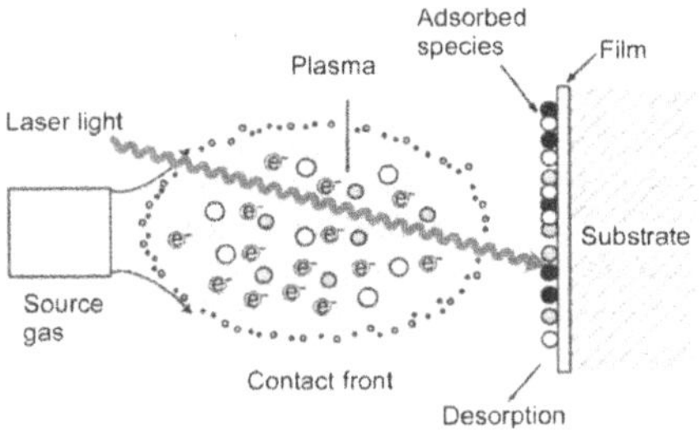

Figure 7. Schematic diagram of laser CVD (LCVD).

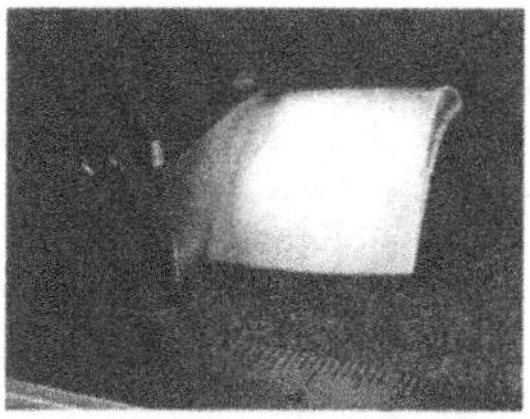

Figure 8. Appearance of YSZ coated turbine blade by LCVD.

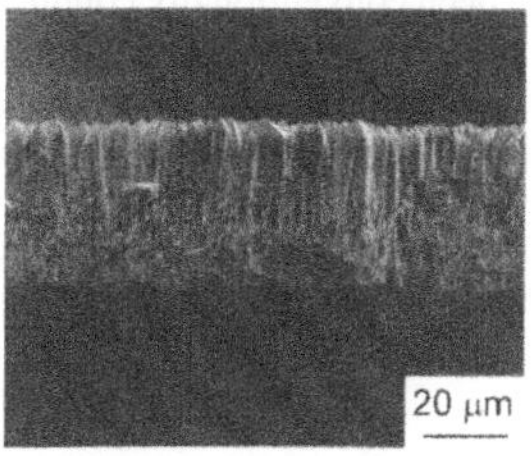

Figure 9. Cross-sectional microstructure of YSZ film prepared by LCVD.

laser CVD has been never applied to thick and wide-area coating particularly to TBC. We have developed new laser CVD using Nd: YAG laser (wave length: 1064 nm), and successfully prepared many kinds of oxide films at significantly high deposition rates ranging from several 100s μm/h to several mm/h [14]. In our laser CVD, a plasma with a bright light appeared around a deposition zone. The plasma should have highly activated source gases. The laser would have also enhanced the diffusion of absorbed atoms or molecules on the substrate surface resulting in high deposition rates. Figure 7 depicts a schematic of the plasma formation around the substrate. Langmuir probe analysis has evidenced the formation of electrons and ionic species in the plasma zone [15]. Since the laser beam has been expanded to about 20 mm in diameter, the temperature of substrates merely increased to 150 to 200°C by laser radiation. However, after introducing source gases the substrate temperature abruptly increased to 800 to 1000 °C by the effect of plasma formation. Figure 8 depicts an appearance of YSZ coated gas turbine blade by the present laser CVD. The YSZ film consisted of significantly (200) oriented columnar grains similar to those of EB-PVD YSZ coatings as depicted in Fig. 9. The highest deposition rate was 660 μm/h by using precursors of $Zr(dpm)_4$

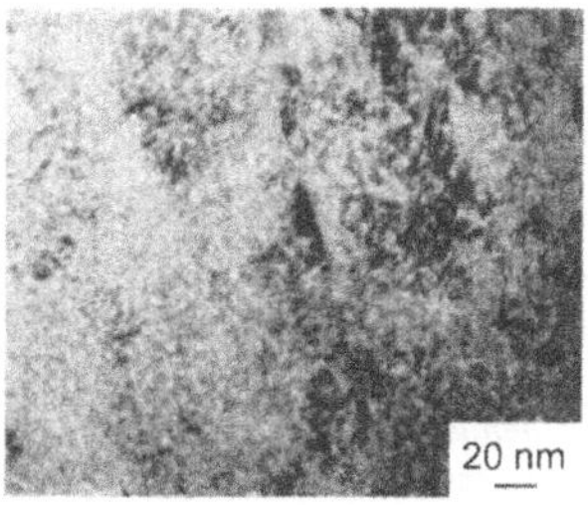

Figure 10. Appearance of nano-pores in YSZ film prepared by LCVD.

and $Y(dpm)_3$. Figure 10 demonstrates nano-pores inside columnar grains of YSZ coating. The porosity of the YSZ film was about 30 % which strongly depended on deposition rate; the higher deposition rate the higher porosity. By changing the precursor gases and deposition conditions, various oxides were able to prepare by the present laser CVD [16]. Rutile and/or anatase TiO_2 films were prepared at most 2300 μm/h. Y_2O_3 films were prepared at 270 μm/h, showing excellent anti-plasma etching properties. α-type (corundum structure) Al_2O_3 films were obtained at most 1000 μm/h, and are applicable to cutting tools due to its high-hardness and thermal stability. SiO_2 film can be prepared at a tremendously high deposition rate of 25 mm/h.

SUMMARY

Thick coatings, typically TBC, have been conventionally conducted by APS and EB-PVD, where the microstructure control is a primary issue to develop high-performance TBC. Since CVD is versatile to prepare various morphologies of films, CVD can be a candidate process for TBC. However, conventional thermal CVD has a serious drawback of low deposition rate. An auxiliary energy of laser in CVD is significantly effective to enhance the deposition rate enabling the application of TBC. The present laser CVD would be also promising in various engineering applications such as Al_2O_3 coating for cutting tools, Y_2O_3 coating for anti-plasma etching and TiO_2 coating for photocatalytic devices.

REFERENCES

[1] D. R. Clarke and C. G. Levi, Materials Design for the Next Generation Thermal Barrier Coatings, *Annu. Rev. Mater. Res.*, **33**, 383-417 (2003).

[2] P. A. Kammer, *Hand Book of Thin Film Technology*, IOP Pub., A4.1: 1.

[3] N. Yamaguchi, K. Wada, K. Kimura and H. Matubara, Microstructure Modification of Yttria-Stabilized Zirconia Layers Prepared by EB-PVD, *J. Ceram. Soc. Jpn.*, **11**, 883-889, (2003).

[4] R. Tu, T. Kimura and T. Goto, Rapid Synthesis of Yttria-Partially-Stabilized Zirconia Films by Metal-Organic Chemical Vapor Deposition, *Mater. Trans.*, **43**, 2354-2356 (2002).

[5] G. Whal, W. Nemetz, M. Giannozzi, S. Rushworth, D. Baxter, N. Archer, F. Cernuschi and N. Boyle, Chemical Vapor Deposition of TBC: An Alternative Process for Gas Turbine Components, *Trans. ASME* **123**, 520-524 (2001).

[6] Y. Akiyama, T. Sato and N. Imaishi, Reaction Analysis for ZrO_2 and Y_2O_3 Thin-Film Growth by Low-Pressure Metalorganic Chemical-Vapor-Deposition Using Beta-Diketonate Complexes, *J. Cryst. Growth*, **147**, 130-146 (1995).

[7] N. Bourhila, F. Felten, J. P. Senateur, F. Schuster, R. Madar and A. Abrutis, Deposition and Characterization of ZrO_2 and Yttria-Stabilized ZrO_2 Films Using Injection-LPCVD, *Proc. 14th Conf. EUROCVD*-**11**, 417-424 (1997).

[8] M. Pulver, W. Nemetz and G. Wahl, CVD of ZrO_2, Al_2O_3 and Y_2O_3 from Metalorganic Compounds in Different Reactors, *Surf. Coat. Tech.*, **125**, 400-406 (2000).

[9] T. Goto, T. Kimura, R. Tu, High-speed Deposition of Nano-pore Dispersed Zirconia by CVD and Improvement of Thermal Barrier Performance, *J. Jpn. Soc. Powder & Powder Metall.*, **51**, 821-828 (2004).

[10] R. Tu and T. Goto, Thermal Cycle Resistance of Yttria Stabilized Zirconia Coatings Prepared by MO-CVD, *Mater. Trans.*, **46**, 1318-1323 (2005).

[11] B. Preauchat and S. Drawin, Properties of PECVD-Deposited Thermal Barrier Coatings, *Surf. Coat. Tech.*, **142**, 835-842 (2001).

[12] D. Bauerler, *Laser Processing and Chemistry*, Springer (2000).

[13] C. Duty, D. Jean and W. J. Lackey, Laser Chemical Vapour Deposition: Materials, Modelling, and Process Control, *Inter. Mater. Rev.*, **46**, 271-287 (2001).

[14] J. R. V. Garcia and T. Goto, Thermal Barrier Coatings Produced by Chemical Vapor Deposition, *Sci. Tech. Adv. Mater.*, **4**, 397-402 (2003).
[15] H. Miyazaki, T. Kimura and T. Goto, Acceleration of Deposition Rates in a Chemical Vapor Deposition Process by Laser Irradiation, *Jpn. J. Appl. Phys.*, **42**, L316-L318 (2003).
[16] T. Goto, High-Speed Deposition of Zirconia Films by Laser-Induced Plasma CVD, *Solid State Ionics*, **172**, 225-229 (2004).

A REVIEW OF NANOCRYSTALLINE DIAMOND/β-SiC COMPOSITE FILMS

Vadali. V. S. S. Srikanth, Thorsten Staedler and Xin Jiang*
Institute of Materials Engineering, University of Siegen
Siegen, NRW, 57076, Germany

ABSTRACT

The idea behind synthesis of nanocrystalline diamond/β-SiC composite films is to obtain films those posses a whole range of combined properties of diamond and β-SiC to serve tribological, thermal barrier, electronics and biological applications. The diamond/β-SiC nanocomposite films are designed in such a way that the resultant material contains required volume of nanometer sized grains of both the components such that the availability of large volume of grain boundaries can be controlled based on the application requirement. In this paper a review of nanocrystalline diamond/β-SiC composite film system with regards to its controlled synthesis, characterization, mechanical properties, and coefficient of friction will be discussed. Microwave plasma enhanced chemical vapor deposition technique was used to carry out the nanocomposite film depositions with the aid of H_2-CH_4-$Si(CH_3)_4$ gas mixtures. Based on the micro-structural analyses these nanocomposite films are classified as granular type composite films that contain diamond and β-SiC components as nanocrystalline grains distributed contiguously and laterally throughout the thickness of the film in a desired volume fraction combinatorial form. Deposition of gradient natured diamond/β-SiC nanocomposite films and novel diamond/β-SiC composite films containing (001) diamond rounded or square faceted surfaces will also be discussed. β-SiC content and diamond microstructure in the films are identified as the compositional and structural factors respectively that influence the mechanical and friction properties.

INTRODUCTION

In general, a composite film obtained from two different materials is useful for applications only when the properties of both the components are resonably incorporated into the resultant film. However, in the case of composite films containing micron sized grains, the overall properties of the film are mostly influenced by only one component, hindering the structure control and thereby, the property control. It is therefore necesarry to develop a composite film that contains nanometer sized grains in such a way that the availability of large volume of grain boundaries can be controlled which in turn will help in controlling the film properties. The motivation behind synthesizing nanocrystalline diamond/β-SiC composite thin films is exactly the same; it is to obtain films those posses a whole range of combined properties of diamond and β-SiC to serve tribological, thermal barrier, electronics and biological applications. In this paper, diamond/β-SiC nanocomposite film system will be reviewed with regards to its structural, mechanical and tribological properties.

EXPERIMENTAL

Diamond/β-SiC nanocomposite films were synthesized by using microwave plasma enhanced chemical vapour deposition (MWCVD) technique. H_2, CH_4, and $Si(CH_3)_4$ (tetramethylsilane, TMS) gas mixtures were employed during the film depositions. In all the film depositions, H_2 and CH_4 gas flow rates were 400 and 2.5 sccm respectively, with the total gas pressure kept constant at 25 Torr. The deposition temperature and microwave power used were 700 °C and 700 W respectively.

(100) Si wafers (that have undergone a bias enhanced nucleation (BEN)[1] pre-treatment step), Mo, W, and WC-6wt.% Co (that have undergone a manual pre-treatment step[2]) were used as the substrates. Other deposition related details have been discussed in the listed references.[3-5]

Field emission scanning electron microscopy (FESEM), glancing incidence x-ray diffraction (GIXRD), transmission electron microscopy (TEM), micro Raman scattering, Fourier transform infrared (FTIR) spectrometry, Rutherford back scattering (RBS) studies and electron probe micro analysis (EPMA) have been carried out to obtain micro-structural and compositional properties of the diamond/β-SiC nanocomposite films. Atomic force microscopy (AFM) and indentation studies have been carried out to obtain film properties on the tribological and mechanical front.

RESULTS AND DISCUSSION

FESEM surface morphology images of diamond/β-SiC nanocomposite thin films deposited on pre-treated Si and WC-6wt.% Co substrates are shown in Fig. 1(a) and Fig. 1(b) respectively. A clear phase contrast, a bright phase, which is diamond and a dark phase, which is β-SiC can be clearly observed in Fig. 1. SEM cross-sectional morphology also showed a similar microstructure. Based on the SEM images, the nanocomposite films are categorized as granular type films that contain both diamond and β-SiC components as nanocrystalline grains distributed contiguously and laterally in the same layer of the film. The nanometer sized grains in the films are due to the presence of TMS in the gas phase during the deposition. TMS not only disturbs the diamond grain growth but also aids in the incorporation of the second phase namely β-SiC.

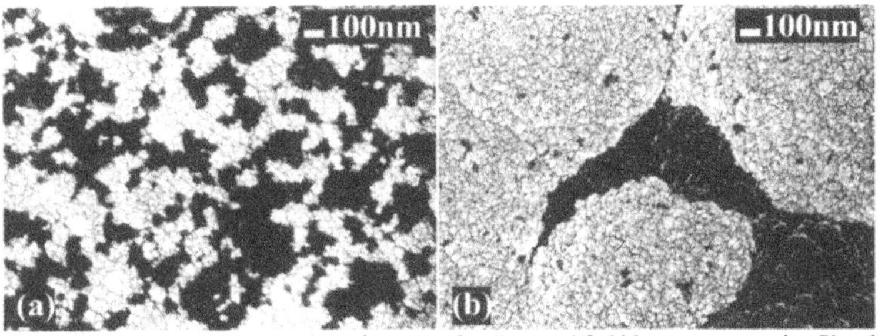

Figure 1. FESEM surface morphology images of a diamond/β-SiC nanocomposite film deposited (a) on Si and (b) on WC-6wt.% Co substrates by using a TMS flow rate of 15 sccm.

The GIXRD patterns of the composite films grown on different substrates showed that both diamond and β-SiC co-exist in the films.[5,6] The XRD results not only showed that higher TMS flow rates result in films that are dominated with β-SiC but also showed that both the components in the film are nanocrystalline in nature. Both phase mixture and nanocrystalline natures of the composite films are further confirmed by TEM analysis. TEM plane view micrograph shown in Fig. 2(a) corresponds to a film deposited using a TMS flow rate of 5 sccm. Both (111) diamond and (111) β-SiC reflexes are seen in the same selected area electron diffraction (SAED) pattern [Fig. 2(b)] obtained from the film in discussion. Similar SAED patterns are obtained for other composite films. The diffraction rings in all the cases are diffused, indicating nanocrystallinity whilst the uniformity of the rings was indicative of the homogeneous distribution of the crystallites. Additionally, in the high resolution (HR) TEM plane view image [Fig. 2(c)] of the film shown in Fig.2 (a), {111} lattice fringes of both diamond and β-SiC can be clearly observed; these are marked with parallel lines including the inter-atomic distance. The areas of amorphous phase are marked with the letter **A**.

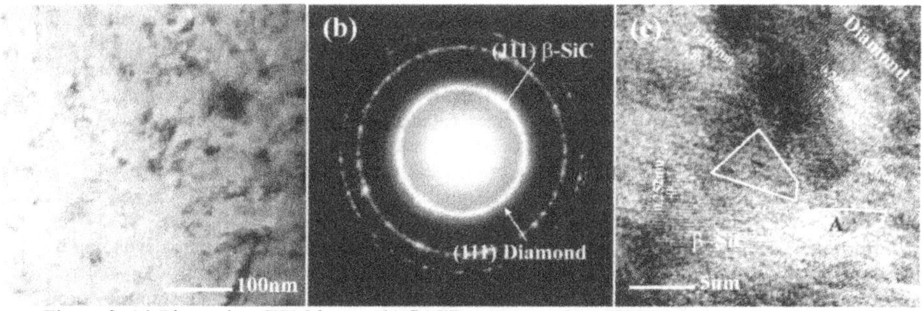

Figure 2. (a) Plane view TEM image (b) SAED pattern and (c) HRTEM plane view image of diamond/β-SiC nanocomposite film deposited using a TMS flow rate of 5 sccm.

Raman spectra showed that the diamond phonon line broadening started to show up even at 5 sccm of TMS flow rate indicating the influence of increasing β-SiC Volume% in the films with an increase in TMS flow rate.[5,6] FTIR measurements illustrated that greater transverse optic phonon (TO) band intensity obtained from the samples deposited with greater TMS concentration showed qualitatively the presence of larger volume of β-SiC in the films. As an example, FTIR spectra obtained from two different diamond/β-SiC nanocomposite films deposited on W substrates are shown in Fig. 3. Additionally, quantitative compositional analysis (RBS measurements & EPMA)[5] showed that the content of β-SiC in the films corresponds almost linearly to the TMS concentration in the gas phase during the film deposition.

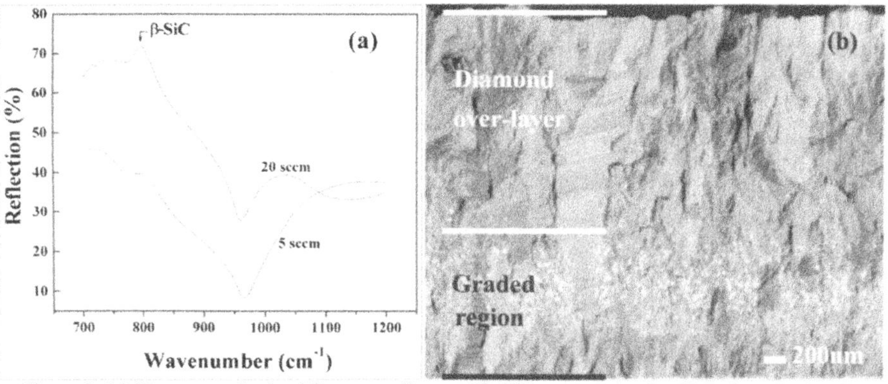

Figure 3. (a) IR spectra obtained from two diamond/β-SiC nanocomposite films deposited on W substrates by using different TMS flow rates. The transverse optical phonon band around 800 cm^{-1} corresponds to the presence of β-SiC. (b) Backscattered electron cross-sectional micrograph of a gradient natured diamond/β-SiC nanocomposite film deposited on BEN pre-treated (100) Si substrate. The bright spots indicate β-SiC phase.

The knowledge on β-SiC composition control has led to the synthesis of gradient natured composite film [Fig. 3(b)] in a single process step. The gradient is β-SiC dominated composite on the substrate side to a pure diamond top layer. Gradient composite films are deposited at the same experimental conditions that were used in the case of the homogeneous composite films (Fig. 1) except for the TMS gas concentration which is continuously decreased with time during the deposition, from a higher value to a lower value TMS gas situation with a subsequent diamond deposition for few hours. With the aid of micro Raman scattering experiments low residual stress values in the range of -0.1 to -0.5 GPa have been calculated for various diamond top layers on different gradient inter layers (on Si substrates). Similar reduction was also observed in the case of such films deposited on other substrates. This proves that the diamond/β-SiC nanocomposite film when used as an interlayer, accommodates the thermal stresses that should have appeared in the diamond top layer in the absence of the inter layer.

RMS roughness values ranging from 30 nm to 300 nm are measured for the films depending on the diamond pre-treatment method (except BEN method) and also on the substrate material. However, a low RMS roughness of 12 ± 1 nm was measured for all the nanocomposite films deposited on BEN pre-treated Si substrates. Additionally, a general trend of linearly increasing friction coefficients with increasing TMS flow rate, i.e. increasing amount of β-SiC, was observed for these films. The friction coefficient decreased from 0.28 to 0.19 for films prepared with 20 and 5 sccm TMS flow rate. With such low friction co-efficient values in addition to low RMS roughness values, these films are potential candidates as protective coatings on mirror polished Si based ceramic tool applications.[7]

Microhardness values (as evaluated[8] from Vickers indentation data) of the diamond/β-SiC nanocomposite films deposited on ultrasonically pre-treated Si substrates as a function of β-SiC volume fraction are shown in the Figure 4. The hardness graph reflects a phase transition in the films from a nanodiamond dominated to a β-SiC dominated diamond/β-SiC nanocomposite complimenting well with the compositional analyses[5] on the same samples. With increasing TMS concentration, microhardness of the diamond/β-SiC composite film decreases slowly in the beginning, then falls down rapidly and levels off at a hardness value of 20 GPa for TMS flow rate above 10 sccm, a value representative of pure β-SiC. This indicates the dominance of β-SiC phase in the composite film deposited with greater TMS flow rates. On the other hand, the nanoindentation results obtained (by using a Berkovich indenter) from diamond/β-SiC nanocomposite films deposited on BEN pre-treated Si substrates, showed a linear decreasing indentation modulus and hardness with increasing TMS flow rate (0 to 20sccm) from 400 to 200

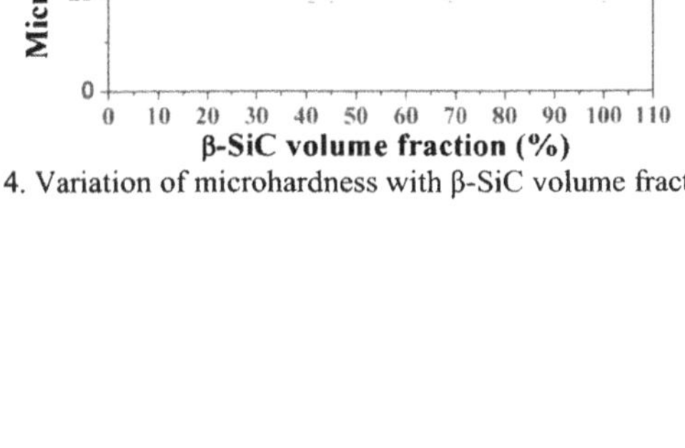

Figure 4. Variation of microhardness with β-SiC volume fraction.

GPa and from 70 to 30 GPa, respectively.[9] The hardness values obtained for the diamond/β-SiC composite films at lower and moderate TMS flow rates can be directly related to the high density of the interfaces or grain boundaries present in the films owing to the nanocrystallinity of both the phases.[10,11] Frictional and mechanical properties of the diamond/β-SiC nanocomposite films clearly indicate that β-SiC volume fraction can be considered as an important compositional factor to determine any physical properties of the nanocomposite film system.

As for the hardness values of the nanocomposite films, it has been observed that diamond phase acts as a load bearing component, whereas β-SiC acts as a binder providing the structural flexibility. At higher TMS flow rates the roles seemed to have reversed. An optimum content of about 40% β-SiC in the nanocomposite seems to be a good compromise meeting both the requirements of low friction and relatively high hardness. Diamond/β-SiC nanocomposite films deposited on refractory metals showed improved fracture toughness than that of pure diamond film;[6,12] this is plausible due to the continuous (graded) or discrete (as in diamond on homogeneous composite) variation of composition, structure, and mechanical properties of the diamond/β-SiC nanocomposite film system with depth beneath the indented surface. Additionally, contributions for improved toughness come from mechanisms such as reduced residual stresses and difficult crack deflection. The crack deflection was interpreted by analyzing the crack propagation (due to Brinell indentation) by using SEM; the nanocomposite films showed lesser lateral crack lengths indicating greater toughness.

Apart from the nanocomposite films discussed above, novel composite films with nanocrystalline diamond and β-SiC phases along with clear (001) diamond rounded or square faceted surfaces have also been deposited.[13] The growth of the novel composite films has been attributed to the reactivity dependent selective deposition of nano-β-SiC on diamond surfaces and its effect in controlling the diamond grain morphology. Marked increase in the measured modulus of the composite film containing faceted diamond structures was observed. The values obtained for a composite film without the diamond faceted structures (Fig. 1) fell within a narrow range whilst the values obtained for a composite film with faceted diamond structures are scattered over a broad range. This shows that diamond microstructure in the composite films acts as a structural factor while determining the mechanical properties. Figure 5 shows the SEM surface morphology of the novel composite film in the discussion along with its comparative modulus values.

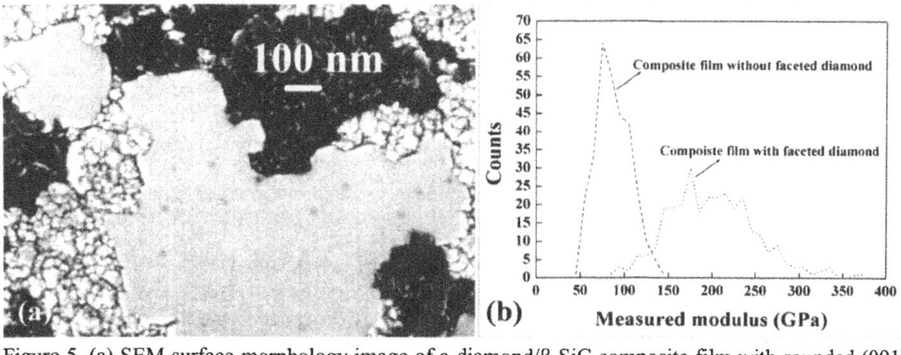

Figure 5. (a) SEM surface morphology image of a diamond/β-SiC composite film with rounded (001) diamond facets (large bright regions) along with nanocrystalline diamond (bright) and β-SiC (dark) phases. This film is deposited on BEN pre-treated Si substrates with a TMS addition of 0.0506% to H_2 and CH_4. (b) Indentation modulus values obtained for different diamond/β-SiC composite films.

CONCLUSION

A review of diamond/β-SiC nanocomposite film system with regards to its synthesis, structural, tribological and mechanical properties has been presented. β-SiC content and diamond microstructure in the films are identified as the compositional and structural factors respectively that influence the mechanical and friction properties. The potential of the diamond/β-SiC nanocomposite film system in thermal, electronic and biological applications is being explored. A thorough understanding of such two component nanocomposite films can not only lead to the new design but also to the improvement of other diamond nanocomposite films[14,15] and multi-component nanocomposite films[16] depending on the application specifications.

ACKNOWLEDGEMENTS

This work was supported by Deutsche Forschungsgemeinschaft. We would also like to thank our project (DFG, JI 22/11-1) partners for the useful discussions.

*Invited speaker and corresponding author; electronic email: xin.jiang@uni-siegen.de

REFERENCES

[1]X. Jiang, K. Schiffmann and C.-P. Klages, Nucleation and initial growth phase of diamond thin films on (100) silicon, *Phys. Rev. B*, **50**, 8402-10 (1994).

[2]C. P. Chang, D. L. Flamm, D. E. Ibbotson and J. A. Mucha, Diamond crystal growth by plasma chemical vapor deposition, *J. Appl. Phys.*, **63**, 1744-48 (1988).

[3]X. Jiang and C.-P. Klages, Synthesis of diamond/β-SiC composite films by microwave plasma assisted chemical vapor deposition, *Appl. Phys. Lett.*, **61**, 1629-31 (1992).

[4]Vadali. V. S. S. Srikanth, M. H. Tan and X. Jiang, Initial growth of nanocrystalline diamond/β-SiC composite films: A competitive deposition process, *Appl. Phys. Lett.*, **88 (7)**, 073109 (2006).

[5]Vadali. V. S. S. Srikanth, T. Staedler and X. Jiang, Structural and compositional analyses of nanocrystalline diamond/β-SiC composite films, *Appl. Phys. A*, **91 (1)**, 149-155 (2008).

[6]Vadali. V. S. S. Srikanth, H. A. Samra, T. Staedler and X. Jiang, Nanocrystalline diamond/β-SiC composite interlayers for the deposition of continuous diamond films on W and Mo substrate materials, *Surf. Coat. Technol.*, **201(22-23)**, 8981-85 (2007).

[7]A. V. Sumant, A. R. Krauss, D. M. Gruen, O. Auciello, A. Erdemir, M. Williams, A. F. Artiles, and W. Adams, Ultrananocrystalline diamond film as a wear-resistant and protective coating for mechanical seal applications, *Tribology Transactions*, **48 (1)**, 24–31 (2005).

[8]M. F. Doemer and W. D. Nix, A method for interpreting the data from depth-sensing indentation instruments, *J. Mater. Res.*, **1 (4)**, 601-09 (1986).

[9]T. Staedler, Srikanth Vadali and X. Jiang, Diamond/carbide nano-composite gradient films: a route to solve the adhesion issues of diamond films, *Mater. Res. Soc. Symp. Proc.*, **890**, 0890-Y01-04 (2006).

[10]S. Yip, Nanocrystals: The strongest size, *Nature*, **391**, 532-33 (1998).

[11]S. Barnett and A. Madan, Superhard superlattices, *Phys. World*, **11**, 45-48 (1998).

[12]G. Dinger, Vadali. V. S. S. Srikanth, H. A. Samra, C. Friedrich, X. Jiang, H. Hoche and S. Gross, Indentation loading behaviour and simulation of nanocrystalline diamond-composite films, submitted to *Plasma Processes & Polymers*, (2008).

[13]X. Jiang, Vadali. V. S. S. Srikanth, Y. L. Zhao and R. Q. Zhang, Facet dependent reactivity and selective deposition of nanometer sized β-SiC on diamond surfaces, *Appl. Phys. Lett.*, **92**, 243107 (2008).

[14]H. A. Samra, R. J. Hong, and X. Jiang, The preparation of diamond/tungsten-carbide composite films by microwave plasma-assisted CVD, *Chem. Vap. Deposition*, **13 (1)**, 17–20 (2007).
[15]F. Z. Ding and Y. L. Shi, The study of diamond/TiC composite film by a DC-plasma–hot filament CVD, *Surf. Coat. Technol.*, **9-11**, 5050-53 (2007).
[16]Stan Veprek and A. S. Argon, Towards the understanding of mechanical properties of super- and ultrahard nanocomposites, *J. Vac. Sci. Tech. B*, **20 (2)**, 650-64 (2002).

EFFECT OF TEMPERATURE FIELD ON DEPOSITION OF BORON CARBIDE COATING FORM BCl_3-CH_4-H_2 SYSTEM

Yongsheng Liu*, Litong Zhang, Laifei Cheng, Wenbin Yang, Weihua Zhang, Yongdong Xu
National Key Laboratory of Thermostructure Composite Materials, Northwestern Polytechnical University, Xi'an Shaanxi 710072, People's Republic of China

ABSTRACT

Boron carbide was prepared by low pressure chemical vapor deposition from BCl_3-CH_4-H_2 system. Firstly, the temperature distributions of field A and B were tested. The results showed that the temperature distribution of field B is more uniform than that of field A. The effects of temperature field on deposit characteristic and deposition mechanism were investigated. The results showed that the temperature field had an important effect on the morphologies, phases, microstructure and compositions of deposits. Under the temperature field A, the morphologies were crystalline-like, the boron concentration was 87.11at.%, and the phase of deposit was $B_{13}C_2$ as confirmed by XRD and TEM. Under the temperature field B, the morphologies were cauliflower-like, the boron concentration was 75.34at.%, the phase of deposit was amorphous boron carbide as confirmed by XRD and TEM. The above differences were attributed to the different reaction mechanism during the deposition process under field A and field B. The early reactions between BCl_3 and CH_4 before deposition process were critical to the formation of amorphous boron carbide.

KEYWORDS: effect; temperature field; CVD; boron carbide; deposition mechanism

1. INTRODUCTION

Boron carbide plays an important role to improve the oxidation resistance of ceramic matrix composites, such as multilayer self-healing silicon carbide matrix composites[1], SiC-B_4C oxidation protective coatings[2], and oxidation protection matrix in C/C composites[3]. These composites have received increasing interest in the military and aerospace industries due to the self-healing and oxidation resistant functionality, in addition to high strength, low density and high melting point.

CVD methods have been extensively studied and developed for boron carbide ceramics. Many types of reactive gas mixtures were used for the B_4C deposition. Some of the most commonly adapted mixtures are listed as follows: $BCl_3/CH_4/H_2$[4-6], $B_2H_6/CH_4/H_2$[7-8], $BBr_3/CH_4/H_2$[9] and $BCl_3/C_2H_4/H_2$[10].

The recent research focus of CVD B_4C is to reveal the deposition mechanism under different deposition conditions and establish the relationship between deposition parameter and deposition mechanism, for which thermodynamic, mass transfer and kinetic modeling attempts have been studied by several research groups[11-13]. The CVD B_4C from $BCl_3/CH_4/H_2$ precursor is a very complex chemical reaction process, and the B_4C can be deposited by different mechanisms. Some reasonable deposition mechanisms have been established, such as Thomas S. Moss et al[14], and Mustafa Karaman et al[15], their experiments were performed under a small variation of deposition parameters and thus all of the B_4C deposits had similar microstructure and phase composition, which probably suggest the B_4C coatings were deposited by a single mechanism.

In addition to the deposition parameters (such as deposition temperature, total pressure, gas flow and gas ratio), the furnace state (for example structure of furnace and temperature field in furnace)

* Corresponding author. Tel.: +86-29-8848-6068-823; fax: 86-29-8849-4620.
E-mail address: liuys99067@163.com; yongshengliu@nwpu.edu.cn

would play a very important role on deposit characteristic. However, no report can be found about the effect of temperature field on deposition of boron carbide.

In this work, the effect of two kinds of temperature field on boron carbide deposition was investigated. Firstly, two kinds of temperature field were compared, which were named field A and field B respectively. Then, morphologies, phase, composition and structure of boron carbide under field A and field B were characterized and compared. Finally, different deposition mechanisms were proposed to explain the difference of deposits under field A and field B.

2. EXPERIMENT PRODUCEDURE

Firstly, the temperature distribution of field A and field B were tested before deposition experiments using RTC (Rings for Temperature Check), which was produced by Xiamen Quantum Star Technology Co., Ltd in China. The used type of RTC is AQS type, which measured range was from 700°C to 1100°C. In order to know the effect of gas flow on temperature distribution, temperature distribution of field A was tested under three kinds gas flow, which were vacuum, Ar with 600mL/min flow and Ar with 1500mL/min flow. Temperature distribution of field B was tested only under 1500mL/min Ar.

The experimental system has been reported in previous paper[16]. Boron trichloride ($BCl_3 \geq 99.99$vol.% and iron$\leq$10ppm) was used as boron source for CVD boron carbide. The carbon source was provided by the methane ($CH_4 \geq 99.95$vol.%) gas. Hydrogen ($H_2 \geq 99.999$vol.%) was used as a dilution gas of BCl_3. High purity graphite slices (30×15×2mm) was used as substrate material for CVD SiC. Then the graphite+CVD SiC slices were used as substrate for deposition of B_4C. The deposition parameters for SiC were from CH_3SiCl_3-H_2-Ar system. Methyltrichlorosilicane (Called MTS, content $CH_3SiCl_3 \geq 98.0$ wt. %) was a precursor of SiC. Hydrogen (content $H_2 \geq 99.99$ %) was a carrier gas of MTS. Argon (content $H_2 \geq 99.9$ %) was used as dilution gas. The deposition conditions were as follows: (MTS)/H_2=1/10 for 80 h at P=3 kPa, Ar=350mL/min, and T=1000 °C. The process parameters for different temperature fields were fully same, which were summarized in table 1.

Table 1 Process Parameters of LPCVD Boron Carbide

Substrate	BCl_3 (ml.min^{-1})	CH_4 (ml.min^{-1})	H_2 (ml.min^{-1})	T (°C)	t (h)	Pressure (Pa)
Graphite+CVDSiC	500	100	500	950	40	1000

The morphology of the coatings were examined by a Leo 1530 SEM and a XL30 ESEM-TMPSEM with an attachment of EDX, X-ray diffraction (XRD) was made using a Panalytical X'pert PRO at glancing incidence (DIXRD, glancing angle of 1°) with Cu-Kα radiation. Transmission Electron Microscopy (TEM) observations and high resolution images were preformed using a TECNAI F30 operated at 300 kV.

3. RESULTS AND DISCUSSION

3.1 COMPARSION OF TEMPERATURE DISTRIBUTION OF FIELD A AND FIELD B

In order to know the effect of temperature fields on the deposition of boron carbide, temperature distribution of field A and field B were tested first as shown in Fig.1. For field A, the temperatures at different position are very non-uniform from top to bottom of the furnace. The position of highest temperature was at 20cm from the furnace top, which was 940°C. Then the temperature decreased rapidly along the furnace from top to bottom. The temperature was only 780°C at 60cm position. On the other hand, the temperature distributions were very similar with different gas flows such as vacuum, 600mL/min Ar or 1500mL/min Ar, which indicated that gas flows have no influence on temperature distributions. Therefore, the temperature distribution of field B was tested only with 1500mL/min Ar. It was obvious that the temperature distribution were more uniform for field B than that for field A. The isothermal regions existed at position from 15cm to 40cm and from 50cm to 60cm. The temperature gradient was smaller than that of field A. Based on the above results, we can conclude that temperature distributions of field B are more uniform than that of field A.

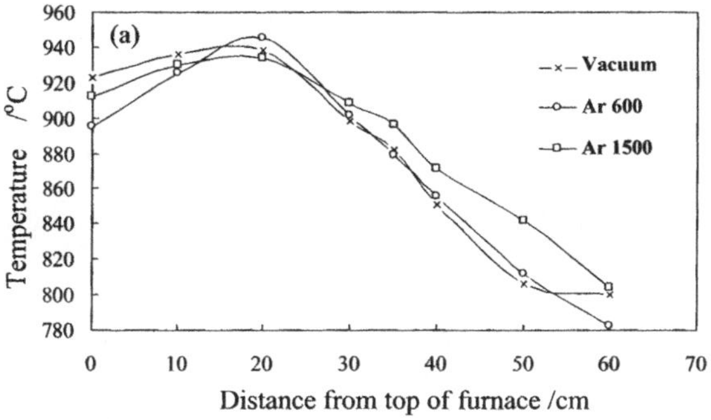

(a) field A

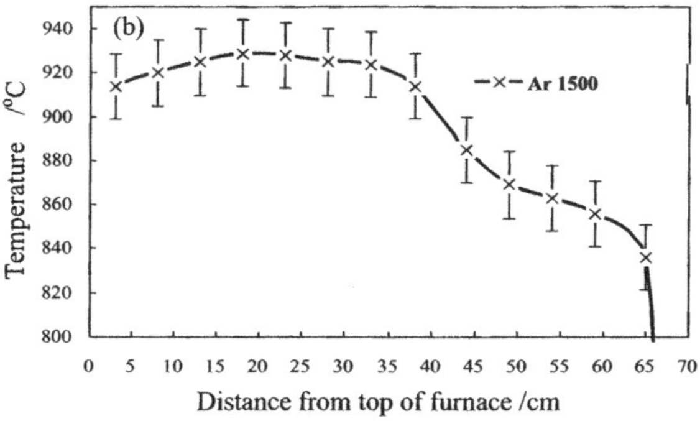

(b) field B

Figure.1 Temperature distributions of field A and field B in deposition furnace

3.2 COMPARSION OF MORPHOLOGIES, COMPOSITION AND MICROSTRUCTURE WITH DIFFERENT FIELDS

The B_4C surface morphologies of field A and field B were shown in **Fig.2**. It can be seen that the temperature fields have important effects on the surface morphologies of deposits. Under the role of field A, the surface morphology was crystal-like and coarse. The edges of surface particles were very obvious. However, the surface morphology was cauliflower-like under the role of field B. The cauliflower-like particles consisted of small particles.

The B_4C cross-section morphologies of field A and field B were shown in **Fig.3**. The temperature fields also have an important influence on cross-sections of deposits. Under the role of field A, the cross-section morphologies were dense, coarse and crystal-like. The fracture surface could be found in the cross-section. The thickness of boron carbide coating was about 17μm. However, the cross-section morphology was dense, smooth, and glass-like under the role of field B. The thickness of boron carbide coating was about 19μm.

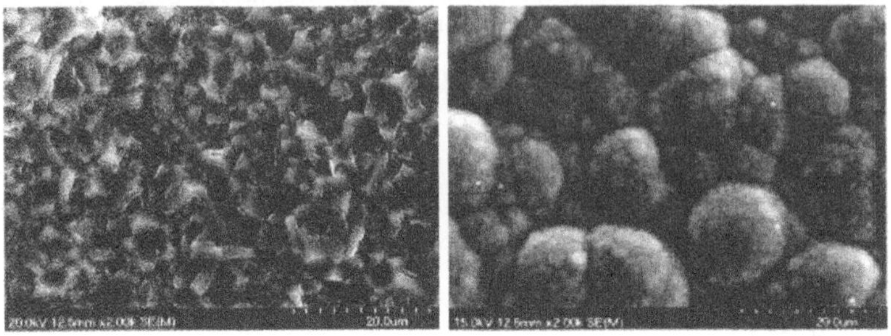

(a) field A (b) field B

Figure.2 SEM photos of surface morphologies for boron carbide deposited at different temperature fields

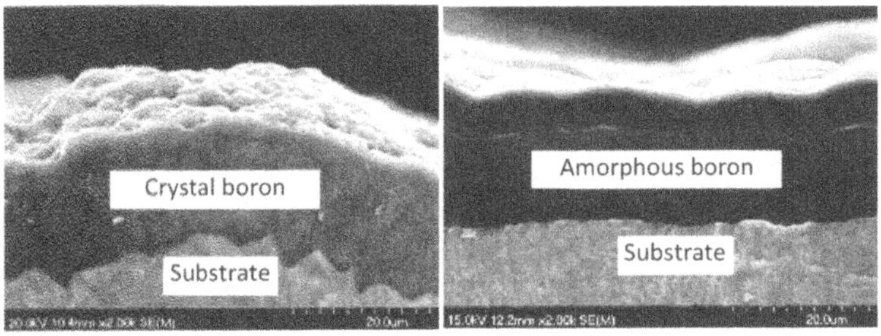

(a) field A (b) field B

Figure.3 SEM photos of cross-section morphologies for boron carbide deposited at different temperature fields

The element compositions of deposits under different temperature fields can be seen in Fig.4. The temperature fields have some effect on the element compositions of deposits. Under the field A, the B concentration was 87.11at% and the C was concentration was 12.89at.%. Under the field B, the B concentration was 75.74at% and the C was concentration was 24.66at.%. Therefore, there are higher B and lower C in deposit under the field A than that under the field B.

According to previous results[16], the deposit was $B_{13}C_2$. The XRD patterns of deposits under field B were shown in Fig.5. Only SiC and PyC peaks were observed in the XRD pattern of field B. No boron carbide peaks could be detected by XRD, which was consistent with J. Berjonneau's results[11, 17-18]. They results showed that the deposit under field B is amorphous.

To know B-C bond state, the FTIR was used to characterize the deposits. The FTIR spectrums of boron carbide deposited at different temperature fields were shown in Fig.6. The main absorb peaks of deposits under two kinds temperature fields were 1082.26 cm^{-1}, 792.13 cm^{-1} and 485.36

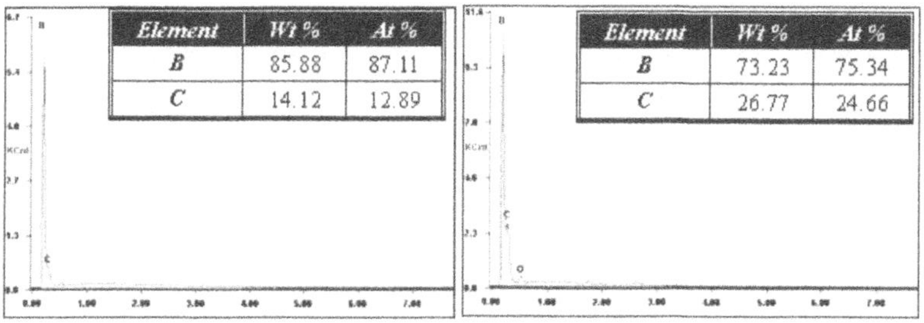

(a) field A (b) field B

Figure. 4 EDS spectrums of boron carbide deposited at different temperature fields

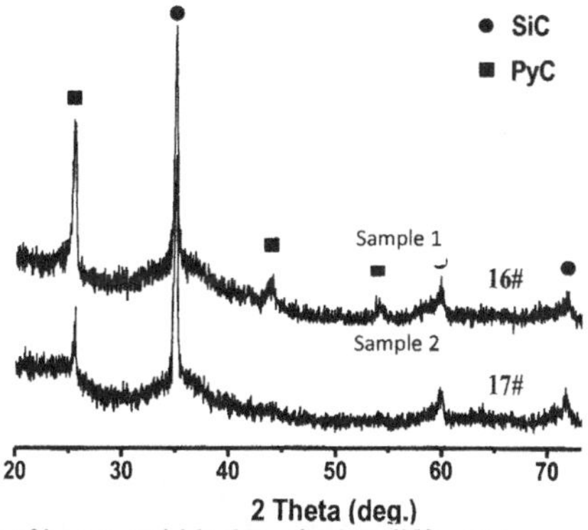

Figure.5 XRD patterns of boron carbide deposited at different temperature fields

cm^{-1}. The standard absorb peak of B_4C were 1100 cm^{-1}, 800 cm^{-1} and 470 cm^{-1}. Therefore, the B-C bond state in deposits under different temperature fields are the same.

In order to investigate the microstructure of boron carbide under field A and field B, TEM studies were performed. Fig.7 showed the microstructure of deposits under different temperature fields. Under field A, the space of (021) plane were 0.24nm, which showed the deposits were crystal $B_{13}C_2$. Under field B, no crystal phase can be found, which showed the deposits were amorphous.

3.3 COMPARSION OF DEPOSITION MECHANISM

The above results showed that the morphology, composition and microstructure were different under two different temperature fields. Under field A, the deposits were crystal $B_{13}C_2$ with high boron concentration and crystalline-like morphologies. Under field B, the deposits were amorphous boron carbide with low boron concentration and cauliflower-like morphologies. The characteristics of deposits depended on deposition mechanism. It was apparent that there were different deposition mechanisms for boron carbide since the characteristics of deposits were very great different under

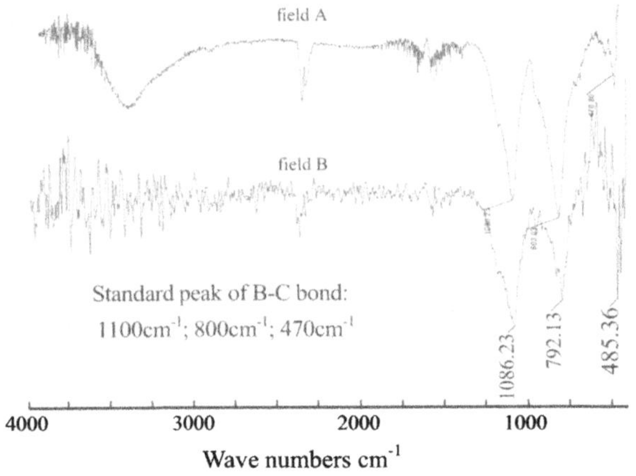

Figure.6 FTIR spectrums of boron carbide deposited at different temperature fields

field A and field B.

Under field A, the deposition mechanism has been discussed in reference[16]. The reaction pathways were as follows:

$$BCl_3+H_2=BHCl_2+HCl \tag{1}$$

$$CH_4+H^* = CH_3+H_2 \tag{2}$$

$$BHCl_2+CH_3 = B_{13}C_2+2HCl+H_2 \tag{3}$$

Under field B, the deposition mechanism can be hypothesized as follows. According to cauliflower-like morphology and amorphous phase, BCl_3 and CH_4 might be reacted before deposition. Therefore, the following reaction might have occurred during deposition process expect for reaction (1) and (2).

$$BHCl_2+CH_3 = B_xC_yH_zCl+HCl \tag{4}$$

$$B_xC_yH_zCl = BC_x+HCl \tag{5}$$

The early reaction between BCl_3 and CH_4 might be the key to the formation of cauliflower-like boron carbide.

4. SUMMARY

We have demonstrated the effects of temperature fields on deposit characteristics. The temperature distribution of field B is more uniform than that of field A. The temperature fields have important effect on the morphologies, phases, microstructure and compositions of deposits. Under the temperature field A, the morphologies were crystalline-like, the boron concentration was 87.11 at.%, the phase of deposit was $B_{13}C_2$ according to XRD and TEM examinations. Under the temperature field B, the morphologies were cauliflower-like, the boron concentration was 75.34 at.%, the phase of deposit was amorphous boron carbide based XRD and TEM. The early reactions between BCl_3 and CH_4 before deposition process might be the key to the formation of amorphous boron carbide.

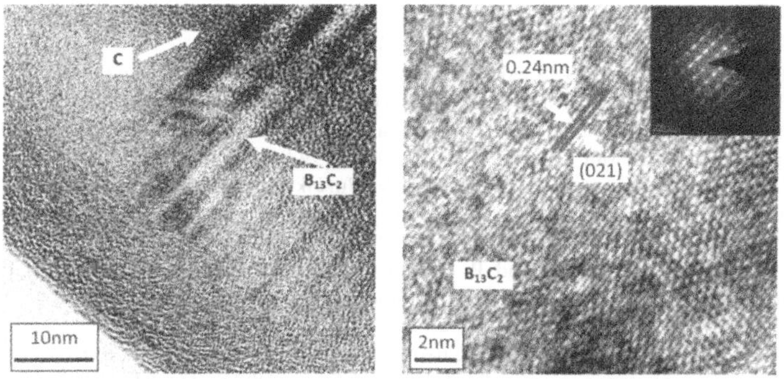

(a) field A

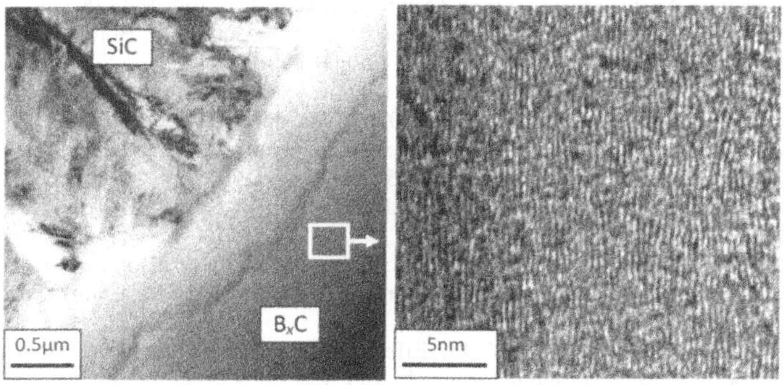

(b) field B

Figure.7 TEM photos of boron carbide deposited at different temperature fields[16]

ACKNOWLEDGMENTS

This work was supported by the National Science Foundation in China (No.90405015, No.50672076, No.50425208, No.50642039). This work was also supported by the Doctorate Foundation of Northwestern Polytechnical University (CX200505).

REFERENCES

1. Q. Ludovic, R. Francis, G. Alain, T. Henri, and L. Caroline, Self-healing mechanisms of a SiC fiber reinforced multi-layered ceramic matrix composite in high pressure steam environments, J. Eur. Ceram. Soc. 27[4], 2085-2094(2007).

2. Q. Liu, X. Xu, Q. Huang, and B. Huang, Anti-oxidation mechanism of SiC-B_4C-C composites, Trans. Nonferrous Met. Soc. China, 15[6], 1346-1350(2005)

3. S. Goujard, L. Vandenbulcke, and H. Tawil, Oxidation behavior of 2D and 3D carbon/carbon thermostructural materials protected by CVD polylayer coatings, Thin Solid Films, 252[2], 120-130(1994).
4. K. Lee, and S. Harris, Boron carbide films grown from microwave plasma chemical vapor deposition, Diamond Relat. Mater., 7[10], 1539-1543(1998).
5. U. Jansson, J. Carlsson, B. Stridh, S. Soederberg, and M. Olsson, Chemical vapour deposition of boron carbides I: Phase and chemical composition, Thin Solid Films, 172[1], 81-93(1989).
6. J. Oliveira, and O. Conde, Deposition of boron carbide by laser CVD: a comparison with thermodynamic predictions, Thin Solid Films, 307[1-2], 29-33(1997).
7. S. Veprek, S. Rambert, M. Heintze, F. Mattenberger, M. Jurčik-Rajman, W. Portmann, D. Ringer, and U. Stiefel, Development of plasma CVD and feasibility study of boron carbide in-situ coatings for tokamaks, J. Nucl. Mat. 162–164, 724-731(1989).
8. H. Küenzli, P. Gartenbein, R. Steiner, and P. Oelhafen, Influence of B_2H_6/CH_4 and $B(CH_3)_3$ as process gas on boron carbide coatings: an in situ photoelectron spectroscopy study, J. Nucl. Mater., 196–198, 622-626(1992).
9. V. Cholet, R. Herbin, and L. Vandenbulcke, Chemical vapour deposition of boron carbide from BBr_3---CH_4---H_2 mixtures in a microwave plasma, Thin Solid Films, 188[1], 143-155(1990).
10. M.J. Santos, A.J. Silvestre, and O. Conde, Laser-assisted deposition of r-B_4C coatings using ethylene as carbon precursor, Surface and Coatings Technology, 151 –152, 160–164(2002).
11. J. Berjonneau, G. Chollon, and F. Langlais, Deposition Process of Amorphous Boron Carbide from $CH_4/BCl_3/H_2$ Precursor, J.Electrochem.Soc. 153[12], C795-C800(2006).
12. S. Noyan Dilek, H. Özbelge, N. Sezgi, and T. Doğu, Kinetic Studies for Boron Carbide Formation in a Dual Impinging-Jet Reactor, Ind. Eng. Chem. Res. 40[3], 751-755(2001).
13. L. Vandenbulcke, Theoretical and experimental studies on the chemical vapour deposition of boron carbide, Ind. Eng. Chem.: Process Des. Dev., 24[4], 568-575(1985).
14. T. Moss, W. Lackey, and K. More, Chemical vapour deposition of $B_{13}C_2$ from BCl_3-CH_4-H_2-Argon mixtures, J. Am. Ceram. Soc., 81[12], 3077-3086 (1998).
15. M. Karaman, H. Özbelge, N. Sezgi, and T. Dogu, Mechanism and characterization studies on boron carbides deposited by chemical vapor deposition technique, Mater. Res. Soc. Symp. Proc., 886, 455-460(2006)
16 Y. Liu, L. Cheng, L. Zhang, W. Yang, and Y. Xu, Effect of carbon precursors on the microstructure and bonding state of a boron–carbon compound grown by LPCVD, Int. J. Appl. Ceram. Technol., 5 [3], 305–312 (2008)
17 J. Berjonneau, F. Langlais, G. Chollon, Understanding the CVD process of (Si)-B-C ceramics through FTIR spectroscopy gas phase analysis. Sur. Coat. Techno., 201[16-17], 7273-7285 (2007)
18 J. Berjonneau, G. Chollon, F. Langlais. Deposition process of Si–B–C ceramics from $CH_3SiCl_3/BCl_3/H_2$ precursor. Thin Solid Films, 516[10], 2848-2857(2008)

EFFECT OF DEPOSITION RATE ON MICROSTRUCTURE AND THERMAL CONDUCTIVITY OF YSZ FILMS PREPARED BY MOCVD

Rong Tu and Takashi Goto
Institute for materials research, Tohoku University, Sendai, 980-8577, Japan

ABSTRACT

Yttria stabilized zirconia (YSZ) films were prepared by a vertical cold-wall type CVD apparatus using $Zr(dpm)_4$ and $Y(dpm)_3$ as source precursors and the correlation among deposition rate, microstructure and thermal conductivity were investigated. The deposition rate (R_{dep}) of YSZ films was controlled in between 2 and 108 μm h^{-1} by changing total pressure in reactor (P_{tot}), deposition temperature (T_{dep}) and precursor concentration (C_{Zr}). The microstructure of YSZ films changed from dense to columnar with increasing R_{dep}. The YSZ films prepared at a deposition rate higher than 50 μm h^{-1} at 1073 K had a columnar structure and each column was consisted of polycrystals. Two kinds of nano-pores around 100 and 10 nm were observed at the grain boundaries and within grains, respectively. The thermal conductivity (κ) of YSZ films decreased with increasing deposition rate (R_{dep}) and then kept at a constant of 0.8 W m^{-1} K^{-1} over R_{dep} = 40 μm h^{-1}, which is about 1/3 of YSZ bulk.

INTRODUCTION

Yttria stabilized zirconia (YSZ) has been employed as thermal barrier coatings (TBCs) on nickel-based superalloys for its chemical inertness, low thermal conductivity and high thermal expansion coefficient compatible with metals [1]. Since the TBCs should be several 100 μm in thickness to realize the thermal barrier effect, the YSZ films intended for TBCs have been mainly fabricated by atmospheric plasma spray (APS) [2] or electron-beam physical vapor deposition (EB-PVD) [3, 4], which can provide thick coatings at a high deposition rate of several 100 μm h^{-1}. On the other hand, although chemical vapor deposition (CVD) is capable of preparing high quality YSZ films with excellent conformal coverage [5~9], deposition rates of conventional CVD were usually too slow to obtain thick coatings.

In CVD process, metal halide precursors such as $ZrCl_4$ have been commonly employed to prepare ZrO_2 films [5]. However, the reaction temperatures for halides were usually higher than 1200 K and the deposition rates were insufficient to obtain thick TBCs. Furthermore, since corrosive by-products such as HCl gas would degrade the metal substrates, halide CVD is not suitable to prepare TBCs. On the other hand, metal-organic complex precursors, such as $Zr(dpm)_4$ (dpm = dipivaloylmethanate) [6] and $Zr(thd)_4$ (thd = 2,2,6,6,-tetramethyl-3,5,-heptanedionate) [7~9] can yield high deposition rates at relatively low temperatures due to their high vapor pressures and reactivities. Since $Zr(dpm)_4$ has much stable and high vapor pressures (1.3 kPa at 553 K) [10], in the previous study, we have selected $Zr(dpm)_4$ and $Y(dpm)_3$ as precursors and constructed a cold-wall type CVD apparatus to achieve high deposition rates of YSZ films [11,12]. In the present study, YSZ films were prepared at different deposition rate by changing precursor concentration and deposition temperature and the correlation among the deposition rate, microstructure and thermal conductivity was investigated.

EXPERIMENTAL

YSZ films were prepared a vertical cold-wall type CVD apparatus [11]. Source precursors of $Zr(dpm)_4$ and $Y(dpm)_3$ were vaporized at 483 to 593 and 393 to 473 K, respectively. The source vapors were carried with Ar gas into the CVD reactor. O_2 gas was separately introduced by using a double tube nozzle, and mixed with precursor vapors in a mixing chamber placed above a substrate holder. The molar concentration (C_i) of Zr and Y precursor vapors was defined as equation (1),

$$C_i = M_i/(M_{Zr} + M_Y + M_{O2} + M_{Ar}) \qquad (1)$$

where the amount of precursors (M_{Zr}, M_Y) were calculated from the mass loss of precursors after experiments. The values of C_{Zr} and C_Y were changed from 0.0005 to 0.02 and 0 to 0.002, respectively. The Y_2O_3 contents in the YSZ films were fixed at 4 mol% by changing the evaporation temperature of $Zr(dpm)_4$ and $Y(dpm)_3$ precursors. The total gas flow rate was fixed at 3.33×10^{-6} m^3 s^{-1}. The total pressure in the CVD reactor (P_{tot}) was controlled between 0.4 to 2.0 kPa. Fused quartz glass plates (10 by 15 by 1 mm) were used as substrates. The crystal structure was studied by X-ray diffraction (XRD). The microstructure and thickness of films was examined by scanning electron microscopy (SEM) and the nanostructure was observed by transmission electron microscopy (TEM). The deposition rate of films was calculated by thickness and deposition time.

The thermal conductivity of YSZ films was measured at room temperature by a LaserPIT system (Ulvac Riko). Figure 1 shows the schematics of the laserPIT system. YSZ films were prepared on a half of an alumina plates (30 by 3 by 0.1 mm). In the laser-heating AC method, thermal diffusivity parallel to specimen surface can be obtained by a laser-irradiated AC calorimetric technique [13]. When a laser-heating AC heat flux is partially applied to the sample surface, AC temperature waves propagate in the specimen. The logarithmic amplitude ($\ln Am$) of the AC thermal wave changed linearly with the position (x). Thermal conductivity was calculated by equation (2).

$$\kappa = \left\{\left(\frac{d_1}{d_2}+1\right)\left(\frac{k_{12}}{k_1}-1\right)+\frac{C_2}{C_1}\right\}\times\frac{\pi f C_1}{k_1^2} \qquad (2)$$

where, d_1 and d_2 are the thickness of substrate and film, C_1, C_2 are for thermal capacity per unit volume of substrate and film, f is frequency of laser pulse.

RESULTS AND DISCUSSION

Figure 2 shows the effect of total pressure (P_{tot}) on the deposition rate (R_{dep}) of YSZ films. The R_{dep} slightly increased with increasing P_{tot} up to 0.8 kPa, showing a maximum, and then decreased significantly with increasing P_{tot}. The decrease of R_{dep} could be resulted from condensation of the precursor or insufficient mixing of precursor and O_2 gases due to viscous gas flow around the substrate. On the other hand, the R_{dep} increased with increasing deposition temperature (T_{dep}).

Figure 3 shows the effect of molar concentration of Zr precursor (C_{Zr}) on the deposition rate at P_{tot} = 0.8 kPa and T_{dep} = 923 and 1073 K. The R_{dep} increased with increasing C_{Zr}. In a chemical reaction limited process, R_{dep} should be independent of molar concentration of precursors. Therefore, the deposition at lower than 1073 K may be a diffusion limited process. The similar relationship between R_{dep} and precursor concentration was also reported in MOCVD of ZrO_2 [8] and SiO_2 [14]. In the present study, a high R_{dep} of 3.0×10^{-8} m s^{-1} (108 μm h^{-1}) was obtained at 1073 K [12], which represents the highest value for YSZ by MOCVD reported in the literature. This value may be competitive to those of APS and EB-PVD (2.7 to 6.8×10^{-8} m s^{-1}) [15].

Figure 4 shows the surface and cross-sectional microstructure of the YSZ films prepared at various R_{dep}. The YSZ film prepared at R_{dep} = 3 μm h^{-1} (P_{tot} = 0.8 kPa, T_{dep} = 923 K, C_{Zr} = 0.14 mol%) showed a fine granular surface texture (Fig. 4(a)) and a dense cross section (Fig. 4(b)). The YSZ film prepared at R_{dep} = 12 μm h^{-1} (P_{tot} = 0.8 kPa, T_{dep} = 1073 K, C_{Zr} = 0.13 mol%) showed a faceted surface texture (Fig. 4(c)) and a columnar cross section (Fig. 4(d)). The YSZ film prepared at R_{dep} = 50 μm h^{-1} (P_{tot} = 0.8 kPa, T_{dep} = 1073 K, C_{Zr} = 1.3 mol%) showed a cauliflower-like surface

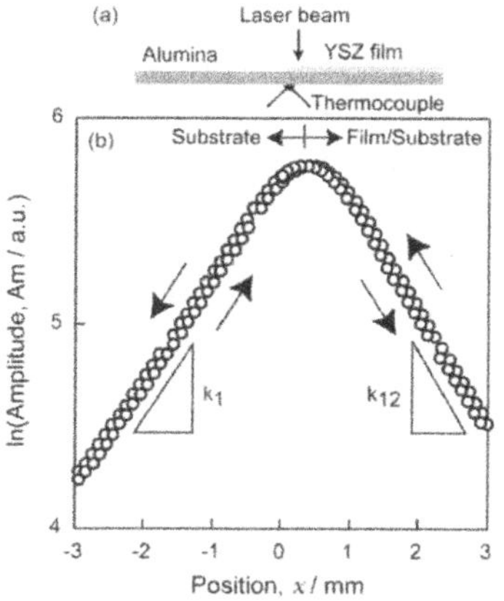

Figure 1. Schematics of thermal conductivity measurement by a laserPIT system.

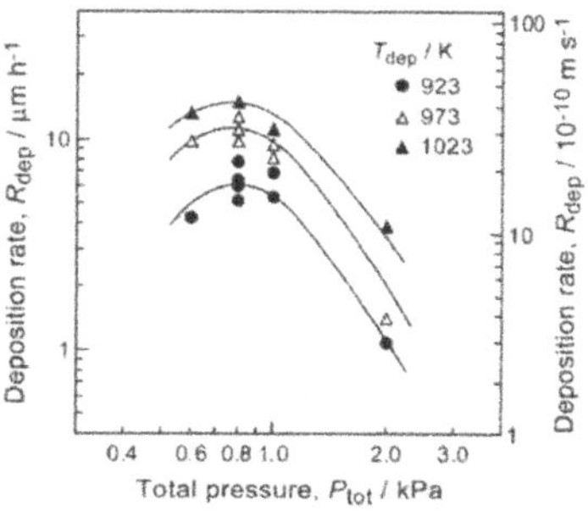

Figure 2. Effect of T_{dep} and P_{tot} on R_{dep} of YSZ films.

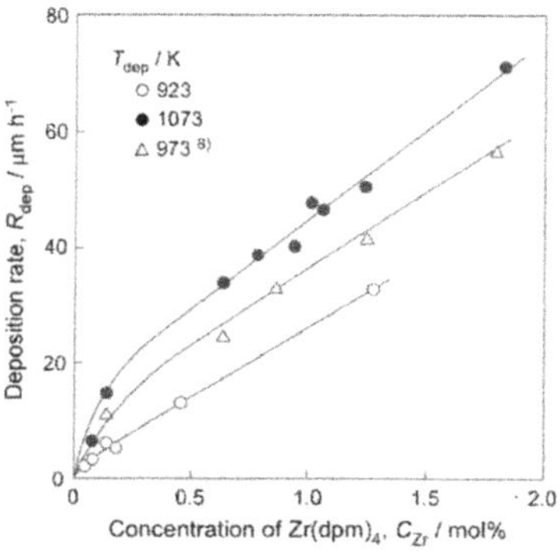

Figure 3. Effect of molar concentration of Zr precursor (C_{Zr}) on R_{dep} at P_{tot} = 0.8 kPa and T_{dep} = 923 to 1073K.

texture (Fig. 4(e)) and a columnar cross section (Fig. 4(f)). The films with a columnar cross section showed a (200) orientation. It is generally understood that the texture changes from dense to columnar to cauliflower-like with increasing T_{dep} and saturation of precursors in the gas phase [16]. In the present study, the change of microstructure is consistent with the trend.

Figure 5 shows the TEM images of the YSZ films prepared at R_{dep} = 5 and 10 μm h^{-1}. The YSZ film prepared at R_{dep} = 5 μm h^{-1} (P_{tot} = 0.8 kPa, T_{dep} = 923 K, C_{Zr} = 0.15 mol%) showed a dense nanostructure several nm in grain size (Fig. 5(a)). The YSZ film prepared at R_{dep} = 10 μm h^{-1} (P_{tot} = 0.8 kPa, T_{dep} = 1073 K, C_{Zr} = 0.12 mol%) showed a typical columnar nanostructure about 100 nm in width and several μm in length (Fig. 5(b)). Each column was almost a crystalline grain.

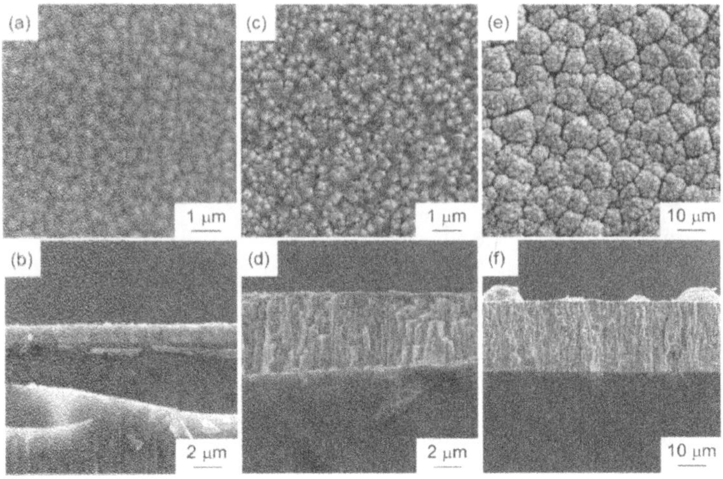

Figure 4. Surface and cross-sectional microstructure of YSZ films prepared at R_{dep} = 3 (a, b), 12 (c, d) and 50 μm h^{-1} (e, f).

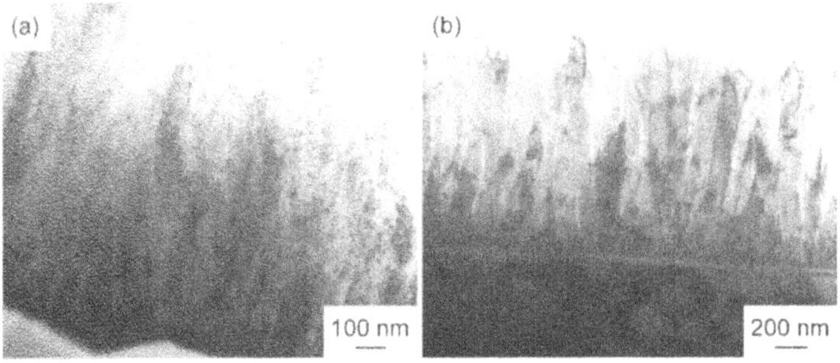

Figure 5. TEM images of YSZ films prepared at R_{dep} = (a) 5 and (b) 10 μm h^{-1}.

Figure 6 shows the TEM images of the YSZ film prepared at R_{dep} = 52 μm h^{-1} (P_{tot} = 0.8 kPa, T_{dep} = 1073 K, C_{Zr} = 1.2 mol%). A columnar nanostructure accompanying with many pores in nanosize was observed (Fig. 6(a)). The nanopores around 100 nm in diameter were dispersed near the substrate and aligned along the growth direction. Each column composed of a great deal of polycrystals about 200 nm in grain size and the nanopores were located at the grain boundaries (Fig. 6(b)). Furthermore, numerous nanopores about 10 nm in diameter were observed within the grains (Fig. 6(c)). We have reported that nano-pores about 10 nm in diameter dispersed in the columnar YSZ films prepared by a laser-assisted CVD [17, 18]. The nano-pores were observed in the case of high precursor concentration and high deposition rate around several 10 to several 100 μm h^{-1}. In the present study, the high deposition rate may result in the formation of nanopores.

Figure 7 shows the relationship between the thermal conductivity (κ) and the R_{dep} of YSZ films. The κ of YSZ films prepared at 5 μm h^{-1} was 1.2 $Wm^{-1}K^{-1}$ and decreased to 0.8 $Wm^{-1}K^{-1}$ with increasing R_{dep} to about 40 μm h^{-1}. The κ at high R_{dep} was about 1/2–1/3 of a bulk YSZ, may be due to the columnar structure to impede the transfer of heat flux and nanopores to scatter the propagation of phonons.

Figure 8 shows the relationship between the porosity (P) and the R_{dep} of YSZ films. The porosity was evaluated by equation (3).

$$P = 1 - \Delta m/(t * A)/D \qquad (3)$$

where Δm is mass gain of substrate after deposition, t is thickness of film, A is surface area of substrate and D is the theoretic density of YSZ. The porosity of YSZ films increased with increasing R_{dep} and kept at 20 to 30% even increasing R_{dep} further. The high porosity in YSZ films prepared at high R_{dep} may be also an effective factor to lower the thermal conductivity. The nanopores in the CVD YSZ films had an excellent thermal stability that will be discussed in another report. The low thermal conductivity and the thermal stability of nanopores in the YSZ films may be advantageous to improve the effect of thermal barrier coatings.

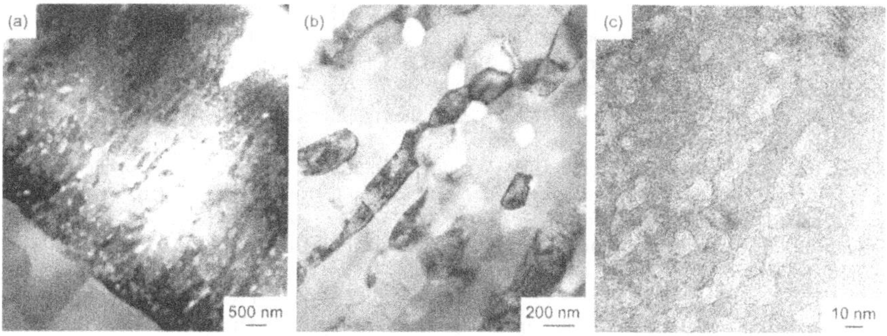

Figure 6. TEM images of YSZ films prepared at R_{dep} = 52 μm h^{-1}.

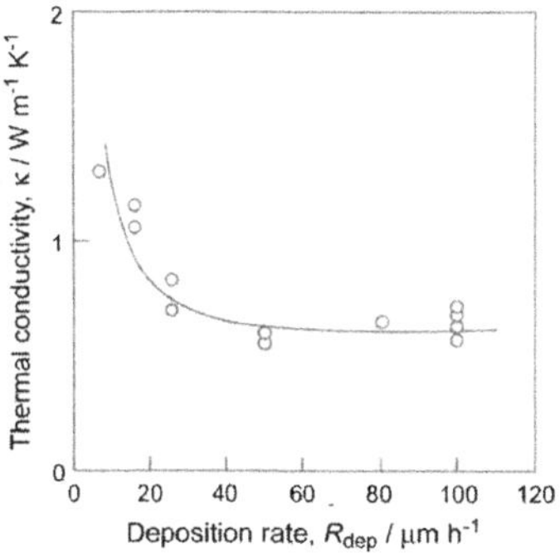

Figure 7. Relationship between thermal conductivity and deposition rate of YSZ films.

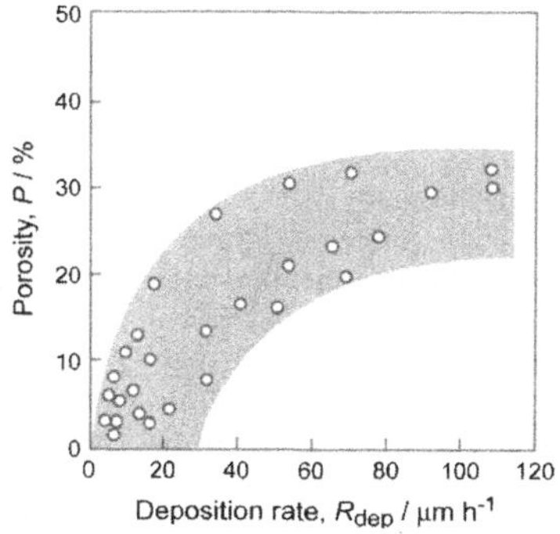

Figure 8. Relationship between porosity and deposition rate of YSZ films.

CONCLUSIONS

The deposition rates (R_{dep}) of the YSZ films prepared by MOCVD increased with increasing deposition temperature (T_{dep}) and precursor concentration (C_{Zr}). The microstructure of YSZ films changed from a dense texture to a single-crystal-like columnar structure and then polycrystalline columnar structure with increasing R_{dep}. In the YSZ film prepared at a high R_{dep} more than 50 μm h^{-1}, a great deal of nanopores about 100 nm in diameter and that about 10 nm in diameter were observed at the grain boundaries and within grains, respectively. The thermal conductivity decreased with increasing R_{dep}, may be related to the phonon scattering effect of nanopores and the increment of porosity.

ACKNOWLEDGEMENTS

This work was financially supported by the Global COE Program of Materials Integration in the International Center of Education Research, and by the Asian CORE Program supported by the Japan Society for the Promotion of Science (JSPS).

REFERENCES

1) W. Beele, G. Marijnissen and A. van Lieshout, The evolution of thermal barrier coatings - status and upcoming solutions for today's key issues, *Surf. Coat. Tech.*, **120-121**, 61-67 (1999).
2) S. Sharafat, A. Kobayashi, Y. Chen, and N. M. Ghoniema, Plasma spraying of micro-composite thermal barrier coatings, *Vacuum*, **65**, 415–425 (2002).
3) U. Schulz and M. Schmücker, Microstructure of ZrO_2 thermal barrier coatings applied by EB-PVD, *Mater. Sci. Eng. A*, **276**, 1-8 (2000).
4) C. Leyens, U. Schulz, B. A. Pint and I. G. Wright, Influence of electron beam physical vapor deposited thermal barrier coating microstructure on thermal barrier coating system performance under cyclic oxidation conditions, *Surf. Coat. Tech.*, **120**, 68-76 (1999).
5) H. Yamane and T. Hirai, Preparation of ZrO_2-film by oxidation of $ZrCl_4$, *J. Mater. Sci. Letters*, **6**, 1229-1230 (1987).
6) Y. Akiyama, T. Sato and N. Imaishi, Reaction Analysis for ZrO_2 and Y_2O_3 Thin-Film Growth by Low-Pressure Metalorganic Chemical-Vapor-Deposition Using Beta-Diketonate Complexes, *J. Cryst. Growth*, **147**, 130-146 (1995).
7) N. Bourhila, F. Felten, J. P. Senateur, F. Schuster, R. Madar and A. Abrutis, Deposition and Characterization of ZrO_2 and Yttria-Stabilized ZrO_2 Films Using Injection-LPCVD, *Proc. 14th Conf. EUROCVD*-**11**, 417-424 (1997).
8) M. Pulver, W. Nemetz and G. Wahl, CVD of ZrO_2, Al_2O_3 and Y_2O_3 from Metalorganic Compounds in Different Reactors, *Surf. Coat. Tech.*, **125**, 400-406 (2000).
9) G. Whal, W. Nemetz, M. Giannozzi, S. Rushworth, D. Baxter, N. Archer, F. Cernuschi and N. Boyle, Chemical Vapor Deposition of TBC: An Alternative Process for Gas Turbine Components, *Trans. ASME* **123**, 520-524 (2001).
10) J. A. Belot, R. J. McNeely, A. Wang, C. J. Reedy and T. J. Marks, Expedient route to volatile zirconium metal-organic chemical vapor deposition precursors using amide synthons and implementation in yttria-stabilized zirconia film growth, *J. Mater. Res.*, **14**, 12-15, (1999).
11) R. Tu, T. Kimura and T. Goto, Rapid synthesis of yttria-partially-stabilized zirconia films by metal-organic chemical vapor deposition, *Mater. Trans.*, **43**, 2354-2356 (2002).
12) R. Tu, T. Kimura and T. Goto, High-speed deposition of yttria stabilized zirconia by MOCVD, *Surf. Coat. Tech.*, **187**, 238-244, (2004).
13) F. Takahashi, T. Mori, Y. Hamada, I. Hatta, AC calorimetric thermal diffusivity measurement in relatively thick samples by a distance-variation method, *Jpn. J. Appl. Phys.*, **38**, 4741-4744, (2001).
14) J. Arndt, G. Wahl, Kinetic study of low pressure chemical vapor deposition of SiO_2 using

tetraethoxysilane (TEOS), *Chemical Vapor Deposition. Proceedings of the Fourteenth International Conference and EUROCVD-11*, 147-154, (1997).

15) B. A. Movchan, EB-PVD technology in the gas turbine industry: Present and future, JOM-J. Miner. Metal. Mater. Soc., **48**, 40-45, (1996).

16) C. E. Morosanu, *Thin Films by Chemical Vapor Deposition*, ELSEVIER, p.101 (1990).

17) T. Kimura and T. Goto, Thermal conductivities of yttria-stabilized zirconia films measured by a laser-heating AC method, *Surf. Coat. Tech.* , **198**, 129-132 (2005).

18) T. Kimura, R. Tu and T. Goto, Nano-structure of YSZ films prepared by laser CVD, *J. Jpn. Institute Metals*, **69**, 12-16 (2005).

VIII. Novel Processing of Ceramics

PREPARATION OF Na-β"-Al_2O_3 GREEN BODIES THROUGH NONAQUEOUS GEL-CASTING PROCESS

Xiaogang Xu, Zhaoyin Wen*, Ning Li, Xiangwei Wu, Jiu Lin, Zhonghua Gu
Shanghai Institute of Ceramics, Chinese Academy of Sciences, Shanghai 200050, P. R. China
*Corresponding author: Tel: +86-21-5241-2044 E-mail: zywen@mail.sic.ac.cn

ABSTRACT:

A nonaqueous gel-casting process was applied to the forming of Na-β"-Al_2O_3 green bodies. Ethanol and n-butanol were used as solvents, methacrylamide (MAM) was used as a monomer, and N,N'-methylenebisacrylamide (MBAM) as a crosslinker. The precursor powder of Na-β"-Al_2O_3 with a mean particle size of 4.3 μm and a specific area of 4.1 m^2/g was prepared via solid state reaction. The effects of dispersant content and solid loading on the rheological properties of the ceramic slurries were studied systematically. It was found that most of the slurries exhibited a shear thinning behavior at low shear rates and a shear thickening behavior at high shear rates. Only the slurries with solid loadings no more than 35 vol% displayed a shear thinning behavior over the entire measuring shear rate range. Microstructure of the green bodies was investigated using scanning electron microscopy.

KEYWORDS: Na-β"-Al_2O_3; gel-casting; forming; slurries; rheology; SEM

INTRODUCTION:

The Gel-casting process, which is a combination of polymer chemistry, colloidal chemistry and ceramic technology, has drawn worldwide attention for the past few decades [1, 2]. In such a process, a stable and flowable slurry is first prepared by dispersing ceramic powders into a premixed monomer solution. After mixing for several hours, an initiator is added and the slurry is casted into a mould. The monomer and crosslinker copolymerize in situ at a certain temperature and the ceramic green parts are therefore obtained [3, 4]. Compared to the conventional ceramic forming techniques such as dry pressing, slip-casting and injection molding, gel-casting is a new ceramic forming technique with several distinctive advantages: near-net-shape forming, homogeneous green parts and machinability of the green because of high strengths [5, 6].

Na-β"-Al_2O_3 is a well-known excellent sodium ion conductor. It exhibits a conductivity over $1*10^{-2}$ s/cm at room temperature, making it the best solid electrolyte material for the sodium-sulfur battery [7]. This kind of battery has been recognized as a potential candidate for the energy storage applications, because of its various significant advantages such as high power and energy density, low material cost and low rate of self-discharge [8]. The Na-β"-Al_2O_3 electrolyte is the central part of the sodium-sulfur battery, and plays a particularly important role in the whole system. For such an application, specifically shaped Na-β"-Al_2O_3 ceramics need to be prepared through the ceramic forming processes. In this study, we try to prepare the Na-β"-Al_2O_3 green bodies with the nonaqueous gel-casting technique.

Since the early 1990's when gel-casting was first reported by Omatete et al, lots of studies have been conducted on aqueous gel-casting, which use deionized water as the solvent [1, 9]. However, the precursor powder of Na-β"-Al_2O_3 hydrolyze intensely and produce large amounts of OH^- anions in an aqueous system, making it difficult to prepare the water based slurries. For this reason, organic solvents are first introduced into the gel-casting of Na-β"-Al_2O_3 green parts in present work. Compared to the deionized water, organic solvents have several advantages, e.g. low surface tension and surface energy, excellent wetting property, etc, and these characteristics are considered to be beneficial for the slurry preparation and solidification during the whole process [10].

EXPERIMENTAL

The precursor powder of the Na-β"-Al_2O_3, i.e. $Li_2O{\cdot}5Al_2O_3$ and $Na_2O{\cdot}5Al_2O_3$, was prepared via a solid state reaction method. Stoichiometric amounts of $Li_2C_2O_4$ and $Na_2C_2O_4$ were mixed with α-Al_2O_3 by ball-milling for 24 h. The slurry was dried in air, and then calcined at 1250 °C for 2 h. The resultant powder was ball-milled for another 1 h in order to break the agglomerates. The mean particle size and specific area of the powder were determined by a laser particle size analyzer (Mastersize 2000, Malvern Ins. Ltd., UK). The hydrolyzation behavior of the powder in an aqueous system was studied.

In the nonaqueous gel-casting process, ethanol and n-butanol were used as solvents, monofunctional methacrylamide (MAM) was used as a main monomer, difunctional N,N'-methylenebisacrylamide (MBAM) as a coupling agent, and triethanolamine (TEA) as a dispersant. The precursor solution contained 10 wt% organic chemicals, and the mass ratio of MAM to MBAM was set as 3:1. The slurries of the precursor powder, with solid loadings ranging from 30 to 45 vol%, were prepared by ball-milling. The rheological properties of the slurries were studied by a rotary rheometer (RVDVIII+, Brookfield, USA). After casting, gelling and drying, the resultant green bodies with a size of 8 × 8 × 40 mm were obtained. The microscopic morphology was observed by scanning electron microscopy (SEM, JSM-6700F, JEOL, Japan).

RESULTS AND DISCUSSION

Fig.1 shows the XRD patterns of the precursor $Li_2O{\cdot}5Al_2O_3$, $Na_2O{\cdot}5Al_2O_3$ and the resulting Na-β"-Al_2O_3. It was indicated that the $Li_2O{\cdot}5Al_2O_3$ presented a spinel structure, and the $Na_2O{\cdot}5Al_2O_3$ exhibited a structure similar to the resultant Na-β"-Al_2O_3. The pure Na-β"-Al_2O_3 was obtained after the mixture of $Li_2O{\cdot}5Al_2O_3$ and $Na_2O{\cdot}5Al_2O_3$ was calcined at 1600 °C.

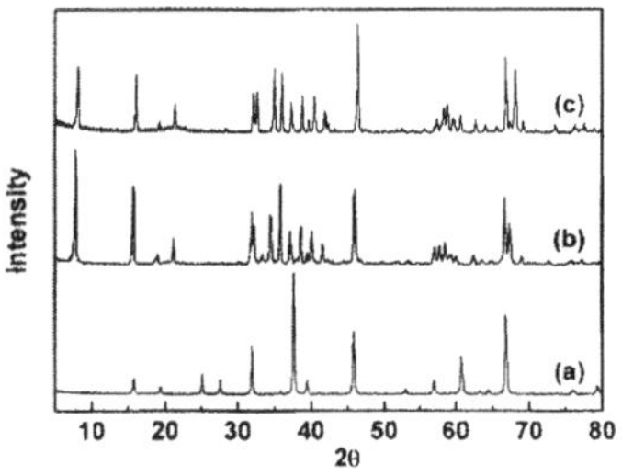

Fig.1. XRD patterns of (a) $Li_2O{\cdot}5Al_2O_3$, (b) $Na_2O{\cdot}5Al_2O_3$ and (c) Na-β"-Al_2O_3

It is well known that the starting materials, including the ceramic powder, the solvent, the monomer and crosslinker, etc, all have a significant impact on the forming process. C. G. Ha et al investigated the effects of the particle size on the rheological properties and gelation behaviors of Al_2O_3 slurries during the gel-casting process [11]. Generally, micro and sub-micro sized ceramic powders are believed to be favorable for obtaining stable and homogeneous slurries. The particle size distribution of the precursor powder is given in Fig. 2. A bimodal particle size distribution with the first peak centering around 0.56 μm and the second peak at about 4.0 μm was dispalyed. The minimum primary particle had a size smaller than 0.05 μm and the maximum agglomerate larger than 30 μm. The mean particle size and specific area were 4.3 μm and 4.1 m^2/g, respectively.

The pH value as a function of the powder content in an aqueous system is shown in Fig. 3. It was clear that, with the increase of the powder content, the alkalinity of the suspension became higher. The pH value reached as high as 11.7 as the concentration of the powder rose to 10 g/L. A further study indicated that the OH^- anions in the suspension were mainly produced by the hydrolyzation of the $Na_2O{\cdot}5Al_2O_3$. If only $LiAl_5O_8$ was dispersed in the water, the hydrolysis behavior was not observed. The different performances between $Na_2O{\cdot}5Al_2O_3$ and $Li_2O{\cdot}5Al_2O_3$ were thought to be relevant to their different crystal structures. The $Li_2O{\cdot}5Al_2O_3$ presented a spinel structure, while the $Na_2O{\cdot}5Al_2O_3$ was a multiphase mixture.

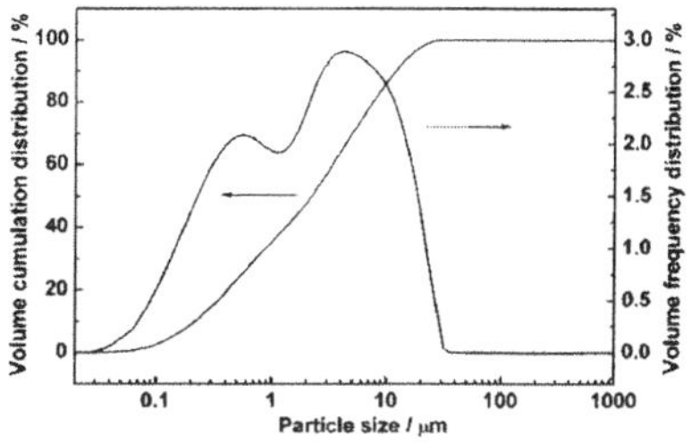

Fig.2. Particle size distribution of the precursor powder

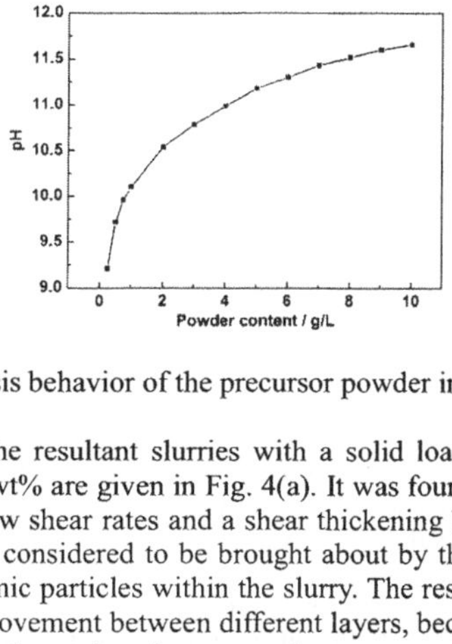

Fig. 3. Hydrolysis behavior of the precursor powder in aqueous system

The rheological curves of the resultant slurries with a solid loading of 40 vol% and dispersant contents of 1.7, 2.0 and 2.3 wt% are given in Fig. 4(a). It was found that all the slurries exhibited a shear thinning behavior at low shear rates and a shear thickening behavior at high shear rates. The shear thinning behavior was considered to be brought about by the forming of a two-dimensional layered structure of the ceramic particles within the slurry. The resistance to flow, primarily caused by the particle and solvent movement between different layers, became lower with the shear action. Therefore, a decrease of the shear viscosity versus shear rate was observed. However, the layered arrangement of the particles was unstable, and could be disrupted above a critical shear rate. The ensuing random arrangement of the ceramic powder induced an increase of the viscosity, and the shear thickening behavior appeared [12]. As shown together in Fig.4, the slurry with a dispersant content of 2.0 wt%, compared to the other two samples, exhibited the lowest viscosity and shear stress over the whole measuring range. This suggested that 2.0 wt% is an appropriate dispersant amount for the slurries with ethanol as the solvent.

For the gel-casting process, the solid loading of a slurry has a direct influence on the density of the green parts. On the other hand, a low viscosity of the slurry is beneficial for both milling and casting processes. Therefore, it is important to achieve a solid loading as high as possible while maintaining a proper fluidity of the slurry. Fig. 4(b) shows the effects of solid loading on the rheological properties of the slurries. As expected, both the apparent viscosity and the shear stress increase obviously with the solid loading rose from 40 to 45 vol%. The shear viscosity curves displayed similar trends, i.e. a shear thinning behavior at low shear rates and shear thickening at high shear rates. Furthermore, the shear stress versus shear rate depicted in Fig. 4 had also been analyzed using the Herschel-Buckley model [13]:

$$\tau = \tau_0 + k\,\gamma^{n}$$

where τ is the shear stress, τ_0 the yield stress, k a consistency coefficient, γ the shear rate and n the flow behavior index. The values of the flow behavior index (n), determined by the software provided with the equipment, were all greater than unity, which was a typical characteristic of a dilatant fluid that always showed a shear thickening behavior.

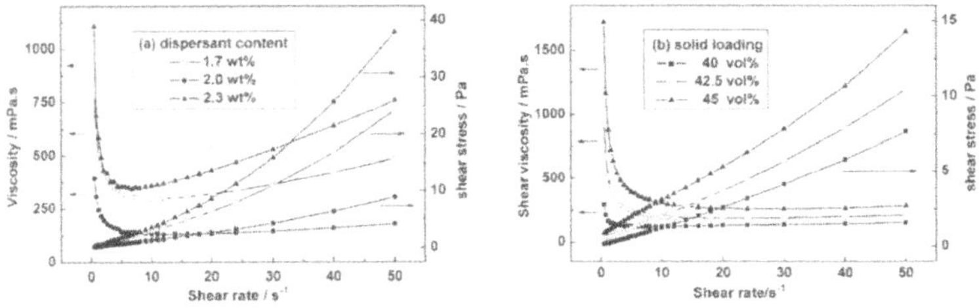

Fig. 4. Effects of (a) dispersant content and (b) solid loading on the shear viscosity and shear stress of the slurries with ethanol as the solvent

For the slurry preparation, n-butanol was also used as the solvent. Compared with ethanol, it has a higher boiling point and a lower vapor pressure, which are good for the gelling and drying of the wet green bodies. Fig. 5(a) illustrates the shear viscosity and shear stress as a function of shear rate for the slurries with dispersant concentrations of 1.0, 1.5 and 2.0wt% respectively. It was found that the viscosity of the slurries varied in the same manner as those with ethanol as the solvent. The slurries containing 1.0 and 1.5 wt% dispersants exhibited nearly the same viscosity and shear stress, especially in the low shear rate region. However, an increase of the dispersant content to 2.0 wt% resulted in poor rheological performance, the fluidity property of the slurry was damaged by the excess dispersant. Consequently, a value of 1.5 wt% was chosen as an optimum dispersant amount for the slurries with n-butanol as the solvent.

The influence of solid loading on the apparent viscosity and shear stress for the n-butanol based ceramic slurries are shown in Fig. 5(b). Different from the one with a solid loading of 40 vol%, the slurries containing 30 and 35 vol% ceramic powders showed only a shear thinning behavior over the whole measuring shear rate range. The flow behavior index (n), calculated according to the Herschel-Buckley model, was less than unity. The results suggested that both the slurries belonged to the family of pseudoplastic fluid. Besides, it was proven extremely difficult to achieve a 45 vol% slurry. The precursor solvent of n-butanol could barely wet all the ceramic powders, and the slurry was paste and nonflowable.

Fig. 6 shows the SEM micrographs of the resultant green bodies prepared respectively with ethanol and n-butanol as the precursor solvent. Both of the green samples were prepared from the slurry with a solid loading of 40 vol%, and the dispersant amounts for the ethanol and n-butanol based slurries were 2.0 wt% and 1.5 wt%, respectively. The polymer network, usually formed by the copolymer of MAM and MBAM, was hardly visible in either of the pictures. Distinctive differences were obviously observed between the two green samples. In Fig. 6(a), the ceramic particles within the green parts were clear and distinguishable, a few pores were found and the green bodies displayed a lower density. And in Fig. 6(b), the particles were held very close to each other, a homogeneous and compact microstructure without apparent agglomerates and pores was therefore obtained.

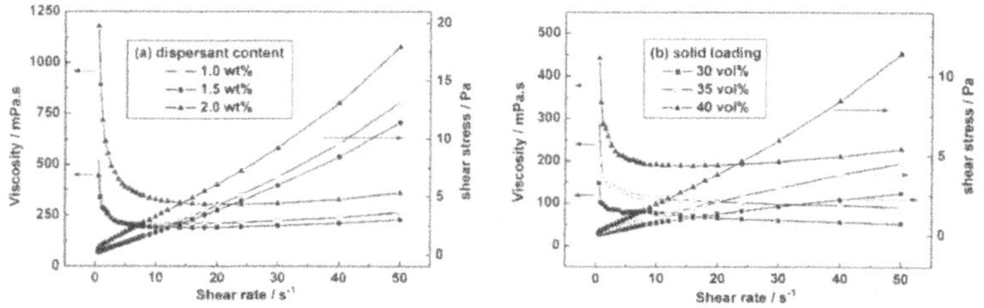

Fig. 5. Influences of (a) dispersant content and (b) solid loading on the rheological properties of the slurries with n-butanol as the solvent

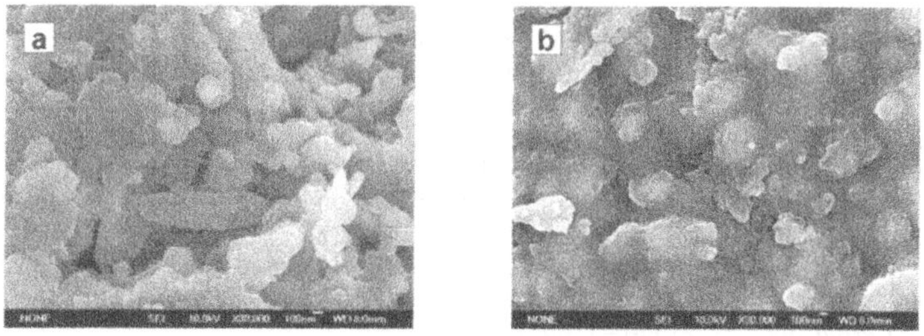

Fig. 6. SEM micrographs of the green parts prepared with (a) ethanol and (b) n-butanol as the precursor solvent

CONCLUSIONS

A nonaqueous gel-casting process using ethanol and n-butanol as precursor solvent was successfully applied to the forming of Na-β"-Al_2O_3 green bodies. The precursor powder of Na-β"-Al_2O_3, generally prepared by the solid state reaction, possessed a bimodal particle size distribution with mean particle size of 4.3 μm and specific area of 4.1 m^2/g. When added into water, it displayed strong basic characteristics, which made it difficult to prepare an aqueous slurry with a high solid loading. It was found that, 2.0 and 1.5 wt% were considered as the proper dispersant amounts for ethanol and n-butanol based slurries respectively. The slurry using ethanol as the solvent could achieve a higher solid loading than the one with n-butanol. However, different performances were observed in the SEM micrographs. The green parts with n-butanol as the precursor solvent showed a compact microstructure with few agglomerates and pores.

ACKNOWLEDGEMENT

This work is financially supported by the "973" Project (2007CB209700) of the Ministry of Science and Technology of China and key project of Natural Science Foundation of China (NSFC, No.50730001).

REFERENCES

[1]O. O. Omatete, M. A. Janney, and R. A. Strehlow, Gelcasting—A new ceramic forming process, *Ceram. Bull.*, 70, 1641-9 (1991).

[2]A. C. Young, O. O. Omatete, M. A. Janney, and P. A. Menchhofer, Gelacsting of alumina, *J. Am. Ceram. Soc.*, 74, 612-8 (1991).

[3]I. Santacrus, C. Baudin, M. I. Nieto, and R. Moreno. Improved green properties of gelcast alumina through multiple synergistic interaction of polysaccharides, *J. Eur. Ceram. Soc.*, 23, 1785-93 (2003).

[4]F. Li, H. Y. Chen, R.Z.Wu, and B.D.Sun, Effect of polyethylene glycol on the surface exfoliation of SiC green bodies prepared by gelcasting, *Mater. Sci. Eng. A*, 368, 255-9 (2004).

[5]O. O. Omatete, M. A. Janney, and S. D. Nunn, Gelcasting: From laboratory development toward industrial production, *J. Eur. Ceram. Soc.*, 17, 407-13 (1997).

[6]J. Tong, and D. Chen, Preparation of alumina by aqueous gelcasting, Ceram. Int., 30, 2061-6 (2004).

[7]K. Terabe, S. Yamaguchi, and Y.Iguchi, Characterization of sodium β-alumina prepared by sol-gel method, *Solid state ionics,* 40-41, 111-4 (1990).

[8]Z. Wen, J. Cao, Z, Gu, X, Xu, F, Zhang, and X, Lin, Research on sodium sulfur battery for energy storage, *Solid state ionics*, 179, 1697-701 (2008).

[9]S. L. Morissette, and J. A. Lewis, Chemorheology of aqueous based alumina-poly(vinyl alcohol) gelcasting suspension, *J. Am. Ceram. Soc.*, 82, 521-8 (1999).

[10]A. Muller, F. Yu, and M. Willert-Porada, Cellulose acetate gelcasting process for Gd-containing ceramic bodies, *J. Eur. Ceram. Soc.*, 26, 2743-51 (2006).

[11]C. H. Ha, Y. G. Jung, J. W. Kim, C. Y. Jo, and U. Paik, Effect of particle size on gelcasting process and green properties in alumina, *Mater. Sci. Eng. A*, 337, 212-21 (2002).

[12]H. A. Barnes, J. F. Hutton, and K. Walers, An introduction to rgeology, *Elsevier science publishers B. V.*, 1989.

[13]D. T. Beruto, A. Ferrari, F. Barberis, and M. Giordani, Dispersions of micrometric powders of molybdenum and alumina in liquid paraffin: role of interfacial phenomena on bulk rheological properties, *J. Eur. Ceram. Soc.*, 22, 2155-64 (2002).

ROD-LIKE β-SIALON POWDER PREPARED BY A NEW N_2-ASSISTED CARBOTHERMAL REDUCTION OF CARBON AND ALUMINUM NANOCASTED MESOPOROUS SILICA

Tongping Xiu[1,2], Qian Liu[1]*, Minghui Wang[1,2], and Qiang Yan[1,2]
[1] State Key Laboratory of High Performance Ceramics and Superfine Microstructure, Shanghai Institute of Ceramics, Chinese Academy of Sciences, Shanghai 200050, P. R. China.
[2] Graduate School of the Chinese Academy of Sciences, Beijing 100049, P. R. China

ABSTRACT

A kind of β-Sialon powder was prepared by the N_2-assisted carbothermal reduction of carbon and aluminum nanocasted mesoporous silica, where the mesoporous silica was prepared by the hydrothermal method using triblock copolymer as the template. For the nanocasting procedure, by using sucrose and aluminum chloride as the starting precursor of carbon and aluminum, the mesoporous silica was impregnated with carbon and aluminum species in its nano-sized channels. After carbonization in nitrogen at 800 °C for 6 hours, the composite was heated again at 1450 °C in nitrogen for another 6 hours to undergo a carbothermal process and finally at 600 °C in air for 6 hours to remove residual carbon. X-ray diffraction and high resolution transmission electron microscope analysis results showed that a β-Sialon ($Si_3Al_3O_3N_5$) phase was formed in the resulting powder and the powder was of high crystalline state. Scanning electron microscopy confirmed that the morphology of the powder was rod-like. In addition to the rod-like β-Sialon, some SiC whiskers were also found as a by-product.

INTRODUCTION

The β-Sialon is a solid solution of β-Si_3N_4 that has some of Si and N atoms in Si_3N_4 replaced by Al and O, respectively. The solid solution has a formula of $Si_{6-z}Al_zO_zN_{8-z}$ (β-Sialon, $0<z\leq4.2$). Meanwhile, Sialon ceramics exhibit high strength and hardness, and are highly resistant to oxidation and corrosion. The development of high-quality Sialon powders is essential if the structural reliability of the corresponding ceramics is to be improved. Therefore, Sialon powders have been largely synthesized through carbothermal reduction–nitridation (CRN) of silicate based oxides for many years. There are several reports which have shown that β- Sialon can be synthesized from mixed powders of kaolinite ($Al_2Si_2O_5(OH)_4$),[1] or powder mixture of SiO_2–$Al_2O_3{\cdot}2H_2O$,[2] or powder mixtures of SiO_2–Al_2O_3,[3] or an alkoxide-derived SiO_2–Al_2O_3 co-precipitate,[4] or SiO_2–Al_2O_3 gel,[5] and Y-zeolite,[6] all in the presence of carbon in flowing N_2 gas. In this study, we focus our attention on using mesoporous silica (SBA-15) as the prime source for preparing β-Sialon powder.

SBA-15 has larger pore size (normally 4 to 6 nm) and high surface areas, and is widely studied.[7] Due to its larger pore size, carbon and aluminum species can be easily introduced into its nanosized channels in SBA-15. It is also possible for the impregnated aluminum species to penetrate into the whole framework of silica at high temperature,[8] because the mesoporous silica framework is also in the nanoscale (3 to 4 nm). All these advantages make SBA-15 a useful starting material for the fabrication of SiAlON powders. So herein we report the synthesis of β-Sialon using SBA-15 as the starting materials by a CRN method.

EXPERIMENTAL

Preparation of SBA-15: The pure silica SBA-15 mesoporous material was made according to the literature[7] using tri-block poly(ethylene oxide)–poly(propylene oxide)–poly(ethylene oxide), $(EO)_{20}(PO)_{70}(EO)_{20}$, as a template in acidic conditions. For example, 4 g of $[(EO)_{20}(PO)_{70}(EO)_{20}]$ was dissolved into the solution of 30 mL of deionized water and 120 ml of HCl (2 mol L^{-1}), followed by the addition of 9.15 ml of tetraethyl orthosilicate (TEOS, >98%) slowly. Then the mixture was maintained at 35–40 °C for 24 h under a stirred condition, later it was loaded into an autoclave and heated at 100 °C for 48 h. The solid product was recovered by filtration, washed with a large amount of deionized water. After drying at 80 °C for 24 h, the solid powder material was calcined in air atmosphere at 500 °C for 4 h to remove the template.

Nanocasting of SBA-15: The impregnation of carbon and aluminum species into the channels of SBA-15 was carried out as follows. 1 g of SBA-15 was added to a solution obtained by dissolving 2.25 of anhydrous aluminum chloride,1.25 g of sucrose and 0.14 g of H_2SO_4 in 5 g of H_2O. The mixture was placed in a drying oven for 6 h at 80 °C , and subsequently the oven temperature was increased to 160 °C and maintained there for 6 h. The sample turned dark brown or black during the treatment in the oven. The silica sample, containing partially polymerized and carbonized sucrose at the present step, was treated again at 80 and 160 °C using the same drying oven after the addition of 0.8 g of sucrose, 0.09 g of H_2SO_4 and 5 g of H_2O. The carbonization was completed by pyrolysis with heating to typically 800 °C in nitrogen flow.

Carbothermal reduction–nitridation: The process was carried out by heating the impregnated SBA-15 slowly in a carbon tube furnace to 1400-1450 °C for 4-6 h under N_2 at a flow rate of 600 ml/min. The products are denoted as S-T-t (where T refers to the temperature, t refers to the holding time). Excess carbon was removed by calcining the product in air at 600 °C for 6 h.

RESULTS AND DISSCUSSIONS

Figure 1 shows the XRD patterns of the final product calcined at different temperatures. In S-1400-4, a broad amorphous peak can be seen together with several crystalline peaks, which means there still exist some unreacted amorphous silica in the final product. Those crystalline peaks can be assigned to the diffraction peaks of β-$Si_3Al_3O_3N_5$ (JCPDF:79-0483). While prolonging the holding time, the crystallization is much better, with a β-$Si_3Al_3O_3N_5$ phase the main phase in S-1400-6. At higher temperature, almost no other phases except the β-$Si_3Al_3O_3N_5$ phase can be detected in S-1450-6, which indicates that the product is totally crystalline. In addition, the molar ratio of Si/Al got from β-$Si_3Al_3O_3N_5$ is in accordance with the starting materials.

Figure 2 shows the HRTEM image of S-1450-6. Figure 2 also shows a very good crystalline structure of hexangular β-$Si_3Al_3O_3N_5$ along the [212] direction. EDS results show that the Si/Al is 1.1 in the product, which is almost the same with the starting materials.

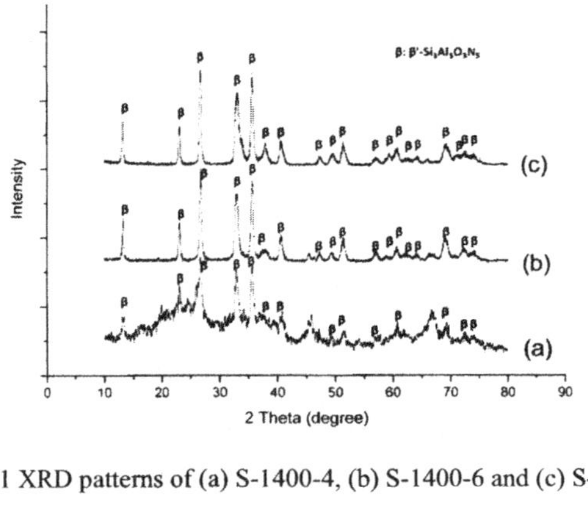

Figure 1 XRD patterns of (a) S-1400-4, (b) S-1400-6 and (c) S-1450-6.

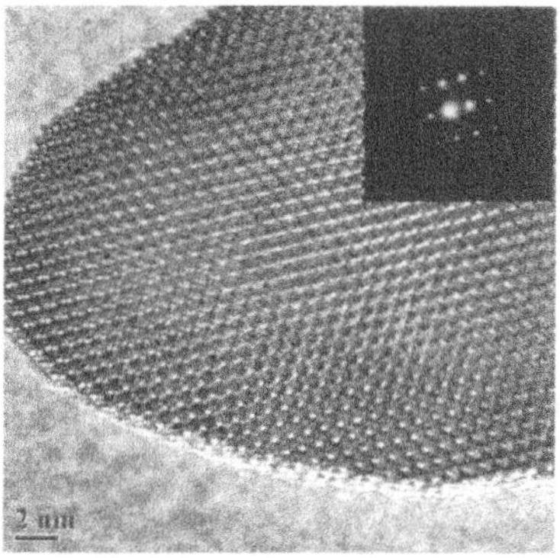

Figure 2 HRTEM image and SEAD pattern (inset) of S-1450-6.

Morphology of the powders can be seen in the SEM observation, which is shown in Figure 3. Some rod-like powders can be seen in the picture. The diameter of the rod is around 10 μm and the length is about 50 um. In addition to the rod-like morphology, some irregular powders are also observed. It is well known that the mother SBA-15 powder has a rod-like morphology,[7] so the existence of some irregular morphology suggests the collapse of the framework of the some SBA-15 at higher carbothermal reduction temperature.

It should be mentioned that, some crystal whiskers can also be observed in the SEM picture (Figure 3). HRTEM image confirms that the whisker is 4H-α-SiC grains (not shown). However, the further research is still needed to confirm the whisker formation mechanism.

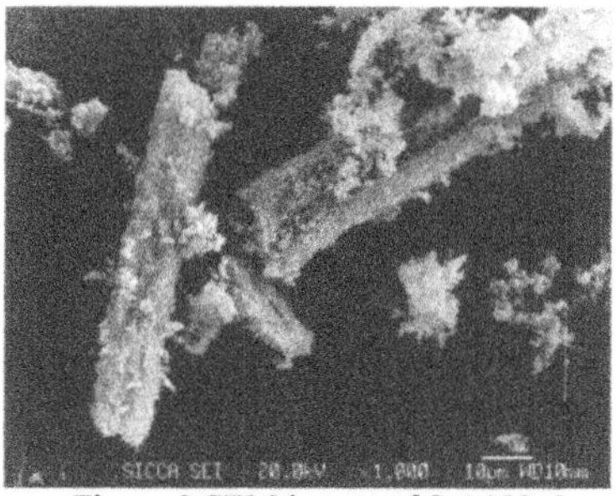

Figure 3 SEM image of S-1450-6

CONCLUSION

In conclusion, a kind of rod-like β-Sialon (β-$Si_3Al_3O_3N_5$) powder was prepared through a carbothermal reduction–nitridation method using mesoporous silca as the starting material, combined with a carbon and aluminum nanocasting procedure. The resultant β-Sialon powder can be expected to find its use in preparing high performance Sialon ceramics.

FOOTNOTES

Corresponding author *E-mail: qianliu@sunm.shcnc.ac.cn

REFERENCES

[1]H. Yoshimatsu, M. Mitomo and H. Mihashi , The Preparation of Sialon Powder from Kaolinite, *J. Ceram. Soc. Jpn.*, 91, 24–31 (1983).

[2]H. Yoshimatsu, M. Mitomo and H. Mihashi, Preparation of Sialon Powder from Mixture of SiO_2 and Al_2O_3-$2H_2O$ by Thermal Carbon Reduction, *J. Ceram. Soc. Jpn.*, 95, 28–32 (1987).

[3]M. Sopicka-Lizer, R. A. Terpstra and R. Metselaar, Carbothermal Production of β-Sialon from Alumina Silica and Carbon Mixture. *J. Mater. Soc.*, 30, 6363–9 (1995).

[4]M. Mitomo, T. Shiogai, H. Yoshimatsu and Y. Kitami, Preparation of Sialon Powder from Alkoxides, *J. Ceram. Soc. Jpn.*, 93, 364–9 (1985).

[5]O. Yamamoto, M. Ishida, Y. Saitoh, T. Sasamoto and S. Shimada, Influmence of Mg^{2+} on the Formation of β-Sialon by the Carbothermal Reduction–nitridation of Homogeneous gel. *Int. J. Inorg. Mater.*, 3, 715–9 (2001).

[6]F. J. Li, T. Wakihara, J. Tatami, K. Komeya, and T. Meguro, Synthesis of β-SiAlON Powder by Carbothermal Reduction-Nitridation of Zeolites with Different Compositions, *J. Eur. Ceram. Soc.*, 27, 2535–40 (2007).

[7]D. Y. Zhao, J. L. Feng, Q. S Huo, N. Melosh, G. H. Fredrickson, B. F. Chmelka, G. D. Stucky, Triblock Copolymer Syntheses of Mesoporous Silica with Periodic 50 to 300 Angstrom Pores, *Science*, 279, 548-52 (1998).

[8]S. Jun, S. H. Joo, R. Ryoo, M. Kruk, M. Jaroniec, Z. Liu, T. Ohsuna, O. Terasaki, Synthesis of New Nanoporous Carbon with Hexagonally Ordered Mesostructure, *J. Am. Chem. Soc.*, 122, 10712-3 (2000).

CERIA-STABILIZED ZIRCONIA/ALUMINA NANOCOMPOSITE SUITABLE FOR ELECTROPHORETIC DEPOSITION IN THE FABRICATION OF DENTAL RESTORATIONS

Takashi Nakakmura[1], Hisataka Nishida[2], Tohru Sekino[3], Xuehua Tang[1], and Hirofumi Yatani[1]
[1]Department of Fixed Prosthodontics, Osaka University Graduate School of Dentistry, Suita, Osaka 565-0871, Japan
[2]Department of Operative Dentistry, Osaka Dental University, Hirakata, Osaka 573-1121, Japan
[3]Institute of Multidisciplinary Research for Advanced Materials, Tohoku University, Sendai, Miyagi 980-8577, Japan

ABSTRACT

The aim of this study was to prepare the ceria-stabilized zirconia (Ce-TZP)/alumina nanocomposite powder suitable for the electrophoretic deposition (EPD) in the fabrication of dental restorations. The mixtures of Ce-TZP and γ- Al_2O_3 powder were calcined at 1200, 1300, or 1400 °C and were ball-milled in a wet and dry condition. Then, EPD was performed and the weight of the deposited material was determined.

In case of the specimens calcined at 1300 and 1400 °C composite particles were observed with a transmission electron microscope. However, no composite particles were found in the specimens prepared at 1200 °C. The maximum deposit weight was 115.2 mg/cm^2 (1200 °C), followed by 70.0 mg/cm^2 (1300 °C), and 66.6 mg/cm^2 (1400 °C). The deposit weight tended to decrease as the calcination temperature increased and the particle size became larger.

INTRODUCTION

Zirconia-based materials are very strong and fracture-resistant[1,2]. In addition, these materials bond well to resin cement when the surface is properly treated or primers are used[3,4]. Thus in recent years zirconia-based materials have come to be widely used for making the frameworks of all-ceramic dental restorations.

It is said that the yttria-stabilized zirconia (Y-TZP) generally used will degrade if it remains for a long period of time under low-temperature and moist conditions, such as in the human body. Indeed, Piconi et al.[5] have reported on a case of fracture of an artificial hip joint made of Y-TZP.

On the other hand, ceria-stabilized zirconia (Ce-TZP) is highly resistant to fracture and won't degrade at low temperature, but is inferior in strength and hardness to yttria-stabilized zirconia (Y-TPZ). It has become possible, however, to solve this problem by compounding Ce-TPZ with alumina (Al_2O_3), a material with the same degree of strength as zirconia, thus producing a material that is both strong and highly resistant to fracture[6]. The ceria-stabilized zirconia/alumina (Ce-TZP/Al_2O_3) nanocomposite thus produced is not only as strong and fracture-resistant as metal[6], but also resists degradation[7] because it does not undergo phase changes in the temperature range and moisture level that degrade Y-TZP. Therefore, a Ce-TZP/Al_2O_3 nanocomposite appears to be suitable for use in the human body, for example as artificial hip joints[8] and artificial bone[9]. A Ce-TZP/Al_2O_3 nanocomposite has been processed by mechanically milling a block of dense ceramic with a CAD/CAM system[10]. However, CAD/CAM systems are usually expensive and the strength of Ce-TZP/Al_2O_3 nanocomposite might be impaired by milling[11].

We have studied the use of electrophoretic deposition (EPD), which is an easy means of processing ceramics. The results of our previous study showed that a powder of a mixture of Ce-TZP and Al_2O_3 can be deposited by electrophoresis[12]. However, the material deposited by this means must be sintered at a high temperature to compound the components completely, and then it inevitably shrinks a great deal when sintered at a high temperature.

If Ce-TZP/Al_2O_3 powder that has been compounded in advance using a powder metallurgy could be deposited by electrophoresis, high-temperature sintering would no longer be required and thus there would be less problem with material shrinkage. This study was conducted to explore the production of a Ce-TZP/ Al_2O_3 nanocomposite that can be used for EPD.

MASTERIALS AND METHODS

Ce-TZP powder (Matsushita Electric Works, Ltd., Osaka, Japan) and γ- Al_2O_3 powder (Nano Tek, C.I. Kasei Co., Tokyo, Japan) were used as the starting materials. The γ- Al_2O_3 was added to the Ce-TZP at 30 vol%. The mixture was ground and mixed further in ethanol using a wet ball mill for 24 hours, using zirconia balls with a diameter of 2 mm (YTZ Ball, Nikkato Co., Osaka, Japan) as the grinding media. After drying, the mixture was ground further using a dry ball mill for 24 hours, to prepare a completely mixed powder. Samples of the mixed powder thus prepared were calcined for two hours in the ordinary atmosphere at 1200, 1300, and 1400 °C, to prepare three types of mixed powder. Each of these was ground again and mixed using wet and then dry ball mills, 24 hours for each grinding, yielding three types of sample powder. The sample powders were observed under a transmission electron microscope (TEM). In addition, they were subjected to elementary analysis using EDX (Energy Dispersive X-ray spectroscopy), a standard TEM tool for microanalysis.

EPD was performed using each of the sample powders, prepared at three different calcination temperatures, and the weight of the deposited material was determined. The electrodes were carbon plates measuring 40 mm long, 10 mm wide and 1.5 mm thick. The back sides and edges were coated with silicone rubber leaving the areas (10 mm x 30 mm) facing each other bare. The electrodes were set up with a 15 mm space between them. For each type of sample powder, 250 ml of slurry with a powder to solvent ratio of 20 wt% was prepared, using ethanol as the solvent. Each 250 ml aliquot of slurry was then dispersed and mixed for three minutes, using an ultrasonic homogenizer (U200S, IKA Labortechnik, Stanfen, Germany). In keeping with the results of our previous study[12], the slurries were adjusted to a pH level of about 7.0, and a voltage was 100 V. EPD was applied for one to five minutes. The weight of the powder deposited on the electrodes was determined using an electronic scale.

RESULTS

TEM pictures of samples deposited by EPD revealed that some particles started necking in the sample prepared at a calcination temperature of 1200 °C, but no composite particles were found (Figure 1).

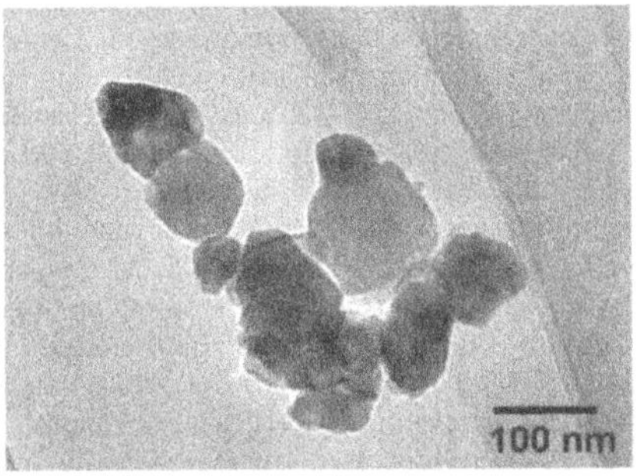

Figure 1. TEM micrograph of Ce-TZP/Al_2O_3 mixture calcined at 1200 °C. Some particles started necking but no composite particles were observed.

However, the powders prepared as calcination temperatures at 1300°C and at 1400°C were composite particles with black particles incorporated into the white particles (Figs 2 and 3). EDX revealed that the white particles were alumina and the black particles were zirconia. This demonstrated that when the powders were calcinated at 1300°C or 1400°C, the zirconia became incorporated with the alumina to form composite particles.
In addition, the higher the calcination temperature, the greater the particle size tended to become: When calcined at 1200 °C, the sample powder contained some particles with a diameter of 100 nm or more; at 1300 °C, it contained some particles with a diameter of 300 nm or more; and at 1400 °C, some particles had a diameter of 400 nm or more.

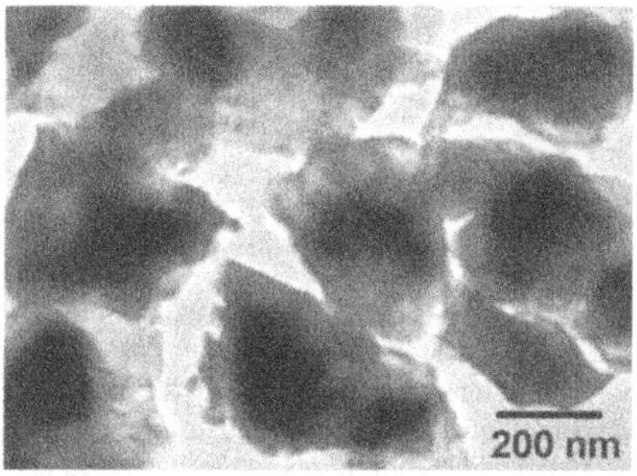

Figure 2.TEM micrograph of Ce-TZP/Al_2O_3 mixture calcined at 1300 °C. A TEM micrograph showed composite particles with black particles (zirconia) incorporated into the white alumina particles.

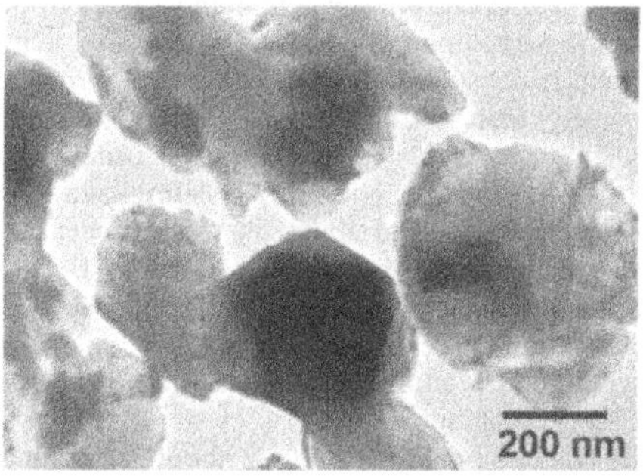

Figure 3. TEM micrograph of Ce-TZP/Al_2O_3 mixture calcined at 1400 °C. Composite particles were also observed and the size of the particles was greater than that observed in the specimens calcined at 1300 °C.

The weight of the sample powders deposited by EPD tended to increase as the migration time was longer, regardless of the calcination temperature at which the sample powders were prepared. The maximum deposit weight was 115.2 mg/cm^2, with the sample prepared at a calcination temperature of 1200 °C, followed by 70.0 mg/cm^2 with the sample powder prepared at 1300 °C, and 66.6

mg/cm^2 with the one prepared at 1400 °C. The deposit weight tended to decrease as the calcination temperature increased and the particle size became larger (Figure 4).

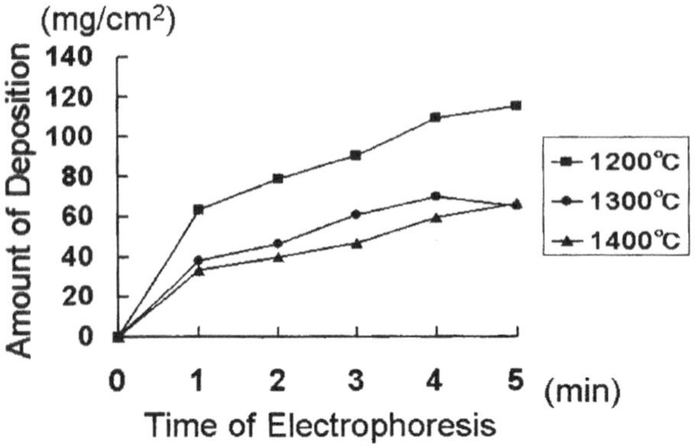

Figure 4. Amount of deposition of the sample powders with different clacination temperatures. The weight of the sample powders deposited by EPD tended to increase as the migration time was longer, but tended to decrease as the calcination temperature increased and the particle size became larger.

DISCUSSION

The results of our previous study demonstrated that a granulated powder mixture of Ce-TZP and Al_2O_3 could be used for EPD[12]. However, both materials must be high-density sintered[6] at a high temperature (1450 °C or more) to compound them sufficiently so that the resultant composite delivers great strength and fracture resistance. Like other types of zirconia-stabilized materials, the Ce-TZP/Al_2O_3 composite usually shrinks 20 to 30 percent during high-density sintering. If a CAD/CAM system is used to cut the composite, a framework is usually cut from a composite block to a somewhat larger size than required, in view of the shrinkage that might take place after sintering[10]. EPD may be conducted, using an abutment tooth model as an electrode, instead of carbon plates[13]. However, if the composite material deposited by EPD shrinks very much when subject to sintering, it would not fit the abutment tooth model, making it difficult to fabricate a precise framework in the fabrication of dental restorations. For this reason, the shrinkage of the composite material must be minimized, and thus the possibility of compounding Ce-TZP powder and Al_2O_3 powder prior to EPD was examined.

Ce-TZP can be sintered easily and compounded simply without performing gas-pressured sintering. It is understood that the material is suitable for compounding by powder-metallurgical methods. In this study, Ce-TZP and Al_2O_3 were compounded using the ratios of both materials obtained in previously reported studies[6, 14].

In a preliminary experiment, an attempt was made to prepare a compound powder by using a granulated powder mixture of Ce-TZP and α-Al_2O_3, the raw material used for ceramic blocks in the CAD/CAM system. Contrary to our expectations, both types of powder were not sufficiently compounded, even through a range of calcination temperatures from 1300 to 1600 °C.

Al_2O_3 is available in four types, α, γ, δ, and θ, classified according to crystal morphology. The γ-Al_2O_3 type is used chiefly as a catalyst, because it is highly catalytic[15]. In this experiment, a combination of Ce-TZP and γ-Al_2O_3 was therefore used to prepare a composite powder, with the expectation that the high catalytic activity of γ-Al_2O_3 would work favorably in compounding.

When γ-Al_2O_3 was used to prepare sample powders, it could be confirmed that almost all the particles in the samples prepared at 1300 or 1400 °C were completely compounded; that is to say, the zirconia was completely incorporated into the alumina particles.

This result suggests that it is effective to use γ-Al_2O_3 during the dispersion phase because of its high catalytic activity, in order to nano-compound Ce-TZP and Al_2O_3 in the powder stage. In addition, the result of our X-ray diffraction of the powders prepared in this study showed that most of the zirconia crystals were tetragonal at any calcination temperature and they did not become monoclinic after being ground by a ball mill, suggesting that they existed in a stable state.

EPD was performed on the sample powders prepared using γ-Al_2O_3. When an EPD session was performed using the sample powder prepared at a calcination temperature of 1200 °C, which was not sufficiently compounded so that particle size was small, the deposit thickness was 500 μm or more. By contrast, when the sample powders prepared at calcination temperatures of 1300 and 1400 °C were used, the deposit did not become sufficiently thick. It is known that in EPD using alumina and zirconia the deposit weight tends to decrease as the particle size becomes larger[16, 17]. It appeared that the reason for the deposit weight of the compound materials in this study was small because the particles of the compound materials were hardly reduced in size by grinding.

After completing this study, we attempted to grind a composite powder consisting of Ce-TZP and γ-Al_2O_3 calcined at a temperature of 1300 °C, using a planetary ball mill[18] capable of grinding material to a very high degree of fineness, with good success. That is, the particle size could be reduced to about 1/2 to 2/3 of the particle size attained by an ordinary ball mill. When EPD sessions were performed using sample powders finely ground by using a planetary ball mill, the maximum deposit weight could be increased about 40 percent over when sample powders prepared using an ordinary ball mill were used, with a deposit thickness of 400 μm or more. Therefore, it is likely that the weight of deposits obtained by EPD can be increased, if composite powders can be ground more finely.

The material deposited by EPD during this study was porous and its physical properties are not acceptable if used as it is. It may be possible to ensure strength and fracture-resistance without causing severe shrinkage if this porous material is impregnated with glass[19]. We are planning to report the experimental results of impregnating this material with glass in the next article.

ACKNOWLEDGEMENTS

The authors would like to special thanks to Dr. Masahiro Nawa (Matsushita Electric Works) for his assistance in preparing experimental materials. This study was supported in part by a Grant-in-aid for Scientific Research (18592119) from the Japan Society for the Promotion of Science.

REFERENCES

[1]A. J. Raigrodski. Contemporary materials and technologies for all-ceramic fixed partial dentures: a review of the literature. *J. Prosthet. Dent.*, 92, 557-62 (2004).

[2]M. Guazzato, M. Albakry, S. P. Ringer, and M.V. Swain. Strength, fracture toughness and microstructure of a selection of all-ceramic materials. Part 2. Zirconia-based dental ceramics. *Dent. Mater.*, 20, 449-56 (2004).

[3]M. Uo, G. Sjogren, A. Sundh, M. Goto, F. Watari, and M. Bergman. Effect of surface condition of dental zirconia ceramic (Denzir) on bonding. *Dent. Mater. J.*, 25, 626-31 (2006).

[4]Y. Tsuo, K. Yoshida, and M. Atsuta. Effects of alumina-blasting and adhesive primers on bonding between resin luting agent and zirconia ceramics. *Dent. Mater. J.* 2006; 25: 669-74 (2006).

[5]C. Piconi, G. Maccauro, L. Pilloni, W. Burger, F. Muratori, and H. G. Richter. On the fracture of zirconia ball head. *J. Mater. Science. Mater. Med.*, 17, 289-300 (2006).

[6]M. Nawa, S. Nakamoto, T. Sekino, and K. Niihara. Tough and strong Ce-TZP/Almina nanocomposite. *Ceram. Int.*, 24, 497-506 (1998).
[7]T. Tanaka, J. Tamura, K. Kawanabe, M. Nawa, M. Uchida, T. Kokubo, and T. Nakamura. Phase stability after aging and its influence on pin-on disk wear properties of Ce-TZP/Al_2O_3 nanocompopsite and conventional Y-TZP. *J. Biomed. Mater. Res. A*, 67, 200-07 (2003).
[8]T. Tanaka, J. Tamura, K. Kawanabe, M. Nawa, M. Oka, M. Uchida, T. Kokubo, and T. Nakamura. Ce-TZP/Al_2O_3 nanocompopsite as a bearing material in total joint replacement. *J. Biomed. Mater. Res.*, 63, 262-70 (2002).
[9]M. Takemoto, S. Fujibayashi, M. Neo, J. Suzuki, T. Kokubo, and T. Nakamura. Bone-bonding ability of a hydroxyapatite coated zirconia-alumina nanocompopsite with a microporous surface. *J. Biomed. Mater. Res. A* , 78, 693-701 (2006).
[10]J. Fischer and B. Stawarczyk. Compatibility of machined Ce-TZP/Al_2O_3 nanocomposite and a veneering ceramic. *Dent. Mater.*, 23, 1500-05 (2007).
[11]R. G. Luthardt, M. Holzhuter, O. Sandkuhl, V. Herold, J. D. Schnapp, E. Kuhlish, and M. H. Walter. Reliability and properties of ground Y-TZP-Zirconia ceramics. *J. Dent. Res.*, 81, 487-91 (2002).
[12]T. Nakamura, H. Nishida, T. Sekino, K. Wakabayashi, Y. Mutobe, and H. Yatani. Electrophoretic deposition behavior of ceria-stabilized zirconia/alumina powder. *Dent. Mater. J.*, 26, 623-27 (2007).
[13]T. Moritz, D. Linaschke, and W. Eiselt. Development of ceramic crowns and bridges using electrophoretic deposition. *Key Eng. Mater.*, 314, 207-12 (2006).
[14]M. Nawa, N. Bamba, T. Sekino, and K. Niihara. The effect of TiO_2 addition on strengthening in intragranular type of 12Ce-TZP/Al_2O_3 nanocomposite. *J. Europ. Ceram. Soc.*, 18, 209-19 (1998).
[15]K. Sohlberg, S. Pennycook, and S. Pantelides. Hydrogen and the structure of the transition aluminas. *J. Am. Ceram. Soc.*, 121, 7493-99 (1999).
[16]T. Uchikoshi, S. Furumi, T. Suzuki, and Y. Sakka. Electrophoretic deposition of alumina on conductive polymer-coated ceramic substrates. *J. Ceram. Soc. Japan,* 114, 55-58 (2006).
[17]S. Hayashi and Z. Nakagawa. Electrophoretic deposition of different yttria-stabilezed zirconia powder on stainless steel and carbon electrodes. Key Eng. Mater., 314, 75-80 (2006).
[18]H. Mio, J. Kano, and F. Saito. Scale-up method of planetary ball mill. *Chemical Engineering Science* , 59, 5909-16 (2004).
[19]X. J. Sheng, H. Xu, Z. H. Jin, Y. L. Wang. Preparation of glass-infiltrated 3Y-TZP/Al_2O_3/glass composite. *Materials Letters*,. 58, 1750-53 (2004).

PREPARATION OF POROUS ALUMINA BY GEL-CASTING PROCESS USING COMMERCIAL STARCHES AS A GELLING AGENT

Vorrada Loryuenyong[1, 2,*], Ajcharaporn Aontee [1], Daruni Kaeoklom[1], and Adisorn Sridej[1]
[1]Department of Materials Science and Engineering, Faculty of Engineering and Industrial Technology, Silpakorn University, Nakhon Pathom, Thailand
[2]National Center of Excellence for Petroleum, Petrochemicals and Advanced Materials, Bangkok, Thailand

ABSTRACT

In this study, porous alumina has been prepared by gelcasting method, using biopolymers instead of conventional toxic monomers as a gelling agent. The feasibility of this method and the factors (such as the type and the amount of starch used - 5 wt.%, 10 wt.%, 15 wt.% and 20 wt.%) affecting the porosity and mechanical properties of the alumina samples have been studied. The green bodies were sintered at 1500°C for 6 hours. The results showed that with increasing amount of added biopolymers, the bulk density and the modulus of rupture of the samples would decrease. In addition, samples prepared with tapioca starch were found to have better overall properties than those with corn starch at low starch content. Alumina ceramics with maximum porosity of ~61% and those with maximum theoretical relative density of ~83% were made by the present method.

INTRODUCTION

Porous ceramics usually have unique characteristics such as large surface area, low density and high specific strength. As a consequence, the development of such porous ceramics has recently become a new challenge to a variety of applications and industries. Since pore structure may have an effect on the properties of final products, porous ceramics with controlled pore structures are then required for any particular applications. Several methods have been reported for the fabrication of porous ceramics including the polymeric sponge, the foaming and the gelcasting.

Nowadays, gelcasting technique has received increasing attention as a new ceramic forming process, especially for products with highly porous and complex shapes [1]-[4]. Unlike slip casting, gelcasting technique relies on the gelation mechanism of ceramic slurry containing a gelling-agent. Slurry is poured into a mold, and upon polymerization or gelation process, a solid part containing ceramic powder is formed. This process is simple and economical, giving it advantages over other conventional processes.

Originally, the gelling agent is acrylamide-based system, which is high in toxicity [2]. Previous researchers have then investigated alternative gelling agents such as agarose- and starch-based systems for low-toxicity gelcasting of ceramic foams [3]-[4].

This work has developed the new process to prepare porous ceramics by using commercial starches as a gelling agent in gelcasting process. The study reports the influence of starch type and starch content on the physical and mechanical properties of the alumina specimens.

EXPERIMENTAL PROCEDURE

Commercial native tapioca and corn starches from Thailand (Tesco) were purchased and used as received. Alumina powder (99.5% purity, Amarin, Thailand), wherein the particle size distribution is shown in Figure 1, was used as the ceramic raw material. Ceramic suspensions were prepared to different % solid loadings with sodium silicate (Amarin, Thailand) as a dispersant, having a fixed concentration of 3 wt.% on a dry alumina basis. The wt.% starch powders were added in different quantities; 5 wt.%, 10 wt. %, 15 wt.% and 20 wt.% on a dry alumina basis, which corresponded to 69%, 66%, 63% and 60% solid loadings, respectively. Homogenization was carried out by ball

milling in high-density polyethylene (HDPE) bottles using alumina milling media.
After ball milling, alumina green bodies were obtained by pouring the suspensions into the metallic molds. The molds were then heated to 80-120°C, followed by subsequent cooling in ambient condition. After cooling, the samples were removed from the molds. Sintering was performed for 6 hours at 1500°C using a heating rate of 2°C min^{-1}

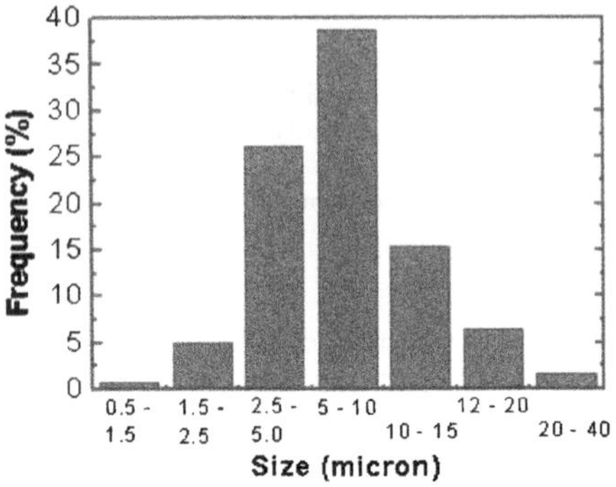

Figure 1. Particle size distribution of alumina powder.

The density of sintered bodies was measured by Archimedes' method. The theoretical density of fully-dense alumina (3.9 g cm^{-3}) was used as a reference to calculate the total volume fraction of porosity and relative density (R.D). The microstructure of ceramic bodies was observed by scanning electron microscopy (SEM). The fractured samples were first coated with a thin layer of gold. The flexural strength of sintered bodies was determined from three-point bending using universal testing machine (Lloyd, LR50K). The span width and the applied loading rate were 50 mm. and 1 mm./min, respectively.

RESULTS AND DISCUSSION

Figure 2 show the total porosity of the samples, determined by Archimedes' method. While both types of starch had the total porosity of the sintered samples in the range of 51-61%, the tapioca starch seemed to cause a lower amount of porosity. This might be corresponded to the different characteristics as a gelling agent from different kinds of starches. Unlike corn starch, tapioca starch normally becomes as a gel-like structure at lower temperature and at higher concentration [5].
Figure 3 illustrates the density properties of the porous alumina. The bulk density of the porous alumina ranged between 1.52 – 1.90 g/cm3, while the apparent density was in the range of 2.85 – 3.22 g/cm3 (R.D. = 73% - 83%). The highest value (83% of theoretical value) of apparent density was observed for alumina containing the lowest amount of corn starch. The bulk density tended to decrease with decreasing % solid loading due to an increase in the calculated total porosities.
The trends in apparent density of porous bodies, on the other hand, varied between the samples. For samples containing tapioca starch, apparent densities were high when the starch content is 5 wt.% or 20 wt.%. It was expected that a decrease in the density at the first step is due to a decrease in % solid loading and an increase in total porosities. When the starch content is as high as 20 wt.%, the starch solution becomes very concentrated. The resultant high viscous suspension can cause inhomogeneous mixing and particle agglomeration in the starting slurry, producing poor packing of particles, increasing opened porosity and hence increasing the apparent density. For the corn starches, it was observed that the suspension was even more viscous, and as a consequence,

apparent density increased with increasing starch content. Nevertheless, further increase in the corn starch content caused a decrease in the apparent density, which might be due to a considerable increase in total porosity from a high amount of added starch content. From Figure 2, with 20 wt.% of the starch content, 57% and 61% total porosity could be obtained for tapioca and corn starches, respectively.

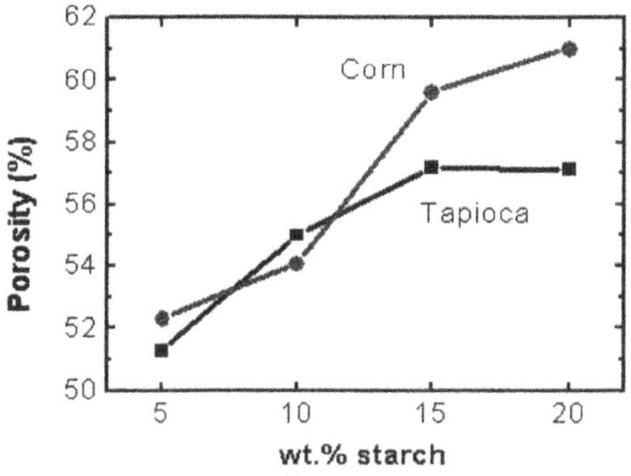

Figure 2. Total porosity of alumina samples prepared from tapioca and corn starches.

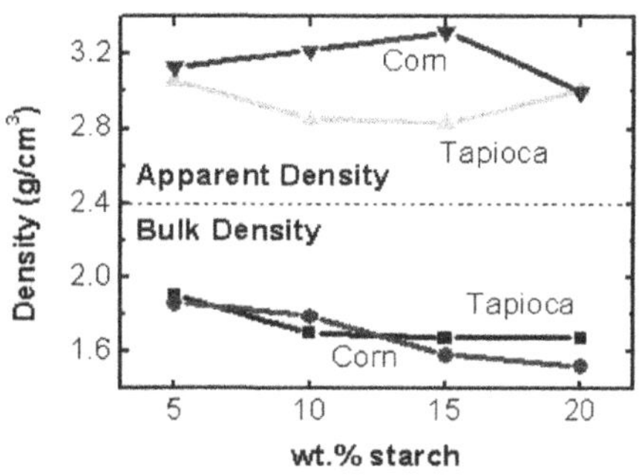

Figure 3. Density of alumina samples prepared from tapioca and corn starches.

Figure 4 shows modulus of rupture of the samples as a function of the starch contents. In general, pores can act as fracture origins, and pore size is the key factor affecting the flexural strength. At low starch content, the strength of samples with tapioca starch as a gelling agent was found to be higher than those with corn starches. This can be explained as follows. The better dissolution of tapioca starch allowed better uniform packing of particles, producing small pores (at similar % porosity and higher apparent density for corn starch) and hence higher strength structures. However, the strengths decrease with increasing starch content, and the curves become close to each other at high starch content when % porosity of both types of sample is very high.

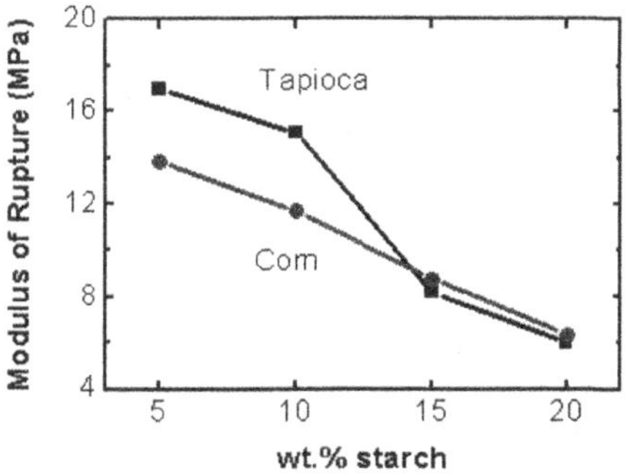

Figure 4. Modulus of rupture of alumina samples prepared from tapioca and corn starches.

As one would expect, decreasing solids loading results in a lower bulk density, higher porosity and lower strength. Additionally, samples with better physical and mechanical properties can be prepared from tapioca starch at low starch content. There is a little difference in the strength of samples prepared from tapioca and corn starches at high starch content. Alumina ceramics with porosity of ~61% and those with maximum theoretical relative density of ~83% can be obtained in this experiment.

The microstructure of the sintered, highly porous alumina (20 wt.% starch content) is presented in Figure 5. The green sample (Figure 5a) shows the fractured surface of a porous alumina. It was shown that a number of pores can be formed by the burnout of starches. The other figures (Figure 5b - 5c) show the packing of sintered-alumina grains with the interconnected pores between them. These grains are bigger than those of green sample and were approximately microns in size.

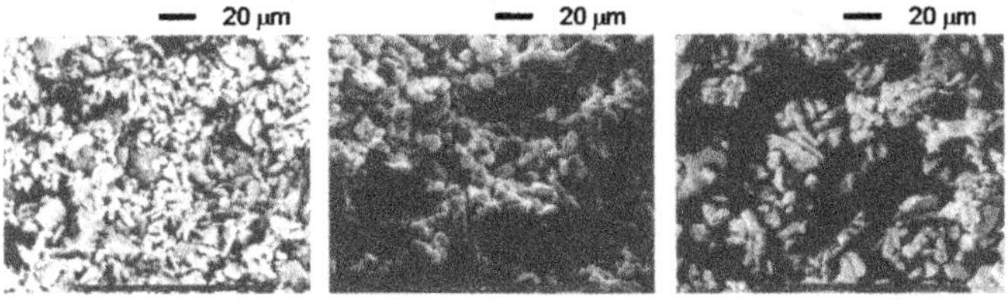

(a) Green body (b) 20 wt.% tapioca starch (c) 20 wt.% corn starch

Figure 5. SEM micrographs of gelcast green body and sintered alumina samples: (a) green body, (b) 20 wt.% tapioca starch and (c) 20 wt.% corn starch.

CONCLUSION

A method for fabricating a highly porous alumina ceramic which has a high porosity of not less than 50% has been proposed, using gelcasting method with tapioca and corn starches as a gelling agent. The high-concentrated suspension containing corn starches was found to cause higher porosity and as a consequence lower strength at low starch content. Porous ceramics (51%-61%) were prepared in this study. The use of different starch content affected the physical and mechanical properties of the samples. Depending on the % porosity, the moduli of rupture of alumina foams were measured to be 6-17 MPa.

REFERENCES

[1]O. O. Omatete, M. A. Janney and R. A. Strehlow, *Am. Ceram. Soc. Bull.*, **70**(10), 1641-1649 (1991).
[2]G. Meng, H. Wang, W. Zheng, X. Liu, *Mater. Lett.*, **45**, 224-227 (2000).
[3]I. Santacruz, M. I. Nieto and R. Moreno, *Ceram. Inter.*, **31**, 439-445 (2005).
[4]X. Mao, S. Wang, S. Shimai, *Ceram. Inter.*, **34**, 107-112 (2008).
[5]S. Mishra, T. Rai, *Food Hydrocolloids*, **20**, 557-566 (2006).

ACKNOWLEDGEMENT

This work has been supported by Faculty of Engineering and Industrial Technology, Silpakorn university. The authors also wish to thank Department of Materials Science and Engineering, Silpakorn university to provide equipment and some of materials for this investigation and National Center of Excellence for Petroleum, Petrochemicals and Advanced Materials for supporting and encouraging this investigation.

THE EFFECT OF POLYVINYL ALCOHOL ON THE MICROSTRUCTURE OF THE POROUS TiO_2 SHEETS FABRICATED BY FREEZE TAPE-CASTING

Linlin Ren [1, 2], Yu-Ping Zeng [1] *, Dongliang Jiang [1]
[1] Shanghai Institute of Ceramics, Chinese Academy of Sciences, Shanghai 200050, China
[2] Graduate School of the Chinese Academy of Sciences, Beijing 100049, China
* yuping-zeng@mail.sic.ac.cn

ABSTRACT

In order to investigate the effect of polyvinyl alcohol (PVA) on the microstructure of porous TiO_2 sheets fabricated by freeze tape-casting process, well-dispersed aqueous TiO_2 slurries with different PVA content were prepared. After casting onto the aluminum substrate, the slurries were then frozen and freeze-dried. The porous TiO_2 sheets with different porosities were obtained after sintering at 1100°C. The experimental results showed that the crystal phase of the TiO_2 sheets sintered at 1100°C was rutile. With increasing PVA content, the bamboo-like pores disappeared and uniform pores formed in the TiO_2 sheets. The porosity decreased from 87% to 53% as the PVA content increased from 2% to 5%. The effect of PVA content on the microstructure of the porous TiO_2 sheet was attributed to PVA gelation and the growth behavior variation of the ice crystals during the slurries freezing.

INTRODUCTION

Porous materials have potential applications in many fields, such as catalyst supports, separation filters, and absorbents etc. [1, 2] The performance of such materials highly depends on their pore structure, pore morphology as well as pore size. In traditional processing, pore formers are usually needed to prepare porous materials, the porosity and the pore size can be tuned by adjusting the amount of the additives and their size. However, it is still a challenge to control the pore morphology easily. Freeze casting technique is a useful and fascinating method to fabricate porous materials with controllable structure and open interconnected pores. [3-6] It involves solidification of the solvent and ice sublimation at a reduced vacuum. For the water-based slurry system, the resultant pore morphology is the replica of the ice crystals, so the microstructure of porous materials can be modulated by the parameters influencing the growth manner of the ice crystals, like slurry additives, freezing temperature and cooling rate etc. Tape casting is a traditional processing for preparation of large-area 2-dimensional sheets, and nowadays aqueous tape casting is top-priority than organic system in terms of its environmental friendliness. So in the present study, freeze-casting combined with tape casting was utilized to fabricate porous TiO_2 sheets, and the effect of polyvinyl alcohol (PVA) on the microstructure of porous TiO_2 sheets was mainly investigated.

EXPERIMENT PROCEDURE

The commercially available TiO_2 powder (Shanghai Xinrun Co. Ltd., Shanghai, china) with particles size about 100-200 nm was used as the starting material. The TiO_2 aqueous slurries were obtained by addition of dispersant ammonium polyacrylate (NH_4PAA), binder polyvinyl alcohol (PVA), and plasticizer polyethylene glycol (PEG) into the TiO_2 suspensions. After ball-milled, the TiO_2 slurries were de-aired and then casted onto the aluminum substrate, with the doctor blade controlling the thickness of the sheets. Then the tapes along with the substrate were put into the freezing zone to be solidified. After frozen completely, the sheets were transferred into a lyophilizer to freeze-drying till the ice crystals are totally sublimated. Then the dried sheets were heated in a muffle furnace to 600°C with a heating rate of 5°C/min in air, holding for 1h for binder removal, finally sintered at 1100°C for 1h. For investigating the effect of the PVA content on the microstructure of the TiO_2 sheets, the slurries with 2%, 3%, 4% and 5% PVA were prepared. Fig.1 shows the schematic illustration of the freeze-tape casting process.

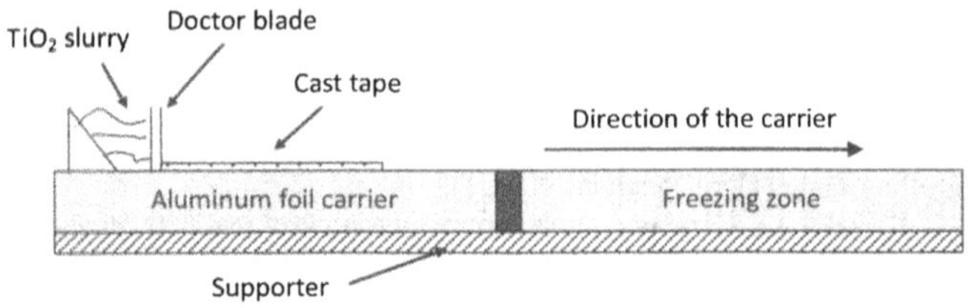

Fig.1. Schematic illustration of the freeze-tape casting process

The microstructures of the TiO_2 sheet cross-section were observed by scanning electron microscopy (SEM), and the phase composition was confirmed by X-ray diffraction (XRD). The porosities of the samples were determined by Archimedes method.

RESULTS AND DISCUSSION

Fig.2 shows the XRD patterns of the initial powder and sintered TiO_2 sheets, the initial phase of the powder is anatase, after sintering at 1100°C for 1h, the crystal phase of the TiO_2 sheets has totally transformed to rutile, and this result can be confirmed by the main peaks at 25.3° and 27.4°, which are corresponding to the characteristic peak of anatase and rutile respectively. High temperature sintering favorably induces the phase transformation from anatase to rutile.

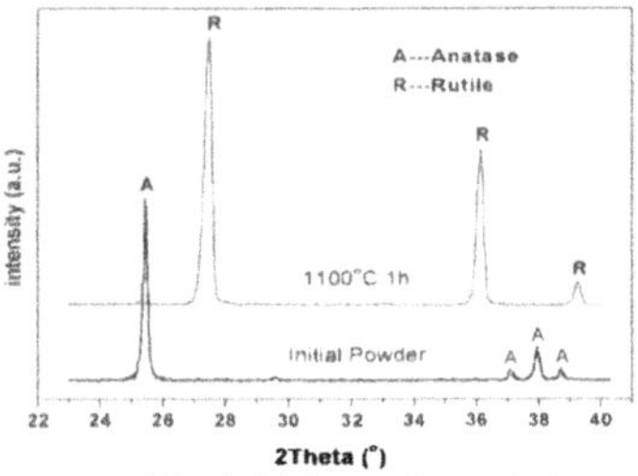

Fig.2. XRD patterns of the initial powder and sintered TiO_2 sheets

In the freezing process, thermally induced phase separation which occurs in aqueous slurries results in various types of microstructure due to *in situ* ice crystal growth. The ice crystals nucleate and grow along the direction of the temperature gradient. At the same time, the ceramic particles and PVA were expelled from the moving ice front. The resultant pore morphologies are obtained after sublimation of the ice crystals. The microstructures of the TiO_2 sheet cross-section are shown in Fig.2. From the images, it can be observed that the pore structure of the TiO_2 sheets remarkably changes with the increase of the PVA content. For the low PVA content, the macroporous property is distinct, and the parallel bamboo-like pores distribute uniformly in the whole sample, which are replica of bamboo-like ice crystals. The pores in ceramic walls are originally from the partial sintering of the ceramic particles or the removal of the organic additives. When the PVA content is up to 4% and 5%, the porous TiO_2 sheets become very homogeneous and the bamboo-like pores has disappeared, which suggests that PVA addition prevents from water development to macroscopic ice crystals. The variation of the microstructure of these TiO_2 sheets can be attributed to the fact that the PVA content affects the growth behavior of the ice crystal in the slurry freezing process.

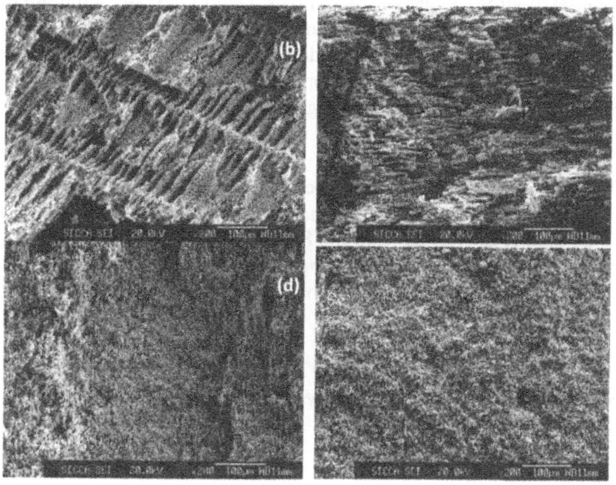

Fig.3. SEM micrographs of the TiO_2 sheets prepared by the slurries with different PVA content (a) 2% (b) 3% (c) 4% (d) 5%

In the process of nucleation and growth of the ice crystal, ceramic particles together with PVA were expelled from the freezing front, and phase separation occurs. Since PVA had polymeric long chain configuration, at relative high temperature, the PVA chains moved freely in the water. With the temperature decrease, PVA chains lost fluidity and became gelation, accordingly, the viscosity of the slurry increased. Once the temperature of slurry was below 0°C, the phase separation will happen, one was ice crystal, and the other was PVA-wrapped TiO_2 particles. The growth of the ice crystal is carried out through the water molecular nuclei. Therefore, the migration of the water molecular had great influence on the size and morphology of the ice crystals. With the increase of the PVA content, the water migration resistance increase due to PVA gelation, which resulted in the small ice crystals and the ice crystals morphology was also modified.

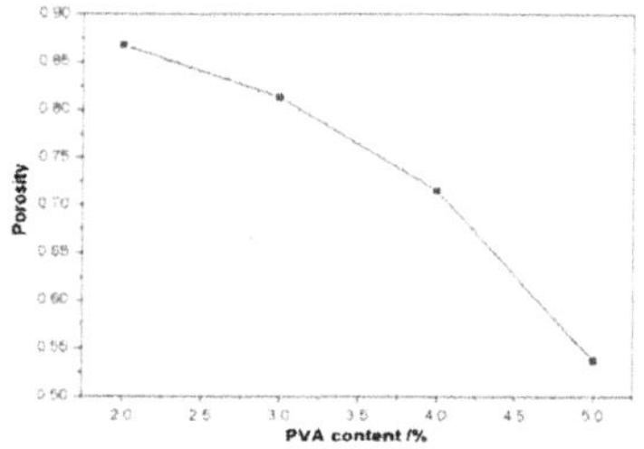

Fig.4. Relationship between the porosity of the TiO_2 sheet and the PVA content

Fig.4 displays the relationship between the porosity of the TiO_2 sheet and the PVA content. The porosity decreases from 87% to 53% as the PVA content changes from 2% to 5%. The general trend is that the porosity of the TiO_2 sheet decreases with the increase of the PVA content.

Actually, the freeze casting method which strategically uses ice crystals as templates provides a facile way to modulate the microstructure of the samples. Due to the pore structure being a replica of the ice crystal's morphology, so the microstructure of ceramics can be tuned by controlling the growth manner of the ice crystal. Besides the additives, other means can also vary the pore morphology, for example, quick freezing of such precursors produces a sponge-like morphology, which contains three-dimensional interconnected regular/irregular pores. [7, 8]

CONCLUSION

Highly porous TiO_2 sheets were fabricated by freeze tape-casting method. As the PVA content increased from 2% to 5%, the bamboo-like pores gradually transformed to the uniform pores in the whole sample, and the porosity decreased from the 87% to 53%. The results indicate that the PVA content in the slurries has significant effect on the microstructure of the TiO_2 sheet, which is attributed to the PVA gelation and the growth behavior variation of the ice crystals during the slurries freezing.

ACKNOWLEDGMENT

This work was supported by Science and Technology Commission of Shanghai Municipality under Contracts No. 07JP14093 and No.08JC1420300.

REFERENCES

[1]U. Soltmann, H. BÖttcher, D. Koch, and G. Grathwohl, Freeze gelation: a new option for the production of biological ceramic composites (biocers), *Mater. Lett.*, 57, 2861-65 (2003).

[2]K. Shqau, M.L. Mottern, D. Yu, and H. Verweij, Preparation and Properties of Porous α-Al_2O_3 Membrane Supports, *J. Am. Ceram. Soc.*, 89, 1790-94 (2006).

[3]S. Deville, E. Saiz, R.K. Nalla, and A.P. Tomsia, Freezing as a Path to Build Complex Composities, *Science*, 311, 515-18 (2006).

[4]T. Fukasawa, Z.Y. Deng, M. Ando, T. Ohji, and Y. Goto, Pore Structure of Porous Ceramics Synthesized from Water-Based Slurry by Freeze-Dry Process, *J. Mater. Sci.*, 36 2523-27 (2001).

[5]S. Deville, E. Saiz, and A.P. Tomsia, Freeze Casting of Hydroxyapatite Scaffolds for Bone Tissue Engineering, *Biomaterials*, 27, 5480-89 (2006).

[6]Y.H. Koh, J.H. Song, E.J. Lee, and H.E. Kim, Freezing Dilute Ceramic/Camphene Slurry for Ultra-High Porosity Ceramics with Completely Interconnected Pore Networks, *J. Am. Ceram. Soc.*, 89, 3089-93 (2006).

[7]J. Fukasawa and K. Tsujii, *J. Colloid Interface Sci.*, 125, 155 (1988).

[8]S. W. Sofie and F. Dogan, *J. Am. Ceram. Soc.*, 84, 1459 (2001).

PRECERAMIC PAPER DERIVED FIBRILLAR CERAMICS

Cynthia M. Gomes, Bjoern Gutbrod, Nahum Travitzky, Tobias Fey and Peter Greil
Department of Materials Science (Glass and Ceramics), University of Erlangen-Nuernberg, D-91058 Erlangen, Germany and Centre for Advanced Materials and Processes (ZMP), University of Erlangen-Nuernberg, D-90762 Fuerth, Germany

ABSTRACT

Preceramic paper may serve as a preform to manufacture light weight as well as multilayer ceramic products. Fiber reinforced preceramic paper was prepared from bioorganic pulp fibers, alumina fibers and ceramic fillers. After filtration, pressing and drying the preceramic paper sheet was processed into single sheet and multilayer components. Applying well established paper processing technologies including laminated object manufacturing lightweight ceramic structures of variable shape, size, and complexity may easily be achieved offering a high potential for economical production process based on preceramic paper.

INTRODUCTION

Stimulated by conversion of natural plant fibers [1,2] and wood tissue [3-5] into ceramic ma-terials of various phase composition, ligno-cellulosic fiber preforms in the form of paper and fiber boards were used to fabricate biomorphous ceramics. Approaches to convert biopolymeric materials into non-oxide as well as oxide ceramic products include pyrolytic decomposition resulting in a porous carbon replica (template). The template may subsequently be reacted to form carbide phases or it may be infiltrated with gaseous or liquid organometallic and metalorganic precursors followed by oxidation to remove the free carbon phase to yield oxide reaction products [5]. Highly porous SiC high-temperature filter, for example, were prepared by chemical vapour infiltration of silanes into carbonised paper preforms [6]. Processing of dense composites involves infiltration of a metal melt into the porous fiber network to fill up the inter-fibrillar space and eventually react with the fibers to form new reaction phases [7]. Laminated paper structures and corrugated cardboard were infiltrated with different preceramic polymer-filler (Si, Al, SiC) suspensions and converted into low density Si-Al-C-O ceramic composites [8].

Compared to common writing paper novel preceramic paper is distinguished by a substantially higher loading of inorganic fillers up to 90 wt.% and a grammage (weight per area) exceeding 300 g/m^2. While the cellulose fiber network generated upon paper making provides good flexibility and machinability, filler loading controls the paper-to-ceramic conversion via sintering or reaction infiltration. In order to improve the mechanical properties reduction of density variation at high filler fractions as well as tailoring of fiber templated porosity in the paper sheets are key factors to be controlled carefully [9]. Reduction of fiber templated porosity may be achieved by substituting part of the bioorganic pulp fibers by inorganic ceramic fibers. Furthermore, active fillers providing specific physical (optical, electrical, thermal, magnetical) or chemical (catalytic) properties [10-12] may be incorporated in the preceramic paper. In the following, key issues of the processing of preceramic paper, their properties and potential applications will be presented.

PAPER PROCESSING

Preceramic paper is a multi-scale composite material that is formed in the wet state from a suspension of discrete ceramic fibers and ceramic filler powder. Major steps in the preparation include coagulation of the fiber and filler in the suspension by means of retention and flocculating agents followed by dewatering of the feedstock. Figure 1 shows the composition triangle of preceramic paper.

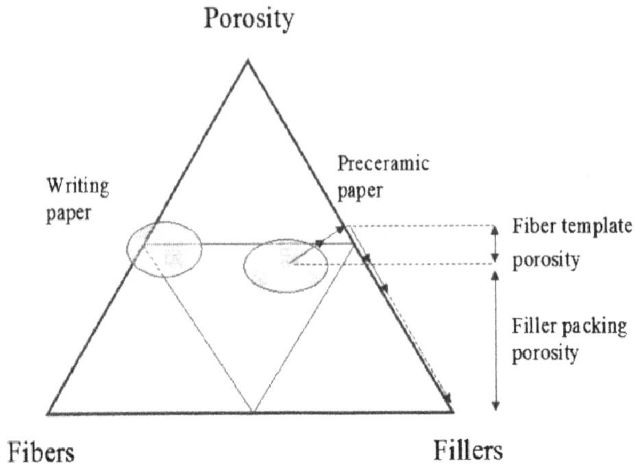

Fig. 1 Compositional diagram (volume fractions) of preceramic paper. Arrows indicate compositional change upon paper-to-ceramic conversion; initial fiber fraction may transform to fiber template porosity upon sintering in air.

For an alumina filled preceramic paper, for example, cellulose based pulp fibers (average length of 1 mm), and PVA-coated-chopped alumina fibers (Nextel™ 610, 3M, USA) with an average diameter of 12 μm were dispersed in a water based suspension. Submicron size Al_2O_3 powder (mean particle size d_{50} = 0.5 μm, CT 3000SG, Almatis GmbH, Germany) was added under vigorous stirring. Prior to filtration appropriate amounts of retention (based on vinylamine and N-vinylformamide) and flocculation aids (acrylonitrile) were added. Wet paper sheets with a typical thickness of 300 μm were produced in a laboratory paper making machine (Haage BBS-2, Estanit, Germany), using a Rapid Köthen-like aqueous handsheet forming process.
The tensile stress-strain behaviour of the preceramic papers prepared was measured on a testing device (Instron 5565, Instron Corp., Canton, MA, USA) equipped with an optical extensometer. Commonly, the total strain ε is split into elastic ε^{elast} and plastic ε^{plast} parts described by the Ramberg–Osgood relation [13]

$$\varepsilon = \varepsilon^{elast} + \varepsilon^{plast} \qquad \text{and} \qquad \varepsilon = \frac{\sigma}{E} + \left(\frac{\sigma}{E_0}\right)^n \qquad (1)$$

where E is Young's modulus and E_0 and $n \geq 1$ are the hardening modulus and the hardening exponent, respectively. For the Al_2O_3 filler loaded preceramic paper prepared from bioorganic pulp fibers: n = 2.0 - 2.8 for loading along the fiber sheet direction and E_0 = 1.1 - 1.9 GPa. In preceramic paper containing a substantial amount of particle filler > 40 vol. %, tensile strength is based mainly on fibre pull-out of the fibers as the dominating mechanism [14]. The tensile strength generally scales with a frictional strength, σ_{fric}, the number of fibers per unit area n_f, the critical fiber length l_f and the fiber strength σ_f [15]

$$T_s \propto \sigma_{fric} n_f l_f^{\,2} + \sigma_f n_f{}^* \qquad (2)$$

The left term represents the tensile strength of weakly bonded sheets additional contribution given by the fibre strength, σ_{f},. The number of fibres of embedded length greater than half the ciritical length. n_f* in the right term takes into account strong bonding conditions. While except for the lowest pulp/alumina – fiber ratio tensile strength remained on the same level of 2 (- 6) MPa Young's modulus and strain-to-failure were significantly influenced by substitution of pulp fibers by alumina fibers, fig. 3.

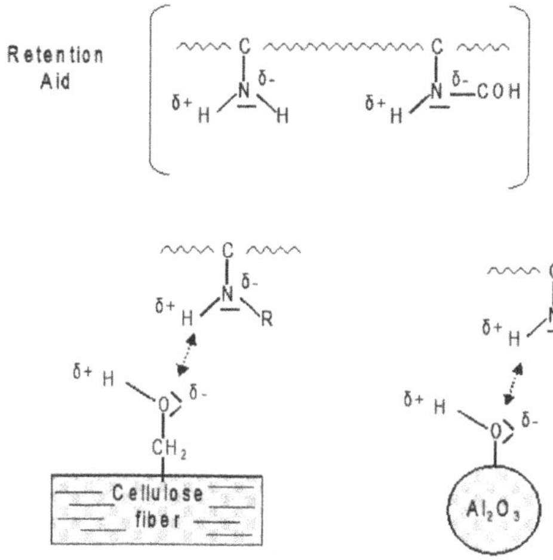

Fig. 2 Retention of fibers (cellulose) and filler (alumina) during preceramic paper formation.

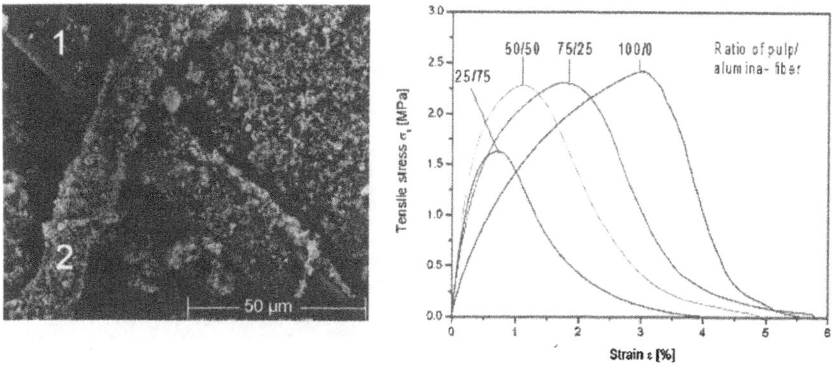

Fig. 3 Left: Preceramic paper microstructure (1 – alumina fiber, 2 – cellulose (pulp) fiber); right:Stress-strain behaviour of paper with various cellulose/alumina-fiber ratio.

PAPER TO CERAMIC CONVERSION

A high tensile strength combined with a high plasticity is necessary for preceramic paper shaping. Preceramic paper may be processed into a variety of single sheet and multi-sheet structures applying well established and versatile shaping procedures of paper technology including cutting, folding, embossing, lamination, or laminated object manufacturing, Fig. 4.

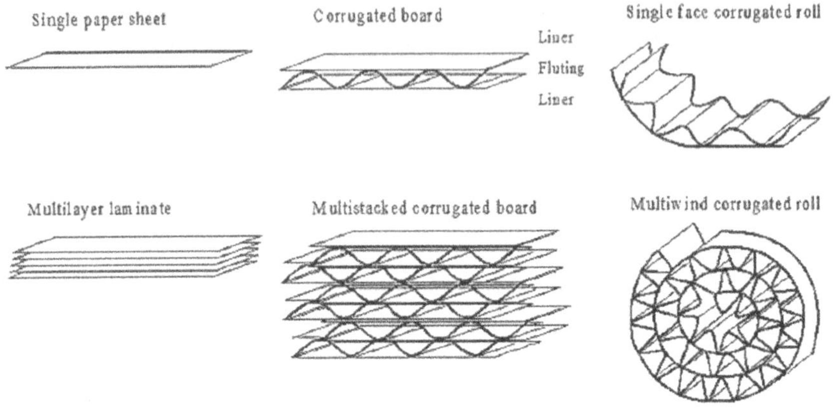

Fig. 4 Multilayer and corrugated preform structures build up from single sheet and corrugated preceramic paper.

Conversion of the preceramic paper preform into a ceramic pro-duct involves removal of the bio-organic pulp fibers and consolidation of the inorganic filler powder compact. Oxide ceramics are formed by annealing oxide filled preceramic paper preform in air to decompose and oxidise the fibers in the temperature range of 300 – 800 °C followed by sintering at elevated temperatures (1200 – 1600 °C); non-oxide composite ceramics involve formation of a highly porous biocarbon template preform into which a liquid or gas phase is infiltrated and final consolidation during

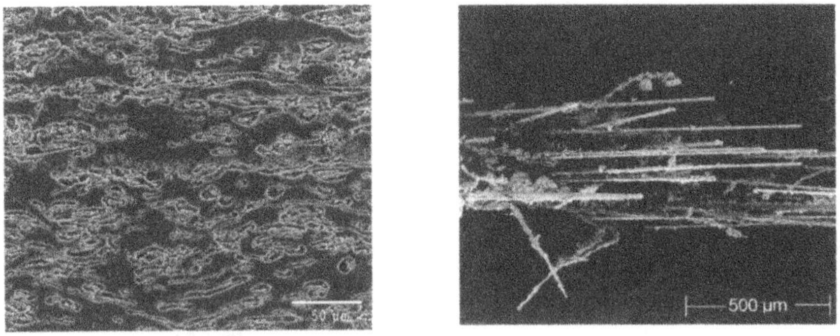

Fig. 5 Microstructures of alumina filler loaded preceramic paper: left – pulp fiber paper sintered at 1600 °C (porous); right – fracture surface of sintered alumina fibre paper.

cooling (melt) or decomposition (gas phase); Since the fluid infiltrant fills up remaining pore space without changing the overall preform dimensions a near net shape composite can be obtained which is distinguished by low residual porosity.

PROPERTIES AND APPLICATIONS

Experimentally a Young's modulus of E = 155 – 175 MPa (E/E_0 = 0.41 – 0.46) was measured by ultrasonic vibration analysis for sintered Al_2O_3 having a porosity of 31 %. The porosity resulting from pulp fibers is characterized by elongated shape and exhibits a preferential in-plane orientation.. A preferential fibre texture generated by controlled machining of paper sheet might be favourable for preceramic derived multilayer ceramics subjected to in-plane loading parallel to the initial fibre orientation e.g. pore texture.

Orientation dependence of mechanical behaviour will have a pronounced influence for design and manufacturing of corrugated lightweight structures. Stacks and rolled structures of corrugated board can easily be fabricated into three dimensional bodies containing directed macroscopic pore channels with a minimum diameter down to approximately 3-5 times the single layer thickness (200 – 300 μm). Based on FE modelling (MARC/MENTAT) of local stress and strain distribution optimization of macroscopic honeycomb structure design with respect to minimizing of local tensile stress loading may be achieved. Fig. 6 shows the distribution of principal stress (σ_{xx}) calculated for loading a multistack corrugated structure derived from preceramic paper sheets applying a loading stress of 10 MPa (y-direction).

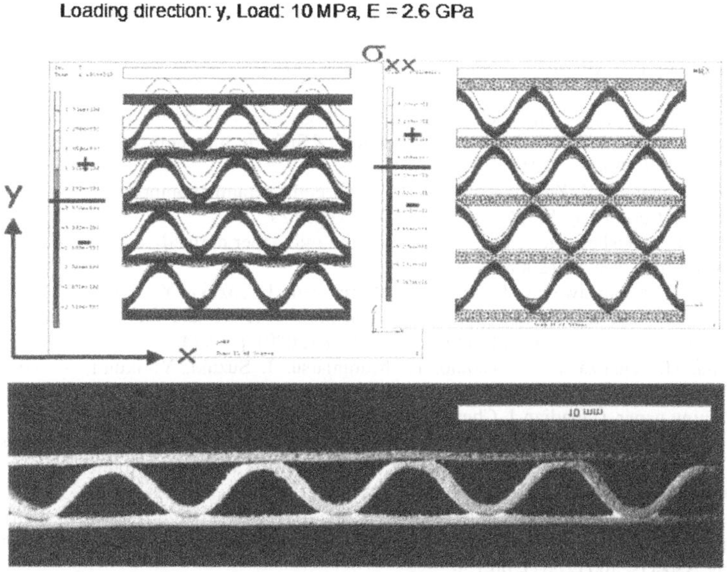

Fig. 6 Honeycomb lightweight structures derived from preceramic paper sheets: top – FEM calculation of principal stress (σ_{xx}) distribution for two different wave positions showing significantly lower tensile stress loading to occur in the right structur model; bottom – corrugated stack of alumina manufactured from preceramic paper.

The novel class of preceramic paper might offer a versatile and economic approach to process light-weight ceramics with tailored macro- and microscopic porosities for a broad field of applications in transportation (particle filtration, friction sheets in clutches), energy (porous burner, hat exchangers, solar radiation receivers, photovoltaic substrates), and environment (water cleaning, exhaust gas purification, catalytic reactor inserts).

CONCLUSIONS

Preceramic paper may serve as a preform to manufacture lightweight fibrillar ceramics of variable composition. Applying well established paper processing technologies, including Laminated Object Manufacturing, ceramic structures of variable shape, size, and complexity may easily be achieved offering a high potential for economical production process. Thus, a novel class of preceramic paper might offer a versatile and economic approach to process lightweight ceramics with tailored macro- and microscopic porosities for a broad field of applications.

REFERENCES

[1] S. Dasgupta, S. K. Das, Paper pulp waste—A new source of raw material for the synthesis of a porous ceramic composite, *Bull. Mater. Sci.*, Vol. 25, (2002) pp. 385.
[2] M. Patel., B.K. Padhi, Production of alumina fibre through jute fibre substrate, J.*Mat. Sci.* Vol. 25 (1990) pp. 1335.
[3] T. Ota, M. Takahashi, T: Hibi, M. Ozawa, H. Suzuki, Biomimetic process for producing SiC wood, *J.Am.Ceram.Soc.* Vol. 78 (1995) pp. 3409.
[4] C.E. Byrne, D.C. Nagle, Carbonization of wood for advanced materials applications,*Carbon* Vol. 35 (1997) pp. 259.
[5] P. Greil, Biomorphous ceramics from lignocellulosics, *J.Europ.Ceram.Soc.* Vol. 21(2001) pp. 105.
[6] Y. Ohzawa, H. Hshino, M. Fujikawa, K. Nakane, K. Sugiyama, Preparation of high-temperature filter by pressure-pulsed chemical vapour infiltration of SiC into carbonized paper preforms, *J.Mat.Sci.* Vol. 33 (1998) pp. 1211.
[7] H. Sieber, H. Friedrich, Z. Zeschky, P. Greil, Light-weight ceramic composites from laminated paper structures, *Ceram.Eng.Sci.Proc.* Vol. 21 (2000) pp. 129.
[8] N. Travitzky, H. Windsheimer, T. Fey, P. Greil, Preceramic paper derived ceramics, (Feature), to be publ. J.Am.Ceram.Soc. 91 (2008)
[9] N. Kinoshita, H. Katsuzawa, S. Nakano, H. Muramatsu, J. Suzuki, Y. Ikumi, Y. Toyotake, Influence of fibre length and filler particle size on pore structure and mechanical strength of filler-containing paper, Canadian J. Chem. Eng., vol. 78 (2000), pp. 974.
[10] N. Kinoshita, H. Katsuzawa, S. Nakano, H. Muramatsu, J. Suzuki, Y. Ikumi, Y. Toyotake, Influence of fibre length and filler particle size on pore structure and mechanical strength of filler-containing paper, Canadian J. Chem. Eng., 78 (2000), pp. 974.
[11] H. Kogaa, S. Fukahoria, T. Kitaokaa, M. Nakamurab, H. Wariishi, Paper-structured catalystnext term with porous fiber-network microstructure for autothermal hydrogen production, Chemical Engineering Journal, 139 (2008), pp. 408.
[12] J. C. Leea, H. C. Kwona, Y. P. Kwona, J. Leeb, S. Park, Porous ceramic fiber glass matrixnext term composites for solid oxide fuel cell seals, Colloids and Surfaces A: Physicochemical and Engineering Aspects, 300 (2007), pp.150.
[13] P. Mäkelä, S. Östlund, Orthotropic elastic-plastic material model for paper materials, *Int.J.Sol.Struct.* Vol. 40 (2003) pp. 5599.
[14] L. Li, A.Collis, R. Pelton, A New Analysis of Filler Effects on Paper Strength, J.Pulp Paper Sci. Vol. 28 (2002) pp. 267.

[15] M.W. Kane, The effect of beating on fiber length distributions and tensile strength –part 2, *Pulp and Paper Mag.Can.* Vol. 60 (1959) pp. T 359.

IX. Composites

IN-SITU SYNTHESYS AND PROPERTIES OF TiB_2/Ti_3SiC_2 COMPOSITES

Wei Gu, Jian Yang, Tai Qiu
Coll. of Mater. Sci. and Eng., Nanjing Univ. of Tech., China
Nanjing, JiangSu, 210009, China

ABSTRACT

Based on thermodynamic analysis, TiB_2/Ti_3SiC_2 composite ceramics were in-situ prepared by hot-pressing in an Ar atmosphere. The phase composition and microstructure of the materials were characterized by XRD and SEM. Effect of temperature on phase composition, sintering performance, microstructure and bending strength of the materials was investigated. The results showed that for the materials sintered at 1350°C ~1500°C, the main phases were Ti_3SiC_2 and TiB_2. However, due to the volatilization of raw material Si at high temperature, small amount of TiC remained in all materials. With the increase of sintering temperature, the density and bending strength of the materials increased. Higher temperature was favorable to the grain growth of reinforcing phases and matrix. The materials sintered at 1500°C showed optimal microstructure and the highest bending strength of 741MPa. When sintering temperature is higher than 1550°C, the density and strength of the materials deceased gradually as a result of the decomposition of matrix Ti_3SiC_2.

KEYWORDS: TiB_2/Ti_3SiC_2, in-situ synthesis, hot-pressing, bending strength

1. INTRODUCTION

Ti_3SiC_2 ceramics has been promisingly employed in many high-temperature structural and functional applications for their combined merits of metals and ceramics, such as high thermal and electrical conductivity, high machinability, excellent thermal shock resistance, self-lubricity and oxidation resistance, and so on[1~4]. However, their relatively poor mechanical strength restricts their potential use as a structural material. Introduction of second-phase has been considered as an effective enhancement for materials. A number of works have been reported on improving the mechanical properties of Ti_3SiC_2 by adding second-phase into it. Zhang et al[5] synthesized SiC/Ti_3SiC_2 nano composites by spark plasma sintering-reactive synthesis method (SPS-RS) with Ti, Si, C and Al powers. The bending strength of the material with 20vol%(volume fraction) SiC reached to 501 ± 50 MPa. Wang et al[4] prepared Al_2O_3/Ti_3SiC_2 composite with bending strength of 600MPa by SPS using Ti, TiC, Si and Al_2O_3 powers as starting materials. Konoplyuk et al[6] prepared TiC/Ti_3SiC_2 composites by pulse discharge sintering (PDS) from TiH_2, SiC and TiC powders. The materials show a bending strength as high as 550±60MPa. Zhu et al[7] prepared TiB_2/Ti_3SiC_2 composites by hot pressing(HP) method, in which TiB_2 was introduced by directly mixing TiB_2 power with other starting materials. The highest bending strength of 485.1MPa was obtained for the composite with 10vol% TiB_2 added.

In-situ synthesis is a novel preparation technology for composites, in which reinforcing phases are formed by in-situ reaction occurred during the preparation process of the composites. Comparing with mechanical mixing, in-situ synthesis has many advantages such as small grain size and stable thermodynamic performance of reinforcing phases, clean grain boundary, high bonding strength between reinforcing phases and matrix, etc.[8] TiB_2 is an important boride ceramics with many distinguished properties such as high hardness and modulus, good electrical conductivity, etc. Moreover, TiB_2 has a similar crystal structure of hexagonal system to Ti_3SiC_2 and they also possess similar coefficient of thermal expansion. Therefore, TiB_2 can be expected as an ideal candidate of reinforcing phases for Ti_3SiC_2 ceramics.

In this work, an attempt on preparation of TiB_2/Ti_3SiC_2 composite ceramics by in-situ hot-pressing synthesis was made and the microstructure and mechanical properties of the materials were investigated.

2. EXPERIMENTAL PROCEDURE

Powers of Ti (purity>99.99%, average particle size 50μm), Si (purity>99.99%, average particle size 50μm), graphite(purity>99.99%, average particle size 10μm) and B_4C (purity>99.99%, average particle size 2.5μm) were used as starting materials. These powers were precisely weighed with a certain ratio and mixed in a polyethylene mill for 24h. Then the mixed powers were hot-pressed at 1350°C ~1600°C for 2h under the pressure of 25MPa in Ar atmosphere.

Density of the sintered materials was determined by the Archimedes method, and the phase identification of the materials was performed by X-ray diffractometer (ARL X'TRA, Switzerland) with CuKα radiation. The sintered materials were cut, ground and polished into strips with the size of 3mm×4mm×50mm and three point bending method was applied to measure room temperature bending strength using a span of 40mm with a crossing speed of 0.5mm/min. Fracture microstructure of the specimens was investigated by scanning electron microscope (SEM) (JSM-5900, JEOL, Japan).

3. RESULTS AND DISCUSSION

3.1 Thermodynamic analysis

The formation of Ti_3SiC_2 from Ti, Si and C can be generalized in the following

$$3Ti + Si + 2C = Ti_3SiC_2 \qquad (1)$$

In addition, another reaction between Ti and B_4C powder may also occur in the system[9].

$$3Ti + B_4C = 2TiB_2 + TiC \qquad (2)$$

The thermodynamic calculation[10] indicated that the ΔG^0 of reaction (2) is $-634412 J\cdot mol^{-1}$~$-622414 J\cdot mol^{-1}$ in the temperature range from 1350°C to 1600°C, in which reaction (1) can proceed sufficiently and Ti_3SiC_2 can exist stably[3]. This means that reaction (2) is thermodynamically feasible in this temperature range. The by-product TiC of reaction (2) also can react again with Ti and Si as follows[11]

$$Ti + 2TiC + Si = Ti_3SiC_2 \qquad (3)$$

Therefore, the general reaction in the present system can be described as follows

$$(3+3.5\alpha)\ Ti + (1+\frac{\alpha}{2})Si + 2C + \alpha B_4C = (1+\frac{\alpha}{2})Ti_3SiC_2 + 2\alpha TiB_2 \qquad (4)$$

in which α represents TiB_2 molar content in the composites. In this work, an α value of 0.17 was applied for all specimens.

3.2 Phase analysis

Fig.1 shows the XRD patterns of the materials sintered at different temperatures. When sintered at 1350°C, TiC and Ti_3SiC_2 are the main crystal phases and a small amount of Ti_5Si_3 and TiB_2 are also identified, which means that reactions (1) and (2) did not proceed completely as Ti_5Si_3 has been believed to be the intermediate product of reaction (1)[12,13]. With the increase of sintering temperature, the content of Ti_5Si_3 decreases and the content of Ti_3SiC_2 and TiB_2 increases dramatically. When sintered at 1500°C, Ti_5Si_3 disappears and only Ti_3SiC_2, TiB_2 and TiC remains in the material, which suggests that reactions (1) and (2) have proceeded completely. With sintering temperature increased to 1550°C Ti_3SiC_2 begins to decompose. For the materials sintered at 1600°C, a thoroughly different phase composition was obtained. In this sample, the content of Ti_3SiC_2 decreases dramatically and TiB_2, SiC, TiC become the main phases as a result of the further decomposition of Ti_3SiC_2. So in the following research, the material sintered at 1600°C was not considered. It should be noted that a small amount of TiC remains in all materials. This is due to the volatilization of raw material Si at high temperature[14], which led to the excess of TiC. Therefore, it can be deduced that TiB_2/Ti_3SiC_2 composite ceramics would be obtained by adjusting the ratio of starting materials and sintering at 1500°C. On the other hand, as TiC is also an effective reinforcing phase for Ti_3SiC_2 ceramics, $(TiB_2+TiC)/Ti_3SiC_2$ materials, which was prepared at 1500°C as mentioned above, may be expected as another novel composite with integrated multivariant enhancement mechanisms.

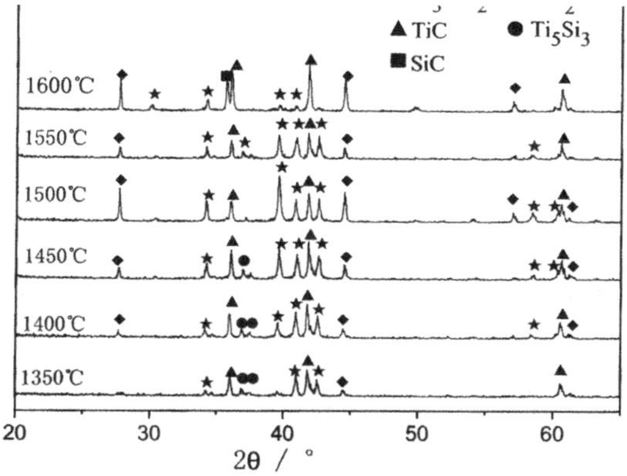

Fig.1 XRD patterns of the materials sintered at different temperature

3.3 Sintering property

Fig.2 shows the sintering properties of the materials sintered at different temperatures. It can be seen that before 1500°C, with the increase of sintering temperature, apparent porosity decreases and bulk density increases dramatically. The apparent porosity of the materials sintered at 1350°C, 1400°C and 1450°C was still in at a high level. When sintered at 1500°C, a very dense material with an apparent porosity of 0.07% was obtained. With sintering temperature increased to 1550°C, the apparent porosity increases and the material becomes incompact again because of the decomposition of Ti_3SiC_2.

3.4 Bending strength

Fig.3 shows the bending strength of the materials sintered at different temperatures. With the increase of sintering temperature, the bending strength of the materials increases and then decreases, which shows a reverse variation rule compared with apparent porosity. The highest bending strength of 741MPa was obtained at 1500°C, which is much higher than the strength value reported for Ti_3SiC_2 ceramics and other Ti_3SiC_2-matrix composites. This result suggests that a very significant reinforcing effect has been obtained in the materials.

Fig.4 shows the micrographs of fracture surface for the materials sintered at 1350°C and 1500°C. When sintered at 1350°C, the grains of the material did not developed well and obvious pores can be found in the materials. With sintering temperature increased to 1500°C, the material shows very dense microstructure with well-developed grains. As seen in Fig.4(b), plate like Ti_3SiC_2 grains, columnar grains of TiB_2 and equiaxed grains of TiC can be easily identified and the bonding between them was very tight. It is the ideal phase composition and optimal microstructure characteristics which result in the distinguished enhancement effect and high bending strength for the material.

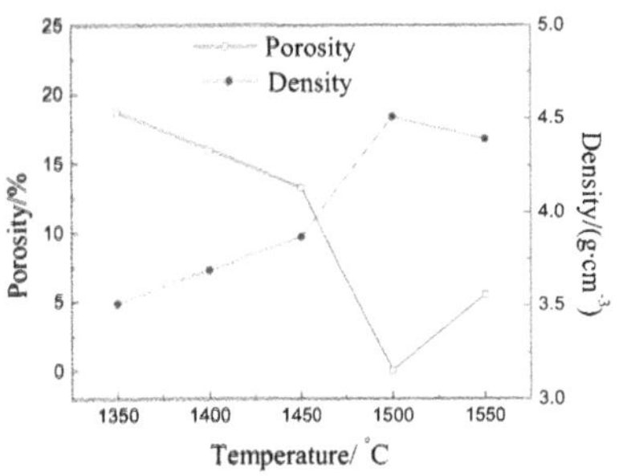

Fig. 2 Sintering properties of the materials sintered at different temperatures

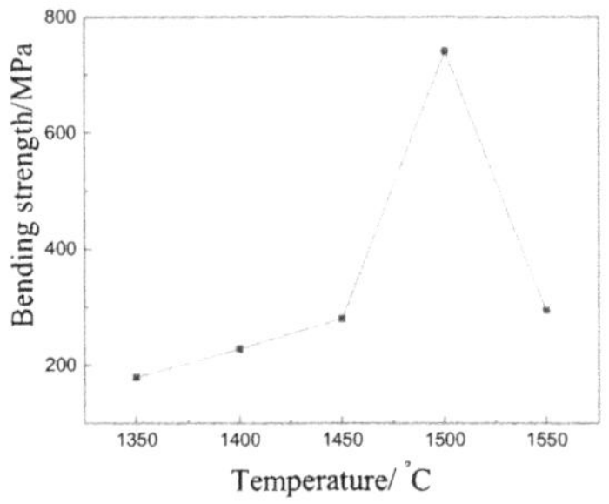

Fig.3 Bending strength of the materials sintered at different temperatures

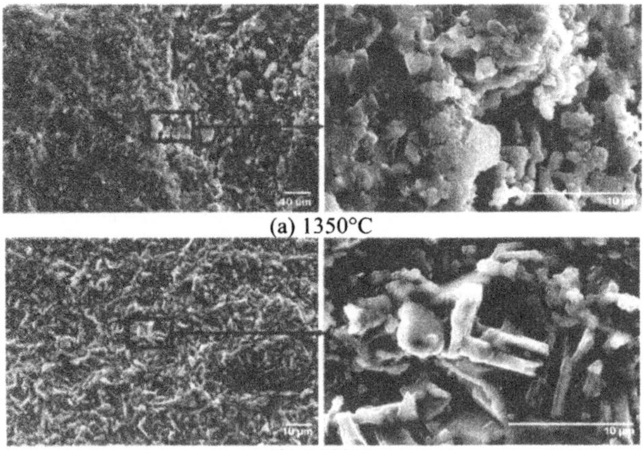

(a) 1350°C

(b) 1500°C

Fig.4 SEI micrographs of fracture surface for the materials sintered at 1350°C and 1500 °C

4 CONCLUSION

(1) The main phases of the materials sintered at 1350°C ~1500°C were Ti_3SiC_2 and TiB_2. Owing to the volatilization of raw material Si at high temperature, small amount of TiC remained in all materials. TiB_2/Ti_3SiC_2 composite ceramics would be obtained by adjusting the ratio of starting materials and sintering at 1500°C.

(2) With the increase of sintering temperature, the density and bending strength of the materials increased. Higher temperature was favorable to the grain growth of reinforcing phases and matrix. The materials sintered at 1500°C showed the highest bending strength of 741MPa owing to the ideal phase composition and optimal microstructure characteristics.

(3) When sintering temperature is higher than 1550°C, the density and strength of the materials deceased gradually as a result of the decomposition of matrix Ti_3SiC_2.

ACKNOWLEDGE

This work was supported by Natural Science Foundation of Jiangsu Province Education Commission under Grant No. 07KJB430039, the Opening Project of State Key Laboratory of High Performance Ceramics and Superfine Microstructure under Grant No. SKL200705SIC, the Key Laboratory of New Inorganic Materials and Its Composites, Jiangsu Province.

REFERENCE

[1] M. W. Barsoum, T. El-Raghy. Progress Report on Ti_3SiC_2, Ti_3GeC_2, and the H Phases, M_2BX. *J. Mater. Synth. Proc.*, 5, 197–216(1997)

[2]J. Travaglini, M. W. Barsoum. The Corrosion Behavior of Ti_3SiC_2 in Common Acids and Dilute NaOH, *Corros. Sci.*, 45, 1313–27(2003).

[3]R. Radhakrishnan, J. J. Williams, M. Akinc. Synthesis and High-temperature Stability of Ti_3SiC_2. *J. Alloys Compd.*, 285, 85–88(1999).

[4]H. J. Wang, Z. H. Jin, Y. Miyamotob. Effect of Al_2O_3 on Mechanical Properties of Ti_3SiC_2/Al_2O_3 Composite, *Ceram. Int.*, 28, 931–34 (2002).

[5]J. F. Zhang, L. J. Wang, L. Shi, W. Jiang and L. D. Chen. Rapid Fabrication of Ti_3SiC_2–SiC Nanocomposite Using the Spark Plasma Sintering-reactive Synthesis (SPS-RS) Method, *Scripta Mater.*, 56, 241–44(2007).

[6]S. Konoplyuk, T. Abe, T. Uchimoto, T. Takagi. Synthesis of Ti_3SiC_2/TiC Composites from TiH_2/SiC/TiC Powders, *Mater. Lett.*, 59 , 2342–46(2005).

[7]D. Y. Zhu, J. Q. Zhu, B. C. Mei, W. B. Zhou. Fabrication, Microstructure and Mechanical Properties of Ti_3SiC_2/TiB_2 Composite, *J. WuHan Univ. Tech.*, 27, 1-4(2005).

[8]G. J. Zhang. In-situ Reaction Synthesis of Structural Ceramic Nanocomposites, *The Chin. J. Proc. Eng.*, 2, 380-84(2002)

[9]D. Brodkin, S. R. Kalidindi, M. W. Barsoum and A. Zavaliangos. Microstructural Evolution during Transient Plastic Phase Processing of Titanium Carbide-Titanium Boride Composites, *J. Am. Ceram, Soc.*, 79, 1945-49(1996)

[10]Y. J. Liang, Y. C. Che. Handbook of Thermodynamic Properties for the Inorganic Substances. *Shenyang: Northeastern University Press*, 1993.

[11]H. Li, D. Chen, J. Zhou, J. H. Zhao, L. H. He. Synthesis of Ti_3SiC_2 by Pressureless Sintering of the Elemental Powders in Vacuum. *Mater. Lett.*, 58, 1741– 44(2004).

[12]Y. Zou, Z. M. Sun, S. Tada. Effect of Al Addition on Low-temperature Synthesis of Ti_3SiC_2 Powder, *J. Alloys Compd.*, 461, 579–84(2008).

[13]E. Wu, D. P. Riley, E. H. Kisi, R. I. Smithc. Reaction Kinetics in Ti_3SiC_2 Synthesis Studied by Time-Resolved Neutron Diffraction, *J. Eur. Ceram.Soc.*, 25, 3503–08(2005).

[14]R. Radhakrishnan, J. J. Williams, M. Akinc. Synthesis and High-temperature Stability of Ti_3SiC_2, *J. Alloys. Compd.*, 285, 85-88(1999)

EFFECT OF La2O3 ADDITIVE ON MICROSTRUCTURE AND PROPERTIES OF Si3N4-AlN COMPOSITE CERAMICS

Peng Xu,* Jian Yang, Tai Qiu
Col. of Mater. Sci. and Eng., Nanjing Univ. of Tech., China
Nanjing , JiangSu, 210009, China

ABSTRACT

As a novel microwave-transparent ceramics with low dielectric loss, high thermal conductivity and high strength, Si_3N_4-AlN(30wt%) composite ceramics were prepared by hot-pressing at 1800 °C using La_2O_3 as additive. The ceramics were characterized by XRD and SEM. Effect of La_2O_3 on phase composition, sintering performance, microstructure, bending strength, dielectric loss and thermal conductivity of Si_3N_4-AlN ceramics was investigated. The results showed that increasing La_2O_3 content promoted the phase transformation of α-Si_3N_4 to β-Si_3N_4 and the formation of more lanthanum melilites in the Si_3N_4-AlN system. With the increase of La_2O_3 content from 4% to 10%, the dielectric loss decreased and then increased, while for thermal conductivity, the case was on the contrary. The lowest dielectric loss (4.5×10^{-3}) and highest thermal conductivity ($11.7W\cdot m^{-1}\cdot K^{-1}$) were obtained at the level of 6% La_2O_3 content. Furthermore, the increase of La_2O_3 content played a negative effect on bending strength which decreased from 574MPa to 480MPa.

KEY WORDS: Si_3N_4-AlN composites, bending strength, thermal conductivity, dielectric loss

INTRODUCTION

Microwave-transparent ceramics (MTC) has been used widely as output windows of traveling wave tube (TWT) in the field of microwave communication. Commonly, excellent mechanical property and thermal shock resistance are also required for MTC to release the heat caused by dielectric loss in time and prevent the thermal breakdown, besides low dielectric loss and high thermal conductivity. Recently, with the increment of output power and broadening of wave band, a higher mechanical property is desired in response to the miniaturization design of output windows.

As we know, aluminum nitride (AlN) has been an excellent candidate for MTC due to its high thermal conductivity and low dielectric loss[1,2]. However, AlN possesses relative low mechanical strength, which limits its applications on high-power output windows. Silicon nitride (Si_3N_4) is considered as excellent structural ceramic for engine components and wear parts owing to its excellent mechanical properties at both room and elevated temperatures, good resistance to thermal shock and chemical attack. Recent investigations indicated that thermal conductivity of $100\text{-}155W\cdot m^{-1}\cdot K^{-1}$ or even more could be obtained for Si_3N_4 ceramics[3-5], which suggests a potential application of Si_3N_4 as microwave-transparent ceramics. The Si_3N_4 ceramics with low dielectric loss of less than 5×10^{-4} at 10GHz, which is suitable for use as output windows, has been prepared by Taijima et al.[6]. Therefore, based on the excellent properties of AlN and Si_3N_4, Si_3N_4-AlN composite ceramics might be a novel microwave-transparent ceramics with low dielectric loss, high thermal conductivity, high strength and thermal shock resistance.

In this study, Si_3N_4-AlN composite ceramics were prepared by hot-pressing at 1800 °C using La_2O_3 as additive. The effect of La_2O_3 on phase composition, sintering performance, microstructure, bending strength, dielectric loss and thermal conductivity of the materials was also investigated.

EXPERIMENTAL PROCEDURE

Commercial α-Si_3N_4 (α-phase content >93%, average particle size 0.5μm, oxygen content <2.0wt%) and AlN (purity>98.6%, average particle size 0.5μm) powder were selected as starting materials. In all specimens, the mass ratio of Si_3N_4: AlN is fixed at 70:30 and certain amount of La_2O_3

(purity >99.9%) was added as a sintering additive. The amount of La_2O_3 ranged from 4 to 10 wt%. After ball-milling in ethanol for 6h, the slurry was dried and sieved. Then the mixed powders were hot-pressed at 1800 °C for 1h under the pressure of 30 MPa in N_2 atmosphere.

The density was determined by the Archimedes method, and the phase identification was performed by X-ray diffractometer (ARL X'TRA, Switzerland) with CuKα radiation. The specimens were cut, ground and polished into strips with the size of 3mm×4mm×40mm and three point bending method was applied to measure room temperature bending strength using a span of 30mm with a crossing speed of 0.5 mm/min.

The thermal conductivity was determined by laser–flash method (LFA447, Netzsch, German) with billets dimension of φ12.7mm×2.5mm. The dielectric loss (tanδ) was measured at 1MHz by the perturbation method using a cavity resonator and a vector network analyzer (HP-4294A).

The Microstructure was investigated by SEM (JSM-5900, JEOL, Japan). Prior to observation by SEM, heat-etching at 1700 °C in N_2 atmosphere was conducted for the polished specimens.

RESULTS AND DISSCUTION

Si_3N_4 and AlN are difficult to be fully densified during sintering due to their strong covalent bonding, and sintering additives are often introduced to realize densification by liquid phase sintering mechanism. Y_2O_3 and Al_2O_3-Y_2O_3 are effective additives widely used in Si_3N_4 and AlN ceramics. However, with these additives, it is liable to induce α-sialon phase in the Si_3N_4-AlN system that two lattice vacancies existing in α-Si_3N_4 unit cell can be packed by Y^{3+} (ionic radius is 0.0893nm) to equilibrate the electrical valence unbalance generated by the replacement of Si^{4+} by Al^{3+}[7]. It was reported that in Si_3N_4-AlN-Al_2O_3-Y_2O_3 system, the structure of α-sialon can be stabilized by many kinds of metallic ions, which includes Li^{+}, Mg^{2+}, Ca^{2+}, Y^{3+} and lanthanide ions(excluding La^{3+}, Ce^{3+}, Pr^{3+})[8], which have ionic radius larger than the size of lattice vacancies in α-Si_3N_4 cell. This has beenproved by our experiment, i.e., distinct diffraction peaks of α-sialon was identified with XRD pattern (Fig.1) of the Si_3N_4-AlN system using Y_2O_3-Al_2O_3 as additive. Since α-sialon has poor dielectric loss and thermal conductivity compared with β-Si_3N_4, in this work, La_2O_3 was selected as the sintering additive for obtaining as-desired materials due to its largest ionic radius (0.106nm) in lanthanide ions.

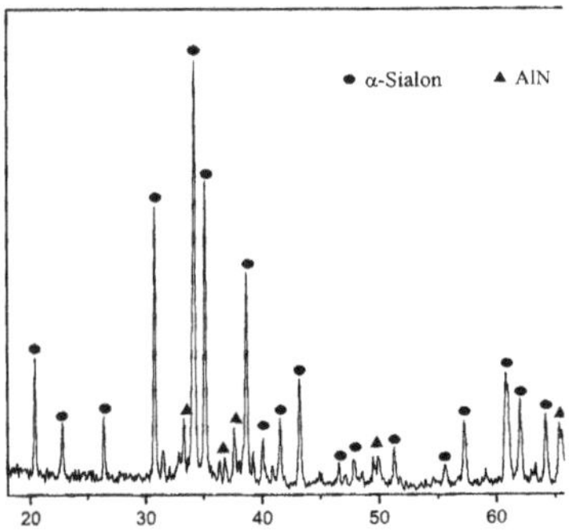

Fig.1 XRD pattern of Si_3N_4-AlN composite ceramics with Y_2O_3-Al_2O_3 as additive

Phase composition and density

Fig.2 shows the XRD patterns of the specimens with different La_2O_3 content. Both β-Si_3N_4 and AlN were identified as the main phases in all specimens, which indicated that the desired Si_3N_4-AlN composite ceramics had been prepared using La_2O_3 as sintering additive. For the specimens with 4-8wt% La_2O_3 added, small amounts of α-Si_3N_4 were also found, which suggested an incomplete transformation of Si_3N_4 from α-phase to β-phase. In addition, when La_2O_3 content was higher than 6wt%, lanthanum melilites ($La_2Si_6N_8O_3$) appeared in the specimens and its content increased gradually with increasing La_2O_3 addition. The formation of $La_2Si_6N_8O_3$ could be explained by the reaction between La_2O_3 and Si_3N_4 as follows:

$$La_2O_3+2Si_3N_4=La_2Si_6N_8O_3 \qquad (1)$$

Therefore, the conclusion could be drawn that increasing La_2O_3 content promotes the phase transformation of α-Si_3N_4 to β-Si_3N_4 and the formation of $La_2Si_6N_8O_3$ phase in the system.

Fig.3 shows the density and apparent porosity of Si_3N_4-AlN composites with different La_2O_3 content. As can be seen, La_2O_3 additive with concerned content effectively promoted the sintering densification and all samples showed high density(apparent porosity<0.5%). With the increase of La_2O_3 addition, the porosity reduced and then increased slightly, which reached the minimum value of 0.07% at the level of 8wt% La_2O_3 content.

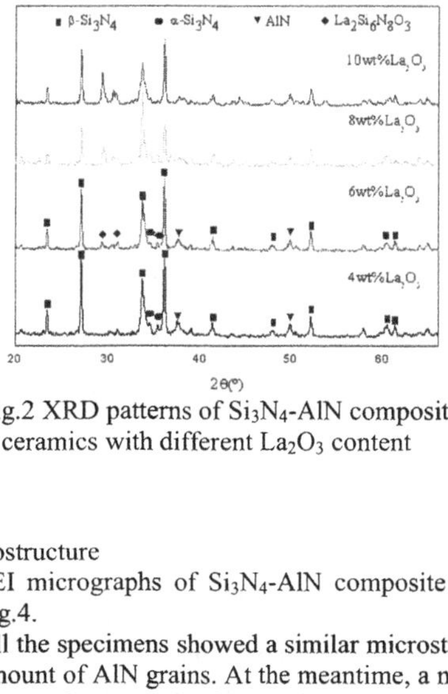

Fig.2 XRD patterns of Si_3N_4-AlN composite ceramics with different La_2O_3 content

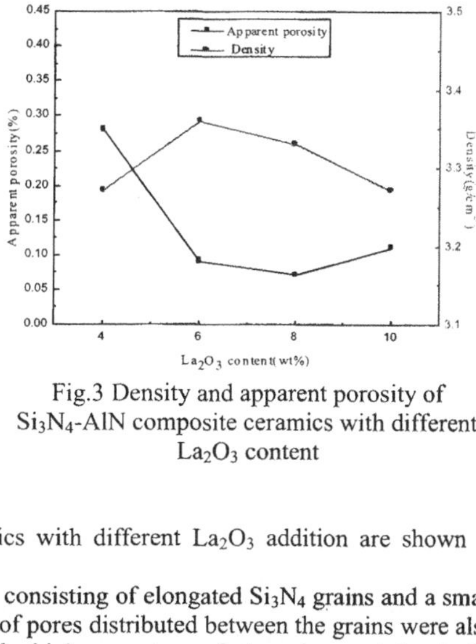

Fig.3 Density and apparent porosity of Si_3N_4-AlN composite ceramics with different La_2O_3 content

Microstructure

SEI micrographs of Si_3N_4-AlN composite ceramics with different La_2O_3 addition are shown in Fig.4.

All the specimens showed a similar microstructure consisting of elongated Si_3N_4 grains and a small amount of AlN grains. At the meantime, a number of pores distributed between the grains were also observed as a result of hot-etching, which indicated a high percentage of glass phase existing in the system. As can be seen in Fig.4, with La_2O_3 content increased from 4wt% to 10wt%, the amount of glass phase increased, which dramatically promoted the solution-precipitation process and grain growth during sintering process.

Bending strength

Fig.5 shows the bending strength of Si_3N_4-AlN composite ceramics with various La_2O_3 content. It can be seen that all materials exhibited high room-temperature strength, however, with the increase of La_2O_3 content from 4wt% to 10wt%, the bending strength reduced from 574MPa to 480MPa gradually. This could be attributed to more glass phase and $La_2Si_6N_8O_3$ formation. Fig.6 shows the SEI micrograph of the fracture surface of Si_3N_4-AlN composite ceramics with 8wt% La_2O_3. As can be seen, intergranular fracture was the leading fracture mode due to the weak bonding between grains and intergranular phase. Therefore, the more La_2O_3 was added, the more intergranular phase was formed in the systems, which led to the dramatic decrease of bending strength.

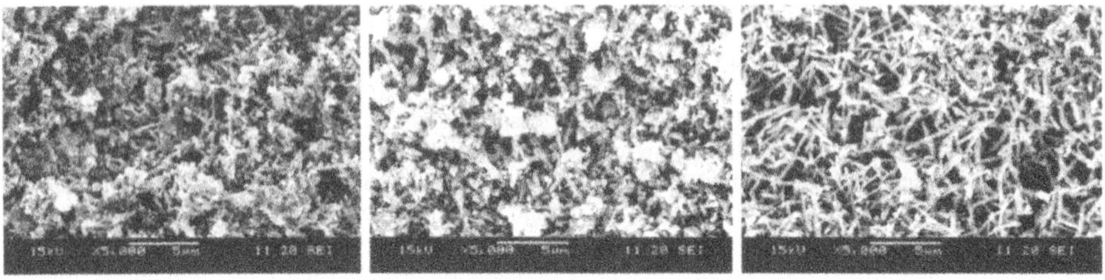

(a) (b) (c)

Fig.4 SEI micrographs of the etched surfaces of Si_3N_4-AlN composite ceramics with various La_2O_3 content: (a) 4wt%, (b) 8wt%, and (c) 10wt%.

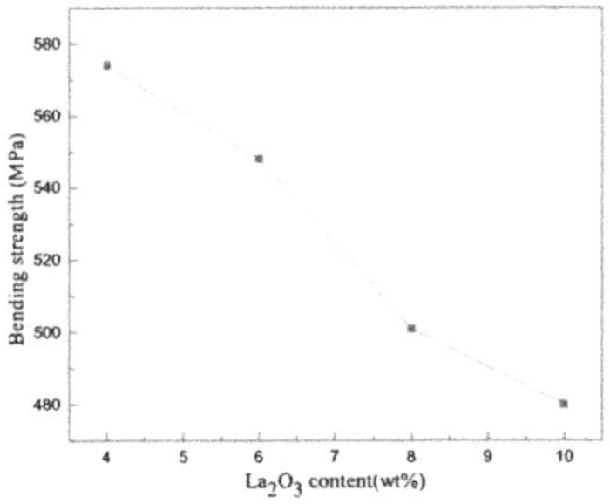

Fig.5 Bending strength of Si_3N_4-AlN composite ceramics with different La_2O_3 content

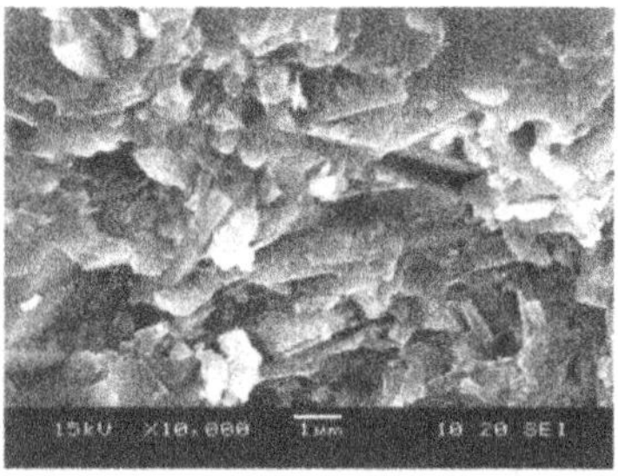

Fig.6 SEI micrograph of the fracture surface of Si_3N_4-AlN composite ceramics with 8wt% La_2O_3

Dielectric loss and thermal conductivity

Fig.7 demonstrates the relationship between the dielectric loss of Si_3N_4-AlN composite ceramics and La_2O_3 addition. It can be seen that with the increase of La_2O_3 content, the dielectric loss decreased rapidly and then increased slightly, which reached the minimum value of 4.55×10^{-3} at 6wt% La_2O_3. The dielectric loss is influenced greatly by the microstructure characteristics such as porosity, grain size and glass phase content. The decrease of dielectric loss at first can be explained by the increment of density. With the increment of La_2O_3 content from 6wt% to 10wt%, there was no remarkable variation in density. So, the increase of dielectric loss can be ascribed to the grain growth and increasing amount of glass phase, which is in agreement with other works reported[9,10].

Effect of La_2O_3 content on thermal conductivity of Si_3N_4-AlN composite ceramics is shown in Fig.8. With the increase of La_2O_3 addition, the thermal conductivity increased firstly and then decreased, exhibiting a reverse variation trend compared with tanδ. However, the variation was not remarkable and the highest thermal conductivity of only $11.7 W\cdot m^{-1}\cdot K^{-1}$ was obtained at 6% La_2O_3

content. Obviously, the thermal conductivity of Si_3N_4-AlN composite ceramics is much lower than that of pure AlN and Si_3N_4 ceramics. Such a low thermal conductivity is believed to be caused by the relative large amount of glass phase existing in the materials, which has very low thermal conductivity. As can be observed in SEM results, the glass phase distributed continuously around AlN and Si_3N_4 grains and therefore greatly decreased the thermal conductivity.

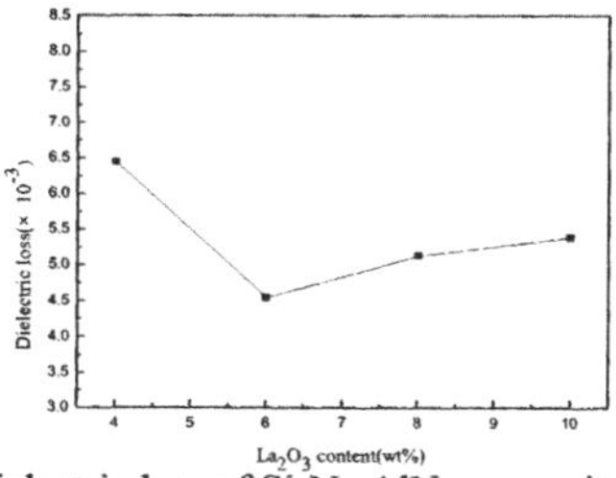

Fig.7 Dielectric loss of Si_3N_4-AlN composite ceramics with different La_2O_3 content

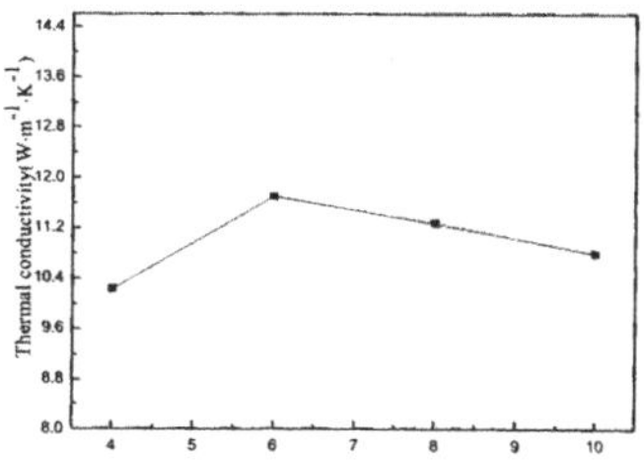

Fig.8 Thermal conductivity of Si_3N_4-AlN composite ceramics with different La_2O_3 content

CONCLUSION

(1).Si_3N_4-AlN(30wt%) composite ceramics were prepared by hot-pressing at 1800 °C using La_2O_3 as additive. With the increase of La_2O_3 content, the phase transformation of α-Si_3N_4 to β-Si_3N_4 was promoted and more lanthanum melilites ($La_2Si_6N_8O_3$) existing in the materials.

(2).The increase of La_2O_3 content showed a negative effect on bending strength of the materials, which decreased from 574MPa to 480MPa with La_2O_3 addition increasing from 4wt% to 10wt%. The leading fracture manner of the materials was intergranular fracture, owing to the existence of relative large amount of intergranular phases.

(3).With the increase of La_2O_3 content from 4% to 10%, dielectric loss of the materials decreased and then increased, while for thermal conductivity, the case was on the contrary. The lowest dielectric loss(4.5×10^{-3}) and highest thermal conductivity($11.7W\cdot m^{-1}\cdot K^{-1}$) were obtained at 6% La_2O_3 content.

ACKNOWLEDGEMENT

This work was supported by the Commission of Science, Technology and Industry for National Defence.

REFERENCE

[1]G. A. Slack , R. A. Tanzilli, R. O. Pohl, J. W. Vandersande. The Intrinsic Thermal Conductivity of AlN. *J. Phys Chem Solids*., 48. 641-47 (1987).

[2]K. Watari, H. J. Hwang, M. Toriyama, S. Kanzaki. Effective Sintering Aids for Low-temperature Sintering of AlN Ceramics. *J. Mater Res*., 14.1409-17 (1999).

[3]K. Watari, Kiyoshi Hirao, Motohiro Toriyama, Kozo Ishizaki. Effect of Grain Size on the Thermal Conductivity of Si_3N_4. *J. Am. Ceram. Soc*., 82. 777-79 (1999).

[4]K. Hirao, K. Watari, M. E. Brito, M. Toriyama, S. Kanzaki. High Thermal Conductivity in Silicon Nitride with Anisotropie Microstructure. *J. Am. Ceram. Soc*., 79.2485-88 (1996).

[5]K.Watari, K. Hirao, M. E. Brito, M. Toriyama, S. Kanzaki. Hot Isostatic Pressing to Increase Thermal Conductivity of Si_3N_4 Ceramics. *J. Mater. Res.*, 14. 1538-541 (1999).

[6]Tajima, Kenichi, Uchimura, Hideki, Tanaka, Koichi, Kohsaka, Shoji, Maruyama, Hiroshi. Dielectric Material Having a Low Dielectric Loss Factor for High-frequency Use. *U. S. Patent*, 5885916,

3.23.1997.
7S. Hampshire, H.K. Park, D. P. Thompson, α'-Sialon Ceramics. *Nature*, 274: 880-82(1978).
[8]V. A. Izhevskiy, L. A. Genova, J. C. Bressiani and F. Aldinger. Progress in SiAlON Ceramics. *J. Eur Ceram. Soc.*, 20. 2275-95 (2000).
[9]M.K. Park, H.N. Kim, K.S. Lee, S.S. Baek, E.S. Kang, Y.K. Baek, D.K. Kim. Effect of Microstructure on Dielectric Properties of Si_3N_4 at Microwave Frequency. *J. Key Eng. Mater.*, 287. 247-52 (2005).
[10]H. Miyazaki, Y. Yoshizawa, K. Hirao. Effect of Crystallization of Intergranular Glassy Phases on the Dielectric Properties of Silicon Nitride Ceramics. *Mater. Sci. Eng.*, 148.257-60 (2008).

VAPOR SILICON INFILTRATION FOR FIBER REINFORCED SILICON CARBIDE MATRIX COMPOSITES

Qing Zhou[1], Shaoming Dong[1], Haijun Zhou[1] and Dongliang Jiang[1]
[1] Shanghai Institute of Ceramics, Chinese Academy of Sciences, Shanghai, 200050, China

ABSTRACT

At high temperature, there is very low silicon vapor pressure acting as an ideal gas. It can infiltrate into open pores with any pore size, and deliver much smaller exothermal heat. Using this way, C_f/SiC composites were fabricated. When the infiltration temperature increased, the mass gain increased, accompanied with density increase and open porosity decrease. Strength increased and then kept nearly the same, but the work of fracture decreased with temperature above 1923K. At 1973K, its density and open porosity were 2.25g/cm^3 and 6%, respectively. Its strength and work of fracture reached about 240MPa and 9.54 kJ·m^{-2} at 1973K. There were no α-SiC formed during the VSI process because its a less violent reaction. However, the silicon content was large (14.5vol. %) and its distribution was heterogeneous with large size. Stress-displacement curve showed its non-brittle fracture behavior.

INTRODUCTION

Silicon carbide (SiC) matrix composites have been fabricated by chemical vapor infiltration (CVI), polymer impregnation and pyrolysis (PIP), and reaction sintering (RS). The RS process can be recognized as an attractive technique, because it offers a high density and good thermal conductivity, compared to those of CVI and PIP process. In general, the fabrication of fiber reinforced SiC matrix composites by reaction sintering involves melt infiltration (MI) or liquid silicon infiltration (LSI). However, the fabrication of continuous fiber reinforced SiC matrix composites by RS focused in melt infiltration (MI) such as liquid silicon infiltration (LSI)[1,2,3]. Vapor silicon infiltration was rarely used for SiC matrix composites.

In this paper, the vapor silicon pressure and reaction to C were calculated. C_f/SiC composites were also manufactured by VSI at high temperature and properties were investigated.

CALCULATION OF VSI

It is supposed to reach chemical equilibrium when liquid and gaseous silicon transforms to each other above melting point of silicon (1414 °C). The saturated gas pressure can be expressed by Clapeyron equation:

$$\frac{dP}{dT} = \frac{\Delta_{vap} H_m}{T \Delta_{vap} V} \tag{1}$$

Where P is vapor pressure (Pa), T is temperature (K), $\Delta_{vap}H_m$ is enthalpy of vaporization and $\Delta_{vap}V$ is the change of volume during vaporization.

At this high temperature, it can be written by Clausius-Clapeyron equation when considering the gaseous silicon to be ideal gas.

$$\frac{d \ln P}{dT} = \frac{\Delta_{vap} H_m}{RT^2} \tag{2}$$

Where R is a constant, 8.31J/mol·K. Then the vapor pressure of gaseous silicon is evaluated to be

90-100Pa at 1700 °C, by referring the data of $\Delta_{vap}H_m \approx 392$kJ/mol.
At high temperature, gaseous silicon can be considered to be an ideal gas due to its vapor pressure of ~100Pa. The amount of gaseous silicon can be derived with vapor silicon infiltration conducted in a sealed crucible with a diameter of 180mm and a height of 150mm.

$$P=nkT \\ =NkT/V \tag{3}$$

Where k denotes Boltzmann constant and N is the amount of gas (mol). When T is about 1727 °C(2000K),

$$N=4.44\times10^{-51}\text{mol}$$

The enthalpy and Gibbs free energy calculated according to their chemical reaction at 2000K:

Si(g) +	C(s) =	SiC(s)		(4)
69.111kJ/mol	-46.069kJ/mol	-192.143kJ/mol	ΔG=-215.185kJ/mol	
486.235kJ/mol	35.323kJ/mol	7.499kJ/mol	ΔH=-514.059 kJ/mol	

This means that SiC can de derived at this temperature, due to ΔG=-215.185kJ/mol<0. The root-mean-square velocity of gaseous silicon can be expressed according to statistical thermodynamics as:

$$\sqrt{\bar{v}^2}=\sqrt{\frac{3RT}{\mu}} \tag{5}$$

where μ is 28.085×10^{-3}kg/mol for silicon so the root-mean-square velocity of gaseous silicon can be figured out.

$$\sqrt{\bar{v}^2}=1.33\times10^{3}\text{m/s}$$

The amount of gaseous silicon can be derived from its velocity, and then the heat delivery in the unit time can be obtained.

$$\begin{aligned} Q&=\Delta H\times\frac{1}{t}\times N \\ &=\Delta H\times\frac{\sqrt{\bar{v}^2}}{l}\times N \\ &=514.059\ \text{kJ/mol}\times\frac{1.33\times10^{3}}{150\times10^{-3}}/\text{s}\times4.44\times10^{-51}\text{mol} \\ &=2.02\times10^{-44}\text{kJ/s} \end{aligned}$$

For liquid silicon infiltration, Capillary force is the driving force. It can be stated

$$\Delta P=\frac{2\sigma\cos\theta}{r} \tag{6}$$

Where σ is the interfacial energy of solid-liquid, θ is wetting angle and r is radius of pore. The length of liquid silicon in the pore can be expressed by writing

$$l = (\frac{\sigma \cos\theta}{2\eta} r\tau)^{\frac{1}{2}} \quad (7)$$

Where η is viscosity of liquid silicon and τ is time. It is reported that θ is 0°for C-Si system and it is ~8°for SiC-Si. σ (mN/m) can be denoted in 1459-1845K

$$\sigma(T) = 721.13 - 0.0615 \times (T - T_m) \quad (8)$$

And viscosity of molten silicon in 1635-1845K is

$$\eta(T) = 0.5572 - 5.39 \times 10^{-4}(T - T_m) \quad (9)$$

With a hypothesis of a porous carbon with a porosity of 30%, a diameter of 2μm, and a specific surface area of $300m^2 \cdot g^{-1}$. Thus the length of silicon in pore reaches 25.58mm per second.
When liquid silicon infiltration are conducted at 1427 °C (1700K),

Si(l) +	C(s) =	SiC(s)		(10)
-69.477kJ/mol	-34.4393kJ/mol	-163.460kJ/mol	ΔG=-59.544 kJ/mol	
86.597kJ/mol	28.021kJ/mol	-8.426kJ/mol	ΔH=-123.044 kJ/mol	

The heat given off is about

Q=N×ΔH
=680.43kJ/s

From the above calculation, it is found that there is much less heat delivery per unit time during vapor silicon infiltration than liquid silicon infiltration. It is also noted that the driving force of molten silicon is greatly affected by the pore size (diameter), which might result in problem (clogging, etc.). However, gaseous silicon, acting like ideal gas, can enter into any open pore.

FABRICATION OF SiC MATRIX COMPOSITES BY VSI

PyC-SiC bi-layer interphase was deposited on 3D M40JB carbon fiber (Toray, Tokyo, Japan) preform as interfacial layer with methane and MTS/H_2 precursors by FP-CVI[4]. The coated fabrics were impregnated by the slurry, followed by pyrolysis at 2073K to form a porous body. After pyrolyzation, the pyrolyzed body contained a carbon matrix and a distribution of micropores was put in a sealed graphite crucible and enough silicon powder was put below the sample. The gaseous Si, which was nearly similar to perfect gas, penetrated into the porous channel and concurrently facilitated the reaction to form SiC. The temperature was raised to the target value at a heating ramp rate of 10K/min, and VSI process was conducted at high temperature (1873K-1973K) for 3 hours inside a vacuum furnace (~1Pa).

The density was measured by Archimedes' method. The flexure strength were tested by three point bending way using an INSTRON 5566 test machine (Canton, MA, USA) with a span of 24 mm and a crosshead speed of 0.5 mm/min. The work of fracture was obtained from the characteristic area under the stress–displacement curve divided by the cross section of the specimen. In order to determine the work of fracture effectively, we defined the characteristic area which started from initial point to the 60% drop of the curve. After the test, the silicon volume content was calculated according to the mass change before and after immersed in 70wt%HF+30wt%HNO_3 acid. The weight loss tests were conducted at 1273K in muffle furnace in atmosphere. The heating rate was 10K/min.

In Table 1, some properties of C_f/SiC composites before and after VSI were shown. It was noted that before the VSI process, the preform before VSI has nearly the same density, which is between 1.52-1.56g/cm^3. After VSI at different temperature, the mass gain was different. It increased with the temperature and it reached 57.8% at 1973K. It clarified that the VSI temperature resulted in different mass gain at different temperature.

Table 1 Properties of C_f/SiC composites before and after VSI

VSI temperature/K	1873	1923	1973
Density of the perform before VSI/g·cm^{-3}	1.51	1.56	1.52
Mass gain of the perform/%	10.7	18.6	57.8
Flexure strength/MPa	149.0±13.3	239.5±35.6	238.9±41.2
Work of Fracture/ kJ·m^{-2}	--	12.13	9.54

The bulk density and the open porosity of the C_f/SiC composites are shown in Figure 1 as a function of infiltration temperature. It was also found that trend of density. As the infiltration temperature increased from 1873K to 1973K the bulk density of the composites increased from 1.68g/cm^3 to 2.25g/cm^3 while the open porosity decreased from 28% to 6%. It is known[5] that the reaction between silicon vapor and carbon depend on the concentration of silicon vapors and the infiltration temperature. In this case it was found that at 1873K, the amount of reactive vapor silicon was not enough to react with the carbon in the matrix, resulting in the composites with the lowest density. At higher temperatures, a larger amount of vapor silicon reacted with the carbon and resulted in denser composite with better mechanical behavior. When the temperature was 1973K, the silicon vapor became abundant. Some reacted with carbon forming SiC matrix and the other remained in the matrix.

Table 1 showed that the strength increased with VSI temperature at first, but the strength kept nearly the same above 1923K.It was about 240MPa. Work of fracture (WOF), which was indicated by area under the curve, also increased from the 1873 to 1923K but it decreased when T increased to 1973K. Table 1 showed the work of fracture (WOF) at 1923K and 1973K, which were 12.13 and 9.54 kJ·m^{-2}. It can be ascribed to the large silicon (14.5vol.%) content at 1973K, which results in the volume increase (~8vol.%) during cooling, making the strength and WOF decrease.

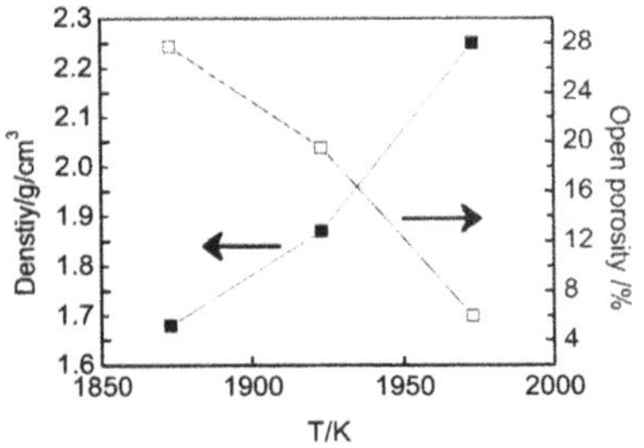

Figure 1 Density and porosity of the composites

Under TEM, the PyC and SiC interphase was observed, as shown in Figure 2 (a). As a result of the interphase, the crack was deflected during the propagation. Figure 2 (b) showed that the crack changed from the transverse direction to longitudinal direction, which bring about larger work of fracture and better non-brittle fracture of VSI composites. It can be said that the fiber protected from gaseous silicon during the VSI process as a result of the interphase.

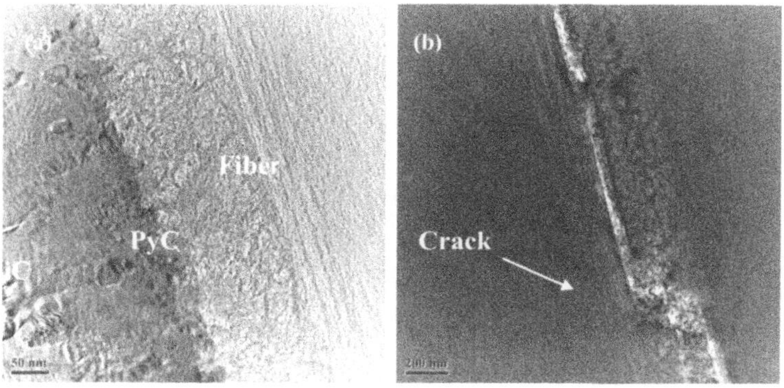

Figure 2 PyC-SiC interphase (a) and the crack propagation (b) in the fiber surface

Figure 3 (a) showed the cross section of composites after VSI for 3h at 1973K. It was observed that dense matrix was also received by VSI process even in the intra-fiber area. However, micropores were also observed in the intra-fiber area. It was regardless of VSI process but resulted from other reasons. After fiber coating process, fibers adhered to each other tightly and became accumulated tows. They were difficult to be separated and prone to result in the formation of micropores. During VSI, the amount of gas silicon increased with temperature and it resulted in larger mass increase after reaction with C at high temperature. The density was dependent on vapor silicon infiltration temperature and the higher temperature was more advantageous to make dense matrix. Especially at 1973K, the dense matrix were received in the inter- and intra- bundle areas, which demonstrated that VSI was an appropriate solution to fabricate dense SiC matrix and form dense C_f/SiC composites.

Figure 3 (b) showed the RS SiC and residual Si distribution in the matrix. The RS SiC (grey area) was not continuous and distributed in the white Si area. The SiC area was more than 20 μm in size. It can be concluded that the pyrolyzed carbon from resin distributed discontinuously and it mainly distributed around the fiber bundles. At 1973K, the larger amount of vapor silicon (more than 100Pa) infiltrated and made the C template become SiC so the VSI SiC kept discontinuous just like the carbonized resin template. However, there was large gap area without RS, which was filled by the Si during the cooling process. In the end, the high density composites was received and the residual silicon content also high about 14.5 vol. %.

CONCLUSION

Vapor silicon has a very low pressure, facilitating its superior infiltration ability and low exothermal heat during reaction. Based on theory calculations, VSI was conducted for the fabrication of C_f/SiC composites. As the infiltration temperature increased from 1873K to 1973K the bulk density of the composites increased from 1.68g/cm^3 to 2.25g/cm^3 while the open porosity decreased from 28% to 6%. Strength increased and then kept nearly the same, but the work of fracture decreased with temperature above 1923K. Its strength and work of fracture reached about 240MPa and 9.54 kJ·m^{-2} at 1973K. However, the silicon content was large (14.5vol. %) and its distribution was heterogeneous with large size. The crack was deflected during propagation as a result of PyC-SiC interphase, making the composites have larger work of fracture. Stress-displacement curve showed its non-brittle fracture behavior.

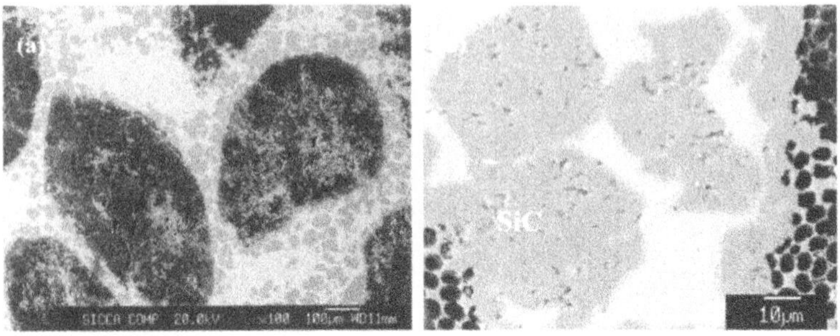

Figure 3 Cross section (a) and Si distribution in the matrix (b) of VSI composites made at 1973K

ACKNOWLEDGEMENT

This work was supported by the National Natural Science Foundation Program of China under Grant No. 50472015 and the National High Technology Research and Development Program of China (863 Program) under project No. 2006AA03Z565.

REFERENCES

1.H. Lin and M. Singh, "Evaluation of Lifetime Performance of Hi-Nicalon™ Fiber-reinforced Melt-infiltrated SiC Ceramic Composites," *Ceram. Trans.*, 144, 207-20 (2002).

2.G. Morscher, "Stress-dependent Matrix Cracking in 2D Woven SiC-fiber Reinforced Melt-infiltrated SiC Matrix Composites," *Compos. Sci. Technol.*, 64 [9] 1311-9 (2004).

3.E. Tani, K. Shobu, and K. Kishi, "Two-dimensional-woven-carbon-fiber-reinforced Silicon Carbide/Carbon Matrix Composites Produced by Reaction Bonding," *J. Am. Ceram. Soc.*, 82 [5] 1355-57 (1999).

4.Qing Zhou, Shaoming Dong, Xiangyu Zhang, Yusheng Ding, Zhengren Huang and Dongliang Jiang, "Carbon Fiber Surface Coating by Forced Pressure-pulsed CVI",J. Inorg. Mater, 21 [6], 1378-1384 (2006).

5.Y. Wang, "The preparation of reaction-formed SiC/Si composite from controllable porous carbon by Si infiltration," Ph. D. Thesis, Institute of Graduate, Chinese Academy of Sciences, China (in Chinese), 2004.

TAILING PROPERTIES OF C_f/SiC COMPOSITES VIA MODIFICATION OF MATRIX COMPOSITION

Shaoming Dong*, Zhen Wang, Yusheng Ding, Xiangyu Zhang, Ping He and Le Gao
Structural Ceramics Engineering Research Center, Shanghai Institute of Ceramics, Chinese Academy of Sciences
Shanghai 200050, China

ABSTRACT

C_f/SiC-BN composites and C_f/SiC-ZrC composites were fabricated to improve the oxidation resistance and high temperature performance of C_f/SiC composites through the modification of matrix. SiC-BN matrix was formed through an *in-situ* reaction of active filler boron and protective gas N_2 in the active-filler-controlled polymer pyrolysis (AFCOP). The oxidation performance of C_f/SiC-BN composites was greatly improved when oxidized at 1000°C compared to that of C_f/SiC composite. Meanwhile, SiC-ZrC matrix was fabricated using the ZrC particles as inert filler. Both C_f/SiC-BN composites and C_f/SiC-ZrC composites show non-catastrophic fracture behavior. The microstructures were also characterized by SEM and EDS. It was shown that the fiber reinforcement hindered the impregnation of solid particles into the fiber bundles so that most of the fillers remained in the inter-bundle matrix and most of the intra-bundle matrices were composed of SiC that resulted from the decomposition of polycarbosilane (PCS).

KEY WORDS: matrix modification, microstructure, filler

INTRODUCTION

Due to the excellent high-temperature performance of SiC-matrix composites, much attention in the past decades was focused on the development of composites consisting of SiC matrix, such as C_f/SiC and SiC_f/SiC composites[1,2]. It is well-known that the properties of CMCs strongly rely on the consolidation of fiber, matrix and their interphase (and/or interface). Sound microstructures such as those with low porosity, and with proper thickness of interphase will provide high performance to the composites. High fracture toughness of CMCs is achieved through a proper design of the fiber/matrix (FM) interface to arrest and deflect cracks formed under load in the brittle matrix and prevent the early failure of the fibrous reinforcement[3]. Therefore, some interphases such as PyC, h-BN, $(PyC/SiC)_n$ and $(h\text{-}BN/SiC)_n$ have been developed to improve the performance of continuous fiber reinforced ceramic matrix composites[4]. However, to improve the properties of C/SiC composites such as oxidation resistance or high temperature performance, modification of the matrix composition is also an effective method.

As a result of microcracks from the mismatch of coefficient of thermal expansion and residual pores formed in the fabrication processes, the in-depth diffusion of oxidative gases will be favored[3], which would lead to the easy oxidation and degradation of PyC interphase and carbon fibers. Therefore, the principal obstacle for long-term applications of carbon-containing materials is known to be their relatively poor oxidation resistance[5].The improvement of oxidation resistance is thus one of the main problems to solve for the application of carbon fiber reinforced CMCs to long-term applications [6]. Boron-bearing species are efficient to improve the oxidation resistance of C/SiC composites at relatively low temperatures. An important property of all boron-bearing species is the formation of fluid oxide phases (B_2O_3 or B-M-O ternary phase) over a broad temperature range (600-1200°C for B_2O_3) when heated in an oxidizing atmosphere[7]. The formed B_2O_3 fluid will seal the cracks and improve the diffusion barrier to oxidative gases.

The maximum use temperature of silicon based ceramics is limited to about 1600°C due to the onset of active oxidation and lower temperatures in water vapor environments. The development of structural materials for use in oxidizing and rapid heating environments at temperatures above 1600°C is therefore of great importance[8]. However, hypersonic flight vehicles including sub-orbital and earth-to-orbit vehicles need sharp nose-caps and leading edges, which are projected to require reusable materials capable of operating at 2000 to 2400°C in air[9,10]. The introduction of UHTCs into C_f/SiC composites allows these materials as potential candidates for extreme environments associated with hypersonic flight and rocket propulsion due to their retained strength at high temperatures[11]. Zirconium compounds have exceptional properties. The melting points of ZrC and ZrB_2 are as high as 3540°C and 3040°C, respectively, and their oxide can effectively reduce the diffusion rate of oxygen. The formation of a ZrO_2 coating may be a good method to reduce the ablation of carbon materials at ultra-high temperatures[12]. UHTC particles have been introduced into C/C composites to improve their high temperature performances[11].

The purpose of this work was to study the effects of filler addition on the properties and microstructues of both C_f/SiC-BN composites and C_f/SiC-ZrC composites.

EXPERIMENTAL

Sample fabrication

Boron (Tangshan WeiHao Magnesium Powder Co., Ltd, Tangshan, China) and ZrC(Kai'er Nanometer Technology Development Co. Ltd., Hefei, China) powders were applied as active fillers and inert fillers, respectively, to fabricate C_f/SiC-BN composites and C_f/SiC-ZrC composites. Fillers were ball milled with PCS (National University of Defense Technology, Changsha, China) for 48h to form the homogenously dispersed slurries, using xylene as solvent. Carbon fibers (T700SC, 12K and M40JB, 6K, Toray, Japan) for C_f/SiC-BN composites and C_f/SiC-ZrC composites respectively, deposited with interphases by CVI were impregnated into the aforementioned slurries and stacked in a graphite die and dried. Pressures were applied to control their thicknesses to achieve a fiber volume of ~45% in the hot-pressing furnace at ~200°C. Then the samples were densified by several cycles of PIP using PCS as the polymer precursor. The pyrolysis temperatures were 800°C and 1100°C for C_f/SiC-BN composites and C_f/SiC-ZrC composites, respectively. Furthermore, composites with boron were heat-treated at 1800°C in N2 to favor the *in-situ* formation of h-BN.

Oxidation of C_f/SiC-BN composites

After being cut and ground into 5mm×2mm×20mm specimens, a SiC coating of ~10μm thickness was deposited to isolate the sample cross-sections from the air for oxidation. Oxidation tests were performed in a muffle furnace. The specimens were pushed into the furnace at 1000°C and then taken out quickly to weigh the mass after each 2h oxidation.

Sample characterization

The composites were cut and ground into bars for three-point-bending test in an Instron-5566 machine with a crosshead speed of 0.5mm/min and a span of 24mm. Both the polished cross section and fracture surface were observed by Electron Microprobe Analysis (EPMA; JXA-8100, JEOL, Tokyo, Japan) equipped with EDS.

RESULTS AND DISCUSSION

C_f/SiC-BN composite

Fig.1 shows the back-scattered electron micrographs of the polished cross-section of C_f/SiC-BN composite. It is clear that though the inter-bundle matrix is dense, there are still some pores remaining in the intra-bundle areas. Such pores formed because of hindrance to impregnation of solid particles in the slurries by the fiber tows, which led to a higher solid concentration in the inter-bundle areas and a higher PCS concentration in the intra-bundle areas. It has been proved that the incorporation of fillers can decrease the volume shrinkage accompanying the polymer pyrolysis process, so there are more pores left in the intra-bundle areas. It can also be observed that there are some black pots in both inter-bundle matrix and intra-bundle matrix, which can correspond to h-BN due to the lower atom weight of both boron and nitrogen atom.

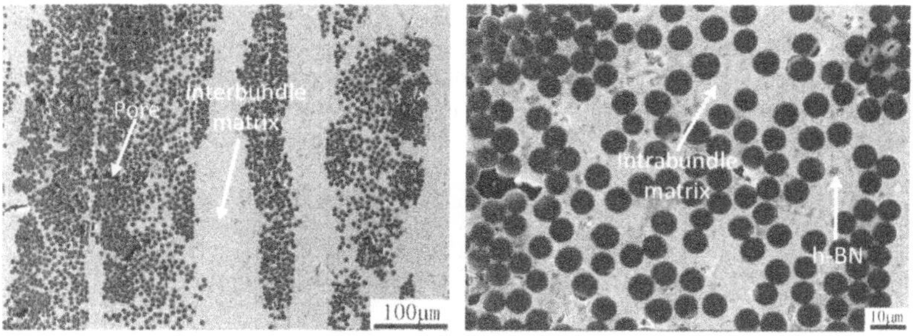

Fig1. Back-scattered electron images for the polished cross-sections of C_f/SiC-BN composites

The weight loss curves of composites oxidized at 1000°C are plotted in Fig.2. The weight loss curves of C_f/SiC-BN composite were compared with those of C_f/SiC composite which was just oxidized for 10h considering the high weight loss. It is obvious that the weight losses of the C_f/SiC-BN composite are much smaller than those of C_f/SiC composite; only about 5% weight loss was evident for C_f/SiC-BN after being oxidized for 20h, which means that with the introduction of boron as active filler, the oxidation resistance of carbon fiber reinforced CMCs can be greatly improved.

Fig.3 shows the bending stress/displacement curves of C_f/SiC-BN composites. Typical non-catastrophic fracture behaviors were observed both before and after oxidizing at 1000°C for 20h but there are some small decreases in both bending stress and elastic modulus after oxidation.

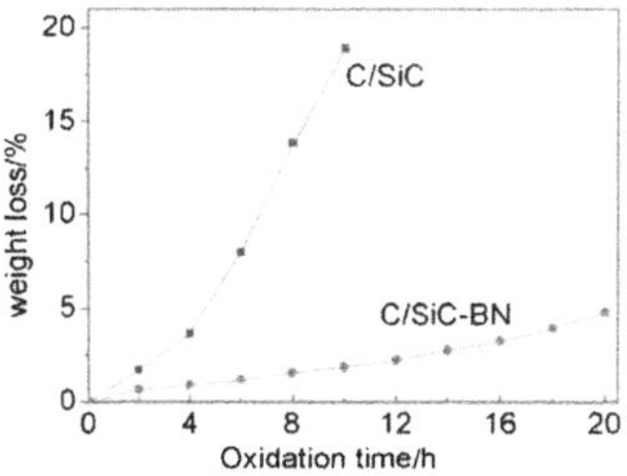

Fig.2 Weight loss curves of composites oxidized at 1000°C

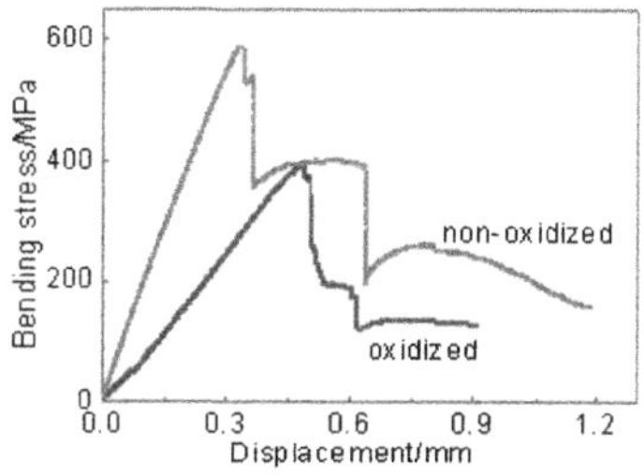

Fig.3 Bending stress/displacement curves for composites before and after oxidation at 1000°C for 20h

C_f/SiC-ZrC composite

Micrographs of the polished cross-section of C_f/SiC-ZrC composite are shown in Fig.4. Some matrix-rich areas existing between the fiber bundles are almost fully dense except for a few partially filled large cracks. However, when the intra-bundle matrix is considered, a large number of pores are noted to exist; this may be attributed to the same reason mentioned above. It can also be found that there are three phases with different backgrounds in the matrix. The results of EDS shown in Fig.5 indicate that the white spots were actually large ZrC particles, the gray areas are a combination of ZrC and SiC resulting from the first infiltration and pyrolysis of slurry, and the black zones are almost completely SiC resulting from the PIP process that followed. Moreover, the oxygen content increased with the increase of zirconium, which can be attributed to the easy oxidation of ZrC [13], especially particles in nano-scales. After the mold-pressing process, there may be some cracks and some pores left, so in the subsequent PIP process, PCS were infiltrated in and transferred to SiC. As a result of the closure of open pores after several cycles of infiltration and pyrolysis, and the large shrinkage of PCS during pyrolysis, some of the cracks and pores can only be partially filled.

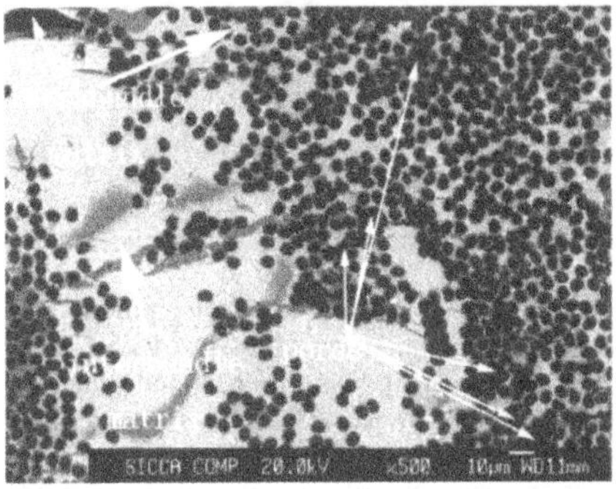

Fig.4 SEM micrograph of polished cross-section of the C_f/SiC-ZrC composite

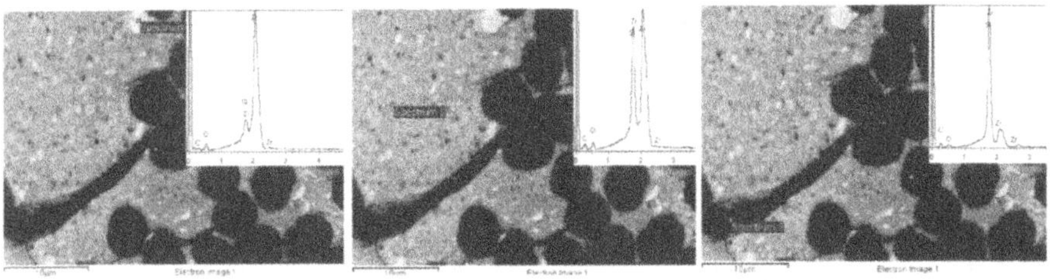

Fig.5 EDS analysis of Cf/SiC-ZrC composite

The bending stress/ displacement curve from three-point-bending test plotted in Fig.6 reveals that the composite shows non-brittle fracture behavior and the bending stress reaches 493.7±36.7MPa. Consistent with the non-brittle fracture behavior, some fiber pull-out can be observed from the fracture surface of C_f/SiC-ZrC composite and the pulled-out fiber surface is smooth as a result of the weak bonding between the fiber and the matrix due to the existence of an interphase.

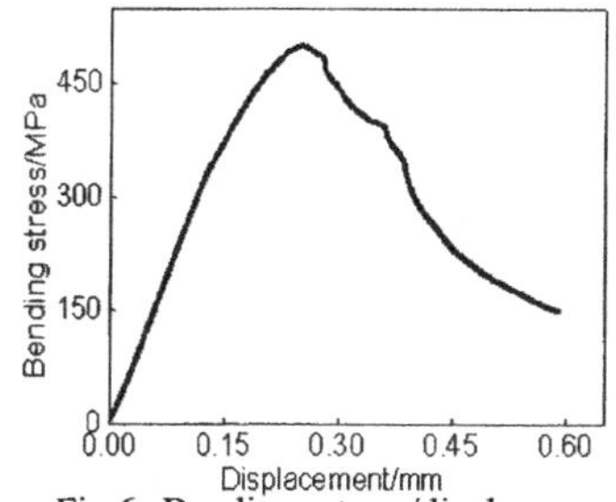

Fig.6 Bending stress/displacement curves for Cf/SiC-ZrC composite

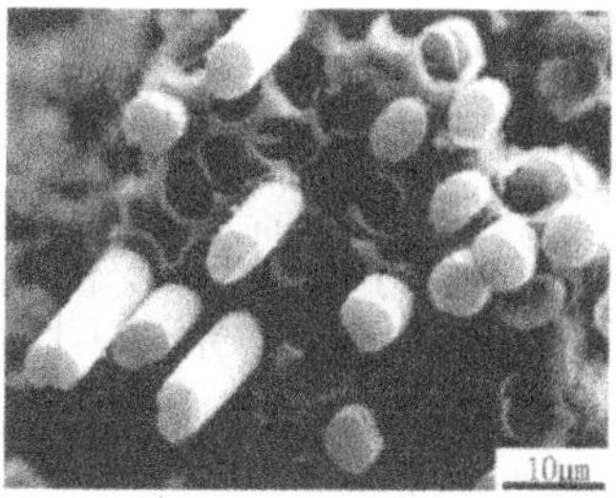

Fig.7 Scanning electron microscope micrographs on the fracture surface of the

CONCLUSION

Both C_f/SiC-BN composite and C_f/SiC-ZrC composite show non-brittle fracture behavior in the three-point-bending test. Due to the hindrance to the impregnation of filler particles by carbon fiber tows, there is a higher filler concentration in the inter-bundle area. The inter-bundle matrix is dense and some pores remain in the intra-bundle matrix. With the introduction of boron as active filler, the oxidation resistance of the as-derived composite increased greatly. Only a small decrease occurred in bending stress and elastic modulus after oxidation at 1000°C for 20h. The bending stress of C_f/SiC-ZrC composite reaches 493.7±36.7MPa. Three phases with different background (white, gray and black) in representative micrographs of the matrix correspond to ZrC, ZrC-SiC mixture and SiC, respectively, and the oxygen content increases with the increase of ZrC.

ACKNOWLEDGEMENT

This work was supported by the National High Technology Research and Development Program of China (863 program) under Project No. 2006AA03Z565.

REFERENCES

1 R.Naslain, SiC-matrix composites: nonbrittle ceramics for thermostructural application, Int. J. Appl. Ceram. Technol., 2(2005), 75-84

2 S.M. Dong, Y. Katoh, A. Kohyama, Preparation of SiC/SiC composites by Hot pressing, using Tyranno-SA fiber as reinforcement, J. Am. Ceram. Soc., 86 (2003), 26-32

3 R Naslain, Design, preparation and properties of non-oxide CMCs for application in engines and nuclear reactors: an overview, Compos. Sci. Technol., 64(2004) 155-170

4 R Naslain, The design of the fibre- matrix interfacial zone in ceramic matrix composites. Compos. Pt. A-Appl. Sci. Manuf. Part A, 29(1998) 1145-1155

5 Y.J. Lee, L.R. Radovie, Oxiadtion inhibition effects of phosphorus and boron in different carbon fabrics, Carbon 41 (2003) 1987-1997

6 F. Lamouroux, S. Bertrand, R. Pailler, R, Naslain, M. Cataldi, Oxidation-resistant carbon-fiber-reinforced ceramic-matrix composites, Compos. Sci. Technol 59(1999) 1073-1085

7 R. Naslain, A. Guette, F. Rebillat, R. Pailler, F. Langlais, X. Bourrat, Boron-bearing species in ceramic matrix compistes for long-term aerospace applications, J. Solid State Chem. 177(2004), 449-456

8 E. Wuchina, M. OPeka, A. Causey et al, Designing for ultrahigh-temperature applications: The mechanical and thermal properties of HfB_2, HfC_x, HfN_x and aHf(N), J. Mater. Sci. 39(2004) 5939-5949

9 M.M. Opeka, I.G. Talmy and J. A. Zaykoski, Oxidation-based materials selection for 2000°C + hypersonic aerosurfaces: Theoretical considerations and historical experience, J. Mater. Sci., 39(2004) 5887-5904

10 D.M. Van wie, D.G. Drewry JR., D.E. King and C.M. Hudson, The hypersonic environment: Required operating Conditions and design challenges, J. Mater. Sci., 39(2004) 5915-5924

11 S.F. Tang, J.Y. Deng, S.J. Wang, W.C. Liu and K. Yang, Ablation behaviors of ultra-high temperature ceramic composites, Mater. Sci. Eng. A-Struct. Mater. Prop. Microstruct. Process. A, 465(2007) 1-7

12 X.T. Li, J.L. Shi, G.B. Zhang, H. Zhang, Q.G. Guo and L. Liu, Effect of ZrB_2 on the ablation properties of carbon composites, Mater. Lett., 60 (2006) 892-896

13 Zirconia growth on zirconium carbide single crystals by oxidation, Surf. Coat. Technol., 197(2005):294-302

STATUS AND CRITICAL ISSUES OF SiC/SiC COMPOSITES FOR FUSION APPLICATIONS

Zhou Xingui, Yu Haijiao, Cao Yingbin, Liu Rongjun, Wang Honglei, Zhao Shuang, and Luo Zheng
State Key Laboratory of Advanced Ceramic Fibers and Composites,
National University of Defense Technology,
Changsha, Hunan, 410073, P.R. China

ABSTRACT

Owing to their outstanding properties, silicon carbide fiber reinforced silicon carbide matrix (SiC/SiC) composites have been specified in several applications especially in recent fusion power plant design studies and have been considered internationally in several power plant studies. These characteristics include high-temperature properties and stability, corrosion resistance, as well as low induced radioactivity, quick decay of activity, low afterheat, low atomic number, good fracture resistance and more.

This paper summarizes the recent progress in design and R&D status of the SiC/SiC composites for fusion energy applications. Meanwhile, several technologically critical issues that will be solved for fusion applications, such as thermal properties, Pb-17Li compatibility, hermeticity, joining technique and protective coatings are also identified.

INTRODUCTION

SiC/SiC composites are being considered as candidates for structural materials for fusion reactors because of their excellent high-temperature properties and stability, corrosion resistance, thermal conductivity, as well as their low-induced radioactivity by 14 MeV neutron irradiation, quick decay of activity, low afterheat, low atomic number and good fracture resistance.

In recent years, experimental activities on industrial SiC/SiC composites have highlighted their main features under irradiation and provided important guidelines for further development of radiation compliant materials. Equal efforts have focused on design studies, advanced fiber fabrication, new methods for matrix processing and joining technology, and testing of properties.

Though the above efforts have led to improved properties in many aspects, critical issues related to the SiC/SiC composites for nuclear applications are still present. These issues are mainly connected to the *He* bubble effects under fusion reactor environment, poor thermal properties, the residual porosity, and to the Pb-17Li compatibility. Meanwhile, some technology issues must be addressed, for example, joining methodology and coating for hermeticity.

This paper briefly reviews the recent progress in design and R&D status of SiC/SiC composites for fusion reactor applications in the initial section. It also overviews the remaining critical issues related to nuclear applications of SiC/SiC composites, such as transmutation gases, thermal properties, Pb-17Li compatibility, hermeticity, joining techniques and protective coatings. Finally, emphases of future work on SiC/SiC composites for fusion applications are prospected.

PROGRESS IN DESIGN AND R&D STATUS OF SiC/SiC COMPOSITES

Proposed fusion reactor blanket design concepts

SiC/SiC composites were considered several years ago in the ARIES-I [1], ARIES-IV [2]and PROMETHEUS [3]power plant studies in the US and more recently in other international studies. Examples also include the TAURO [4], a self-cooled Pb-17Li blanket (SCLL), whose latest reference design assumes a surface heat flux of 0.5 MW/m^2, and shows a Pb-17Li outlet temperature of 950℃ with an estimated conversion efficiency of ~55%; the *He* cooled pebble bed (HCCB) concept—DREAM [5]reactor study in Japan, with assumed inlet/outlet *He* coolant temperatures of ~600/900°C and a gross thermal efficiency of ~50%; and the latest SCLL conceptual designs, EU Power Plant Conceptual Study (PPCS) Model D [6], with assumed *He* inlet/outlet temperatures of ~600/700°C and efficiency > 50%, and US ARIES-AT [7,8], with the assumed lowest/highest operating temperatures for SiC/SiC structures of ~700/1000°C yielding a power conversion efficiency of ~60% for the blanket circuit [5].

R&D of the SiC/SiC composites has made remarkable progress in many aspects during the last few decades. Firstly, many kinds of high performance SiC fiber, such as Hi-Nicalon S, Sylramic and Tyrano-SA, have been developed. Table 1 presents the composition and characteristics of some advanced SiC fibers [9]. Secondly, both various new technologies and modifications of the existing processes were developed to obtain SiC/SiC composites with improved properties. For example, RS (Reaction Sintering) method for improving both strength and thermal conductivity, CVI-PIP (Chemical Vapor Infiltration — Polymer Infiltration and Pyrolysis) routine for effectively controlling the densification and the accessory shape, HP (Hot-Pressing) technology for higher density, CVR (Chemical Vapor Reaction) process for high thermal conductivity across the thickness and lower cost, and NITE (Nano-Infiltrated Transient Eutectoid) process for pseudo-ductile properties and nearly full-dense SiC matrix production [10]. Finally, the efforts to make appropriate fiber-matrix interphase layer with various kinds of materials and microstructure, by methods such as PIP, CVD (Chemical Vapor Deposition) and EPD (Electronic Physical Deposition) were successful, resulting in the production of high strength and high fracture toughness SiC/SiC composites.

Table I. Composition and Characteristics of Advance SiC Fibers

Properties		Nippon Carbon		Ube Industries			Dow Corning	Bayer AG
		Nicalon		Tyranno			Sylramic	Siboramic
		Hi-Nicalon	Hi-Nicalon S	LoxE	ZE	SA		
Composition	Si	63.7	68.9	~56	~61	67.8	66.6	~34
	C	35.8	30.9	~37	~35	31.3	28.5	~12
	O	0.5	0.2	5.0	2.0	0.3	0.8	1.0
	N	-	-	-	-	-	0.4	~40
	B	-	-	-	-	-	2.3	11.6
	Al	-	-	-	-	0.6	-	-
	Ti	-	-	2.0	-	-	2.1	-
	Zr	-	-	-	2.0	-	-	-
Density /g •cm^{-3}		2.74	3.10	2.55	2.55	3.1	>3.1	1.85
Diameter /μm		14	12	11	11	11	10	12-14
Tensile strength/GPa		2.8	2.6	3.4	3.5	>2.5	2.8-3.4	4.0
Yong's modulus/GPa		270	420	206	2.33	>300	386	290
HHR* / ℃		1600	>1800	1300	1400	>1900	>1800	>1800

*The highest heat-resistance (under inert atmosphere).

Flow channel insert application

For the dual coolant design, to reduce the MHD (magnetohydrodynamic)-induced pressure drop in the flowing Pb-17Li, an electrically and thermally insulating component called a Flow Channel Insert (FCI) is located within each steel channel. Therefore, it must satisfy the following requirements:

- Low transverse electrical conductivity (1-50S/m) and thermal conductivity (<2W/m·K)
- Chemically compatible with Pb-17Li at temperature up to ~800℃ or even ~1000℃ in a flowing system and with FS (ferritic steel) at lower temperatures (400-500℃)
- Near 100% dense 'sealing layers' or alternate strategy to insure its impermeability to the liquid Pb-17Li
- Good thermal shock and thermal cycling behavior (in-plane E~100GPa) (That means resistance to

high temperature gradients and primary stresses caused by normal and off-normal tokamak operation events like disruption.)

- Irradiation resistance and acceptable changes in thermal and electrical conductivity, dimensional and structural integrity at projected operating temperatures of 500-800°C
- Non-magnetic
- Fabrication to match flow channel geometry [11]

SiC/SiC has low electrical conductivity, allowing for sufficient reduction of the MHD pressure drop and heat loss, and good compatibility with Pb-17Li flow at temperatures approaching 800℃. So it has been proposed as a promising material for the construction of FCIs in several advanced Pb-17Li blanket concepts [12]. This idea was proposed initially in the EU PPCS Model C DCLL(Dual-Cooled Lead-Lithium) blanket concept, of which the design inlet/outlet Pb-17Li temperatures are 480/700℃, and the US ARIES-ST blanket design[13]. It is also a candidate for the proposed DCLL US Test Blanket Module (TBM) for ITER [14,15].

In recent years, study on the thermal conductivity of SiC/SiC composites for FCI application has significantly increased. It seemed that 2D architecture was more suitable for such an application. Through analysis of EC and TC for various SiC and SiC/SiC forms, Youngblood et al [16]also suggested that an architectural or "engineering" composite design would be necessary to achieve the desired transverse EC and TC goals for 2D SiC/SiC. Typical through-thickness non-irradiated thermal conductivity for CVI 2D SiC-matrix composites with radiation-resistant SiC fibers at 500℃ is ~15 W/m·K [17]. Under neutron irradiation at the same temperature, the thermal conductivity decreases to a saturation value of 2-4 W/m·K at 1 dpa, which is in the range required for the FCI.

Katoh et al [18]used the linear thermal resistance model and the temperature dependent defect thermal resistance data for the composites' constituents to predict the effect of neutron irradiation on temperature dependent thermal conductivity of the composites. The prediction suggested that the maximum and minimum post-irradiation through-thickness conductivity was 10-15 W/m·K at 800-1000℃for 3D architecture, and less than 5 W/m·K at <800℃ for 2D architecture.

The electrical conductivity for SiC/SiC is strongly affected by the matrix conductivity; and in the fiber direction, it is usually dominated by conduction through the PyC (Pyrolitic Carbon) interphase [19,20]. By 2-probe DC (direct current) and AC (alternating current) and 4-probe DC methods, Youngblood et al [21]measured the EC-values of 2D Hi-Nicalon S/PyC interphase/ICVI SiC matrix composites, and composites with the PyC interphase layer removed by oxidation (in plane) at temperatures up to 500℃. It was found that by removing the PyC interphase, in plane EC of the composites decreased ~ 30%.

Electrical conductivity of chemically vapor deposited or infiltrated SiC is most commonly determined by the nitrogen and other impurity concentrations, and typically has a range of 100-1000 S/m at 500℃. A readily achievable through-thickness electrical conductivity for CVI SiC/SiC with impurity controlled matrix is in the range of 10-100 S/m [22]. Thus, many of the assumed basic requirements are already satisfied or advanced to a substantial extent.

REMAINING CRITICAL ISSUES

Transmutation gases

Extensive production of *He* occurs in materials as a result of nuclear transmutation due to fusion neutrons. *He* is an insoluble element in almost all the solid materials. It stabilizes vacancy type clusters, interacts with the radiation defects and impurities, and consequently may alter various irradiation effects on physical or mechanical properties. Therefore, its effect must be taken seriously.

Calculation of the total amount of transmuted *He* was carried out by Noda [23]. At the first wall region of the solid breeding blanket, the *He*/dpa ratio is predicted to be ~100 appm/dpa. Sawan and El-Guebaly [24]also indicated production ratios of 80-170 appm *He*/dpa and 30-70 appm H/dpa at the peak radiation regions of the first wall for various blanket concepts.
In experiments, transmuted *He* gas effects are simulated using *He* implantation or simultaneous *He*-ion and heavy-ion irradiations. Single-, dual-, or triple-beam ion irradiation experiments and TEM were used to investigate *He* effects on microstructural evolution. The *He* diffusion behavior and microstructural observation of SiC/SiC composites, SiC fiber and monolithic SiC were reported by Hasegawa [25]. Results suggested that *He* might be mobile in C and the amorphous like Si-C phase in the composite in the range of the expected reactor operating temperature. Microstructural observations of *He*-implanted SiC/SiC composites were also reported by Hasegawa [26]. *He* bubbles were neither observed at the interface between matrix/interphase (C) or between fiber/interphase (C), nor in the SiC fiber after about 10000 appm *He* implantation at room temperature and annealing at 1400℃. Bubbles were observed only at grain boundaries in SiC. *He* desorption results can explain this microstructure. More research is required to clarify the mechanical property degradation of SiC fibers and composites caused by *He*. Also, further modifications in experimental facilities and investigations of *He*/*H* effects are needed [27].

Thermal properties

In order to be used as fusion structural materials, quite the contrary to the FCI case, enhanced thermal conductivity will be mandatory for SiC/SiC composites. The value of the thermal conductivity should comply with design requirements for heat removal and minimization of thermal stresses. Thermal property of SiC is highly dependent on both composition and microstructure. At room-temperature, it can vary from 1 W/m·K for nano crystalline fibers up to 490 W/m·K for high-purity single crystal SiC, according to impurity content, lattice defect density, average grain size, porosity and the presence of amorphous and/or interfacial phases [25]. In the same way, for current industrial Nicalon CG fiber /SiC (CVI) composites, the thermal properties are relatively modest, for example, 12 W/m·K for current 3-D Nicalon CG fiber /SiC (CVI). This is due to the low thermal conductivity of Nicalon CG fiber, considerable matrix porosity and micro-cracks, and small grain size. Therefore, it is advisable to stress the microstructure improvement of both the fiber and the matrix.
As fiber is a primary component in continuous fiber reinforced ceramic matrix composites, its characteristic is an important factor that confines the thermal conductivity of the composites. The ideal SiC fiber should be highly crystalline, oxygen-free, and stoichiometric. As shows in Table 1, many kinds of new SiC fiber such as Hi-Nicalon S [28], Sylramic and Tyrano-SA [29]are all oxygen-free and stoichiometric, and they all have higher thermal conductivities compared to other SiC fibers. On the other hand, densification of the matrix is also a key aspect to improve the thermal conductivity of SiC/SiC composites. Yamada [30]showed the density dependence of thermal conductivity of the SiC/SiC composites fabricated by CVI process. Thermal conductivity was shown to increase with sample density, and it was suggested that localized large cavities had a worse effect on the thermal diffusivity than smaller size pores distributed in the whole volume. To improve thermal conductivity, matrix processing for porosity control and densification are required for fusion applications [25]. Many new technologies were developed to improve the thermal properties of the composites, such as RS, F-CVI, modified PIP, CVI-PIP, HP, CVR and NITE. 31,32,33.
As is shown in Table 2 [18,31-33], for one thing, stoichiometric SiC fibers have higher thermal conductivity; for another thing, SiC/SiC composites processed by CVR [34], RS [35], or NITE [36]show relatively higher thermal conductivity compared to the PIP and CVI ones because of their higher bulk density. Meanwhile, architecture and configuration of the fiber perform are important to

thermal properties of the SiC/SiC composites as well. Furthermore, accession of medium with fairly high thermal conductivity is also an attractive way.

Although efforts to meet high thermal conductivity goals are continuing, there's a new goal to develop or design a radiation-resistant SiC/SiC-FCI with low thermal conductivity. In fact, the new value desired for the FCI-application is <2W/m·K. Until the last decade, little effort has been made in reducing the thermal conductivity, but circumstances have now changed with the advent of the SiC/SiC-FCI concept.

Table II. Some Typical Through-thickness Thermal Conductivity Data of SiC/SiC Composites [18,31-33]

Method	Fiber	Configuration of fiber texture	Fiber volumes (%)	Bulk density (g/cm^3)	Thermal conductivity (W/m·K)
CVI (DuPont)	Hi-Nicalon	2D	40	2.63	12.1(RT) 10.3 (1000℃)
CVI (DuPont)	Nicalon	2D	36	2.3	2.5(RT)
CVI	Tyranno SA	3D	34	2.82–2.88	45 (RT) 24 (1000℃)
CVI	Tyranno SA	3D Orthogonal, x:y:z = 1:1:1	45	2.8	46.1 (RT)
CVI	Tyranno SA	3D Orthogonal, x:y:z = 1:1:4	45	2.8	58.2 (RT)
CVI	Tyranno SA/P120S (Graphite fiber)	3D Hybrid, x:y:z = 1:1:1.8	45	2.2	53 (RT)
CVI	Hi-Nicalon Type S	3D	40	2.9	36 (RT) 20 (1000℃)
CVI	Hi Nicalon	2D, Plain-weave		2.55	13 (RT) 10 (1000℃)
CVI	Tyranno SA	2D, Plain-weave	40	2.5	18.9 (RT)
CVI	Hi-Nicalon Type S	2D, Plain-weave	40	2.6	18.6 (RT)
CVI	Hi-Nicalon Type S	2D, 5-Harness satin-weave	44	2.5.	18.1 (RT)
PIP	Hi-Nicalon	3D Satin, x:y:z = 1:1:0.2			3.6 (RT)
PIP	Hi-Nicalon Type S	3D Satin, x:y:z = 1:1:0.1			4.9 (RT)
PIP	Hi-Nicalon Type S	3D Satin, x:y:z = 1:1:0.2			5.9 (RT)
PIP/CVI	Tyranno SA	3D	34	2.71–2.81	38 (RT) 18 (1000℃)
PIP/CVI	Tyranno SA/ Graphite fiber	2D Hybrid, Plain-weave		2.55	86 (RT) 28 (1000℃)
CVR	CVR （T-300 1K 8HS fabric）	2D, Plain-weave		2.65	75 (RT) 35 (1000℃)
RS	Hi-Nicalon	UD	30	3.0	50 (RT) 30 (1000℃)
NITE	Tyranno SA	UD			29 (RT) 20 (1000℃)

Thermal conductivity of irradiated SiC decreases due to the accumulation of point defects that effectively scatter phonons in SiC. In the 500-800℃ range of interest, the thermal conductivity of conventional SiC/SiC composites at 1 dpa could be in the range required for the FCI [17]. As the point defects accumulate, meeting the requirement of low conductivity becomes much easier. So as far as the FCI application is concerned, the critical time for SiC/SiC composites is the beginning of life (BOL) for irradiation exposure. Unfortunately, SiC/SiC composites derived from advanced SiC fibers and crystalline SiC matrix (to preserve radiation stability) likely will have thermal conductivity values even greater than the Hi-Nicalon reference values of 10-14 W/m·K, which are well above the required ones for a SiC/SiC-FCI at BOL. Therefore, to provide the desired extremely low thermal conductivity for the FCI, advanced 2D SiC/SiC composites will have to be specially designed and fabricated [37].

Pb-17Li Compatibility

Liquid Pb-17Li is considered as the coolant and breeding material for the TAURO and ARIES-AT reactor designs. Fenici and Scholz [38]reported that CVI SiC/SiC composites were stable in a static solution of Pb-17Li at 800℃ for up to 1500h. They concluded that SiC should be very stable in this environment because the free energy change for the following reaction is about +99 kJ/mol over the temperature range of 700-900 ℃:

$$2SiC \oplus 2Li \longrightarrow Li_2C_2 \oplus 2Si \quad (1)$$

Terai et al [39]reported that SiC/SiC composites and monolithic SiC exhibited excellent stability in Pb-17Li at 300 and 500℃ for 666 h exposure. The largest weight loss reported was 1.5% for SiC/SiC composites with Hi-Nicalon fibers, PyC fiber/matrix interface and matrix produced by the PIP process. Two other composite materials showed a factor of 10 less weight loss. In contrast, samples immersed in Li at 427℃ for the same periods of time were totally dissolved with the exception of a high-purity, monolithic CVD SiC. Reactions with other phases such as residual Si and C are a likely reason for the high reactivity of the other samples tested. In another experiment, SiC was unchanged after 1500h exposure in the alloy at 1073℃ [40]. In a review of temperature limits for fusion reactors, Zinkle and Ghoniem [41]concluded the limits for SiC in Li and Pb-17Li.
Pint et al [42]studied the Pb-17Li compatibility of high-purity CVD β-SiC. The initial 1000h exposures at 800℃ and 1100℃ showed no mass change and no increase in the Si content (30 ppma detection limit) of the Pb-17Li. After 5000h at 800℃, no increase in the Si content was measured. But after 2000h at 1100℃ and 1000h at 1200℃, increased levels of Si were detected in the Pb-17Li (185 and 370 ppma, respectively) [43]. Based on these results, the maximum use temperature of SiC composites in Pb-17Li appears to be ≤1100℃. However, the current corrosion results in Pb-17Li are limited to static immersion tests [44]. To fully determine the compatibility of these materials, future work will need to include testing in a flowing system with a thermal gradient.

Coating

Hermeticity of composites is an important issue for first wall and blanket structure applications that require a pressure boundary and/or gas or fluid containment. It is widely considered to be a weakness of common SiC composites that they are not fully dense, for the matrix micro-cracks, existing more or less in the as-fabricated CVI, PIP or MI SiC/SiC. That may easily cause unacceptable *He* leaks [45]. A hermetic seal coating likely will be required for such applications [46,47]. Exactly, it is both for hermeticity considerations [48]and for corrosion protection of the fine-grained SiC fibers and of the fiber/matrix interphase material [49,50].
In the ARIES-I design, CVD-SiC coatings[51] were considered. Glass-ceramics can be versatile joining or coating materials with tailorable thermal and mechanical properties; in addition, they are not affected by oxidation and can be self-sealant at temperatures above the glass softening point. A eutectic composition in the binary phase $CaO\text{-}Al_2O_3$ glass has been studied for joining and coating materials because it has a high characteristic temperature with B and Li. The coated composites did not seem to be affected after 140h at 800℃ in contact with a ceramic breeder [52]. PIP SiC coatings

also have been studied as a gas permeation barrier of CVI-processed composites [53].

Joining

Further, the SiC/SiC component only can be fabricated into simple shapes that require development of attachment technologies in order to achieve more complex configurations. Nevertheless, limitations occur in assembling complex geometry SiC/SiC elements due to the fact that: (1) Welding is not possible due to SiC fibers partially losing their mechanical properties as a result of the presence of tramp properties (for example oxygen) before reaching the melting temperature; and (2) Diffusion bonding does not seem suitable as the inter-diffusion of SiC is very low, even at high temperature. For these reasons the availability of a reliable joining technique is fundamental to realize fusion reactors structures. Several joining techniques are under development, which include: reaction-bonding, homogeneous joining by preceramic polymers [54,55], brazing with evolved special system and joining by glass and glass ceramic [56]. Besides, the materials utilized should be with low activation elements, operate at high temperature (800-1000℃), retain radiation stability even at high temperature and be chemically compatible with the coolant and breeder. Thus, the corresponding candidate joining materials are: (1) melt infiltrated and reaction-formed SiC, (2) preceramic-polymer derived SiC, (3) in situ reinforced silicides, (4) high-temperature braze, and (5) low activation, high-temperature glasses.

By preceramic polymers, several joints were obtained through PIP, RS, and new NITE processes [48,50]. The NITE SiC/SiC composites were produced to demonstrate high-performance hot-pressed joining [57,58], and its tensile strengths is > 200 MPa; however, the joining process requires pressurization at high temperatures in a controlled environment. A homogeneous joining technique has been developed by ENEA and Padua University [59]. Relevant shear strength, measured by means of an almost pure shear test was obtained for a joining sintered α-SiC (40 MPa). The use of a high-temperature adhesive based on Ceraset preceramic polymer was reported by Fareed [60]. Bend strength of joints using Ceraset was about 40-60 MPa at 1100℃ and at room temperature, respectively. A well known difficulty with preceramic polymers is the mass loss, which can exceed 50%, on conversion to a ceramic phase. A slightly different approach uses a linear chain of polyhydridomethylsiloxane (PHMS) as a precursor to a highly crosslinked polysiloxane, by applying a catalytic chemistry approach developed at SRI International. That has some advantages, such as much lower mass loss [61,62,63] on ceramic conversion compared to other systems and pyrolysis occurs at temperatures as low as 600℃.

For the high-temperature braze technique, a brazing alloy series for joining 2D and 3D SiC/SiC composites has been developed at CEA Grenoble and named Brasic® [64]. This brazing compound was composed of low activation elements and was conceived to work at elevated temperatures. In particular the brazing system contains a sufficient amount of silicon to prevent reaction with the SiC substrate and promote good wetting and to induce appropriate infiltration in the composites. Using a Brasic® Ti-Si system and a brazing temperature close to 1350℃, an effective joint has been obtained with limited infiltration. On this joint, tensile tests up to 1200℃ are in progress. Using Brasic V3® and carrying out the joining at 1300℃ in a neutral atmosphere a sound joint was obtained with a perfect filling of the joint gap but no infiltration of the composite. For this joint a shear strength of 174 MPa was obtained at room temperature and about 100 MPa at 800℃. This system is currently the most promising. However, high open porosity of the composites and free silicon content have been shown to be the main limitations for brazing [65].

Encouraging results were also obtained using a glass ceramic phase to join SiC CMCs: Singh [66] has reported on reaction-forming methods of SiC/SiC composites. The flexural strength of the joints was about 150 MPa at RT, and about 217 MPa at 1350℃. For calcia-alumina glass-ceramics joints, which have been studied for joining and coating materials [52], the strength of the joint was 28 MPa at room temperature.

There are reports on the use of solid state displacement reactions, too. The joints were made from TiC and Si powder processed at 1400℃, and their shear strength at room temperature reached 50 MPa [67]. Nevertheless, these joints were weak compared to SiC/SiC composites, so strength improvement is required for their applications to gas/liquid cooled reactor systems. Additionally, irradiation behavior of these joints will also require to be studied [25].

CONCLUSION

The recent progress in design and R&D status and several technologically critical issues of SiC/SiC composites for fusion applications were reviewed. Future work should put the emphases on:

1. The remaining technological issues of SiC/SiC composites for fusion applications such as transmutation effect, thermal properties, especially through the thickness, Pb-17Li compatibility, hermeticity should be further studied.
2. To fulfill the goal of final fusion application, advanced test methods and facilities for test fusion application materials that can create circumstances much closer to the real operation environment of the fusion reactor should be developed. The clarification of deformation and fracture mechanisms and the models modification at fusion reactor environment are also required.
3. Nowadays, R&D of SiC/SiC composites for fusion applications should be in collaborations with non-fusion application programs on a more extensive scale.

REFERENCES

1 F. Najmabadi, R. W. Conn, P. I. H. Cooke, S. P. Grotz, M. Z. Hasan, E. Ibrahim, T. Kunugi, T. K. Mau, R. C. Martin, and S. Sharafat, The ARIES-I Tokamak Reactor Study- The Final Report, UCLA Rep., UCLA-PPG-1323, UCLA, CA, USA, 1991.

2 F. Najmabadi, R.W. Conn and The Aries Team, The ARIESII and ARIES-IV Second-stability Tokamak Reactors, TMTFE-10, La Grange Park, IL, USA, 1992.

3 L.M. Waganer, Innovation Leads the Way to Attractive IFE Reactors-Prometheus-L & Prometheus-H, The 5th IAEA Technical Committee Meeting and Workshop on Fusion Reactor Design and Technology, CA, USA, 1993.

4 H. Golfier, G. Aiello, and L. Giancarli, Progress on the TAURO Blanket System, CEA internal report, SERMA/LCA/RT/00-2837/A, 2000.

5 S. Nishio, S. Ueda, I. Aoki, R. Kurihara, T. Kuroda, H. Miura, T. Kunugi, Y. Seki, T. Nagashima, M. Ohta, J. Adachi, S. Yamazak, I. Kawaguchi, T. Hashimoto, K. Shinya, Y. Murakami, H. Takase, and T. Nakamura, Improved Tokamak Concept Focusing on Easy Maintenance, Fusion Eng. Des., 41[1], 357-64 (1998).

6 D. Maisonnier, I. Cook, P. Sardain, R. Andreani, L.Di Pace, R. Forrest, L. Giancarli, S. Hermsmeyer, P. Norajitra, N. Taylor, and D. Ward, A Conceptual Study of Commercial Fusion Power Plants: Final Report of the European Power Plant Conceptual Study (PPCS), EFDA Rep., EFDA-RP-RE-5.0, 2004.

7 A.R. Raffray, L. El-Guevaly, S. Gordeev, S. Malang, E. Mogahed, F. Najmabadi, I. Sviatoslavsky, D.-K. Sze, M.S. Tillack, X. Wang, and A. Team, High Performance Blanket for ARIES- AT Power Plant, Fusion Eng. Des., 58-59, 549-53 (2001).

8 A.R. Raffray, L. El-Guebaly, S. Gordeev, S. Malang, E. Mogahed, F. Najmabadi, I. Sviatoslavsky, D.K. Sze, M.S. Tillack, X. Wang, and the ARIES Team, High Performance Blanket for ARIES-AT Power Plant, The 21st Symposium on Fusion Technology, Madrid, Spain, 2000.

9 H.M. Yun, and J.A. Di Carlo, Comparison of the Tensile, Creep, and Rupture Strength Properties of Stoichiometric SiC Fibers, Ceram. Eng. Sci. Proc., 20[3], 259-72 (1999).

10 Y. Katoh, S.M. Dong, and A. Kohyama, Thermo-mechanical Properties and Microstructure of

Silicon Carbide Composites Fabricated by Nano-infiltrated Transient Eutectoid Process, Fusion Eng. Des., 61-62, 723-31 (2002).
11 C.P.C. Wong, V. Chernov, A. Kimura, Y. Katoh, N. Morley, T. Muroga, Y.C. Wu, and M. Zmitko, ITER-Test Blanket Module Functional Materials, J. Nucl. Mater., 367-370, 1287-92 (2007).
12 C. Wong, S. Malang, M. Sawan, S. Smolentsev, S. Majumdar, B. Merrill, D.K. Sze, N. Morley, S. Sharafat, P. Fogarty, M. Dagher, P. Peterson, H. Zhao, S. Zinkle, and M. Youssef, Assessment of Liquid Breeder First Wall and Blanket Options for the DEMO Design, The 16th ANS TOFE Meeting, Madison, 2004.
13 S. Smolentsev, M. Abdou, N.B. Morley, M. Sawan, S. Malang, and C. Wong, NumericalAnalysis of MHD Flow and Heat Transfer in a Poloidal Channel of the DCLL Blanket with a SiC/SiC Flow Channel Insert, Fusion Eng. Des., 81, 549-53 (2006).
14 M. Abdou, D. Sze, C. Wong, M. Sawan, A. Ying, and N. Morley, US Plans and Strategy for ITER Blanket Testing, The 16th ANS TOFE Meeting, Madison, 2004.
15 C.P.C. Wong, S. Malang, M. Sawan, M. Dagher, S. Smolentsev, B. Merrill, M. Youssef, S. Reyes, D.K. Sze, N.B. Morley, S. Sharafat, P. Calderoni, G. Sviatoslavsky, R. Kurtz, P. Fogarty, S. Zinkle, and M. Abdou, An Overview of Dual Coolant Pb-17Li Breeder First Wall and Blanket Concept Development for the US ITER-TBM Design, Fusion Eng. Des., 81, 461-67 (2006).
16 G.E. Youngblood, R.J. Kurtz, and R.H. Jones, Development of SiC/SiC for Application as a Flow Channel Insert, Fus. Mater. Semi-Annual Prog. Rep., DOE/ER-0313/37, 3-8 (2004).
17 G.E. Youngblood, D.J. Senor, and R.H. Jones, Effects of Irradiation and Post-irradiation Annealing on the Thermal Conductivity/Diffusivity of Monolithic SiC and F-SiC/SiC Composites, J. Nucl. Mater., 329-333, 507-12 (2004).
18 Y. Katoh, T. Nozawa, L.L. Snead, T. Hinoki, and A. Kohyama, Property Tailorability for Advanced CVI Silicon Carbide Composites for Fusion, Fusion Eng. Des., 81, 937-44 (2006) .
19 L.L. Snead, Limits on Irridation-induced Thermal Conductivity and Electrical Resistivity in Silicon Carbide Materials, J. Nucl. Mater., 524, 329-33 (2004).
20 R. Scholz, F. Dos Santos Marques, and B. Riccardi, Electrical Conductivity of Silicon Carbide Composites, J. Nucl. Mater., 307-311, 1098-101 (2002).
21 G.E. Youngblood, E. Thomsen, and G. Coffey, Electrical Conductivity of SiC/SiC, Fus. Mater. Semi-Annual Prog. Rep., DOE/ER-0313/39, 46-51 (2005).
22 L.L. Snead, S.J. Zinkle, and D.P. White, Thermal Conductivity Degradation of Ceramic Materials Due to Temperature Low Dose Neutron Irradiation, J. Nucl. Mater., 340, 187-202 (2005).
23 T. Noda, H. Araki, S. Ito, M. Fujita, and K. Maki, Proc. 2nd IEA/JUPITER Joint Int. WS on SiC/SiC Ceramic Composites for Fusion Applications, Sendai, Japan, 244-54 (1997).
24 T. Tanabe, Reconsideration of Hydrogen Recycling Process at the First Wall, Proc. The 4th Int. WS on Tritium Effects in Plasma Facing Components, Santa Fe, New Mexico, 47-49 (1998).
25A. Hasegawa, A Kohyama, R.H. Jones, L.L. Snead, B. Riccardi, and P. Fenici, Critical Issues and Current Status of SiC/SiC Composites for Fusion, J. Nucl. Mater., 283-287, 128-37 (2000).
26 A. Hasegawa, M. Saito, S. Nogami, K. Abe, R.H. Jones, and H. Takahashi, Helium Bubble Formation Behavior of SiCf/SiC Composites After Helium Implantation, J. Nucl. Mater., 264, 355-58 (1999).
27 Y. Katoh, L.L. Snead, C.H. Henager Jr., A. Hasegawa, A. Kohyama, B. Riccardi, and H. Hegeman, Current Status and Critical Issues for Development of SiC Composites for Fusion Applications, J. Nucl. Mater., 367-370, 659-71 (2007).
28 H. Ichikawa, K. Okamura, and T. Seguchi, Oxygen-free Ceramic Fibers From Organo-silicon Precursors and E-beam Curing, Proc. Second International Conference on High-temperature Ceramic Matrix Composites, Santa Barbara, CA, USA, 65-74 (1995).
29 T. Nakayasu, M. Sato, T. Yamamura, K. Okamura, Y. Katoh, and A. Kohyama, Recent Advancement of Tyranno/SiC Composites R&D, Ceram. Eng. Sci. Proc., 20, 301-08 (1999).

30 R. Yamada, T. Taguchi, N. Igawa , and J. Nakano, Thermal Conductivity of CVI and PIP SiC/SiC Composites, Ceram. Eng. Sci. Proc., 20[3], 273-80 (1999).
31 D.B. Fischbach, The Kinetics and Mechanism of Graphitization, Chemistry and Physics of Carbon (Edited by P L. Walker, Jr.), New York: Marcel Dekker, 7, 1-105, (1971).
32 R. Yamada, T. Taguchi, and N. Igawa, Mechanical and Thermal Properties of 2D and 3D SiC/SiC Composites, J. Nucl. Mater., 283-287, 574-78 (2000).
33 R. Yamada, N. Igawa, and T. Taguchi, Thermal Diffusivity/Conductivity of Tyranno SA fiber and Hi-Nicalon Type S Fiber-reinforced 3-D SiC/SiC Composites, J. Nucl. Mater., 329-333, 497-501 (2004).
34 G.E. Youngblood, D.J. Senor, T.H. Jones, and N.D. Witold Kowbel, Composites for Fusion Applications, The 9th International Conference on Fusion Reactor Materials, Colorado Springs, CO, USA, 1999.
35 A. Sayano, F. Sutoh, S. Suyama, Y. Itoh, and S. Nakagawa, Development of a Reaction-sintered Silicon Carbide Matrix Composite, J. Nucl. Mater., 271-272, 467-71 (1999).
36 J.-S. Park, A. Kohyama, T. Hinoki, K. Shimoda, and Y.-H. Park, Efforts on Large Scale Production of NITE-SiC/SiC Composite, J. Nucl. Mater., 367-370, 719-24 (2007).
37 G.E. Youngblood, D.J. Senor, R.H. Jones, and S. Graham, The Transverse Thermal Conductivity of 2D-SiC/SiC Composites, Compos. Sci. Technol., 62, 1127-39 (2002).
38 P. Fenici, and H.W. Scholz, Advanced Low-activation Materials Fibre-reinforced Ceramic Composites, J. Nucl. Mater., 212-215, 60-68 (1994).
39 T. Terai, T. Yoneoka, and S. Tanak, Compatibility Test of SiC with Liquid Metal Breeders, International Town Meeting on SiC/SiC Design and Material Issues for Fusion Systems, ORNL, Tennessee, USA, 2000.
40 V. Coen, H. Kolbe, L. Orechia, and M. Della Rossa, High Temperature Corrosion of Technical Ceramics, Elsevier Applied Science, Amsterdam, 1989.
41 S.J. Zinkle, and N.M. Ghoniem, Operating Temperature Windows for Fusion Reactor Structural Materials, Fuusion Eng. Des., 51-52, 55-71 (2000).
42 B.A. Pint, L.D. Chitwood, and J.R. Di Stefano, Compatibility of CVD SiC with Pb-17Li at 800°-1100°C, Fus. Mater. Semi-Annual Prog. Rep., DOE/ER-0313/35, 13-17 (2003).
43 B.A. Pint, J.L. Moser, and P.F. Tortorelli, Investigation of Pb–Li Compatibility Issues for the Dual Coolant Blanket Concept, J. Nucl. Mater., 367, 1150-54 (2007).
44 R.H. Jones a LG, A. Hasegawa, Y. Katoh, A. Kohyama, B. Riccardi, L.L. Snead, and W.J. Weber, Promise and Challenges of SiC/SiC Composites for Fusion Energy Applications, J. Nucl. Mater., 307-311, 1057-72 (2002).
45 T. Hino, T. Jinushi, Y. Hirohata, M. Hashiba, Y. Yamauchi, Y. Katoh, and A. Kohyama, Helium Gas Permeability of SiC/SiC Composite Developed for Blanket Component, Fusion Sci. Technol., 43, 184-90 (2003).
46 C. A. Lewinsohn, M. Singh, and C. H. Henager, Jr., Brazeless Approaches to Joining of Silicon Carbide-based Ceramics for High Temperature Applications, Ceram. Trans., 138, 201-12 (2003).
47 B. Riccardi, C. A. Nannetti, T. Petrisor, and M. Sacchetti, Low Activation Brazing Materials and Techniques for SiC/SiC Composites, J. Nucl. Mater., 307-311, 1237-41 (2002).
48 T. Hino, Y. Hirohata, Y. Yamauchi, M. Hashiba, A. Kohyama, Y. Katoh, Y. Lee, T. Jinushi, M. Akiba, K. Nakamura, H. Yoshida, S. Sengoku, K. Tsuzuki, Y. Kusama, K. Yamaguchi, and T. Muroga, Plasma Material Interaction Studies on Low Activation Materials Used for Plasma Facing or Blanket Component, J. Nucl. Mater., 329-333, 673-77 (2004).
49 C.H. Henager Jr., Y. Shin, Y. Blum, L.A. Giannuzzi, B.W. Kempshall, and S.M. Schwarz, Coatings and Joining for SiC and SiC-composites for Nuclear Energy Systems, J. Nucl. Mater., 367-370, 1139-43 (2007).
50 B.A. Pint, K.L. More, H.M. Meyer, and J.R. Di stefano, Recent Progress Addressing Compatibility

Issues Relevant to Fusion Environments, Fusion Sci. Technol., 47[4], 851-55 (2005).
51 L. Giancarli, G. Aiello, A. Caso, A. Gasse, G. Lemarois, Y. Poitevin, J.F. Salay, and J. Szczepanski, R&D Issues for SiCf/SiC Composites Structural Material in Fusion Power Reactor Blankets, Fusion Eng. Des., 48, 509-20 (2000).
52 M. Ferraris, M. Salvo, C. Isola, M.A. Montorsi, and A. Kohyama, Glass-ceramic Joining and Coating of SiC/SiC for Fusion Applications, J. Nucl. Mater., 258-263, 1546-50 (1998).
53 S.D. Connery, L.L. Snead, and D. Steiner, Hermeticity of SiC/SiC Composites Applied Stress, The 9th International Conference on Fusion Reactor Materials, Colorado Springs, CO, USA, 1999.
54 C.A. Lewinsohn, R.H. Jones, T. Nozawa, M. Kotani, Y. Katoh, A. Kohyama, and M. Singh, Silicon Carbide Based Joining Materials for Fusion Energy and Other High-temperature Structural Applications, Ceram. Eng. Sci. Proc., 22, 621-25 (2001) .
55 P. Colombo, B. Riccardi, A. Donato, and G. Scarinci, Joining of SiC/SiCf Ceramic Matrix Composites for Fusion Reactor Blanket Applications, J. Nucl. Mater., 278, 127-35 (2000).
56 B. Riccardi, P. Fenici, A. Frias Rebelo, L. Giancarli, G. Le Marois, and E. Philippe, Status of the European R&D Activities on SiCf/SiC Composites for Fusion Reactors, Fusion Eng. Des., 51-52, 11-22 (2000).
57 T. Hinoki,and A. Kohyama, Current Status of SiC/SiC Composites for Nuclear Applications, Ann. Chim. - Sci. Mat., 30[6], 659-71 (2005).
58 T. Hino, E. Hayashishita, Y. Yamauchi, M. Hashiba, Y. Hirohata, and A. Kohyama, Helium Gas Permeability of SiC/SiC Composite Used for In-vessel Components of Nuclear Fusion Reactor, Fusion Eng. Des., 73[1], 51-56 (2005).
59 B. Riccardi, A. Donato, P. Colombo, and G. Scarinci, Development of Homogeneous Joining Technique for SiC/SiCf Composites, Fusion Technology 1998-20th SOFT, Marseilles, France, 1998.
60 A.S. Fareed, C.C. Cropper, and B.R. Rossing, Joining Techniques for Fiber-reinforced Ceramic-matrix Composites, Ceram. Eng. Sci. Proc., 20[4], 61-70 (1999).
61 Y.D. Blum, D.B. and MacQueen, Modifications of Hydrosiloxane Polymers for Coating Applications, Surf. Coat. Int., B: Coat. Trans., 84[1], 27-33 (2001).
62 S.M. Johnson, Y.D. Blum, C. Kanazawa, and H.-J. Wu, Low-cost Matrix Development for an Oxide–oxide Composite, Met. Mater., 4[6], 1119-25 (1998).
63 Y.D. Blum, D.B. Mac Queen, and H.-J. Kleebe, Synthesis and Characterization of Carbon-enriched Silicon Oxycarbides, J. Eur. Ceram. Soc., 25, 143-49 (2005).
64 A. Gasse, F. Saint Antonin, and G. Coing Boyat, Specific Non Reactive BraSiC Alloys for SiC/SiC Joining, CEA-Grenoble Rep., DEMN.DR 25/97, 1997.
65 L. Giancarli, J.P. Bonal, A. Caso, G.L. Marois, N.B. Morley, and J.F. Salavy, Design Requirements for SiC/SiC Composites Structural Material in Fusion Power Reactor Blankets, Fusion Eng. Des., 41, 165-71 (1998).
66 M. Singh, A Reaction Forming Method for Joining of Silicon Carbide-based Ceramic, Scripta Mater., 37, 1151-54 (1997).
67 R.H. Jones, L.L. Snead, A. Kohyama, and P. Fenici, Recent Advances in the Development of SiC/SiC as a fusion Structural Material, Fusion Eng. Des., 41, 15-24 (1998).

PREPARATION AND CHARACTERIZATION OF C/SiC-ZrB_2 COMPOSITES VIA PRECURSOR INFILTRATION AND PYROLYSIS PROCESS

Jun Wang[1], Haifeng Hu[1], Yudi Zhang[1], Qikun Wang[1], and Xinbo He[2]
[1] CFC Lab, College of Aerospace and Materials Engineering, National University of Defense Technology, Changsha 410073, Hunan Province, P.R.China
[2] Institute of Powder Metallurgy, University of Science and Technology Beijing, Beijing 100083, P.R.China

ABSTRACT

Ultra-high temperature ceramic matrix composites (C/SiC-ZrB_2) are prepared by slurry infiltration and precursor infiltration and pyrolysis method. C/SiC-ZrB_2 composites with ZrB_2 volume content from 10% to 24.6%, have balanced performance of fracture toughness (17.7-8.1MPa-$m^{1/2}$), flexural strength at room temperature (366.7-162.8MPa) and at high temperature (strength retention 74.1% at 1800°C and over 31.8% at 2000°C), better anti-oxidation, anti-ablation under oxyacetylene torch environment (recession rate 0.01mm/s).

INTRODUCTION

Next generation hypersonic re-entry vehicles need ultra-high temperature materials as leading edges and nose caps to maintain sharp bodies to increase the lift-to-drag ratio so as to improve the vehicles' performance in many ways.[1] Solid rocket engines, with ever-increasing higher chamber pressure and aluminum content, also need zero-erosion throat materials which must withstand temperature over 3000°C. Ultra-high temperature ceramics (UHTC) are the most promising materials that stands up for these requirements and the most widely investigated systems. Numerous papers about components, processing, oxidation, and arc-jet ablation are published, and also flight experiment testified the application of UHTC.[2]

Usually UHTC are prepared by hot-press sintering,[3] non-pressure sintering,[4] spark plasma sintering,[5] reaction-sintering,[6] etc., so these methods face the same difficulty of all bulk ceramics, that is, low fracture toughness (2-4MPa·$m^{1/2}$) which limits thermal shock resistance property and also reliability. Furthermore, sintering process also limits preparation of large, complicated, thick articles.

Introduction of ceramic fiber to make ultra-high temperature ceramic matrix composites (UHTCMC) will greatly improve fracture toughness, reliability of composites and articles. Previous papers reported carbon fiber reinforced composites with ZrB_2 or HfC as matrix, but with rather low fracture strength or low fracture toughness. Sayir[7] reported C/HfC composite preparation by chemical vapor infiltration process, but with flexural strength of only 26MPa. Tang[8] reported C/SiC-ZrB_2 fabrication by powder infiltration and CVI process, in which fracture toughness is about 5MPa·$m^{1/2}$ and ablation property is greatly improved compared with C/SiC composites under oxyacetylene torch ablation.

In this paper, carbon fiber cloth reinforced SiC-ZrB_2 composites (2D C/SiC-ZrB_2) were prepared by slurry infiltration of ZrB_2 powders and precursor infiltration and pyrolysis (PIP) process to infiltrate SiC matrix, and accordingly mechanical property, anti-oxidation property, and torch ablation property were investigated.

EXPERIMENTAL PROCEDURE

3K PAN-based plain carbon fiber cloth (Jilin Carbon Corporation, China) was used as reinforcement. Polycarbosilane (PCS), with molecular weight ~1300 and softening point ~210°C, was synthesized in our lab. ZrB_2 powder with particle size ~2.5μm and purity over 99.0% was used.

The sample preparation is as follows. Firstly carbon fiber cloth was cut into 60 mm×90 mm pieces, and then 12 pieces were vacuum infiltrated with the slurry (PCS/ZrB_2/divinyl benzene), stacked into a graphite mold, pressed to about 7MPa, and cured in an oven at 150°C for 2 hours. Finally the sample together with the graphite mold was pyrolyzed up to 1200°C in a furnace under the protection of flowing nitrogen atmosphere. Further densification was continued by repetition of vacuum infiltration of PCS/divinyl benzene (DVB) solution, cure and pyrolysis. The C/SiC-ZrB_2 samples were labeled as ZB-0, ZB-20, ZB-30, ZB-40, ZB-50 and ZB-60 according to the volume ratios of ZrB_2 powder in slurry (the numbers indicated the volume ratios).

The apparent density was measured by Archimedes's method. Flexural strength (σ_f) was determined using a three-point-bending test on specimens of 4.0 mm×4.5 mm×60 mm with 50 mm span and 0.5 mm/min crosshead speed. A single-edge-notched-beam (SENB) test was applied on notched specimen of 4.0 mm×8.0 mm×60 mm (notch with 0.3 mm in width and 4.0 mm in depth) with 0.05 mm/min crosshead speed and 30 mm span to determine fracture toughness (K_{IC}). The high temperature mechanical properties were tested on YKM2200 high temperature instrument (Northwestern Polytechnical University, China) with specimen size of 5 mm×3.5 mm×70 mm and span of 60mm and crosshead speed of 0.597mm/s.

Oxidation experiment was performed in a muffle furnace at 1200°C. The sample was held for different time (20min and 40min) and then cooled down naturally to room temperature in air. Anti-ablative property test was carried out in a flowing oxyacetylene torch environment, with approximately 4187kW·m^{-2} heat flux and ~3100°C flame temperature. During the test, the specimen with a size of 30 mm×30 mm×5 mm, was vertically exposed to the flame. The distance between the nozzle tip and the surface of the specimen was 10 mm and the inner diameter of the nozzle tip was 2.0 mm.

The microstructure and composition of the samples were examined by scanning electron microscopy (JSM-5600LV, JEOL), and X-ray Diffraction (Siemens D-500).

RESULTS AND DISCUSSION

Microstructure and Composition

Since the components (carbon fiber, ZrB_2 powder, and SiC matrix) were introduced step-by-step, and during preparation there exist weight gain and weight loss, so it is important to determine exactly the contents of various components in the composite.

Suppose the ceramic yield of PCS is 65%, and that of DVB is 0 (small amount in slurry and low ceramic yield ~20%), so the following equations are presented to calculate the volume percentages of carbon fiber (V_f), ZrB_2 powder (V_{ZrB2}), and SiC (V_{SiC}) in the composites.

$$V_f = \frac{m_f}{V_0 \cdot \rho_f} \times 100\%$$

$$V_{ZrB_2} = \frac{(m_1 - m_f) \cdot \dfrac{m_{ZrB2}}{m_{ZrB2} + m_{PCS} \cdot \eta}}{\rho_{ZrB_2} \cdot V_0} \times 100\%$$

$$V_{SiC} = \frac{\rho_0 - V_f \cdot \rho_f - V_{ZrB_2} \cdot \rho_{ZrB_2}}{\rho_{SiC}} \times 100\%$$

V, ρ, m means volume, density, mass, respectively, and subscript of 0, f, ZrB_2, SiC, PCS means sample, fiber, ZrB_2 powder, SiC matrix, PCS, respectively. m_1 means sample mass after first pyrolysis, and η means ceramic yield of PCS.

Table 1. Contents of Different Components in C/SiC-ZrB_2 Samples

sample	ρ(g/cm^3)	V_{ZrB2}(vol%)	V_{SiC}(vol%)	V_f(vol%)	*porosity*(vol%)
ZrB-0	1.87	0	48.7	37.1	14.2
ZrB-20	2.06	10.0	33.8	35.9	20.2
ZrB-30	2.35	16.9	33.6	30.0	19.5
ZrB-40	2.46	19.4	35.4	25.7	19.6
ZrB-50	2.50	21.5	35.1	21.2	22.1
ZrB-60	2.59	24.6	33.2	18.9	23.3

From Table 1, the volume percentage of ZrB_2 increases with the increase of ZrB_2 content in the original slurry, and this value reaches 24.6% when ZrB_2 content in the slurry is 60%. When ZrB_2 content is too high, e.g., over 60%, the viscosity of the slurry is too high and slurry-infiltration of the carbon fiber cloth is impossible, so ZrB_2 content in the slurry, and thus ZrB_2 content in the composites, is limited.

It is also obvious that with ZrB_2 increase in the composites, the volume percentage of carbon fiber, V_f, decrease from 37.1% (without ZrB_2 introduction) to 18.9% (with 24.6% ZrB_2), and V_f decrease will decrease mechanical properties of the composites. The reason for V_f decrease comes from the thicker matrix layer between two pieces of carbon fiber cloth due to higher content of ZrB_2 in the slurry, and at the same time, V_{SiC} remains almost the same level (33%) in all C/SiC-ZrB_2 samples, except 48.7% for C/SiC (ZB-0) sample.

The density of the composites also increases with ZrB_2 increase, and reaches 2.59g/cm^3 when ZrB_2 content in the slurry (ZrB-60) is 60%, compared with 1.87g/cm^3 of C/SiC (ZB-0) composites. The difference would be much larger if C/SiC-ZrB_2 and C/SiC samples had the same porosity because the density of ZrB_2 (5.8g/cm^3) is much higher than that of SiC (3.2g/cm^3) as matrix. The porosity of C/SiC-ZrB_2 composites (~20%), are significantly higher than that of C/SiC (14.2%), indicating that ZrB_2 introduction is not beneficial for densification. Another trend is also observed, that higher content of ZrB_2 will cause higher porosity in the composites, which would be ascribed as the agglomerates of ZrB_2 particles and hindrance of PCS infiltration.

Mechanical Property

The flexural strength and modulus of C/SiC-ZrB_2 composites are shown in Fig. 1. The flexural strength and modulus of the samples decrease almost linearly with ZrB_2 content increase, in which the flexural strength and modulus decrease from 366.7MPa and 55.5GPa of sample ZB-0 to 162.8MPa and 24.5GPa of sample ZB-60, respectively. This trend may be caused by (1) ZrB_2 introduction and/or (2) V_f decrease. Comparison of C/SiC (ZB-0) and C/SiC-ZrB_2 (ZB-20) indicates that small amount of ZrB_2 introduction will not influence fiber strength and fiber/matrix inter-phase, because two samples have almost the same flexural strengths, e.g., 366.7MPa vs 365.5MPa.

Further analysis of carbon fiber content in the composites (Fig. 2) shows that V_f decreases linearly with V_{ZrB2} increase, and this decrease causes decrease of flexural strength and modulus, showing that in the composites both SiC matrix and ZrB_2 matrix contribute only integral of the composites, and carbon fiber contributes mainly both strength and modulus.

Introduction of carbon fiber improves greatly fracture toughness of the samples (see Table 2), that is, 8.1~17.67MPa·m$^{1/2}$, which is almost one magnitude higher than that of sintered ZrB_2 and ZrB_2-SiC composites (2~4MPa·m$^{1/2}$).[9] This improvement shows the significant advantages of C/SiC-ZrB_2 over sintered ZrB_2 ceramics, in that excellent thermal shock resistance under super-high ramping rate during re-entry, and also in the higher reliability of the composites and articles, and also in the process availability of thick, complicated articles. Further study testified the excellent thermal shock resistance of the composites under ablation tests.

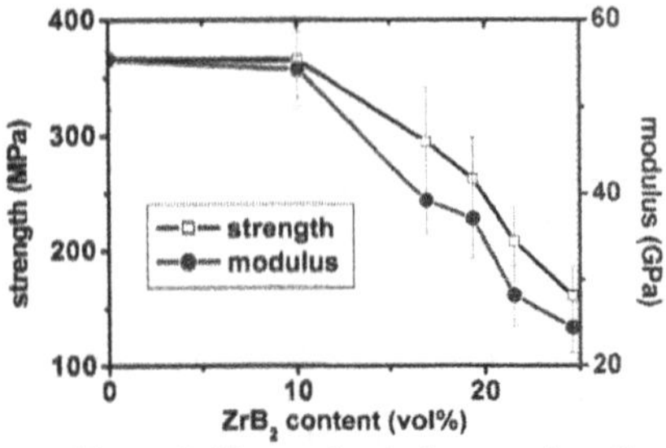

Figure 1. The mechanical properties of C/SiC-ZrB_2 samples with different ZrB_2 contents

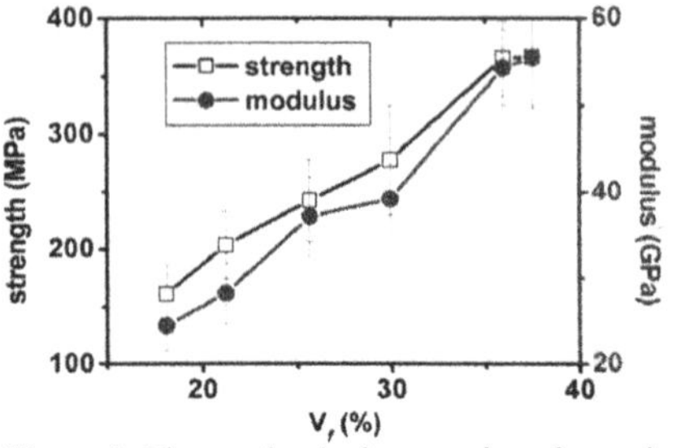

Figure 2. The mechanical properties of samples with different carbon fiber content

Table 2. Fracture toughness of C/SiC-ZrB_2 composites

Sample	ZB-0	ZB-20	ZB-30	ZB-40	ZB-50	ZB-60
K_{IC} ($MPa{\cdot}m^{1/2}$)	23.63	17.67	12.92	10.51	9.47	8.06

Oxidation Property

The samples of C/SiC-ZrB_2 (ZB-40) and C/SiC were oxidized at 1200°C for 20 and 40 minutes, after which the mass loss rate and flexural strength were determined (Table 3). Sample C/SiC-ZrB_2 after oxidation for 20min, showed 4.49% mass loss, while C/SiC showed 10.62%. Further oxidation for 40min sample C/SiC-ZrB_2 showed 11.45% mass loss. The mass change of C/SiC-ZrB_2 and C/SiC comes from two reasons. One is oxidation mass loss of carbon fiber, which can be described as follows.

$C_f + 0.5O_2 \rightarrow CO\uparrow$, mass loss 12g/molar

The other is oxidation mass gain of ZrB_2 and SiC matrix, which is described as follows.

$ZrB_2 + 2.5O_2 \rightarrow 2ZrO_2 + B_2O_3$, mass gain 80g/molar

$SiC + 1.5O_2 \rightarrow SiO_2 + CO\uparrow$, mass gain 44g/molar

Both mass gain and loss contribute total mass loss of the sample, indicating severe carbon fiber oxidation.

Besides the difference of mass loss, the flexural strength retention after oxidation of C/SiC-ZrB_2 and C/SiC is different, that is, after 20min oxidation, the flexural strength retention of sample C/SiC-ZrB_2 is 69.2%, significantly higher than that of C/SiC (20.2%), indicating that introduction of ZrB_2 improves greatly the anti-oxidation property of the composites. After exposition to oxidation for 40min, the flexural strength retention of C/SiC-ZrB_2 remains 52.8%, still higher than that of C/SiC.

Ablation Property

During the ablation test with oxyacetylene torch, the surface temperatures of the samples may reach as high as 2300°C after 8-10 seconds, and the mass loss rate and linear recession rate are shown in Table 4. When ZrB_2 content in the samples is raised from 0 to 20%, the mass loss rate is reduced from 0.013g/s to 0.007g/s, and linear recession rate is reduced from 0.03mm/s to 0.019mm/s, showing that introduction of ZrB_2 remarkably improves the ablation performance. And further increase of ZrB_2 content in the composites leads to slight decrease of mass loss rate and recession rate. With the highest content of ZrB_2 (24.5%) in the composites the mass loss rate and recession rate decrease accordingly to 0.005g/s and 0.01mm/s, respectively.

Table 3. Mass loss and flexural strength of C/SiC-ZrB_2 (ZB-40) and C/SiC before and after oxidation at 1200°C

sample	Time (Min)	Mass loss ratio (%)	σ_f (MPa) Before (after) oxidation
C/SiC-ZrB_2	20	4.49	266.4 (184.3, 69.2%)
	40	11.45	266.4 (140.6, 52.8%)
C/SiC	20	10.62	351.0 (71.0, 20.2%)

Table 4 Ablation properties of C/SiC-ZrB_2 samples

sample	ZB-0	ZB-20	ZB-30	ZB-40	ZB-50	ZB-60
Recession rate (mm/s)	0.03	0.019	0.016	0.015	0.014	0.010
Mass loss (%)	0.0128	0.0066	0.0058	0.0054	0.0052	0.0051

The ablation of the samples is mainly composed of three aspects: high temperature decomposition (destroy), oxidation, and gas denudation. The improvement of ablation property after ZrB_2 introduction is because (1) ZrB_2 matrix can withstand higher temperature due to high melting point (3240°C), thus reducing high temperature decomposition (destroy). Considering that the surface temperature of the samples is as high as nearly 2400°C, SiC matrix will tend to decompose, and (2) ZrB_2 matrix has much better anti-oxidation property than SiC matrix, as is proved both in the pure oxidation experiment at 1200°C and in the analysis of ablation test. Further study of surface morphology after torch test shows that sample of C/SiC-ZrB_2 forms dense, glassy coating, while C/SiC cannot form glassy coating and shows priority of carbon fiber ablation and leaves hollow pores around tip-sharp carbon fiber.

The dense, glassy coating is analyzed by XRD measurement, and mainly composed of ZrO_2 and small amount of B_2O_3, obviously from ZrB_2 oxidation. It is much strange that no SiO_2 exists in the surface coating, though existence of SiC matrix. The absence of SiO_2 maybe due to the high temperature gas stream flowing away SiO_2 layer, which is low viscosity at high temperature over 1800°C (note that melting point of SiO_2 is 1710°C). Same phenomenon is observed by other researchers in ablation tests of ZrB_2-SiC composites, which is depletion of SiC in the outer surface of the sample.[10] At the same time, ZrO_2 (and small amount of B_2O_3) forms viscous coating on the surface, thus decreasing the recession rate and protecting the fiber from further oxidation.

CONCLUSION

Ultra-high temperature ceramic matrix composites (C/SiC-ZrB_2) were prepared by slurry-infiltration and precursor infiltration and pyrolysis method. Introduction of carbon fiber enhances fracture toughness up to 8.1-17.7MPa·m$^{1/2}$, and introduction of ZrB_2 matrix improves greatly anti-oxidation property at 1200°C, and anti-ablation property under oxyacetylene torch environment, compared with C/SiC composites. This method also has advantages in preparation of thick, complicated articles, compared with sintered bulk ceramics.

ACKNOWLEDGMENT

The work was financially supported by the National Natural Science Foundation of China (90816020).

REFERENCES

[1] M. J. Gasch, D. T. Ellerby and S. M. Johnson, "Ultra High Temperature Ceramic Composites"; pp. 197-224, In Handbook of Ceramic Composites, Edited by N. P. Bansal. Kluwer Academic Publishers, Boston, USA, 2005.

[2] W. G. Fahrenholtz, G. Hilmas, I. G. Talmy, and J. A. Zaykoski, "Refractory Diborides of Zirconium and Hafnium," *J. Am. Ceram. Soc.*, 90 [5] 1347-1364 (2007).

[3] M. Gasch, D. Ellerby, E. Iiby, S. Beckman, M. Gusman, and S. Johnson, "Processing, Properties and Arc Jet Oxidation of Hafnium Diboride/Silicon Carbide Ultra High Temperature Ceramics," *J. Mater. Sci.*, 39, 5925-5937 (2004).

[4] M. Quabdesselam and Z. A. Munir, "The Sintering of Combustion Synthesized Titanium Diboride," *J. Mater. Sci.*, 22 [5] 1799-807 (1987).

[5] V. Medri, F. Monteverde, A. Balbo, and A. Bellosi, "Comparison of ZrB_2-ZrC–SiC Composites Fabricated by Spark Plasma Sintering and Hot-Pressing," *Adv. Eng. Mater.*, 7 [3] 159-63 (2005).

[6] F. Monteverde, "Progress in the Fabrication of Ultra-High-Temperature Ceramics: "In Situ" Synthesis, Microstructure and Properties of a Reactive Hot-Pressed HfB_2–SiC Composite," *Comp. Sci. Tech.*, 65, 1869–79 (2005).

[7] A. Sayir, "Carbon Fiber Reinforced Hafnium Carbide Composite," *J. Mater. Sci.*, 39, 5995-6003 (2004).

[8] S. Tang, J. Deng, S. Wang, and W. Liu, "Fabrication and Characterization of an Ultra-High-Temperature Carbon Fiber-Reinforced ZrB_2-SiC Matrix Composite," *J. Am. Ceram. Soc.*, 90 [10] 3320-3322 (2007).

[9] F. Monteverde, A. Bellosi, and S. Guicciardi, "Processing and Properties of Zirconium Diboride-Based Composites," *J. Europ. Ceram. Soc.*, 22 [3] 279-288 (2002).

[10] W. G. Fahrenholtz, "Thermodynamic Analysis of ZrB_2–SiC Oxidation: Formation of a SiC–Depleted Region," *J. Am. Ceram. Soc.*, 90 [1] 143-148 (2007).

FABRICATION OF C_f/SiC-BN COMPOSITES USING POLYCARBOSILANE(PCS)- BORON-SiC FOR MATRIX DERIVATION

ZhenWang[1,2], Shaoming Dong[1], Le Gao[1], Haijun Zhou[1,2], Jinshan Yang[1,2], Dongliang Jiang[1]
[1] Shanghai Institute of Ceramics, Chinese Academy of Sciences,
Shanghai, 200050, PR China
[2] University of Graduate, Chinese Academy of Sciences
Beijing, 100049, PR China

ABSTRACT

C_f/SiC-BN composite was fabricated via in situ reaction by applying boron as active filler. The influence of pyrolysis temperatures on ceramic yields and linear shrinkages were studied. At 1200°C B_4C was synthesized and above 1300°C h-BN appeared as a result of the reaction between boron and N_2. There are some fiber pull-outs in both C_f/SiC composite and C_f/SiC-BN composite on the fracture surface, otherwise, the layered structure of h-BN can be observed on the fracture surface of C_f/SiC-BN composite.

Key words: Active filler, in situ, C_f/SiC-BN composite

INTRODUCTION

Carbon fiber reinforced ceramic-matrix composites are attractive materials for applications requiring low specific weight, high strength and high toughness at elevated temperatures.[1] Chemical vapor infiltration (CVI), liquid silicon infiltration (LSI) and precursor impregnation and pyrolysis (PIP) techniques have been most frequently reported as the fabrication methods of fiber-reinforced composites. The general procedure of PIP technique is to impregnate woven fiber fabrics with a liquid precursor, cross-link the polymer precursor by heating or UV lightening, and pyrolyze the assembly at temperature in the range of 1000-1400°C in a controlled atmosphere. [2] However, a major drawback is low ceramic yield of usually less than 70wt% and the large shrinkage of up to 60vol%. Therefore, several infiltration-cure-pyrolysis cycles are required to prepare dense fiber reinforced ceramic-matrix composites by this method, which makes the process both expensive and time-consuming[3,4,5]. Filler materials are often applied to fill up the spaces between the fiber yarns to inhibit excessive shrinkage and pore formation of FRC matrix during pyrolysis and reduce the necessary number of impregnation and pyrolysis cycles to decrease the processing time. When investigating the near-net shaping of composites, Greil [6] added fillers to the precursors used in order to control the volume change of components during processing. There are two kinds of fillers, namely active fillers and passive (inert) fillers. When active filler particles are applied, near-net-shape conversion can be achieved because of the compensation of the polymer shrinkage by filler expansion.

In this work, boron was chosen as the active filler due to its considerable volume expansion ($\Delta V/Vo$) on nitridation ($B+0.5N_2=BN$, $\Delta V/Vo$ =142%). Influence of heat-treatment temperatures on the compositions, ceramic yields and linear shrinkages as well as effects of boron on the microstructure evolution and properties of composites were studied.

EXPERIMENTAL PROCEDURE

Matrix preparation

Boron , PCS and α-SiC powders (15wt%:30wt%:55wt%) were mixed by ball-milling for 48h using 120# gasoline as solvent to form homogenous slurries followed by evaporation of the solvent in vacuum. The well-dispersed mixtures were screened through a sieve to obtain the small particles. Mixtures with α-SiC were molded at room temperature under a pressure of ~5MPa then further compacted by cold isostatic pressing at 200MPa. The composition of the mixture is 55wt%α-SiC, 15wt% boron and 30wt% PCS. These green compacted bars were pyrolyzed in N_2 at different temperatures for 1h (nitridation). These bars were heated to 800°C at 5°C/min and then 10°C/min to the desired temperatures. The weights and lengths of bars were measured before and after pyrolysis respectively. The ceramic yields (ω) and linear shrinkages (ε) were calculated according to the formulas (1) and (2):

$$\omega = m_{fin} / m_{ini} \times 100\% \qquad (1)$$

$$\varepsilon = (L_{ini} - L_{fin}) / L_{ini} \times 100\% \qquad (2)$$

Mixture without SiC was pyrolyzed at different temperatures without molding and the pyrolysis residues were ground for composition characterization. The composition of mixture without SiC is 67wt% PCS and 33wt% boron.

Composite preparation

Boron powders were mixed with PCS by ball-milling for 48h (45wt% boron and 55wt% PCs), using xylene as solvent to form homogenous slurries for impregnation of 2D woven carbon fiber plain fabrics (Mitsubishi Rayon, $200g/m^2$) deposited with PyC interphase. Then the impregnated fabrics were stacked in a graphite die after drying pressures were applied to control their thicknesses according to a fiber volume of ~45% in the hot-pressing furnace at ~200°C and then pyrolyzed at 800°C. Before nitridation, two cycles of PIP were performed using PCS as precursor with the pyrolysis temperature of 800°C. Then the samples were nitrided at 1800°C in N_2 for 1h. After nitridation, PIP was performed again to fill the open pores and the pyrolysis temperature was 1100°C. For comparison, composite without boron was also prepared by the same method using SiC as inert filler.

Sample Characterization

The compositions of both matrix powders pyrolyzed at different temperatures and composites were characterized by XRD. Mechanical properties of composites were measured by three-point-bending tests with 5mm×2mm×40mm specimens in an Instron-5566 machine, operated at a crosshead speed of 0.5mm/min and a span of 24mm. The microstructures of composites both with and without boron were observed by electron probe microanalyzer (EPMA, JXA-8100, JEOL, Tokyo, Japan).

RESULTES AND DISCUSSION

Fig.1 shows the XRD patterns of the residues of powder mixtures pyrolyzed at different temperatures. It can be found from these curves that at 1200°C, there is no obvious diffraction peaks except a broad peak for β-SiC with very low intensity, the diffraction peaks of boron also disappeared. There are some small diffraction peaks of B_4C when pyrolysed at 1300°C. Above 1400°C the intensities of diffraction peaks for β-SiC increased and the peak for h-BN at ~26° also appears. When the samples are heat-treated at high temperatures, several reactions may occur. At temperatures below 800°C, PCS will decompose to some hydrocarbons and solid residues. When the temperature increases to temperatures above 1000°C, the solid residues will gradually convert into amorphous SiC plus excessive carbon., then at temperatures above 1200°C, active filler (boron) will react with hydrocarbons and carbon to form B_4C. At temperatures above 1300°C, the formed B_4C and the unreacted boron will react with protective gas (N_2) to form h-BN. When the temperature reaches 1400°C, SiOC phase in the materials begin to decompose into SiO and CO, which will result in weight loss and volume shrinkage.

It can be concluded from the ceramic yield and linear shrinkage versus pyrolysis temperature curves shown in Fig.2 that both ceramic yields and linear shrinkages increase with the increase of pyrolysis temperature in the temperature range of 1200°C and 1400°C. This can be caused by the volume changes in the B_4C formation process and β-SiC conversion process. The volume changes accompanied in the reaction of boron with amorphous carbon and hydrocarbons are -7% and 20% respectively [6]and there is also some volume shrinkage in the β-SiC conversion process[7]. Though the reaction between boron and N_2 takes place above 1300°C and there is weight gain and volume expansion accompanied in this reaction, however, the reaction rate is relatively slow. The volume expansion and weight gain caused as a result of formation of h-BN is not enough to compensate the volume shrinkage and weight loss induced by the decomposition of SiOC phase. At temperature higher than 1600°C, the weight gain and volume expansion resulted from the reaction between boron and N_2 are larger than the weight loss and volume shrinkage caused by the decomposition of SiOC, so the ceramic yield increases and the linear shrinkage decreases.

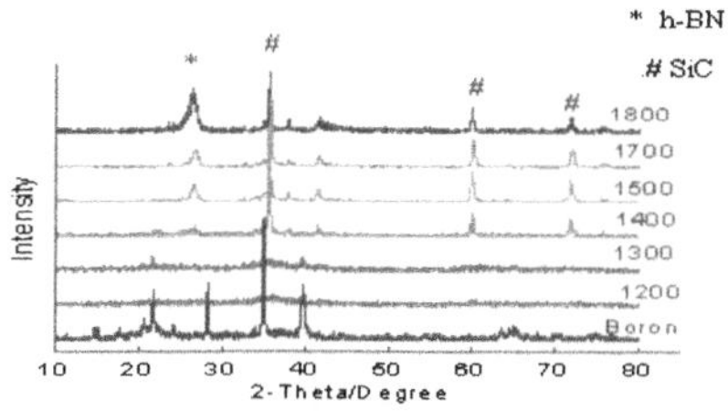

Fig. 1 XRD patterns of boron/PCS mixtures pyrolyzed at different temperatures

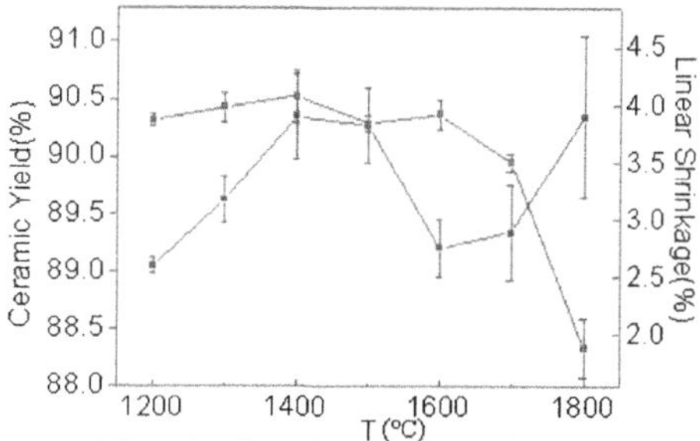

Fig. 2 Ceramic yields and linear shrinkages pyrolyzed at different temperatures.

X-ray diffraction patterns of composites with and without boron(S* and S respectively) are compared in Fig.3. As it can be found from these pattern curves, with boron introduced, the intensities of diffraction peaks for β-SiC weakened, meanwhile those for h-BN enhanced, which may be induced by the formation of h-BN as a result of reaction between boron and N_2. It is also shown in the XRD patterns that the diffraction peak at ~26° for composite S* is much sharper than that for S. The diffraction peaks for h-BN and graphite are overlapped as they have the similar

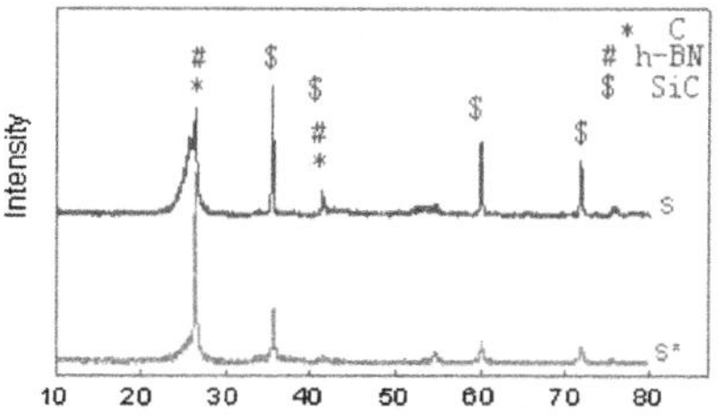

Fig. 3 XRD patterns of composites with and without boron nitrided at 1800°C(S without boron and S* with boron)

layered crystal structures. The diffraction peak at ~26° in XRD pattern for S* is the combination of carbon fiber and h-BN while that for S mainly belongs to carbon fibers. Meanwhile, h-BN formed at 1800°C has a very high crystallinity which corresponds to sharper diffraction peaks.

Fig4. shows the microstructures of the fracture surfaces. Some holes can be observed in both composites as a result of the fibers-pull-out, however, the lengths of the pulled-out fibers are very short. Compared to the nitridation temperature of 1800°C, the carbonization temperature of carbon fibers in the preparation process is relatively low, therefore, in the nitridation process, some damages may be done to the fibers and lead to the degradation in the strengths of fiber reinforcements and result in the low strength of composites. Some phases with layered structures can also be observed in the microstructures of composite with boron, which corresponds to h-BN in situ formed in the nitridation process. When the cracks propagate to h-BN, they will deflect.

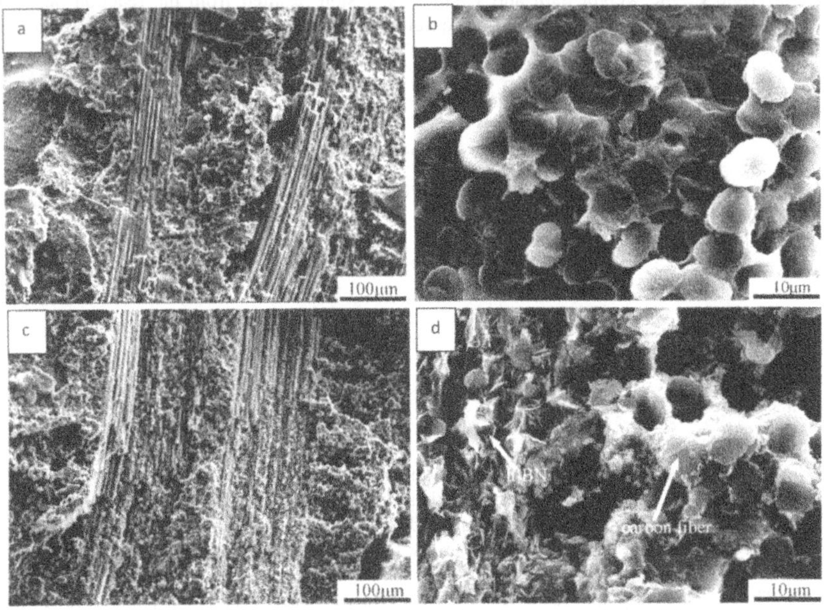

Fig.4 Scanning electron microscope graphs on the fracture surface of the composites. (a,b: C_f/SiC composites; c,d: C_f/SiC-BN composites)

CONCLUSION

h-BN was successfully introduced into the matrix of composite. At 1200°C, boron will react with hydrocarbons and carbon to form some B_4C and h-BN appears at temperatures above 1300°C due to the reaction between boron and protective gas N_2. From 1200°C to 1400°C, both ceramic yields and linear shrinkages increase with the increase of pyrolysis temperatures and above 1600°C the ceramic yield increases while the linear shrinkage decreases. Pulled-out fibers can be observed on fracture surfaces in both C_f/SiC and C_f/SiC-BN composites, moreover, the layered structures of h-BN can be found as well in the matrix of C_f/SiC-BN composite.

REFERENCES

1 R. Naslain, Design, preparation, and properties of non-oxide CMCs for application in engines and nuclear reactors: An overview," Compos. Sci. Technol., 64[2] 155-70(2004)

2 R. riedel, G. Passing, H. Schonfelder, R.J. Brook, Synthesis of dense silicon-based ceramics at low-temperatures, Nature, 355, 714-717(1992)

3 R.M. Rocha, C.A.A. Carro, M.L.A. Graca, Formation of carbon fiber-reinforced ceramic matrix composites with polysiloxane/silicon derived matrix, Materials Science and Engineering A 437(2006) 268-273

4 G. Ziegler, I. Richter, D.Suttor, Fiber-reinforced composites with polymer-derived matrix: processing, matrix formation and properties, Composites: part A, 30,411-417(1999);

5 K. W. Chew, A. Sellinger, R.M. Laine, processing aluminium nitride-silicon carbide composites via polymer infiltration and pyrolysis of polymethyl silane, a precursor to stoichiometric silicon carbide, J. Am. Ceram. Soc., 82, 857-866 (1999)

6 P.Greil, Active-filler-controlled pyrolysis of preceramic polymers, J. Am. Ceram. Soc., 78,835-48(1995)

7 Y.S. Ding, S.M. Dong, Z.R. Huang, and D.L.Jiang, "Preparation of C/SiC composites by hot pressing, using different C fiber content as reinforcement," J. Am. Ceram. Soc., 89[4] 1447-9 (2006)

SINTERABILITY, THERMAL CONDUCTIVITY AND MICROWAVE ATTENUATION PERFORMANCE OF AlN-SiC SYSTEM WITH DIFFERENT SiC CONTENTS

Wenhui Lu, Xiaoyun Li, Weihua Cheng, Tai Qiu
College of Material Science and Engineering, Nanjing University of Technology
Nanjing, Jiangsu, 210009, China

ABSTRACT

AlN-SiC composites with TiO_2 and Y_2O_3 as sintering additives were prepared by hot-pressing in a nitrogen atmosphere. The effects of different SiC contents on the sinterability, thermal conductivity and microwave attenuation performance of AlN-SiC composites were studied by XRD, SEM and network analyzer. The composites reach a high relative density (99%) with SiC below 70wt%, while the content of SiC is increased continually, the relative density is decreased. When the SiC content is lower than 40wt%, the AlN-SiC composites show narrow frequency band attenuation performance and have low attenuation. When increasing SiC content continually (40wt %-80wt %), the composites show wide frequency band attenuation performance. The relationship between the content of SiC and the thermal conductivity is also presented.

Key words: aluminum nitride, silicon carbide, sinterability, thermal conductivity, microwave attenuation

1. INTRODUCTION

Electrically lossy materials are used extensively in vacuum electronic devices for preventing instabilities, reducing reflections, adjusting cavity quality factors, and controlling electromagnetic dispersion and growth rates. With the increased frequency of microwave devices, new lossy materials with high thermal conductivity are urgently needed. Recently, emerging types of AlN-based lossy materials, especially AlN-SiC composites, are attracting great attention, as an effective alternative for the traditional, toxic BeO-based ceramics and low thermal conductivity Al_2O_3-based lossy ceramics[1-2]. Because of its excellent physical properties and electronic characteristics, AlN has been recognized as a promising wide band-gap semiconductor material for high-temperature, high-frequency, high-power and radiation-hardened devices[1-3]. Meanwhile, SiC materials, as the lossy additive investigated considerably, have been found to have not only strong microwave absorption but also high temperature resistance and high strength properties[4]. Landon et al. obtained AlN-50wt%SiC and AlN-70wt%SiC with a high thermal conductivity of 73 $W \cdot m^{-1} \cdot K^{-1}$ and found that thermal behavior was correlated with the formation of SiC polytypes and with the phase transformations[5]. Bu et al. reported that a single AlN-SiC solid solution could be completed at 1900 °C, and the formation of partial AlN-SiC solid solution was helpful to improve the wide frequency band attention performance[6-7]. AlN-40wt%SiC ceramics were prepared by hot-pressing, exhibiting good wideband attenuation property with the dielectric loss changing from 0.19 to 0.26 within the frequency range of 8~12GHz[8].

The primary objective of this work is to prepare and characterize AlN-SiC composites by hot-pressing and explain the relationship between the content of attenuator SiC and the sinterability, thermal conductivity, microwave attenuation performance of AlN-SiC composites.

2. EXPERIMENTAL

Commercially available AlN powders (average particle size of 0.5 μm, N≥33%) and SiC powders (average particle size of 0.5~0.7μm, SiC≥99.2%) were used as raw materials, with compound additives containing TiO_2 (99.0% pure) and Y_2O_3 (99.95% pure).The mixtures contained SiC ranging from 0~80wt% and AlN ranging from 92wt%~12wt%, doping of 4wt%TiO_2 and

4wt%Y_2O_3.

The mixtures were ball-milled by planetary milling for 4 h using ethanol as a mixing medium. After drying, the mixed powders were put into a graphite die (40×50mm), and then hot pressed at 1900°C for 1 h in a flowing nitrogen atmosphere under 30MPa pressure.

The bulk density was measured by Archimedes principle. The theoretical density was determined by the rule of mixtures, using the starting compositions to calculate the relative densities. The identification of the phase composition was performed by using an X-ray diffraction (XRD, ARL X'TRA, Switzerland). The microstructure was investigated on fracture surfaces by means of scanning electron microscopy (SEM, JSM—5900, JEOL, Japan). The sintered samples were machined to disks (12.7 mm in diameter, 2.5 mm in thickness) for thermal diffusivity measurement by a flash thermal constant analyzer (LFA 447, Germany Netzsch) and the heat capacities were calculated using a rule of mixtures relationship. The thermal conductivity at room temperature was calculated from the equation $\lambda = C \cdot \alpha \cdot \rho$, where C is heat capacity, α is thermal diffusivity, and ρ is density of the sample. The microwave attenuation performance was studied by network analyzer (WILTRON360B, Britain), combining with the theory of microstrip line.

3. RESULTS AND DISCUSSION

3.1. Sinterabilites of AlN-SiC composites with different SiC contents

Fig.1 shows the apparent porosity and relative density of AlN-SiC composites with 0wt%~80wt% SiC hot-pressed at 1900℃ for 1 hour.

When the SiC content is lower than 70wt%, it is obvious that the apparent porosity is low, no more than 0.1%, indicating that the samples are densified. While the content of SiC is increased continually, the apparent porosity suddenly increases to 13.86%. The relative density of composites reaches a high value(99%) with SiC below 70wt%, agreement with the apparent porosity, a turning point appeared at 70wt%, and after that the relative density begin to decrease sharply. Results indicate that the addition of more than 70wt% SiC brings disadvantages on densification, leading to a quite high apparent porosity 13.86% and a very low relative density 84.9% when the SiC content is up to 80wt%.

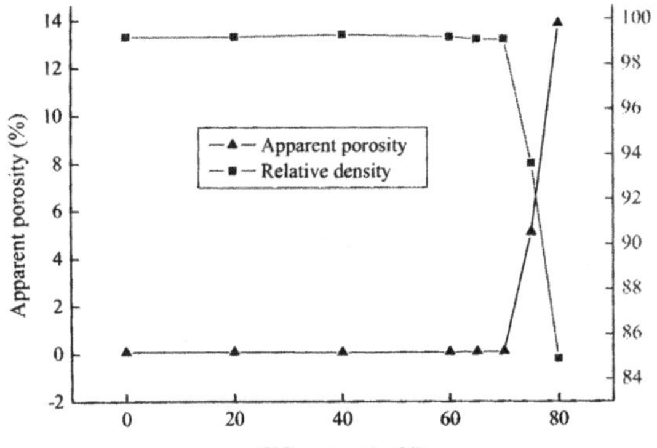

Fig.1 Apparent porosity and relative density of samples with different SiC contents

3.2 Phase analysis and Microstructure

Fig.2 shows XRD patterns of AlN-SiC composites with 20wt%~60wt%SiC, which indicate that the three samples have similar phases. The major phases are AlN and SiC, and the minor phases are TiC and $Y_3Al_5O_{12}$. It is estimated that a reaction occurs as follows: $5Al_2O_3 + 3Y_2O_3 \rightarrow 2Y_3Al_5O_{12}$, leading to the formation of $Y_3Al_5O_{12}$ phase. The appearance of small amount of TiC is due to the reaction between the sintering additive TiO_2 and carbon, infiltrated from the graphite die at high temperature.

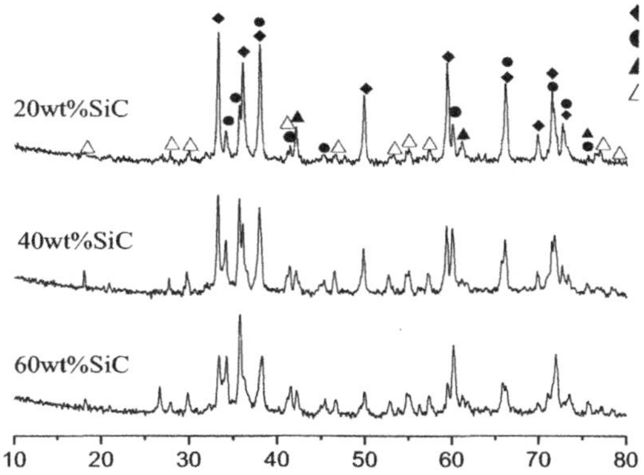

Fig. 2 XRD patterns of composites with different SiC contents

Typical microstructures of fracture surfaces from AlN-SiC composites with 0wt%~75wt%SiC are shown in Fig.3. Results of these micrographs are all in good agreement with the results of the apparent porosity and relative density discussed in previous section 3.1(Fig.1). The sample without SiC addition contains few visible pores. The grains grow very well, close to each other, and have uniform grain size distributions(5~10μm) (see Fig.3 (1)).

But with increasing SiC addition, especially when the content of SiC is up to 75wt% (see Fig.3 (4)), the structure of samples display more and bigger pores, the crystals also change obviously from uniform size to different size. Moreover, as shown in EDS analysis, the grey bulk areas (b) are the major phases AlN and SiC, as compared with the bright areas (a), which are the minor phases (see Fig.3 (2) and Fig(4).

3.3 Effects of SiC content on the Thermal Conductivity of AlN-SiC composites

Table I. shows the thermal diffusion parameters and thermal conductivity values of AlN-SiC composites with 0wt%~80wt%SiC.

As can be seen in Table I, the trend in thermal diffusivity is carried into the thermal conductivity data. Despite SiC additions, thermal diffusivity and, thus, conductivity are lower for AlN-SiC ceramics than AlN ceramic. With SiC addition, thermal conductivity decrease first from 56.5 $W \cdot m^{-1} \cdot K^{-1}$ for 40wt% SiC to 50.0 $W \cdot m^{-1} \cdot K^{-1}$ for 65wt% SiC, and then increase to 67.7 $W \cdot m^{-1} \cdot K^{-1}$ for 75wt% SiC. The thermal conduction mechanism of AlN is phonon transmission. There would be a continuous path in AlN ceramic without SiC. With the increase of SiC addition the AlN grains would be isolated gradually, resulting in a decrease in thermal conductivity. When the SiC content is high enough, grains of SiC connect with each other and SiC ceramics have high thermal conductivity. Therefore, the composite with 75wt% SiC shows an increased thermal conductive value. The sample with 80wt% SiC has a low thermal conductivity, because it is non-densification and has large quantity of pores. There may be some other factors influencing thermal conductivity. For example, increases of grain boundary reduce thermal conductivity, mainly due to the increase of scattering sites. Another effect controlling the thermal conductivity, namely the solid solution formed by AlN and SiC, are known to decrease thermal diffusivity, by increasing the number of phonon scattering sites[10].

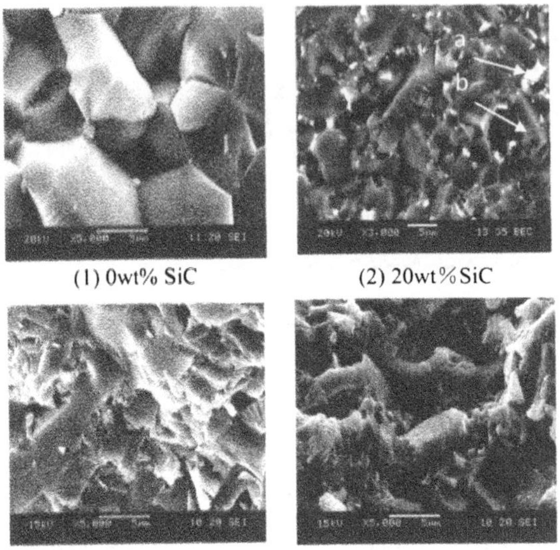

(1) 0wt% SiC (2) 20wt%SiC

(3) 70wt% SiC (4) 75wt% SiC

Fig.3 SEM micrographs of the fracture surface of samples with different SiC contents

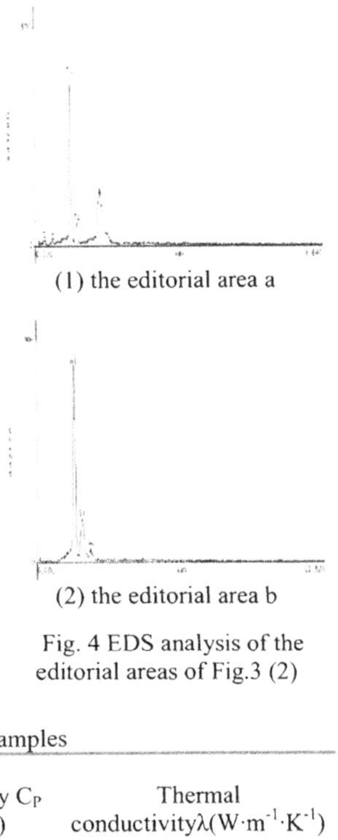

(1) the editorial area a

(2) the editorial area b

Fig. 4 EDS analysis of the editorial areas of Fig.3 (2)

Table I. Thermal diffusion coefficient and thermal conductivity of samples

SiC content (wt%)	Thermal diffusivity $\alpha(mm^2 \cdot s^{-1})$	Bulk density $\rho(g \cdot cm^{-3})$	Heat capacity C_P $(J \cdot g^{-1} \cdot K^{-1})$	Thermal conductivity$\lambda(W \cdot m^{-1} \cdot K^{-1})$
0	38.2	3.23	0.79	97.7
40	23.2	3.29	0.74	56.5
60	22.6	3.28	0.71	51.7
65	21.6	3.27	0.71	50.0
70	22.3	3.27	0.70	51.2
75	31.5	3.09	0.70	67.7
80	19.51	2.80	0.69	37.6

3.4 Microwave attenuation property

Fig. 5 shows the microwave attenuation characteristics of AlN-SiC composites within the range of 2-20GHz. Samples with less than 20wt%SiC show narrow frequency band attenuation performance and have low attenuation. When the SiC content is increased (40wt%~80wt%), the compositions show wide frequency band attenuation and the attenuations rise obviously with the SiC addition. The attenuation of 0wt%SiC is 1.1dB and it is 2.4dB for 80wt%SiC at 18.65GHz.

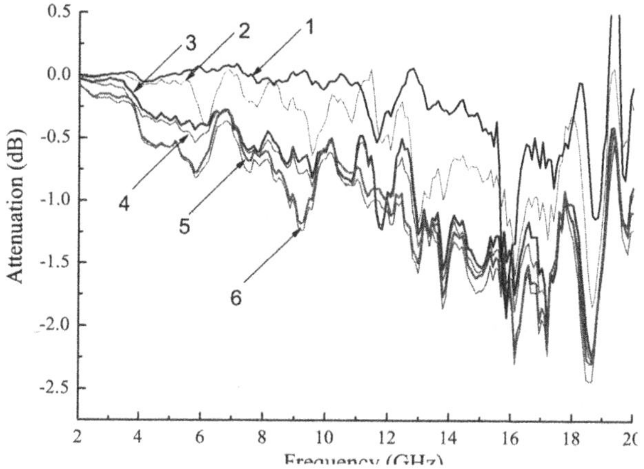

Fig. 5 Microwave attenuation curves of AlN-SiC composites with different SiC content
1-0wt% SiC 2- 20wt% SiC 3-40wt% SiC 4-60wt% 5-70wt% SiC 6-80wt% SiC

The wave absorption energy of multiphase materials is calculated by the following formula[9]:

$$Q = \omega(\varepsilon'' E^2 + \mu'' H^2)/4\pi \tag{1}$$

Where Q is energy release of 1 cm^3 material within 1s, E and H are electric intensity and magnetic intensity, ωis a microwave frequency. As seen in this formula, the wave absorption energy is determined by the electric loss (ε'') and magnetic loss (μ''). Because AlN and SiC are non-magnetic material either, the main microwave attenuation of the AlN-SiC composites comes from dielectric loss[11]. As shown in Fig.5, attenuation of the sample without SiC only takes place at special frequencies, most probably caused by graphite infiltrated from the die at high temperature. With the increasing of SiC addition, SiC begins to make the main contribution to the wave attenuation, exhibiting wide frequency band attenuation. When the semiconductor phrase SiC is dispersed in the insulating medium, electrons would move urgently under an external electrical field, and are partly blocked and accumulated at the interface, producing space charges. These space charges can form many macro-dipoles which would oscillate with the external field. Thus a space charge relaxation process is created. These oscillations can occur at any frequency, so the attenuation characteristic exhibits wideband attenuation. The loss increases as electrical conductivity increases. Therefore, with the SiC content increasing, the attenuation band is broadened and the attenuation is increased. In addition, there are many defects in the ceramics hot-pressed at high temperature. These defects, such as vacancies, antiphase boundary, dislocation, can also make the microwave energy lose.

4. CONCLUSIONS

AlN-SiC compositions with SiC below 70wt% are all densification to ~99% relative density by hot pressing in this work. The main crystalline phases in the composites are AlN and SiC as well as small amount of TiC and $Y_3Al_5O_{12}$. The trend in thermal diffusivity is carried into the thermal conductivity data, with thermal conductivities values of 97.7 $W \cdot m^{-1} \cdot K^{-1}$ to 50.0 $W \cdot m^{-1} \cdot K^{-1}$ from 0wt% to 65wt%SiC, but increasing to 67.7 $W \cdot m^{-1} \cdot K^{-1}$ at 75wt% SiC. The attenuation characteristic is divided into two regions due to the different SiC content: one is narrowband attenuation at certain frequencies (0wt% and 20wt%SiC), the other is wideband attenuation (40wt%~80wt%SiC). With the SiC addition from 0wt% to 80wt%, the attenuation rise from 1.1dB to 2.4dB at 18.65GHz.

REFERENCE

[1]G. Q. Chen, X. Y. Li, and T. Qiu, Research Progress of AlN-based Microwave Attenuation Composite Ceramics with High Thermal Conductivity, *J. Mater. Sci. Eng.*, 25(2), 321-324 (2007).

[2]B. Mikijelj, D. K. Abe, and Ron. Hutcheon, AlN-based Lossy Ceramics for High Average Power Microwave Devices: Performance-property Correlation, *J. Eur. Ceram. Soc.*, 23(14), 2705-2709 (2003).

[3]T. B. Jackson, A. V. Virkar, K. L. More, R. B. Dinwiddie, and R. A. Cutler, High-thermal-conductivity Aluminum Nitride Ceramics: The Effect of Thermodynamic, Kinetic, and Microstructural Factors, *J. Am. Ceram. Soc.*, 80(6), 1421-1435 (1997).

[4]E. Mouchon, and P. H. Colomban, Microwave Absorbent: Preparation, Mechanical Properties and r. f.-microwave Conductivity of SiC (and/or mullite) Fibre Reinforced Nasicon Matrix Composites, *J. Mater. Sci.*, 31(2), 323-334 (1996).

[5]M. Landon, and F. Thevenot, Thermal Conductivity of SiC-AlN Ceramic Materials, *J. Eur. Ceram. Soc.*, 8(5), 271-277 (1991).

[6]W. B. Bu, T. Qiu, and J. Xu, Preparation and Microwave Attenuation Performance of AlN-SiC Composites, *J. Chin. Ceram. Soc.*, 31(9), 828-831 (2003).

[7]W. B. Bu, J. Xu, T. Qiu and X. Y. Li, Investigation of Reaction Synthesis of AlN-SiC Solid Solution, *J. Mater. Sci. Lett.*, 21, 731-732 (2002).

[8]J.P. Calame, Broadband Microwave and W-band Characterization of BeO-SiC and AlN-based Lossy Dielectric Composites for Vacuum Electronics, *IEEE Vacuum Electro. Conf.*, 37-38 (2006).

[9]G. Q. Chen, X. Y. Li, and T. Qiu, Effect of Tungsten on Microwave Attenuation of AlN-Based Composites, *Chinese Journal of Rare Metals*, 30(6), 813-817 (2006).

[10]A. A. Buchheita, G. E. Hilmasa, W. G. Fahrenholtz, D. M. Deason, and H. Wang, Mechanical and thermal properties of AlN-BN-SiC ceramics, *Mater. Sci. Eng.*, 494, 239-246 (2008).

[11]Q.F. Huang, S.F. Yoon, Rusli, Q. Zhang and J. Ahn, Dielectric Properties of Molybdenum-containing Diamond-like Carbon Films Deposited Using Electron Cyclotron Resonance Chemical Vapor Deposition, *Thin Solid Films*, 409, 211-219 (2002).

EFFECT OF ALKALINE EARTH OXIDES ON DIELECTRIC PROPERTIES OF POLYCRYSTALLINE $BaTi_2O_5$ PREPARED BY ARC MELTING

Xinyan Yue[1,2], Rong Tu[2], Takashi Goto[2] and Hongqiang Ru[1]
1 Key Laboratory for Anisotropy and Texture of Materials (MOE), School of Materials and Metallurgy, Northeastern University, Shenyang, 110004, China
2 Institute for materials research, Tohoku University, Sendai, 9808577, Japan

ABSTRACT

Dielectric properties of SrO, CaO and MgO substituted $BaTi_2O_5$, $Ba_{1-x}A_xTi_2O_5$ (A = Sr, Ca and Mg) prepared by arc melting were investigated. The lattice parameters of $Ba_{1-x}A_xTi_2O_5$ showed different changes depended on the increase of SrO, CaO and MgO substitutions. $Ba_{1-x}A_xTi_2O_5$ showed a significant *b*-axis orientation of (010). The Curie temperature of SrO and CaO substituted $BaTi_2O_5$ decreased from 750 to 703 and 685 K with increasing x up to 0.10, respectively, whereas that of MgO substituted $BaTi_2O_5$ decreased to 666 K with increasing x only to 0.01. The permittivity of $Ba_{1-x}Sr_xTi_2O_5$ and $Ba_{1-x}Ca_xTi_2O_5$ showed a sharp peak at Curie temperature while that of $Ba_{1-x}Mg_xTi_2O_5$ showed a flat peak at $x \geq 0.03$. The maximum permittivity of $Ba_{1-x}Sr_xTi_2O_5$, $Ba_{1-x}Ca_xTi_2O_5$ and $Ba_{1-x}Mg_xTi_2O_5$ showed the highest values of 3300, 4950 and 3250 at $x = 0.01$, 0.03 and 0.005, respectively.

INTRODUCTION

Barium titanate ($BaTiO_3$, BT) is a well-known ferroelectric material with a sharp peak of dielectric permittivity (ε_r) at the Curie temperature (T_c) about 400 K.[1-2] In order to modify the temperature dependence of ε_r and T_c, intensive studies on the substitution of Ba^{2+} and Ti^{4+} sites by various foreign elements have been conducted.[3-8] By choosing the appropriate substitution elements, the ε_r of BT can be enhanced and the T_c was changed significantly in a wide-range.[3] Since alkaline-earth ions could have similar characteristics and be easily substituted with Ba^{2+}, the substitution of alkaline-earth ions in BT has been studied by many researchers.[9-13] Zhou *et al.* reported that the T_c of $Ba_{1-x}Sr_xTiO_3$ decreased with increasing SrO substitution and the peak permittivity around the T_c increased from 3000 to 30000 by optimizing SrO substitution.[14-16] Jeong *et al.* reported that the T_c of BT shifted to 360 K with increasing MgO concentration below 1 mol%.[10] Chen *et al.* reported that the permittivity of BT decreased with increasing the Ca substitution.[11]

In the BaO-TiO_2 system, several TiO_2-rich compounds such as $BaTi_4O_9$ and $Ba_2Ti_9O_{20}$ are promising dielectric materials at microwave frequencies due to a significant small dielectric loss. $BaTi_2O_5$ (BT_2), on the other hand, has not been focused on since BT_2 has been misunderstood as a common paraelectric compound.[17] Recently, our research group and Akishige *et al.* independently discovered ferroelectricity of BT_2 and reported that single crystalline BT_2 has a high ε_r of 13000 to 30000 at the T_c (750 K) in the *b*-direction.[18-19] We have also revealed that the crystal structure of BT_2 does not belong to a monoclinic *C2/m* space group[20] but to a *C2*.[21] BT_2 could be expected as a new lead-free high-T_c ferroelectric material.[22] However, in practical applications, polycrystalline material are more preferable. We have first prepared (020) orientated polycrystalline BT_2 by arc melting.[23] In order to modify the dielectric properties of polycrystalline BT_2, substitution in Ba^{2+} site may be an effective approach. Therefore, in the present study, alkaline earth oxides substituted BT_2 were prepared by arc melting and the effect of SrO, CaO and MgO substitutions on the dielectric property of polycrystalline BT_2 was investigated.

EXPERIMENTAL PROCEDURES

TiO_2, $BaCO_3$, $SrCO_3$, $CaCO_3$ and MgO (99.9% in purity) powders were mixed in compositions of $Ba_{1-x}A_xTi_2O_5$ (A = Sr, Ca and Mg, $x = 0 \sim 0.10$). The mixed powders were pressed into pellets 20 mm in diameter at 10 MPa, and calcined at 1223 K for 43 ks in air. The pellets were melted on a water-cooled copper plate by arc melting in Ar. The specimens were heat-treated at 1323 K for 43 ks in air. The crystal orientation was measured by X-ray diffraction (XRD). A program based on the least squares analysis was used to calculate the lattice parameters and the standard deviation of d-values was less than 0.02%. The microstructure was observed by scanning electron microscopy (SEM) and field-emission scanning electron microscopy (FESEM). The specimens for dielectric measurements were coated plane electrode having 3 mm × 3 mm dimension parallel to the copper plate, and was set to be 1 mm in thickness. Gold paste as an electrode was painted on the two sides of all specimens, and was fired at 1123 K for 300 s. The dielectric properties were measured using an AC impedance analyzer (Hewlett Parckard 4194A) at frequencies (f) from 10^2 to 10^7 Hz and temperatures from 293 to 1073 K in air.

RESULTS AND DISCUSSION

Figure 1 shows the powder XRD patterns of polycrystalline BT_2 with alkaline earth oxides at the composition that the second phases began to be identified in the specimens. For polycrystalline BT_2 with CaO substitution, second phases were not identified at any composition by XRD analysis. For polycrystalline BT_2 with SrO and MgO substitutions, second phases began to be identified at the composition about $x = 0.15$ and 0.05, respectively.

Figure 2 shows the effect of alkaline earth oxides on lattice parameters of polycrystalline BT_2. The lattice parameters of polycrystalline $Ba_{1-x}Sr_xTi_2O_5$ and $Ba_{1-x}Ca_xTi_2O_5$ decreased first and then kept constant, while the lattice parameters of $Ba_{1-x}Mg_xTi_2O_5$ increased first and then decreased, finally kept constant. For the lattice parameters of $Ba_{1-x}Sr_xTi_2O_5$, the a-axis decreased from 1.6895 to 1.6874 nm, while b-axis decreased from 0.3934 to 0.3928 nm and c-axis decreased from 0.9411 to 0.9394 nm with increasing x up to 0.10. For the lattice parameters of $Ba_{1-x}Ca_xTi_2O_5$, the a-axis decreased from 1.6895 to 1.6877 nm, while the b-axis decreased from 0.3934 to 0.3923 nm and the c-axis decreased from 0.9411 to 0.9387 nm with increasing x up to 0.10. For the lattice parameters of $Ba_{1-x}Mg_xTi_2O_5$, the a, b and c-axis first increased from 1.6895, 0.3934 and 0.9411 nm to 1.6912, 0.3935 and 0.9414 nm with increasing x up to 0.01, then decreased to 1.6903, 0.3932 and 0.9410 nm with increasing x up to 0.03, and finally became almost a constant at $x > 0.03$. According to the change of the lattice parameters originating from the different ionic radius ($Ba^{2+} = 0.16$ nm, $Sr^{2+} = 0.14$ nm, $Ca^{2+} = 0.13$ nm and $Mg^{2+} = 0.072$ nm), it could be confirmed that the solubility limits of SrO and CaO in BT_2 were higher than that of MgO in BT_2.

Figure 3 shows the bulk XRD patterns of polycrystalline BT_2 with alkaline earth oxides. Polycrystalline BT_2 with SrO and CaO all had an good orientation of (020). Polycrystalline BT_2 with MgO also showed the orientation of (020) at $x<0.10$, however the peaks of secondary phases appeared when the MgO substitution content was at $x = 0.10$. The non-orientation of $Ba_{1-x}Mg_xTi_2O_5$ at $x = 0.10$ could be caused by the increasing of second phases.

Figure 4 presents the SEM micrographs of $Ba_{1-x}A_xTi_2O_5$ at $x = 0.05$: $Ba_{1-x}Sr_xTi_2O_5$ (a), $Ba_{1-x}Ca_xTi_2O_5$ (b) and $Ba_{1-x}Mg_xTi_2O_5$ (c). Several cracks were observed on the surfaces perpendicular to the growth direction. It might have been caused by two reasons: one was a significant difference of thermal expansion (α) among the a-, b- and c- directions of BT_2 ($\alpha_a = 5.14$, $\alpha_b = 0.86$ and $\alpha_c = 12.5 \times 10^{-6}$ K^{-1} at 900 K),[24] the other could be the difference of thermal expansion between BT_2 matrix and second phases.[25] In $Ba_{1-x}Sr_xTi_2O_5$ almost no second phase was observed according to the SEM image. For $Ba_{1-x}Ca_xTi_2O_5$ except cracks a number of crack-like phases were observed on the surface of the specimens that were confirmed as second phases by

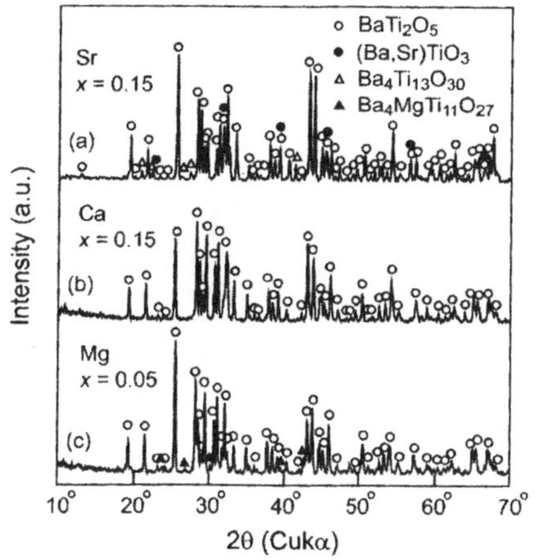

Fig. 1 Powder XRD patterns of $Ba_{1-x}A_xTi_2O_5$.

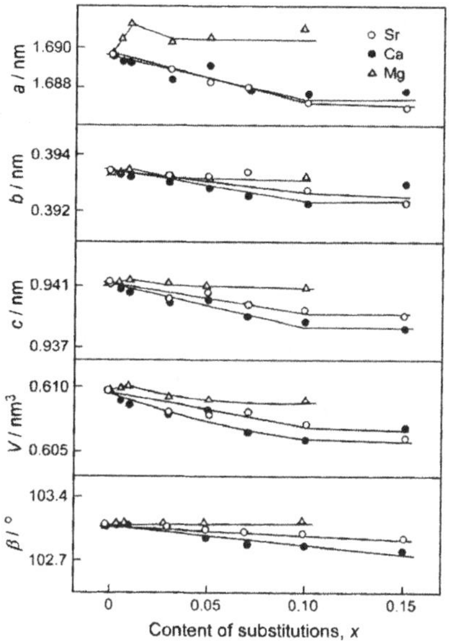

Fig. 2 Lattice parameters of $Ba_{1-x}A_xTi_2O_5$.

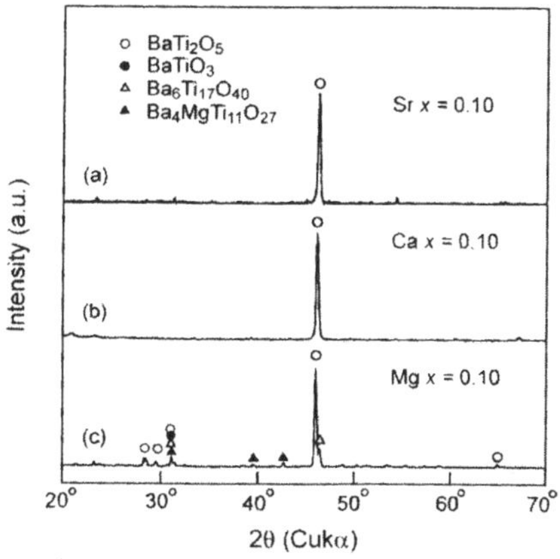

Fig. 3 Bulk XRD patterns of $Ba_{1-x}A_xTi_2O_5$.

FESEM in a higher magnification. In $Ba_{1-x}Mg_xTi_2O_5$ the second phases in dark contrast dispersed in the matrix. A small amount of secondary phases of BT in bright contrast and B_6T_{17} in dark contrast located in BT_2 matrix were identified although the XRD patterns showed no secondary phases. The sensitivity of XRD would be insufficient to detect the small amount of second phases in $Ba_{1-x}A_xTi_2O_5$ as that reported for polycrystalline BST_2 (Sr substituted BT_2).[26] According to a recent equilibrium phase diagram of BaO-TiO_2,[27] a melt with the composition of BT_2 would be first solidified to a mixture of BT and B_6T_{17}. So BT and B_6T_{17} might be precipitated as secondary phases in a small amount. However, the BT_2 phase could be obtained due to the narrow temperature range of the BT + B_6T_{17} mixture zone and the quench of the melt in arc melting process.

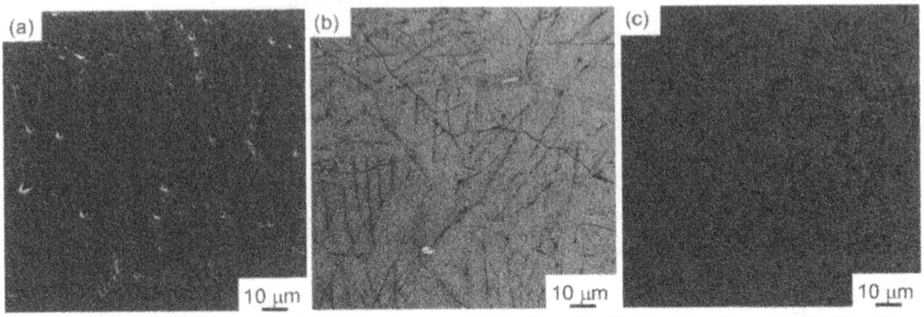

Fig. 4 SEM photos of $Ba_{1-x}A_xTi_2O_5$ at x = 0.05: $Ba_{1-x}Sr_xTi_2O_5$ (a), $Ba_{1-x}Ca_xTi_2O_5$ (b) and $Ba_{1-x}Mg_xTi_2O_5$ (c).

Figure 5 shows the temperature dependence of ε_r of $Ba_{1-x}A_xTi_2O_5$ at f = 100 kHz: $Ba_{1-x}Sr_xTi_2O_5$ (Fig. 5 (a)), $Ba_{1-x}Ca_xTi_2O_5$ (Fig. 5 (b)) and $Ba_{1-x}Mg_xTi_2O_5$ (Fig. 5 (c)). The permittivity of $Ba_{1-x}Sr_xTi_2O_5$ and $Ba_{1-x}Ca_xTi_2O_5$ showed a sharp peak at Curie temperature. The peak value of permittivity (ε_{max}) at the T_c changed depending on x. The permittivity of $Ba_{1-x}Mg_xTi_2O_5$ showed flat peaks at $x \geq 0.03$ around 400 to 700 K. The flat peak of permittivity indicated a distribution with more than one peak maximum caused by the inhomogeneity of micro regions of the specimens.[28] All the responses of the different micro regions to the different T_c could compose the broad peaks.

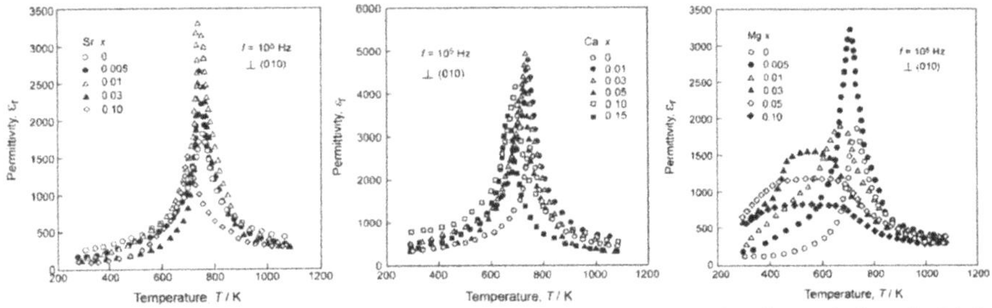

Fig. 5 Temperature dependence of $Ba_{1-x}A_xTi_2O_5$ at f = 100 kHz: $Ba_{1-x}Sr_xTi_2O_5$ (a), $Ba_{1-x}Ca_xTi_2O_5$ (b) and $Ba_{1-x}Mg_xTi_2O_5$ (c).

Figure 6 shows the effect of alkaline earth oxides on the maximum permittivity of polycrystalline BT_2. The maximum permittivity of SrO substituted polycrystalline BT_2 specimens showed the highest value of 3300 at x = 0.01. The maximum permittivity of CaO substituted polycrystalline BT_2 specimens showed the highest value of 4950 at x = 0.03. The maximum permittivity of MgO substituted polycrystalline BT_2 specimens showed the highest value of 3250 at x = 0.005. The Ca element was the most effective to increase the permittivity of polycrystalline BT_2 which may be caused by the longer balling time than that of other specimens.

Figure 7 shows the effect of alkaline earth oxides on the Curie temperature of polycrystalline BT_2. The Curie temperature decreased with increasing substitution content. The Curie temperature of SrO substituted polycrystalline BT_2 specimens decreased from 750 to 703 K with increasing x up to 0.10. The Curie temperature of CaO substituted polycrystalline BT_2 specimens decreased from 750 to 685 K with increasing x up to 0.10. The Curie temperature of MgO substituted polycrystalline BT_2 specimens decreased from 750 to 666 K with increasing x up to 0.01. The Mg element was the most effective to decrease the Curie temperature of polycrystalline BT_2.

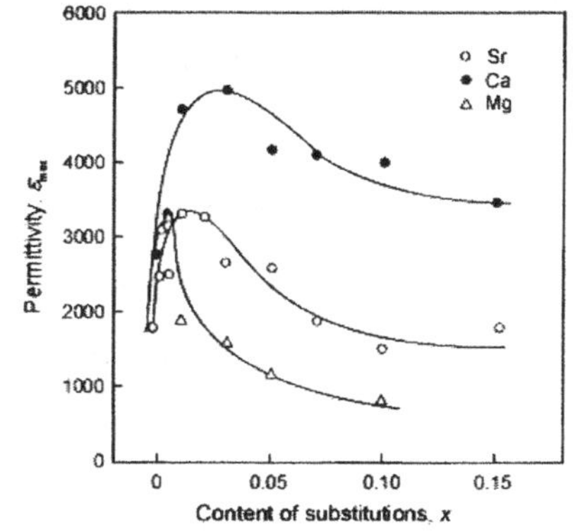

Fig. 6 Maximum permittivity of $Ba_{1-x}A_xTi_2O_5$.

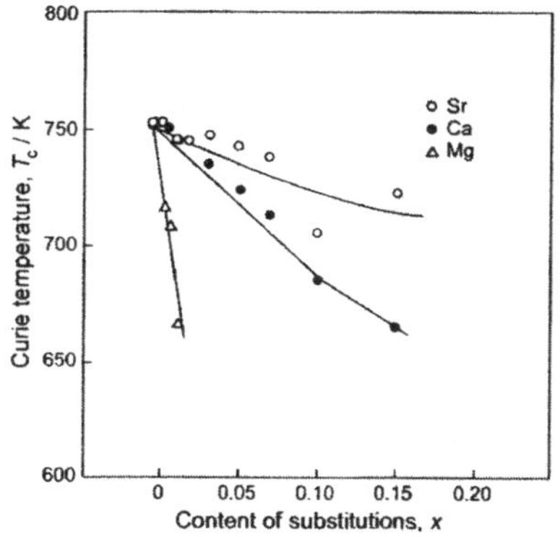

Fig. 7 Curie temperature of $Ba_{1-x}A_xTi_2O_5$.

CONCLUSION

SrO, CaO and MgO substituted $BaTi_2O_5$, $Ba_{1-x}A_xTi_2O_5$ (A = Sr, Ca and Mg) were obtained by arc melting method. The lattice parameters of polycrystalline $Ba_{1-x}Sr_xTi_2O_5$ and $Ba_{1-x}Ca_xTi_2O_5$ decreased first and then kept as a constant, while the lattice parameters of $Ba_{1-x}Mg_xTi_2O_5$ increased first and then decreased, finally kept as a constant. At $x = 0.05$ the amount of secondary phases in $Ba_{1-x}Mg_xTi_2O_5$ was more than that in polycrystalline $Ba_{1-x}Sr_xTi_2O_5$ and $Ba_{1-x}Ca_xTi_2O_5$ observed by SEM. The polycrystalline $Ba_{1-x}A_xTi_2O_5$ had an orientation of (010) plane along *b*-axis. The Curie temperature of SrO and CaO substituted polycrystalline $BaTi_2O_5$ decreased from 750 to 703 and 685 K with increasing x up to 0.10, respectively. The Curie temperature of MgO substituted polycrystalline $BaTi_2O_5$ decreased from 750 to 666 K with increasing x up to 0.01. The permittivity of $Ba_{1-x}Sr_xTi_2O_5$ and $Ba_{1-x}Ca_xTi_2O_5$ showed a sharp peak at Curie temperature while the permittivity of $Ba_{1-x}Mg_xTi_2O_5$ showed a flat peak at $x \geq 0.03$. The permittivity of $Ba_{1-x}Sr_xTi_2O_5$, $Ba_{1-x}Ca_xTi_2O_5$ and $Ba_{1-x}Mg_xTi_2O_5$ showed the highest values of 3300, 4950 and 3250 at $x = 0.01$, 0.03 and 0.005, respectively.

REFERENCES

[1] H. Beltrán, B. Gómez, N. Masó, E. Cordoncillo, P. Escribano and A. R. West, Electrical properties of ferroelectric $BaTi_2O_5$ and dielectric $Ba_6Ti_{17}O_{40}$ ceramics, J. Appl. Phys., **97** 084104-1-6 (2005).

[2] A. Hushur, H. Shigematsu, Y. Akishige and S.Kojima, Observation of relaxation mode in ferroelectric barium dititanate by micro-Brillouin scattering, Jpn. J. Appl. Phys., **43**, 6825-6828 (2004).

[3] A. D. Hilton and B. W. Ricketts, Dielectric properties of $Ba_{1-x}Sr_xTiO_3$ ceramics, J. Phys. D: Appl. Phys., **29**, 1321-25 (1996).

[4] A. Kirianov, T. Hagiwara, H. Kishi and H. Ohsato, Effect of Ho/Mg ratio on formation of core-shell Jpa. J. Appl. Phys., **41**, 6934-6937 (2002).

[5] Z. Yu, R. Guo and A. S. Bhalla, Dielectric behavior of $Ba(Ti_{1-x}Zr_x)O_3$ single crystals, J. Appl. Phys., **88**, 410-15 (2000).

[6] Y. S. Jung, E. S. Na, U. Paik, J. Lee and J. Kim, A study on the phase transition and characteristics of rare earth elements doped $BaTiO_3$, Mater. Res. Bull., **37**, 1633-40 (2002).

[7] D. Hennings and A. Schnell, Diffuse Ferroelectric Phase Transitions in $Ba(Ti_{1-y}Zr_y)O_3$ Ceramics, J. Am. Ceram. Soc., **65**, 539-44 (1982).

[8] R. Varatharajan, S. Madeswaran and R. Jayavel, Nb:BST:Crystal growth and ferroelectric properties, J. Crystal Growth, **225**, 484-88 (2001).

[9] B. Su, J. E. Holmes, B. L. Cheng and T. W. Button, Processing Effects on the Microstructure and Dielectric Properties of Barium Strontium Titanate (BST), J. Electroceramics., **9**, 111-16 (2002).

[10] J. Jeong and Y. H. Han: Electrical properties of MgO doped $BaTiO_3$, Phy. Chem. Chem. Phys., **5**, 2264-67 (2003).

[11] X. M. Chen, T. Wang and J. Li: Dielectric characteristics and their field dependence of $(Ba, Ca)TiO_3$ ceramics, Materials Science and Engineering B **113**, 117-20 (2004).

[12] L. Szymczak, Z. Ujma, J. Handerek and J. Kapusta, Sintering effects on dielectric properties of $(Ba,Sr)TiO_3$ ceramics Ceramics, International, **30**, 1003-08 (2004).

[13] C. B. Samantaray, H. Sim and H. Hwang, Electronic structure and optical properties of bariumstrontium titanate ($Ba_xSr_{1-x}TiO_3$) using first-principles method, Physica B: Condensed Matter, **351(1-2)**, 158-62 (2004).

[14] X. Wei and X. Yao, Nonlinear dielectric properties of barium strontium titanate ceramics, Materials Science and Engineering, B, **99**, 74-78(2003).

[15] L. Zhou, P. M. Vilarinho and J. L. Baptista, Dependence of the Structural and Dielectric Properties of $Ba_{1-x}Sr_xTiO_3$ Ceramic Solid Solutions on Raw Material Processing, J. Euro. Cera. Soci., **19**, 2015-20 (1999).

[16] M. Valant and D. Suvorov, J. Am. Ceram. Soc., Low-Temperature Sintering of $(Ba_{0.6}Sr_{0.4})TiO_3$, **87(7)**, 1222-1226 (2004).

[17] K. W. Kirby and B. A. Wechsler: Phase relations in the Barium Titanate—Titanium Oxide System, J. Am. Ceram. Soc. **74(8)**, 1841-47 (1991).

[18] T. Akashi, H. Iwata and T. Goto, Preparation of $BaTi_2O_5$ single crystal by a floating zone method, Mater. Trans., **44**, 802-04 (2003).

[19] Y. Akishige, K. Fukano and H. Shigematsu, Crystal growth and dielectric properties of new ferroelectric barium titanate: $BaTi_2O_5$, J. Electroceramics., **13**, 561-65 (2004).

[20] F. W. Harrison, The crystal structure of barium dititanate, Acta. Cryst., **9**, 495-500 (1956).

[21] T. Kimura, T. Goto, H. Yamane, H. Iwata, T. Kajiwara and T. Akashi, A ferroelectric barium titanate: $BaTi_2O_5$, Acta. Cryst., **C59**, i128-30 (2003).

[22] Y. Akishige: Synthesis and physical properties of single crystals and ceramics of new ferroelectric $BaTi_2O_5$, Jpn. J. Appl. Phys. **44(9B)** 7144-47 (2005).

[23] R. Tu and T. Goto, Dielectric properties of poly- and single-crystalline $BaTi_2O_5$, Mater. Trans., **47**, 2898-2903 (2006).

[24] Y. Akishige, and H. Shigematsu, A Kitahara and I Takahashi, Phase transition of new ferroelectric $BaTi_2O_5$, J. Korean Phys. Soc., **46**, 24-28 (2005).

[25] X. Y. Yue, R. Tu and T. Goto, Dielectric property of polycrystalline Ta_2O_5 substituted $BaTi_2O_5$ prepared by arc melting, J. Ceram. Soc. Jpn, **116(3)**, 436-40 (2008).

[26] X. Y. Yue, R. Tu and T. Goto, A. C. Impedance Analysis on *b*-axis Oriented $Ba_{1-x}Sr_xTi_2O_5$ Prepared by an Arc-melting Method, J. Ceram. Soc. Jpn, **115(10)**, 648-53 (2007).

[27] S. Lee and C. A. Randall, Modified Phase Diagram for the Barium Oxide–Titanium Dioxide System for the Ferroelectric Barium Titanate, J. Am. Ceram. Soc., **90(8)**, 2589-94 (2007).
[28] N. Maso, H. Beltran, E. Cordoncillo, A. S. Foores, P. Escribano, D. C. Sinclair and A. R. West: Synthesis and electric properties of Nb-doped $BaTiO_3$, J. Mater. Chem. **16,** 3114-19 (2006).

JOINING AND INTEGRATION OF ADVANCED CARBON-CARBON AND CARBON-SILICON CARBIDE COMPOSITES TO METALLIC SYSTEMS

M. Singh[1] and R. Asthana[2]

[1]Ohio Aerospace Institute
22800 Cedar Point Road
Cleveland, OH 44142

[2]Department of Engineering and Technology
University of Wisconsin-Stout
Menomonie, WI 54751

ABSTRACT

Carbon-carbon composites with CVI and resin-derived matrices, and C-SiC composites reinforced with T-300 carbon fibers in a CVI SiC matrix were joined to Cu-clad Mo using two Ag-Cu active braze alloys, Cusil-ABA (1.75% Ti) and Ticusil (4.5% Ti). The brazed joints revealed good interfacial bonding, preferential precipitation of Ti at the composite/braze interface, and a tendency toward delamination in the case of resin-derived composites. Extensive braze penetration of the inter-fiber channels in the CVI C-C composites was observed. The Knoop microhardness distribution across the joints revealed hardness gradients at the interface, and a higher hardness in Ticusil than in Cusil-ABA. For the C-SiC composites, the effect of composite surface preparation revealed that joints made using ground samples did not crack whereas un-ground samples cracked due conceivably to amplification of residual stress at surface imperfections. Theoretical predictions of the effective thermal resistance suggest that composite-to-Cu-clad-Mo joints may be promising for lightweight thermal management applications.

INTRODUCTION

Carbon-carbon (C–C) composites are used in the nose cone and leading edges of the space shuttle, rocket nozzles, exit cones, heat shield, aircraft braking systems and other components. One promising area to utilize C–C composites is in thermal management applications. Over the years, a number of materials such as Al, Cu-clad-Mo, Cu-clad-Invar, B/Al and SiC/Al have been considered for thermal management applications, each exhibiting distinct benefits and limitations. For example, Cu-clad-Mo and Cu-clad-Invar have high density, Al has large CTE mismatch with ceramics, and B/Al and SiC/Al are relatively costly. Joining of monolithic materials to obtain desired conductivity, CTE, and specific gravity has also been explored in systems such as graphite/Cu-clad-Mo[1] and more recently, C-C/Cu-clad-Mo[2]. Joining of Cu-clad-Mo and C–C composites containing high-conductivity carbon fibers is particularly attractive for thermal management applications because of increased functionality at reduced weight which is important to a number of thermal management applications. Acting in combination, Cu-clad Mo and C–C can provide excellent heat dissipation capability and weight advantage. Additionally, by controlling the clad layer thickness in Cu-clad-Mo, the CTE mismatch between C-C and Cu-clad Mo can be designed to mitigate residual stresses during joining and service while maintaining acceptable levels of thermal conductivity.

Besides C-C composites, C fiber reinforced SiC (C-SiC) composites are being evaluated for applications in industrial gas turbine engines, combustor liner components, shrouds, expansion nozzles of rocket propulsion systems, as well as exhaust cones, engine flaps and flame holders of jet engines. Table 1 gives representative properties of C-SiC, C-C and SiC-SiC composites. Recently, there has been considerable interest in joining C-SiC composites[3-6] to high-temperature metals and alloys; however, studies on joining of C-SiC to molybdenum were not found.

In the present work, we report on the brazing and joint characterization of C-C composites and C-SiC composites to Cu-clad-Mo using two Ti-containing Ag-Cu active braze alloys (Ticusil and Cusil-ABA). These alloys have good thermal conductivity ($K_{Cusil-ABA}$=180 W/m.K and $K_{Ticusil}$=219 W/m.K) and good ductility (42% and 28% for Cusil-ABA and Ticusil, respectively). Copper as a cladding on Mo and as an alloying additive in braze is expected to promote the wetting and metallurgical bonding. The brazed joints were characterized using optical microscopy, field emission scanning electron microscopy (FESEM), energy dispersive spectrometry (EDS), and by microhardness measurements.

Table I. Representative Properties of C-C, C-SiC and SiC-SiC Composites[7-9]

Composite	UTS, MPa	E*, GPa	Flexural Strength, MPa	ILSS, MPa	CTE, $\times 10^{-6}$/K	K, W/m.K
CVI C-SiC (42-47% fiber)[8]	350	90-100	500-700	35	3.0[a] 5.0[b]	14.3-20.6[a] 6.5-6.9[b]
LPI C-SiC[8]	250	65	500	10	1.16[a] 4.06[b]	11.3-12.6[a] 5.3-5.5[a]
GE's HiPerComp SiC-SiC (22-24% fiber)[9]	--	285	--	135[c]	3.5[a] 4.07[b]	33.8[a] 24.7[b]
CVI C-C composite (3-D)[7]	99[i] 105[j]	56[i] 58[j]	--	--	~2.0-4.0	115[d] 95[e] 60[f]
C-C[7]	600[a] 4[b]	125[a][m] 56[k]	1250-1600[a] 20[b]	--	~2.0-4.0	140[g] [a] 12[g] [b] 105[h] [a] 6[h] [b]

* Strongly depends on fiber type and architecture, matrix structure, and heat treatment; [a]in-plane value; [b]through-thickness value; [c]from fast fracture strength test; [d]pierced weave (HM fiber) at 500C; [e]3-D fine weave (orthogonal) at 500C; [f]pierced weave (LM fiber) at 500C; [g]pitch-derived at 500C; [h]resin-derived at 500C; [i]orthogonal (16% fiber); [j]pierced weave (25% fiber; [k]3-D orthogonal (room temperature); [m]1-D C-C.

EXPERIMENTAL PROCEDURE

The 3-D C-C composites were obtained from Goodrich Corp., Santa Fe, CA. Some joints were made of 2-D C-C composites with T-300 fibers in a resin-derived matrix, which were obtained from Carbon-Carbon Advanced Technology (C-CAT Composites) Inc., Fort Worth, TX. The C-SiC composites were obtained from GE Power Systems Composites, Newark, DE. These composites are reinforced with T-300 carbon fibers (1K tow; PW; 3.2 mm thick) in an amorphous CVI SiC matrix. The C-SiC composites were used in both as-received (unground) and ground conditions. Grinding was done using 320#, 400# and 600# grit silicon carbide papers to obtain macroscopically smooth composite surfaces; the as-received C-SiC had rough, uneven surface. The surface roughness was not measured and not used as a quantifiable variable in the study.

Copper-clad Molybdenum (Cu-Mo-Cu) plates from H.C. Starck, Inc., Newton, MA, were used as the metal substrate. The Cu-to-Mo-to-Cu layer thickness ratio was 13%-74%-13%. The material combines the high conductivity of Cu with the CTE of Mo; the CTE of the material is tailored by changing the clad ratio of Cu-Mo-Cu. Powders of active metal brazes Cusil-ABA (63Ag-35.3Cu-1.75Ti, T_L=815C) and Ticusil (68.8Ag-26.7Cu-4.5Ti, T_L=900C) were obtained from Morgan Advanced Ceramics, Hayward, CA. Selected properties of the braze alloys are given in Table I.

The C-C, C-SiC, and Cu-clad-Mo plates were sliced into 2.54 cm x 1.27 cm x 0.25 cm pieces, and ultrasonically cleaned in acetone for 15 min. prior to brazing. Braze powders were mixed with glycerine in a viscous paste with dough-like consistency, and a small (~0.2 g) quantity of this paste was applied to the joint region between the composite and metal. The sandwiched composite/braze/metal structure was placed under ~6.0 kPa pressure (2.0 N load) in a vacuum furnace which was heated to the brazing temperature (~20°C above T_L) under a pressure of ~10^{-6} torr. After a 5 min. soak at the brazing temperature, the assembly was slowly cooled at a controlled rate (~5°C per min.) to 400°C, followed by furnace cooling.

The brazed joints were mounted in epoxy, ground, polished, and examined using Field Emission Scanning Electron Microscopy (FESEM) (model: Hitachi 4700) coupled with energy dispersive x-ray spectroscopy (EDS). Microhardness scans were made with a Knoop indenter across the joint interfaces on a Struers Duramin-A300 machine under a load of 200 g and loading time of 10 s. Multiple (4 to 6) hardness scans were made across each joint to check the reproducibility.

RESULTS AND DISCUSSION

Joint Microstructure and Composition

C-C/Cu-Clad-Mo Joints: The microstructure of the composite/braze interface (Fig. 1) reveals braze infiltration of the inter-fiber regions to several hundred micrometer distance in 5 min. This is consistent with the sessile-drop wettability test results [10] on Cu-Ti/porous C in which the sessile drop volume continuously decreased due to the reactive infiltration of open porosity in graphite, and sessile drops of high Ti content (e.g., Cu-28Ti) rapidly and completely disappeared into the graphite substrate. The reaction of carbon with Ti in the braze forms the wettable compound titanium carbide which facilitates self-infiltration and sound bonding.

Figures 2 through 4 show the SEM images of C-C/Cu-clad-Mo joints. All joints display intimate physical contact, and are free of common imperfections. The braze matrix exhibits a two-phase eutectic structure with Ag-rich light-grey areas (e.g., point 3, Fig. 2) and Cu-rich dark areas (point 4, Fig. 2). In the Ag-Cu-Ti system, intermetallics such as AgTi, Ti_2Cu_3, and $TiCu_2$ may also form. The C-C/Cusil-ABA interface is rich in Ti and the Ti concentration decreases with increasing distance from the interface (9.2 atom%, 4.2 atom% and 1.8 atom% at points 2, 4 and 5, respectively, in Fig. 2b). Small amounts of Ag and Cu from braze are detected within the C-C composite region (point 1, Fig. 2b). The Cusil-ABA/Cu-clad-Mo interface (Fig. 2c) displays evidence of good wetting and somewhat diffuse interface character. The Cu cladding at the braze/Cu-clad-Mo interface remains untransformed because the joining temperature (830°C) is below the melting point of Cu (1086°C). Some dissolution probably occurred at the Cu-cladding/braze interface.

In C-C/Cu-clad-Mo joints made using Ticusil (Fig. 3), a small amount of Cu is detected within the composite region (points 5 and 6, Fig. 3b). The normal two-phase structure with a characteristic acicular morphology (Fig. 3b & c) is observed within the braze region. Some carbon has dissolved in the molten braze (points 1 and 2, Fig. 3b), possibly because of the higher brazing temperature (915°C) of Ticusil which led to C diffusion in the eutectic micro-constituents. In addition, carbon is detected within the Cu-clad-Mo region (points 3-6, Fig. 3c).

Figure 4 shows joint interfaces between resin-derived C-C composite (C-CAT Composites) and Cu-clad-Mo made using Ticusil. Microstructurally sound joints formed but there was some cracking within the C-C composite (Fig. 4a) presumably due to the low inter-laminar shear strength of C-C composites. Ag- and Cu-rich phases formed in the braze matrix with the Ag-rich phase preferentially precipitating onto C-C (point 2, Fig. 4b) and Cu-clad-Mo surface (point 2, Fig. 4c). A small amount of Cu was detected within the composite (point 4, Fig. 4b).

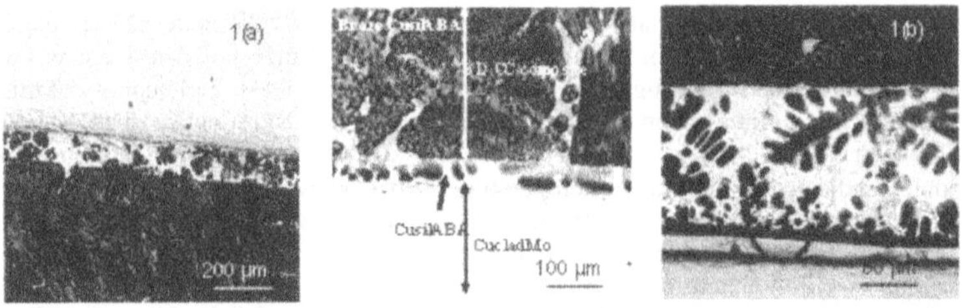

Fig. 1 A 3-D C-C composite joined to Cu-clad-Mo using (a) Ticusil and (b) & (c) Cusil-ABA. Copious infiltration of inter-fiber channels by molten brazes, and dissolution of Cu cladding have occurred.

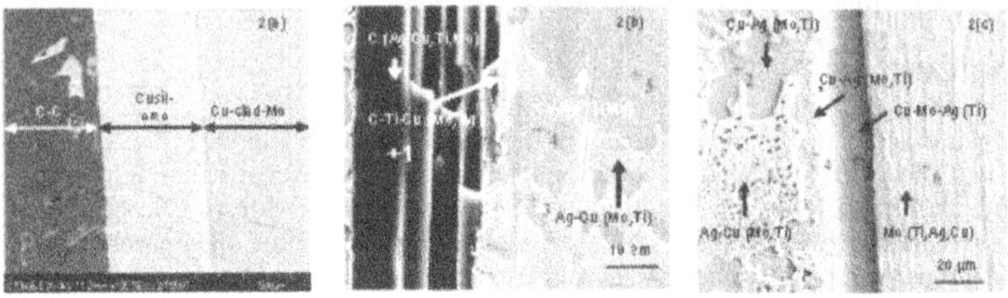

Fig. 2. A 3-D C-C composite/Cusil ABA/Cu-clad-Mo joint showing (a) overall view of the joint, (b) C-C/Cusil-ABA interface, and (c) Cusil-ABA/Cu-clad-Mo interface.

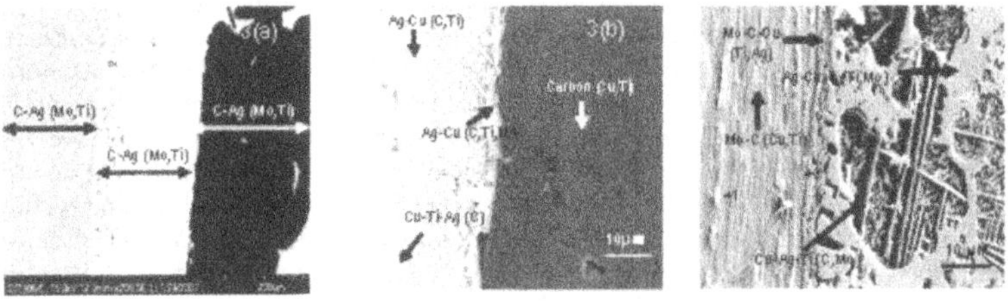

Fig. 3 A 3-D C-C composite/Ticusil/Cu-clad-Mo joint showing (a) overall view of the joint, (b) C-C/Ticusil interface, and (c) Ticusil/Cu-clad-Mo interface.

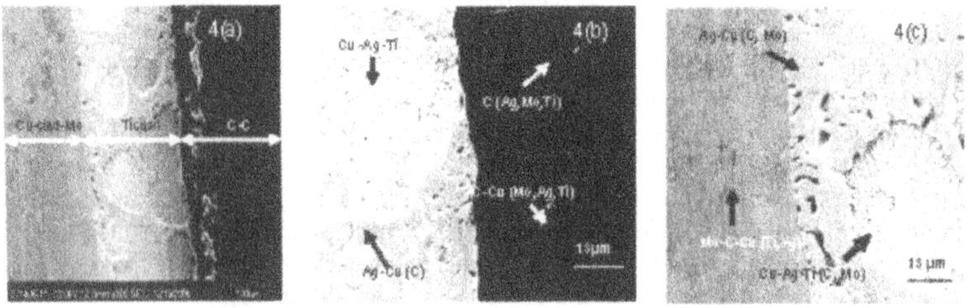

Fig. 4 A C-C (resin-derived) composite/Ticusil/Cu-clad-Mo joint showing (a) overall view of the joint, (b) C-C/Ticusil interface, and (c) Ticusil/Cu-clad-Mo interface.

In summary, whereas extensive chemical interactions did not occur and observable interfacial reaction layers did not form at the C-C/CuAgTi interfaces, some redistribution of alloying elements occurred. Large Ti concentrations occurred at the C-C/braze interface, indicating favorable surface modification due to a carbide-forming reaction that promoted bonding.

C-SiC/Cu-Clad-Mo Joints: Figure 5 shows the microstructure of a ground C-SiC/Cu-clad-Mo joint made using Cusil-ABA braze. Intimate physical contact at both CMC/braze (Fig. 5b) and braze/Cu-clad-Mo (Fig. 5c) interfaces is noted. The braze matrix displays the characteristic two-phase eutectic microstructure comprised of Cu(Ag) and Ag(Cu) phases. The EDS analysis showed that small amounts Ti and Si dissolved in the braze (points 2, 3 and 4 in Fig. 5b). Within the C-SiC composite, minute amounts (2.5 atom %) of Ag are detected (point 5, Fig. 5b).

The grinding operation prior to joining removed the SiC coating from the composite's surface (Fig. 5b). The ground C-SiC/Cusil-ABA interface (point 4, Fig. 5b) is enriched in Ti (45.8 atom %) and Si (9.6 atom %), suggesting possible formation of a wettable and well-bonded titanium silicide interface layer. The Cusil-ABA/Cu-clad-Mo interface (Fig. 5c) is metallurgically sound but there was little indication of interdiffusion of alloying elements.

Figure 6 shows the joint interfaces in an un-ground C-SiC/Cu-clad-Mo joint with Cusil-ABA braze interlayer. Whereas the interface between Cu-clad-Mo and Cusil-ABA (Fig. 6c) is sound the C-SiC/braze interface (Fig. 6b) is cracked. The crack exists between the SiC coating on the unground substrate and braze (Fig. 6a & b). Minute quantities (~2.0-3.0 atom %) of the braze constituents, Cu and Ag, are detected to a distance of ~50-100 μm within the composite (points 4, 5 and 6, Fig. 6b). This suggests that interfacial de-cohesion had probably resulted from the residual stresses during post-braze cooling and not because the wetting of the un-ground composite was poor. Good, physical contact between molten braze and the composite would permit diffusion of Cu and Ag into the composite across the interface. For the Ag-Cu alloys containing Ti, contact angle measurements[10,11] reveal that wetting on C and SiC surfaces should be excellent. Because in a wettable system, surface roughness promotes rather than inhibits the wetting, most probably the interfacial de-cohesion noted in Fig. 6b occurred due to residual stresses resulting from the large CTE-mismatch during post-braze cooling.

Figure 7 shows the joint interfaces in ground C-SiC/Cu-clad-Mo joints made using Ticusil. There is evidence of good braze/composite interaction (Figs. 7a & b), and relatively large quantities of Ti (18.6 atom%), Mo (36.4 at%) and Ag (45 at%) are detected within the C-SiC composite (point 1, Fig. 7b). The SiC coating on the composite surface has been removed by grinding and an intimate composite-to-braze contact established. Silicon is detected at ~15-20 μm distance within the braze region near the interface (point 4, Fig. 7b). As before, the braze matrix displays the Ag-rich and Cu-rich two-phase eutectic structure with the Ag-rich phase preferentially segregating at the C-SiC surface (Fig. 7b). The Ag-rich phase has also preferentially deposited at the interface on the Cu-clad-Mo side (Fig. 7c). Interestingly, there is some carbon dissolution and diffusion in braze (points 1 & 2, Fig. 7c) and also in Mo (point 5, Fig. 7c) to a depth of ~30 μm. Additionally, some Cu (10.6 at%) from the clad layer was detected within the Mo substrate (point 5, Fig. 7c).

The un-ground C-SiC/Ticusil/Cu-clad-Mo joints (Fig. 8) show good bonding unlike the un-ground C-SiC/Cusil-ABA/Cu-clad-Mo joints of Fig. 6 where interfacial de-cohesion had occurred at the composite/braze interface. In Fig. 8, the CVI SiC layer on the unground composite is intact. With a higher joining temperature for Ticusil (920°C) than Cusil-ABA (835°C), the resulting thermal strain ($\Delta\alpha\Delta T$) from CTE mismatch will be higher in Ticusil than in Cusil-ABA joints. However, a higher Ti content in Ticusil (4.5% Ti) than in Cusil-ABA (1.75% Ti) may have contributed to better wetting and stronger bonding[11] with Ticusil than with Cusil-ABA. It is possible that the negative effects of a large thermal strain in Ticusil may have been partly offset by

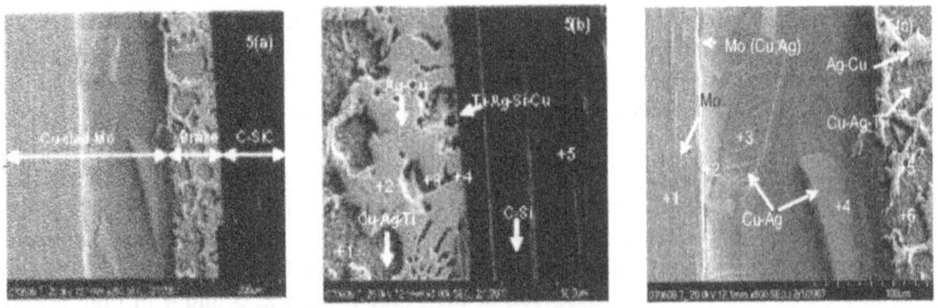

Fig. 5 C-SiC (ground)/Cusil-ABA/Cu-Clad-Mo joint: (a) overall view of the joint, (b) C-SiC/Cusil-ABA interface, and (c) Cusil-ABA/Cu-Clad-Mo interface.

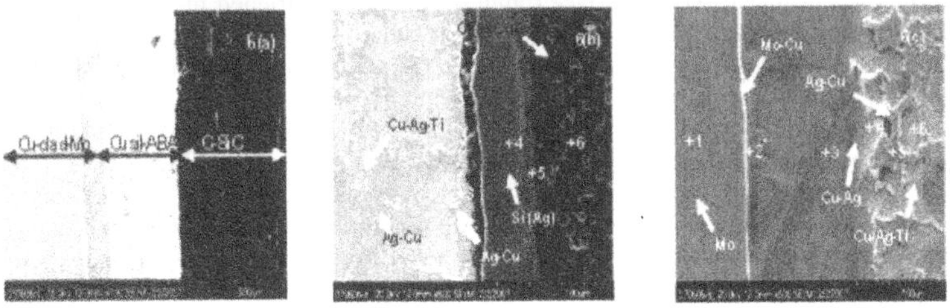

Fig. 6 C-SiC (unground)/Cusil ABA/Cu-Clad-Mo joint: (a) overall view of the joint region, (b) C-SiC/Cusil-ABA interface, and (c) Cusil-ABA/Cu-clad-Mo interface.

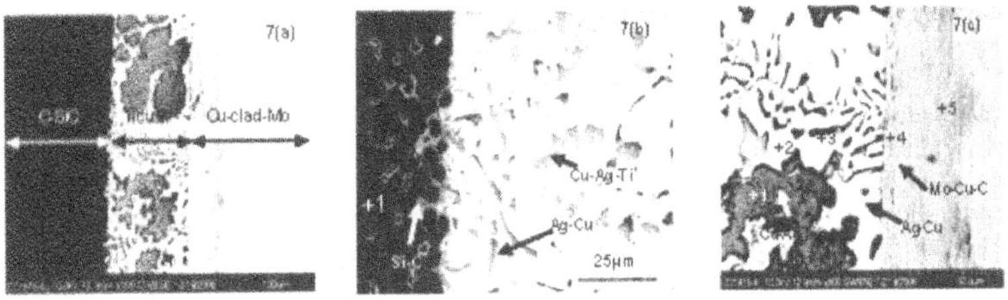

Fig. 7 C-SiC (ground)/Ticusil/Cu-Clad-Mo: (a) overall view of the joint region, (b) C-SiC/Cusil-ABA interface, and (c) Cusil-ABA/Cu-clad-Mo interface.

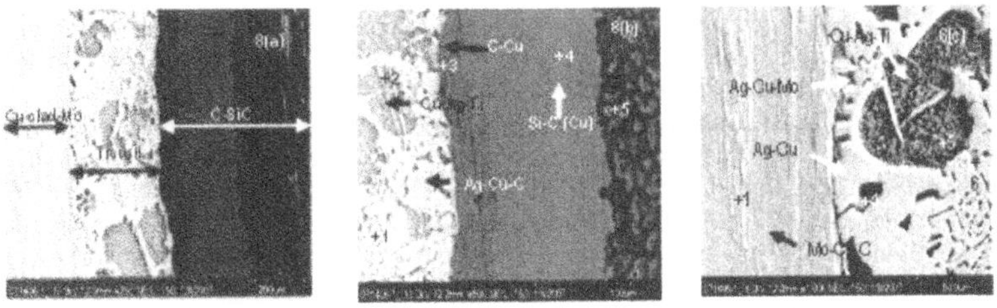

Fig. 8 C-SiC (unground)/Ticusil/Cu-Clad-Mo joint: (a) overall view of the joint region, (b) C-SiC/Ticusil interface, and (c) Ticusil/Cu-clad-Mo interface.

the formation of a stronger joint in Ticusil than Cusil-ABA.
Some Cu was detected to ~100 μm distance within the composite (points 4 and 5, Fig. 8b). In addition, the Ag-rich phase in braze preferentially precipitated on to both the composite surface (Fig. 8b) and on the Mo substrate (points 2 and 3, Fig. 8c).

Microhardness

C-C/Cu-Clad-Mo Joints: Knoop microhardness (HK) profiles across the C-C/Cu-clad-Mo joints made using Cusil-ABA and Ticusil are shown in Fig. 9(a-c). There was no effect of the composite type (CVI *vs* resin-derived) on the HK values within the braze region. The hardness of the Mo substrate is ~200-330 HK and the hardness of braze depends on braze type; Ticusil (Fig. 9b & c) exhibited higher hardness (~85-200 HK) than Cusil-ABA (~50-180 HK). This is consistent with the somewhat greater Ti-induced hardening expected in Ticusil (4.5%Ti) than in Cusil-ABA (1.75%Ti), and with the somewhat larger residual stresses expected with Ticusil because of its higher liquidus temperature ($T_L \sim 920°C$) than Cusil-ABA ($T_L \sim 815°C$).

C-SiC/Cu-Clad-Mo Joints: Figures 9(d-g) show the distribution of Knoop hardness (HK) across the C-SiC/Cu-clad-Mo joints made using Cusil-ABA and Ticusil. The data show that there is no effect of composite surface preparation on hardness profiles; both ground and un-ground composite substrates led to similar distributions across the joint. The hardness of the Mo substrate is 250-350 HK. The braze regions display a drop in hardness. The hardness of the composite is sensitive to the actual path traversed by the indenter, with the hardness values rising to ~1,500 HK when hard SiC matrix regions were encountered by the indenter between C fibers. Additionally, residual stresses due to CTE mismatch also possibly influenced the hardness value.

Residual Stress at the Joint

Upon cooling a brazed joint, large residual stresses arising from CTE mismatch may lower the fracture strength. A model due to Eager and coworkers[12] analyzes residual stress relief by metal interlayers taking into account the CTE mismatch and interlayer plasticity. Their models permit estimation of the strain energy in the ceramic for well-bonded ceramic-metal joints. For a small CTE mismatch between the ceramic (C) and the metal substrate (M), but with a large CTE mismatch between the ductile interlayer (I) and the base materials the elastic strain energy, U_{eC}, in the non-metallic substrate can be calculated using equations (1)–(3) of ref[12] (Note: powder-based braze interlayers in this work were substantially thicker than the braze foils used in ref. [12] where Ni, W, Mo etc were considered as the interlayers. Here we consider the thick braze layers themselves to be the interlayer materials). The required material properties are yield strength of interlayer (σ_{YI}), radius of the joint, elastic modulus of the ceramic (E_C), elastic modulus of the interlayer (E_I), temperature change (ΔT), and the CTE (α) of metal (M), ceramic (C) and interlayer

(I). The configuration analyzed by Eager et al[12] is a cylindrical disc-shaped joint whereas our joints are rectangular in cross-section (2.54 cm × 1.25 cm). For calculation purpose, we take an effective joint radius to be the minimum distance to the edge of our samples (0.625 cm). The calculations are made in terms of the dimensionless parameters Π_I and Φ defined in ref.[12], where Π_I is the ratio of the thermal residual strain at the interface to the yield strain of the braze interlayer, and Φ specifies the relative difference in CTE's between the ceramic (C), braze interlayer (I), and metal substrate (M).

C-C/Cu-Clad-Mo Joints: The strain energy in the C-C/Cu-clad-Mo joints was computed using the generic property data summarized in Table 1. The average values of E_{C-C} = 56 GPa and α_{C-C} =

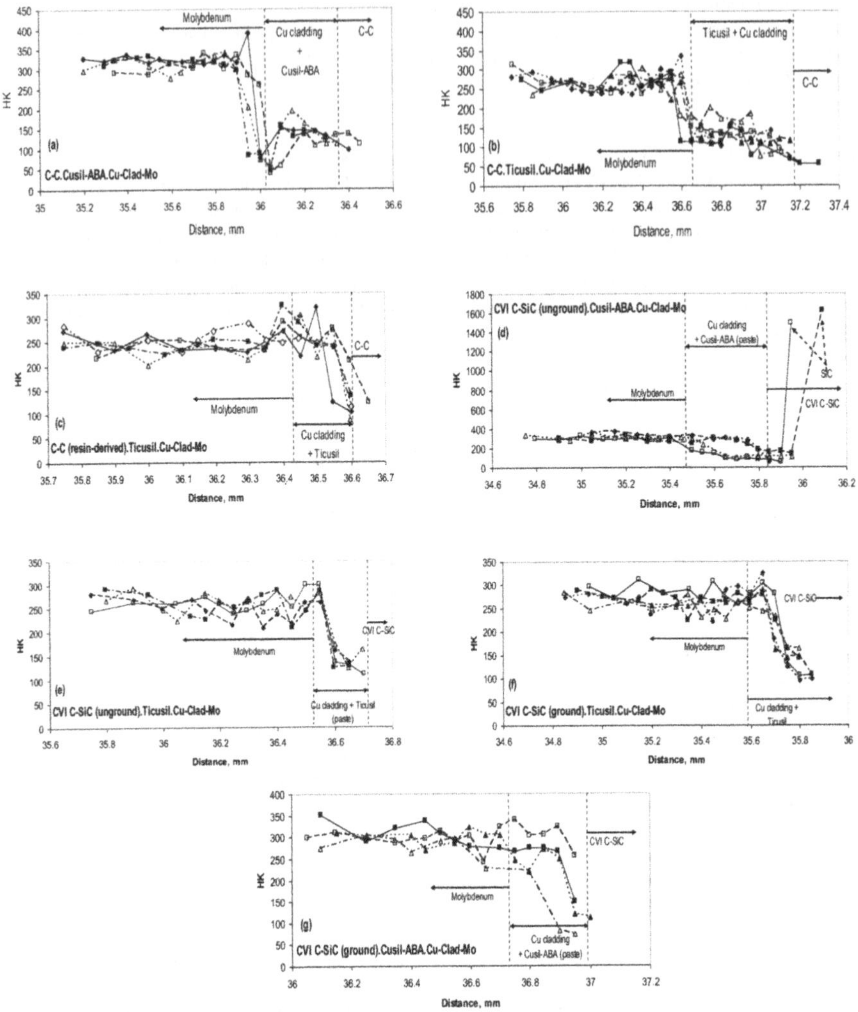

Fig. 9 Knoop hardness (HK) distribution across joints in (a) CVI C-C/Cusil-ABA/Cu-clad-Mo, (b) CVI C-C/Ticusil/Cu-clad-Mo, (c) C-C (resin-derived)/Ticusil/Cu-clad-Mo, (d) C-SiC (un-ground)/Cusil-ABA/Cu-clad-Mo, (e) C-SiC (un-ground)/Ticusil/Cu-clad-Mo, (f) C-SiC (ground)/Ticusil/Cu-clad-Mo, and (g) C-SiC (ground)/Cusil-ABA/ Cu-clad-Mo. Multiple scans across each joint are identified with symbols.

3×10^{-6} C^{-1}. For the braze interlayer, the following data from the manufacturer were used: $E_{Ticusil}$ = 85 GPa, $E_{Cusil\text{-}ABA}$ = 83 GPa, $\sigma_{Ticusil}$ = 292 MPa, $\sigma_{Cusil\text{-}ABA}$ = 271 MPa, and $\alpha_{Ticusil} = \alpha_{Cusil\text{-}ABA}$ = 18.5×10^{-6} C^{-1}. The thermal property data for Cu-clad-Mo as a function of clad layer thickness were taken from[13]. These data yield U_{eC} as ~38-190 mJ for clad layer thicknesses ranging from 0 to 40% (Fig. 10), with U_{eC} increasing at a progressively faster rate with an increase in Cu layer thickness. In addition, less strain energy develops in Cusil-ABA joints than in Ticusil joints.

C-SiC/Cu-Clad-Mo Joints: Two types of C-SiC composite substrates listed in Table 1 were considered: CVI C-SiC and LPI C-SiC (for comparison, HiPerComp SiC-SiC were also considered). The values of α_C and E_C used in the calculations are average values for each type of composite from Table 1. The calculated strain energies are shown in Fig. 10 for C-SiC/Cu-clad-Mo joints as a function of clad layer thickness. With regard to composite type, the HiPerComp SiC-SiC is the best candidate for joining to Cu-clad-Mo followed by CVI C-SiC and LPI C-SiC (least favorable; largest U_{eC}). For all types of joints, U_{eC} is slightly lower for Cusil-ABA than Ticusil suggesting that Cusil-ABA joints are probably more favorable; the greater ductility and lower Ti content of Cusil-ABA are beneficial from strain energy considerations although the higher Ti content of Ticusil is better from the standpoint of braze flow and spreading (larger Ti content causes greater drop in the contact angle). Thus a tradeoff between chemical affinity and thermoelastic compatibility probably exists.

Considering the fact that the calculations strictly apply to a cylindrical joint configuration and monolithic ceramics rather than anisotropic materials such as composites, and the fact that chemical interactions and solute segregation will irrevocably and unpredictably modulate the joint properties, the calculated strain energy values are probably reasonable.

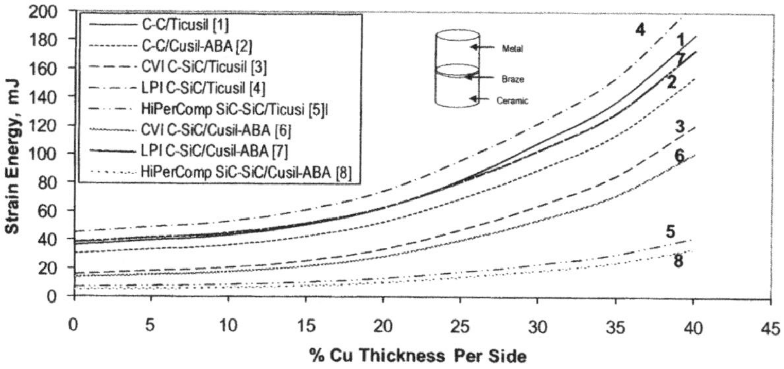

Figure 10 Calculated strain energy in C-C/Cu-clad-Mo and C-SiC/Cu-clad-Mo joints made using Ticusil and Cusil-ABA based on a model due to Eager et al[12].

Thermal Conduction in Brazed Joints

The thermal conduction in composite-to-Cu-clad-Mo joints is important for thermal management applications. For 1-D steady-state heat conduction, the joined materials form a series thermal circuit with an effective thermal resistance, $R_{eff} = \Sigma(\Delta x_i/K_i)$, where Δx_i and K_i are the thickness and thermal conductivity, respectively, of the ith layer.

C-C/Cu-Clad-Mo Joints: For our joints, $\Delta x_{C\text{-}C} = \Delta x_{Cu\text{-}Mo} = 0.25\times10^{-2}$ m, $\Delta x_{Ticusil} \sim 100\times10^{-6}$ m, $K_{Ticusil} = 219$ W/m-K and $K_{Cusil\text{-}ABA} = 180$ W/m-K. For composites, even though $K_{C\text{-}C}$ is anisotropic and variable, we take an average value as 125 W/m-K for 2D and 3D composites. The conductivity of Cu-clad-Mo, $K_{Cu\text{-}Mo}$, varies with the clad layer thickness, and is taken from ref.[13]; $K_{Cu\text{-}Mo}$ varies from 138 W/m-K to 235 W/m-K for 0 to 30% clad layer thickness. The effective thermal resistance of the joints is shown in Fig. 11(a). R_{eff} varies in the range 31.5 to 38.5×10^{-6} m^2.K/W, and that there is insignificant (<1%) difference between Ticusil and Cusil-ABA. Because the difference in the R_{eff} of the joints with the two brazes is insignificant, there may be considerable flexibility in selecting brazes to satisfy other criteria such as ductility and wetting characteristics without impairing the thermal conductivity and weight advantages of the joined materials.

Figure 11(a) also compares the R_{eff} values of the joints to the R_{eff} values of C-C and Cu-clad-Mo substrates of the same total thickness (5.1×10^{-3} m) as the joined materials; the thermal resistance of the C-C block is about 40.8×10^{-6} m^2.K/W and that of Cu-clad-Mo substrate decreases with increasing clad layer thickness. The decrease in the conductivity of our joints (Cu:Mo:Cu thickness ratio of 13%:74%:13%) relative to an un-joined Cu-clad-Mo substrate is compensated by 39%

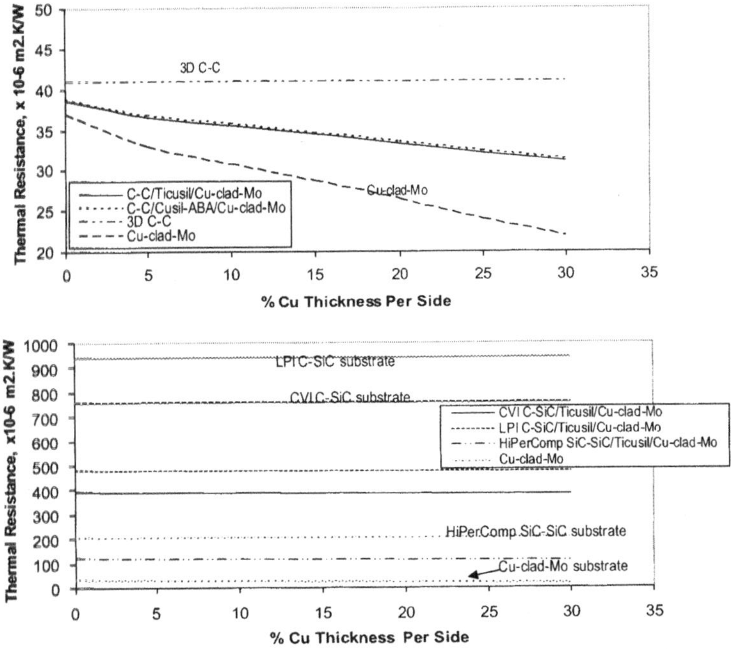

Figure 11 Thermal resistance of C-C/Cu-clad-Mo joint (top) and C-SiC/Cu-clad-Mo joint (bottom). The decrease in R_{eff} is small (<10%) for clad layer thickness up to 30% and is not visually resolvable in the figure. The lower figure also shows the projections for HiPerComp SiC-SiC composite.

ORPHAN HOLD
DO NOT SHIP

H1–000000000374

ISBN 9780470408421

CLIENT	Alternate Sourcing
STATION	TPP Receiving Station 01
HELD	2026–10–08 11:44

SHELF

If its order arrives, scan this label at a receiving station.

decrease in weight. The Rule-of-Mixtures (ROM) density (ρ) of our joints is ~5,919 kg.m^{-3} (with $\rho_{C\text{-}C}$ = 1,900 kg.m^{-3}) compared to a density of 9,937 kg.m^{-3} for Cu-clad-Mo (ignoring the thin braze interlayer does not introduce any sensible error in the density calculations). This points to a beneficial effect of joining Cu-clad-Mo to C-C composite.

C-SiC/Cu-Clad-Mo Joints: For the joints created in this work, $\Delta x_{C\text{-}SiC} = \Delta x_{Metal} = 0.25\times10^{-2}$ m, $\Delta x_{Braze} \sim 100\times10^{-6}$ m, $K_{Ticusil}$ = 219 W/m-K, $K_{Cusil\text{-}ABA}$ = 180 W/m-K, and from Table 1, $K_{C\text{-}SiC}$ = 6.7 W/m.K (CVI C-SiC), 5.4 W/m.K (LPI C-SiC), and 24.7 W/m.K (HiPerComp SiC-SiC); the anisotropy of the $K_{C\text{-}SiC}$ values is neglected in the simplified 1-D thermal calculations. Figure 11(b) shows the effective thermal resistance of the various C-SiC/metal joints made using Ticusil; the computations with Cusil-ABA differ by less than 1% relative to Ticusil. Braze composition can therefore be chosen to satisfy criteria related to the wettability and other properties without significant detriment to the thermal properties. Figure 11(b) shows the effect of clad layer thickness on R_{eff}, which decreases as clad layer thickness increases; however, the maximum decrease is less than 7% at 30% clad thickness. The results are shown for the two types of C-SiC (and HiPerComp SiC-SiC) joined to Cu-clad-Mo; the effect of clad layer thickness on the thermal resistance of bare (un-joined) Cu-clad-Mo substrate of the same total thickness (5.1 mm) as the joined materials is also displayed. For comparison, the thermal resistance of the un-joined composite substrates is also displayed. Different composite substrates lead to different reductions in the thermal resistance; the lowest thermal resistance is achieved with the use of HiPerComp SiC-SiC composite joined to Cu-clad-Mo.

CONCLUSIONS

Brazed joints of C-C composites to Cu-clad-Mo using active braze alloys, Cusil-ABA and Ticusil led to large-scale braze penetration of the inter-fiber spaces in the composites, good interfacial bonding, and segregation of Ti at the composite/braze interface. The distribution of microhardness across the joints was reproducible and revealed gradients at the Cu-clad-Mo/braze interface. The projected benefits of reduced thermal resistance suggest that C-C composite/Cu-clad-Mo joints may be considered for potential thermal management applications. The C-SiC composites in ground and un-ground conditions were also joined to Cu-clad-Mo using Cusil-ABA and Ticusil. The C-SiC/braze interfaces were enriched in Ti and Si and some diffusion of braze constituents occurred in the composite. The composite surface preparation did not affect the hardness profiles. The theoretical projections of strain energy and thermal resistance in C/SiC joints highlight a pronounced effect of the C-SiC composite type, with the CVI C-SiC composite joints exhibiting lower tendency for joint fracture and lower thermal resistance than the LPI composites.

REFERENCES

[1] S.A. McKeown, R.D. LeVasseur, CH 3030-4/91/0000-0153, IEEE, 153-157 (1991)
[2] M. Singh, R. Asthana, and T.P. Shpargel, *Mater. Sci. Eng. A*, **452-453**, 699-704 (2007)
[3] G. Lin and J. Huang, *Powder Metallurgy*, **49(4)**, 345-8 (2006)
[4] J. Xiong, J. Li, and F. Zhang, *Scripta Materialia*, **55(2)**, 151-4 (2006)
[5] T. Qiao-Ying, C. Lai-Fei, and Z. Li-Tong, *J. Aero. Mater.*, **24(1)**, 53-6 (2004)
[6] M. Salvo, M. Ferraris, P. Lemoine, M.A. Montorsi and M. Merola, *J. Nuclear Mater.*, **233-235**, 949-953 (1996)
[7] R. Taylor, in *Comprehensive Composite Materials*, Elsevier Science, Boston, **4**, 387-426 (2000)
[8] G.S. Corman and K.L. Luthra, in *Handbook of Ceramic Composites*, N.P. Bansal (ed.), Kluwer, 99-115 (2005)
[9] W. Krenkel, in *Handbook of Ceramic Composites*, N.P. Bansal (ed.), Kluwer, 117-147 (2005)
[10] N. Sobczak, J. Sobczak, P. Rohatgi, M. Ksiazek, W. Radziwill and J. Morgiel, in *Proc. Int. Conf.*

High-Temperature Capillarity, Krakow, eds. N. Eustathopoulos, N. Sobczak, 145-51 (1997)
[11] N. Eustathopoulos, M.G. Nicholas, and B. Drevet, *Wettability at High Temperatures*, Pergamon, Boston, 281-282 (1999)
[12] J. -W. Park, P. F. Mendez and T.W. Eagar, *Scripta Mater.*, **53(7)**, 857-861 (2005)
[13] C.A. Harper, *Electronic Materials and Processes Handbook*, McGraw-Hill (2003)

JOINING OF ZIRCONIUM DIBORIDE-BASED CERAMIC COMPOSITES TO METALLIC SYSTEMS FOR HIGH-TEMPERATURE APPLICATIONS

M. Singh[1] and R. Asthana[2]

[1]Ohio Aerospace Institute
22800 Cedar Point Road
Cleveland, OH 44142

[2]Department of Engineering and Technology
University of Wisconsin-Stout
Menomonie, WI 54751

ABSTRACT

Hot-pressed ZrB_2-SiC (ZS) composites were joined to Cu-clad-Mo using two AgCuTi brazes (T_L~1073-1173°K) and two Pd-base brazes (T_L~1493-1513°K). The SEM and EDS examinations revealed crack-free joints in both sets of braze alloys but greater chemical interaction and elemental redistribution in joints made using Pd brazes. Calculations of strain energy reveal a pronounced dependence on clad layer thickness with the strain energy increasing with increasing clad layer thickness. Joints made using Pd-base brazes displayed higher microhardness than joints made using AgCuTi brazes. Projected reductions in the joint's thermal resistance highlight the potential benefits of joining the composite to Cu-clad-Mo.

INTRODUCTION

Refractory metal borides such as ZrB_2 have high melting points (~3493°K), chemical inertness, and good resistance to oxidation upon exposure to extreme environments all of which impart these materials capability to be used at ultra-high temperatures (~2150-2770°K). A number of ZrB_2-based composites have been developed[1-6] to enhance the heat tolerance in nose cap and leading edge of space vehicles reentering the Earth's atmosphere. Additives to ZrB_2 such as SiC improve the oxidation resistance and strength and permit design for properties. One area of special interest is the development of joining and integration technology for structural applications of these composites. Studies on joining of ZrB_2–based composites are very few[7,8] even though this represents an area of considerable practical importance.

Here we report the joining of a ZrB_2-SiC (ZS) composite to Cu-clad-molybdenum, a good heat sink material[9] using two AgCuTi brazes (T_L~1073-1173°K) and two Pd-base brazes (T_L~1493-1513°K). Acting in combination with the composite, Cu-clad Mo can potentially improve the heat dissipation capability at high temperatures. In contrast to Ag- and Ni-base brazes used in recent studies[7,8], Pd brazes offer higher use-temperatures and good oxidation-resistance. The coefficient of thermal expansion (CTE) of ZrB_2-SiC composite is $\sim 7.5\times 10^{-6}$/K[4,8] and the CTE of Cu-clad-Mo varies with the clad layer thickness (e.g., in the range 5.6 to 11.6 /K for 5 to 40% clad thickness per side of Mo[9]). In a ZS/Cu-clad-Mo joint, the negative effect of a large temperature excursion, ΔT, on residual strain, $\Delta T\Delta\alpha$, can be offset by decreasing the CTE mismatch, $\Delta\alpha$, by judiciously selecting the clad layer thickness on Mo to tailor the CTE. This shall mitigate the residual stress and preserve joint integrity. In addition, Cu as a cladding on Mo shall serve as a stress-absorbing compliant layer for the joint.

Unfortunately, most ceramics are poorly wet by molten metals, which presents a challenge to achieving good flow and spreading characteristics in joints. The wettability data for Pd on ZrB_2 could not be found but contact angle (θ) values for the transition metals Co and Ni on ZrB_2 are 39° and 42° at 1773K[10-12], respectively, which suggests that Pd-base brazes probably shall also wet this ceramic. Silver (base metal in our AgCuTi brazes) does not wet ZrB_2 (θ ~114° at 1373K); however,

Ti, Zr or Hf in Ag improve the wetting, and θ drops to 20-80°[12]. Therefore, the presence of Ti in AgCuTi brazes shall facilitate reactive wetting of ZrB_2. Additionally, these brazes contain Cu which wets ZrB_2 (θ ~80° at 1413K[11]).

EXPERIMENTAL PROCEDURE

The ZrB_2-SiC_p (ZS) composite were fabricated by hot pressing in a graphite die at Materials and Machines, Inc, Tucson, AZ, and contained 20 v/o SiC particles. Cu-clad-Mo plates, obtained from H.C. Starck, Inc., Newton, MA, had Cu-to-Mo-to-Cu layer thickness ratio of 13%-74%-13%. The commercial brazes, Palco, Palni, Cusil-ABA and Ticusil were obtained in foil form (thickness ~50 μm) from Morgan Advanced Ceramics. The compositions, liquidus and solidus temperatures, and selected properties of the braze alloys are given in Table I.

Table I. Composition and Selected Properties of Brazes used

Braze (composition, %)	Density Kg.m^{-3}	T_L, °C	T_S, °C	E, GPa	YS, MPa	UTS, MPa	CTE, ×10^{-6} C^{-1}	% El.	K, W/m.K
Cusil-ABA® (63Ag-35.3Cu-1.75Ti)	18,500	815	780	83	271	346	18.5	42	180
Ticusil® (68.8Ag-26.7Cu-4.5Ti)	18,500	900	780	85	292	339	18.5	28	219
Palco® (65Pd-35Co)	-	1219	1219	--	341	661	--	43	35
Palni® (60Pd-40Ni)	15,000	1238	1238	--	772	978	15	23	42

E: Young's modulus, YS: yield strength, UTS: tensile strength, CTE: coefficient of thermal expansion, %El: percent elongation, K: thermal conductivity. ®registered trademark of Morgan Advanced Ceramics, Hayward, CA

The composite, metal substrate, and braze foils (two layers, ~100 μm total thickness) were sliced into 2.54 cm x 1.25 cm x 0.25 cm pieces and ultrasonically cleaned in acetone for 15 min. The braze foil was sandwiched between metal and composite, and a normal pressure of 1.2-4.7 kPa (0.38-1.5 N) was applied to the assembly. The assembly was heated in a furnace to ~15-20 °C above the braze liquidus under vacuum (10^{-6} torr), soaked for 5 min., and slowly cooled to room temperature. The brazed joints were prepared for metallography and examined with a Scanning Electron Microscope (SEM) coupled with energy dispersive x-ray spectroscope (EDS) on a JEOL-840 A unit. Microhardness scans were made with a Knoop micro-indenter on a Struers Duramin A-300 machine under a load of 200 g and loading time of 10 s.

RESULTS AND DISCUSSION

Microstructure

ZS/Cu-clad-Mo Joints using AgCuTi Brazes

Figure 1 shows the microstructure of the ZS composite (Fig. 1a) and ZS/Cusil-ABA joint (Figs. 1b-d). The ZS composite comprises of nearly equi-axed ZrB_2 (light gray) particles (6-12 μm in size) and tabular/plate-like SiC (dark phase) particulates with an average size of ~3-11 μm major axis and 1.5-3 μm thickness. The ZS/Cusil-ABA/Cu-clad-Mo interfaces show intimate physical bonding (Fig. 1b). The braze is well-bonded to ZS and Mo, contains large amounts of dissolved silicon, and the interfaces are crack- and void-free. The different elements at various locations across the joint are identified in Figs. 1c & d. Some near-interfacial compositional changes due to limited dissolution and interdiffusion have occurred although there is no evidence of reaction layer formation. The ZS/Ticusil/Cu-clad-Mo joint (Fig. 2) also shows excellent bonding, and some segregation of Ti at the interface (e.g., 7.9at% Ti at point 6, Fig. 2a).

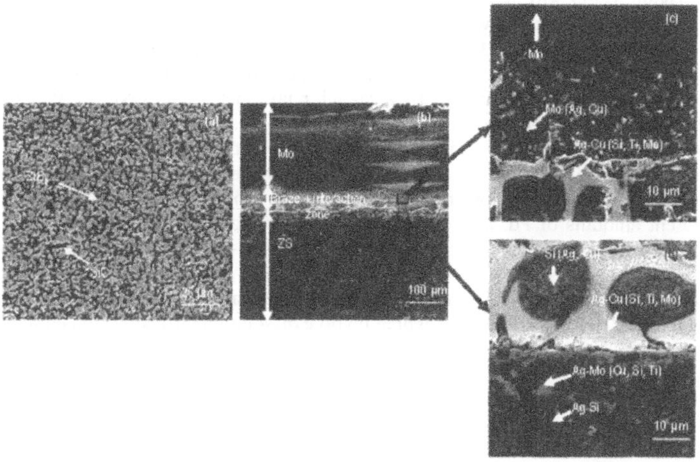

Fig. 1 (a) Microstructure of ZS, (b) ZS/Cusil-ABA/Cu-clad-Mo joint, (c) & (d) show the microstructure and elements at the (c) braze/Cu-clad-Mo interface, and (d) braze/ZS interface.

ZS/Cu-clad-Mo Joints using Pd-Base Brazes

Evidence of strong chemical interaction is seen in ZS/Palco/Cu-clad-Mo joints (Fig. 3a). A multilayer transition zone (Fig. 3a & b) separates the ZS from molybdenum. A change in the microstructure occurs upon traversing from the ZS side through the braze region (Fig. 3b-d) to the braze/Cu-clad-Mo interface (Fig. 3e & f) and into the Cu-clad-Mo region (Fig. 3f). The EDS analysis across the joint (Fig. 3b) showed that Co, Cu, Mo and Pd had penetrated within ZS to a

Fig. 2 (a) SEM view of ZS/Ticusil/Cu-clad-Mo joint, (b) shows the inset from (a), and (c) shows Ticusil/Cu-clad-Mo interface.

distance on the order of ~100μm (points 1 & 2, Fig. 3b). Considerable dissolution of Si from ZS has occurred within the braze region (points 7 & 9, Fig. 3b).

The ZS/Palni/Cu-clad-Mo joint (Fig. 4a) shows a complex multi-layer interaction zone indicative of significant chemical interaction. The joint, however, is sound and defect-free. The ZS side shows a very thick interaction zone (Fig. 4a). The lighter, plate-like phase in the lower half of Fig. 4b is a Zr-rich phase (point 5, Fig. 4b) that also contains relatively large amounts of Cu and some Ni, Pd and Si. The boundary (Fig. 4c) between the interaction zone of Fig. 4b and the braze exhibits a Zr-rich light-gray phase (points 1, 3, 4 and 6, Fig. 4c) and a darker plate-like Zr-Cu phase that contains Pd, Ni and Si (point 2, Fig. 4c). The eutectic-type two-phase structure in the intergranular regions in the lower half of Fig. 4c is actually a Zr-rich phase that contains Ni, Cu, Pd and Si. The transition structure near the braze/Cu-clad-Mo interface is shown in Fig. 4e and a higher magnification view is shown in Fig. 4d. The light region and the dark round spots (Fig. 4d) are Cu-rich phases that contain Pd, Si and Zr (the dark spots have slightly higher Pd content). The entire cross-section of the joint region shown in Fig. 4d from point 1 through point 6 is rich in Cu but with different amounts of Pd, Si and Ni at different locations. At the boundary between the Mo-rich interphase layer and pure Mo (substrate), a thin Cu-rich layer (from Cu-cladding) is observed.

In summary, the commercial AgCuTi and Pd-base braze alloys exhibit complex chemical interactions with the ZrB_2-SiC composites which involve dissolution, diffusion, and compositional

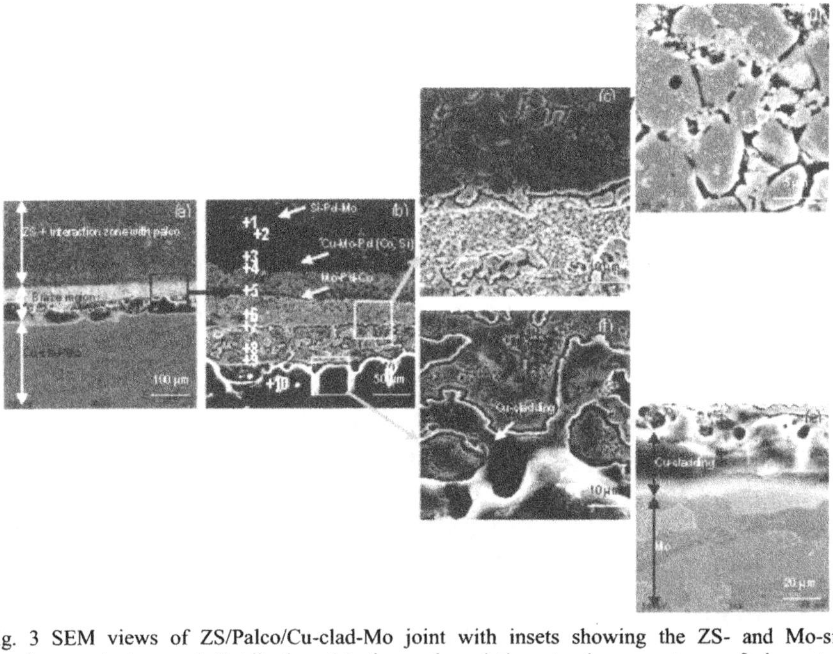

Fig. 3 SEM views of ZS/Palco/Cu-clad-Mo joint with insets showing the ZS- and Mo-side interfaces and elemental distribution; (g) shows the relative atomic percentages of elements at points marked in (b).

modulations, particularly in Pd-base joints. New phase and compound formation in joints is particularly likely. The Gibb's free energy change, ΔG, per mole of the ceramic for a number of reactions (Fig. 5) was calculated using the Software HSC Chemistry version 4.1 (Outokumpu Ra, Oy, Finland). The calculations show that while formation of TiC, Ni_2Si and $TiSi_2$ from the reaction of SiC with Ti and Ni may be thermodynamically possible, the formation of borides (TiB_2, NiB, Ni_2B, Ni_3B, CoB) and some carbides (Ni_3C) and silicides ($CoSi_2$) is unlikely although the formation of Ni_2B above 1260°C has been reported in the literature[7]. Additionally, the Pd-Zr equilibrium diagram reveals that intermetallics such as Pd_3Zr, Pd_2Zr, PdZr, and $PdZr_2$ could form at 800-1200°C. Similarly, with Co in Pd (Palco braze), CoZr, Co_2Zr, and $CoZr_2$ could form. Besides these binaries, other reaction products and phases may also be possible.

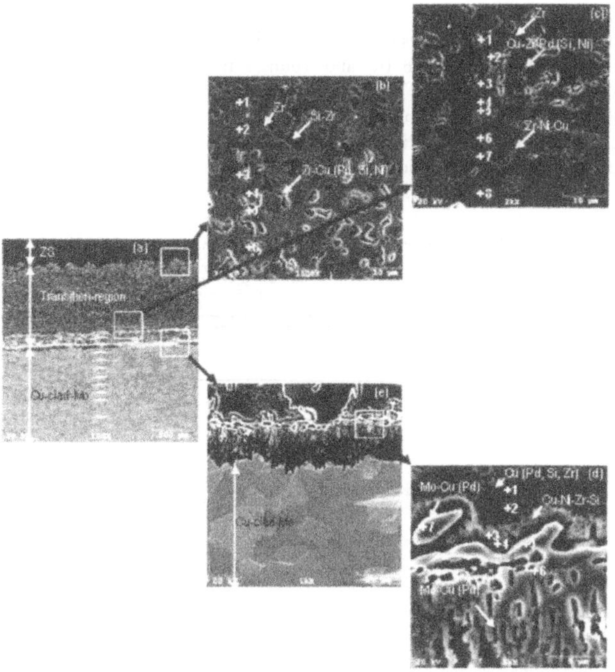

Fig. 4 SEM views of ZS/Palni/Cu-clad-Mo joint showing regions in the vicinity of the braze/ZS interface and braze/Cu-clad-Mo interface.

Microhardness

For all joints, at least three-to-six Knoop hardness profiles were generated as a function of distance across the interface. In the ZS/Cusil-ABA/Cu-clad-Mo joints (Fig. 6a), the hardness of the ZS composites, molybdenum substrate, and Cu-cladding plus the braze region are 1900-2280 HK, 270-350 HK and 65-80 HK, respectively. Steep hardness gradients exist at the braze/ZS interface but no decohesion occurred because of good ductility (42%) of Cusil-ABA. With Ticusil braze in place of Cusil-ABA, the hardness profiles were similar to Cusil-ABA (Fig. 6b). The HK of ZS was 1860-2465. As with Cusil-ABA, the region comprised of Ticusil plus Cu cladding exhibited low

hardness (65-170 HK); there was no significant difference in the hardness of the two brazes.
In ZS/Palco/Cu-clad-Mo joints (Fig. 6c) the HK values of ZS and Palco braze regions are 1878-2110 and 1025-1365, respectively, and the HK values of Mo substrate and Cu-clad regions are appreciably lower (225-250 and 75-90, respectively). The ZS/Cu-clad-Mo joints made using Palni (Fig. 6d) display a high hardness in both ZS composite (2200 HK) and the Palni (1000-1365 HK) regions. The HK of Pd-brazes is significantly higher than that of AgCuTi brazes. Relatively large interaction zones had formed at the ZS/braze interface in Pd-base brazes, which led to shallower hardness gradients in joints.

Residual Stress

In ceramic/metal joints where the CTE's of substrates may be appreciably different and processing and service temperatures relatively high, residual stresses could significantly affect the joint integrity. The magnitude of residual stresses determines the joint strength; joints with large residual stresses fracture at low stresses. The CTE (α) and Young's modulus (E) of ZS composite, hot pressed to a relative density of 98%, are 7.5×10^{-6}/C and 421 GPa[1], respectively[4]. The CTE of Cu-clad-Mo depends upon the clad layer thickness, and is ~5.7×10^{-6}/K for 5% thickness of Cu per side of Mo substrate[9]; this value is only slightly different from the CTE of ZS. However, the CTE of our brazes is large ($15\text{-}18.5\times10^{-6}$/K, Table 1).

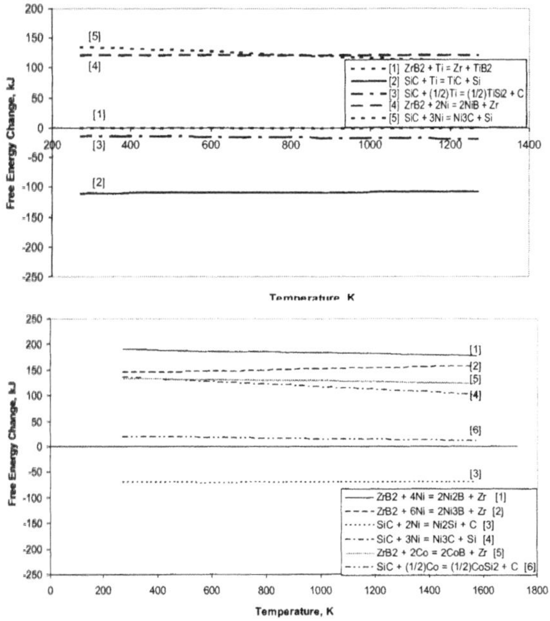

Fig. 5 Gibb's free energy change for reactions of ZrB_2 and SiC with Ti and Ni calculated using the software HSC Chemistry version 4.1 (Outokumpu Ra, Oy, Finland).

[1] The value quoted is for ZrB_2 + 20v/o SiC + sintering aids[7].

A model due to Eager et al[13] estimates the strain energy in the ceramic in well-bonded ceramic-metal joints. For a small CTE mismatch between the ceramic (C) and the metal substrate (M), but with a large CTE mismatch between the interlayer (I) and the base materials, the elastic strain energy, U_{eC}, in the ceramic for a disc-shaped joint is calculated in terms of the yield strength (σ_{YI}) of the braze, the radial distance from the center of the joint, and the elastic moduli of the ceramic (E_C) and braze (E_I). Eager et al[13] proposed analytical expressions to calculate U_{eC} as asymptotic approximations (to 1% accuracy) to their finite element calculations; these analytical expressions (equations [1]-[3] in ref.[13]) are used here to estimate the strain energy.
The CTE of Cu-clad-Mo (α_M) depends on clad layer thickness and is taken from ref. [9]. The elastic strain energy, U_{eC}, in the ZS composite was calculated as a function of the clad layer thickness for the four brazes used in the study. The elastic moduli of Palco and Palni are not reported by the vendor. The dependence of E on composition of eutectic alloys without terminal solid solubility (true for Pd-Co and Pd-Ni equilibrium diagrams) follows a linear (rule-of-mixtures) relationship when the moduli of the constituents differ by a small amount, and the modulus exhibits a negative deviation from linearity when the phases are randomly distributed and have a large difference in their moduli[20]. For Palco and Palni, E_{Pd} (121 GPa) is significantly smaller than E_{NI} (200 GPa) and E_{co} (209 GPa), which indicates a negative deviation from linearity. Using the ROM, approximate lower bounds on E are obtained as E_{Palco} = 151.8 GPa and E_{Palni} = 152.6 GPa.

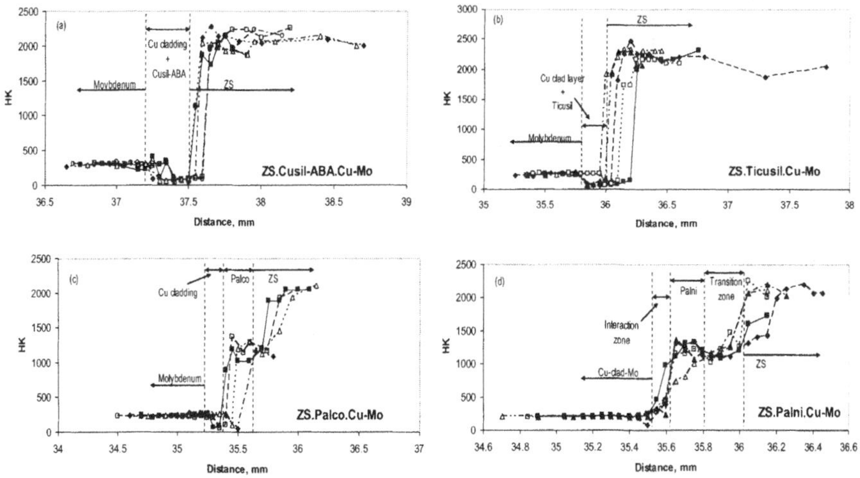

Fig. 6 Distribution of Knoop microhardness in ZS/Cu-clad-Mo joints made using (a) Cusil-ABA, (b) Ticusil, (c) Palco and (d) Palni brazes.

The CTE of Palco needed in strain energy calculations is obtained from Turner's model[12] which predicts the CTE of Pd-Ag, Pb-Sb, Be-Al and other alloys. According to Turner's model, the volumetric coefficient of expansion may be obtained from $\beta = \sum \frac{\beta_i P_i K_i}{\rho_i} / \sum \frac{P_i K_i}{\rho_i}$, where β, P, K and ρ denote the volumetric expansion coefficient, weight percent, bulk modulus and density, respectively, i denotes the ith component in the alloy, and $\beta \sim 3\alpha$, (α: linear CTE). The following data for properties of pure Pd, Co and Ni were used to estimate β_{Palni} and β_{Palco}: $\alpha_{Pd} = 11.8 \times 10^{-6}$/K,

α_{Ni} = 13.4 x 10^{-6}/K, α_{Co} = 13 x 10^{-6}/K, $K_{Pd} = K_{Ni} = K_{Co}$ = 180 GPa, ρ_{Pd} = 12,023 kg.m^{-3}, ρ_{Ni} = 8,908 kg.m^{-3}, ρ_{Co} = 8,900 kg.m^{-3}. These values in conjunction with the Turner equation yield, β_{Palni} = 37.67 x 10^{-6}/K and α_{Palni} = 12.6 x 10^{-6}/K. Noting that the calculated α_{Palni} is about 16% less than the CTE value (15 x 10^{-6}/K) quoted by the manufacturer (Table 1), and assuming the same margin of error for Palco as for Palni, the CTE of Palco, α_{Palco}, is obtained as 14.27 x 10^{-6}/K.

The results of strain energy calculations (Fig. 7) show that U_{eC} is negative up to ~23% clad layer thickness; above this value, $U_{eC} > 0$, and it increases sharply with increasing clad layer thickness. Whereas 'negative' energy cannot be assigned physical meaning, here it reflects the difference relative to a reference or stress-free state (analogous to thermodynamic free-energy). In essence, therefore, negative strain energy can be taken to represent increased fracture stress. There is a small (max. ~15%) difference in the U_{eC} values for Ticusil and Cusil-ABA joints, with Cusil-ABA exhibiting marginally lower strain energy than Ticusil above 23% clad layer thickness. For most ceramic-to-metal joints, $\alpha_M > \alpha_C$ and U_{eC} is positive and in the range 0.5-80 mJ[13]. However, for the ZS/Cu-clad-Mo joints (<23% clad layer thickness), $\alpha_M < \alpha_C$, and $U_{eC} < 0$. Even at larger clad layer thicknesses, the maximum values of strain energy (at 40% thickness) are only 15.1 mJ for Ticusil and 12.7 mJ for Cusil-ABA; the slightly lower U_{eC} for Cusil-ABA is beneficial from the standpoint of joint integrity.

Chemical interactions which lead to compositional changes and diffusion processes in joints can influence the joint properties. Models for residual stress that take into account compositional changes from such interactions are scant and perhaps prohibitively difficult to construct but highly desirable to understand the behavior of reactive systems.

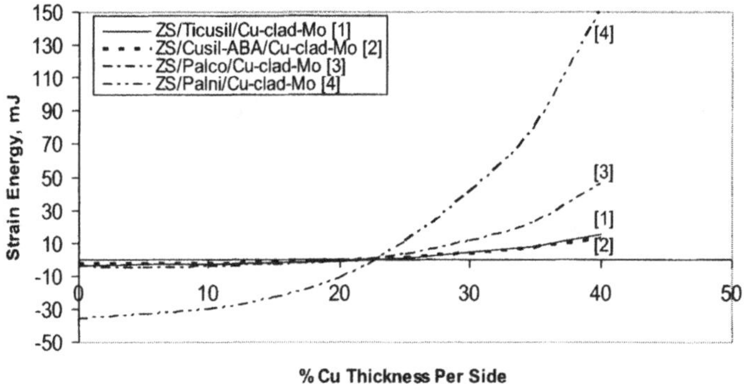

Fig. 7 Calculated strain energy in the ceramic in ZS/Cu-clad-Mo joints with Ticusil, Cusil-ABA, Palco and Palni brazes as a function of % Cu cladding thickness per side on Mo substrate.

Thermal Considerations

For 1-D steady-state heat conduction, the joined materials form a series thermal circuit with an effective resistance, $R_{eff} = \Sigma(\Delta x_i/K_i)$, where Δx_i and K_i are the thickness and the thermal conductivity, respectively, of the ith layer. Figure 8 shows the projected thermal resistance of ZS/Cu-clad-Mo joints made using the four brazes as a function of % clad layer thickness. This figure also shows the thermal resistance of the ZS composite and Cu-clad-Mo of the same total thickness (5.1 mm) as the joined assembly. For calculation: $\Delta x_{ZS} = \Delta x_{Cu\text{-}Mo} = 0.25\times10^{-2}$ m, $\Delta x_{Ticusil} = 100\times10^{-6}$ m, and K of Cu-clad-Mo with different Cu layer thicknesses is from ref.[9]. The conductivity of ZS (K_{ZS}) is calculated from the Maxwell equation for spherical particles homogeneously dispersed in and well-bonded (zero contact resistance) to the matrix, according to which

$$K_{ZS} = \frac{K_{ZrB2}[2\{(K_{SiC}/K_{ZrB2})-1\}V_{SiC}+(K_{SiC}/K_{ZrB2})+2]}{[\{1-(K_{SiC}/K_{ZrB2})\}V_{SiC}+(K_{SiC}/K_{ZrB2})+2]} \quad ......[1]$$

where V_{SiC} is the volume fraction of SiC in the ZS composite. For V_{SiC} = 0.20 in ZS, and with K_{ZrB2} = 58 W/m-K[7] and K_{SiC} = 425 W/m.K (average value, based on K_{SiC} values of 360-490 W/m.K for different SiC polytypes), the conductivity of ZS, K_{ZS} = 85.3 W/m-K. The calculated R_{eff} is shown in Fig. 8. It decreases when the clad layer thickness increases (e.g., by ~15% when thickness increases from 0 to 30%). The values of R_{eff} for Ticusil and Cusil-ABA joints are nearly identical and vary from 47.9×10^{-6} m^2.K/W (without Cu cladding) to 40.4×10^{-6} m^2.K/W (with 30% Cu cladding). The corresponding values of R_{eff} for Palco and Palni joints vary from $\sim50.3\times10^{-6}$ m^2.K/W to $\sim42.8\times10^{-6}$ m^2.K/W. Because of its miniscule thickness, the interlayer makes a negligible contribution to R_{eff}. The above changes in R_{eff} should be viewed in light of an appreciably greater change in the strain energy (Fig. 7) with clad layer thickness. There is thus considerable flexibility in selecting the clad layer thickness to design a low CTE to mitigate residual stresses with minimal impact on the effective thermal resistance of the joined assembly. These calculations, however, disregard any thermal discontinuity (interface resistance) at the joint.

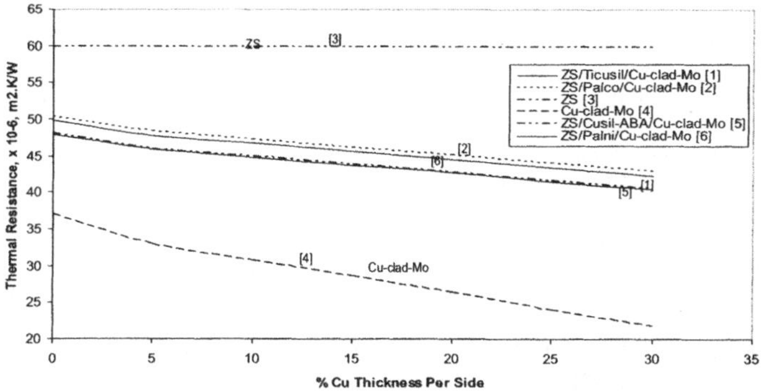

Fig. 8 Projected effective thermal resistance of ZS/Cu-clad-Mo joints made using Cusil-ABA, Ticusil, palco and Palni brazes as a function of % Cu cladding thickness per side on Mo substrate.

The calculated R_{eff} may be compared to the thermal resistance ($\Delta x/K$) of just the ZS composite of the same total thickness (5.1×10^{-3} m) as the joined materials to visualize the potential benefits of joining ZS to Cu-clad-Mo. As one extreme, consider a Mo substrate with a ~27% thick Cu cladding (K_{Cu-Mo}~224 W/m.K[9]) joined to ZS using Ticusil. The R_{eff} of the joint is 40.9×10^{-6} m^2.K/W and the thermal resistance of just the ZS substrate of the same total thickness is 59.8×10^{-6} m^2.K/W (i.e., a 32% increase in R_{eff} compared to the joint). Similar calculations for ZS/Cu-Mo joints made using the lowest conductivity Palco braze (K_{Palco} = 35 W/m-K, Table I) show that the thermal resistance of the joint will be 43.32×10^{-6} m^2.K/W (~28% decrease relative to ZS). Reducing the braze layer thickness will only marginally change these percentages; this provides for considerable flexibility in selecting braze on the basis of other requirements (e.g., strength, ductility and braze wettability) without compromising the thermal conductivity.

What emerges from the above discussion is that a reduction in the thermal resistance may justify joining the ZS to Cu-clad-Mo for enhanced heat transfer even though this reduction in thermal resistance shall be accompanied by an increase in density. The Rule-of-Mixtures (ROM) density of the ZS composite is 5,492 kg.m^{-3} (with ρ_{ZrB2} = 6,090 kg.m^{-3}, ρ_{SiC} = 3,100 kg.m^{-3}), and the ROM density of our ZS/Cu-Mo joint is 7,638 kg.m^{-3} (the contribution of braze layer is negligible and ignored). This represents a ~22% weight increase for the ZS/Cu-clad-Mo joints relative to the ZS composite. The percentage gains in R_{eff} presented above shall not change as the component size is increased provided the thickness ratio of ZS and Cu-clad-Mo remains fixed (~50% each as in our joints).

CONCLUSIONS

A ZrB_2-SiC composite was successfully joined to Cu-clad-Mo using Cusil-ABA, Ticusil, Palco and Palni brazes. The extent of chemical interaction as revealed by the formation of interaction zones at the braze/ZS interface depended upon the braze used; joints made using Palco and Palni exhibited considerably greater interaction than joints made using Cusil-ABA and Ticusil although sound, defect-free joints formed in all cases. Braze regions in Pd-base brazes displayed greater hardness than in AgCuTi brazes. Calculations revealed an increase of fracture stress of the ceramic up to a certain clad layer thickness. Potential thermal benefits of joining the ZS to Cu-clad-Mo for enhanced heat transfer were highlighted.

References

[1]S.R. Levine, E.J. Opila, M.C. Halbig, J.D. Kiser, M. Singh, J.A. Salem, Evaluation of ultra-high temperature ceramics for aeropropulsion use, *J. Eur. Ceram. Soc.*, **22**, 2757-2767 (2002).

[2]D. Sciti, S. Guicciardi, A. Bellosi, G. Pezzotti, Properties of a pressureless-sintered ZrB_2-$MoSi_2$ ceramic composite, *J. Amer. Ceram. Soc.*, **89(7)**, 2320 (2006).

[3]F. Monteverde, A. Bellosi and S. Guicciardi, Processing and properties of zirconium diboride-based composites, *J. Eur. Ceram. Soc.*, **22(3)**, 279-288 (2002).

[4]A. Bellosi, and F. Monteverde, Fabrication and properties of zirconium diboride-based ceramics for UHT applications, in *Proc. 4th European Workshop, 'Hot Structures and Thermal Protection Systems for Space Vehicles*, Palermo, Italy, Nov. 2002 (ESA SP-521, April 2003).

[5]I. G. Talmy, J. A. Zaykoski, and M. M. Opeka, Properties of Ceramics in the ZrB2/ZrC/SiC System Prepared by Reactive Processing, *Ceram. Eng. Sci. Proc.*, **19[3]**, 105–112 (1998).

[6]G.-J. Zhang, Z. -Y. Deng, N. Kondo, J. -F. Yang, and T. Ohji, Reactive Hot Pressing of ZrB_2–SiC Composites, *J. Am. Ceram. Soc.*, **83[9]**, 2330–2332 (2000).

[7]M.L. Muolo, E. Ferrera, L. Morbelli, C. Zanotti and A. Passerone, Joining of zirconium diboride based refractory ceramics to Ti6Al14V, in *Proc. of the 9th International Symposium on Materials in a Space Environment*, June 2003, Noordwijk, The Netherlands, compiled by K. Fletcher, ESA SP-540, Noordwijk, Nethelands: ESA Publications Div., 467-472 (2003).

[8]M. Singh and R. Asthana, Joining of advanced ultra-high-temperature ZrB_2-based ceramic composites using metallic glass interlayers, *Mater. Sci. Eng. A*, **460-461**, 153-162 (2007).

[9]C.A. Harper, *Electronic Materials and Processes Handbook*, McGraw-Hill, 10.67 (2003).

[10]M.L. Muolo, E. Ferrera, A. Passerone, A., Wetting and spreading of liquid metals on ZrB_2-based ceramics, *J. Mater. Sci.*, **40**, 2295-2300 (2005).

[11]A. Passerone, M.L. Muolo and D. Passerone, Wetting of Group IV diborides by liquid metals, *J. Mater. Sci.*, **41**, 5088-5098 (2006).

[12]A. Passerone and M.L. Muolo, Metal-ceramic interfaces: wetting and joining processes, *International Journal of Materials and Product Technology*, **20(5-6)**, 420-439 (2004).

[13]J.-W. Park, P.F. Mendez and T.W. Eager, Strain energy release in ceramic-to-metal joints by ductile metal interlayers, *Scripta Mater.*, **53(7)**, 857-861 (2005).

X. Bioceramics

PREPARATION AND CHARACTERISATION OF PLGA-COATED POROUS BIOACTIVE GLASS-CERAMIC SCAFFOLDS FOR SUBCHONDRAL BONE TISSUE ENGINEERING

Timothy Mark O'Shea and Xigeng Miao
Institute of Health and Biomedical Innovation, and School of Engineering Systems, Queensland University of Technology
Brisbane, Queensland, 4059, Australia

ABSTRACT

The present work involved the development of a poly(lactic-co-glycolic acid) (PLGA)-coated, porous 58S bioactive glass-ceramic scaffold for the bone-forming layer of a future bilayered scaffold for osteochondral tissue engineering. Scanning electron microscopy (SEM), mechanical testing and preliminary cell culture investigations were performed to determine the suitability of the scaffold for the subchondral bone tissue engineering application. The bioactive-glass ceramic scaffolds possessed excellent pore interconnectivity. Mechanical testing of the PLGA-coated bioactive glass-ceramic scaffolds demonstrated a compressive strength of 0.25±0.05 MPa. The capacity of the bioactive glass-ceramic scaffold to proliferate human bone marrow stromal cells (BMSCs) was also assessed through *in vitro* culturing. SEM morphology investigations revealed that the BMSCs had a relatively high affinity for the PLGA-coated bioactive glass-ceramic scaffold and significant attachment and proliferation towards confluence occurred over a 21 day period.

INTRODUCTION

Osteochondral tissue engineering, a hybrid of bone and cartilage regeneration has emerged as a tangible alternative for promoting superior cartilage integration as well as a treatment for osteochondral defects, which pertain to joint damage that extends the articular cartilage and penetrates the underlying subchondral bone [1-3]. Bilayered scaffolds can be used to encourage simultaneous bone and cartilage growth to form integrated repair tissue. The incorporation of bilayered scaffolds assists in the establishment of tissue specific biological environments for each respective layer via variations in the physiochemical properties of the comprising scaffold layers.

Scaffolds for bone tissue engineering have been produced through a surfactant foaming and casting process incorporating sol-gel derived bioactive glass [4], as well as porous dry powder pressing and polymer replication using melt-derived 45S5 Bioglass® [5,6]. Scaffolds produced by polymer replication have been shown to possess structural properties similar to those of trabecular bone [7]. Sol-gel derived 58S bioactive glass has been shown to facilitate a faster rate of ionic dissolution and associated hydroxyl-carbonated apatite (HCA) layer formation compared to that of the melt-derived 45S5 Bioglass® alternative [8]. Thus, it was thought that 58S sol-gel bioactive glass-ceramic scaffolds when toughened by a biodegradable polymer (e.g. PLGA) would possess the physiochemical properties required to promote the bone formation for the repair of the subchondral bone of an osteochondral defect.

The work reported in this paper was about the scaffolds only suitable for the subchondral bone tissue engineering. The ultimate aim of our research is the formation of an integrated bilayered scaffold consisting of the current bioactive glass-ceramic scaffold layer and an additional biopolymer scaffold layer for entire osteochondral (i.e. bone-cartilage) tissue engineering. The bilayered scaffolds may be used to additionally deliver growth factors to promote cartilage growth, while ensuring improved engineered tissue integration via bone growth into the subchondral scaffold layer at the osteochondral defect site.

MATERIALS AND METHODS

Scaffold fabrication

58S bioactive glass, a ternary system of composition SiO_2 (58 wt%)-CaO (33 wt%)-P_2O_5 (9 wt%), was produced using reagent grade chemicals: tetraethoxysilane (TEOS, $Si(OC_2H_5)_4$; 131903, Aldrich); triethylphosphate (TEP, $PO(OC_2H_5)_3$; 538728, Aldrich); and calcium nitrate (CN, $Ca(NO_3)_2.4H_2O$; 237124, Sigma-Aldrich) by a sol-gel method as described elsewhere [9,10]. Briefly, an HCl-water solution of pH 0.5 was prepared by combining a measured quantity of 1N HCl to a beaker of demineralized water. A ratio of mols of water to mols of TEOS of 8 was used for the present study. TEOS was added to the acid-water solution and magnetically stirred for thirty minutes to facilitate partial hydrolysis. TEP was then subsequently combined before being stirred for an additional twenty minutes. Finally, a suitable quantity of CN was progressively added to the solution and allowed to mix for one hour to permit further hydrolysis and complete dissolution of reagents. The resultant solution was then covered and sealed with aluminum foil and allowed to age in a temperature control chamber (Votsch VCL 4006) at 60°C and 55% relative humidity for 54 hours. Following aging the supernatant was removed and the sol gel was allowed to dry under near equilibrium drying conditions in the same chamber up to a temperature of 130°C at 95 % relative humidity for 72 hours. A final stabilization (calcination treatment) phase was performed in a furnace up to 700°C. The resultant bioactive glass particles were ground and sieved and then ball milled for 24 hours to produce fine bioglass powder.

Porous bioactive glass-ceramic scaffolds were produced using the polymer replication technique as articulated previously [11]. Specifically, polyurethane (PU) foams were cut into a variety of sizes and immersed in a dilute sodium hydroxide (NaOH) solution before being rinsed clean with water. The PU foams were dried and stored in a vacuum desiccator until use. Bioactive glass slurry was prepared through the measured addition of 40 grams of bioactive glass to a beaker of demineralised water to form a 45% w/v mixture. The mixture was stirred using a magnetic stirrer for one hour. Aqueous solutions of 25 wt% sodium polymethacrylic acid and 10 wt% polyvinyl alcohol (PVA) were used as a dispersant and a binder respectively and added to the slurry mixture at a concentration of 1 wt% on the basis of bioactive glass weight. The resultant solution was stirred for periods of thirty minutes and one hour following the addition of the dispersant and the binder. A porous green construct was produced by immersing the PU foams into the bioactive glass slurry ensuring complete penetration of slurry throughout the entire foam structure. The foams were manually retrieved and the excess slurry was squeezed out by hand. A compressed air gun was used to eliminate any pore blockages ensuring a relatively homogenous coating on the struts of the foam. The green ceramic foams were allowed to dry on a smooth flat surface for twenty-four hours before being sintered. A two stage heat treatment regime was used to sinter the scaffolds. A temperature of 600°C held constant for one hour was used to burn off the sacrificial polymer template, with a subsequent sintering period of one hour at 1000°C used to promote densification of the scaffold struts. The rate of temperature increase and decrease during sintering was 2°C/min and 5°C/min respectively. The scaffolds were finally coated with a thin layer of poly(lactic-co-glycolic acid) (PLGA). For coating with the PLGA, PLGA pellets (Sigma–Aldrich; PLA:PGA = 75:25; mol. wt = 90,000–126,000) were dissolved in dimethyl carbonate (DMC, ≥ 99%, 517127, Sigma Aldrich) at a ratio of 1:5 wt/vol. The polymer-solvent solution was sonicated for 15 minutes just prior to use. Then the bioactive glass-ceramic scaffolds were immersed into the polymer solution, drained by taking out of the scaffolds and then centrifuged at 1000 rpm for 1 minute to remove the excess polymer solution. The coating action was repeated to ensure a homogenous coating of the struts throughout the entire scaffolds. The scaffolds were then allowed to dry for 12 hours in a fume hood and then for 24 hours in a vacuum desiccator to assist solvent evaporation.

Scaffold characterisation

Sintered and PLGA-coated bioactive glass-ceramic scaffolds were examined by SEM to assess important macro- and micro- structural characteristics. Each specimen was mounted onto a separate aluminium stub using carbon tape and coated with gold film on a sputter coater (BioRad SC500) in an argon atmosphere. The individual samples were then analyzed using a scanning electron microscope (FEI QUANTA 200) and viewed under an acceleration operation voltage at 20 kV.

The total porosity (P) of the bioactive glass-ceramic scaffold (as sintered) as well as the scaffold coated with PLGA was determined using the relationship between the bulk density (ρ_B) and the theoretical density of the scaffold:

$$P_{\text{as sintered}} = 1 - \frac{\rho_{B1}}{\rho_{BG}} \tag{1}$$

$$P_{\text{coated}} = P_{\text{as sintered}} - \left(\frac{\rho_{B2} - \rho_{B1}}{\rho_{PLGA}} \right) \tag{2}$$

where, ρ_{B1} is for the as sintered scaffold, ρ_{BG} is the theoretical density of the bioactive glass-ceramic, ρ_{B2} is the bulk density of the scaffold coated PLGA, and ρ_{PLGA} is the theoretical density of PLGA. The theoretical density of the bioactive glass-ceramic sintered at 1000 °C was estimated to be approximately 3.125 g/cm^2, as obtained through aprevious study while for PLGA a theoretical density value of 1.25 g/cm^2 was used. The bulk densities were calculated from measurements of the weights and volumes of individual scaffolds.

The compressive strengths of the PLGA-coated bioactive glass-ceramic scaffolds were measured individually using a Hounsfield testing machine with a load cell of 500 N. Cylindrical samples with a diameter of 6 mm were prepared for all compressive tests with a height of 20mm being used for the bioactive glass-ceramic scaffolds. Prior to testing, the distal ends of the bioactive glass-ceramic were immersed and fixed in a liquid paraffin wax up to 5 mm, which was used to ensure uniform loading of the scaffold cross section during testing and prevent any shearing force being applied at the extremities. A loading rate of 0.5 mm/min was used for all compressive testing with 5 specimens from each category being assessed. During testing, the compressive load was applied until densification of the bioceramic struts and compaction of microspheres was observed.

In vitro evaluation by cell culture

PLGA coated-bioactive glass-ceramic scaffolds were used for *in vitro* evaluation by cell culture. For cell morphology studies, scaffolds with dimensions 4 mm x 4 mm x 4 mm were placed into individual 96 well plates. All samples were sterilised in 70% ethanol for thirty minutes with samples subsequently rinsed twice with potassium phosphate buffer solution (PBS) before being allowed to dry in a fume hood overnight. Before cell seeding, the scaffolds were incubated in standard Dulbecco's modified Eagle medium (DMEM) culture medium for a 48 hour period.

BMSCs were obtained by the method developed early [12]. Before seeding of the BMSCs, the incubating medium was removed from the scaffold well plates through aspiration. Approximately 10 μL of DMEM standard medium was directed onto the scaffold as a means of wetting the scaffold to prevent drying out of cells following incubation as well as to provide nutrients to these cells during initial seeding. Then 1×10^4 cells were used for scaffold morphology investigations. Following seeding, the cells were allowed to adhere for 3 to 4 hours before being topped up with standard medium. 1 mL was added to the 96 well plates used for morphology investigations. The scaffolds were then incubated at 37°C in a humidified atmosphere containing

5 % (v/v) carbon dioxide for the designated time periods with the medium changed every 3 days during culture.

Cell morphology and attachment onto the bioactive glass-ceramic scaffold was observed using SEM. Scaffolds were cultured for 3 days, 5 days and 21 days in standard culture medium. Following culture, the medium was removed through aspiration and scaffolds were rinsed with PBS before being fixed with a 3% glutaraldehyde solution. The fixed scaffolds were stored at 4 °C until further processing. Before SEM observation, the scaffold samples were prepared using a standardized process. The preparation procedure involved an initial buffer wash using 0.1 M cacodylate buffer followed by a dehydration process facilitated through a series of increasing grade concentration ethanol solutions with a repetition at each stage. The dehydration process was completed through a repeat incubation in 100% amyl acetate. Final drying was performed using a supercritical point dryer of carbon dioxide (Denton Vacuum critical point dryer). The scaffolds were then mounted on aluminum stubs and subsequently coated with gold using the same sputter coater as before and stored in a desiccator until SEM viewing. For cell morphology assessment a voltage of 10 kV was used.

RESULTS AND DISCUSSION

Scaffold structure

The porous 58S bioactive glass-ceramic scaffold, produced by polymer replication maintained the comprehensive pore interconnectivity of its PU foam template. The total porosity of the bioactive glass ceramic scaffold without PLGA coating was calculated to be approximately 88.21 ± 0.846 %. This porosity value accounted for the macro- and micro- porosity that presented in the as sintered samples and was consistent with previous studies which developed a hydroxyapatite/tri-calcium phosphate scaffold using the same PU foam template[11]. The addition of the PLGA coating did not appear to significantly modify the macroporous structure of the scaffold. In contrast, the micropores apparent on the individual struts of the as sintered scaffold were infiltrated with PLGA following coating, which probably accounted for the approximate 2% reduction in overall porosity. The polymer coating also appeared to repair crack-like defects that presented in the as sintered scaffold due to processing and handling as well as any microvoids that were created as a result of the burning-off of the original PU template during scaffold sintering. The open macropore size distribution of the PLGA coated bioactive glass-ceramic scaffold, characterized from SEM micrographs (Figure 1), was within the range of approximately 220-500 μm, with the thickness of scaffold struts being around 70-95 μm. The observed pore dimensions were considered to be appropriate to facilitate cellular migration and tissue ingrowth for possible *in vivo* applications [13]. The high degree of porosity and large pore size would seem ideal for enhanced vascularisation in vivo and possibly promote more rapid and direct osteogenesis formation from progenitor cells such as BMSCs.

Mechanical testing

The bioactive glass-ceramic scaffolds demonstrated a mechanical response characteristic of open cell ceramic foams. The response included an initial linear elastic region followed by a plateau section created through the brittle crushing of the struts, with final densification accompanying a stress increase. The compressive strength of the glass-ceramic scaffolds was 0.12±0.03 MPa. On the other hand, the compressive strength of the PLGA-coated bioceramic scaffolds was 0.25±0.05 MPa, which was much higher than that of the as sintered scaffolds and was within the range of cancellous bone (i.e. 0.2-4 MPa) [5]. The addition of the PLGA coating acted to enhance the structural integrity of the scaffold and therefore provided it with a significant increase in compressive strength as well as a smoother and more stable stress-strain curve. The currently obtained compressive strength for the PLGA coated scaffolds does however lie toward the lower end of the defined scale and may, in its present form, be incapable of functioning appropriately as an implant for load bearing sites in synovial joints. High sintering temperatures would promote improved the mechanical properties, however the increased presence of the crystalline phase would limit the bioactivity of the scaffold somewhat. Additional slurry coating to increase the strut thickness would enhance the compressive strength, but care must be taken at the same time not to restrict the pore interconnectivity of the scaffold. In this study, no coupling agent was used to enhance the interfacial bonding strength between the PLGA coating and the glass-ceramic strut. Thus, improvement of the interfacial bonding will need to be done in the future.

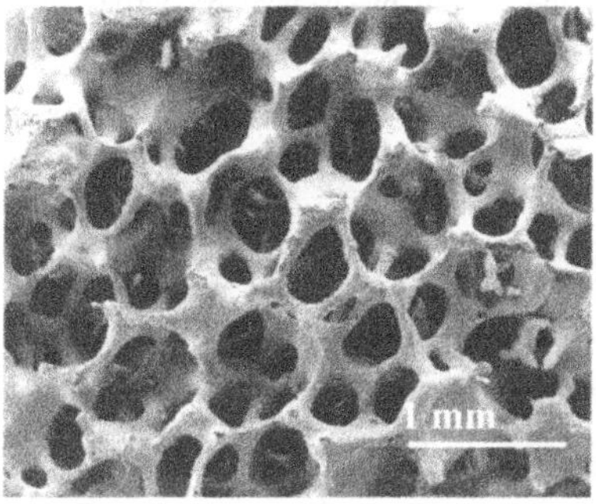

Figure 1. SEM micrographs of the macrostructure of the PLGA-coated bioactive glass-ceramic scaffold.

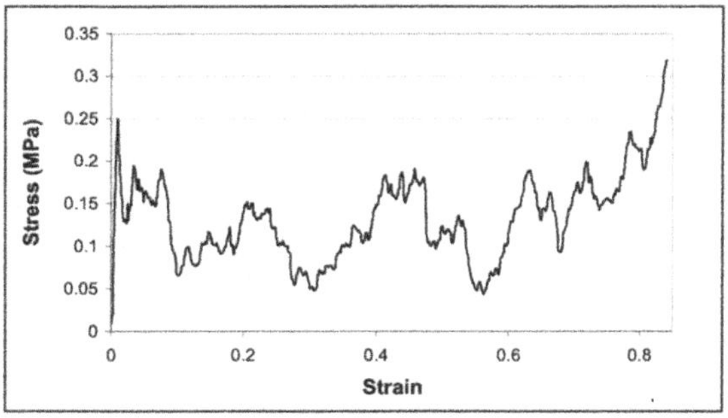

Figure 2. Stress-strain curve for the PLGA-coated bioactive glass-ceramic scaffold.

In vitro cell culture

Cell culturing activities were performed on the PLGA coated bioactive glass-ceramic composite scaffold to assess whether it possessed sufficient biocompatibility. It was apparent through observations that BMSCs had a high affinity for the surface of the scaffold and that cell migration, attachment, and proliferation readily occurred. After only 3 days in culture, cell attachment was evident on this scaffold type and while the original spindle characteristic of the cells was maintained, initial cross linking between cells appeared to take place. For culture time of 5 days significant proliferation of cells was apparent characterised by cell penetration and migration into the internal regions of the scaffold. The morphology of cells at day 5 transitioned from the original spindle shape to a more flattened appearance suggesting a near complete attachment and integration of the cell with the scaffold. An increase in cross liking between cells was also observed at this 5 day interval. At 21 days proliferation was substantial and approaching confluence, with cells of a flattened morphology appearing to line the surface of the scaffold (Figure 3). Additional cells looked to cross link transversely from ridges and grooves of the scaffold at positions elevated above the material surface as well as progressing across the large marcopores of the scaffold. The cells also seemed to actively migrate to, and cover, the cracks and defects present in the scaffold. The observation of this phenomenon was encouraging as it could potentially act to improve the overall mechanical properties of the scaffold as well as allowing for the transition of load to repair tissue as the material continues to resorb and these defects become larger.

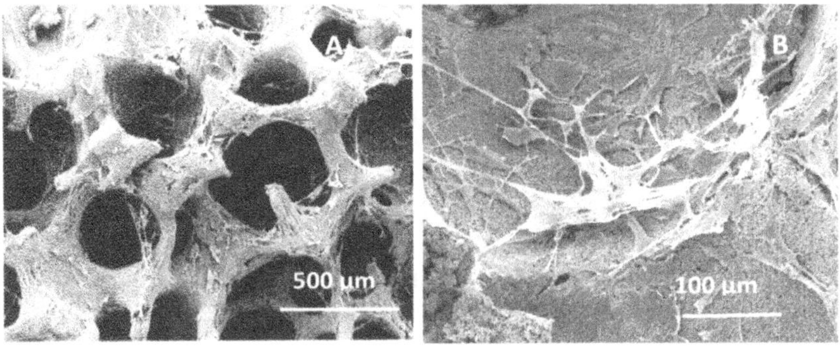

Figure 3. SEM micrographs with different magnifications (A and B) of BMSCs cultured onto the PLGA-coated bioactive glass-ceramic scaffold cultured at day 21.

CONCLUSION

This paper outlined the development and characterisation of a novel porous PLGA-coated 58S bioactive glass-ceramic scaffold for subchondral bone tissue engineering. Structural characterisation, mechanical testing and preliminary *in vitro* cell culture work were conducted. Results suggested that the PLGA-coated bioactive glass-ceramic scaffolds possessed respectable pore interconnectivity, a high porosity of ~88%, and a pore size range of 220-500 m. Complementary cell culture investigations showed that the PLGA-coated bioactive glass-ceramic scaffolds could facilitate the proliferation of BMSCs derived from patients with osteoarthritis with near confluence achieved by 21 days in culture. While the PLGA coating was able to increase the mechanical integrity of the sintered porous 58S glass-ceramic scaffolds, the limited mechanical strength of ~0.25 MPa suggests that iterative refinement of the fabrication protocol or an alternative method will be required to optimise the mechanical properties without significantly compromising the overall biological properties.

ACKNOWLEDGEMENT

The contribution of Dr. Shobha Mareddy in the processing of the cell culture work is gratefully acknowledged.

REFERENCES

[1]I. Martin, S. Miot, A. Barbero, M. Jakob, and D. Wendt, Osteochondral Tissue Engineering, *Journal of Biomechanics*, **40**(4), 750-765 (2007).
[2]S.N. Redman, S.F. Oldfield, and C.W. Archer, Current Strategies for Articular Cartilage Repair, *European Cells and Materials*, **9**, 23-32 (2005).
[3]J.F. Mano, and R.L. Reis, Osteochondral Defects: Present Situation and Tissue Engineering Approaches, *Journal of Tissue Engineering and Regenerative Medicine*, **1**(4), 261-273 (2007).
[4]M.M. Pereira, J.R. Jones, L.L. Hench, Bioactive Glass and Hybrid Scaffolds Prepared by Sol-Gel Method for Bone Tissue Engineering, *Advances in Applied Ceramics: Structural, Functional & Bioceramics*, **104**(1), 35-42 (2005).
[5]Q.Z. Chen, I.D. Thompson, A.R. Boccaccini, 45S5 Bioglass(R)-Derived Glass-Ceramic Scaffolds for Bone Tissue Engineering, *Biomaterials*, **27**(11), 2414-2425 (2006).
[6]T. Livingston, P. Ducheyne, and J. Garino, *In vivo* Evaluation of a Bioactive Scaffold for Bone Tissue Engineering, *Journal of Biomedical Materials Research*, **62**(1), 1-13 (2002).
[7]Q.Z. Chen, A. Efthymiou, V. Salih, A.R. Boccaccini, Bioglass®-Derived Glass-Ceramic Scaffolds: Study of Cell Proliferation and Scaffold Degradation *in vitro*, *Journal of Biomedical Materials Research Part A*, **84A**(4), 1049-1060 (2008).
[8]P. Sepulveda, J.R. Jones, L.L. Hench, *In vitro* Dissolution of Melt-Derived 45S5 and Sol-Gel Derived 58S Bioactive Glasses, *Journal of Biomedical Materials Research: Part A*, **61A**(2), 301-311 (2002).
[9]R. Li, A.E. Clark, and L.L. Hench, An Investigation of Bioactive Glass Powders by Sol-Gel Processing, *Journal of Applied Biomaterials*, **2**(4), 231-239 (1991).
[10]J. Zhong, and D.C. Greenspan, Processing and Properties of Sol-Gel Bioactive Glasses, *Journal of Biomedical Materials Research: Applied Biomaterials*, **53**(6), 694-701 (2000).
[11]X. Miao, D.M. Tan, J. Li, Y. Xiao, and R. Crawford, Mechanical and Biological Properties of Hydroxyapatite/Tricalcium Phosphate Scaffolds Coated with Poly(lactic-co-glycolic acid), *Acta Biomaterialia*, **4**(3), 638-645 (2007).
[12]S. Mareddy, R. Crawford, G. Brooke, and Y. Xiao, Clonal Isolation and Characterization of Bone Marrow Stromal Cells from Patients with Osteoarthritis, *Tissue Engineering*, 13, 819-829 (2007).
[13]V. Karageorgiou, and D. Kaplan, Porosity of 3D Biomaterial Scaffolds and Osteogenesis, *Biomaterials*, **26**(27), 5474-5491 (2005).

CERAMIC MATERIALS FOR BONE TISSUE REPLACEMENT AND REGENERATION

W. Swieszkowski[1], Z. Jaegermann[2], D.W. Hutmacher[3,4], K.J. Kurzydlowski[1]
[1] Faculty of Materials Science and Engineering, Warsaw University of Technology, Warsaw, Poland;
[2] Institute of Glass, Ceramics, Refractory and Construction Materials, Warsaw, Poland;
[3] Division of Bioengineering, National University of Singapore, Singapore;
[4] Institute of Health and Biomedical Innovation, Queensland University of Technology, Brisbane, Australia

ABSTRACT

The aim of this study is to show how ceramic materials could be used in bone tissue replacement and regeneration. The possibility of treatment of bone defects using porous alumina grafts, calcite porous scaffolds, polymer/ceramic biocomposite scaffolds and ceramic coatings will be shown.

Physical properties as well as architecture of the porous structure of porous alumina bone grafts formed by gel casting will be described. The long term clinical results of using porous alumina implants will be presented.

The calcite based porous ceramics will be discussed as a material for bone regeneration. The influence of chemical compositions and sintering parameters on the physical properties of sintered calcite ceramics as well as forming of porous material by polymeric sponge method will be shortly described.

A novel material made of biodegradable polymer reinforced with modified calcium phosphates (TCP) particles will be proposed to be used in fabrication of novel constructs for the repair of critical-sized bone defects. Several composite materials made of PLLA/PDLA or PCL reinforced with TCP micro and nanoparticles will be discussed.

To make the metallic implants more osteogenic, hydroxyapatite or TiO_2 coatings might be applied. The different methods of coating of the porous and non porous titanium implants with ceramic layers will be described.

INTRODUCTION

The development of contemporary medicine implies the necessity of improvements of biomaterials that have been already applied in medicine, as well as research on possibilities of the introduction of new materials, including, among others, bioceramics.

Alumina, being one of the best known and most widely used group of bioceramics have been already a part of orthopaedic practice. They have been implemented in the form of non-porous ceramics as well as in the porous form - with the porosity varying, according to the application given. Excellent chemical stability, perfect mechanical properties and good biocompatibility are the main advantages of this group of materials.

In the search for bioresorbable bone graft substitutes a variety of synthetic materials have been investigated. The usefulness for bone grafting of calcium carbonate-containing biomaterials has been demonstrated by a number of research groups[1,2]. Natural coral has been studied for over 25 years[3-5] and has been used as a filling bone defects material for more than 20 years. Bioresorbable bone substitutes also might be used to replace harvested autogenous bone at the donor site, making it possible to reharvest bone later at the same site, if necessary. The calcite based porous ceramics is a result of research on new bioresorbable calcium-containing biomaterial.

A bone regeneration by using tissue engineering approach is a quickly developing treatment in orthopedics. The main goal of the research in this field consists in elaboration of porous scaffolds for adding the process of a new bone formation. One of the potential material used for scaffold fabrication is composite materials consisting of polymeric matrix and ceramic fillers.

The aim of this study is to show how the ceramic materials could be used in bone tissue replacement and regeneration. The possibility of treatment of bone defects using the porous alumina grafts, calcite porous scaffolds, biocomposite (polymer/ceramic) scaffolds and ceramic coatings will be shown.

POROUS ALUMINA BONE GRAFTS

Porous alumina bone grafts - PABG (Fig.1) were prepared by mixing powdered substrate (96,2wt.% Al_2O_3, 2,1wt.% MgO and 1,7wt.% $CaCO_3$) with gelling binder of water solution of aluminum polyoxychloride $(AlOCl)_n$. As soon as the foamed solution was ready it was poured into elastic moulds. After the gelling process terminated ready blocks of porous castings were dried and sintered up to the temperature of 1730°C.

The grafts were characterized by the open porosity of 65÷75% (with the 80% of pores ranging between 100 and 1500 m). Open porosity as well as the pores sizes made a structure analogous, although not identical as the mineral structure of the cancellous bone. The pores within the ceramic material were spherically shaped and the walls between them were built of well sintered polycrystalline alumina. The diameter of connective ducts between pores are usually close to 150 m (Fig.2).

The compressive strength, depending on the material porosity ranged between 15÷40 MPa (Fig.3), making it easy to handle the material during the surgical procedures. After the overgrowing, the composite improved considerably its mechanical strength (up to 70%) and the modulus of elasticity (close to 30 GPa) was equal as the one characterizing the compact bone, guaranteeing proper biomechanical correlation between the bone and the graft.

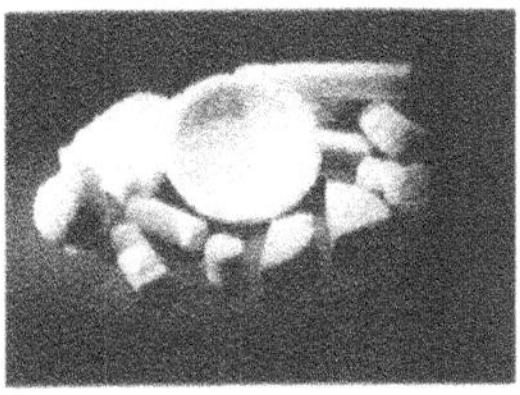

Figure 1. Porous alumina bone grafts - PABG

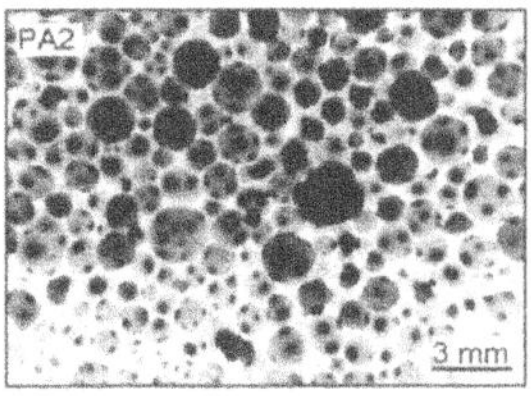

Figure 2. Macroporous structure of PABG

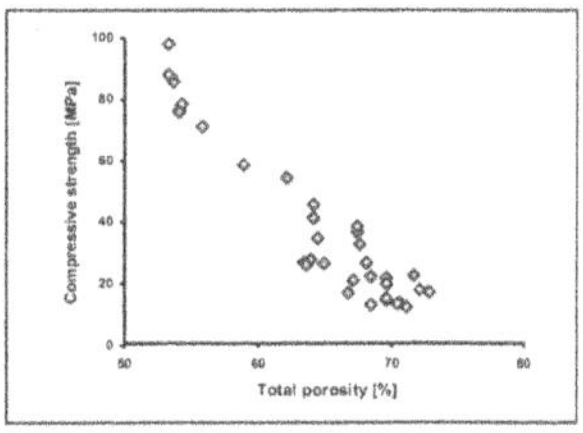

Figure 3. Compressive strength vs. total porosity of PABG

In vivo" tests (sheep model) have proved that after 16 weeks overgrowing and the bone mineralization are completed - Havers ducts and osseous lacunae can be distinguished within the tissue filling the pores[6]. It has been demonstrated as well that the pores overgrow by the kind of bone tissue that remains in contact with the graft[7]. Electron microscope tests have shown that the ingrowing bone tissue is closely connected to the pores surface, and that aluminium (Al^{3+}) from the ceramics does not penetrate the bone in contact[8]. Histochemical and histopathologic analysis demonstrated the correct callus development and the normal wound healing process. Clinical tests on the porous alumina ceramics have been carried on in several Polish renowned clinical centres[9]. Starting from 1998 the number of surgeries is close to three thousand now, and the observation period of some of the cases is close to 20 years (Fig.4). In the great majority of cases the patient treatment has been successful. It seems important to stress that surgical wounds after the operations usually healed by primary adhesion and the implants were overgrown. It has been also observed that after the surgeries the necessary period of limbs fixation was by 1/3 shorter then in similar cases where bone grafts were applied. This feature reduces harm to the patient's organs and thanks to it the intense rehabilitation treatment can be initiated earlier.

According to the results of diverse physical, chemical and biological analysis, porous alumina grafts are totally biocompatible, are being successfully overgrown by well mineralized bone tissue, are characterized by mechanical strength sufficient for many kinds of operations as well as are solid enough for free manipulation during surgical procedure and could be sterilized by all the methods accepted in medicine.

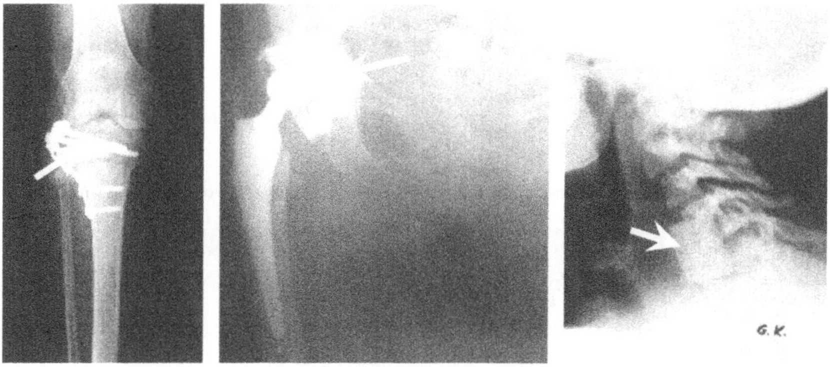

a) b) c)

Figure 4. Examples of applications of porous alumina bone grafts: a) PABG wedge used for subcondylar osteotomy, b) acetabular cup superstructure for pelvis reinforcement, c) PABG implants for vertebrae stabilization

CALCITE BASED POROUS MATERIAL

The purpose of our work was to create and evaluate the properties of synthetic bioresorbable calcite based porous material (CBPM) for bone regeneration.

Three kinds of calcite ceramics based on pure $CaCO_3$, containing 1wt%, 5wt% and 10wt% of pure lithium fluoride LiF were evaluated. Testing samples were formed by uniaxial pressing and sintered in temperatures of 450, 470, 490, 510 and 530°C. Green density and apparent density were determined by geometrical method, relative density was calculated on the basis of calcite theoretical density (3,156 g/cm^3) and mechanical properties were evaluated on the basis of compression tests.

The results of our studies show that the materials containing 1 and 5wt% of LiF behave in a similar way during sintering. They achieve the highest apparent density and compressive strength in the temperature of 510°C, but in 530°C a rapid decrease of both parameters can be observed. The material containing 10wt% of LiF sinters well only in the temperature of 530°C (fig.5). Analysis also demonstrated that both chemical composition and thermal treatment highly influence physical and mechanical properties of calcite materials.

Polymeric sponge method were used for forming CBPM, characterised by total porosity ranging from 65 to 90% and compression strength not exceeding 2,5 MPa. Few examples of porous structure of materials formed on the base of 30, 45 and 60 pore per inch density polyurethane sponges are presented in Figure 6.

Biological "in vitro" and "in vivo" tests show good biocompatibility of synthetic calcite based porous scaffolds.

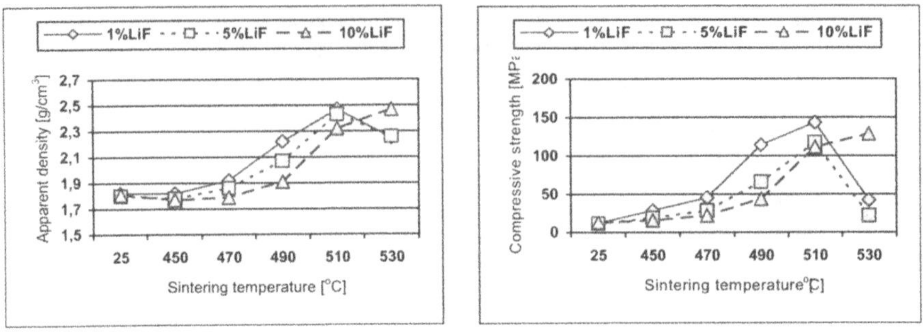

Figure 5. Apparent density and compressive strength vs sintering temperature of calcite ceramics

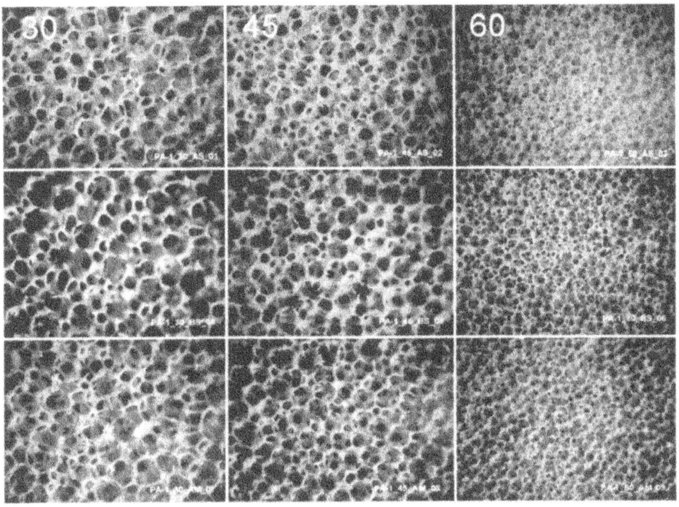

Figure 6. Structures of CBPM

POLYMER/CERAMIC BIOCOMPOSITES

The above mentioned scaffolds were made completely of the ceramic materials. Other potential materials which could be used to fabricate a novel construct for the repair of critical-sized bone defects is a novel material made of biodegradable polymer reinforced with ceramics particles. The properties of such a composite depend on: 1) properties of the polymer used for the matrix and properties of the ceramics used for the reinforcement, 2) composition of the composite (i.e. content of ceramic particles) and 3) size, shape and arrangement of the particles in the matrix. Several polymer-composite composites have been used for scaffolds fabrication including polylactide (PLA) and polycaprolacton (PCL) reinforced with calcium phosphate (CaP) micro and nanoparticles. Authors proposed a novel composite material by blending copolymer - Poly(L-lactide-co-D,L-lactide) (PLDLLA) a copolymer with a ceramic – Tri-Calcium Phosphate

(TCP). Poly(L-lactide-co-D,L-lactide) with intermediate strength, degradation period of 6–36 months was purchased from Boehringer Ingelheim GmbH and -Tri-Calcium Phosphate from Progentix BV. The final composition of the composite was PLDLLA (90%wt) and TCP (10%wt). Using this composite material, the computational modeling together with computer controlled extrusion deposition were applied to develop and then fabricate the 3D scaffold with high open porosity. Scaffold architecture and its porosity were analyzed by scanning electron microscopy and nano-computed tomography (Figure 7). The TCP particles were homogenously distributed in the polymeric matrix (this will improve mechanical performance of the scaffold). The fabricated scaffold possessed mechanical properties similar to the trabecular bone[10]. Cell culture study showed that PLDLLA/TCP scaffolds promotes proliferation and osteogenic differentiation of BMSC in vitro. These results demonstrated the high potential of using this composite scaffold in bone tissue engineering.

a) 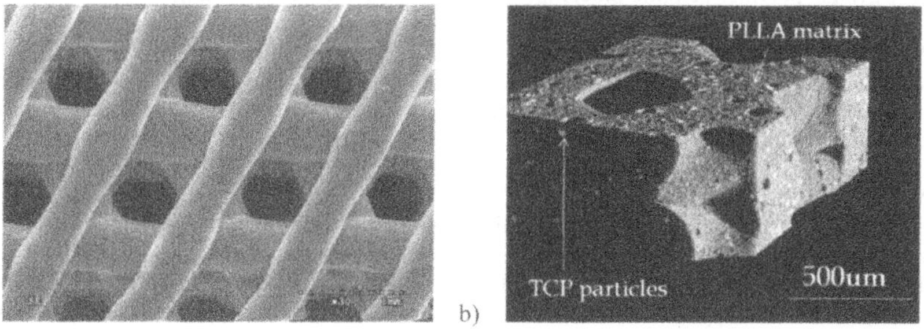

Figure 7. Composite scaffold for bone tissue engineering: a) SEM image and b) NanoCT image

CERAMIC COATINGS

The ceramics such as hydroxyapatite (HAp) or titanium oxide have been used to make the metallic implants more osteogenic. Endoprostheses coated with HAp are widely used to improve implant fixation to the bone. The different methods of coating of the porous and non porous titanium implants with the ceramic layers. A electrochemical deposition method was used to obtain HAp layer on the pure highly porous titanium scaffolds (total porosity – 75 %) fabricated by Powder

a) 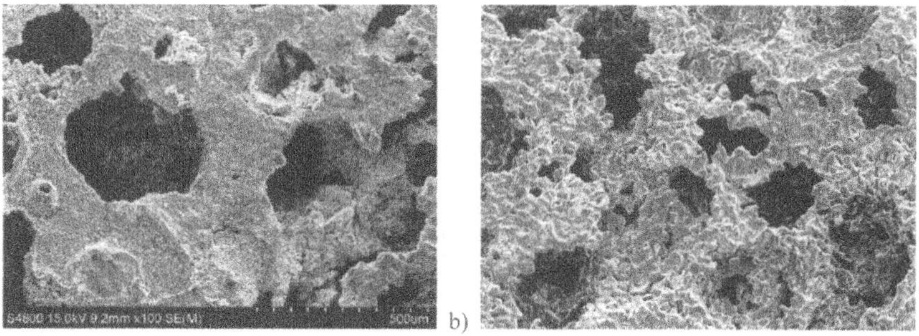

Figure 8. No-modified porous Ti scaffold (a) and Ti scaffold coated with HAp layer (b)

Metallurgy (PM)[11]. Already after few hours a dense HAp layer could be formed on the Ti scaffold imbedded in theelectrochemical solution consisted of $Ca(NO_3)_2 \cdot 4H_2O + NH_4H_2PO_4$ (Figure 8). Another method to form a osteoinductive ceramic coating on the Ti substrate material could be a electrochemical anodization process. When a potential ramp from open circuit potential to 25 V followed by holding the applied potential at 25 V for 1500 s, and NH_4F + DI water + glycerol as a electrolyte were used the titanium oxide nanotubes layer has been grown on the surface of the pure Ti [12]. It was already proved that such layer of TiO_2 nanotubes promote osteoblast differentiation and matrix production, and enhance short- and long-term osseointegration in vitro[13].

CONCLUSION

There is a high potential for using bioceramics in bone tissue replacement and regeneration. The study shows possibility of treatment of bone defects using: the porous alumina grafts, calcite porous scaffolds, polymer/ceramic biocomposite scaffolds, and ceramic coatings.

ACKNOWLEDGEMENT

This work was partially supported by funds provided by the Polish Ministry of Science and Higher Education and A*Star, Singapore for the project: A Composite Material Technology Platform for Bone Engineering and COST533.

REFERENCES

1. G. Guillemin, J.L. Patat, J. Fournie, M. Chetail - The use of coral as a bone graft substitute, *J. Biomed. Mat. Res.* 21, 557-567 (1987).
2. G. Guillemin, A. Meunier, P. Dallant, P. Christel, J.C. Pouliquen, L. Sedel - Comparison of coral resorption and bone apposition with two natural corals of different porosities, *J. Biomed. Mat. Res.* 23, 765-779 (1989).
3. J.P. Ouhayoun, A.H.M. Shabana, S. Issahakian, J.L. Palat, G. Guillemin, M.H. Sawaf, N. Forest - Histological evaluation of natural coral skeleton as a grafting material in miniature swine mandible, *J. Mat. Sci. Mat. Med.* 3, 222-228 (1992).
4. C. Voigt, C. Merle, C. Müller-Mai, U. Gross - Substitution of natural coral by cortical bone and bone marrow in the rat femur (Part I), *J. Mat. Sci. Mat. Med.* 5, 688-691 (1994).
5. B.F. Shahgaldi - Coral graft restoration of osteochondral defects, *Biomaterials,* 19, 205-213 (1998).
6. J. Kotz, J. Bieniek and A. Bieniek, Application of porous bioceramics in experimental therapy in bone injuries. I: Morphological and histological studies in the control animals, *Arch. Immunol. Ther. Experim.*, 36, 89-96 (1988).
7. J. Bieniek, Z. Swiecki, G. Rosiek and C. Miksiewicz, Porose Korundkeramik als Biomaterial, *Deutsche Geselschaft fur Orthopadie und Traumatologie,* 3, 86 (1983).
8. Z. Swiecki, G. Michalska and J. Bieniek, Bioceramics for orthopedic purposes. Part III. Porous alumina in living tissue, *Szklo i Ceramika (Acta Ceramica)* 6, 155-158 (1980).
9. J. Kiwerski, A. Ogonowski, J. Bieniek and M. Krasuski, The use of porous corundum ceramics in spinal surgery, *International Orthopaedics (SICOT),* 18, 10-13 (1994).
10. C.X.F. Lam, R. Olkowski, W. Swieszkowski, K.C. Tan, I. Gibson, D.W. Hutmacher. Composite PLDLLA/TCP scaffolds for Bone Engineering: Mechanical and In Vitro Evaluations. *Proceedings of the ICBME08, Singapore, 2008. (in press)*
11. B. Dabrowski, W. Swieszkowski W. S. Hiromoto, K.J. Kurzydlowski. *Paper in preparation.*
12. A. Roguska, M. Pisarek, M. Dolata, M Lewandowska, K.J. Kurzydlowski, M. Janik-Czachor, *Paper in preparation*
13. KC. Popat, L. Leoni, CA. Grimes, and TA Desai. Influence of engineered titania nanotubular surfaces on bone cells. *Biomaterials. 2007 Jul;28(21):3188-97.*

CHEMICAL INTERACTION BETWEEN HYDROXYAPATITE AND ORGANIC MOLECULES IN BIOMATERIALS

K. Tsuchiya, T. Yoshioka, T. Ikoma and J. Tanaka
Department of Metallurgy and Ceramic Science, Tokyo Institute of Technology
2-12-1, Ookayama, Meguro-ku, Tokyo, 152-8550 Japan.

ABSTRACT

In this study, we tried to elucidate chemical interactions between hydroxyapatite and organic molecules using an ab initio calculation method, i.e. discrete variational Xαmethod. Two cluster models were investigated. One cluster model consisted of hydroxyapatite and methyl acetate: the later corresponds to a simplified molecule of poly-L-lactide that is a typical biodegradable material. Another cluster is the model of hydroxyapatite and glycine that is the simplest molecule in 20 amino acids. Although the glycine molecule has three types of ionization states, two types of interface models, cation model and zwitter ion model were adopted. An overlap population between a Ca ion of hydroxyapatite and COO^- in glycine was larger than that between a Ca ion of hydroxyapatite and CO in methyl acetate. Further, in the case of the hydroxyapatite/glycine model, the zwitter ion model had larger chemical interaction than that of the cation model.

INTRODUCTION

The chemical interaction of organic-inorganic interface is very important for the development of novel biomaterials. Especially the interfacial interactions contribute to the fabrication of higher-order structure through self-assemble process and are closely related to the mechanical properties of organic-inorganic composites. Despite of many researches performed on organic-inorganic interfaces, the detailed mechanism of the chemical interactions has not been sufficiently understood yet. For example, the main components of bone and tooth are hydroxyapatite (HAp) and collagen[1]. In bone and tooth, the collagen molecules interact with HAp nano-crystals, resulting in the formation of higher-order structure through biomineralization process. Since chemical interaction of organic molecule / inorganic crystal interface influences the mechanical property of the composite, bone and tooth have sufficiently high mechanical toughness as hard tissues, although a pure HAp is hard and fragile.

Many researches have synthesized calcium phosphate / organic molecule composites[2,3]; in general, as experimental evaluation on organic-inorganic interfaces is so difficult and complicated, the detailed interfacial interactions have not been elucidated yet. In this study, we adopted a HAp / poly-L-lactide (PLLA) composite and a HAp / glycine composite as model materials to elucidate the interfacial chemical interactions using ab initio calculation method, *i.e.* discrete variational (DV)-Xα method.

CALCULATION METHOD

The DV-Xα method assumes comparatively small clusters. Therefore its calculation time is quite shorter than that of the other ab initio calculation methods. The geometry optimization of the organic molecules was performed using HF/6-31G in Gaussian03. We adopted the DV -Xα method to the following three cluster models.

(1) Surface model of HAp

A surface model of HAp was the ($Ca_8(PO_4)_5$) cluster obtained from X-ray structure data, shown in Fig. 1. In order to reproduce the coulomb interaction of HAp crystal, 2376 Madelung potential ions were put into a bulk side of the cluster and no potential ion was put on a surface side.

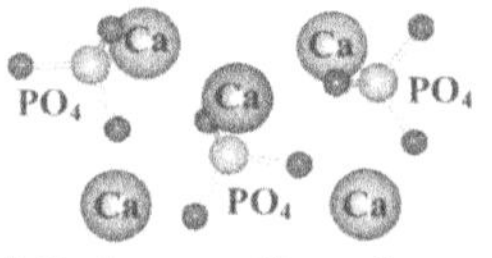

Fig. 1 Hydroxyapatite surface model

(2) Interface model of HAp/methyl acetate

In this study, a methyl acetate molecule (CH_3COOCH_3) was adopted as a simplified model for PLLA: both compounds have the same C=O double bonds. Ca ions on the HAp surface are positively charged to interact with the C=O bond in methyl acetate.

The cluster model of HAp/methyl acetate interface was shown in Fig.2; overlap population analysis was applied to this model. Using Monte Carlo method, 300 sampling points were put around each atom in the cluster. Molecular orbitals in the cluster were constructed by a linear combination of atomic orbitals (LCAO). Atomic orbitals used in this model were 1s-2p for C, 1s-2p for O, 1s for H, 1s-3d for P and 1s-4p for Ca, which were numerically calculated for atomic Hartree-Fock method. Overlap population was evaluated by Mulliken's population analysis.

The methyl acetate molecule was moved along the HAp surface to fix a bonding site between HAp and methyl acetate. Then, the methyl acetate molecule was vertically approached to the HAp surface to investigate a bond length and a chemical interaction between them.

(3) Interface model of HAp/glycine

Glycine was adopted as a model material of collagen, since glycine was the simplest molecule in amino acid and main components of collagen. In this study, we supposed that the carboxyl groups in collagen molecule interact with HAp surface. The ionization states of glycine molecules were transformed, when the surround pH was changed. Thus we adopted two types of cluster model; zwitter ion model ($CH_2NH_3^+COO^-/Ca_8(PO_4)_5$), and cation model ($CH_2NH_3^+COOH/ Ca_8(PO_4)_5$).

The zwitter ion model and the cation model were shown in Fig.3 and Fig.4, respectively. Overlap population analysis was performed for the cluster models. For the Monte Carlo method, 300 of sampling points were put on each atom in the cluster. Molecular orbitals of this cluster were constructed by LCAO. Atomic orbitals used in this model were 1s-2p for C, 1s-2p for O, 1s for H, 1s-2p for N, 1s-3d for P and 1s-4p for Ca, which were numerically calculated for atomic Hartree – Fock method. The electron density was evaluated by Mulliken's population analysis.

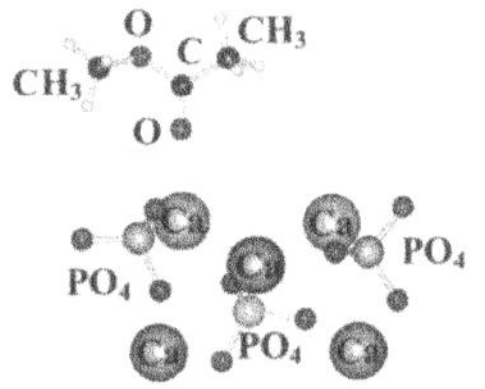

Fig.2 HAp/ Methyl acetate model

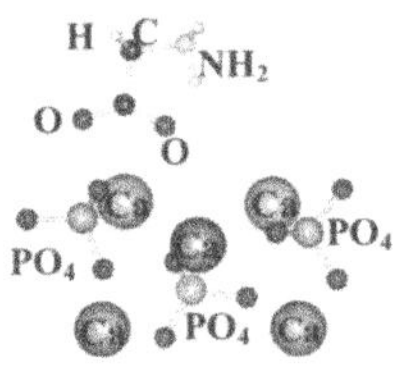

Fig.3 Zwitter ion model

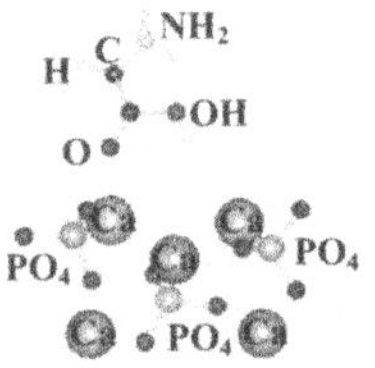

Fig.4 Cation model

RESULTS AND DISCUSSION

(1)Chemical interaction in the HAp/methyl acetate model

Fig.5 shows the change of the overlap population between the Ca ion of HAp and the O atom in the methyl acetate molecule along the HAp surface. This result indicates that the overlap population became maximal, when the methyl acetate molecule lied directly on the Ca ion. Then, the methyl acetate molecule was vertically approached to the Ca ion of the hydroxyapatite surface. Fig.6 shows the overlap population between the Ca ion and the O atom as a function of Ca-O distance. This result indicates that the chemical bond between the Ca ion and the O in C=O bond was formed. The overlap population between the Ca and the O becomes maximal at the Ca-O distance of 0.23nm.

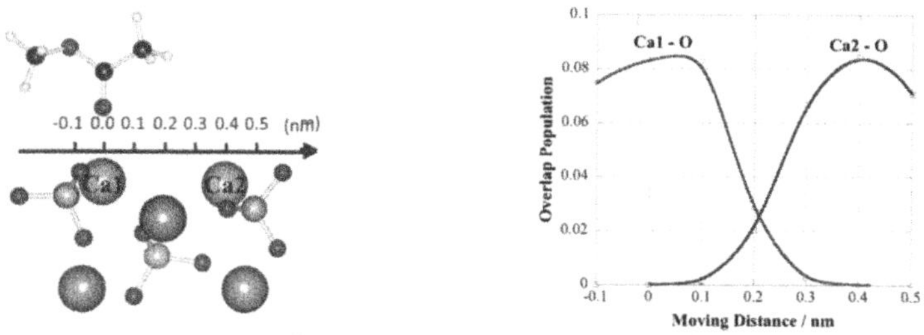

Fig.5 Overlap population between Ca ion and O atom (horizone)

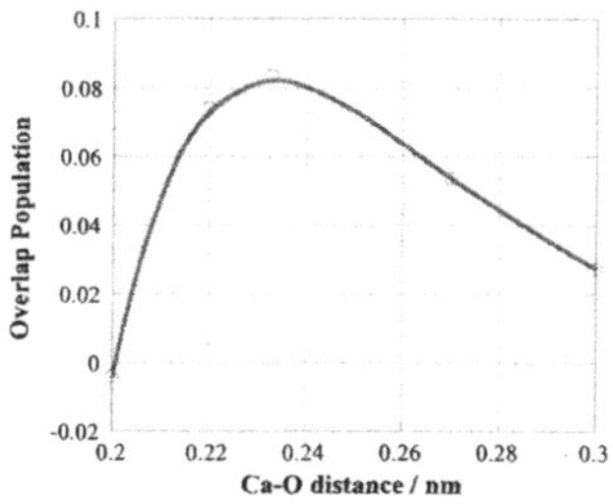

Fig.6 Overlap population between Ca ion and O atom as a function of Ca-O distance

(2)Chemical interaction in the HAp/glycine model

In cation model, we supposed that C=O double bond in un-ionized carboxyl group interacts with Ca ion on the HAp surface. Fig.7 shows the overlap population between Ca ion and O atom of cation of glycine molecule. When the cation of the glycine molecule approaches the HAp surface, the overlap population between the Ca ion and the O atom increased to Ca-O distance d=0.23nm, and reversely decreased with the decrease of the distance d less than d=0.23nm.

On the other hand, on the zwitter ion model(Fig.8), the chemical bonds were formed between the Ca ion and the O in COO- group: O1 and O2, when the glycine was approached to Ca on the HAp. The overlap populations between the Ca and the O1, the Ca and the O2 became maximal at the Ca-O1 distance of 0.23 nm and Ca-O2 distance of 0.25nm respectively.

(3)Comparison of interfacial interaction

Fig.9 shows the comparison of interfacial interaction among these cluster models. As the result of

the simulation, the interaction of HAp/glycine interface was larger than that of HAp/methyl acetate interface. On the other hand, the interaction of zwitter ion model was larger than that of cation model. These results indicates that ionized carboxyl group mainly interacts with the Ca of HAp.
The bond distance between Ca ion and O atom in organic molecule calculated by DV-Xα method was 0.23 nm. On the other hand, the ion radius of Ca^{2+} ion and O^{2-} ion were 0.114 nm and 0.126 nm respectively. Total bond length between Ca^{2+} ion and O^{2-} ion is 0.240nm. It was suggested that the chemical interactions of these interfaces are similar to ion bonding rather than covalent bonding.

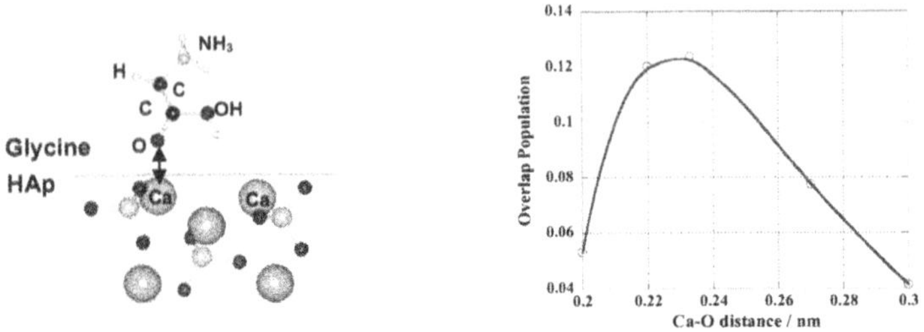

Fig.7 Overlap population between Ca ion and O atom of cation of glycine molecule

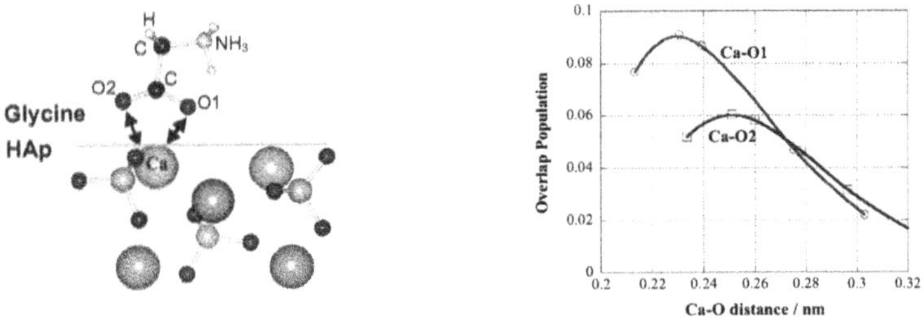

Fig.8 Overlap population between Ca ion and O atom of zwitter ion of glycine molecule

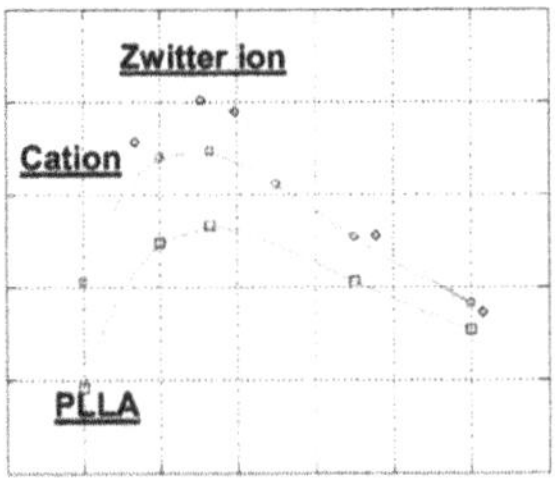

Fig.9 Comparison of interfacial interaction among three cluster models

CONCLUSIONS

Three interface models consisting of HAp surface and organic molecules were analyzed using DV-Xα method. In the methyl acetate, the overlap population between a Ca ion of HAp and an O atom in methyl acetate became maximal at the Ca-O distance of 0.23nm. In order to elucidate the chemical interaction of the HAp/gycine interface, we adopted two models, i.e. cation and zwitter ion models. In the cation model, the overlap population between a Ca ion of HAp and an O atom was also maximal at the Ca-O distance of 0.23 nm. On the other hand, in the Zwitter ion model, the overlap populations between a Ca ion and O1 and O2 atoms in a functional group COO^- became maximal at the Ca-O distance of 0.23-0.25nm. The overlap interaction between a functional group (COO^-) and a HAp surface in the HAp/glycine system was larger than that in the HAp/methyl acetate system. Further, the chemical interaction in the zwitter ion model was larger than that in the cation model. Such qualitative features may be expanded to other molecular systems involving proteins, biomolecules etc.

REFERENCES

[1]M. Kikuchi, T. Ikoma, S. Itoh, H. N. Matsumoto, Y. Koyama, K. Takakuba, K. Shinomiya and J. Tanaka, Biomimetic Synthesis of Bone-like Nanocomposites using the Self-organization Mechanism of Hydroxyapatite and Collagen, Composites Science and Technology, 64, 819-825(2004)

[2]M. Kikuchi and J. Tanaka, Chemical Interaction in b-Tricalcium Phosphate/Copolymerized Poly-L-Lactide Composites, J. Ceram. Soc. Jpn., 108, 642-645 (2000).

[3]Shaobing Zhou, Xiaotong Zheng, Xiongjun Yu, Jianxin Wang, Jie Weng, Xiaohong Li, Bo Feng, Ming Yin, Hydrogen Bonding Interaction of Poly(D,L-Lactide)/hydroxyapatite Nanocomposites, Chem. Mater., 19, 247-253 (2007)

CONCLUSION

POROUS Al_2O_3 PREPARED VIA FREEZE CASTING AND ITS BIOCOMPATIBILITY

Jing Li[1], Kaihui Zuo[2], Wenjuan Liu[1],Yu-Ping Zeng[2], Fu-Qiang Zhang[1]*, Dongliang Jiang [2]
1. Department of prosthodontics, No.9th Hospital, Shanghai Jiaotong University , School of Medicine, Shanghai 200011, China
2. Shanghai Institute of Ceramics, Chinese Academy of Sciences, Shanghai 200050, China

Abstract:

The porous Al_2O_3 ceramics with macro lamellar pores were prepared via freeze casting process received due to the phase separation of the water- based Al_2O_3 ceramic slurry. The mechanical properties results showed that the porosity and the strength of the porous Al_2O_3 ceramics were 37.04 % and 31.0 MPa, respectively. The cytotoxicity, acute hemolysis, as well as skin sensitization tests were employed to characterize the biocompatibility of the porous Al_2O_3 ceramics, the results indicated that the porous Al_2O_3 ceramics have good biocompatibility and can be used as bone scaffolds in the future.

Keywords: Porous alumina; Mechanical properties; freeze casting; pore; biocompatibility

1 Introduction

Al_2O_3 is a bio-inert ceramic, which can be used as a bone scaffold. Normally, bone scaffolds should have biocompatibility and several physical characteristics, such as, high porosity, large pore size, uniform pore distribution as well as sufficient strength etc. Many methods are employed to prepare the porous ceramic scaffolds, for example, porogen leaching, polymer sponge, gel casting, freeze casting and freeze gelation etc 1-4. Among of them, freeze casting is an ideal fabrication technique, which can be used to prepare complex shape scaffolds with controlled pore size and pore morphology. Several literatures have reported the microstructure and mechanical properties of porous bioceramics fabricated by freeze casting 4-8.

In this work, porous Al_2O_3 ceramics were fabricated by freeze casting from Al_2O_3 aqueous ceramic slurries, and its biological properties, such as, in vitro cytotoxicity, acute hemolysis tests, and skin sensitization reactions were also investigated.

2 Experimental

2.1 Preparation of porous Al_2O_3

Alumina powder (Shanghai Wusong Fertilizer Factory, China, d50=0.6 μm) was used as the starting material. Kalium polyacrylate (Lopon 895, BK Guiulini Chemic Representative Office, Germany) was used as the dispersant and deionized water was used a solvent. Firstly, Al_2O_3 powder was stirred via the magnetic bar in diluted HNO_3 (5 wt%) for 24 h to remove possible metal ions, and then washed with deionized water until pH=7. The pretreated Al_2O_3 powder was used to prepare Al_2O_3 slurry after drying in oven at 80°C for 24 hours. The treated Al_2O_3 powder was dispersed in deionized water with 2 wt% Lopon 895 (based on the powder) by ball milling for 24 h using Al_2O_3 balls as balling medium. The solid loading of the Al_2O_3 slurry was 70 wt%.

Al_2O_3 slurry was poured into a rectangular silicone rubber container with the dimension of 40 ×6 ×8 mm. The container was then frozen in a chamber, where the temperature was -18°C. After the slurries completely solidified, the samples were taken out from container and moved into a lyophilizer (TLG-A, Zhongke Biologic Co. Ltd, China). After freeze drying, the samples were sintered in muffle furnace at 1600°C for 1h.

2.2 Mechanical properties and microstructure characterization

Microstructure of porous Al_2O_3 was observed by scanning electron microscopy (SEM, JSM-6700F, JEOL, Japan). Open porosity and density were conducted by the Archimedes method with distilled water as liquid medium. The compressive strength of the sintered specimens was measured using Instron 5500R universal testing machine (Instron Ltd., Norwood, MA, USA) at a crosshead speed of 2.0 mm/min, the size of specimen was Φ8.5*12.5 mm and 6 specimens were used to get the average value.

2.3 Biological properties characterization

Preparation of the extract of porous Al_2O_3 ceramics

The disinfected ceramics were immersed in culture solution DMEM (Dulbecoo's Modified Eagle's Medium) for cytotoxicity test and physiological saline was also used to get the extract for hemolysis test. The extract experiment was conducted under 37°C for 24 h and 10 ml physiological saline for each 0.2 g porous Al_2O_3 was used.

Cytotoxicity test

The mouse fibroblast cell line L929 was routinely cultivated in DMEM with 10% fetal bovine serum at 37°C in 5% CO_2 contained air atmosphere. In 96-well plates, 3000 cells were seeded in each well and incubated for 24 hours. The cells were divided into three groups: sample group, positive group, and negative group. Each group has eight wells. The DMEM with 10% fetal bovine serum medium was then replaced with the extract of porous ceramics (sample group), 0.64% phenol (positive group), and DMEM (negative group), respectively. After 2, 4, 7 days incubation, 20μl 5% MTT (3-(4,5)-dimethylthiahiazo (-z-yl) -3,5-di- phenytetrazoliumromide) was added into each well and incubated for another 4 h. The incubation liquid in each well was let off, 350 μl dimethyl sulfoxide (DMSO) was then added to every well. After jolted by a shaker for 15 min at room temperature, the supernatant was gathered and their optical density value (OD) was measured at 490 nm using the spectrophotometric microplate reader. The relative growth rate (RGR) was then calculated according to the OD values of MTT tests (RGR= $\overline{X}$ treated/ $\overline{X}$ negative×100%) and was graded into 6 grades according to Table 1.

Acute hemolysis test

In order to get fresh anti-coagulated rabbit blood, 0.5 ml of 20 g/L potassium oxalate solution was added into each 10 ml fresh rabbit blood. Each 8 ml of anti-coagulated rabbit blood was then diluted with 10 ml physiological saline. 0.2 ml diluted rabbit blood was mixed with 10 ml of the extract of porous Al_2O_3 ceramics (sample group), distilled water (positive group), and 0.9% physiological saline solution (negative group), respectively. There are five specimens in each group. After centrifugal separation for 5 min at 1000 rpm, the optical density (OD) of the supernatant was measured by a spectrophotometer at 545 nm. The final data of each group was an average of 5 samples. Hemolytic rate was calculated using the formula as below:

$$\text{Hemolytic rate} = (\overline{OD}\,\text{sample} - \overline{OD}\,\text{negative}) / (\overline{OD}\,\text{positive} - \overline{OD}\,\text{negative}) \times 100\%.$$

Skin sensitization test (Buehler test)

Healthy young adult albino guinea pigs are preferred for the test. The animals are randomized and assigned to the treatment groups. 20 animals were used in the treatment group and 10 animals were used in the control group. The hair of animals on the back and abdominal sites was removed 24 hours ago before test. The abdominal site of the guinea was parted to the induction area and the challenge area equally from the middle line. The animals were divided into three groups in this test: the extract of ceramics (sample group); 5% formaldehyde (positive group); physiological saline (negative group). The experiment can be divided into three steps.

Induction phase: The test patches fully loaded with the liquid of the sample group, the negative group, and the positive group, respectively. The patches were then affixed the induction areas and removed after 6 hours. The same procedure was done repeatedly every one week and totally three times.

Challenge phase: After 14 days, the patches were held in contact for 6 hours on the challenge areas of animals and the patches was then removed.

Observation and grading phase: 24 hours later, skin reactions in challenge areas were observed and recorded according to the grades shown in table 2.

3 Results and discussion

3.1 Mechanical properties of the porous Al_2O_3 ceramics:

Figure 1 is the SEM micrograph of the porous Al_2O_3 ceramics. The Al_2O_3 ceramics show the lamellar porous microstructure. The low magnification Fig 1(a) indicates that the pores are unidirectional. The deep channel areas in Fig 1(b) are macro lamellar pores, which are caused by sublimation of the macro columnar ices. In the experiment, since the rubber container has a low coefficient of thermal conductivity, the ice crystals grow in the direction from the top of slurry to the bottom and form unidirectional lamellar ice crystals. Simultaneously, Al_2O_3 particles were expelled in the front of ice crystals and formed the particles walls. By controlling the slurry concentration, the composition of slurry, freezing rate and sintering condition *etc*, the pore's micrograph and pore size can be tuned 4-8. Therefore, this kind of porous ceramics has potential uses as scaffolds for human beings due to the controlled pore structure and morphology.

The relative density, porosity and the strength of the porous Al_2O_3 ceramic is about 61.59 %, 37.04 % and 31.0 MPa, respectively. The porosity calculated from the relative density is almost the same as the date given by the Archimedes method. Therefore, it is reasonable to suppose that most of the pores are open, which is useful as the scaffolds. The porosity and strength of the porous Al_2O_3 ceramic are both suitable for hard tissue repair applications.

3.2 In vitro biocompatibility evaluation

As to the biomaterial for human tissue replacement, it is necessary to demonstrate if the material has any effect on the biological properties of the tissue. Bioceramics exhibit some possible toxic reactions due to metal ions leaching from the ceramics, resulting in the tissue dying or heavy reactions. In this experiment, cytotoxicity test, hemolysis test as well as skin irritation were conducted to value the biocompatibility of the porous Al_2O_3 ceramics.

Cytotoxicity test: The L929 fibroblast cell line is the most widely used for cytotoxicity assessment 9-11. The cell metabolic activity was evaluated by succinc acid dehydrogenase (SDH) of mitochondria in living cell, which can deoxidized MTT salt into insoluble purple crystal. These crystals can be resolved by dimethyl sulfoxide (DMSO), and measured by the spectrophotometric microplate reader at 490 nm. To some extent, the optical density value (OD) of supernatant raised with the number of the living cell. Therefore, the measurement of MTT crystal OD can reflect the relative growth rate and the activity of cells, which can also leads to a reliable assessment to the cytotoxicity of the sample.

The experimental results are shown in Table 3. It can be seen that the RGR of both porous Al_2O_3 and negative group (NG) increased steadily with time. However the RGR of positive group (PG) is very low and some L929 cells are dead. The cytotoxicity grade of the porous Al_2O_3 ceramic samples is grade 1 at the 2, 4, 7 days, which indicate that the porous Al_2O_3 ceramics is not toxic for cell.

The acute hemolysis test: The results of the hemolysis test are presented in table 4. The absorbance is based on the hemoglobin solubility from the erythrocytes. Thus, the absorbance corresponds to the number of hemolyzed erythrocytes. The data directly reflected if newly prepared porous Al_2O_3

ceramic can hemolyze the erythrocyte 11-13. Theoretically, if the hemolytic rate is less than 5%, the result is considered to be satisfactory. In the experiment, the hemolytic rate of porous Al_2O_3 ceramics was 3.22%, which indicated that the porous Al_2O_3 is safe for the red blood cells.

Skin irritation test: The skin irritation reaction is listed in table 5. Skin sensitization is an immunologically mediated reaction to a substance. In a human, the response can be characterized by pruritis, erythema, edema, papules, vesicles, bullae, or a combination of these. In other species, the reactions are different and only erythema and edema can be seen 14-15. The skin irritation reactions showed that there are no allergic responses, which indicated that the porous Al_2O_3 ceramics did not induce any immune reactions to guinea pigs.

According to the results of cytotoxicity test, acute hemolysis test, skin irritation test, the in vitro biological reactions of the porous Al_2O_3 ceramics prepared via freeze casting is biosafe and can be further used in implant experiments.

4 Conclusions

Al_2O_3 ceramics with lamellar pores and ceramic walls can be fabricated by a freeze-drying process using aqueous ceramic slurries. The relative density, porosity and strength of the porous Al_2O_3 ceramic is about 61.59 %, 37.04 % and 31.0 MPa, respectively.

The cytotoxicity, acute hemolysis, skin sensitization results suggest that Al_2O_3 porous ceramics prepared via freeze casting in vitro biological reactions is biosafe and the porous Al_2O_3 ceramics have a potential application for implants. Further experiments, such as, direct cells contact in vitro, long-term implantation test in vivo will be conducted to meet the practical application requirements.

Acknowledgements

The authors acknowledge financial support of the Shanghai of Committee Science and Technology under the object "07jp14093" and Shanghai Leading Academic Discipline Project (T0202).

References

1. L. Freed, G. V. Noyokovic, R. J. Biron, D. B. Eagles, D. C. Lesnoy, S. K .Barlow and R. Langer, "Biodegradable polymer scaffolds for tissue engineering," *Biotechnology.*, 12 689-693 (1994).
2. H. R. Ramay and M.Q. Zhang, "Preparation of porous hydroxyapatite scaffolds by combination of the gel-casting and polymer sponge method," *Biomaterials.*, 24[19] 3293-3302 (2003).
3. M. H. Ho, P. Y. Kuo, H. J. Hsieh, T. Y. Hsien, L. T. Hou, J.Y. Lai and D. M. Wang, "Preparation of porous scaffolds by using freeze-extraction and freeze-gelation methods," *Biomaterials.,* 25 129-138 (2004).
4. D. Silvain, S. Eduardo, P. T. Antoni, "Freeze casting of hydroxyapatite Scaffolds for bone tissue engineering," *Biomaterials.*, 27 5480-5489 (2006).
5. S. Deville, E. Saiz, R. K. Nalla, and A. P. Tomsia, "Freezing as a Path to Build Complex Composites," *Science.*, 311 [27] 515-518 (2006).
6. T. Fukasawa and M. Ando, "Synthesis of Porous Ceramics with Complex Pore Structure by Freeze-dry Processing," *J. Am. Ceram. Soc.*, 84 [1] 230-232 (2001).
7. T. Fukasawa, Z. Y. Deng, and M. Ando, "Synthesis of Porous Silicon Nitride with Unidirectionally Aligned Channels Using Freeze-drying Process," *J. Am. Ceram. Soc.*, 85 [9] 2151-2155 (2002).
8. K. H. Zuo, Y. P. Zeng and D. L. Jiang, "Properties of microstructure controllable porous YSZ Ceramics fabricated by freeze casting," *Inter. J. Appl. Ceram. Tech.,* in press.
9. B. J. Nablo and M. H. Schoenfisch, "In vitro cytotoxicity of nitric oxide-releasing sol-gel derived materials," *Biomaterials.*, 26 [21] 4405-4415 (2005).
10. ISO 10993-5: 1999(E), Biological Evaluation of Medical Devices, Part 5: Tests for in vitro

cytotoxicity
11. R. Quan, D. Yang, X. Wu, H. Wang, X. Miao, W. Li, "In vitro and in vivo biocompatbility of graded hydroxyapatite-zirconia composite bioceramic," *J Mater Sci: Mater Med.,* inpress
12. A. Nakahira, M. Tamai, H. Aritani, S. Nakamura, K. Yamashita, "biocompatibility of dense hydroxyapatite prepared using an SPS process," *J Biomed Mater Res.*, 62 [4] 550-557 (2002).
13. L. Montanaro, C. R. Arciola, E. Cenni, "Cytotoxicity, blood compatibility and antimicrobial activity of two cyanoacrylate glues for surgical use," *Biomaterials.*, 22 [1] 59-66 (2001).
14. S. Kitajima, J. Momma, T. Inoue, "Reactivities of the skin-sensitization test in guinea pig (GPMT) as a function of three parameters: induction doses (MID), challenge doses (SCD), and direct exposures (DED)," *Ann. N. Y. Acad. Sci.,* 919 312-314 (2000).
15. H. Sueki, A. M. Kligman, "Cutaneous toxicity of chemical irritants on hairless Guinea pigs ," *J. Dermatol.*, 30 [12] 859-870 (2003).

Figure captions

Fig. 1. SEM micrograph of the porous Al_2O_3 ceramics sintered at 1600°C for 1h: (a) low magnification (b) high magnification

Table 1. Grading of cytotoxicity

Table 2. Degree of skin irritation assay

Table 3. The cytotoxicity results (OD, RGR, cytotoxic grade)

Table 4. The results of the acute hemolysis test

Table 5. Degree of irritation

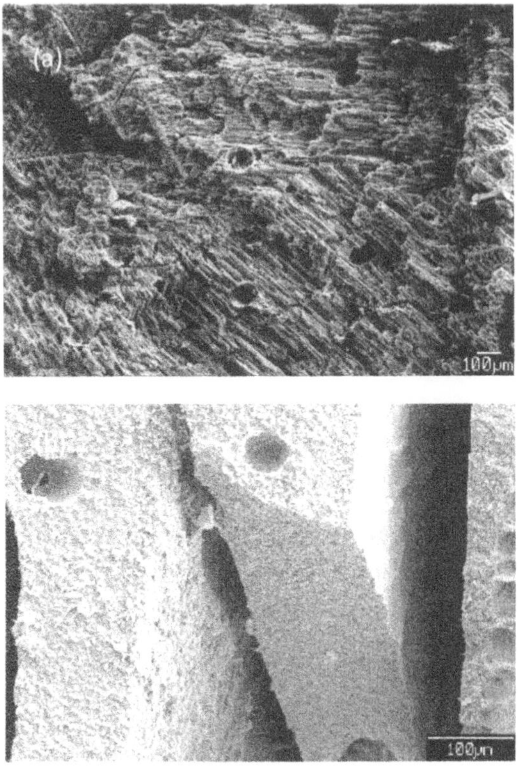

Fig. 1. SEM micrograph of the porous Al_2O_3 ceramics sintered at 1600°C for 1h: (a) low magnification (b) high magnification

Table 1. Grading of cytotoxicity

Relative growth rate(RGR)	Cytotoxicity rate
>=100	0
75-99	1
50-74	2
25-49	3
1-24	4
0	5

Table 2. Degree of skin irritation assay

Score	Reaction Description	Degree of Irritation
0	No erythema/ edema	No irritation
1	(Barely perceptible)Very slight erythema/ edema	very slight irritation
2	Well defined erythema/ slight edema	Slight irritation
3	Moderate to severe erythema/ edema	Moderate irritation
4	Severe erythema/ edema	Severe irritation

Table 3. The cytotoxicity results(OD, RGR, cytotoxic grade)

group	2d			4d			7d		
	OD	RGR	Grade	OD	RGR	Grade	OD	RGR	Grade
Al2O3	0.18±0.04	76.5	1	0.29±0.04	78.6	1	0.40±0.04	81.6	1
NG	0.24±0.03	100	0	0.37±0.04	100	0	0.49±0.02	100	0
PG	0.08±0.01	35.3	3	0.08±0.01	21.3	4	0.09±0.01	18.4	4

Table 4. The results of the acute hemolysis test

group	Sample 1	Sample 2	Sample 3	Sample 4	Sample 5	Hemolysis rate%
Al_2O_3	0.0391	0.0383	0.0385	0.0386	0.0384	3.22
NG	0.022	0.021	0.023	0.021	0.020	
PG	0.561	0.553	0.555	0.553	0.557	

Table 5. Degree of irritation

group	Irritation number	Irritation rate	Degree of irritation
Al_2O_3	0/10	0	0
NG	0/10	0	0
PG	10/10	100	4

XI. Laser Ceramics

PREPARATION OF TRANSPARENT CERAMIC Nd:YAG WITH MgO AS ADDITIVE

Yongchao Li[1], Tiecheng Lu[1,2,*], Nian Wei[1], Ruixiao Fang[1], Benyuan Ma[1] and Wei Zhang[1]
[1]Department of Physics and Key Laboratory for Radiation Physics and Technology of Ministry of Education, Sichuan University, Chengdu 610064, P. R. China
[2]International Center for Material Physics, Chinese Academy of Sciences, Shenyang, 110015, P. R. China

ABSTRACT

In this paper, high quality Nd:YAG nanopowders with different amounts of MgO (0~0.4 wt %) as additive were synthesized by alcohol-water solvent co-precipitation method. Transparent Nd:YAG ceramics were successfully fabricated by vacuum sintering at 1780°C for 10 hours using the powders. Surfaces morphologies of thermally etched Nd:YAG ceramics was observed by means of SEM. Optical transmittance of the transparent Nd:YAG ceramics was measured over the wavelength region from 400nm to 850nm. Few pores and defects of transparent Nd:YAG ceramics were observed and the optical transmittance of them reached 80% in the visible-near infrared range. The effect of MgO on the sintering of transparent ceramic was investigated and the sintering mechanism was discussed. The results showed that MgO as sintering additive can restrain abnormal grain growth and reduce pores in grains and consequently enhance optical transmittance of ceramics. The optimal weight percentage of added MgO can be determined as 0.2 wt %.

INTRODUCTION

A lot of efforts have been made for synthesizing transparent YAG ceramics. In 1995, Ikesue et al. fabricated Nd:YAG transparent ceramic by solid-state reaction and accomplished laser output for the first time [1]. In 2002, Nd:YAG transparent ceramic was pumped by LD and obtained an output power as high as 1460W [2]. Because of excellent physical properties, superior optical and laser characteristics, Nd:YAG transparent ceramic is a promising material as a very good substitute to YAG single crystal as laser material, and will be used widely in various areas, such as military affairs, communications, metal processing and medical operations, etc.

Powder synthesis and sintering additive selection are essential for transparent ceramic fabrication. In this study, high quality Nd:YAG powders were prepared by alcohol-water solvent co-precipitation method [3]. The method of using MgO as additive for fabrication of YAG ceramics was proposed by De With et al for the first time and has not been studied deeply since then[4]. In this paper, transparent Nd:YAG ceramics were successfully fabricated using MgO as additive, the effect of MgO on the sintering of transparent ceramic was investigated and the sintering mechanism was discussed.

EXPERIMENTAL

Nd:YAG precursors were synthesized via alcohol-water solvent co-precipitation method. The mixed solution containing Al^{3+} ,Y^{3+} and Nd^{3+} in molar ratio 5:2.94:0.06 was prepared by dissolving $NH_4Al(SO_4)_2 \cdot 12H_2O$ (purity≥99.99%), $Y(NO_3)_3 \cdot 6H_2O$ (purity≥99.99%) and $Nd(NO_3)_3 \cdot 6H_2O$ (purity≥99.99%) in distilled water. Different amounts of MgO (0~0.4 wt %) was added into mixed solution with $MgSO_4 \cdot 7H_2O$ (purity≥99.99%) as raw material. Concentration of the mixed solution was 0.12M for Al^{3+}. The precipitant solution was prepared by dissolving NH_4HCO_3 (purity≥98%) in mixed solvent of alcohol and distilled water. The mixed solution was dripped into the precipitant solution at a dripping speed of 200ml h^{-1} under stirring at room temperature. The resultant suspensions, after aging for 20h, were filtered and washed with distilled water and alcohol, respectively. Then precursors were produced after the precipitate was dried at 70°C for 24h. The precursors were calcined to powders in air at 1100°C for 2h. Calcined powders were uniaxially pressed into pellets and then isostatically pressed at 250 MPa. Then these pellets were sintered at 1780°C for 10h in a vacuum-sintering furnace.

Phase identification was performed via X-ray diffraction (XRD, model D/max-rA, using nickel-filtered Cu-Kα radiation). A transmission electron microscope (TEM, JEM-100CX II) was used to observe the microstructure of YAG powders. Surface microstructures of the thermally etched ceramics were observed by a scanning electron microscopy (SEM, model HITACHIS-3000N). The optical transmittance of each ceramics mirror-polished on both surfaces, with a dimension of Φ10mm×1mm, was measured over the wavelength region from 400nm to 850nm using an ultraviolet spectrophotometer (model SCAN683). Densities of ceramics with different amounts of additive MgO were measured by the Archimedes draining method.

RESULTS AND DISCUSSION

Fig.1 shows the XRD pattern of Nd:YAG powders calcined at different temperature for 2h. The powders were found to be amorphous until about 800 °C. The precursor crystallized to pure YAG at 900 °C without the formation of any intermediate phases, indicating higher cation homogeneity of the precursor synthesized via alcohol-water solvent co-precipitation method. Above 900 °C, continued refinement of perk shapes and intensities were observed, indicating crystallite growth of the YAG powders with temperature increasing.

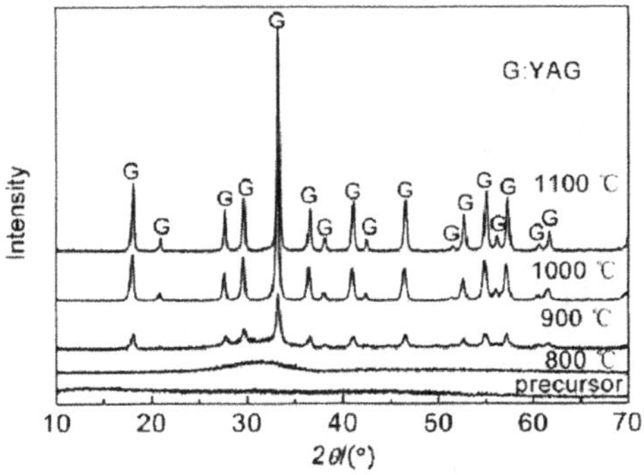

Fig.1 XRD pattern of Nd:YAG powders calcined at different temperature for 2h

Fig.2 shows TEM image of YAG powders calcined at 1100 °C for 2h. TEM observation indicates that YAG powders are well dispersed and have a small particle diameter, about 40nm. Alcohol-water solvent is beneficial to the preparation of high quality YAG powders. The reason is that the dielectric constant of alcohol is smaller than that of water, it is easier to gain precipitate using alcohol-water solvent. Furthermore, alcohol has steric effect, which makes it more difficult for particles to approach to each other, to reduce powder agglomeration.

Fig.3 shows the photos of ceramics prepared with different weight percentage of MgO as additive. All specimens are transparent, those with MgO as additive take on pink. The letters under specimen A is misty, while the letters under specimens B and C is legible. The specimen B with 0.2wt % MgO as additive exhibits the best transparency among them.

Fig.4 shows the optical transmittance spectra of ceramics in wavelength ranging from 400nm to 850nm. The maximum transmittance of specimens A, B and C over the wavelength region from 400 to 850 nm is 47%, 83% and 53%, respectively. It is obvious that MgO as sintering aid can improve optical transmittance of ceramics. The in-line transmittance of Nd:YAG ceramics A and C increase with increasing wavelength, while this phenomenon can not be observed in single crystal. This may be attributed to the differences in microstructure between single crystal and polycrystalline ceramics. There are some scattering centers including grain boundaries and pores in ceramics. If the grain boundaries are narrow and no grain-boundary phases exist, optical scattering is mainly caused by the pores [1,5]. When the optical scattering center is significantly smaller than the wavelength, the scattering coefficient can be determined by Rayleigh's equation, the scattering coefficient increases proportionally λ^{-4}, where λ is the wavelength. Thus, the scattering intensity increases with the decrease of wavelength [6].

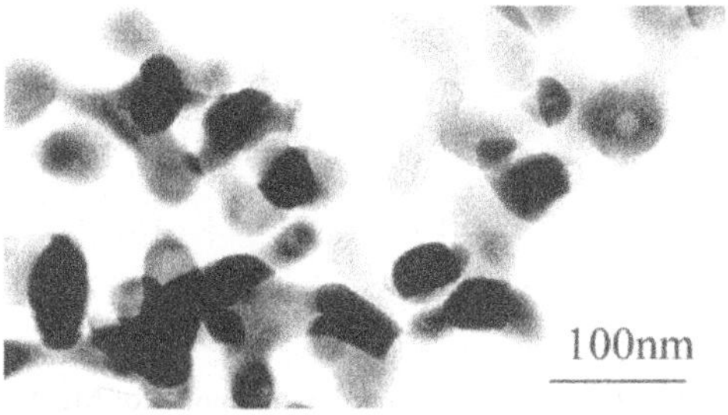

Fig.2 TEM image of Nd:YAG powder calcined at 1100 °C for 2h

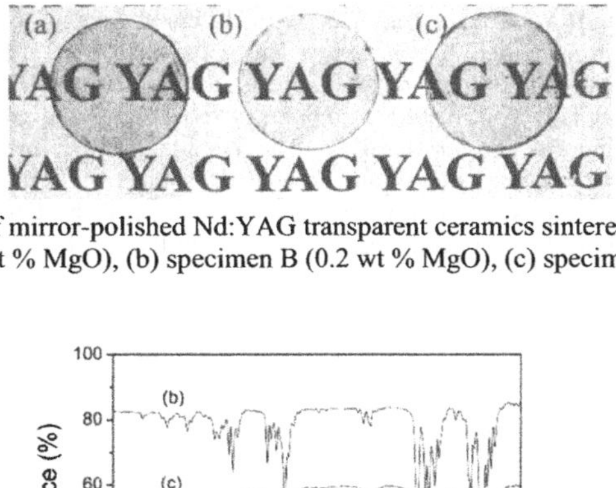

Fig.3 Appearance of mirror-polished Nd:YAG transparent ceramics sintered at 1780°C for 10 h (a) specimen A (0.0 wt % MgO), (b) specimen B (0.2 wt % MgO), (c) specimen C (0.4 wt % MgO)

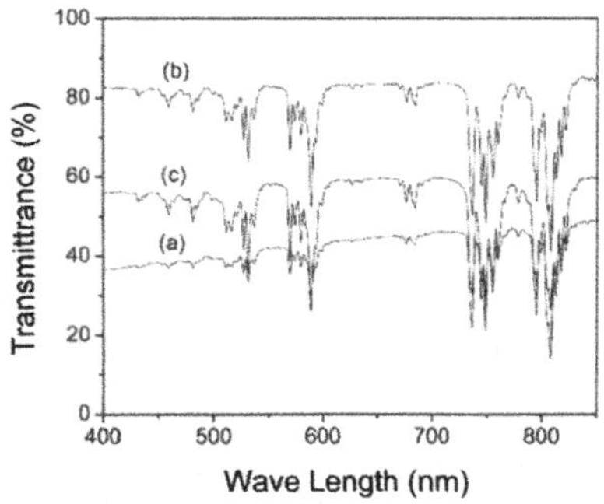

Fig.4 Optical transmittance of Nd:YAG ceramics (1mm thick) sintered at 1780°C for 10h (a) specimen A (0.0 wt % MgO) (b) specimen B (0.2 wt % MgO) (c) specimen C (0.4 wt % MgO)

Fig.5.shows relative density of Nd:YAG ceramics, with different amount of MgO, sintered at 1780°C for 10h. The relative density of all specimens is above 99.5%. As the weight percentage of MgO increases to 0.2 wt %, the density increases to full density and highly transparent specimen is obtained.

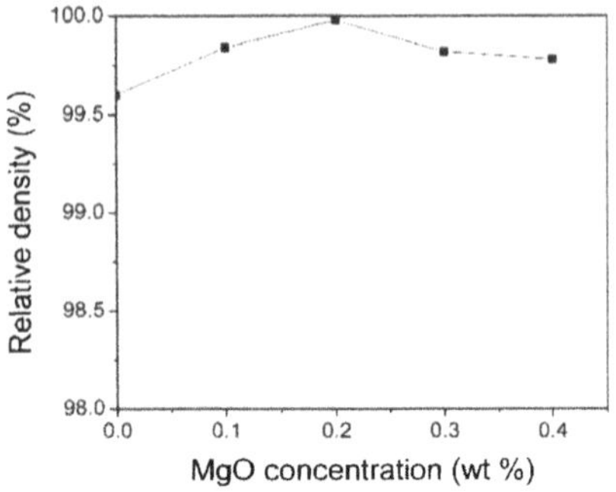

Fig.5.Relative density of Nd:YAG ceramics, with different amount of MgO, sintered at 1780°C for 10h

Fig.6. shows surface microstructure of ceramics, with different amounts of MgO, after thermal etching. It is obvious that specimen A is not very dense, and that there are many pores in grain boundaries and in grains. So its optical transmittance is low, as shown in Fig.4 (a). The average grain size is about 200μm. However, there are no pores in specimen B and impurity and the second phase in clear grain boundaries are not observed. The average grain size is about 7μm. Because microstructure of specimen B was greatly ameliorated by adding MgO as sintering aid, optical transmittance of specimen B was greatly improved as shown in Fig.4 (b). Observed from Fig.6 (c), there are many surface defects in specimen C, optical transmittance of specimen C is lower than 60%, as shown in Fig.4 (c).

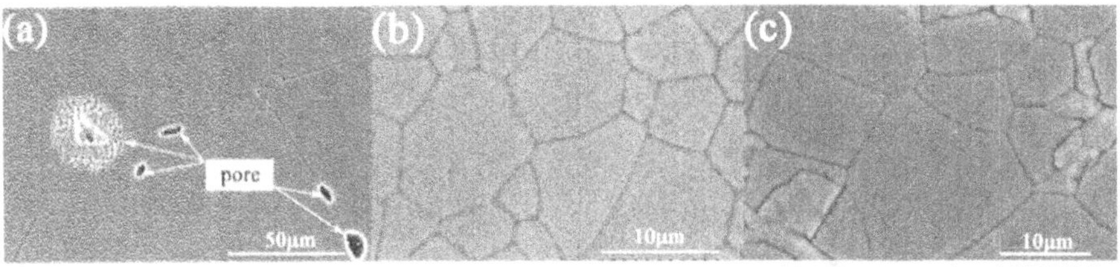

Fig.6. Surface morphologies of the ceramics, with different mount of MgO, sintered at 1780°C for 10h. (a) specimen A (0.0 wt % MgO), (b) specimen B (0.2 wt % MgO), (c) specimen C (0.4 wt % MgO)

There are scattering centers in ceramics, such as grain boundary, pore and so on. It has been shown that the main factor influencing beam scattering is not grain boundary, but the residual pore from Ikesue previous' study [5,6]. Normally, without sintering aid, it is very difficult to prepare transparent oxide ceramics technically. The reason is that at the final stage of sintering, pores declining, grain boundary moving more quickly, abnormal grain growth always occurs, resulting in a breakaway of grain boundary-pores [7]. When this happens, pores would be entrapped in grains and be dispelled with difficulty. MgO as grains growth inhibitor transforms into liquid phase at grain boundaries during vacuum sintering, enwrapping grains and delaying the grain boundary's quick move[8] and consequently resulting that pores at the grain boundaries will be dispelled during sintering and abnormal grain growth will be restrained. So the average grain size of YAG ceramics with MgO as sintering aid is far smaller than that of ceramics without MgO, as shown in Fig.6 (a) and (b). On the other hand, when Mg^{2+} ions displace Al^{3+} ions in YAG lattice, it is most possible that oxygenic vacancy (Vo)¨ appears as charge compensation. Appearance of defect (Vo)¨ increases the diffusion coefficient of grain-boundary area, which enhances the sintering of YAG ceramics.

Furthermore, MgO as sintering aid can enhance optical transmittance of ceramics. The percentage of MgO is very important for sintering Nd:YAG transparent ceramics. If MgO is not enough, it is not sufficient for forming liquid phase to enwrap grains and delay the grain boundary's move. Pores at grain boundaries would be entrapped inner grains at the final stage of sintering and be not dispelled. If MgO is excessive, exceeding the ultimate solubility of MgO in YAG phase at sintering temperature, large amount of liquid will be yielded. It is easy for excrescent MgO to react with Al_2O_3, resulting forming $MgAl_2O_4$ phase at grain boundaries. It is detrimental for optical properties of ceramics. So an appropriate amount of MgO is a key factor for sintering transparent Nd:YAG ceramics. Pores have been removed almost completely and the ceramic has uniform microstructure, as shown in Fig.6 (b), and a full transparent Nd:YAG ceramic can be obtained. The SEM results are in agreement with these analyses and indicate that the optimal weight percentage of MgO can be determined as about 0.2wt%.

CONCLUSION

High quality Nd:YAG powders were synthesized via alcohol-water solvent co-precipitation method. Transparent Nd:YAG ceramics with MgO as additive were fabricated. Few pores and defects of the Nd:YAG ceramic with 0.2 wt % MgO have been observed and the optical transmittance of that reaches almost 80% in the visible-near infrared range. Additive MgO can restrain abnormal grain growth and reduce pores in grains and consequently enhance optical transmittance of ceramics. The optimal percentage of added MgO can be determined as 0.2wt%.

ACKNOWLEDGEMENTS

This research was supported by NSFC of P. R. China under grant No. 50742046 and No. 50872083.

REFERENCES

[1] A. Ikesue, T. Kinoshita, K. Kamata and K. Yoshida, Fabrication and Optical Properties of High-Performance Polycrystalline Nd:YAG Ceramics for Solid-State Lasers, *J. Am. Ceram. Soc.*, **78**, 1033-40 (1995).

[2] J. Lu, K. Ueda, H. Yag, et al, Neodymium doped yttrium aluminum garnet ($Y_3Al_5O_{12}$) nanocrystalline ceramic- a new generation of solid state laser and optical materials, *J .Alloy. Comp.*, **341**, 220-225 (2002).

[3] S. Tong, T. Lu and G. Wang, Synthesis of YAG Powder by Alcohol-water Co-precipitation Method, *Mater Lett.*,**61**, 4287-4289 (2007).

[4] G. De With and HJA, Van Dijk, Translucent $Y_3Al_5O_{12}$ Ceramics, Mater. Res. Bull., **19**, 1669-1674 (1984).

[5] A. Ikesue, Polycrystalline Nd:YAG ceramics lasers, *Opt. Materials.*,19, 183-187 (2002)

[6] A. Ikesue, K. Yoshida, T. Yamamoto, et al, Optical scattering centers in polycrystalline Nd:YAG laser, *J. Am. Ceram. Soc.*, **80**, 1517-1522 (1997).

[7] L. Wen, X.D. Sun, Z.M. Xiu, et al, Synthesis of nanocrystalline yttria powder and fabrication of transparent YAG ceramics, *J. Eur. Ceram. Soc.*, **24**, 2684-2688 (2004).

[8] Jorgensen P J and Westbrook J H, Role of solute segregation at grain boundaries during final-state sintering of alumina, *J. Am. Ceram. Soc.*, **47**, 332-338 (1964).

SYNTHESIS OF La, Yb CODOPED Y_2O_3 POWDER AND LASER CERAMICS

Yihua Huang[1,2] , Dongliang Jiang[1*] , Jingxian Zhang[1], Qingling Lin[1]
[1] The State Key Laboratory of High Performance Ceramics and Superfine Microstructure, Shanghai Institute of Ceramics, Shanghai 200050, China
[2] Graduate University of the Chinese Academy of Sciences, Beijing 100049, China

ABSTRACT

Yb, La:Y_2O_3 nanocrystalline powder was prepared by co-precipitation method. The transparent ceramic was fabricated by vacuum sintering at 1700 °C for 4h in vacuum. The structural, morphological and luminescence properties of the sample were characterized by BET, SEM, XRD, and fluorescence analyzer. The results showed that Yb^{3+} and La^{3+} dissolves completely in the Y_2O_3 cubic phase and the lattice parameters became bigger. After calcining at 1100℃ for 4h, the particles were nearly spherical with narrow size distribution and the average diameter of the particles was in the range of 60-80 nm. Transparent polycrystalline ceramics with uniform lanthanum distribution was obtained. The grain size was around 21±7 μm. The relative density of the transparent ceramics reached 99.8%, the in-line transmittance of the transparent ceramics exceeded 75% at 800 nm for 2 mm thick samples. All the absorption bands were broadened. The absorption cross section is 4.02×10^{-21} cm^2 and its fluorescent lifetime is 0.84 ms, which was bigger than that of Yb:Y_2O_3. The emission cross section was 2.20×10^{-20} cm^2.

INTRODUCTION

Ytterbium ion is the most hopeful ion that can be used in a non-Nd laser in the same range of emission wavelength, and a very attractive dopant for efficient diode pumped solid-state lasers due to its very simple energy level[1]: the $^2F_{7/2}$ ground state and the $^2F_{5/2}$ exited state. There is no excited state absorption reducing the effective laser cross section, no up conversion, no concentration quenching. Thus, Yb^{3+} shows high quantum efficiency, weak non-radiative transitions. Yttria is one of the promising host materials for solid-state laser[2], because of its high thermal conductivity (twice than YAG), which makes it capable to endure more energy from laser radiation. However, yttria single crystals are very difficult to grow. Now it is possible to fabricated highly transparent ceramics for laser application[3], which is easier to fabricate than single crystal. Many cubic garnet and sesquioxide transparent polycrystalline ceramics have shown efficient laser emission properties and got laser output[4].

Lu[5] et al developed Yb^{3+}:Y_2O_3 solid state laser with continuous wave laser output of 0.47 W, and the slope efficiency is about 32% Highly efficient continuous-wave operation at 1030 and 1075 nm wavelengths of LD pumped Yb^{3+}:Y_2O_3 ceramic laser(1.4 W) was obtained by Takaichi [6] in 2004. They applied pressureless sintering technologies to obtain homogeneous grain size. Qi [7] et al reported 10.5 W output power for Yb^{3+}:Y_2O_3 laser ceramic in 2007, and the corresponding slope efficiency was 37.5%.

Sintering with additives is one of the traditional sintering methods for transparent yttria ceramics. Some additives can form liquid in the sintering of yttria. Lefever[8] et al fabricated the first transparent yttria ceramic in 1967. They used hot pressing at 950 °C with 70 MPa using LiF as sintering additive. The addition of LiF was responsible for the formation of liquid phase in sintering. SrO[9] was found to be a good candidate for accelerating densification process. The achievement of transparency was attributed to the creation of point defects, especially oxygen vacancies, which helped to accelerate pore removal during the sintering process. In 2002, Ikegami 10 et al fabricated transparent yttria ceramics by low temperature synthesis of yttrium hydroxide. Sulfate ion was used as additive in the progress and liquid phase sintering may be occurred to eliminate considerable pores in the final sintering.

Some additives can inhibit the grain growth in the sintering of transparent yttria ceramics by solute drag mechanism [11]. Greskovich[12] fabricated 1%Nd, 2.5-10%Th co-doped transparent ceramic and got laser output in 1974. The grain size could be kept about 120 μm after sintering at 2170 °C. Belyakov [13] et al introduced HfO_2 to the sintering of yttria, cation vacancies were formed in the inner parts of the crystal, which facilitates diffusion mass transfer and perfects the structure inside the crystal. The segregation of hafnium oxide at the crystal boundaries leads to the formation of oxygen vacancies in the material, which impedes mass transfer between the crystals and facilitates their rapid growth and the capture of intercrystalline pores.

Rhodes [14] used oxalic acid co-precipitation method to produce lanthanum doped yttria powder and sintered by controlled transient solid second-phase sintering method. Materials with near theoretical total transmittances and specular transmittance within 6% of theoretical were obtained.

According to Y_2O_3-La_2O_3 binary phase diagram, limited solid solution can be produced when the mole ratio of La_2O_3 is less than 14%. La^{3+} could be dissolved into the yttria crystalline lattice uniformly[15]. As a trivalent dopant, it can enhance the grain boundary mobility in yttria. The appearance of La^{3+} could change the crystalline field of yttria and change the optical properties. $(Y_{0.95}La_{0.03}Yb_{0.02})_2O_3$ nanopowders and transparent ceramics were fabricated in this paper. The effect of La^{3+} addition on microstructure development during sintering and the optical properties affected by the crystal field was studied.

EXPERIMENTAL PROCEDURE

Powder preparation and compaction

Yttria powder and appropriate amount of lanthanum oxide powder (99.99%, Shanghai YueKai New Materials Co., Ltd., Shanghai) according to $(Y_{0.95}La_{0.03}Yb_{0.02})_2O_3$ were dissolved in nitric acid, and then diluted with suitable deionized water to prepare the 0.3 M nitrate solution, which was used as the mother solution. Some ammonia sulfate was added into the mother solution. Precursors were prepared by adding ammonia solution (2 M) to the mother solution at a rate of 3 ml/min with stirring. When the pH value of the system reached 8, titration was stopped followed by a 3 hour aging. Then the gel was washed by deionized water for several times to remove the byproducts. After washing, the gel was frozen and dried by freeze-drying (Beijing Boyikang Experiment Instrument Co., Ltd., FD-1A-50,Beijing). The precursors were calcined at 1000 °C. The lanthanum doped yttria powder was uniaxially compacted into disks in Φ20 mm steel die at 40 MPa and then isostatically pressed at a pressure of 200 MPa. Specimens were sintered at 1500 or 1700 °C for 4h at a rate of 10 °C /min under vacuum of 5×10-3 Pa in a furnace with molybdenum heating element (Shanghai Chen Rong Co., Ltd., Shanghai). And then specimens were annealed at 1500 °C for 20 h in air.

Characterization

Powder size of initial lanthanum doped yttria was determined by X-ray diffraction (Model D/MAX-2550V, Rigaku, Japan), and field emission scanning electron microscopy (FESEM, JSM-6700F, JEOL, Tokyo, Japan). The content of La and Yb was detected by EDX. BET was used to obtain the specific surface area of the calcined yttria. Guinier-Hagg camera (XDC-1000, Stockholm, Sweden) was used to characterize precisely the lattice parameters of lanthanum doped yttria. Optical microscope was used to determine the grain size, at least 300 grains was measured to get the mean size. The grain sizes were measured by the linear-intercept method and calculated from $G=1.5\ L$, where G is the average grain size and L is the average intercept length. Mirror-polished samples were etched by 20% boiling HCl for 30 s. Archimedes method was utilized for measuring the densities of samples. Mirror-polished samples on both surfaces were used to measure the optical transmittance (Model U-2800 Spectrophotometer, Hitachi, Japan). The fluorescence spectra and fluorescent lifetime of the ceramics were measured by a spectrofluorimeter (Fluorolog-3, Jobin Yvon, France) with 940nm LD exciting.

RESULTS AND DISCUSSION

The microstructures of 3% La, 2%Yb co-doped yttria calcined at different temperatures for 4h were shown Figure 1. The particles of the original precursor in Figure 1(a) were plate like. The powder was in less aggregated state due to freeze drying [16]. The XRD peaks of the nitric precursor can be indexed to $Y_2NO_3(OH)_5 \cdot 1.5H_2O$ JCPDS No.49-1107 in Figure 2. The plate like particles was decomposed with the increase of calcination temperature. In Fig 1 (b), many rod like particles appeared when the precursor was calcined at 600 °C. When the calcination temperature increased to 750 °C, more coral like particles were observed in Fig 1(c). Figure 1 (d) shows the photograph of the 3% Yb doped yttria after calcined at 1100 °C. These coral like powder continually decomposed to spherical particles when the calcination temperature increased to 1100℃ from 750 °C. No big agglomeration could be observed after calcination. The mean particle size was around 60nm. This kind of nanopowder with high surface area and well distribution in particle size satisfied sintering behavior for the subsequent work.

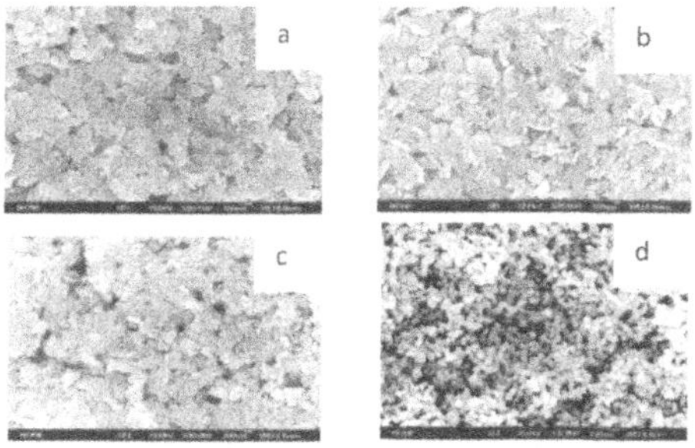

Fig.1. SEM photographs of 3% La, 2%Yb co-doped yttria precursor through freeze drying process (a)and calcined at 600 (b), 750 (c), 1100 °C (d).

Figure 2 shows the XRD spectra of the original precursor and precursors calcined at different temperature (600, 750, 900 and 1100 °C) for 4 h. It was shown that the yttria phase appeared when the calcination temperature over 600 °C. The peaks can be indexed as cubic yttria phase JCPDS No. 65-3178, no peaks for La and Nd were detected. But all of the peaks moved a little uniformly to lower 2θaccording to the JCPDS No. 65-3178. That might be due to the reason that the radius of La^{3+} and Yb^{3+} are larger than that of the Y^{3+} (103.2, 99.4, and 90.0 pm, respectively). When the La and Nd were integrated into the crystal lattice of yttrium oxide, the interplanar spacing (d) would be larger, and then the θ would be smaller according to the Bragg formula. Thus all of the peaks moved a little on the XRD curves. The more the La and Nd added, the bigger the lattice parameter would be. It could be found that with the rise of the calcination temperature, the peaks became sharper. The reason was that the crystal grew bigger at high temperature. According to the Scherrer formula, the peaks will be narrower when the crystal sizes were bigger.

Figure 3 shows the TG-DTA curves of the dried precipitate precursor obtained from ammonia co-precipitation processing. It was observed at the TG curve that there are three stages of weight loss in the temperature ranges from 90 to 250 °C, from 250 to 380 °C and 400 to 620 °C. Respectively, these three stages can be corresponded to three relevant endothermic peaks at 112, 308, and 530 °C in DTA curve. The mass loss in total 3 steps was about 38%. It can be seen that most of the weight loss occurred before 600 °C.

Additional useful information could be deduced from TG-DTA curve is the formula of the precursor. The weight losses from 90 to 180°C, 220 to 380 °C and 420 to 650 °C are 10.9, 11.4 and 17.6%,respectively. To our knowledge, when the rare earth salt was precipitated by ammonia, the ratio of $[OH^-]$ to $[R^{3+}]$ (R=Yb, La, and Y) was between 2.5 to 3. The chemical formula of the precursor can be deduced as $R_2NO_3(OH)_5 \cdot 1.5H_2O$, which was consistent with the result of XRD.

The average particle sizes of the as-synthesized 3% La, 2%Yb co-doped yttria were calculated by two different ways respectively, broadening of diffraction peaks of x-ray patterns according to the Debye-Scherrer equation and specific surface areas by BET method. Figure 4 showed that the specific surface area (S_{BET}) decreased sharply from 40 to 16 m^2/g when the calcination temperature increased from 600 to 1000 °C. A comparison of particle size by the two methods was also presented in Fig 4. When the calcination temperature was 600 °C, the primary grain size calculating by XRD was about 15 nm, while from BET was around 30 nm. When the calcination temperature increased to 1000 °C, the primary grain size measured by XRD was about 36nm. The grain size obtained from BET was still about 2 times larger than that from XRD method, but it was consistent with the particle size from the SEM image, demonstrating that there existed a remarkable primary

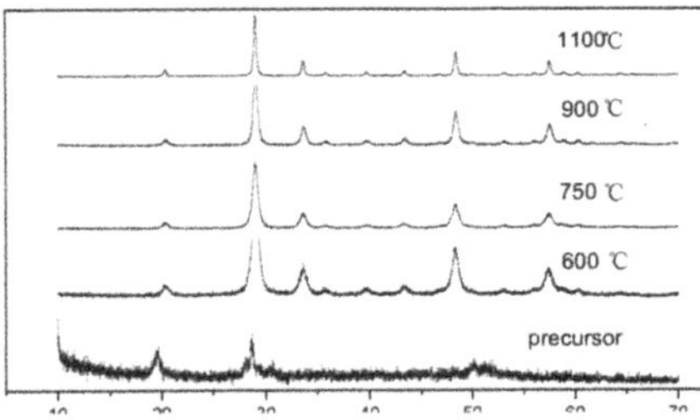

Fig.2. XRD profiles of the 3% La, 2%Yb co-doped yttria precursors before and after calcining at different temperatures.

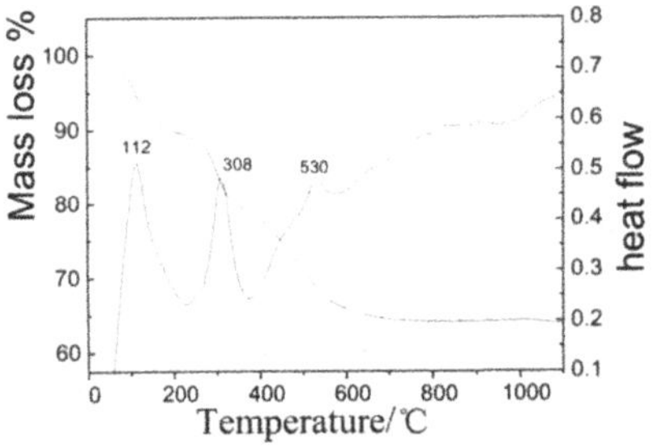

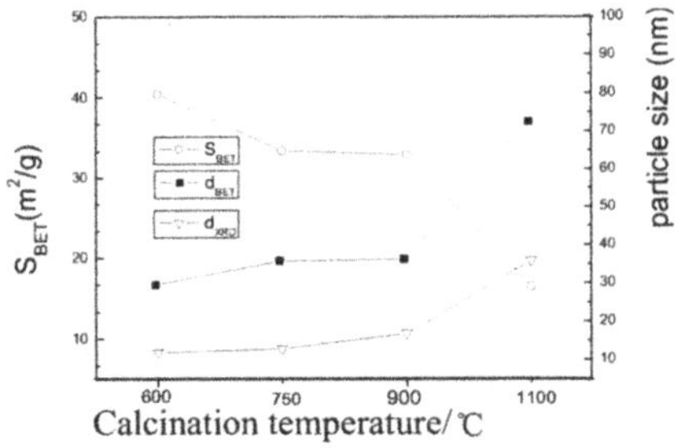

Fig.3. TG-DTA curves of the 3% La, 2%Yb co-doped yttria co-precipitation precursors (10 °C /min).

Fig.4. Dependence of specific surface area and primary grain size of 3% La, 2%Yb co-doped yttria powder as a function of calcining temperature.

grain growth as well as grain coursing. However, the nanoparticles with high BET are still active for the densification process.

Figure 5 shows the microphotograph of 3% La, 2% Yb co-doped yttria samples sintered at 1700 °C. No pores could be detected in the grains on the surface photograph, either in the grain boundary. There were no abnormal grains in the face microphotograph. The average grain size was about 21±7 μm. And the relative density was higher than 99.8%, which was higher than that of undoped $Yb:Y_2O_3$ in the same conditions. This indicates that the appearance of La can enhance the densification progress in the sintering of yttira. Figure 6 shows the photograph of the 3% La, 2%Yb co-doped yttria transparent ceramics sintered at 1700 °C in vacuum for 4 h (2 mm thickness). The words in the paper can be read clearly through the transparent sample.

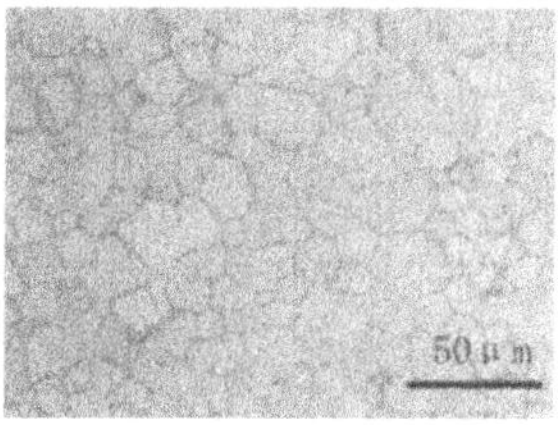

Figure.5. The microphotographs of 3% La, 2% Yb co-doped yttria samples sintered at 1700 °C after etched

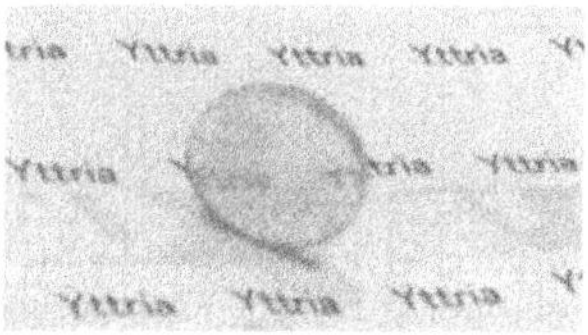

Figure.6. Photograph of transparent ceramics sintered at 1700 °C.

Figure 7 shows the transmittance of 3% La, 2% Yb co-doped yttria transparent ceramics. The highest transmittance reaches 75% at about 850 nm for 2mm thick sample. There are absorption peaks at about 950nm, which are the characteristic absorption peaks for Yb^{3+} ion.

Figure 8 shows the absorption spectrum and absorption cross section of the 3% La, 2%Yb co-doped yttria transparent ceramic. The ground electronic configuration of Yb^{3+} is $4f^{13}$, which corresponds to a single hole in the complete 4f electronic shell. This electronic configuration has a unique spectral term, 2F that is split by the spin-obit interaction in two energy manifolds, $^2F_{7/2}$ (ground manifold) and $^2F_{5/2}$; in crystal fields of symmetry lower than cubic the ground manifold is split in four Stark levels and excited manifold in three levels[17]. So there are three absorption peaks in figure 8(a), which are centered at 904, 948, and 974 nm, all attributed to Yb3+ ion $^2F_{7/2}\rightarrow{}^2F_{5/2}$ transition. The absorption spectrum is similar to those of $Yb:Y_2O_3$. The broad absorption band is benefit for absorbing the pump energy, and it can low the high request for temperature controlling. The absorption cross section can be calculated by:

$$\sigma_{abs}(\lambda)=\frac{2.303k(\lambda)}{N\times L}$$

Where $\sigma_{abs}(\lambda)$ is the absorption cross section at the wavelength, $k(\lambda)$ is the absorption rate at the

wavelength, N is the concentration of Yb^{3+} ion, and L is the thickness of the sample.
Figure 8(b) shows the absorption cross section value according to wavelength. The highest absorption cross section is 4.02×10^{-21} cm^2 at 978 nm, and 3.52×10^{-21} cm^2 at the LD pumped wavelength (940 nm).

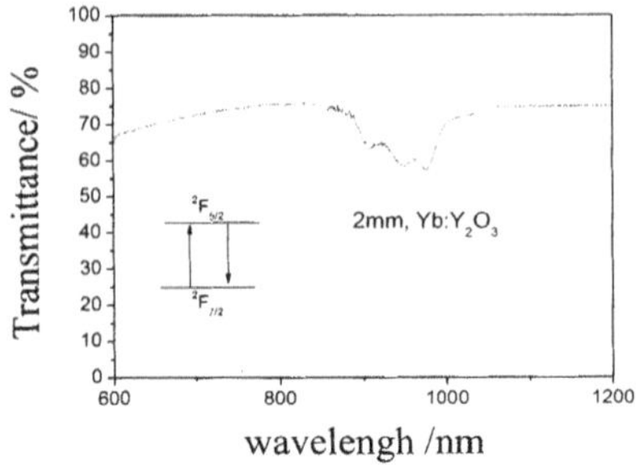

Figure.7. The transmittance of 3% La, 2%Yb co-doped yttria transparent ceramic

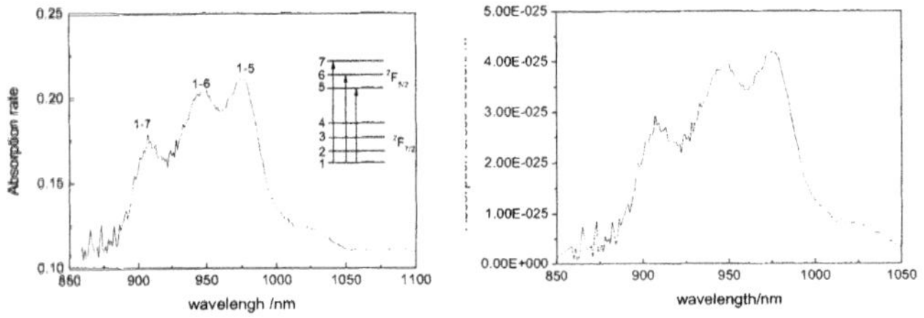

Figure.8. Absorption spectrum and absorption cross section of the 3% La, 2%Yb co-doped yttria transparent ceramic

Figure 9 presents the typical emission spectrum of 3% La, 2%Yb co-doped yttria transparent ceramic and its emission cross sections. There are two main emission peaks at 980 to 1100 nm wavelength in Figure 9(a), centered at 1030 nm and 1078 nm, respectively. According to F-L equation, the emission cross sections are shown in Figure 9(b).

$$\sigma_{em}(\lambda) = \frac{\lambda^5}{8\pi n^2 c\tau} \times \frac{I(\lambda)}{\int \lambda I(\lambda) d\lambda}$$

Where $\sigma_{em}(\lambda)$ is the emission cross section at the wavelength, λ is the wavelength, n is the refractive index, c is velocity of light, τ is the $^2F_{5/2}$ fluorescent lifetime, $I(\lambda)$ is the intensity of fluorescent emission at the wavelength.
The highest emission cross section is 2.20×10^{-20} cm^2 at 1030 nm. This value is suitable for realizing laser oscillation. The full width at half maximum (FWHM) of the 1030 nm emission peak is about 16 nm, and 13 nm at 1078 nm emission peak, which has the potential be used as femtosecond short pulse laser materials by the mode locking technologies. The fluorescent lifetime is 0.84 ms, which is higher than that of yttria crystal. The long fluorescent lifetimes are advantageous for energy storage and make it suitable for high power laser output.

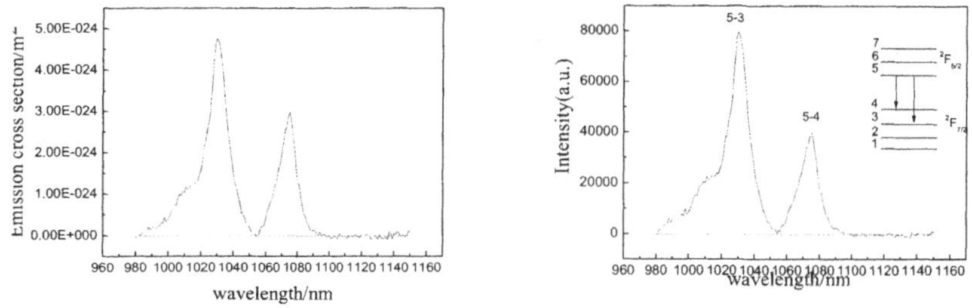

Figure.9. emission spectrum and the emission cross section of the 3% La, 2%Yb co-doped yttria transparent ceramic

CONCLUDING REMARKS

1. High transparency 3% La, 2%Yb co-doped yttria transparent ceramic was fabricated by vacuum sintering technologies. Nano-powders from ammonium precipitation method have high sintering activity. La is a good additive for enhancing the densification progress in the sintering of yttria. The grain size was 21±7 μm, no pores were observed.
2. The broad absorption band is benefit for absorbing the pump energy, and makes it a promising material for miniaturization of laser device applications. The emission cross section of 3% La, 2%Yb co-doped yttria transparent ceramic at 1030 nm is 2.20×10^{-20} cm^2, which is suitable for realizing laser oscillation.

ACKNOWLEDGEMENT

This work was supported by the Shanghai Science and Technology Committee (No. 07DJ14001) and the State Key Laboratory of High Performance Ceramics and Superfine Microstructures.

FOOTNOTES

*Author. Tel.: +86 21 5241 2606; fax: +86 21 5241 3122
E-mail address: dljiang@sunm.shcnc.ac.cn

REFERENCES

[1] A.Brenier, G. Boulon, Overview of the best Yb^{3+}-doped laser crystals, *Journal of Alloys and Compounds.* , 323, 210-3 (2001).

[2] J. Lu, J.H. Lu, T. Murai, K. Takaichi, T. Uematsu, K. Ueda, et al. Nd^{3+}: Y_2O_3 ceramic laser, *Japanese Journal of Applied Physics Part 2-Letters.* 40 (12A), L1277-L9 (2001).

[3] A. Ikesue, Y.L. Aung, Synthesis and performance of advanced ceramic lasers, *9th International Ceramic Processing Science Symposium*, Coral Spring, FL, Jan 08-11, 2006.

[4] V. Lupei, A. Lupei, A. Ikesue, Transparent polycrystalline ceramic laser materials. *Optical Materials,* 30, 1781-1786 (2008).

[5] J.Lu, K. Takaichi, T. Uematsu, A. Shirakawa, M. Musha, K. Ueda, et al, Yb^{3+}:Y_2O_3 ceramics - a novel solid-state laser material, *Japanese Journal of Applied Physics Part 2-Letters,* 41 (12A), L1373-L5 (2002).

[6] K. Takaichi, H. Yagi, J. Lu, J.F. Bisson, A. Shirakawa, K. Ueda, et al, Highly efficient continuous-wave operation at 1030 and 1075 nm wavelengths of LD-pumped Yb^{3+}:Y_2O_3 ceramic lasers. *Applied Physics Letters,* 84, 317-9 (2004).

[7] Y.F. Qi, Q.H. Lou, H.T. Zhu, J. Zhou, J.X. Dong, Y.R. Wei, et al., Optical characteristics of Yb : Y_2O_3 transparent ceramic. *Acta Physica Sinica*, 56, 2657-62 (2007).
[8] R.A. Lefever, J. Matsko, Transparent yttrium oxide ceramics, *Materials Research Bulletin*, **2**, 865-9 (1967).
[9] C. Greskovich, C.R. Oclair, Transparent, sintered $Y_{2-x}Sr_xO_{3-x/2}$ ceramics. *Advanced ceramic materials*, 1, 350-5 (1986).
[10] T. Ikegami, J.G. Li, T. Mori, Y. Moriyoshi, Fabrication of transparent yttria ceramics by the low-temperature synthesis of yttrium hydroxide, *Journal of the American Ceramic Society*, 85, 1725-9 (2002).
[11] A. Ikesue, K. Kamata, K. Yoshida, Effects of neodymium concentration on optical characteristics of polycrystalline Nd:YAG laser materials, *Journal of the American Ceramic Society*, 79, 1921-6 (1996).
[12] C. Greskovich, J.P. Chernoch, Improved Polycrystalline Ceramic Lasers, *Journal of Applied Physics*, 45, 4495-502 (1974).
[13] A.V. Belyakov, D.O. Lemeshev, E.S. Lukin, G.P. Val'nin, E.E. Grinberg, Optically transparent ceramics based on yttrium oxide using carbonate and alkoxy precursors, *Glass and Ceramics*, **63**: 262-4 (2006).
[14] W.H. Rhodes, Controlled Transient Solid Second-Phase Sintering of Yttria, *Journal of the American Ceramic Society*, 64: 13-9 (1981).
[15] C.G. Dou, Q.H. Yang, X.M. Hu, J. Xu, Cooperative up-conversion luminescence of ytterbium doped yttrium lanthanum oxide transparent ceramic, *Optics Communications*, 281, 692-5 (2008).
[16] Y. Rabinovitch, C. Bogicevic, F. Karolak, D. Tetard, H. Dammak, Freeze-dried nanometric neodymium-doped YAG powders for transparent ceramics, *Journal of Materials Processing Technology*, 199, 314-20 (2008).
[17] G. Boulon, V. Lupei, Energy transfer and cooperative processes in Yb^{3+}-doped cubic sesquioxide laser ceramics and crystals, *Journal of Luminescence*, 125 , 45-54 (2007).

MICROCRYSTALLIZATION IN OXYFLUORIDE Nd^{3+} DOPED GLASS DUE TO LASER IRRADIATION

S. González-Pérez[1], P. Haro-González[1] and I. R. Martín[1,2]
[1] Departamento de Física Fundamental y Experimental, Electrónica y Sistemas, University of La Laguna, (CP 38206) Tenerife, Spain.
[2] MALTA Consolider Team Departmento Fundamental y Experimental, Electrónica y Sistemas, Universidad de La Laguna.

ABSTRACT

A local crystalline formation in a Neodymium doped oxyfluoride glass has been obtained using laser irradiation. It has been studied the intense emission around 880 originated from the $^4F_{3/2}$ (Nd^{3+}) level when the glass matrix changes to a glass ceramic structure due to the irradiation of the laser beam. The emission spectra and the lifetime values obtained before and after the irradiation with 1400 mW reveals that the devitrification process made by the laser power beam has been successfully achieved. Moreover, the estimation of the temperature at the local position during irradiation agrees with the temperature of devitrification for these samples. The results confirm that micro-crystals of β-PbF_2 have been created by the laser action and also that the transition from glass to glass ceramic has been completed. These results are in agreement with the bulk spectroscopy measurements.

INTRODUCTION

The lanthanide ion Nd^{3+} is one of the most interesting luminescent ions to be used for their laser applications due to the $^4F_{3/2} \rightarrow {}^4I_{11/2}$ transition centred at about 1064 nm. Moreover in the ranges comprising UV and visible upconversion and optical amplification at 1.3-1.4 μm has been widely investigated according to a four-level scheme by direct excitation or by upconversion processes[1]. Rare-earth doped transparent oxyfluoride glass ceramics, in which rare-earth ions are selectively incorporated into the fluoride nano-crystals embedded among the oxide glassy matrix, possess great potential applications in the field of solid luminescence due to the combination of the advantages of both fluorides and oxides: low phonon energy environment of fluoride crystalline for luminescent ions, and desirable mechanical and chemical properties of oxide glasses[2,3]. This new material has attracted great attention in the continuous research for the novel photoelectric devices, and is usually fabricated by controlled crystallization of fluoride phase in oxide glassy matrix through thermal process using a furnace. Recently, laser irradiation to glass has received much attention as a new tool of micro-fabrication. Compared with current techniques such as photolithography and reactive ion etching, which requires numerous processing steps and fabrication masks, laser induced micro-fabrication has the advantage of being mask-less, allowing single step and very fast processing. The use of this technique has successful results with micro-fabrication of micro holes and micro patterning of refractive index changes, and many studies on laser induced structural modification in glass have been carried out so far[4-6].

In this letter, it has been report the modification of a localized area of an oxyfluoride glass doped with Nd^{3+} ions under continuous Ar laser irradiation. The local transformation into a glass ceramic structure has been made controlling the temperature of the devitrification process through the monotorization of the fluorescence intensity ratio and the time resolved fluorescence.

EXPERIMENTAL

The transparent glass sample was prepared starting with the following composition in mol%: 30 SiO_2, 15 Al_2O_3, 29 CdF_2, 22 PbF_2, 3 YF_3 and 1 NdF_3. The glasses were obtained by melting the components at 1050 °C for 2 hours and finally casting the melt into a slab on a stainless steel plate at room temperature. One of the glasses was heated at 470 °C for 36 h to obtain a transparent glass ceramic for comparison purposes of the irradiated zone.

Measurements of emission spectra of the glass in the wavelength range of 750-980 nm were taken using an Argon laser in multiline mode while increasing the average laser power and detecting with a high resolution spectrometer. The laser beam was focused onto the sample with a 20 mm focal length lens which provided a waist onto the sample of about 2 µm. Luminescence decay measurements were obtained by exciting the samples with light from a Q-switched Nd-YAG laser at 532 nm and the signal was acquired by a digital oscilloscope.

RESULTS

A local area of an oxyfluoride glass doped with 1 mol% of Nd^{3+} has been irradiated using an Argon laser increasing the laser power beam until 1400 mW. We pumped the oxyfluoride glass sample into the $^2G_{9/2}$ energy level and from this level the ions go down non-radiatively to the $^4F_{3/2}$ level decaying then to the fundamental state (Figure 1).

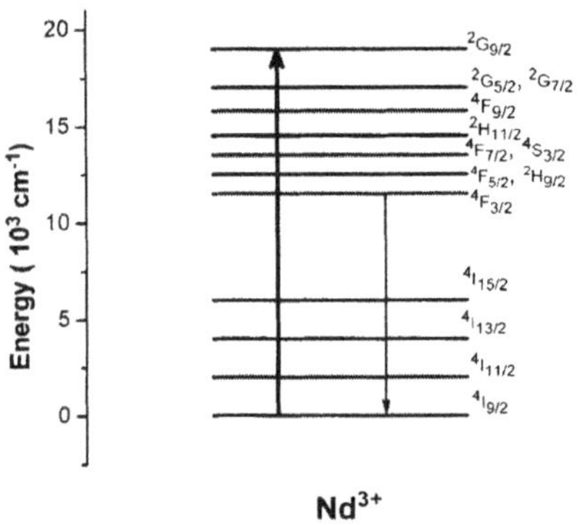

Figure 1. Energy level diagram of Nd^{3+} ions.

Figure 2 shows these emission spectra of the thermalized $^4F_{3/2}$ (Nd^{3+}) and $^4F_{5/2}$ (Nd^{3+}) levels where changes are clearly observed while the pump power increases. Analyzing the ratio between the $^4F_{3/2} \rightarrow {}^4I_{9/2}$ and $^4F_{5/2} \rightarrow {}^4I_{9/2}$ transitions it can be seen how the behaviour change up from 1000 mW of power intensity. From 1200 mW to 1400 mW the ratio changes from 0.21 to 0.8 (factor of 4). Also at this pump power intensity the emission spectrum shows a second peak around 890 nm characteristic of the glass ceramic samples obtained with the furnace method[7]. The emission spectrum of a glass ceramic bulk obtained using a furnace has been also plotted in fig. 2 for comparison. This result indicates that at this intensity power the structure has been modified due to the increase of the temperature.

The temperature of the sample at the irradiated spot for the different laser pump power intensities can be estimated within the theory of the occupation of energy levels in thermal equilibrium of Boltzmann[8]. Using a simple three-level system comprised of the $^4F_{5/2}$ (level 3), $^4F_{3/2}$ (level 2), and $^4I_{9/2}$(level 1), the relation between the ratio of the emission intensities of the excited levels with temperature can be described as follows,

$$\frac{I_3}{I_2} = \frac{p_3^r g_3 h\nu_3}{p_2^r g_2 h\nu_2} \exp\left(-\frac{E_{32}}{KT}\right) \qquad (1)$$

where K is the Boltzman constant, g_3, g_2 are the degeneracies, p^r_3 and p^r_2 are the spontaneous-emission rates of the $^4F_{5/2}$ and $^4F_{3/2}$ levels, respectively, E_{32} is the energy gap between the two excited levels and T the temperature of the sample.

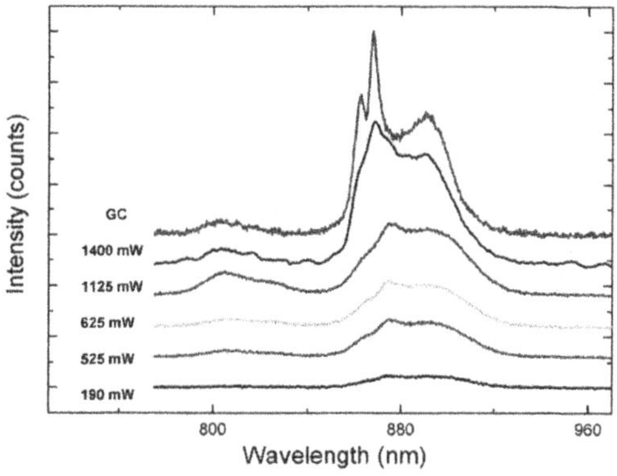

Figure 2. Emission fluorescence spectra obtained at different laser power intensities under excitation with Ar laser in a 1 mol% of Nd^{3+} oxyfluoride glass sample. Emission spectrum of a 1 mol% of Nd^{3+} oxyfluoride glass ceramic sample (GC) obtained using a furnace under Ar laser excitation.

The pre-exponential factor has been calculated with the emission spectrum under a low excitation power when the sample was at room temperature (RT). Therefore, the temperature of the irradiated spot can be calculated using expression (1) introducing the ratio between the intensities from the $^4F_{3/2}$ and $^4F_{5/2}$ thermalized levels to the $^4I_{9/2}$ level for different pump power values. The behaviour of the intensity ratio as function of the temperature is almost linear until the irradiation power intensity exceed 1000 mW. Under an irradiation at 1400 mW of intensity power a second peak appears around 890 nm characteristic of the glass ceramic samples. In this case, the calculated temperature of the irradiated spot increases close to 470 °C that is the temperature used for the thermal treatment of the glass ceramic samples[7]. In order to compare, the emission spectrum of the GC bulk sample obtained using a furnace is also presented in the same graph. As can be seen, the spectrum of the damaged zone is quite similar to the glass ceramic spectrum. Its looks more resolved and with the peaks centred at the same positions that the glass ceramic sample. These facts lead us to conclude that the laser irradiation induced a phase transition on the spot where it has passed from a glassy state to a crystalline phase similar to the glass ceramic bulk samples. Both methods, heating using a furnace or laser irradiation causes an increment on the temperature of the glasses close to its crystallization temperature, the first one at the whole sample and the second in a specific localized point. The heating of the glass in both cases causes a precipitation of fluoride microcrystals of βPbF_2 in the vitreous matrix where the Nd^{3+} ions are incorporated[7].

The temporal evolution from the $^4F_{3/2}$ metastable state of Nd^{3+} fluorescence (monitored at 880 nm) has been obtained under 532 nm excitation before and after the irradiation at 1400mW. As can be observed in Figure 3, its behaviour changes significantly.

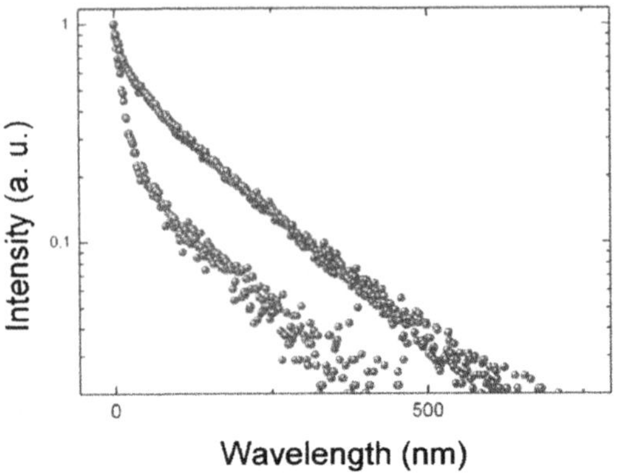

Figure 3. Fluorescence decay of the Nd^{3+} ($^4F_{3/2}$) ions before (blue circles) and after (red circles) irradiation of the 1 mol% of Nd^{3+} oxyfluoride glass sample.

Before the irradiation the curve shows a slow decay characteristic of the glass samples. However the curve shows a slightly non-exponential behaviour due to the energy transfer processes produced by the doping at 1 mol% of Nd^{3+}. In this case, due to the concentration of Nd^{3+} ions, energy transfer between the ions appears and the contributions of nonradiative processes to the decay of the $^4F_{3/2}$ level are significant. After the irradiation at 1400 mW the decay curve changes completely due to the contribution of two different kinds of centers, the fluoride microcrystals (fast initial decay) and the glassy phase (slower tail) of the glass ceramic sample created due to the irradiation[9]. To compare the results after the irradiation we have used a two exponential decay curve to obtain the average lifetime values of the fast decay contribution of both curves. The average lifetime value measured before and after the irradiation changes from 180 μs to 14.5 μs respectively. Notice that these results are in agreement with the lifetimes obtained in the bulk glass (150 μs) and glass ceramic samples (15 μs). Therefore this curve is the result of the contribution of two different kinds of Nd^{3+} centers. As the distances between RE ions are smaller in the microcrystals, mainly Nd^{3+} ions in the crystalline phase are responsible for the fast initial decay and those in the glassy phase give rise to the slower decay curve.

CONCLUSION

In conclusion, a crystalline environment has been created locally in the 1 mol% Nd^{3+} doped glass by laser irradiation at 1400 mW since the laser intensity is high enough to stimulate the formation of a glass ceramic structure. The rise of temperature in the irradiated zone due to the increasing laser power produces a local redistribution of the glass structure and leads to a permanent modification of the microstructure and its properties similarly to the devitrification of the glass samples obtained with the furnace thermal treatment. The optically active rare earth ions are majority hosted in precipitated fluoride microcrystals in both methods. Consequently shortening of distance between the ions inside the microcrystals induces efficient cross relaxation and energy transfer processes.

ACKNOWLEDGEMENTS

This work has been supported by the `Conserjería de Industria, Comercio y Nuevas Tecnologías del Gobierno de Canarias´ and by the National Projects MAT 2007-63319, MAT 2007-65990-C03-02 and MALTA Project.

REFERENCES

[1] W. Höland, G. Beall, Glass-Ceramic Technology, *Am. Ceram. Soc. Bull.*, 81, 1–50 (2002).
[2] M. J. Dejneka, Transparent oxyfluoride glass ceramics, *MRS Bull.*, 23, 57-62 (1998).
[3] M. Mortier, A. Bensalah, G. Dantelle, G. Patriarche, D. Vivien, Rare-earth doped oxyfluoride glass-ceramics and fluoride ceramics: Síntesis and optical properties, *Opt. Mat.*, 29, 1263-1270 (2005).
[4] Patterning of non-linear optical crystals in glass by laser-induced crystallization, *J. Am. Ceram. Soc.* 90, 699-705 (2007).
[5] T. Honma, Y. Benino, T. Fujiwara, Takayuki Komatsu, Line patterning with large refractive index changes in the deepinside of glass by nanosecond pulsed YAG laser irradiation, *Sol. Stat. Com.*, 135, 193-196 (2005).
[6] T. Honma, Y. Benino, T. Fujiwara, R. Sato, T. Komatsu, New optical nonlinear crystallized glasses and YAG laser-induced crystalline dot formation in rare-earth bismuth borate system, *Opt. Mat.*, 20, 27-33 (2002).
[7] M. Abril, J. Méndez-Ramos, I. R. Martín, U. R. Rodríguez-Mendoza, V. Lavín, A. Delgado-Torres, V. D. Rodríguez, P. Núñez and A. D. Lozano-Gorrín, Optical properties of Nd^{3+} ions in oxyfluoride glasses and glass ceramics comparing different preparation methods, *J. Appl. Phys.*, 95, 5271-5279 (2004).
[8] S. González-Pérez, I. R. Martín, D. Jaque and P. Haro-González, Growth of Nanocrystals in Nd^{3+}-Yb^{3+} codoped oxyfluoride glass by laser irradiation, *J. Nanosci. Nanotechnol.*, 8, 1-4 (2008).
[9] S. González-Pérez, I. R. Martín, F. Lahoz, P. Haro-González, J. Herreros, Local crystallization in an oxyfluoride glass doped with Er^{3+} ions using a continuous argon laser, *Appl. Phys. A*, 93, 983-986 (2008).

OPTICAL GAIN BY UPCONVERSION IN Tm-Yb OXYFLUORIDE GLASS CERAMIC

P. Haro-González[1,*], F. Lahoz[1], I. R. Martín[1], S. González-Pérez[1], N.E. Capuj[1]
[1]Dep. de Física Fundamental, Electrónica y Sistemas,
[2]Dep. de Física Básica Universidad de La Laguna, E-38206 La Laguna, Tenerife, Spain

ABSTRACT

Evidence of positive optical gain is observed in Tm^{3+}-Yb^{3+} codoped oxyfluoride glass ceramic in an upconversion pump and probe experiment. The $^1G_4$ level of the Tm^{3+} ions is populated by an upconversion mechanism under excitation of the Yb^{3+} ions at 975 nm with high power pulsed laser and give rise to an intense emission from the $^1G_4$ to the $^3F_4$ levels. The $^1G_4 \rightarrow {}^3F_4$ electronic transition is stimulated with a low signal at 650 nm as a probe.

Under this condition, we reach the population inversion necessary between the Tm^{3+} levels of the transition $^1G_4 \rightarrow {}^3F_4$ and we observed an increase of the emission intensity at the signal wavelength due to the stimulated emission. Transient positive optical gain around 4 cm^{-1} (~17 dB/cm) has been measured in Tm-Yb codoped oxyfluoride glass ceramic.

INTRODUCTION

Interest in obtaining solid state blue and green lasers that can be pumped with available red or near infrared semiconductor laser diodes has stimulated research activity in up-conversion materials. In this sense, oxyfluoride glass ceramics have shown to be an interesting matrix for rare earth (RE) ions due to the lower phonon energy in the fluoride environment which reduces the nonradiative decay rates and shortening of distances between the RE ions, which favours the energy transfer processes[1-2]. Recently, upconversion mechanisms have been reported in Tm^{3+} single doped oxyfluoride glass ceramics and in Tm^{3+}-Yb^{3+} codoped samples[2-3]. These emissions are drastically increased in the codoped samples.

In this context we are investigating the optical amplification properties of trivalent lanthanide ions doped into the oxyfluoride glass ceramics, with the purpose of extending the knowledge and the application perspective of this material[4]. In this paper we present the results of a series of pump and probe experiments carried out on oxyfluoride glass ceramics activated with Tm^{3+}-Yb^{3+}.

EXPERIMENTAL

The samples of this study have been reported in a previous work [2-3].

Measurements of optical amplification were carried out in a pump and probe experimental setup. The pump radiation was provided by an optical parametric oscillator (OPO) sintonized at 975nm with high energy pulses between 50 and 115 mJ/cm^2. The probe beam was obtained by a 1000W lamp, giving a signal power density of 195 $\mu W/cm^2$ at 650 nm.

The incidence of pump and probe beams were normal to the surface of the sample which was situated after a 1 mm diameter pinhole. A dichroic mirror was employed to align both laser beams. In order to cover only the whole area of the pinhole, the pump beam was focused a 20 cm focal length lens.

The detection chain was formed by a TRIAX-180 monochromator with 1 nm resolution and the output of the photomultiplier tube was registered by a digital oscilloscope TEKTRONIX-2430A for temporal analysis of the decay curves.

To determine the optical gain, two kinds of emission spectra were measured. In the first one, the pump and probe lasers were present simultaneously, while the probe laser was blocked for the second one. Both spectra were compared after subtract the continuous background due to the probe.

RESULT AND DISCUSSION

As mentioned in the experimental section, it is recorded two upconversion emission spectra: first with pump and probe laser present and then when the probe is blocked[4-6]. This result is given in Figure 1.

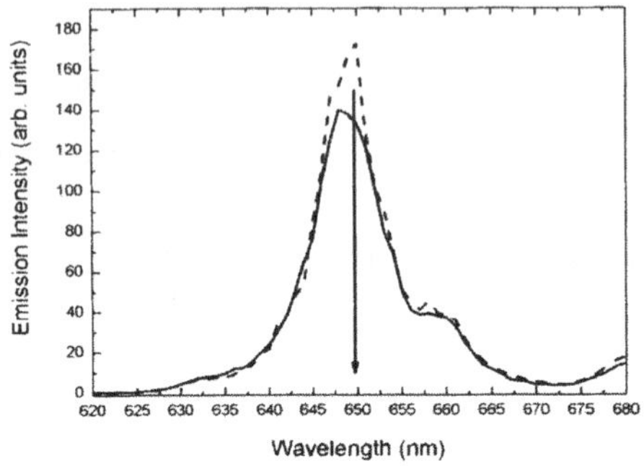

Figure 1- Emission spectrum for Tm^{3+}-Yb^{3+} codoped oxyfluoride glass ceramic associated to the $^1G_4 \rightarrow {}^3F_4$ electronic transition. The solid line gives the emission of the sample under pump excitation and the dashed line shows the emission under pump and probe excitation. The signal wavelength is indicated with an arrow

An increase of the detected intensity at the signal wavelength, 650 nm (indicated by a dotted arrow in the figure), can be clearly appreciated. This increment is due to the stimulated emission associated with the $^1G_4 \rightarrow {}^3F_4$ transition that occurs at the probe wavelength and it is the physical basis of signal amplification.

The Tm^{3+}: $^3H_6 \rightarrow {}^1G_4$ ground state absorption (GSA) is centred about 460 nm. In the pump and probe experiments, the high power pump pulses at 975 nm induce a nearly resonant GSA strongly populating the $^1G_4$ excited level by upconversion mechanism. In these conditions, a probe beam tuned at 650 nm can induce a relaxation process involving the stimulated emission of a photon at the same frequency. The energy levels diagram shows the proposed mechanisms to find optical amplification in this setup (see figure 2).

The optical gain resulting from the stimulated emission process can be determined by evaluating the signal enhancement (SE) when the probe beam passes through the crystal, defined as follows [4-7]:

$$SE = \frac{I_{pp} - I_p}{I_{probe}} \quad (1)$$

Where I_{pp} is the intensity detected at 650 nm in the direction of the probe beam coming out from the sample when it is irradiated simultaneously with the pump and the probe beams, I_p is the spontaneous emission intensity at the same wavelength when the probe is blocked before the sample, and I_{probe} is the intensity of the probe laser. The intensities I_p, I_{pp} and I_{probe} can be experimentally measured and the gain coefficient g is obtained as follow [4-6]:

$$SE = e^{gL} \quad (2)$$

The gain coefficient as a function of the pump energy density is given in figure 3. Continuous growth of the gain coefficient as a function of the pump power density can be observed in the figure and the maximum value has been observed for a pump energy density of 115 mJ/cm^2, corresponding to 4 cm^{-1} (~17 dB).

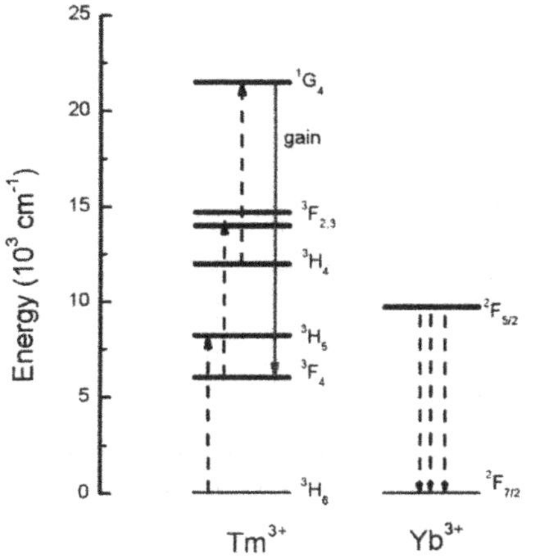

Figure 2. Energy level Scheme of Tm^{3+}-Yb^{3+} codoped oxyfluoride glass ceramic. The Dashed lines show the upconversion process to populate $^1G_4$ level of Tm^{3+} ions by energy transfer from Yb^{3+} ions which are produced by using OPO laser at 975 nm. Solid line shows the optical gain transition.

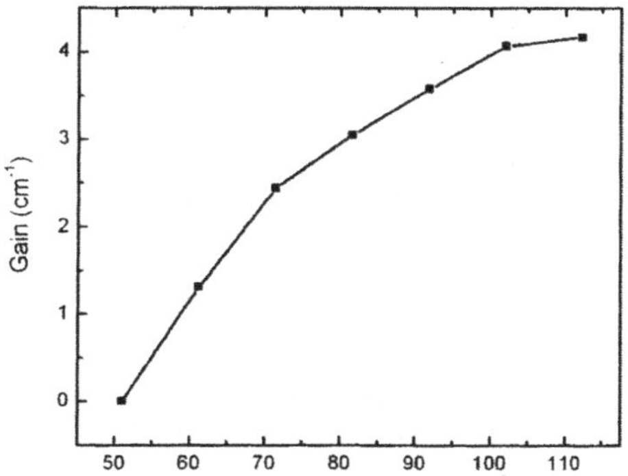

Figure 3. Optical gain as a function of the pump energy density from 50 to 115 mJ/cm^2 with a power probe density of 195 μW/cm^2

Fig. 4 shows the dependence of the g coefficient as a function of the time with pump energy densities of 61.2 mJ/cm^2. As we use a pulsed excitation source, I_p and I_{pp} are experimental curves that decay after the excitation pulse. The maximum of the detected intensity just after the pump pulse was used in order to obtain the g value given in figure 3.

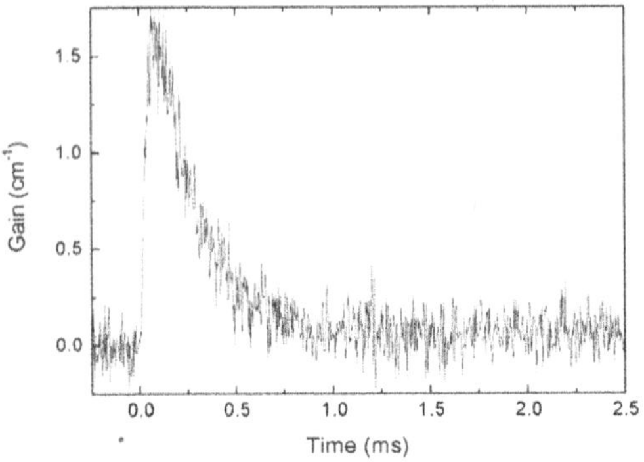

Figure 4. Optical gain as a function of time with pump energy densities of 61.2 mJ/cm^2 with a power probe density of 195 μW/cm^2

CONCLUSION

Positive optical gain at 650 nm has been observed in Tm^{3+}-Yb^{3+} codoped oxyfluoride glass ceramic in an upconversion pump and probe experiment. High power pump laser pulses at 975 nm were used to induce population inversion between the $^1G_4$ and $^3H_6$ levels whereas a 650 nm probe beam from a lamp was used to stimulate the $^1G_4 \rightarrow {}^3F_4$ emission transition. A maximum gain coefficient of about 4 cm^{-1} (~17 dB/cm) was measured for a pump energy density of 115 mJ/cm^2. These new results confirm the remarkable potentialities of in Tm^{3+}-Yb^{3+} codoped oxyfluoride glass ceramic for applications in the solid state laser and optical amplifier technologies.

ACKNOWLEDGMENTS

We would like to thank Comisión Interministerial de Ciencia y Technología (MAT 2007-63319 and MAT 2007-65990-C03-02) and SEGAI Grant for financial support.

REFERENCES

[1] F. Lahoz, J. M. Almenara, U. R. Rodríguez-Mendoza, I. R. Martín, and V. Lavín, Dopant partitioning influence on the near-infrared emisión of Tm^{3+} in oxyfluoride glass ceramics, *J. Appl. Phys.* **99**, 053103 (2006)

[2] F. Lahoz, I. R. Martin, J. Mendez-Ramos, P. Nuñez, Dopant distribution in a Tm^{3+}–Yb^{3+} codoped silica based glass ceramic: An infrared-laser induced upconversion study, *J. Chem. Phys.* **120**, 6180-90 (2004)

[3] J. Mendez-Ramos, F. Lahoz, I.R. Martin, A.B. Soria, A.D. Lozano-Gorrin and V.D. Rodriguez, Optical properties and upconversion in Yb^{3+}-Tm^{3+} co-doped oxyfluoride glasses and glass ceramics, *Molec. Phys.* **101,** 1057-65 (2003)

[4] F. Lahoz, S.E. Hernandez, N.E. Capuj and D. Navarro-Urrios, Optical amplification in Ho^{3+}-doped transparent oxyfluoride glass ceramics at 750 nm, *Appl. Phys. Lett.* **90**, 201117 (2007)

[5] P. Haro-González, F. Lahoz, I.R. Martín, S. González-Pérez, F. Rivera and N.E. Capuj, Optical amplification in Er^{3+}-doped fluoroindate glass at 840 nm and 1550 nm, *Opt. Mater.*

doi:10.1016/j.optmat.2008.10.014
[6] P. Haro-González, I. R. Martín, E. Cavalli, F. Lahoz, N.E. Capuj and S. González-Pérez, Optical amplification in Er^{3+}-doped transparent $Ba_2NaNb_5O_{15}$ single crystal at 850 nm, *Opt. Exp. In Press*
[7] D. Navarro-Urrios, M. Melchiorri, N. Daldosso, L. Pavesi, C. García, P. Pellegrino, B. Garrido, G. Pucker, F. Gourbilleau and R. Rizk, Optical losses and gain in silicon-rich silica waveguides containing Er ions, *J. Lumin.* **121**, 249-255 (2006)

FEMTOSECOND LASER MODIFICATION ON STRONTIUM BARIUM NIOBATE GLASSES DOPED WITH Er^{3+} IONS

P. Haro-González[1], I. R. Martín[1], S. González-Pérez[1], L. L. Martin[1], F. Lahoz[1], D. Puerto[2], J. Solís[2]
[1]Dep. de Física Fundamental, Electrónica y Sistemas, Universidad de La Laguna, E-38206 La Laguna, Tenerife, Spain
[2]Instituto de Optica, CSIC, Serrano 121, E-28006 Madrid, Spain

ABSTRACT

A localized modification of the optical properties in Er^{3+} doped Strontium Barium Niobate (SBN) glasses using a femtosecond laser were carried out. The samples were irradiated with a different number of pulses per spot at two laser fluences.

Confocal micro-luminescent has been developed to analyze the optical changes produced by exciting the sample with an argon laser. The emission of the Er^{3+}: $^4I_{11/2} \rightarrow {}^4I_{15/2}$ and $^4I_{13/2} \rightarrow {}^4I_{15/2}$ transitions are reported and shown structural differences after the femtosecond irradiation. The lifetimes of the levels involved in these transitions are measured inside and outside the damaged area. These measurements are compared with the bulk glass ceramic sample to estimate the optimal condition to produce nanocrystals in a localized area.

INTRODUCTION

The research in glass modification by use of short laser pulses is driven by scientific interest and their applications have been demonstrated for the formation of three dimensional optical memories[1,2] and multicolour images[3], the direct writing of waveguides[4-6], waveguide couplers and splitters[7,8], waveguide optical amplifier[9], and optical gratings[10,11].

The femtosecond laser has two apparent features compared with cw and long pulsed laser[12]:

- Elimination of the thermal effect due to the extremely short energy deposition time.
- Participation of various non-linear process enabled by high localization of laser photons in both time and spatial domains.

When a femtosecond pulse is focused in a transparent material, energy is deposited in a limited volume around the focus due to a combination of multiphonon absorption and avalanche ionization. The photogenerated hot electron plasma transfers its energy to the structure, producing high temperatures and pressures[13]. Structural modification, including crystallization can be induced by the excess energy released from the plasma into the surrounding media[14]. Since the electron plasma is generated only at the focal region where the peak power of the laser beam exceeds a threshold of the non-linear absorption, the crystallization process utilizing a femtosecond-pulsed laser is superior in terms of the internal modification of a transparent material such a glass, compared with crystallization which occurs via linear absorption or heat treatment[15].

In this work, erbium doped strontium barium niobate glasses have been irradiated with a femtosecond laser. The properties of these glasses and the changes induced by a cw laser have been studied in a previous paper[16-18]. Optical measurements show the changes of the local structure in a localized area after the irradiation and they are compared with the bulk glass ceramic sample obtained by thermal treatment with a furnace.

EXPERIMENTAL

The Er_2O_3-SrO-BaO-Nb_2O_5-B_2O_3 glasses were prepared using the melt quenching method[16] with the following composition in mol%: 5 Er_2O_3, 11.25 SrO, 11.25 BaO, 22.5 Nb_2O_5 and 50 B_2O_3.

Commercial powders of reagent grade were mixed and melted in a platinum crucible for 1 h in an electric furnace at 1400°C. The melt was poured between two iron plates and the thickness of the obtained sample was 1.6 mm. The glass ceramic was obtained by thermal treatment of the precursor glass at 620°C for 2 hours. It was used to compare with measurements in the locally damage zone by laser action.

A commercial chirped pulse amplification (CPA) Ti:sapphire laser system (Spectra Physics, Spitfire), providing linearly polarized pulses with pulse duration of 120 fs and at a wavelength of ≈ 800 nm, was used for irradiation. The laser pulse energy was measured by means of a pyroelectric detector (Ophir, PE-9). In the fs-irradiation set-up, the sample was placed at 36° of the normal incidence in the focal plane of a 15 cm lens resulting in an elliptical laser spot on the surface. The samples were irradiated at two laser fluences (2.6 and 5.6 J/cm^2) with different number of pulses (1-50 pulses)

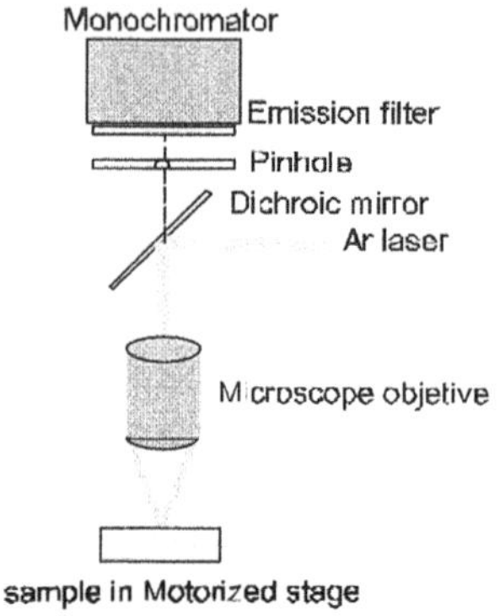

Figure 1. Confocal Micro-luminescence set up.

Confocal micro-luminescent was developed by using the following setup (see figure 1). The sample was situated in the focal plane of a 20X microscope objective (Mitutoyo, M-Plan NIR, numerical aperture (NA) = 0.26) in a motorized stage to displace at different positions. The detection system consists in TRIAX-180 monochromator with a resolution of 0.5 nm and detected with a photomultiplier tube.

The optical measurements were carried out inside and outside the irradiated area under Ar laser excitation for the emission spectra. The lifetimes involved in these transitions were obtained using a mechanical chopper and the signal was recorded by an oscilloscope.

RESULTS AND DISCUSSION

Localized zone of strontium barium niobate glass doped with Er^{3+} were irradiated by using a femtosecond laser at two different fluences and varying the number of pulses per spot. Inside these irradiated areas, the emission spectra of the Er^{3+}: $^4S_{3/2}$ ($^2H_{11/2}$) $\rightarrow$ $^4I_{13/2}$, $^4I_{11/2} \rightarrow {}^4I_{15/2}$ and $^4I_{13/2} \rightarrow {}^4I_{15/2}$ transitions were measured.

The results obtained with a laser fluence of 5.6 J/cm^2 are presented in the figure 2 for the Er^{3+}: $^4I_{11/2} \rightarrow {}^4I_{15/2}$ transition with different number of pulses per spot. As it can see in this figure, the emission corresponding to 1 and 2 pulses are less intense than the rest. These spectra are compared with the emission outside of the irradiated area and there are not differences between both. It can be conclude that with 1 and 2 pulses do not produce or induce any different structure in the sample.

The emission corresponding to 5 and 10 pulses present structural changes. The emission spectra more resolved and the peak at 1005 nm seem to confirm the presence of a new phase. This spectrum is compared with the glass and glass ceramic emission spectra in figure 3. In a previous work, it was found that a fraction of the Er^{3+} ions stay in the glass ceramic environment due to the ceramic process using a thermal treatment at 620°C with a furnace, whereas the rest remains in the glassy phase[16-18].

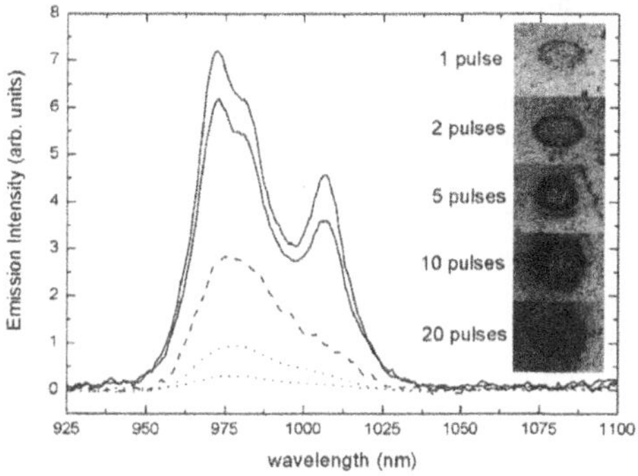

Figure 2. Confocal Micro-luminescence spectra under Ar laser excitation inside the irradiated area of the Er^{3+}: $^4I_{11/2} \rightarrow {}^4I_{15/2}$ transition with different number of pulses at the fluence of 5.6 J/cm^2. The solid lines show the spectra for 5 and 10 pulses, the dashed line for the 20 pulses and the dot line for the 1 and 2 pulses.

By comparing the spectra showed in the figure 3, the emission for the irradiated area is in a good agreement with the glass ceramic sample around the peak of 975 nm. On the other hand, the peak at 1005 nm is not clearly observed in the glass ceramic sample. This result could be explained in basis to radiative transfer processes which change the shape of the emission bands. The emission

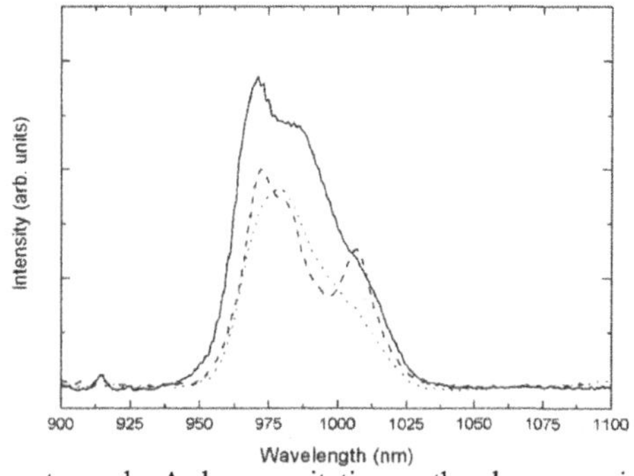

Figure 3. Emission spectra under Ar laser excitation on the glass ceramic samples (solid line), inside the irradiated area (dashed line) and on the glass sample (dotted line) of the Er^{3+}: $^4I_{11/2} \rightarrow {}^4I_{15/2}$ transition.

spectrum for the glass sample is presented to show the changes produced after the irradiation.
It is interesting to note that in Fig. 3 the maximum intensity is obtained after 5 pulses and with 10 pulses the emission decreases. Moreover, after 20 pulses the emission spectrum is less intense and does not present the same structure than 5 and 10 pulses. The appearance of this emission is like the spectrum obtained in the glass samples. It could be conclude that 20 pulses could produce an amorphization of the samples as has been shown in other matrix[19-20].
In the figure 4 are given the emission spectra of Er^{3+}: $^4S_{3/2}$ ($^2H_{11/2}$) → $^4I_{13/2}$ and $^4I_{11/2}$ → $^4I_{15/2}$ transitions for the laser fluence of 2.6 J/cm^2. In this case, the emission corresponding to the irradiated area with 1 and 2 pulses is nearly negligible in similar way to the emission in glass matrix, indicating that with this number of pulses there is not damage in the surface. As can be seen in this figure, the emission spectrum with 5 pulses shows structural differences whereas the emission with 10, 20 and 50 pulses are broader and without structure and less intense.

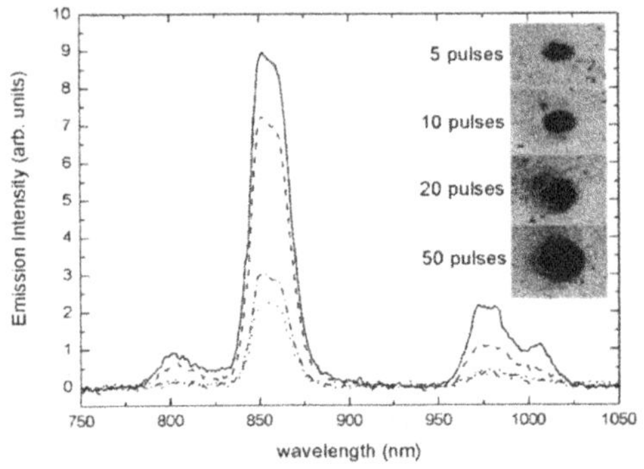

Figure 4. Confocal Micro-luminescence spectra under Ar laser excitation of the Er^{3+}: $^4I_{11/2}$ → $^4I_{15/2}$ and $^4S_{3/2}$ ($^2H_{11/2}$) → $^4I_{13/2}$ transition with different number of pulses at the fluence of 2.6 J/cm^2.The solid line shows the spectrum for 5, whereas the dashed lines correspond to the 10, 20 and 50 pulses.

The emission band at 1550 nm corresponding to the $^4I_{13/2}$ → $^4I_{15/2}$ transitions is measured in the irradiated area at 5 and 10 pulses with a fluence of 5.6 J/cm^2 which is given in the figure 5. There are differences between the emissions inside the irradiated area in comparison with the glass sample. The analysis of the presented results suggest that there are structural changes in the samples after the irradiation with laser fluence of 5.6 J/cm^2 with 5 and 10 pulses in similar way with the results obtained with a laser fluence of 2.6 J/cm^2. Less number of pulses per spot does not affect the structure of the sample and a higher number of pulses causes damages on the surface whereas does not induce the formation of new phases.
In order to investigate if the changes have been obtained due to a desvitrification process on the sample, the lifetime of the $^4I_{11/2}$ level has been obtained. The decay of the luminescence of the $^4I_{11/2}$ level is measured outside and inside of the damage area with 5 pulses at two laser fluences. Inside the irradiated area the decays curves show a double exponential character, while outside there is one single exponential. From the fits of these curves, are obtained the constant decays of the slow and fast components and the values are presented in table 1.

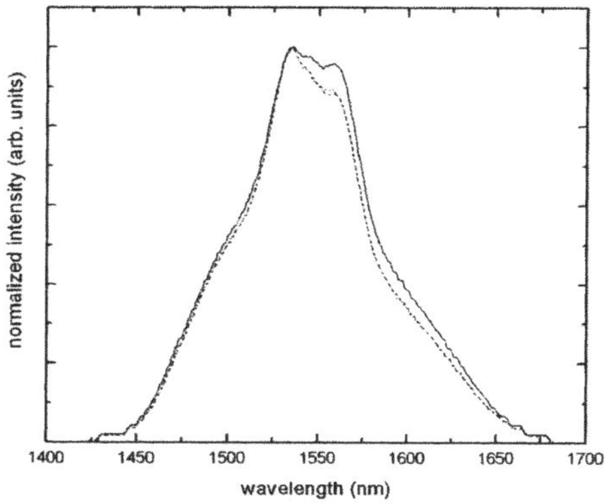

Figure 5. Confocal Micro-luminescence spectra under Ar laser excitation inside the irradiated area of the Er^{3+}: $^4I_{13/2} \rightarrow {}^4I_{15/2}$ transition with different number of pulses at the fluence of 5.57 J/cm^2. The solid line shows the spectra outside the irradiated area and the dashed and dot lines for the irradiated area at 10 and 5 pulses respectively.

Table I. Lifetime of The $^4I_{11/2}$ Level

Sample	Fast component (s)	Slow component (s)
Glass irradiated with 5 pulses at 5.57 J/cm^2	90	398
Glass irradiated with 5 pulses at 2.57 J/cm^2	70	420
Glass	60	
Glass Ceramic	4.5	289

The lifetime of the $^4I_{11/2}$ level for the glass ceramic sample, obtained by a thermal treatment[16] is shown in the table 1 to compare with the values obtained inside the irradiated area and in order to estimate the optimal condition to obtain glass ceramic environment under femtosecond laser excitation. In this sample were found the same double exponential behaviour. The fast component is attributed to the glassy phase of the samples and the slow component to the crystalline phase. The good agreement between the decay constant of the slow component with the lifetime of the glass ceramic samples seem to confirm the presence of a crystalline phase after the femtosecond irradiation. In against, the comparison between the fast components cannot be realized because the glassy phase is too fast for this experimental setup.

CONCLUSIONS

A localized modification of the optical properties in Er^{3+} doped Strontium Barium Niobate glasses using a femtosecond laser has been reported. The samples have been irradiated with a different number of pulses per spot at two laser fluences.

Confocal micro-luminescent measurements have been carried out to spatially select a position inside and outside the irradiated area and to analyze the optical changes produced by exciting the sample with an argon laser. The emission of the Er^{3+}: $^4I_{11/2} \rightarrow {}^4I_{15/2}$ and $^4I_{13/2} \rightarrow {}^4I_{15/2}$ transitions and the lifetimes of these levels have been reported and shown the structural differences after the femtosecond irradiation. As conclusion, using 5 or 10 pulses at two different fluences has been possible to modify the structure of the glass samples and the results seem to confirm the existence of crystalline environment for the Er^{3+} ions in the irradiated area.

ACKNOWLEDGMENTS

We would like to thank Comisión Interministerial de Ciencia y Technología (MAT 2007-63319 and MAT 2007-65990-C03-02) and SEGAI Grant for financial support.

REFERENCES

[1] E.N. Glezer, M. Milosavljevic, L. Huang, R.J. Finlay, T.-H. Her, J.P. Callan and E. Mazur, *Opt. Lett.,* **21** (1996) p. 2023

[2] J. Qiu, K. Miura and K. Hirao, *Jpn. J. Appl. Phys.* **37** (1998) p. 2263

[3] J. Qiu, K. Miura, H. Inouye, Y. Kondo, T. Mutsuyu and K. Hirao, *Appl. Phys. Lett.* **73** (1998) p. 1763

[4] Y. Kondo, T. Suzuki, H. Inouye, K. Miura, T. Mitsuyu and K. Hirao, *Jpn. J. Appl. Phys.* **37** (1998) p. L94

[5] M. Will, S. Nolte, B.N. Chichkov and A. Tunnermann, *Appl. Opt.* **41** (2002) p. 4360

[6] S. Taccheo, G. Della Valle, R. Osellame, G. Cerullo, N. Chiodo, P. Laporta and O. Svelto, *Opt. Lett.*, **29** (2004) p. 2626

[7] D.N. Fittinghoff, C.B. Schaffer, E. Mazur and J.A. Squier, *IEEE J. Sel. Top. Quantum Electron.* **7** (2001) p. 559

[8] K. Minoshima, A.M. Kowalevicz, I. Hartl, E.P. Ippen and J.G. Fujimoto, *Opt. Lett.* **26** (2001) p. 1516

[9] Y. Sikorski, A.A. Said, P. Bado, R. Maynard, C. Florea and K.A. Winick, *Electron. Lett.* **36** (2000) p. 226

[10] K. Miura, J. Qiu, T. Mitsuyu and K. Hirao, *Nucl. Instr. Methods Phys. Res. B* **141** (1998) p. 726

[11] Y. Kondo, K. Nouchi, T. Mitsuyu, M. Watanabe, P.G. Kazansky and K. Hirao, *Opt. Lett.* **24** (1999) p. 646.

[12] Yasuhiko Shimotsuma, Kazuyuki Hirao, Jianrong Qiu and Kiyotaka Miura, *J. Non-Cryst. Solids*, **352** (2006) p. 646

[13] R. Martínez-Vázquez, R. Osellame, G. Cerullo, P. Laporta, R. Ramponi, N. Chiodini, A. Paleari and G. Spinolo, *J. Non-Cryst. Solids*, **351** (2005) p. 1855

[14] S.K. Sundaram, C.B. Schaffer and E. Mazur, *Appl. Phys. A* **76** (2003) p. 379

[15] Yoshinori Yonesaki, Kiyotaka Miura, Ryuhei Araki, Koji Fujita, Kazuyuki Hirao, *J. Non-Cryst. Solids*, **351** (2005) p. 885

[16] P. Haro-González, F. Lahoz, J. González-Platas, J. M. Cáceres, S. González-Pérez, D. Marrero-López, N. Capuj and I. R. Martín, J. Lum., **128** (2008) p. 908

[17] P. Haro-González, I.R. Martín, E. Arbelo-Jorge, S. González-Pérez, J. M. Cáceres, P. Núñez, *J. Appl. Phys.*, **104** (2008) p. 013112

[18] P. Haro-González, S. González-Pérez, I.R. Martin, F. Lahoz, N.E. Capuj, D. Jaque, *Appl. Phys. A*, **93** (2008) p. 977–981

[19] P. Galinetto, D. Ballarini, D. Grando, G. Samoggia, *Appl. Surf. Sci.* **248** (2005) p. 291

[20] D.C. Deshpande, A.P. Malshe, E.A. Stach, V. Radmilovic, D. Alexander, D. Doerr, D. Hirt, J. *Appl. Phys.* **97** (2005) 74316

INFLUENCE OF POWDER TYPE ON THE DENSIFICATION OF TRANSPARENT $MgAl_2O_4$ SPINEL

Adrian Goldstein, Ayala Goldenberg, Meir Hefetz
Israel Ceramic and Silicate Institute
Haifa, 32000, Israel

ABSTRACT

The sinterability of two fine Mg-spinel powders, which have similar characteristics, was investigated. One was prepared by wet-chemistry followed by calcination, while the other – by gas-phase reaction. Marked differences have been observed regarding their densification. The gas-phase powder could be sintered to a bulk density level of 97.5 %TD at a temperature as low as 1280°C. HIPing at 1350°C/3h of the sintered disks generates transparent (total transmission of ~80% at λ=700 nm) specimens, having a grain size of ~1 μm. For the obtainment of transparent disks, the wet-chemistry powder requires sintering and HIPing at temperatures in the 1600 to 1700°C range. The reason(s) for the observed difference in sinterability are not clear. The more compact packing (possibly related to the spheroidal morphology of the basic particles), the lack of large (micron size) voids and the lower calcination temperature (a more disordered lattice, allowing faster diffusion) proper to the gas-phase powder may be among the factors conferring a higher sinterability.

INTRODUCTION

Transparent, polycrystalline $MgAl_2O_4$ spinel (TPSp) can be used in various applications, like: rare-earth and transition element ion host, windows transparent in the visible (VIS) or near-infrared (NIR) domains, etc. [1]

Fine, pure spinel (Sp) powders, necessary for full densification, are prepared by a wide variety of methods, like: coprecipitation, sol-gel processing, combustion synthesis, self-propagating high temperature synthesis, flame-spray pyrolysis, alkoxides decomposition under critical conditions, mechanosynthesis, e.g. [2-8] Many of these powders are quite similar regarding characteristics like basic particles and agglomerates size distribution, specific surface area and impurities content, and yet exhibit markedly different levels of sinterability. For the establishment of economically viable TPSp fabrication technologies, the selection of the most sinterable powder(s) is a critical issue. Despite considerable progress made in the theoretical treatment of the sintering process [9-10], ranking of the powders as to their sinterability still requires comparison of sintering experiment results.

In this work the sintering behavior of two, apparently similar, Sp powders - prepared by, respectively, wet-chemistry and gas-phase reactions- is compared. Possible reasons for the observed differences in sinterability are discussed.

EXPERIMENTAL

Two commercial stoichiometric Sp powders (Mg:Al≅1.0) were used. The first (labeled here: N1) is derived from Al and Mg hydrated sulfate salts, using solution chemistry; it is supplied by Baikowski (La Balme de Silligny, France). The calcination temperature is of ~1100°C. The second, produced by Nanocerox (Ann Arbor, MI, USA) was synthesized by flame-spray pyrolysis from a double Al-Mg alkoxide precursor. The synthesis product is calcined at 650°C.

The as-received powders were pelletized and formed into disk shape (diameter, ϕ≅20 mm and thickness, t=3-6 mm). The powder processing and disk forming specifics constitute ICSI proprietary information, which cannot be disclosed here. Sintering was effectuated in air (AS), using a resistive furnace. The final densification step was performed by the aid of an in-house built hot isostatic press (HIP) in Ar at a pressure of 200 MPa and various temperatures in the 1300 to 1700°C range. The powder characteristics measured included their basic crystallites size and shape by TEM (model 8010F, FEG of JEOL, Tokyo, Japan), agglomerates size distribution (as present in a diluted suspension in ethylene glycol after ultrasonication; with a model LM20 laser scattering sizer produced by NanoSight, Salisbury, UK), specific surface area (nitrogen adsorption curves; BET approximation), phase composition (XRD; model APD2000, Italstructures, Riva del Garda, Italy) and impurity content (glow discharge mass spectroscopy). The green body's bulk density (BD_g) and pores size distribution (mercury intrusion; Macropore 120 porosimeter of Carlo Erba, Torino, Italy) were also measured. The sintered specimens characteristics examined include: the bulk density (BD_A after sintering and BD_H after HIPing; Archimedes technique), the phase composition, Vickers hardness (HV5), grain size (SEM; model Quanta 200, FEI, Eindhoven, NL) and transmission (VIS+NIR; model V-570 of JASCO, Osaka, Japan). For selected specimens the real in-line transmission (RIT) was also measured. This was done with a He-Ne red laser (λ=635 nm). The sensor was located at 80 cm from the source and 73 cm distance from the specimen; an iris with an opening equal to the cross section of the laser beam (2.5 mm) was located between the specimen and the sensor.

RESULTS AND DISCUSSION

The basic crystallites morphology and size is illustrated in Fig. 1 for the two materials. The crystallites are, in both cases, faceted, with a size in the 20 to 70 nm range. Individual crystallites are rare. Most of them form primary clusters in which the basic components are joined by partial sintering. These clusters tend to have a more branch-like shape in the case of the N1 material, being spheroidal in shape in the case of the N2 powder.

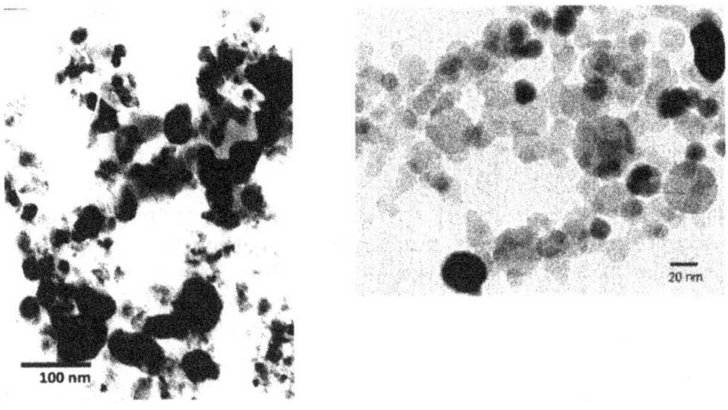

N1 N2

Figure 1. Morphology, size and clustering pattern of Sp powders basic crystallites.

The size distribution of the powder particles (crystallite clusters which are kinetically independent units in a suspension), and other characteristics, are given in Table I. The green body pores size distribution is given in Fig. 2. The sintering shrinkage curves (5°C/min) are shown in Fig. 3, while the evolution of BD_A as a function of the sintering temperature (3 h dwell at peak temperature) is given in Table II.

Table I. Characteristics of the N1 and N2 Powders

Material	Specific surface area (m^2/g)	Size distribution of particles (nm)		Main impurities (ppm)			
N1	30	Mode	45	S	200,	Ca	80
		D_{50}	100	Na	50,	K	35
		D_{90}	300	Fe	20,	Si	20
N2	32	Mode	70	Cl	25,	Si	20
		D_{50}	120	Fe	10,	S	10
		D_{90}	230	Ca	10,	Na	6

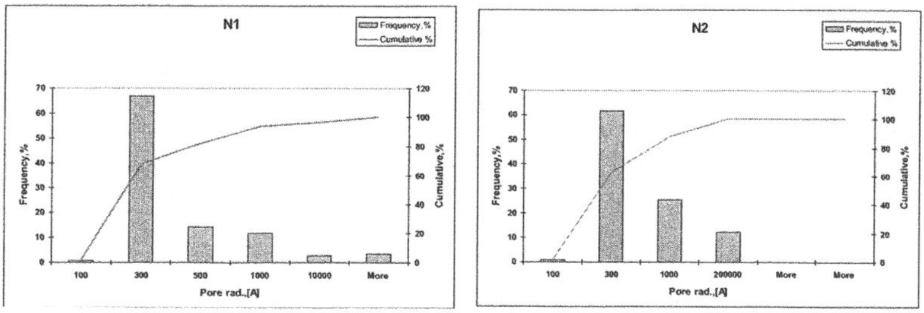

Figure 2. Pore size distribution before sintering (Hg intrusion).

(Voids >>10000 Å are, probably, surface layer, low frequency defects)

As Fig. 3 and Table II show, the N2 powder densifies at significantly lower temperatures than N1. Sintering shrinkage starts, for both materials, around 1100°C, but the densification rate and maximal BD_A values attained differ. The N2 reaches, after 3 h at 1370°C, a $BD_A \approx TD$ (Δl ~17%), while the maximal density attainable by the N1 is of only 99.2 %TD at 1630°C (Δl ~22%). The pores size distribution of the green bodies generated by the two powders is relatively similar. The N1, though, includes some (~3%) micron size voids not present in N2. Such voids may retard densification. The N2 material is also arranged in a more compact way (BD_g of N2 is 2.05 g/cm^3, against 1.77 g/cm^3 for the N1). The difference in BD_g may cause some difference in sinterabililty too; it also explains the higher shrinkage of N1. It is possible that the lower calcination temperature (650°C) of the N2 generates a more disordered lattice (which enhances diffusion) than that exhibited by N1 (calcination at ~1100°C).

Table II Fired State Bulk Density as A Function of Sintering Temperature (3h dwell)

	BD_g (g/cm^3)	BD_A (TD=3.578 g/cm^3) (g/cm^3) 1250°C	1280°C	1320°C	1370°C	1400°C	1500°C	1550°C	1630°C
N1	1.77	3.23	3.50	3.57	~3.58	~3.58	~3.58		
N2	2.05		2.40	2.85	2.95	3.15	3.48	3.51	3.55

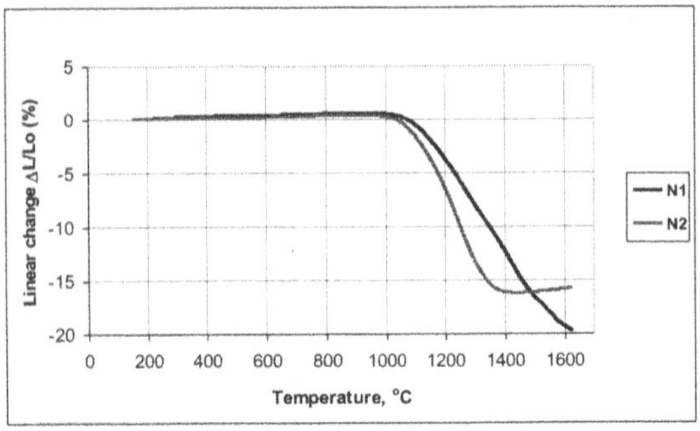

Figure 3 Sintering shrinkage of Sp powder compacts (RT-1620°C).

HIPing at 1350°C/3h is able to transform N2 derived sintered disks, which have a $BD_A \geq 97.5$ %TD, into transparent pieces like those shown in Fig. 4. Obtainment of reasonably transparent parts from N1 sintered disks requires HIPing at 1700°C. The average grain size of the N2 parts is ~1 μm, while that of the N2 disks is ~100 μm.

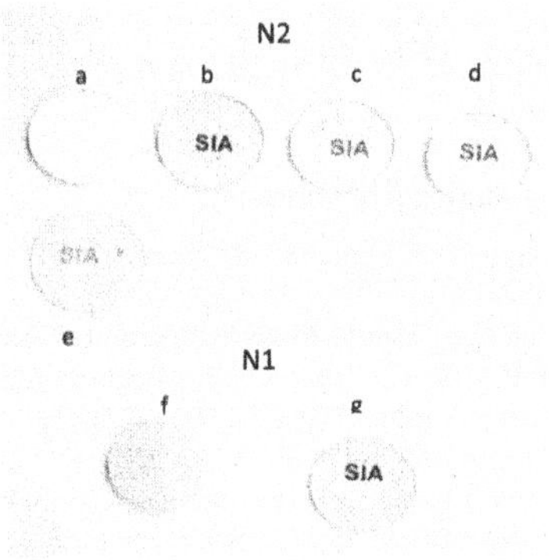

a – AS 1280°C/3h
b – AS 1280°C/3h+HIP 1350°C/3h
c – AS 1320°C/3h+HIP 1350°C/3h
d – AS 1400°C/3h+HIP 1350°C/3h
e – AS 1370°C/3h
f – AS 1630°C/3h
g – AS 1630°C/3h+HIP 1700°C/3h

Figure. 4 Visual aspect of densified Sp disks.

The total forward transmission (TFT) – which includes the RIT and the forward scattered radiation – of HIPed specimens is given in Fig. 5. The transmission of an as-sintered disk (=specimen "e" of Fig. 4) derived from N2 powder is also given. The real transmission is a bit higher than shown in Fig. 5; reflection, by the two major surfaces, is higher than theoretical, owing to the imperfect polishing level that could be achieved.

The level of sinterability exhibited by the N2 powder, when a suitable green-body forming is applied, is exceptionally high. In prior experiments, pressureless sintering (atmospheres like 90% H_2+10% O_2 or Ar) produced Sp disks (t≅1 mm) having TFT of ~40% (λ=700 nm) after firing in the 1800-1850°C range [11,12]. Here, levels of TFT≅45% are attained after AS at 1370°C (see Fig. 5 curve 1). While the level of densification attained by the N2 powder compacts, after AS, is impressive, the level of transmission it allows is modest. These results confirm, once again, that without application of pressure, TPSp usable in practical applications can not be obtained. In the case of most of the powders discussed by the literature, when AS+HIP procedures are applied (without sintering aids), temperatures ≥1550°C (usually ≥1650°C) are required to attain TFT levels ≥80% [1]. The lowest sintering temperature reported yet (in the absence of sintering aids) for the obtainment of good quality TPSp, is of ~1450°C (for both the AS and HIP stages) [13]. Here, TFT levels of ~80% (λ=700 nm) have been attained by combining an AS stage at 1280°C with HIPing at 1350°C.

The RIT value of the N2 disk (AS 1280°C+HIP 1350°C), at 76%, is close to the TFT, indicting that the part has a very low level of residual porosity (much less than 0.01%), offering a quite good image clarity (level of image clarity can not be estimated from Fig. 4). The transparency of the N1 based parts (RIT=65%) is lower (Table III).

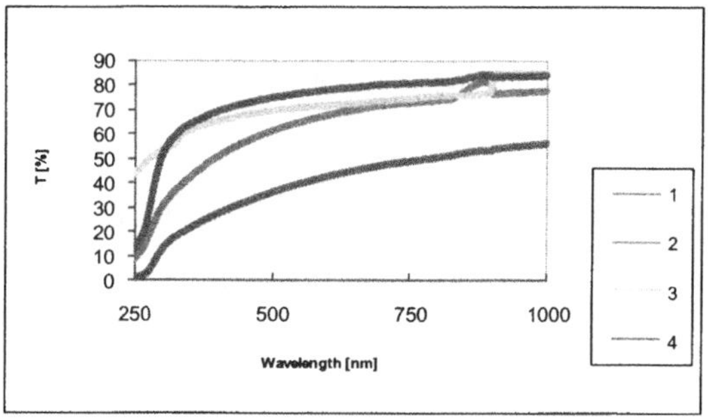

1 – N2; AS 1370°C/3h; t=2mm

2 – N2; AS 1320°C/3h+ HIP 1350°C/3h; t=3.5mm

3 – N1; AS 1630°C/3h+HIP 1700°C/3h; t=2mm

4 – N2; AS 1280°C/3h+HIP 1350°C/3h; t=2mm (Medium quality polishing)

Figure 5. TFT of densified TPSp disks in the 250-1000 nm range.

Table III Characteristics of Sintered Specimens

Material	Fabrication route	Characteristics				
		BD_H (g/cm^3)	Average grain size (μm)	HV5 (GPa)	Transmission (635 nm)	
					TFT (%)	RIT (%)
N2	AS 1280°C/3h+HIP 1350°C/3h	~3.58	~1.0	13.4-14.0	78	76
N1	AS 1630°C/3h+HIP 1700°C/3h	~3.58	~100	12.8-13.6	73	65

For both materials the hardness varies from point to point. The marked increase of HV observed in [13], caused by grain size reduction, was not obtained here, where only a slight effect is seen (Table III).

CONCLUSIONS

Sp powders having similar characteristics differ markedly as to their sinterability. The reasons for the different behavior are not known with certitude yet. The FSP powder can be densified after 3h of sintering in air at 1370°C to a $BD_A \approx$TD. Such highly translucent specimens and, in fact, all specimens which attain a BD_A ≥3.50 g/cm^3 (97.5 %TD) can be converted into highly transparent ones (TFT%≅80 at λ=700 nm) by low temperature, 1350°C, HIPing. The level of transparency attainable before HIP is not high enough for practical applications.

To the best of our knowledge, the firing conditions required for the full densification of the N2 material (level allowing high VIS transparency) are the less energetically reported yet. In such conditions a fine microstructure is also achieved.

REFERENCES

[1] D.C. Harris, History of Development of Polycrystalline Optical Spinel in the US, *Proc. SPIE*, **5786**, 1-22 (2005).

[2] A. Gusmano et al., The Mechanism of Spinel Formation from Thermal Decomposition of Coprecipitated Hydroxides, *J. Eur. Ceram. Soc.*, **7**, 31 (1991).

[3] C.R. Bickmore, K. Waldner, D.R. Treadwell and R.M. Laine, Ultrafine Spinel Powders by Flame Spray Pyrolysis of a Magnesium Aluminum Double Alkoxide, *J. Am. Ceram. Soc.*, **79**, 1419-1423 (1996).

[4] A. Goldstein, L. Geifman and S. Bar-Ziv, Susceptor Assisted MW Sintering of $MgAl_2O_4$ Powder at 2.45 GHz, *J. Mat. Sci. Lett.*, **17(12)**, 977-979 (1998).

[5] J-G Li, T. Ikegami, J.H. Lee and Y. Yajima, A Wet Chemical Process Yielding Reactive $MgAl_2O_4$ Powder, *Ceram. Int.*, **27**, 481-489 (2001).

[6] H. Zang, X. Jia, Z. Liu and Z. Li, The Low Temperature Preparation of Nanocrystalline $MgAl_2O_4$ Spinel by Citrate Sol-Gel Process, *Mater. Lett.*, **58**, 1625-1628 (2004).

[7] S.K. Behera, P. Barpand and S.K. Pratihar, Synthesis of Magnesium-Aluminum Spinel from Autoignition of Nitrate Gel", *Mat. Lett.*, **58**, 1451-1455 (2004).

[8] R. Cook, M. Kochis, I. Reimanis and H.J. Kleebe, A New Powder Production Route for Transparent Windows: Powder Synthesis and Window Properties, *Proc. SPIE*, **5786**, 41-47 (2005).

[9] Y.M. Chiang, D.P. Birnie III, D.W. Kingery, Physical Ceramics, J. Wiley & Sons, N.Y. (1997).

[10] R.M. German, Sintering Theory and Practice, J. Wiley & Sons, N.Y. (1996).

[11] R.J. Bratton, Translucent Sintered $MgAl_2O_4$, *J. Am. Ceram. Soc.*, **57(7)**, 283-285 (1974).

[12] J. Cheng et al, Fabricating Transparent Ceramics by MW Sintering, *Am. Ceram. Soc. Bull.*, **79(9)**, 71-74 (2000)..

[13] A. Krell, T. Hutzler, J. Klimke, Transparent Ceramics for Structural Applications, *Ber. DKG*, **84(4)**, E41-E50 (2007).

SINTERING EVOLUTION OF NOVEL Nd:YAG POWDERS WITH TEOS AS ADDITIVE

Ruixiao Fang[1], Tiecheng Lu[1,2], Nian Wei[1], Yongchao Li[1], Wei Zhang[1], Benyuan Ma[1]
[1] Department of Physics and Key Laboratory for Radiation Physics and Technology of Ministry of Education, Sichuan University, Chengdu, 610064, P. R. China
[2] International Center for Material Physics, Chinese Academy of Sciences, Shenyang, 110015, P. R. China

ABSTRACT

Nd:YAG nanopowders synthesized via a modified co-precipitation method, named as alcohol-water solvent co-precipitation, were used to sinter 2.0 at% Nd:YAG transparent ceramics with 0 wt%, 0.3 wt%, 0.5 wt% and 0.7 wt% TEOS as additive, respectively. The results showed that the optimal amount of TEOS was 0.5wt% for the fabrication of transparent Nd:YAG ceramics. 2.0 at% Nd:YAG ceramics with 0.5 wt% TEOS as additive were sintered from 1550 to 1750 °C for 5h. Densification and microstructure evolution of Nd:YAG transparent ceramics were investigated. The ceramics showed no porosity at 1550°C and obvious grain growth occurred when sintering temperature was increased to ~1650°C. Nearly pore-free microstructured Nd:YAG transparent ceramic with average grain size of ~5μm was fabricated by vacuum sintering at 1750°C for 5h using 0.5wt% TEOS as additive.

INTRODUCTION

Neodymium doped YAG single crystals are widely used as laser host materials in various solid state laser. However, it is hard to fabricate in a large size and with high doping concentration. The first Nd^{3+}:$Y_3Al_5O_{12}$ transparent laser ceramic fabricated using a solid-state reaction among oxide powders and vacuum sintering was reported by Ikesue et al.[1] Since then, YAG transparent ceramic has received much attention as laser host material because of its several advantages, such as low cost, short preparation time, high doping concentration, homogeneity, ease of mass production, etc.[2-5] However, highly reactive powders are crucial for the fabrication of transparent YAG ceramics. In order to achieve fine optical transmittance in YAG ceramics, tetraethyl orthosilicate (TEOS) is normally employed to prevent abnormal grain growth and remove the pores in YAG ceramics sintering.

In this paper, novel Nd:YAG powders were synthesized via a modified co-precipitation method with alcohol-water as the precipitation solvent.[6] Nd:YAG transparent ceramics were fabricated by vacuum sintering using as-prepared ultrafine powders with 0wt%, 0.3wt%, 0.5wt% and 0.7wt% TEOS as additive. The densification and the microstructure evolution of the ceramics were mainly studied.

EXPERIMENTAL

Ammonium aluminum sulfate ($NH_4Al(SO_4)_2 \cdot 12H_2O$, 99.99%), neodymium nitrate ($Nd(NO_3)_3 \cdot 6H_2O$, 99.99%), and yttrium nitrate ($Y(NO_3)_3 \cdot 6H_2O$, 99.99%) were used as starting materials. The ammonium hydrogen carbonate (AHC) is of analytical grade as precipitate. The stock solution of mother salts was made by dissolving $Nd(NO_3)_3 \cdot 5H_2O$, $Y(NO_3)_3 \cdot 6H_2O$ and $NH_4Al(SO_4)_2 \cdot 12H_2O$ into distilled water according to the stoichiometry of 2.0at% Nd:YAG. Concentration of the stock solution was 0.12 M for Al^{3+}. The precipitant solution was prepared as 1 M by dissolving Ammonium hydrogen carbonate into alcohol-water solvents. The mixed solution was then added into the precipitant solution at a speed of 2.5 ml min^{-1} under vigorous stirring at 15 °C. The precipitate slurry was stirred for about 1h after the titration to make the reaction proceed sufficiently. The resultant suspensions, after aging 20h, were filtered and washed with distilled water and alcohol, respectively. Loose precursors were obtained after drying the precipitate at 80 °C for 24h and then calcined at 1100 °C for 2h. Different amount of TEOS was added in the calcined Nd:YAG powders. The mixtures were milled in ethanol for 12h and then calcined at 800°C in oxygen to remove residual organic materials. Calcined powders were uniaxially pressed into pellets and then isostatically pressed at 250 MPa. The compacted pellets were sintered at the temperature from 1450 to 1800 °C for 5h in a vacuum-sintering furnace. After sintering, the specimens were annealed in the oxygen atmosphere at 1400 °C for 10h.

Powder morphology was investigated using a transmission electron microscope (TEM, Model JEM-100CXII). Crystallite size of the powders and grain size of Nd:YAG ceramics calcined at different temperatures were calculated by X-ray diffraction (XRD, model D/maxrA, using nickel-filtered Cu-Kα radiation) patterns from the Scherrer's equation. Microstructures of the fractured and the thermal etched mirror-polished surfaces of Nd:YAG specimens were observed by scanning electron microscopy (SEM, Model S-4800). Densities of the samples were measured by the Archimedes draining method.

RESULTS AND DISCUSSION

Fig.1 shows TEM image of the Nd:YAG powder calcined at 1100 °C for 2h via alcohol-water solvent co-precipitation, using ammonium hydrogen carbonate as the precipitant and alcohol-water as the precipitation solvent. The powder is well-crystallized with an average size of about 40nm, and little agglomeration exits. Alcohol used in the solvent can play the same role as the surfacants and contribute to the well-dispersion of the powder.

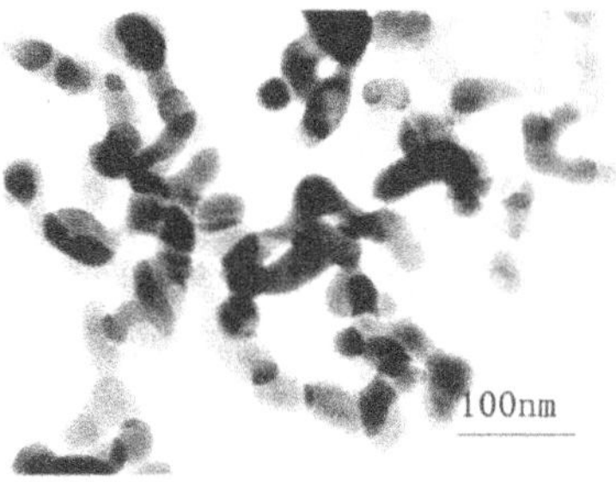

Fig.1. TEM image of the Nd:YAG powder calcined at 1100°C for 2h

Fig.2 exhibits the crystallite size of Nd:YAG powders as a function of calcining temperature. It shows that the particle size increases with increasing temperature. The crystalline grows rapidly as the temperature increasing.

Fig. 3 shows the SEM micrographs of the fractured surfaces of 2.0at% Nd:YAG samples sintered at 1750°C for 5h with different amount of TEOS as additive. The fracture mode of the sample with 0.5wt% TEOS as additive is intergranular . However, the fracture mode of the samples with 0.0wt% and 0.3wt% TEOS as additive are mainly transgranular. When the amount of TEOS is increased to 0.7wt% , the fracture mode is almost transgranular. It can be seen that there is no obvious difference in average grain size with no more than 0.5wt% TEOS as additive. However, nearly pore-free microstructure and uniform grains are just obvious in Fig. 3(c). When the amount of additive is increased to 0.7wt% , the grain size decrease with the increase of TEOS content and the fractured surfaces is impure which can be explained by the excess TEOS. In our experiments, TEOS is used as sintering aid, which decomposes to form SiO_2. A liquid phase begin to form around 1400℃ in SiO_2 doped YAG.[7] According to Lay,[8] if the grains are separated by a liquid phase, the sintering process is enhanced since the material transport involved in grain boundary migration is diffusion through the liquid phase. However, there exists an optimum amount of additive in Nd:YAG ceramics. Inadequate TEOS leads to abnormal grain growth, and numerous pores will be entrapped in the grains. Excess TEOS can result in the formation of impurity phase which is detrimental to optical transmittance. Thus, the optimal amount of TEOS is 0.5wt% for the fabrication of transparent Nd:YAG ceramics.

Fig.4. shows surface morphologies of thermally etched 2.0at% Nd:YAG ceramics, with 0.5wt%

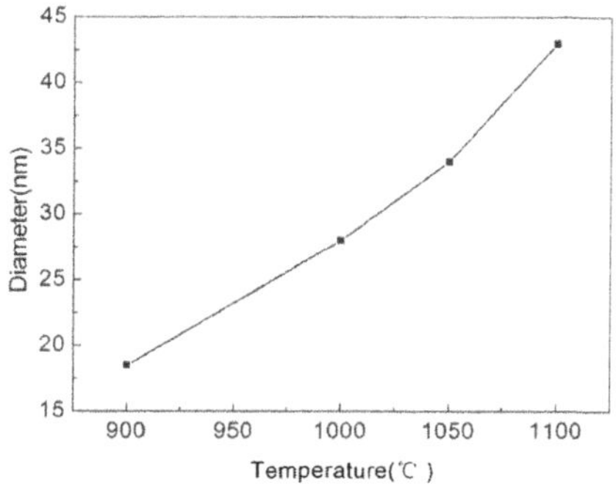

Fig. 2. Crystallite size of the Nd:YAG powder as a function of calcining temperature.

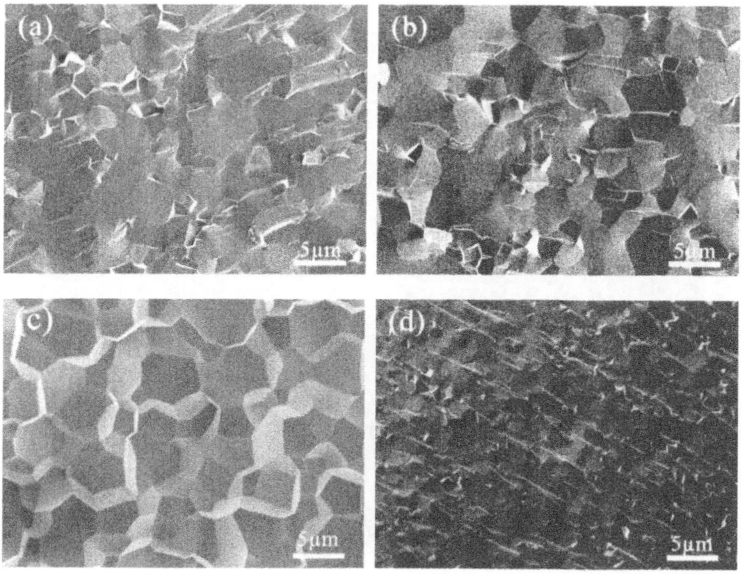

Fig. 3. SEM micrographs of the fractured surfaces of 2.0at% Nd:YAG samples with 0.0wt%(a), 0.3wt% (b), 0.5wt% (c) and 0.7wt% (d) TEOS as additive, respectively

TEOS as additive, sintered at 1550-1750 °C for 5h. It can be seen that grain size increase with increasing sintering temperature. Obvious grain growth occurs when sintering temperature is ~1650 °C. Abnormal grain growth is not observed for the sample sintered at 1750 °C because of an optimum amount of TEOS. The grain boundary of the samples narrows with the increase of sintering temperature. Nearly pore-free microstructure Nd:YAG transparent ceramics with average particle size of ~5μm were fabricated by sintering at 1750 °C for 5h. Note that the sample shows no porosity even at 1550 °C with the average grain size of ~1.5 μm. The temperature obtained no porosity Nd:YAG sample in this paper is lower than that in other reports.[9,10] This phenomenon can be attributed to powder characteristics. In this paper, novel Nd:YAG powders were synthesized via a modified co-precipitation method with alcohol-water as the precipitation solvent. The powders show good dispersity and sinterability. It is a promising way to fabricate Nd:YAG transparent ceramics using the powders as raw materials.

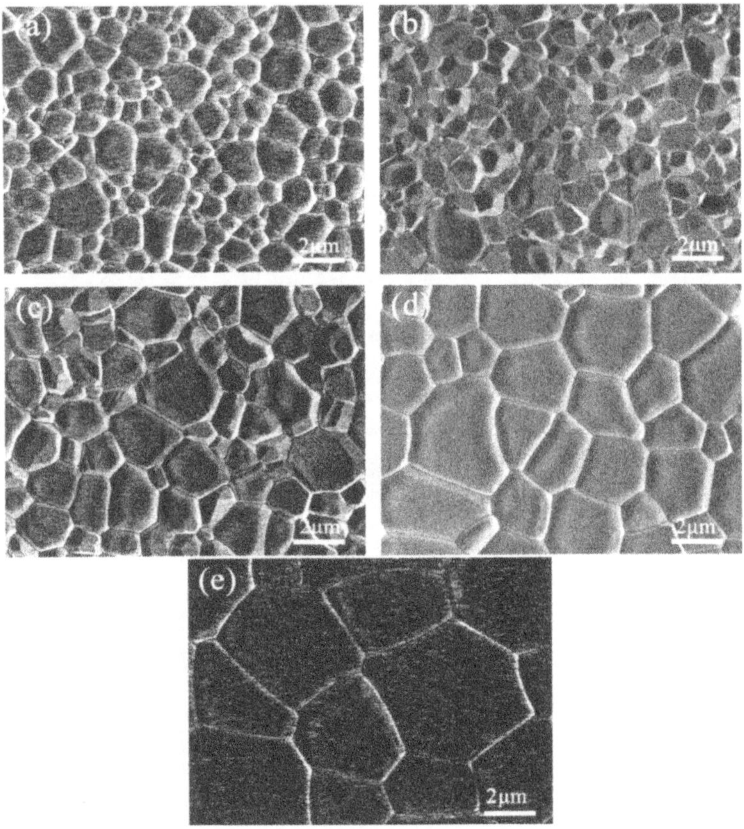

Fig.4. Surface morphologies of thermally etched 2.0 at% Nd:YAG ceramics, with 0.5 wt% TEOS as additive, sintered for 5h at 1550 °C (a), 1600°C(b), 1650 °C (c), 700 °C (d) and 1750°C(e),respectively

Fig.5 shows the densification and grain growth behavior of 2.0 at% Nd:YAG ceramics, with 0.5 wt% TEOS as additive, sintered at 1550-1750 °C for 2h. It can be seen that the relative density and grain size increased as the increase of temperature. Densification occurs mainly at about 1550°C and only increased by ~1% from 1550 °C to 1750 °C. It consists well with SEM micrographs that no porosity Nd:YAG sample is obtained even at 1550 °C. As temperature increasing, relative density increases and grain growth occurs simultaneously which indicate that grain growth is an important process to remove the final pore and achieve full density to obtain Nd:YAG transparent ceramics.

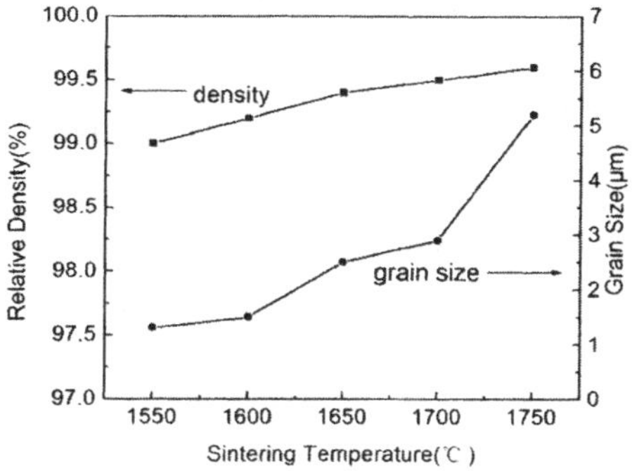

Fig.5. Relative density and grain size of 2.0 at% Nd:YAG ceramics, with 0.5 wt% TEOS as additive, as a function of sintering temperature

CONCLUSION

Nanosized Nd:YAG powder synthesized by alcohol-water solvent co-precipitation showed good dispersion and sinterability. Additive was essential to fabricate transparent ceramics, and the optimal amount of TEOS was 0.5 wt% for the fabrication of transparent Nd:YAG ceramics. Grain size increased with increase of sintering temperature and obvious grain growth occurred when sintering temperature was about 1650 °C. It is interesting to point out that the samples showed no porosity even at 1550 °C with the average grain size of ~1.5 μm. Above 1500 °C, transparent ceramics possess high relative density, moreover, the density of the sample increases as there is increasing sintering temperature.

ACKNOWLEDGEMENTS

This research was supported by the NSFC of P.R.China under grant No. 50742046 and No. 50872083.

REFERENCES

[1] A. Ikesue, T. Kinoshita, K. Kamata, and K. Yoshida, Fabrication and Optical Properties of High-Performance Polycrystalline Nd:YAG ceramics for Solid-State Lasers, *J. Am. Ceram. Soc.*, **78**, 1033-40 (1995).

[2] J. Lu, M. Prahu, J. Xu, K. Ueda, H. Yagi, T. Yanagitani, and A. A. Kaminskii, Highly efficient 2% Nd : yttrium aluminum garnet ceramic laser, *Appl. Phys. Lett.*, **77**, 3707-09 (2000).

[3] J. Lu, T. Murai, K. Takaichi, T. Uematsu, K. Misawa, M. Prabhu, J. Xu, K. Ueda, H. Yagi, T. Yanagitani, A. A. Kaminskii, A. Kudryashov, 72 W Nd: $Y_3Al_5O_{12}$ ceramic laser. *Appl. Phys. Lett.*, **78**,

3586-88 (2001).
[4] A. Ikesue, Y. L. Aung, Synthesis and Performance of Advanced Ceramic Lasers, *J. Am. Ceram. Soc.*, **89,** 1936-44 (2006).
[5] H. Yagi, T. Yanagitani, K. Takaichi, K. Ueda, A.A. Kaminskii, Characterizations and laser performances of highly transparent $Nd^{3+}:Y_3Al_5O_{12}$ laser ceramic, *Opt. Mater.*, **29,** 1258-62 (2007).
[6] S. H. Tong, T. C. Lu, W. Guo, Synthesis of YAG powder by alcohol–water co-precipitation method, *Mater. Lett.*, **61,** 4287-89 (2007).
[7] O. Fabrichnaya, H.J. Seifert, R. Weiland, T. Ludwig, F. Aldinger and A. Navrotsky, Phase equilibria and thermodynamics in the Y_2O_3-Al_2O_3-SiO_2 system, *Z. Metallkd.*, **92** , 1083-97 (2001).
[8] K.W. Lay, Grain growth in UO_2-Al_2O_3 in the presence of a liquid phase, *J. Am. Ceram. Soc.*, **51,** 373-77 (1968).
[9] S. Kochawattana, A. Stevenson, S.H. Lee, M. Ramirez, V. Gopalan, J. Dumm, V. K. Castillo, G. J. Quarles, and G. L. Messing, Sintering and grain growth in SiO_2 doped Nd:YAG, *J. Eur. Ceram. Soc.*, **28,** 1527-34 (2008).
[10] J. Li , Y.S. Wu, Y.B. Pan, W.B. Liu, L.P. Huang, and J.K. Guo. Fabrication, microstructure and properties of highly transparent Nd:YAG laser ceramics, *Opt.Mater.*, **31**, 6-17(2008).

THE EFFECT OF La_2O_3 ON THE PROPERTIES OF Nd^{3+}-DOPED YTTRIUM LANTHANUM OXIDE TRANSPARENT CERAMICS*

Hongxu Zhou
School of Materials Science and Engineering, Shanghai University
Shanghai, 200072, China
Qiuhong Yang**
School of Materials Science and Engineering, Shanghai University
Shanghai, 200072, China
Jun Xu
Shanghai Institute of Ceramics, Chinese Academy of Sciences
Shanghai, 200050, China

ABSTRACT

1%Nd^{3+}:Y_2O_3 and 1%Nd^{3+}:$(Y_{0.9}La_{0.1})_2O_3$ were fabricated and their spectroscopic properties were investigated. The results show that the grain size decreases and becomes more uniform when doped La_2O_3 in Nd^{3+}:Y_2O_3. The absorption and emission spectra show the similar properties. However, the intensity got weaker. Raman spectra show that the structural phases of the samples are unchanged and the phonon energy decreases which is in favor of the quantum efficiency.

INTRODUCTION

With the development of nanocrystalline technology, high quality transparent laser ceramics were successfully obtained by solid sintering method and have attracted more attention because of their advantages of highly doped ion concentration, excellent optical performance, inexpensive and high temperature stability. Many studies on laser diode pumped solid state lasers have focused on Nd^{3+} ion as the active ion, because the strong absorption of Nd^{3+} ion at about 808 nm wavelength matches the emission wavelength of the commercial LD. Neodymium doped YAG ($Y_3Al_5O_{12}$, yttrium aluminum garnet) is one of the best polycrystal laser materials. But it has narrow absorption bandwidths (~1nm) at the LD pump wavelength of 808nm and relative small thermal conductivity which means that it needs the system to control the temperature for LD pump device. The cubic Y_2O_3 is a promising solid state laser material for trivalent lanthanide activators due to its several favorable properties, such as refractory nature, stability, ruggedness, optical clarity over a broad spectral region. The thermal conductivity of Y_2O_3 is twice as large as that of YAG, and their thermal expansion coefficient are very similar. However, it is extremely difficult to fabricate high quality Y_2O_3 single crystal because of its high melting point (~2430°C) and the polymorphic phase from C to a high temperature hexagonal phase H at 2350°C [1-3]. In recent years, the transparent Nd:Y_2O_3 ceramics have been fabricated with the nanocrystalline technology and vacuum sintering method, and the sintering temperature is about 700~800℃ lower than its melting temperature without any influence on the phase and optical transmission[4]. In our previous work, we also found that the sintering temperature can be further decreased with La_2O_3 doped as a sintering additive. Especially, as the radius of Nd^{3+} ion is similar to that of La^{3+} ion, it is easy for Nd^{3+} ion to replace the site of La^{3+} ion in $(Y_{1-x}La_x)_2O_3$ ceramic, which make it possible to dope Nd heavily.

In this paper, Nd^{3+}-doped yttrium lanthanum oxide transparent ceramics Nd:$(Y_{1-x}La_x)_2O_3$ (x=0~0.1) were prepared. The spectroscopic properties of Nd:$(Y_{1-x}La_x)_2O_3$ transparent ceramic at room temperature were investigated.

EXPERIMENTAL

1%Nd:$(Y_{0.9}La_{0.1})_2O_3$ and 1%Nd: Y_2O_3 transparent ceramics were prepared with commercial nanopowders isostatically pressed at 200MPa and sintered at 1700-1750°C for 25h in H_2

atmosphere. The specimens used for spectroscopic studies have been polished to optical quality. The absorption spectrum at room temperature was measured with a spectrophotometer (Model V-570, JASCO) that used Xe light as the pump source. The fluorescence spectra and fluorescence lifetime of the specimens excited with 808nm laser diode were measured with a fluorescence spectrum analyzer (Fluorolog-3, Jobin Yvon Spex, France) at room temperature, of which the resolution is 1 to 2 nm. The Raman spectra were measured by Laser-Raman microspectroscopy (Invia+Plus, Renishaw, England), of which the highest resolution is 0.1nm.

RESULTS AND DISCUSSION

Fig.1 shows the optical microscopic photograph of 1%Nd^{3+}:Y_2O_3 and 1%Nd^{3+}:$(Y_{0.9}La_{0.1})_2O_3$ transparent ceramics sintered in the same condition. It can be seen from the pictures that the specimens display uniform grains and are almost no pores in or between the grain boundaries. The grain average size decreases evidently and the former is two to four times larger than that of the latter when we doped La_2O_3 as an additive. Otherwise, the grain size also becomes more uniform.

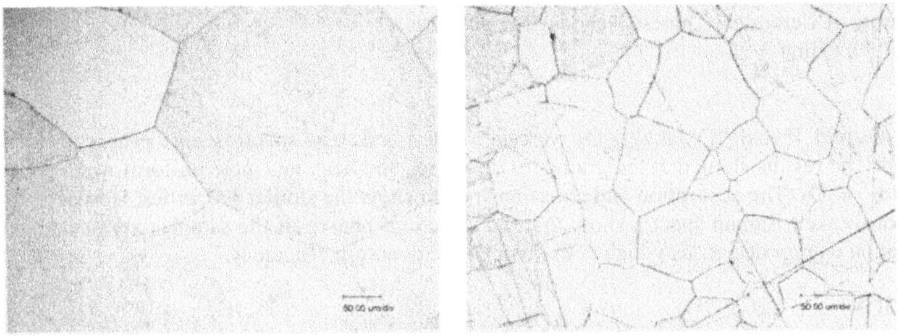

Fig.1 The optical microscopic photograph of (Left)1%Nd^{3+}:Y_2O_3 and (Right) 1%Nd^{3+}:($Y_{0.9}La_{0.1})_2O_3$

The transmittance of the Nd^{3+}:$(Y_{0.9}La_{0.1})_2O_3$ (a) and Nd^{3+}:Y_2O_3 (b) transparent ceramics is presented in Fig. 2. The highest transmittance of Nd^{3+}:$(Y_{0.9}La_{0.1})_2O_3$ reaches 80% which is 4-6 times that of Nd^{3+}:Y_2O_3. So the optical properties can be improved when doped La_2O_3.

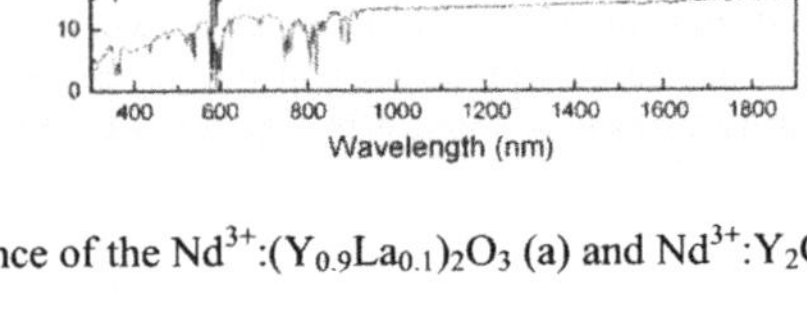

Fig. 2 The transmittance of the Nd^{3+}:$(Y_{0.9}La_{0.1})_2O_3$ (a) and Nd^{3+}:Y_2O_3 (b) transparent ceramics

Fig. 3 shows the absorption spectrum of Nd^{3+}:$(Y_{0.9}La_{0.1})_2O_3$ and Nd^{3+}:Y_2O_3 transparent ceramics at room temperature. It can be seen that the samples have the similar peaks in which the maximum absorption coefficient is at 580nm corresponding to $^4I_{9/2}\rightarrow{}^2G_{7/2}+{}^4G_{5/2}$ transition of Nd^{3+}. The absorption band in the wavelength between 780nm and 850nm corresponds to $^4I_{9/2}\rightarrow{}^4F_{5/2}+{}^2H(2)_{9/2}$ transition of Nd^{3+}ions. The strongest absorption peaks in this area is at 820nm, of which the absorption coefficients are 14.67 cm^{-1} and 3.86cm^{-1} for Nd^{3+}:$(Y_{0.9}La_{0.1})_2O_3$ and Nd^{3+}:Y_2O_3, respectively. It is clearly known that the absorption coefficient decreases remarkably with the adding of La_2O_3 in the Y_2O_3 host.

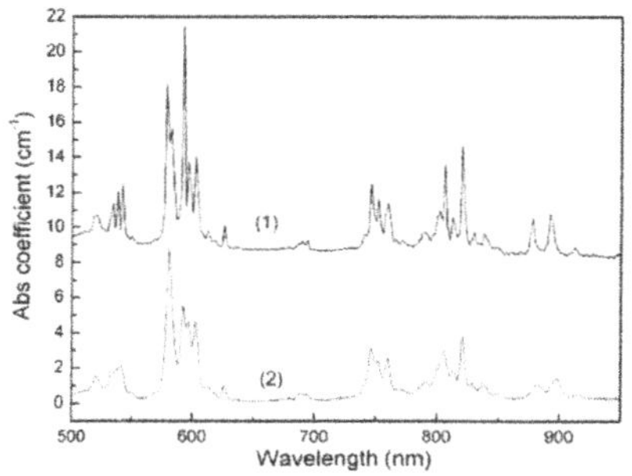

Fig. 3 Room-temperature absorption spectrum of (1) Nd^{3+}:Y_2O_3 and (2) Nd^{3+}:$(Y_{0.9}La_{0.1})_2O_3$ transparent ceramics

The emission spectra of the samples are shown in Fig. 4. The cure of Nd^{3+}:$(Y_{0.9}La_{0.1})_2O_3$ is similar to that of Nd^{3+}:Y_2O_3. There are three groups of emission peaks corresponding to the transitions from the sublevels of $^4F_{3/2}$ to the components of the $^4I_{9/2}$, $^4I_{11/2}$, $^4I_{13/2}$ground states which are located at 890-950nm, 1050-1140nm and 1347-1380nm, respectively. The strongest peak of Nd^{3+}:$(Y_{0.9}La_{0.1})_2O_3$ transparent ceramics is at 1079nm wavelength with the full width at half maximum (FWHM) of about 7.8nm.

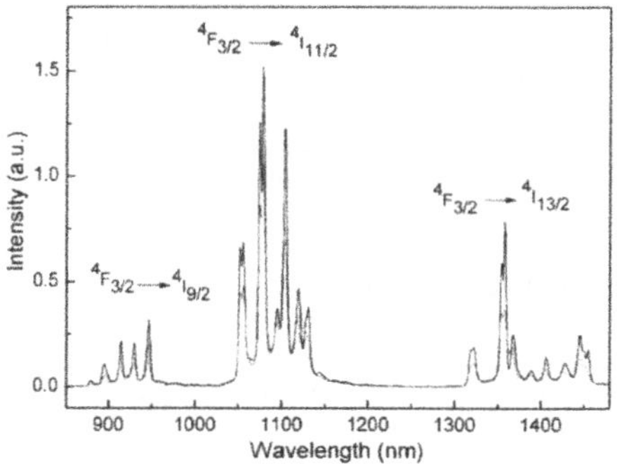

Fig. 4 Room-temperature fluorescence spectra of (red) Nd^{3+}:$(Y_{0.9}La_{0.1})_2O_3$ and (black) Nd^{3+}:Y_2O_3 transparent ceramics

Raman spectra of $Nd^{3+}:(Y_{0.9}La_{0.1})_2O_3$ and $Nd^{3+}:Y_2O_3$ transparent ceramics are measured as shown in Fig. 5 and the insert is the Raman spectra of $Nd^{3+}:(Y_{0.9}La_{0.1})_2O_3$. There are six characteristic peaks for the cube Y_2O_3 phase which are located at 129cm^{-1}, 160 cm^{-1},329 cm^{-1}, 377 cm^{-1}, 468 cm^{-1}, 592 cm^{-1} [5]. The samples show the similar character to that of Y_2O_3 which means that their structural phases are not changed when we doped Nd_2O_3 and La_2O_3 in Y_2O_3. However, there is a red shift for $Nd^{3+}:(Y_{0.9}La_{0.1})_2O_3$ and $Nd^{3+}:Y_2O_3$ transparent ceramics which shows that the phonon energy decreases. The lower phonon is of benefit to the improvement of the quantum efficiency [6].

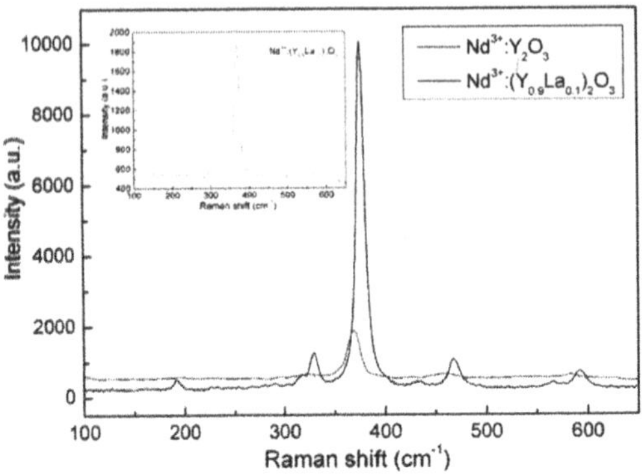

Fig.5 Raman spectra of (red) $Nd^{3+}:(Y_{0.9}La_{0.1})_2O_3$ and (black) $Nd^{3+}:Y_2O_3$ transparent ceramics.

CONCLUSION

$Nd^{3+}:Y_2O_3$ and 1%$Nd^{3+}:(Y_{0.9}La_{0.1})_2O_3$ were fabricated and their spectroscopic properties were investigated. The results show that the grain size decreases and becomes more uniform when doped La_2O_3 in $Nd^{3+}:Y_2O_3$. The highest transmittance of $Nd^{3+}:(Y_{0.9}La_{0.1})_2O_3$ reaches 80% which is 4-6 times that of $Nd^{3+}:Y_2O_3$.

The absorption and emission spectra of the samples show similar properties. However, the intensity got weaker. The absorption band in the wavelength between 780nm and 850nm corresponds to $^4I_{9/2}\rightarrow{}^4F_{5/2}+{}^2H(2)_{9/2}$ transition of Nd^{3+}ions. The emission bands located at 890-950nm, 1050-1140nm and 1347-1380nm correspond to $^4F_{3/2}\rightarrow{}^4I_{9/2}$, $^4F_{3/2}\rightarrow{}^4I_{11/2}$, $^4F_{3/2}\rightarrow{}^4I_{13/2}$ transition.

There is a red shift for its characteristic peaks in Raman spectrum which will induce lower phonon energy for the samples. The full widths of half-maximum of all peaks are increased and peak intensity becomes weak when La_2O_3 was doped as an additive which means that the thermal vibration and the symmetry of the samples weaken. It is proven that the structures of samples present are in disorder and agree with the result that mentioned above. It is possible for $Nd:(Y_{1-x}La_x)_2O_3$ to be a new type of laser material.

FOOTNOTES

*This work was supported by the National Natural Science Foundation of China (Grant No. 60578041).
**The corresponding author: yangqiuhongcn@yahoo.com.cn

REFERENCES:

1 J. Kong, D.Y. Tang, B. Zhao, J. Lu, K. Ueda, H. Yagi, T. Yanagitani, 9.2W diode end pumped $Yb:Y_2O_3$ ceramic laser, *Appl. Phys. Lett.* **86**, 161116-1(2005).

[2] H.X. Ma, Q.H. Lou, Y.F. Qi, J.X. Dong, Y.R. Wei, 5.5W CW Yb^{3+}:Y_2O_3 ceramic laser pumped with 970nm laser diode, *Opt. Commun,* **246**, 465(2005).
[3] J. Kong, D.Y. Tang, J. Lu, K. Ueda, H. Yagi, T. Yanagitani, Passively Q-switched Yb:Y_2O_3 ceramic laser with a GaAs output coupler, *Opt. Exp,* **12**, 3560(2004).
[4] Q.H. Yang, J. Xu, L.B. Su, H.W. Zhang, Spectroscopic characteristics of transparent Yb:$Y_{2-2x}La_{2x}O_3$ laser ceramics, *Act. Phys. Sin.* **55**, 1207(2006).
[5] Schaack G, Koningstein J A, Phonon and Electronic Raman Spectra of Cubic Rare-earth Oxides and Isomorphous Yttrium Oxide, *Optical Society of America,* **13(3),** 284-289(1982).
[6] L.A. Riseberg, The relevance of nonradiative transitions to Solid State Lasers, *Plenum Press, New York*, 369-407(1980).

$Lu_2O_3:Eu^{3+}$ ULTRADISPERSED POWDERS AND TRANSLUCENT CERAMICS

R.P. Yavetskiy, E.A. Vovk, M.B. Kosmyna, Z.P. Sergienko, A.V. Tolmachev, V.M. Puzikov, B.P. Nazarenko, A.N. Shekhovtsov
Institute for Single Crystals of NAS of Ukraine, 60 Lenin Ave.
Kharkov, 61001, Ukraine

ABSTRACT

The $Lu_2O_3:Eu^{3+}$ (1 at. %) ultradispersed powders have been obtained by precipitation method using ammonium hydracarbonate as precipitant. The influence of the precursor calcination temperature on morphology and primary particle size of $Lu_2O_3:Eu^{3+}$ powders has been studied. It has been shown, that calcination temperature of 1000 °C allows one to obtain low-agglomerated monodispersed spherical powders of europium doped lutetium. The translucent $Lu_2O_3:Eu^{3+}$ ceramics (1 mm in thickness) which has optical in-line transmittance of about 20 % in the visible wavelength region and relative density of 99 % has been fabricated by vacuum sintering at T=1800 °C for 10 hours.

INTRODUCTION

Nowadays optical polycrystalline ceramics based on compounds with cubic structure are considered as a novel class of functional optical materials for laser and scintillation techniques. They possess structural, functional and economical advantages over the corresponding single crystals [1]. For example, optical ceramics have excellent optical properties, improved functional characteristics and better mechanical properties in comparison with single crystals with the same compositions. Furthermore, transparent ceramics production process is less expensive compared to single crystal pulling by conventional melt methods. The advantages of optical ceramics over single crystals have been initially realized for yttrium aluminium garnet $Y_3Al_5O_{12}$. At a present transparent ceramics on the basis of different compounds, for example, rare earth sesquioxides RE_2O_3, is actively studied.

Lutetium oxide possesses excellent physicochemical properties, such as high melting temperature (T=2490 °C), low steam pressure, high chemical stability, etc. Lu_2O_3 is a promising material for different spectroscopic application due to transparency in the wide wavelength range and high isomorphic capacity for doping by luminescent rare earth ions. For example, excellent thermo mechanical properties allow one to consider Lu_2O_3 as a promising host for high power solid state lasers, laser fusion drivers etc [2]. $Lu_2O_3:Eu^{3+}$ is a prospective scintillator for medical imaging, for example, for X-ray computed tomography [3].

Recent advances in transparent ceramics fabrication are based on utilization of nanotechnology and pressureless vacuum sintering methods. Nowadays different wet-chemical processes were applied to produce $Lu_2O_3:Eu^{3+}$ nanopowders, including a combustion synthesis [4] and molten salts route [5]. A successful production of optical ceramics by chemical precipitation allows one to consider this technique as one of the most promising. Just recently the light yield of 90000 photons/MeV was achieved with $Lu_2O_3:Eu^{3+}$ (5 at. %) ceramics consolidated using co-precipitated nanopowders [6]. This work is devoted to produce $Lu_2O_3:Eu^{3+}$ highly sinterable nanopowders for optical ceramics using conventional low-temperature co-precipitation route.

EXPERIMENTAL

$Lu_2O_3:Eu^{3+}$ (1 at. %) nanopowders were obtained by co-precipitation technique using ammonium hydrocarbonate NH_4HCO_3 (purity >99.5 %) as a precipitant. Aqueous lutetium and europium nitrate solution was prepared by dissolving of corresponding oxides (purity 99.99 %) in the nitric acid. The precursor precipitate was produced by adding a mother solution to ammonium hydrocarbonate solution (reverse strike method) under mild stirring at the room temperature. The resultant suspension was aged for 24 hours, filtered using suction filter, thoroughly washed several times by deionized water and ethyl alcohol, and finally dried at T=120 °C for 24 hours. Then, precursor was calcinated at T=550, 700, 1000 and 1200 °C for 2 hours to obtain europium doped lutetium oxide. To produce translucent ceramics, Lu_2O_3:Eu^{3+} nanopowders were compacted by dry uniaxial pressure method at the pressure of 330 MPa. Sintering of green compacts was performed in an inductive heating furnace in a vacuum of 10^{-3} Pa at T=1800 °C for 10 hours without any additives.

IR absorbance spectra were obtained using FT-IR spectrometer Spectrum One (Perkin Elmer) with KBr pellets. Differential thermal (DTA) and thermo gravimetrical (TG) analysis of precursor was carried out using a derivatograph MOM Q-1500D (Hungary) with a heating rate of 5 °C /min and α-Al_2O_3 as a reference. Phase identification of precursor and calcinated powders was performed by the X-ray diffraction (XRD) method on diffractometer DRON-4 (Russia) in FeK_α radiation in the range of 2Θ=20-80 degrees. The specific surface area was measured using BET method with home-made setup. Powder morphologies were observed by transmission electron microscopy (TEM-125, Russia). The microstructure of fracture surface was studied by scanning electron microscopy (JSM-6390 LV, JEOL, Japan). The crystallite diameter of thermally etched mirror polished samples was determined using secant method.

RESULTS AND DISCUSSION

IR-spectra of precursor, dried at T=50 °C and Lu_2O_3 powders, calcinated at T=800 °C for 2 hours are presented in Fig. 1. The broad absorption band at 3435 cm^{-1} was attributed to O–H stretching. Wide absorption bands at 1525-1530 and 1385-1400 cm^{-1} were assigned to C–O asymmetric stretch in CO_3^{2-}. Absorption peaks at 1075-1085 cm^{-1} and 840-850 cm^{-1} appearing in the precursor sample are probably connected with stretching and bending of C–O band [7]. These peaks may point out the presence of carbonate groups in the precursor. Precursor calcination at T=800-1200 °C leads to formation of lutetium oxide. The appearance of absorption peaks at 498 and 580 cm^{-1} in the spectra of sample calcinated at 800 °C is the characteristic of Lu–O stretching [8], indicating the crystallization of lutetium oxide. Several residual weak absorption bands were attributed to H_2O and CO_2 absorbed at the surface of the powder in air atmosphere. Their absorption intensity decreases with the increase of calcination temperature.

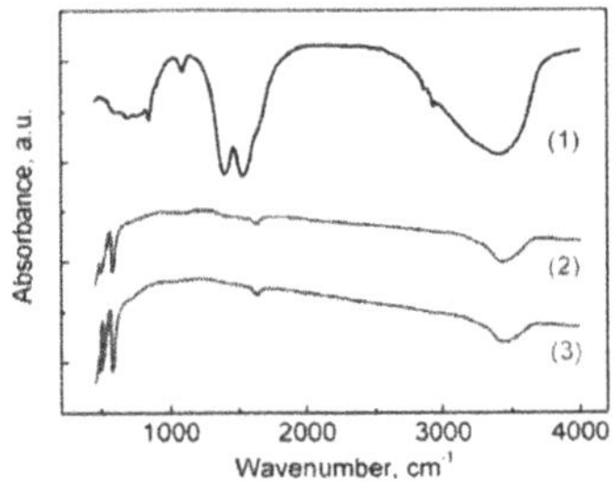

Figure 1. FT-IR spectra of the precipitated precursor (1) and Lu_2O_3:Eu^{3+} (1 at. %) powders, calcinated at T=800 °C (2) and 1200 °C (3).

The DTA-TG traces of precursor after its drying at room temperature are illustrated in Fig. 2. The continuous mass lost by the sample is observed up to T=650 °C and equals to 32 %. Endothermic peaks on the DTA curve at T=130 °C and in the T=200-600 °C temperature range correspond to release of molecular and hydration water, and to decomposition of precursor (which is probably lutetium carbonate). The formation of lutetium oxide occurs at a relatively low temperature and finishes at T=650 °C. Recently it was shown [7], that precipitation of lutetium nitrate by mixed precipitant ($NH_4HCO_3+NH_3 \cdot H_2O$) leads to the formation of the basic lutetium carbonate. The precursor composition has not been studied in this work; however, FT-IR spectra and DTA-TG curves of precipitated precursor are in a good agreement with the data [7]. For this reason we suppose, that precursor's composition is $Lu(OH)_x(CO_3)_y \cdot nH_2O$.

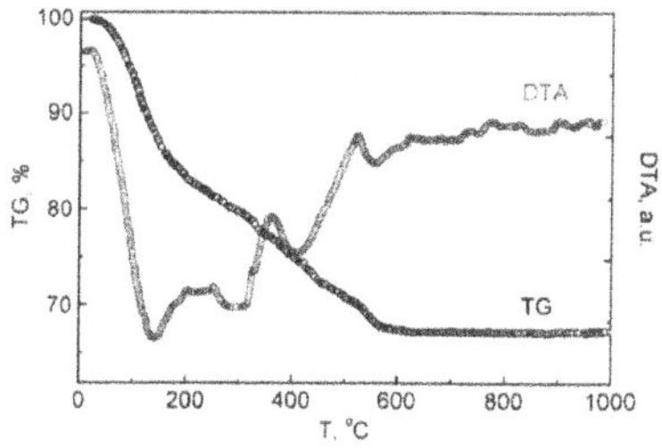

Figure 2. DTA-TG curves of the precursor prepared using NH_4HCO_3.

XRD analysis of $Lu_2O_3:Eu^{3+}$ powders calcinated at different temperatures for 2 hours are presented in Fig. 3. According to the XRD data, the precursor and products of its calcination at T<550 °C were amorphous. The formation of cubic lutetium oxide starts at T=550 °C. Temperature increase from 550 to 1000 °C leads to decrease of half-width of diffraction peaks and to increase of their intensity. This testifies an improvement of $Lu_2O_3:Eu^{3+}$ crystallinity. The average crystallite size estimated by Scherrer's equation also increases from 10.5 to 30.5 Å (fig. 4). All the diffraction peaks on the XRD patterns were attributed to cubic $Lu_2O_3:Eu^{3+}$.

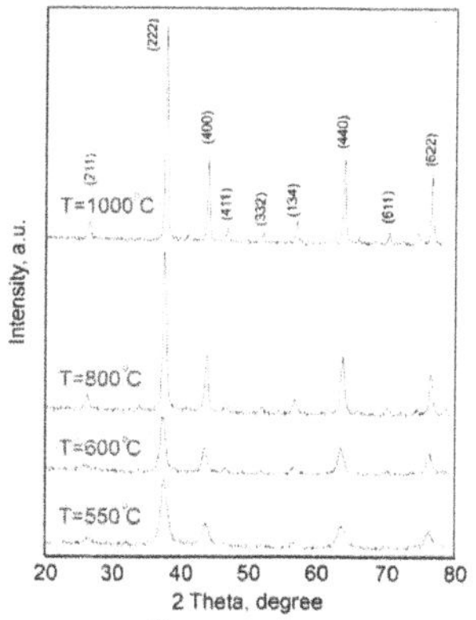

Figure 3. XRD patterns of $Lu_2O_3:Eu^{3+}$ powders, calcinated at T=550 °C (1); T=600 °C (2); T=800 °C (3), and T=1200 °C (4).

In order to produce transparent ceramics, the active non-agglomerated monodispersed spherical powders are required allowing to obtain homogenous green compact. The surface activity and crystallite size of obtained powders were estimated by specific surface area value S_{BET}. Precursor dried at 120 °C was characterized by high specific surface area S_{BET}=45-50 m^2/g (not shown in fig. 4). Temperature increase from 550 to 1200 °C results in decrease of S_{BET} value of $Lu_2O_3:Eu^{3+}$ nanopowders from 18 to 4 m^2/g. It is accompanied by the increase of crystalline size from 28 to 175 nm (fig. 4). TEM images were used to characterize and compare the microstructure of nanopowders calcinated at different temperatures (fig. 5). The powders calcinated at relatively low temperature (T=700 °C and lower), possess high agglomeration degree due to low primary crystalline size. Increase of the crystallite size results in decline of agglomeration degree owing to strongly decrease of interparticle interaction forces. At the optimum calcination conditions (T=1000 °C) all the particles of $Lu_2O_3:Eu^{3+}$ powders have nearly spherical form, the average particle size is 25-30 nm (fig. 5, 2). The particles are only slightly agglomerated, and agglomerates have spherical shape.

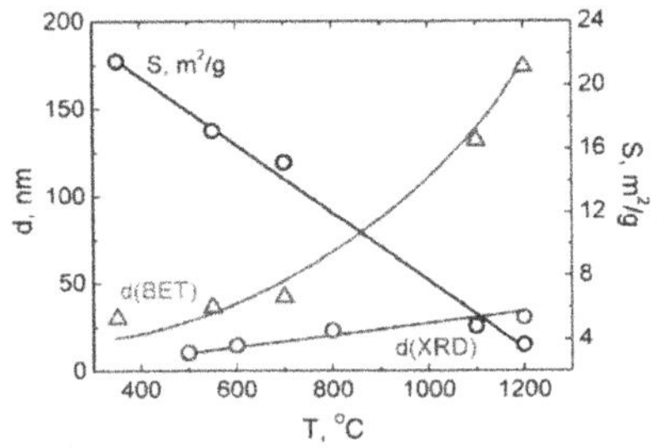

Figure 4. Dependence of specific surface area S_{BET} (1) and average crystalline size, obtained using BET method, d_{BET} (2) and XRD data, d_{XRD} (3) vs. precursor calcination temperature.

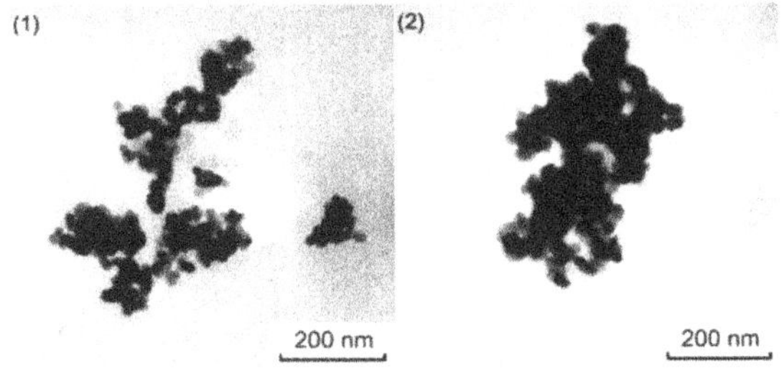

Figure 5. TEM of $Lu_2O_3:Eu^{3+}$ powders calcinated at T=700 °C (1) and T=1000 °C (2) for 2 hours.

To study sinterability of $Lu_2O_3:Eu^{3+}$ nanopowders calcinated at different temperatures, vacuum sintering was performed. $Lu_2O_3:Eu^{3+}$ powders were compacted by conventional uniaxial pressure method. The green body density is ρ=28-47 % for T=550-1200 °C, and is substantially determined by morphology of the powders. The presence of hard agglomerates in the powders considerably interrupts the compaction process. For example, the powders, calcinated at T=550 °C have the highest S_{BET} value and, correspondingly, the highest activity, however, due to low crystalline size they strongly agglomerate. The presence of large and dense agglomerates in the powder prevents the formation of homogenous green compact and is responsible for the formation of inter-agglomerate voids. Such ingomogenities in particles packing lead to the formation of many residual pores resulting in opaque ceramics after vacuum sintering (fig. 6). The powders obtained at T=1200 °C despite the highest green density of compact (47 %) are poorly sintered due to their low surface activity. The best translucency shows $Lu_2O_3:Eu^{3+}$ ceramics obtained from the powders after heat treatment at T=1000 °C – the letters could be read through the pellet. The in-line transmittance coefficient for this specimen was of about 20 % in the visible wavelength region, and the 99 % density was obtained.

(1) (2) (3) (4)

Figure 6. $Lu_2O_3:Eu^{3+}$ (1 at. %) ceramics of 1 mm thickness sintered under vacuum from powders, calcinated at T=550 °C (1); T=700 °C (2); T=1000 °C (3) and T=1200 °C (4).

The SEM images of fracture surfaces of $Lu_2O_3:Eu^{3+}$ ceramics, obtained from nanopowders with different granulometric composition, are given in Fig. 7 The calcination temperature increase from 550 to 1000 °C results in increment of crystallite diameter from 8-10 mkm to 18-20 mkm accompanied by reduction of residual pores quantity. It is known, that even a small fraction of unbroken fragments of agglomerates in the pressed powder causes a substantial decrease in the sintered density. The difference in the porosity of $Lu_2O_3:Eu^{3+}$ sintered pellets was found to be in a good agreement with particle size and agglomeration degree of starting powders. The powders calcinated at T=1000 °C densify to nearly full density and the porosity is almost missing (fig. 7, 3), while in other samples the intergranular and intragranular porosity was observed. The presence of intragranular porosity is a consequence of pore-grain boundary separation followed by pore entrapment under high grain growth rate. Thus, residual porosity in ceramics prepared from high sinterable powders (T=550 and 700 °C) is a consequence of competition between densification and recrystallization processes. Intergranular porosity remains generally in the ceramic samples prepared from powders with low surface activity (T=1200 °C).

The main luminescent properties of $Lu_2O_3:Eu^{3+}$ (1 at. %) nanopowders and ceramics agree well with data reported in [9]. The integral radioluminescence intensity of $Lu_2O_3:Eu^{3+}$ (1 at. %) ceramics under X-ray excitation (E=40 keV) is by an order of magnitude higher than luminescence intensity of starting nanopowders. However, light yield of $Lu_2O_3:Eu^{3+}$ ceramics measured under alpha-particles excitation (^{239}Pu, E=5.15 MeV) was extremely poor (lower than 1000 photons/MeV). Low scintillation yield (and in-line transmittance) is probably connected with application of uniaxial pressure method for powders compaction, which is characterized by extremely non-uniform density distribution in the compact volume. Moreover, non optimized europium concentration can also reduce scintillation response of $Lu_2O_3:Eu^{3+}$ ceramics [6]. We believe, that improvement of optical and scintillation characteristics of $Lu_2O_3:Eu^{3+}$ ceramics can be achieved by utilization of wet compaction methods, namely slip casting in gypsum molds and colloidal casting under pressure [10], which is the subject of our further studies.

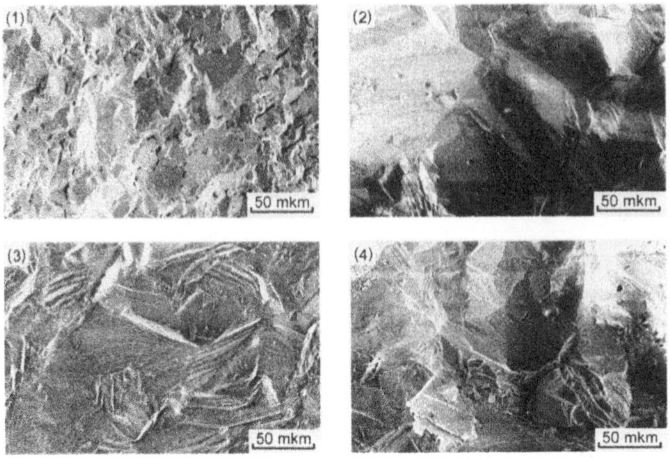

Figure 7. Fracture surface of $Lu_2O_3:Eu^{3+}$ ceramics sintered from powders calcinated at T=550 °C (1); T=700 °C (2); T=1000 °C (3) and T=1200 °C (4).

CONCLUSION

To conclude, the $Lu_2O_3:Eu^{3+}$ (1 at. %) nanopowders were prepared by co-precipitation method using ammonium hydracarbonate as precipitant. It was shown that $Lu_2O_3:Eu^{3+}$ low-agglomerated monodispersed spherical powders with specific surface area of S=14 m^2/g can be obtained by precursor calcination at T=1000 °C. It was determined, that the resultant powders can be used for production of $Lu_2O_3:Eu^{3+}$ translucent ceramics with average crystalline size of 18-20 mkm, nearly full density (99 %), and in-line transmittance coefficient up to 20 % even if the uniaxial pressure method is used for nanopowder compaction.

REFERENCES

[1] V. Lupei, A. Lupei, A. Ikesue, Transparent polycrystalline ceramic laser materials, *Opt. Mater.*, **30**, 1781-86 (2008).

[2] J. Lu, K. Takaichi, T. Uematsu, A. Shirakawa, M. Musha, and K. Ueda, H. Yagi, T. Yanagitani, A.A. Kaminskii, Promising ceramic laser material: Highly transparent $Nd^{3+}:Lu_2O_3$ ceramic, *Appl. Phys. Lett.*, **81**, 4324-26 (2002).

[3] A. Lempicki, C. Brecher, P. Szupryczynski, H. Lingertat, V.V. Nagarkar, S.V. Tipnis, S.R. Miller, A new lutetia-based ceramic scintillator for X-ray imaging, *Nucl. Instrum. Meth. Phys. Res. A*, **488**, 579-90 (2002).

[4] E. Zych, D. Hreniak, W. Strek, Spectroscopy of Eu-doped Lu_2O_3-based X-ray phosphor, *J. Alloys and Compounds*, **341**, 385-90 (2002).

[5] J. Trojan-Piegza, E. Zych, Preparation of nanocrystalline Lu_2O_3:Eu phosphor via a molten salts route, *J. Alloys and Compounds*, **380**, 118-22 (2004).

[6] Y. Shi, Q.W. Chen, J.L. Shi, Processing and scintillation properties of Eu^{3+} doped Lu_2O_3 transparent ceramics, *Optical Materials*, in press. (2008).

[7] Q. Chen, Y. Shi, L. An, S. Wang, J. Chen, J. Shi, A novel co-precipitation synthesis of a new phosphor $Lu_2O_3:Eu^{3+}$, *J. Europ. Ceram. Soc.*, 27, 191-97 (2007).

[8] N.T. McDevitt, W.L. Baun, Infrared absorption study of metal oxides in the low frequency region (700–240 cm^{-1}), *Spectrochim. Acta*, **20**, 799-808 (1964).
[9] Q. Chen, Y. Shi, L. An, J. Chen, and J. Shi, Fabrication and Photoluminescence Characteristics of Eu^{3+}-Doped Lu_2O_3 Transparent Ceramics, *J. Am. Ceram. Soc.*, **89**, 2038-42 (2006).
[10] Y.L. Kopylov, V.B. Kravchenko, S.N. Bagayev, V.V. Shemet, A.A. Komarov, O.V. Karban, A.A. Kaminskii, Development of Nd^{3+}:$Y_3Al_5O_{12}$ laser ceramics by high-pressure colloidal slip-casting (HPCSC) method, *Opt. Mater.*, in press (2008).

FABRICATION AND SPECTROSCOPIC PROPERTIES OF $Nd:Lu_2O_3$ TRANSPARENT CERAMICS FOR LASER MEDIA

Ding Zhou[1], Yan Cheng[2], Yu Ying Ren[1], Ying Shi[1*], Jian Jun Xie[1]
[1]School of Material Science and Engineering, Shanghai University, Shanghai 200072, P. R. China
[2]Shanghai Institute of Optics and Fine Mechanics, Chinese Academy of Sciences, Shanghai 201800, P. R. China

ABSTRACT

Highly transparent $Nd:Lu_2O_3$ ceramics were fabricated from synthesized nanocystalline powder after being sintered in flowing H_2 atmosphere at 1880°C for 8 h. The linear optical transmittance of the polished sample with 1.4mm thickness reaches 75.5% at the wavelength of 1080nm. The absorption band at 807nm has a FWHM of about 4 nm and the absorption cross section σ_{abs} at 807 nm is 2.998×10^{-20} cm^2. According to the Judd-Ofelt theory, the intensity parameters Ω_2, Ω_4 and Ω_6 of 3at.%$Nd:Lu_2O_3$ sample are calculated to be 5.26×10^{-20} cm^2, 13.11×10^{-20} cm^2 and 8.37×10^{-20} cm^2 respectively. The radiative lifetime τ_{rad} of the $^4F_{3/2}$ level attains 165 μs.

INTRODUCTION

Advanced transparent ceramic techniques have attracted more and more attention in the solid-state laser field since Ikesue et al. reported the fabrication of transparent Nd:YAG laser ceramic[1,2]. Lutetia (Lu_2O_3) is one of attractive sesquioxides due to its isotropic cubic crystal structure, extremely high density (9.42g/cm^3), high thermal conductivity (12.5W/mK) and high band gap (~6.4ev), which favor its applications not only as scintillators for radiation detection but also as a promising host material for laser gain media[3-5]. Considering its high melting point up to ~2450°C, it is of great interest to fabricate Lu_2O_3 based polycrystalline transparent materials doped by different rare earth ions. Compared with single crystal, transparent ceramics possess many advantages such as better chemical homogeneity, the flexibility of varying the chemical composition over a wide range, the lower fabrication temperatures and the feasibility of larger size as well [6,7].

In 2002, J. Lu et al. reported a 0.15at.%$Nd:Lu_2O_3$ laser ceramic[4], which generated a 185mw output (pumping power: 1W) by an uncoated square sheet with size of 6mm×6mm×1mm. The laser threshold of this specimen was measured to be 90mW, which was much lower than that of $Nd^{3+}:Y_2O_3$ ceramic laser (200 mW). Spectra properties of Nd^{3+} in the host of Lu_2O_3 ceramic were demonstrated, where the main emission peak was located around 1080 nm.

In this paper, the transparent 3at.%$Nd:Lu_2O_3$ ceramics are fabricated by pressureless sintering under flowing H_2 atmosphere at 1880°C for 8 h without any additives. The spectra properties and parameters are calculated in detail by Judd-Ofelt theory. Based on the spectral characterization results of transparent sample, the relevant spectral parameters such as σ_{abs}, Ω_2, Ω_4, Ω_6, τ_{rad} and the fluorescence branching ratio β for Nd^{3+} in Lu_2O_3 lattice are calculated.

EXPERIMENTAL PROCEDURE

Lutetium and neodymium oxide (purity > 99.995%, Xiyuan International, Shanghai, China) were dissolved in a high purity nitric acid respectively. The as-prepared $Nd(NO_3)_3$ was added to $Lu(NO_3)_3$ solution according to formula $Nd_{0.03}Lu_{1.97}O_3$ to get rare earth mother solution. The precursor for nanosized $Nd:Lu_2O_3$ powder was prepared by adding $NH_4OH+NH_4HCO_3$ mixed precipitant to the upper rare earth mother solution at a rate of 3 ml/min. In the end, pH value of the suspension was kept in the range of 8-9. After aged for 24 h, the amorphous suspension was filtered using a suction filter, washed 4-5 times with deionized water and twice with alcohol, and then dried at 80°C for 24 h. After being crushed and sieved to 200 meshes, the precursor was calcined in a muffle furnace at 1000°C for 2 h with a heating rate of 2°C/min to obtain $Nd:Lu_2O_3$ nanopowders.

Compact disks were dry-pressed in a stainless steel die and then cold-isostatically consolidated under 200MPa pressure without any additives. Densification of Nd:Lu_2O_3 green compacts were performed in a furnace (Model FDB-14-19, NEMS Co., Japan) equipped with a tungsten-mesh heater under flowing H_2 atmosphere at 1880°C for 8 h.
Optical polished Nd:Lu_2O_3 transparent samples were used for the spectroscopic measurement. Linear optical transmittance of 3at.%Nd:Lu_2O_3 transparent ceramic was measured in region of 190-1100 nm on a UV/VIS/NIR spectrophotometer (Lambda 2, Perkin Elmer, U.S.A.). The fluorescence spectrum of the specimen was recorded by a spectrofluorometer (Fluorolog-3, Jobin Yvon, Edision, U.S.A.) equipped with a Hamamatsu R928 photomultiplier tube. A 808nm continuous wave diode laser was used as the excitation source.

RESULTS AND DISCUSSION

Fig. 1 presents the appearance of 3at%Nd^{3+}:Lu_2O_3 transparent ceramic sample (1.4mm thick) which has been polished on both sides. The sintered specimen has a diameter of 15mm and a thickness of 1.4mm. The relative density measured by Archimedes method reached 99.8% theoretical value, indicating a nearly full density. As this photo revealed, the letters under the sample can be seen clearly.

Figure 1. Appearance of transparent Lu_2O_3 ceramic (thickness: 1.4 mm).

Fig. 2 shows the optical linear transmittance of Nd^{3+}:Lu_2O_3 ceramic specimen in the wavelength region from 190 to 1100nm. The relative lower transmittance in visible region may be resulted from the scattering source such as impurities and micro-pores existed in the polycrystalline Nd:Lu_2O_3 ceramic. When the wavelength exceeded 600nm, the linear optical transmittance of the sample increased up to 70%. At the wavelength of 1080nm, as-received Nd:Lu_2O_3 transparent specimen exhibited a linear optical transmittance value of 75.5%, corresponding to a optical attenuation coefficient of 0.0564mm^{-1}.
The absorption spectrum of 3at.%Nd:Lu_2O_3 ceramic was recorded by spectrophotometer at room temperature in the range 700-950nm as shown in Fig. 3. The absorption peaks are originated from the transitions of Nd^{3+} from the ground state $^4I_{9/2}$ to the excited states. The transitions $^4I_{9/2}\rightarrow{}^4F_{7/2}+{}^4S_{3/2}$, $^4I_{9/2}\rightarrow{}^4F_{5/2}+{}^2H_{9/2}$ and $^4I_{9/2}\rightarrow{}^4F_{3/2}$ around 746, 820 and 879 nm are prominent. It can be seen that the strongest absorption peak appears at 822nm, corresponding to the transition of $^4I_{9/2}\rightarrow{}^4F_{5/2}$ of Nd^{3+}. However, considering the unfeasibility of diode emitting at 822nm at present, the preferable emitting wavelength would be 807nm, which matches AlGaAs diode-laser pumping very well. In this spectrum, the absorption band at 807 nm has a FWHM (full width at half maximum) of 4 nm and the absorption cross-section σ_{abs} is calculated to be 2.998×10^{-20} cm^2 by equation (1).

$$\sigma_{abs} = \frac{2.203\log(I_0/I)}{LN} \tag{1}$$

where $\log(I_0/I)$ is optical density, N is the active-ion concentration per unit volume and L is the thickness of the sample.

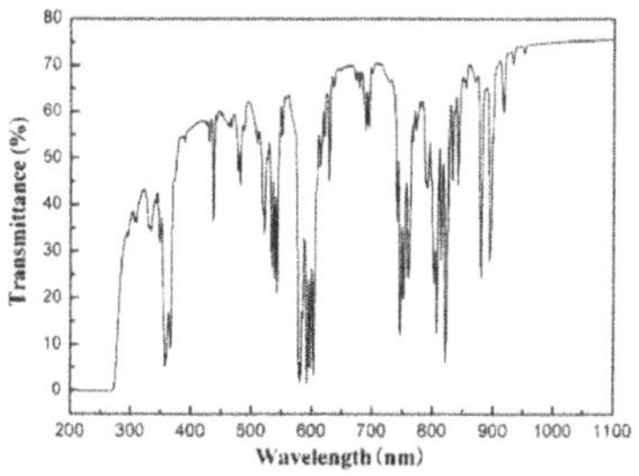

Figure 2. Inline optical transmittance of Nd:Lu_2O_3 transparent ceramic (thickness: 1.4mm).

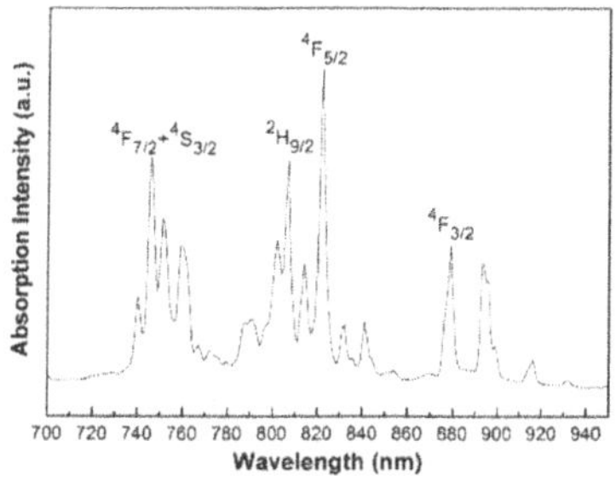

Figure 3. Room temperature absorption spectrum of 3at.% Nd:Lu_2O_3 ceramic.

The Judd-Ofelt theory[8,9] was applied to study the spectral properties of 3at.%Nd:Lu_2O_3 ceramic in this work. For Nd^{3+} ions, the magnetic dipole (md) transitions are much weaker than the electric dipole (ed) transitions and they can be neglected in the *J–O* calculations. The experimental line strength for electric dipole (ed) transitions $S_{ed(exp)}$ can be obtained from the absorption spectrum according to the following equation (2)

$$S_{ed(exp)} = \frac{3hc(2J+1)}{8\pi^3 e^2 \bar{\lambda}} \frac{9n}{(n^2+2)^2} \int \sigma_{abs}(\lambda)d\lambda \tag{2}$$

Where $\sigma_{abs}(\lambda)$ is the absorption cross section of the sample at the wavelength of λ, $\bar{\lambda}$ is the mean wavelength of the absorption band, J is the total angular momentum of the ground state, e is the electron charge magnitude, h is the Planck constant, c is the speed of light in vacuum and n is the refractive index.
The values of $S_{ed(meas)}$ calculated from equation (2) can be substituted into formula (3) to determine the oscillator strength parameter Ω_2, Ω_4 and Ω_6.

$$S_{ed} = \sum_{t=2,4,6} \Omega_t \left| \langle (S,L)J \| U^{(t)} \| (S',L')J' \rangle \right|^2 \qquad (3)$$

Where $\langle (S,L)J \| U^{(t)} \| (S',L')J' \rangle$ are the doubly reduced matrix elements corresponding the transition from J to J' manifold of the U^t unitary tensor operator of rank t, with t = 2, 4, and 6, which were calculated by Carnall[10] *et al.* The values of the intensity parameters Ω_2, Ω_4 and Ω_6 are obtained to be 5.26×10^{-20} cm^2, 7.13×10^{-20} cm^2, and 8.37×10^{-20} cm^2 respectively. Important parameters of absorption and emission spectrum are calculated according to Ω_t (t = 2, 4, and 6) by the following formulas (4)-(8).

The calculated oscillator strength for the transitions from the initial J state to the final J' state is given by

$$f_{cal}^{ed} = \frac{8\pi^2 mc}{3h(2J+1)\bar{\lambda}} \frac{(n^2+2)^2}{9n} S_{ed} \qquad (4)$$

The experimental oscillator strength can be calculated by

$$f_{exp}^{ed} = \frac{mc^2}{\pi e^2 \bar{\lambda}^2} \int \sigma_{abs}(\lambda) d\lambda \qquad (5)$$

The radiative transition rate $A(^4F_{3/2} \rightarrow {}^4I_J)$ from the initial manifold $^4F_{3/2}$ to the terminal manifold $4I_J$ (J = 9/2, 11/2 and 13/2) is given by

$$A(J;J') = \frac{64\pi^4 e^2 n(n^2+2)^2}{3h(2J+1)9\bar{\lambda}^3} S_{ed} \qquad (6)$$

The matrix elements were taken from those determined by Kaminskii[11] *et al.* The radiative lifetime for $^4F_{3/2}$ excited state can be calculated as from the following formula:

$$\tau_{rad} = \frac{1}{\sum_{J'} A(J,J')} \qquad (7)$$

The fluorescence branching ratio $\beta_{JJ'}$ from $J \rightarrow J'$ transition is determined by

$$\beta_{JJ'} = \frac{A(J,J')}{\sum_{J'} A(J,J')} \qquad (8)$$

The absorption spectrum parameters of 3at.%Nd:Lu$_2$O$_3$ sample are listed in Table I and the emission spectrum parameters for Nd^{3+} from the ^{4}F$_{3/2}$ to $^4I_{j'}$ levels calculated according to Ω_t are listed in Table II.

Fig.4 revealed the luminescence spectrum of the 3at.%Nd:Lu$_2$O$_3$ ceramic sample excited by 808 nm. Three emission bands corresponding to the $^4F_{3/2} \rightarrow {}^4I_{9/2}$, $^4I_{11/2}$ and $^4I_{13/2}$ transitions are located at 875–965, 1045–1155 and 1345–1380nm. Among three main emission peaks, the transition of ^{4}F$_{3/2}$ $\rightarrow$ ^{4}I$_{11/2}$ at 1080nm has the largest branching ratio β reaching 52.5%, which suggests the high possibility of photons emission generating from ^{4}F$_{3/2}$ $\rightarrow$ ^{4}I$_{11/2}$. The radiative lifetime τ_{rad} of the ^{4}F$_{3/2}$

level of Nd^{3+} is 165μs, which is a relative longer lifetime to heavily doped 3at.%Nd: Lu_2O_3 ceramic. Therefore, the Nd:Lu_2O_3 ceramic may be regarded as a potential solid-state laser material for diode laser pumping.

Table I. Absorption Spectrum Parameters of 3at.%Nd:Lu_2O_3 Ceramic

Excited states (Ground state $^4I_{9/2}$)	$\bar{\lambda}$ (nm)	f^{ed}_{exp} (10^{-6})	f^{ed}_{cal} (10^{-6})	$\sigma_{abs}(\lambda)$(10^{-20} cm^2)
$^4F_{3/2}$	879	1.35425	3.0515	1.87852
$^4F_{5/2}$	822	2.30001	3.44517	4.21565
$^2H_{9/2}$	807	2.32697	2.24979	2.99798
$^4F_{7/2}$+$^4S_{3/2}$	746	3.51479	1.14564	3.06642

Table II. Emission Spectrum Parameters of Transitions for Nd^{3+} from The $^4F_{3/2}$ to $^4I_{j'}$ Levels in 3at.%Nd:Lu_2O_3 Ceramic

Final State	A (s^{-1})	$\sum$(10^{-18} cm)	β (%)	T_{rad} (μs)
$^4I_{9/2}$	1591	6.76407	26.269	165
$^4I_{11/2}$	4391	13.34466	52.498	
$^4I_{13/2}$	75	0.50783	1.234	

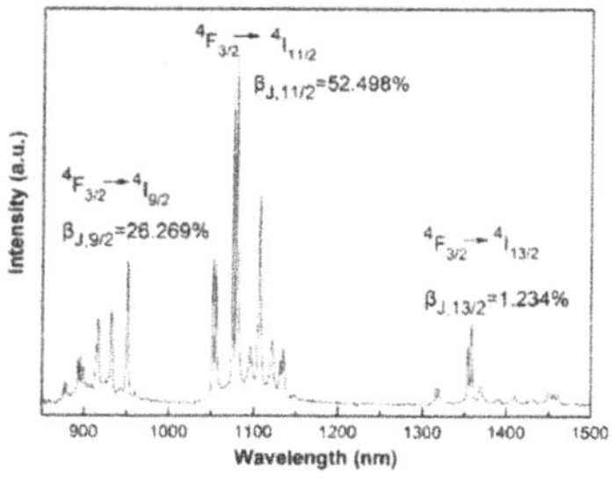

Figure 4. Fluorescence spectrum of 3at.%Nd:Lu_2O_3 ceramic

CONCLUSION

Transparent 3at.%Nd:Lu_2O_3 ceramics were fabricated by pressureless sintering at 1880°C for 8 h under flowing H_2 atmosphere. The polished Nd:Lu_2O_3 ceramic with thickness of 1.4mm achieves inline optical transmittance of 75.5% at the wavelength of 1080nm, corresponding to an optical attenuation coefficient of $0.0564mm^{-1}$. The absorption band at 807nm has a FWHM of 4 nm, which is suitable for AlGaAs diode-laser pumping. The absorption cross section σ_{abs} at 807 nm was 2.998×10^{-20} cm^2. Based on Judd-Ofelt theory, the intensity parameters Ω_2, Ω_4 and Ω_6 are obtained to be 5.26×10^{-20} cm^2, 7.13×10^{-20} cm^2, and 8.37×10^{-20} cm^2. The fluorescence branching ratio of $^4F_{3/2}\rightarrow$ $^4I_{11/2}$ transitions is 52.498% and τ_{rad} is 165μs.

ACKNOWLEDGEMENTS

This work was financially supported by the Natural Science Foundation of China (No. 50572115) and Basic Research Key Project of Shanghai Municipal (06JC14029).

* Corresponding author. Tel.: +86 21 56331793; fax: +86 21 56332694.
E-mail address: yshi@shu.edu.cn (Y. SHI), chengyan130@163.com (Y. CHENG)

REFERENCES

[1] A.Ikesue, I. Furusato and K. Kamata, Polycrystalline Transparent YAG Creamics by a Solid-state Reaction Method, *J. Am. Ceram. Soc.*, 78, 225-8 (1995).

[2] A. Ikesue, Y. L. Kang, T. Taira, T. Kamimura and K. Yoshida, Progress In Ceramic Lasers, *Annu. Rev. Mater. Res.*, 36, 397-429 (2006).

[3] K. Takaichi, H. Yagi, A. Shirakawa, K. Ueda, S. Hosokawa, T. Yanagitani and A. Kaminskii, Lu_2O_3:Yb^{3+} Ceramics-A Novel Gain Material for High-power Solid-state Lasers, *Phys. Stat. Sol. (a)* 202, R1-R3 (2005).

[4] J. Lu, K. Takaichi, T. Uematsu, A. Shirakawa, M. Musha and K. Ueda, Promising Ceramic Laser Material: Highly Transparent Nd^{3+}:Lu_2O_3 Ceramic, *Appl. Phys. Lett.*, 81, 4324-6 (2002).

[5] U. Griebner and V. Petrov, Passively Mode-locked Yb:Lu_2O_3 Laser, *Opt. Express*, 12, 3125-30 (2004).

[6] A. Ikesue, T. Kinoshita, K. Kamata, and K. Yoshida, Fabrication and Optical Properties of High-Performance Polycrystalline Nd:YAG Ceramics for Solid State Lasers, *J. Am. Ceram. Soc.*, 78, 1033–40 (1995).

[7] A.Ikesue, K. Kamata and K. Yoshida, Effects of Neodymium Concentration on Optical Characteristics of polycrystalline Nd:YAG Laser Ceramics, *J. Am. Ceram. Soc.*, 79, 1921-6 (1996).

[8] G.S. Ofelt, Intensities of Crystal Spectra of Rare-earth Ions, *J. Chem. Phys.*, 375, 511-20 (1962).

[9] B.R. Judd, Optical Absorption Intensities of Rare Earth Ions, *Phys. Rev.*, 127, 750-61(1962).

[10] W.T. Carnall, P.R. Fields, and K. Rajnak, Electronic Energy Levels in the Trivalent Lanthanide Aquo Ions. I. Pr^{3+}, Nd^{3+}, Pm^{3+}, Sm^{3+}, Dy^{3+}, Ho^{3+}, Er^{3+}, and Tm^{3+}, *J. Chem. Phys.*, 49, 4424-42 (1968).

[11] A.A. Kaminskii, G. Boulon, M. Buoncristiani, B.D. Bartolo, A. Kornienko, and V. Mironov, Spectroscopy of a new laser garnet $Lu_3Sc_2Ga_3O_{12}$:Nd^{3+}. Intensity luminescence characteristics, stimulated emission, and full set of squared reduced-matrix elements $|< \| >|^2$ for Nd^{3+} ions, *Phys. Stat. sol. (a)*, 141, 471-94 (1994).

FABRICATION AND LASER PERFORMANCE OF $(Yb_{0.05}Y_{0.95-x}La_x)_2O_3$ CERAMICS

Qiuhong Yang, Chuanguo Dou, Hongxu Zhou
School of Materials Science and Engineering, Shanghai University
Shanghai 200072, China
Qiang Hao, Wenxue Li, Heping Zeng
State Key Laboratory of Precision Spectroscopy, and Department of Physics, East China Normal University
Shanghai 200062, China

ABSTRACT

$(Yb_{0.05}Y_{0.95-x}La_x)_2O_3$ (x=0, 0.05, 0.10) transparent ceramics with low phonon energies of 370~376 cm^{-1}, were fabricated by commercial nanopowders and sintered at low temperature of 1600~1700°C under H_2 atmosphere. The lifetimes of $(Yb_{0.05}Y_{0.95-x}La_x)_2O_3$ ceramics are 0.99 ms, 1.12 ms and 1.48 ms with x=0, 0.05 and 0.10, respectively. For $(Yb_{0.05}Y_{0.85}La_{0.1})_2O_3$ ceramics laser, as low as 400 mW pumping threshold and a slope efficiency of 52% were realized. Broadband lasing spectrum up to 68 nm was observed in the tunable experiment.

INTRODUCTION

Transparent ceramic materials have gained much attention as potential solid-state laser materials in recent decades since the first report of laser oscillation in Nd^{3+} doped yttrium aluminum garnet ($Y_3Al_5O_{12}$ or YAG) in 1995[1]. Efficient and high-power laser operation in Nd:YAG and $Yb:Y_2O_3$ ceramic lasers has been demonstrated[2,3]. The investigation of Yb^{3+} doped materials has gained a lot of attention because ytterbium lasers have several advantages over Nd^{3+} doped materials, such as absence of the cross relaxation and excited-state absorption, low thermal loading, long fluorescence lifetime, high quantum efficiency, and so on[4]. Owing to perceptible electron–phonon interaction, Yb^{3+}-doped materials have broad absorption in near-IR which is suitable for laser diode (LD) pumping. The broad luminescence band $^2F_{5/2}$ - $^2F_{7/2}$ is also attractive for mode-locked and ultrashort pulse generation[5]..

The cubic sesquioxide Y_2O_3 has been a promising solid-state laser material because of its excellent thermal, chemical, optical, and mechanical properties. It is, however, extremely difficult to grow large-size high-quality Y_2O_3 single crystals because of their high melting temperature of 2430 °C and the polymorphic phase from C to a high temperature hexagonal phase H at about 2280 °C. Today it is easy to fabricate high quality Y_2O_3 ceramics at a relatively low sintering temperature of 1700 °C, which is about 700 °C lower than its melting temperature by a nanocrystalline and vacuum sintering technology[6]. The sintering temperature could be further decreased by adding La_2O_3 as a sintering aid in $Yb:Y_2O_3$[7].

Lanthanum doped Y_2O_3 transparent ceramics is a good infrared window material[8]. The effect of La_2O_3 on the spectroscopic properties of transparent $Yb:Y_2O_3$ ceramics has been investigated in our previous work and we found Yb-doped $Y_{2-2x}La_{2x}O_3$, a solid solution of Y_2O_3 and La_2O_3, has almost the same optical properties of $Yb:Y_2O_3$. Compared with Yb-doped Y_2O_3 ceramics, Yb-doped $Y_{2-2x}La_{2x}O_3$ has longer lifetime[7,9]. In this work, $(Yb_{0.05}Y_{0.95-x}La_x)_2O_3$ (x=0, 0.05, 0.10) transparent ceramics were fabricated by commercial nanopowders and sintered at low temperature of 1600~1700°C under H_2 atmosphere and its properties were characterized. We also report on low-threshold and broadband continuously tunable $(Yb_{0.05}Y_{0.85}La_{0.1})_2O_3$ ceramics lasers under high-power diode pump.

EXPERIMENTAL

$(Yb_{0.05}Y_{0.95-x}La_x)_2O_3$ (x=0, 0.05, 0.10) transparent ceramics were fabricated with commercial nanopowders. Specimens were synthesized by conventional solid-state processing, calcined at 1100 °C for 5 h in air atmosphere. Disks with 15 mm in diameter and 5~8 mm in thickness isostatically pressed at 200 MPa were sintered at 1600~1700°C for 50 h in H_2 atmosphere, then the specimens were milled and double polished with ϕ10×3.5 mm for spectral analysis and 3×3×1.5 mm for laser performance, respectively. Microstructures were observed with optical microscopy (Model BX60, OLYPMUS). The absorption spectra were measured with a spectrophotometer used Xe light as a pump source (Model V-570, JASCO) at room temperature. The fluorescence spectrum and fluorescence lifetime excited by 940 nm LD were measured with a fluorescence spectrum analyzer (Fluorolog-3, Jobin Yvon Spex, France) at room temperature. The Raman spectra were measured with a micro-laser Raman spectrometer (Model Invia+Plus, Renishaw, Britain) excited by 514.5 nm Ar^+ laser (Model 2000, Spectra Physics, USA) at room temperature.

RESULTS AND DISCUSSION

Fig.1 shows the optical microscopic photographs of $(Yb_{0.05}Y_{0.95-x}La_x)_2O_3$ (x=0, 0.05, 0.10) samples. It reveals that the grains size decreased with the increase of La_2O_3 content and $(Yb_{0.05}Y_{0.95-x}La_x)_2O_3$ (x= 0.05, 0.10) specimen displays uniform grains with average size of about 30-60 μm, and there are almost no pores in or between the grain boundaries.

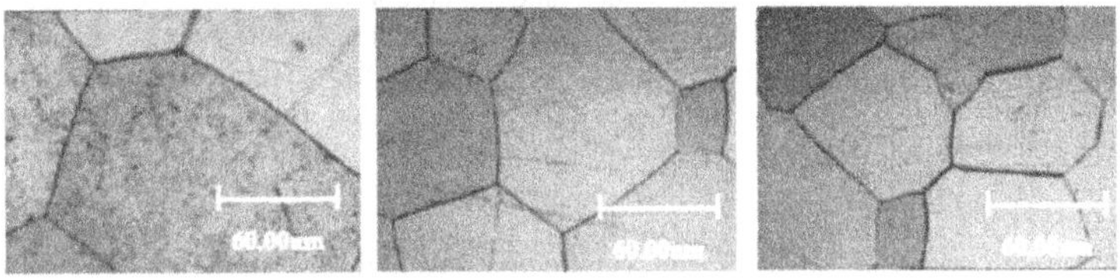

(a) (b) (c)

Fig. 1 Microstructures of $(Yb_{0.05}Y_{0.95-x}La_x)_2O_3$ ceramics. (a) x=0; (b) x=0.05; (c) x=0.1

Fig.2 shows the transmittance of $(Yb_{0.05}Y_{0.95-x}La_x)_2O_3$ (x= 0.05, 0.10) transparent ceramics. The transmittance dramatically increases with the increase of La_2O_3 content. The highest transmittance of $(Yb_{0.05}Y_{0.85}La_{0.1})_2O_3$ transparent ceramics in the wavelength of 1000-1100 nm reaches 81%. It means that the additive La_2O_3 can improved the sintering properties and increasing the transmittance of $Yb:Y_2O_3$ ceramics.

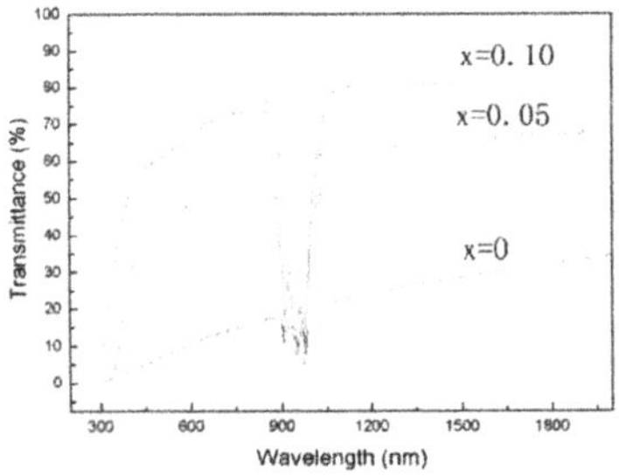

Fig. 2 transmittance of $(Yb_{0.05}Y_{0.95-x}La_x)_2O_3$ transparent ceramics

Fig.3 shows the Raman shift of $(Yb_{0.05}Y_{0.95-x}La_x)_2O_3$ ceramics. The phonon energies of $(Yb_{0.05}Y_{0.95-x}La_x)_2O_3$ ceramics are between 370 and 376 cm^{-1} and the FWHM (the full width of half-maximum) increases with the increase of La_2O_3 content. There exists a red shift for the Raman characteristic peaks after the adding of La_2O_3.

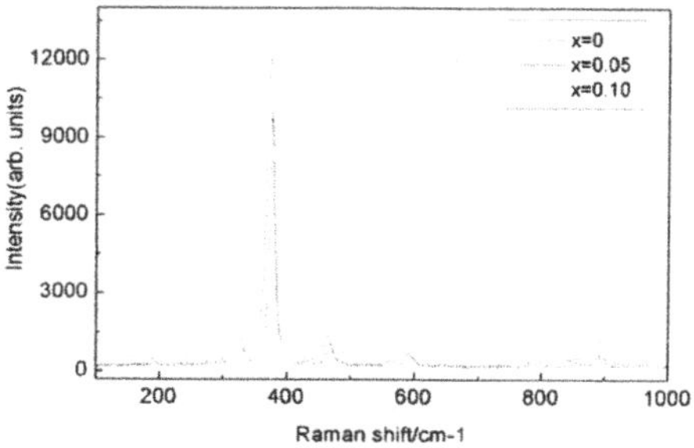

Fig. 3 Raman shift of $(Yb_{0.05}Y_{0.95-x}La_x)_2O_3$ ceramics

Fig.4 is the photograph of $(Yb_{0.05}Y_{0.85}La_{0.1})_2O_3$ transparent ceramics (3.5mm thick). The specimens have high transmittance of 81% in the laser output wavelength near 1100 nm.

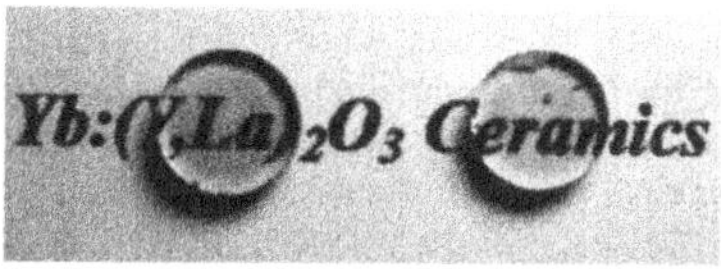

Fig. 4 Photograph of $(Yb_{0.05}Y_{0.85}La_{0.1})_2O_3$ transparent ceramics (3.5mm thick)

Based on the decay curves of the IR $^2F_{5/2}$ manifold emission of $(Yb_{0.05}Y_{0.95-x}La_x)_2O_3$ ceramics at room temperature, the lifetimes of $(Yb_{0.05}Y_{0.95-x}La_x)_2O_3$ ceramics were calculated to be 0.99 ms, 1.12 ms and 1.48 ms when x equals to 0, 0.05 and 0.10, respectively. The lifetime of $(Yb_{0.05}Y_{0.85}La_{0.1})_2O_3$ ceramics is much longer compared with that of $Yb:Y_2O_3$ ceramics. Long lifetime facilitates an enhanced energy storage for a high-power laser output. Since the radius of La^{3+} ion (101.6pm) is larger than that of Y^{3+} ion (89.3pm), the crystal lattice become larger after La_2O_3 doping into Y_2O_3, which makes the strength of the Y_2O_3 crystal field become weaker and resulted in weaker interaction between Yb^{3+} ions and O^{2-} ions. This results in lengthening the lifetime of Yb^{3+} in $Y_{1.8}La_{0.2}O_3$ host.

After optical polishing, the high quality $(Yb_{0.05}Y_{0.85}La_{0.1})_2O_3$ ceramic was cut as 1.5-mm long, 3×3 mm^2 in aperture for lasing test without any antireflection (AR) coatings on both surfaces. It was wrapped with an indium coil and mounted in a water-cooled copper holder to remove the thermal loads. The water temperature was controlled at 14°C. The laser performance was checked with a different fiber-coupled diode laser focused by 1:1 imaging system into the ceramic. A stable three-mirror laser cavity was applied, which consisted of a flat mirror M1 (AR-coated 940-976 nm, high-reflection-coated 1020-1120 nm), a concave mirror M2 (R=300 mm, high-reflection-coated 1020-1120 nm), and an output coupler (OC) with a transmission of 2.5% in 1020-1120 nm. A 980 nm high-brightness diode laser with pigtail fiber core of 50 μm and numerical aperture of 0.22 was used as the pump source. As low as 400 mW pumping threshold was realized at 1080 nm with the pump energy density of 20 KW/cm^2. The maximum output power was 1.0 W at a non-lasing absorbed pump power of 2.4 W (incident pump power of 3.1 W), and the corresponding slope efficiency was as high as 52%, as shown in Fig.5.

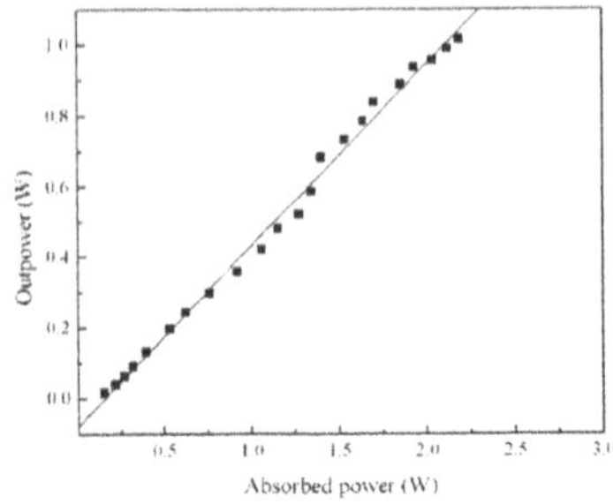

Fig. 5 The output power of a CW $(Yb_{0.05}Y_{0.85}La_{0.1})_2O_3$ ceramic laser

Despite the fact that the emission cross-section around 1030 nm is larger than that around 1070 nm, the ceramic laser oscillated around 1078 nm instead of 1030 nm, due to the strong re-absorption at 1030 nm of the Yb^{3+} ion.

Broad tunable $(Yb_{0.05}Y_{0.85}La_{0.1})_2O_3$ laser was achieved by inserting a silica prism into the stable three-mirror cavity. Pumped by the 940-nm diode laser, the $(Yb_{0.05}Y_{0.85}La_{0.1})_2O_3$ ceramic laser could be tuned from 1018 to 1086 nm, with a continuously tunable range as broad as 68 nm, as shown in Fig. 6. Our experimental results further reveal the promise of ceramic materials as laser gain host materials.

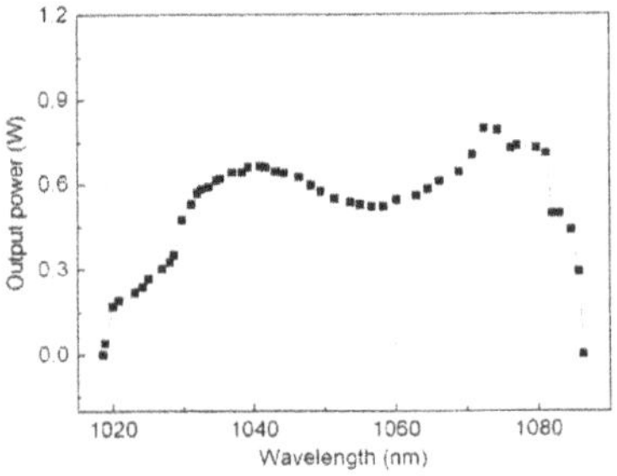

Fig. 6 The tunable curve for a $(Yb_{0.05}Y_{0.85}La_{0.1})_2O_3$ ceramic laser

CONCLUSIONS

$(Yb_{0.05}Y_{0.95-x}La_x)_2O_3$ (x=0, 0.05, 0.10) transparent ceramics were fabricated by commercial nanopowders and sintered at low temperature of 1600~1700°C under H_2 atmosphere. $(Yb_{0.05}Y_{0.95-x}La_x)_2O_3$ ceramics have low phonon energies of 370~376 cm^{-1}. The lifetimes of $(Yb_{0.05}Y_{0.95-x}La_x)_2O_3$ ceramics are 0.99 ms, 1.12 ms and 1.48 ms with x=0, 0.05 and 0.10, respectively. For $(Yb_{0.05}Y_{0.85}La_{0.1})_2O_3$ ceramics laser, as low as 400 mW pumping threshold and a slope efficiency of 52% were realized in the wavelength of 1078 nm. Broadband lasing spectrum up to 68 nm was observed in the tunable experiment from 1018 to 1086 nm.

Acknowledge

The work was supported by the National Natural Science Foundation of China (Grant No. 60578041).

REFERENCES

[1] A. Ikesue, T. Kinoshita, K. Kamata, and K. Yoshida, Fabrication and optical properties of high-performance polycrystalline Nd:YAG ceramics for solid-state lasers, *J. Am. Ceram. Soc.*, **78**, 1033-1040 (1995).

[2] A. Ikesue, and Yan Lin Aung, Synthesis and performance of advanced ceramic lasers, *J. Am. Ceram. Soc.*, **89**, 1936-1944 (2006).

[3] J. Kong, D.Y. Tang, B. Zhao, J. Lu, K. Ueda, H. Yagi, and T. Yanagitani, 9.2-W diode-end-pumped Yb:Y_2O_3 ceramic laser, *Appl. Phys. Lett.* **86**, 161116-18 (2005).

[4] W.F. Krupke, Ytterbium solid-state lasers-the first decade, *IEEE J. Sel. Top. Quantum Electron.*, **6**, 1287-1296 (2000).

[5] K. Takaichi, H. Yagi, J. Lu, J. Bission, A. Shirakawa, K. Ueda, T. Yanagitani, A.A. Kaminskii, Highly efficient continuous-wave operation at 1030 and 1075 nm wavelengths of LD-pumped Yb^{3+}:Y_2O_3 ceramic lasers, *Appl. Phys. Lett.* **84**, 317-319 (2004).

[6] N. Saito, S. Matsuda and T. Ikegami, Fabrication of transparent yttria ceramics at low temperature using carbonate-derived powder, *J. Am. Ceram. Soc.* **81**, 2023-2028 (1998).

[7] Q.H. Yang, J. Ding, H.W. Zhang, and J. Xu, Investigation of the spectroscopic properties of Yb-doped ytrium lanthanum oxide transparent ceramic, *Opt. Commun.* **273**, 238-241(2007).

[8] W.H. Rhodes, G.C. Wei, and E.A. Trickett, Lanthana-doped yttria: a new infrared window material, *SPIE Proc.* **683**, 12-18 (1986).

[9] Q.H. Yang, J. Xu, C.G Dou, H.W. Zhang, J. Ding , Z.F. Tang, Effect of La_2O_3 doping on the spectroscopic properties of transparent Yb:Y_2O_3 laser ceramics, *Acta Phys.Sin.* **56**, 3961-3965 (2007).

A STUDY ON THE $ZnO-Al_2O_3-SiO_2$ SYSTEM NdF_3- DOPED TRANSPARENT FLUORIDE-OXIDE GLASS-CERAMICS

Jing Shao [1], Guohui Feng[1,2], Hongbo Zhang[1],Guangyuan Ma[1] ,Chunhui Su[1,3]
[1]School of Chemistry and Environmental Engineering, Changchun University of Science and Technology
[2]Plastic Institute of Ji Lin Province
[3]Jilin Teacher's Institute of Engineering and Technology, Changchun, Ji Lin,China

ABSTRACT

The system of zinc aluminum silicate ($ZnO-Al_2O_3-SiO_2$) glass doped-NdF_3 and CaF_2 as nucleus was melted in a crucible at 1500°C for 2h, and then two-step heat treatment for the nucleation and the crystal growth was used to prepare transparent glass-ceramics. The characterizations were performed by differential thermal analysis, X-ray diffraction, scanning electron microscopy, UV–Vis-NIR scanning spectrophotometer and fluorescence spectra. When heat-treated below 1000°C, main crystal phase was SiO_2 and the minor crystal phases were $ZnAl_2O_4$, Zn_2SiO_4 and CaF_2. The mean grain size calculated by Sherrer equation was consistent with the result of SEM, the grain size distributes normally. The lattice parameter calculated by the XRD data was a=1.3405nm(±0.0034), which is very closed to that of standard sample. The optical transmittances of glass-ceramics were 80%.The crystallization heat-treated temperature increased with the optical transmittances decreased. The fluorescence spectra of the glass and glass-ceramics were measured. The fluorescent characteristics of Nd^{3+} ion were stronger in the glass-ceramic than in the glass matrix.

INTRODUVTION

Glass-ceramics can be defined as composite materials, lying between glasses and crystals, and combining the good optical and mechanical properties of crystals with easy shaping ability of glasses[1,2]. They are synthesized from melt mixed batches and glasses are annealed for appropriate temperature and time to room temperature. Transparent rare-earth doped glasses-ceramics have been studied for their optical properties. Laser emission has already been observed in Nd^{3+}-doped glass-ceramics[3,4]. Oxyfluoride glass-ceramics are two –phase material (one aluminosilicate glassy phase and another fluoride nanocrystalline phase) that have proved to be interesting matrix for their optical properties as the fluoride environment of the rare-earth ,with its low phonon energy, contributes to reduce the non-radiative de-excitations, leading to an improved quantum efficiency of the radiative emission with respect to an oxide environment[5,6].Nowadays, zinc aluminum silicate glass-ceramics have very promising applications in optical field[7-9]. In this paper, CaF_2 behaved as a nucleating agent and the effect of ceramization on optical transmission of the host and fluorescence of Nd^{3+} ion in the new zinc aluminum silicate system is reported.

EXPERIMENTAL PROCEDURE

Glasses of the 30g batch were prepared by using SiO_2 (99.9%), Li_2CO_3 (99%), Al_2O_3 (99.9%), ZnO (99.9%), CaF_2 (99%), Sb_2O_3 (99.9%) and rare earth oxides 1% NdF_3 (99.99%) used to produce glasses and were as for the base glass preparation. CaF_2 was used as nucleating agents, and Sb_2O_3 was used as the clarifier. The starting materials were fully mixed and melted in a Pt crucible at 1550°C for 2h duration in the $MoSi_2$ electrical resistance heating furnace having PID temperature controller. Then the melt was poured on a stainless steel plate at room temperature, subsequent obtained samples were annealed at 500°C for 2h, and allowed to cool down to room temperature at a rate of 20°C /h to relieve internal stresses. Then two-step heat treatments for the nucleation and the crystal growth were adopted. The objective is to obtain transparent glass-ceramics based on ZnO-Al_2O_3-SiO_2 system.

The study has been carried out using differential thermal analysis (DTA), X-ray diffraction (XRD), FEG-ESEM and UV-VIS-NIR spectrophotometer etc. The nucleation and crystallization temperatures were determined by differential thermal analysis (DTA). Specimens heat-treated at various durations were analyzed by the X-ray diffraction (XRD) to determine the optimum conditions for nucleation and the crystal growth. Scanning Electron Microscopy (SEM) was used to study the glass-ceramics morphology, the grain size and distribution in the residual glass matrix. The fluorescence spectra of the glass and glass-ceramics were measured. The transmittance was measured by UV-VIS-NIR scanning spectrophotometer.

RESULTS AND DISCUSSION

In order to prepare glass-ceramics, differential thermal analysis was performed using Pyris-Diamond (Perkin-Elmer, USA) with a heating rate of 10°C /min recorded over 30~1000 in order to determine the glass transition temperature of the base glass and its stability versus crystallization. Various heat-treatment schedules given in Table I were applied in order to study the nucleation and crystallization processes and the change of microstructure. X-ray powder diffraction system (D/max 2500V, Rigaku, Japan) with Cu Kα radiation (1.54056 Å) has been used for identification of the crystalline phase. Each pattern was scanned from 2θ= 5° to 80° at the rate of 4°/min. The resulting glass-ceramic samples have been investigated by UV-VIS-NIR scanning spectrophotometer (UV-3101PC, SHIMADZU, Japan). The morphology of glass-ceramics samples were studied using Scanning Electron Microscope (FEI Corp., XL30ESEM FEG,USA).

Table I. The Heat-treatment Schedules for Glass Samples

Samples	Temperature/°C×Time/h for Nucleation	Temperature/°C×Time/h for crystallization	Main Crystal phase	Minor Crystal phase	Fluorescence properties
1#	670×1	740×1	SiO_2	-	Better
2#	670×2	740×1	SiO_2	Zn_2SiO_4 CaF_2	Better
3#	670×3	740×1	SiO_2	$ZnAl_2O_4$ Zn_2SiO_4 CaF_2	Better
4#	670×2	740×2	SiO_2	$ZnAl_2O_4$ Zn_2SiO_4 CaF_2	Low
5#	700×1	800×1	SiO_2	$ZnAl_2O_4$ Zn_2S CaF_2	Lower

Typical DTA curve of the ZAS base glass is shown in Figure 1. The DTA curve is characterized by some endothermic peak. The endothermic peak is very obvious at 746°C, 824.6°C, 963°C and 1024°C, which indicates there exists transformation of crystal configuration. The endothermic peak is not very obvious at 963°C and 1024 and the endothermic base line shift at 650°C~740°C gives the beginning of glass nucleation temperature zone.

XRD patterns for samples under present investigations are shown in Figure 2. XRD patterns of the glass-ceramic indicate the evidence of crystallites formed. All the peaks were analyzed with PCPDF files. The diffraction peaks attributable to main crystal phase-SiO_2 observed for all of the heat-treated specimens. The minor crystal phase were $ZnAl_2O_4$, Zn_2SiO_4 and CaF_2 with increase in heat-treatment temperature and time duration. On the other hand, in the glass-ceramics the formation of the CaF_2 phase is favored by the incorporation of Nd^{3+} ions in the crystalline phase, playing the role of heterogeneous nucleating agents and partially replacing the Ca^{2+} ions.

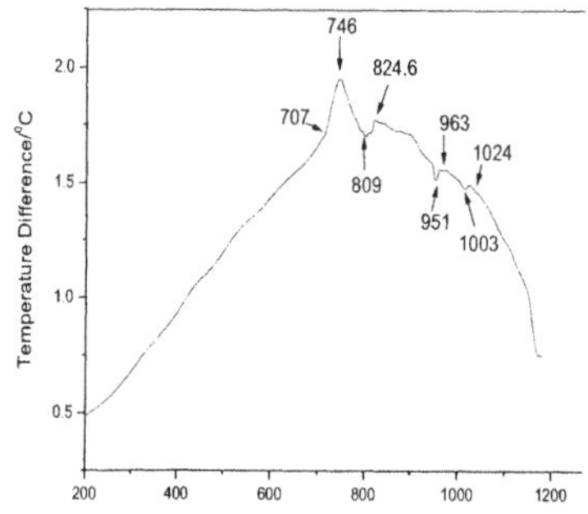

Figure 1. DTA curve for base glass

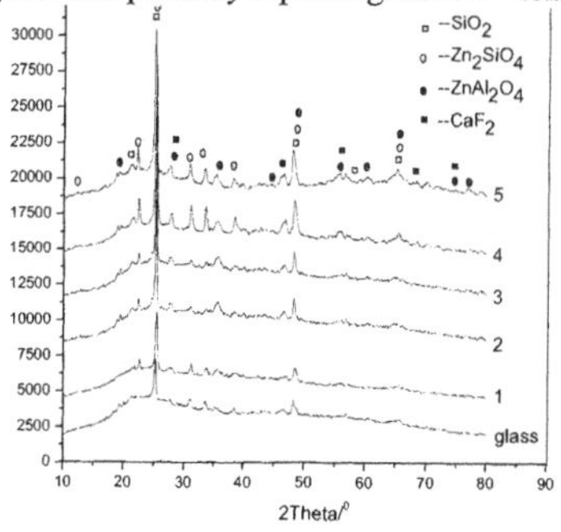

Figure 2. XRD patterns of the base glass and glass-ceramics of ZAS system

The d-spacing of standard sample (JCPDS 71-0968) signed by d_s. The lattice parameters α of the cubic gahnite have been calculated from the experimental lattice spacing with the expression relative to the cubic system:

$$d_{hkl}^2 = \frac{a^2}{(h^2 + k^2 + k^2)} \qquad (1)$$

The high intensity and well defined lines located namely (200), (222), (321), (410), (421), (520), (530), (611), (541), (444), (650) and (741) lines. The a value has been calculated using d_s for these lines respectively. The mean value of the cell parameters of the crystalline phase β-SiO_2 of sample 3# is $\bar{a}$=1.3402nm close to the standard value a=1.3405nm (±0.0034). The mean grain sizes evaluated by Scherrer's equation were 10nm around.

SEM micrograph of glass-ceramic samples 3# are shown in Figure 3 (a), (b), (c) and (d). It can be verified that the grain size calculated by the Scherrer's formula are well in line with the practical sizes measured by SEM.

The UV-VIS-NIR transmittances of glass and glass-ceramic specimens have been measured in the range 240~2000nm. The transmission curves are displayed in Figure 4. The transmittance of glass specimen reach nearly 90% within 700~2400nm, but the transmittance of the glass-ceramics decreases from 80% to 42% within 240~2000 compared with base glass. Heat treatment temperatures and time influence on the microstructure and the transmittance.

Emission measurements showed in Figure 5 performed for the glass-ceramics samples with excitation between 800-1500 nm .The emission spectra show three main fluorescence emission bands centered around 896nm,1062nm and 1329nm corresponding to the electronic transitions of $^4F_{3/2}\rightarrow{}^4I_{9/2}$, $^4F_{3/2}\rightarrow{}^4I_{11/2}$ and $^4F_{3/2}\rightarrow{}^4I_{13/2}$, respectively. It can be observed that the most intense emission band is located at 1062nm for the base glass and glass-ceramics, but peak intensity in glass-ceramics is higher compared to the base glass mainly due to low-energy phonon environments with higher quantum efficiencies.

Figure 3. SEM micrograph of glass-ceramic samples 3#

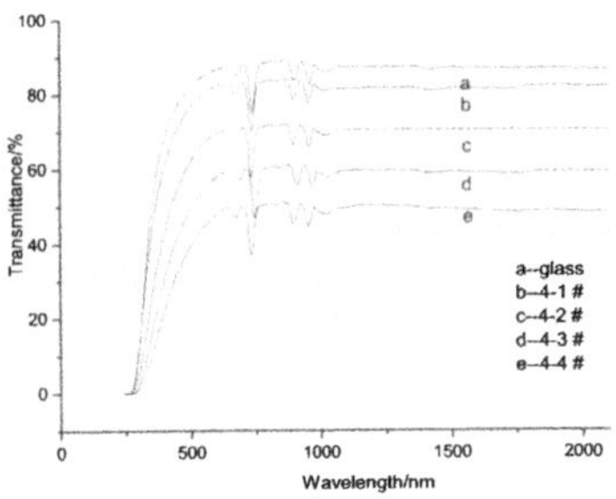

Figure 4. UV-VIS-NIR transmittance of glass and glass-ceramic samples

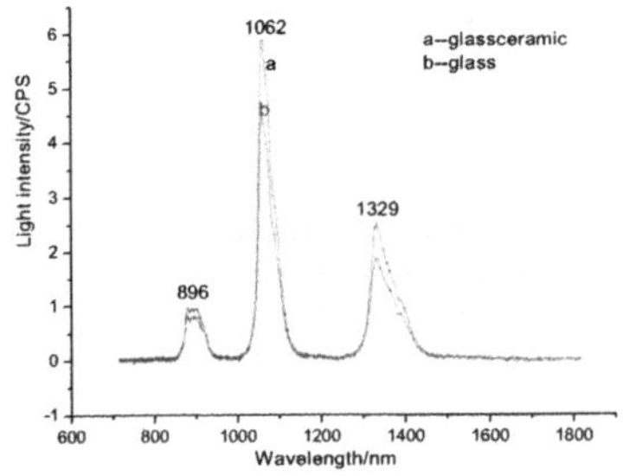

Figure 5. Fluorescence spectra of Nd^{3+} ions in ZAS system between 800-1500 nm

CONCLUSIONS

Zinc aluminum silicate doped-NdF_3 glasses have been prepared by conventional melt and quenching technique, and subsequently converted to transparent glass-ceramics by controlled nucleation and crystallization. Results of XRD indicate that there is an obvious structure change in the glass-ceramics compared to the precursor glass. The transparent ZAS glass-ceramics with main crystal phase-SiO_2 solid solutions and minor crystal phase-Zn_2SiO_4, CaF_2 and the transparent ZAS glass-ceramics with main crystal phase-SiO_2 solid solutions minor crystal phase- Zn_2SiO_4,$ZnAl_2O_4$ and CaF_2 were obtained after ZAS system base glasses were nucleated at 650°C-746°C and crystallized at 740°C-1024°C. The samples lose its transparency when the heat-treatment temperature increased. The effectively induced cross section σ_{in} of Nd^{3+} ions locating in crystal phase-CaF_2 of the transparent glass-ceramics sample at 1064nm advanced than in base glass. The mean grain sizes, calculated by the Scherrer's equation using the results of XRD, keep well with the SEM results. Transparent glass-ceramics were obtained finally.

ACKNOWLEDGEMENT

This work was supported by the Chinese Education Ministry for financial support under Fund Item: KB20026.

REFERENCES

[1] W. Pannhorst, Nucleation and Crystallization in Glass and Liquids, ed. M. Weinberg, *American Ceramics Society, Westerville,OH*.267 (1993).

[2] Wollfgang Pannhorst, Glass ceramics: State-of- the-art, *J.Non-Cryst. Solids*, **219**, 198-204 (1997).

[3] J.del-Castillo, J.Mendez-Ramos,etc.,Gain cross-section of 1.06μm emission in Nd^{3+}-doped SiO_2-LaF_3 glass-ceramics prepared by sol-gel method, *J. Non-Cryst..Solids*, **354**, 2000-2003 (2008).

[4] F.Lahoz, I.R. Martin, Rare earths in nanocrystalline glass-ceramics, *J. Optical Materials*. **27**, 1762-1770 (2005).

[5] S. González-Pérez, I.R. Martín, F. Rivera-López, F. Lahoz, Temperature dependence of $Nd^{3+}\rightarrow Yb^{3+}$-energy transfer processes in co-doped Oxyfluoride glass ceramics, *J. Non-Cryst..Solids*, **353**, 1951-1955 (2007).

[6] D.Q. Chen, Y.S. Wang, etc., Crystallization and fluorescence properties of Nd^{3+}-doped transparent oxyfluoride glass ceramics, *J. Materials Science and Engineering B*, **123**,1-6 (2005).

[7] X.L. Duan, D.R. Yuan, X.F. Cheng, etal. Spectroscopic properties of Co^{2+}: $ZnAl_2O_4$ Nanocrystals in sol-gel derived Glass-Ceramics, *J. Physics and Chemistry of Solids*, **64,** 1021(2003).

[8] X.L. Duan, D.R. Yuan, et al. Preparation and Characterization of Co^{2+}-Doped $ZnO-Al_2O_3-SiO_2$ Glass-Ceramics by the Sol-Gel Method,*J. Materials Research Bulletin*, **38**, 705(2003).

[9] T. Suzuki, K. Horibuchi, Y. Ohishi, Structurl and optical properties of $ZnO-Al_2O_3-SiO_2$ system glass-ceramics containing Ni^{2+}-doped nanoceystal, *J. Non-Crys. Soilds*, **351**, 2304 (2005).

SYNTHESIS OF NANO-SIZED Lu_2O_3 POWDER FOR TRANSPARENT CERAMICS FABRICATION USING CARBONATE DERIVED PRECURSORS

Xiaodong Li, Xudong Sun, Ji-Guang Li, Zhimeng Xiu, Di Huo, and Yan Liu
Key Laboratory for Anisotropy and Texture of Materials (Ministry of Education), Northeastern University, Shenyang 110004, China;

ABSTRACT

Nano-sized Lu_2O_3 powders were synthesized with a precipitation technique using optimized processing parameters. It was found that the pH value of the reaction system after precipitation was among the key factors that influence the sinterability of the powders, while the influences of the aging temperature and aging time on sintering property were relatively insignificant. Cubic Lu_2O_3 phase has formed by calcination at 600°C. The average crystalline sizes were 31, 36, 46 and 58 nm, respectively, for the samples calcined at 800, 900, 1000 and 1100°C. The powder calcined at 800°C showed best sinterability and transparent Lu_2O_3 ceramic (~60% transparency) was fabricated by vacuum sintering at 1700°C for 5h.

INTRODUCTION

The sesquioxide Lu_2O_3 is an attractive host material for high power solid-state lasers, due to its excellent thermo-mechanical properties.[1] It can be doped with all rare earth ions as well as transition metal ions, thus providing a large number of lasing wavelengths. However, till now large crystals of high optical quality are not available. The reason is the high melting point of ~2500°C, which makes the growth from the melt extremely difficult.

In the past two decades, the fabrication of laser-gain crystalline ceramics by using the non-pressing vacuum sintering method has been widely investigated.[2] The vacuum sintering of transparent ceramics rigorously invokes the starting powders of high sinterability. It has been known that the sinterability of a wet-chemically derived oxide powder has a close relationship with the properties of its precursor. In practice, some carbonates precipitated with ammonium hydrogen carbonate (AHC) as the precipitant, proved superior as precursors for less-agglomerated oxide powders.[3] Although synthesis of ultra-fine powders via AHC precipitation techniques have been investigated for a lot of rear earth oxides, research work on fabrication of transparent ceramics of Lu_2O_3 has been rarely found in the literature. This work investigated the fabrication of transparent Lu_2O_3 ceramics using high-sinterability powders synthesized via the precipitation method.

EXPERIMENTAL PROCEDURES

The raw material used was commercially available high purity (>99.99%) Lu_2O_3 powder. Aqueous solution of $Lu(NO_3)_3$ was prepared by dissolving the oxide powder into concentrated aqueous HNO_3 solution. For a typical precipitation, concentrated ammonium hydrogen carbonate (AHC) solution(~2.5 M) was added to the mother solution (0.5 M) at a rate of 3 ml/min. After aging for various times, the white precipitate was collected by centrifugation, which was then washed repeatedly with distilled water, rinsed with acetone, and finally dried at 50°C. The dried precursor powder was lightly pulverized with a mortar and pestle and then calcinated at various temperatures for 4 h with a heating rate of 300°C/h.

Phase identification of the resultant powders were performed by X-ray diffractometry (XRD, Model X'pert PRO MPD, PANalytical, Almelo, the Netherlands) using CuK_α radiation. Crystallite size of the calcined powders was determined by X-ray line broadening and calculated using the Scherrer equation:

$$d_{Crys} = \frac{0.89\lambda}{B\cos\theta},$$

where $B=(B_o^2-B_c^2)^{1/2}$, B_o is the full width at half maximum, B_c is the correction factor for instrument broadening, θ is the angle of the peak maximum, and λ is the Cu K_α weighted average wavelength. Fourier transform infrared (FT-IR) spectroscopy (Model Spectrum IR, Perkin-Elmer) was performed by the standard KBr method. Particle morphology was observed via scanning electron microscopy (SEM, Model SSX-550 Shimadzu, Tokyo).

Constant-rate-of-heating (CRH) sintering of the powder compacts was performed using a thermal mechanical analyzer (TMA, Model SETSYS Evolution 2400, Setaram, France) up to 1600°C in stagnant air at a heating rate of 5°C/min and a cooling rate of 15°C/min. Specimens used for the measurement were compacted by bi-axial pressing at 200 MPa into short cylinders of φ8 mm×5 mm. The green compacts with dimension of φ14 mm×3 mm were used for vacuum sintering. The samples were heated at a rate of 5°C/min to the desired temperatures, and cooled at 30°C/min after holding for 2 h. The vacuum in the furnace was about 5×10^{-3} Pa during the holding period.

RESULTS AND DISCUSSIONS

Optimization of Processing Parameters

A lot of processing parameters are known to influence the sinterability of the oxide powders synthesized via carbonate precipitation method. Through orthogonal design method, selected parameters were optimized firstly for Lu_2O_3 powder synthesis. The factors investigated and their levels adopted are shown in Tab.1. It is found that the pH value of the reaction system after precipitation is among the key factors that influence the sinterability of the powders, while the influences of the aging temperature and aging time on sintering property are relatively insignificant. The effects of pH thus were further investigated and optimized.

Figure 1 shows the pH value versus HCO_3^{-1}/Lu^{3+} molar ratio (R) for the reaction system. The precipitation initiated at pH~4.5, as evidenced by the occurrence of the turbidity and quick reach of a plateau (~4.7) with further drip of the carbonate precipitator. The pH increased gradually again at R>3, followed by a rapid increase after R>3.8, then reached a plateau at pH~7.6. It is interesting to note a rapid decrease of pH at R~6, indicating that some reactions involving consumption of OH^- species might have occurred at this stage. After this, the pH increases gradually and reached the initial pH of AHC solution with a further increase of the R value.

Table 1. The factors and levels

levels	pH	Aging time (h)	Aging temperature (°C)
1	***5.5***	0	20
2	6	14	30
3	7	24	***40***
4	8	***48***	50

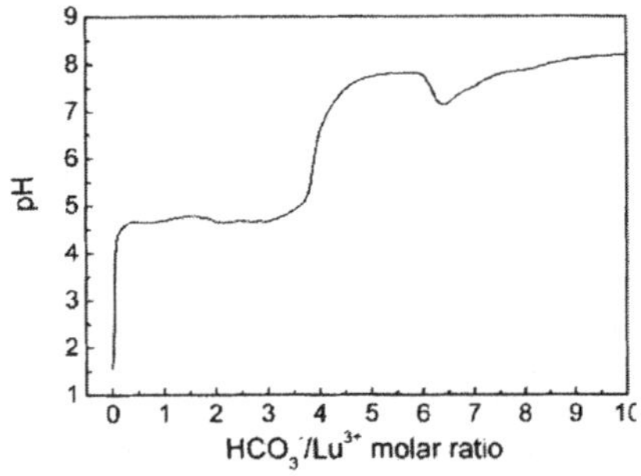

Figure 1. pH value versus HCO_3^{-1}/Lu^{3+} molar ratio (R) for the reaction system

Figure 2 shows the FT-IR absorption spectra of the dried precursor powders synthesized under various pH values. IR spectra of the precursors at pH =5.5, 6, and 7 were almost identical. The broad absorption band at 3390 cm^{-1} is assigned to O-H stretching. The two intense peaks at 1562 cm^{-1} and 1391 cm^{-1} are assigned to the asymmetric stretch of C-O in CO_3^{2-}, while the absorption peaks at 1083 cm^{-1} and 840 cm^{-1} are due to the symmetric stretch of C-O band and deformation vibration of C-O in CO_3^{2-}, respectively. These absorption peaks indicates that the precursors might be hydroxyl carbonate. For the precursor obtained at pH=8, additional adsorptions at 3608 cm^{-1} was observed, suggesting a different compound as compared with precursor of lower pH values. This result is in accordance with subsequent XRD measurement.

Figure 3 shows the XRD pattern of the precursor powders obtained at different pH values. For the precursor prepared at pH=5.5 and 6, nearly identical patterns were observed. The precursors obtained at pH=7, and 8, on the other hand, show quite different XRD patterns in comparison with those of lower pH values. All these patterns could not be indexed to known lutetium compound, which means that the lutetium compounds prepared by our method are new phases that have not been reported in the data-base of JCPDS. Due to limited literature for the lutetium compounds, we cannot decide the definite composition of the precursors at this time.

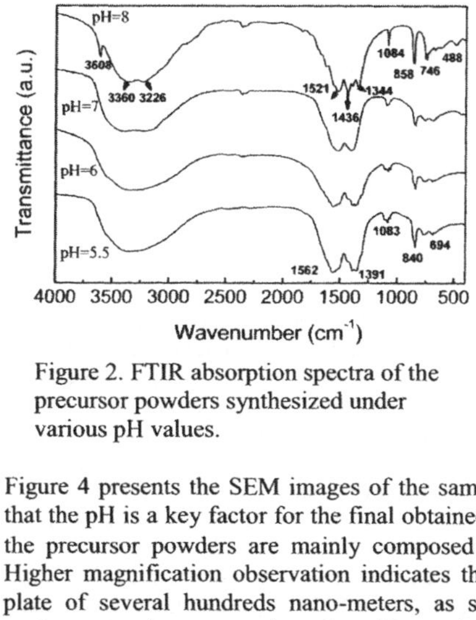

Figure 2. FTIR absorption spectra of the precursor powders synthesized under various pH values.

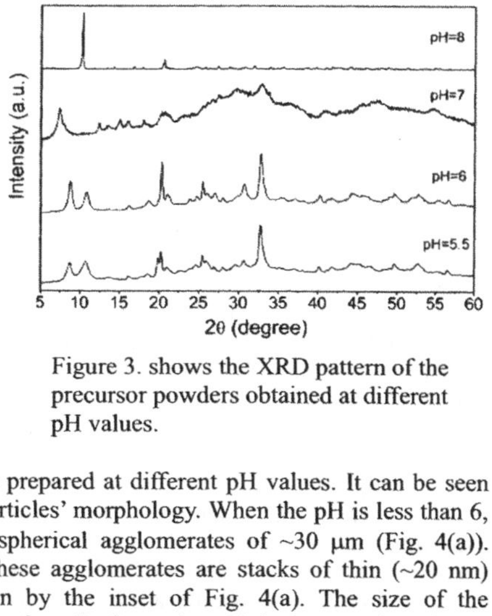

Figure 3. shows the XRD pattern of the precursor powders obtained at different pH values.

Figure 4 presents the SEM images of the samples prepared at different pH values. It can be seen that the pH is a key factor for the final obtained particles' morphology. When the pH is less than 6, the precursor powders are mainly composed of spherical agglomerates of ~30 μm (Fig. 4(a)). Higher magnification observation indicates that these agglomerates are stacks of thin (~20 nm) plate of several hundreds nano-meters, as shown by the inset of Fig. 4(a). The size of the agglomerates increases when the pH was increased to 7. Once the pH is increased to 8, the precursor particles became large-sized plates with rhombus morphologies (Fig. 4(c)).

The sinterability of the powder samples obtained at different pH values were compared by sintering at identical conditions (1500°C×2h). The relative sintered densities were 96.7, 96.4, 94.3, and 89.7%, for the sample of pH=5.5, 6, 7 and 8, respectively. Obviously, the powders obtained from precursors of lower pH values have better sintering behavior. Hereafter, the precursor powder obtained at pH=6, aging time=24 h, and aging temperature=25°C, was further characterized to obtain sinterable oxide powders.

Figure 4. SEM images showing morphologies for the powder prepared at pH=6 (a), 7 (b), and 8 (c), with inset showing high magnification images.

Synthesis of Lu_2O_3 Powder

Figure 5 shows the X-ray diffraction patterns for the precursor powder obtained at pH=6 and its calcined products. Similar to previous result, the precursor could not be indexed to a known compound of lutetium in JCPDS. When calcined at 200°C, the reflections of the precursor were still observable. After calcinating at 400°C, amorphous phase formed due to the decomposition of the precursor. For the sample calcined at 600°C, peaks belonging to the reflections of cubic crystalline Lu_2O_3 occurred, suggesting that the crystal has formed at this stage. Further increasing the annealing temperature from 700 to 1100°C, the peaks become even sharper and stronger, which is due to the increase of crystallinity. The average crystalline sizes were 31, 36, 46 and 58 nm, respectively, for the samples calcined at 800, 900, 1000 and 1100°C.

Figure 6 shows the FT-IR spectra of the precursor and its calcined products. For the precursor, the spectrum shows the absorption peaks of H_2O and OH^- group (near 3373 cm^{-1}) and the CO_3^{2-} anion (near 1581, 1384, 1092, 837, 767, 696 cm^{-1}). These spectra quite similar to that of the rear-earth carbonate reported in previous works.[4] It is known that the solution of ammonium hydrogen carbonate is an alkalescent buffering solution, with three kinds of anions formed in it. Among them, both OH^- and CO_3^{2-} can form precipitates with metal anions. Thus composition of the precursor will be a result of competition between OH^- and CO_3^{2-} in combination with metal cations. Because both the CO_3^{2-} concentration and the CO_3^{2-} to OH^- molar ratio in an AHC solution with pH=6 are very high, and the K_{sp} of hydrate is usually far more than that of carbonate. So the formation of rear earth carbonate in AHC precipitation was always observed. It is thus supposed that $Lu_2(CO_3)_3{\bullet}nH_2O$ will be most likely the precipitation products under present precipitation condition.

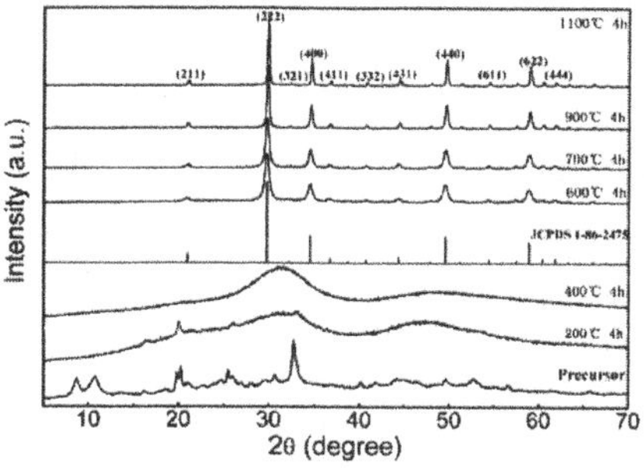

Figure 5. XRD patterns for the precursor powder obtained at pH=6 and its calcined products.

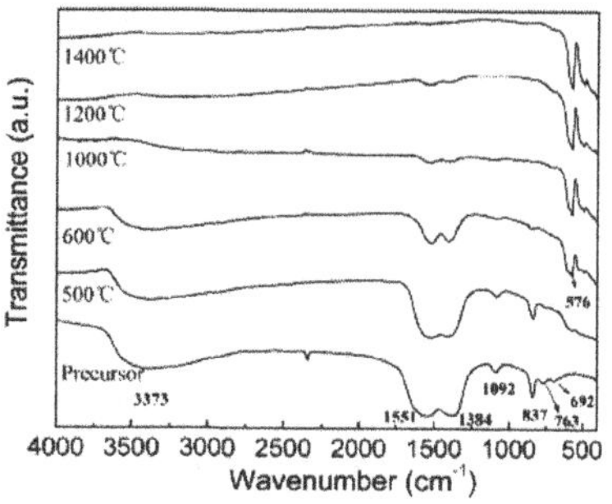

Figure 6. FTIR spectra of the precursor and its calcined products.

For the sample calcined at 500°C, the absorptions due to the carbonate species are declined. In the spectrum of the sample calcined at 600°C, the strong peak near 576 cm^{-1} is associated with the vibration of the Lu-O bond, indicating that Lu_2O_3 has formed by calcination at 600°C. The results coincide with that of XRD patterns. It should be noted that the peaks near 3373 cm^{-1} and 1551, 1384 cm^{-1} are still significant, suggesting the existence of un-decomposed carbonate species. On the other hand, the absorptions due to the carbonate species for samples calcined >1000°C are all weaker than that of the precursor, which is probably due to the residual carbonate or absorption of H_2O and CO_2 from ambient atmosphere.

The sintering behaviors for the powders obtained at different calcination temperatures were characterized by dilatometry, as shown in Fig. 7. Although the powder calcined at 800°C has a lower green density of ~46%, it begins shrinkage at a much lower temperature of ~800°C, due to its finer crystallite size. The relative density of this powder exceeds the other three samples at ~1120°C, and nearly full density was obtained after direct heating to 1600°C, revealing its excellent sinterability. On the other hand, the other three samples showed increased onset temperature of rapid shrinkage with the increase of the calcination temperature. The final densities were 95.0, 91.9, and 87.1%, respectively, for the powders calcined at 900, 1000, and 1100°C, as shown in Fig. 7(a).

The different sintering behaviors of the samples were also clearly revealed on the shrinkage rate curves, as shown in Fig. 7(b). Sintering of the powder calcined at 800°C seems composed of two major events of rapid shrinkage, which peaked at ~1100°C and ~1500°C, respectively. These are ascribed to the sintering of fine particles originally existed in the precursor and the sintering of the crystalline lutetia particles obtained by calcination, respectively. The influence of the first event on the sintering becomes less significant for the sample calcined at 900°C, and then totally disappeared for the samples calcined at 1000 and 1100°C. Consequently, only one maximum of rapid shrinkage was observed for the later two samples, exhibiting shrinkage behaviors for a powder with a mono-dispersed particle size distribution. The two-stage sintering was also observed for YGO powders in our previous works, although it occurred for the powder calcined at a much higher temperature (1000°C).[5]

Transparent ceramics fabrication was performed by vacuum sintering at 1700°C for 5h. All the sintered samples revealed reasonable transparency. Among them the sample calcined at 800°C showed a highest transparency of ~60% in visible light range.

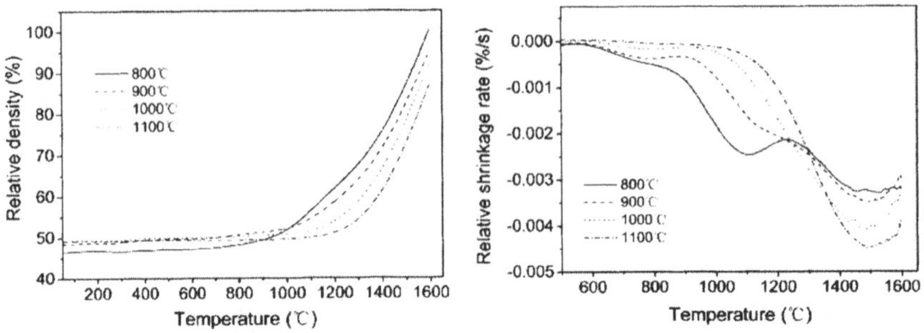

Figure 7. Density (a), and shrinkage rate (b) curves for the powders calcined at different temperatures.

SUMMARIES

The precipitation technique using ammonium hydrogen carbonate as the precipitant was adopted for synthesizing nano-sized Lu_2O_3 powders in this work. It was found that the pH value of the reaction system after precipitation was among the key factors that influence the sinterability of the powders, while the influences of the aging temperature and aging time on sintering property were relatively insignificant. The precursor powder synthesized at pH=6 was further investigated by calcined at different temperatures. Cubic Lu_2O_3 phase has formed by calcination at 600°C. The average crystalline sizes were 31, 36, 46 and 58 nm, respectively, for the samples calcined at 800, 900, 1000 and 1100°C. The powder calcined at 800°C showed best sinterability and reached nearly full density after direct heating to 1600°C.

ACKNOWLEDGMENT

This work was supported by the National Natural Science Foundation of China (50772020, 50672014), National Science Fund for Distinguished Young Scholars (50425413), and Program for New Century Excellent Talents in University (NCET-25-0290).

REFERENCES

[1] L. Fornasiero, E. Mix, V. Peters, K. Petermann, and G. Huber, New Oxide Crystals for Solid State Lasers, *Cryst. Res. Technol.*, **34,** 255-260, (1999).

[2] M. Dubinskii, L. D. Merkle, J. R. Goff, G. J. Quarles, V. K. Castillo, K. L. Schepler, D. Zelmon, S. Guha, L. P. Gonzalez, M. R. Rickey, J. J. Lee, S. M. Hegde, J. Q. Dumm, G. L. Messing, and S.-H. Lee, Processing Technology, Laser, Optical and Thermal Properties of Ceramic Laser Gain Materials, *Proceedings of SPIE*, **5792**, 1-9 (2005).

[3] J-G. Li, T. Ikegami, J.H. Lee, and T. Mori, Well-sinterable $Y_3Al_5O_{12}$ Powder from Carbonate Precursor, *J. Mater. Res.*, **15**, 1514-1523 (2000).

[4] I.Y. Park, D.J. Kim, J.W. Lee, S.H. Lee, and K.J. Kim, Effects of Urea Concentration and Reaction Temperature on Morphology of Gadolinium Compounds Prepared by Homogeneous Precipitation, *Mater. Chem. Phy.* **106,** 149-157 (2007).

[5] X.D. Li, Z.M. Xiu, L.L. Bai, T. Gao, Y.N. Liu, X.Z. Hu, X.D. Sun, Synthesis of $(Y,Gd)_2O_3$:Eu (YGO:Eu) Nano-Powder and Fabrication of Transparent Ceramics, *J. Inorg. Mater.*, **21**, 157-161 (2006).

PREPARATION AND INVESTIGATION OF TRANSPARENT YAG CERAMICS DOPED WITH d[1] IONS

V. B. Kravchenko[1], Yu. L. Kopylov[1], S. N. Bagayev[2], V. V .Shemet[1] , A. A. Komarov[1], L. Yu. Zaharov[1]
[1] Institute of Radio Engineering and Electronics named after V.A.Kotelnikov, Russian Academy of Sciences, 141190 Fryazino, Russia
[2] Institute of Laser Physics, Russian Academy of Sciences, 630090 Novosibirsk, Russia

ABSTRACT

$Y_3Al_5O_{12}$ (YAG) laser ceramics can be produced now as large size samples with excellent quality. We studied YAG ceramics doped with d^1 Ti^{3+} and Zr^{3+} ions as possible broad-band materials for tunable and ultra-short pulses lasers. The procedure of doped YAG ceramics fabrication included chemical co-precipitation, precursors' heat treatment, YAG powder grinding, high pressure colloidal slip-casting for nanopowders compaction and vacuum sintering of performs at 1730-1800°C. Transparent colored samples were obtained. Absorption and luminescence spectra of ceramics samples are similar to the spectra of correspondingly doped YAG single crystals. Zr^{3+} luminescence excited by second harmonic of Nd:YAG laser was observed for the first time. Possibility to obtain laser action is discussed.

INTRODUCTION

There is a certain need in broad-band luminescent laser materials both for tunable lasers and for lasers with ultra-short pulses. Now Ti^{3+}-sapphire Ti^{3+}-Al_2O_3 single crystals are used widely for such lasers. But there are difficulties in production of large-size elements of Ti-sapphire crystals. One possible approach could be fabrication of transparent doped ceramics. It is not possible now to obtain low-loss laser ceramics from birefrigent alumina. On the other hand, yttrium-aluminum garnet $Y_3Al_5O_{12}$ (YAG) ceramics doped with Cr^{4+} ions is a well known material for Q-switching in near IR spectral field and Cr^{3+}-Nd^{3+} YAG laser ceramics was fabricated for efficient lasers with lamp pumping [1].

We tried to obtain YAG optical ceramics doped with other d^1 ions and to investigate properties of the samples. Ti^{3+} ion can be a possible d^1 dopant for YAG, and absorption and luminescence spectra of Ti^{3+} were reported to resemble $Ti3^+$ spectra in sapphire [2, 3]. It was found [4] that Ti distribution in YAG depends strongly on crystal growth atmosphere, with appearance of precipitates in some cases. The situation can change in case of ceramics. So we decided at the beginning to investigate YAG: Ti^{3+} ceramics as well as YAG with less usual d^1 dopant [5] – Zr^{3+}.

EXPERIMENTAL

The procedure of doped ceramics fabrication was similar to one described in [6,7]. The initial nanopowders were produced by chemical co-precipitation using yttrium and neodymium nitrates and $(NH_4)Al(SO_4)_2$ solutions by adding their mixture to NH_4HCO_3 solution as a precipitant. Quasi-spherical particles of the precursor with diameter of 30-100 nm were obtained. After following washing and dewatering the precursor calcining was made at 900-1300°C. X-ray investigation showed only pure YAG peaks. YAG powder mainly with spherical nanoparticles was produced, but, unfortunately, some quantity of hard agglomerates appeared during the heat treatment (Fig.1).

We tried to obtain fully non-agglomerated YAG powder and succeeded to do so together with colleagues from the Institute of Experimental Mineralogy of RAS, Chernogolovka, by modified solvothermal process (Fig. 2) but the quantity of YAG powder obtained till now was not sufficient

for extensive experiments.
Dopants were introduced in YAG powder as appropriate oxides in concentrations 0.1-1.0 w.% during the subsequent milling process for slurry preparation. Zirconium oxynitrate was also used for Zr-doping during precursor preparation process. The milling was made in ball mill with agate balls and deflocculants were added during the process. We used several methods for sample compaction (Fig.3) having ability to self-organization of nanoparticles Among these methods high pressure colloidal slip-casting (2) gives higher uniformity of compact density[6,7].

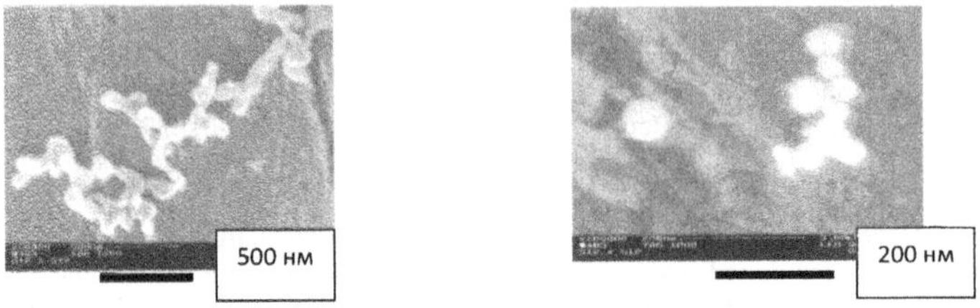

Fig.1. Hard agglomerates (left) and spherical nanoparticles (right) in calcined YAG nanopowders (Taken from the same sample).

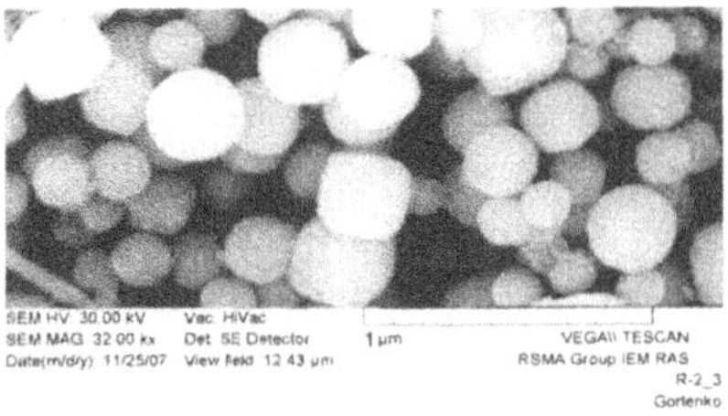

Fig. 2. YAG nanopowders obtained by solvothermal process [8].

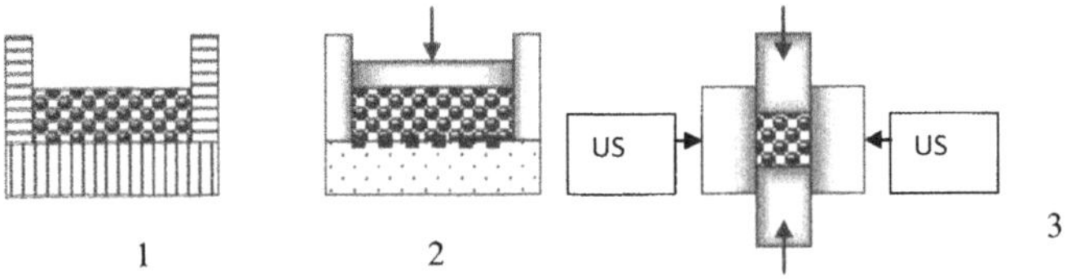

Fig. 3. Methods of YAG nanopowders compaction using nanoparticles self-organization processes. 1-traditional slip casting, 2- high pressure colloidal slip-casting, 3- pressing in the field of high power ultra sound (US)

After air-drying and organic components' removal by 1200 ^{0}C heat treatment the compacts having diameter 27 mm and thickness 2-7 mm were sintered in vacuum furnace with carbon heater at 1750-1800 ^{0}C and transparent ceramics were obtained. They have light- to dark-brown color in case of Ti^{3+} doping and red-to-brown color in case of Zr^{3+} doping (Fig.4). We started investigations of these ceramics with Zr-doped samples and preliminary results are given below. Luminescence was excited by second harmonic of Nd:YAG laser and by Ar and He-Ne lasers.

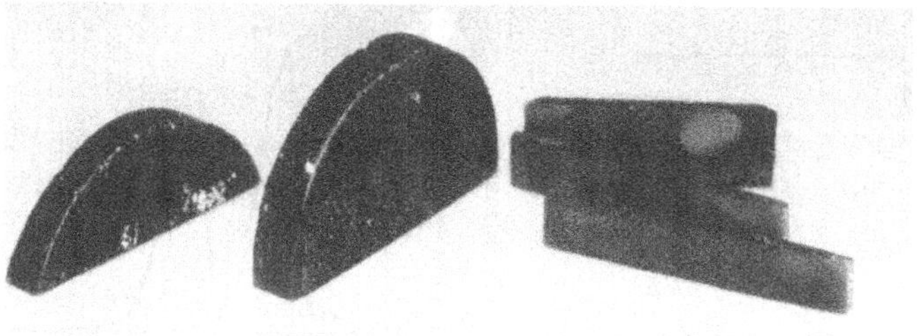

Fig. 4. Samples of Zr3+-YAG ceramics; unpolished (left) and polished (right)

RESULTS

Annealing in air at 1300 °C resulted as a rule in discoloration of the samples owing to transition of 3-valent titanium and zirconium ions into 4-valent state but some samples stayed brown. The real reason for this behavior is not clear at the moment.

Single crystals Zr^{3+}:YAG were grown by A.G. Petrosyan and co-workers[5]. Optical absorption and ESR spectra of these crystals were studied and energy levels diagram shown in Fig.4 was offered. Absorption spectra with strong lines in visible region are quite similar to the spectra of doped single crystals [5] (Fig. 5). Unfortunately absorption intensity was very strong even for the samples with the lowest Zr^{3+} concentration, so the details of the absorption spectra are not seen.

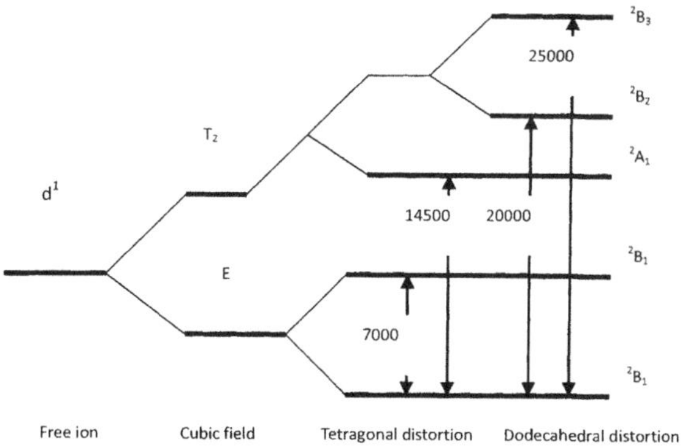

Fig.5. Energy level diagram of Zr^{3+} ion in YAG [5].

YAG:Zr^{3+} where Zr^{3+} ions substitute for Y in dodecahedral sites with D_2 symmetry in YAG lattice gives luminescence band around 780 nm, similar to Ti^{3+} in YAG centered at around 750 nm but we did not find any published data for YAG:Zr^{3+} crystal luminescence to make the comparison. It is noteworthy the shift of luminescence band depending on excitation wavelength is similar to that in Ti^{3+}-Al_2O_3 crystals [9].

Possibility to obtain laser action can be roughly evaluated by comparison of Fig. 5 left and right: there is a window around 800-900 nm where the luminescence falls into lower absorption region, but till now the optical quality of the samples is not good enough for real laser experiments.

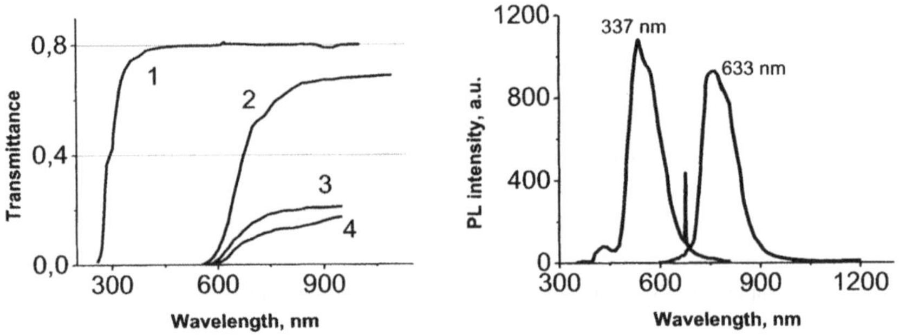

Fig. 6. Transmission spectra of YAG ceramics (left) and luminescence spectra of Zr^{3+}-YAG ceramics (right) at excitation wavelengths 337 and 633 nm. 1 – undoped and 2,3,4 doped samples. Concentration of Zr^{3+}- is increased in samples from 2 to 4.

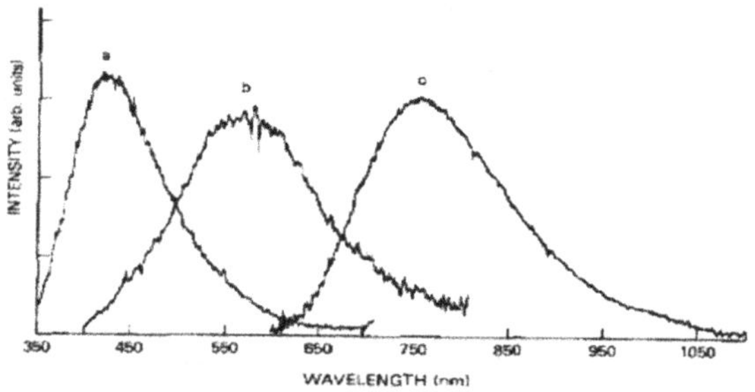

Fig. 7. Fluorescence of Ti^{3+}-Al_2O_3 with excitation at different wavelengths. Curve a – 257,3 nm; curve b - 313 nm; curve c - 454 nm. Intensity scales are different [9].

CONCLUSIONS

Transparent colored YAG ceramics doped with Ti^{3+} and Zr^{3+} ions as possible broad-band materials for tunable and ultra-short pulses lasers were obtained. The procedure of doped ceramics fabrication included chemical co-precipitation, precursors' heat treatment, YAG powder grinding, high pressure colloidal slip-casting for nanopowders compaction and vacuum sintering of performs at 1730-1800°C. Absorption spectra of Zr^{3+} ceramics samples are similar to the spectra of correspondingly doped YAG single crystals. Zr^{3+} luminescence was observed for the first time. Possibility to obtain laser action is discussed.

ACKNOWLEDGEMENTS

This work was supported in part by the Program "Femtosecond physics and new optical materials" of the Russian Academy of Sciences and by the Russian Foundation for Basic Research (grants 07-02-00057, 07-02-12033 and 08-02-12143)

REFERENCES

[1] H. Yagi. Recent advances in transparent ceramics. *3d Laser Ceramics Symp.*, October 8-10, 2007, Paris, 2007.

[2] A.A. Kaminskii, Physics and Spectroscopy of Laser Crystals (Russian), 5-61, Moscow, Nauka. Ed. A.A. Kaminskii (1986).

[3] P. Peshev, V. Petrov, N. Manuilov, Growth and spectral characteristics of $Y_3Al_5O_{12}$:Ti^{3+} single crystals. *Mater. Res. Bull.*, 23, 1193-1198 (1988).

[4] T. Kotani, J.K. Chen, H.L. Tuller, The Dopant Distribution in Ti-, Zr- and Cr-doped $Y_3Al_5O_{12}$ Fibers Grown by the Laser Heated Floating Zone Method. *J. Electroceramics,* 2, 7-20 (1998)

[5] S.R. Asatryan, A.S. Kuzanyan, A.G. Petrosyan, , A.K. Petrosyan, E.G. Sharoyan, Single crystal growth and investigation of optical and ESR absorption spectra of zirconium-doped YAG, *Phys. Stat. Sol. (b),* **135**, 343-352 (1986).

[6] A.A. Kaminskii, V.B. Kravchenko, Y.L. Kopylov, S.N. Bagayev, et al., *Phys. Stat. Sol. (a)*, **204,** 2411-2415 (2007)

[7] Y.L. Kopylov, V.B. Kravchenko, S.N. Bagayev et al., *Optical Materials,* 30, doi:10.1016/j.optmat.2008.03.013., Presented on-line 25. 04. 2008.

[8] M.A. Korjinsky, Y.L. Kopylov., The new hydrothermal synthesis of yttrium aluminum garnet powders., II Russian-French Seminar "Nanotechnology, Energy, Plasma, Lasers (NEPL-2008), Abstracts, 46-47. Tomsk Polytechnic University, Tomsk (2008).

[9] J. F. Pinto, L. Esterowitz, G. H. Rosenblatt, M. Kokta, D. Peressini. *IEEE J. Quantum Electronics,* **30,** 2612-2616 (1994).

PREPARATION AND CHARACTERIZATION OF NEODYMIUM -DOPED LZS TRANSPARENT GLASS-CERAMICS

Hongbo Zhang[1], Yimin Wang[1], Guang Cui[1], Jing Shao[1], Huashan Zhang[1], Chunhui Su *[1 2]
[1]Changchun University of Science and Technology
[2]Jilin Teacher's Institute of Engineering and Technology
[1]Changchun, Jilin, 130022, China
[2]Changchun, Jilin, 130052, China

ABSTRACT

The Neodymium-doped LZS parent glasses were prepared by conventional melt technology at 1400°C for 2 hours. Then the glass-ceramic were precipitated by annealing at the temperature 750°C. Then by the means of X-ray diffraction (XRD), differential thermal analysis (DTA), UV-VIS-NIR spectrophotometer and scanning electron microscope (SEM), the properties of glass and glass-ceramic were studied. From the XRD pattern, a kind of glass-ceramic which contents $Li_8Zn_{10}Si_7O_{28}$ and cristobalite were precipitated. The result of SEM indicate that the grain was very small with the size about 20~50nm. By the UV-VIS-NIR spectrophotometer, the transparence of the glass-ceramic can get 80% which was measured by a spectrophotometer from 400~800nm. Fluorescence spectra showed a maximum peak around 1054nm and with a high transparence, which indicted that the Neodymium-doped glass-ceramic has a better laser property than glass.

INTRODUCTION

Glass-ceramics find various applications in the field of vacuum, sealing, electronics, cook wares, biomedical, etc. because of their superior thermal, mechanical, and other physical properties compare to their parent glass. In the process of conversion of glass to glass-ceramics, controlled crystallization plays the key roll to engineer the different physico-chemical properties[1-2]. Li_2O-ZnO-SiO_2 system has good glass forming ability over a wide range of composition [3-4]. They also has excellent electrical resistivity and chemical durability. LZS glass-ceramics have attracted a great deal of attention because of their adjustable thermal expansion coefficients typically ranging from $36.1\times10^{-7}/°C^{-1}$ to $200\times10^{-7}/°C^{-1}$[5-6]. LZS glass-ceramic also has excellent optical properties which can be used as laser material.

EXPERIMENT

In order to fabricate transparent and high expansion coefficient glass-ceramic, parent glasses having chemical compositions corresponding to Li-Zn-Si. The parent glasses were prepared by conventional melt technology. Analytical grade reagent Li_2CO_3, ZnO, SiO_2, Na_2CO_3, Sb_2O_3 and Na_2SO_4 were used as starting material. After mixing well, the batch was melted at the temperature about 1400°C for 2 hours following by pouring into a steel plate. The poured glass was immediately transferred to a furnace set at about 500°C and held for 2-3 hours for annealing.
A TG-DTA system was employed for recording crystallization temperature. Measurements were done in the temperature from 30~800°C employing a heating rate of 10°C /min. Based on the result of DTA, conversion of glass into glass-ceramics was carried out in a resistance furnace. By the DTA results, the sample was raised to 700°C at a rate of 5°C /min for 2 hours, then the temperature was decreased to room temperature at a rate of 10°C /min.
Identification of various crystalline phase in glass-ceramics sample was carried out using powder X-ray diffractormeter with Cu Kα as X-ray radiation source. The microstructure of the glass-ceramic was observed using SEM, the transmittance of the glass-ceramic was learned by UV-VIS-NIR spectrometer and the emission spectrum was investigated by fluorescence

spectrometry.

RESULTS AND DISCUSSION

Fig. 1 show the DTA curves for LZS parent glass, The endothermic base line shift at 500°C indicates the glass transition temperature and the exothermic peak at about 680°C is a crystallization temperature for this system. Normally speaking, nucleation temperature was about 50°C above the transition temperature. Because the anneal temperature was 500°C, which was very close to the nucleation temperature, so the one-step heat-treatment was adopted.

Fig.2 shows the XRD patter of the LZS glass-ceramics, by the means of XRD, the properties of glass and glass-ceramic were studied. From the JCPDF card, we can see that under the temperature 700°C for 1 hour the major phase of the glass-ceramic was $Li_8Zn_{10}Si_7O_{28}$ and cristobalite. $Li_8Zn_{10}Si_7O_{28}$ belong to the orthorhombic system. From the XRD pattern we also can know that the percentage of the crystal phase can get very high.

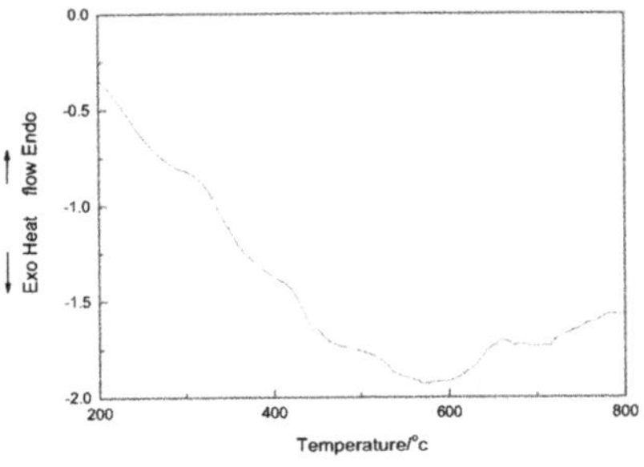

Fig. 1. DTA curve for LZS glass

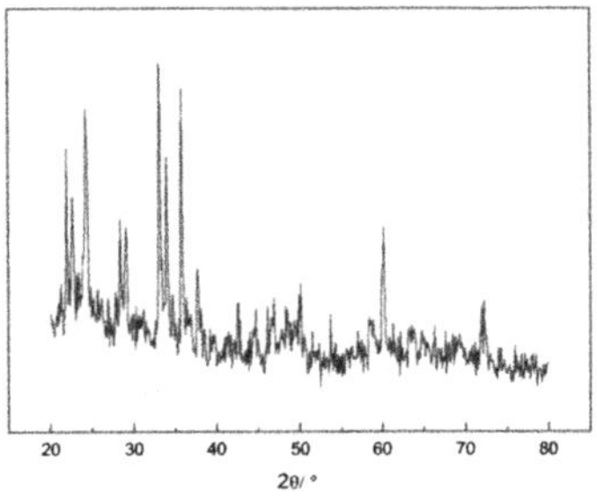

Fig. 2. XRD patterns of the glass-ceramics

Fig.3 shows the transparency of the glass-ceramic, the optical transmittance was about 76-90%. In the visible region the optical transmittance was about 86-90%, which indicated that the

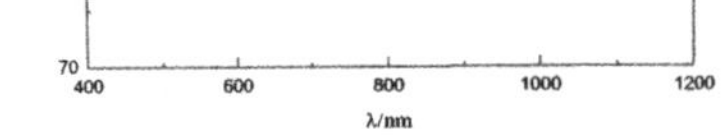

Fig. 3 The transparency of the glass-ceramics

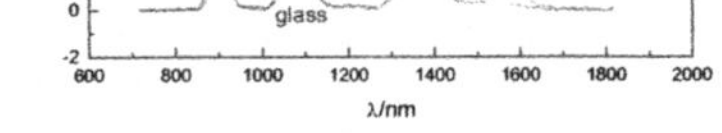

Fig. 4 Nd^{3+}-doped fluorescence spectrometry of the glass and glass-ceramic

glass-ceramic has a very good optical property. During the 300-500nm there is an absorption band which maybe caused by the absorption of the lattice. From the transparency of the glass-ceramic we can know that this kind of material can be used as an optical material.

In Fig.4, the fluorescence spectrometry of the glass and glass-ceramic were investigated, a strong emission was observed at 878nm, 1054nm, 1330nm and 1536nm, which corresponding to $^4F_{3/2}\rightarrow^4I_{9/2}$, $^4F_{3/2}\rightarrow^4I_{11/2}$ and $^4F_{3/2}\rightarrow^4I_{13/2}$[4] of the Nd ion. For the glass-ceramic, its emission peak is much stronger than the parent glass which indicated that the glass-ceramic has better optical character than the parent glass.

Fig.5 shows the SEM of the glass-ceramic. From the picture we can see that the crystal size of the $Li_8Zn_{10}Si_7O_{28}$ was about 20~50nm which can get a very high transparency and optical characters.

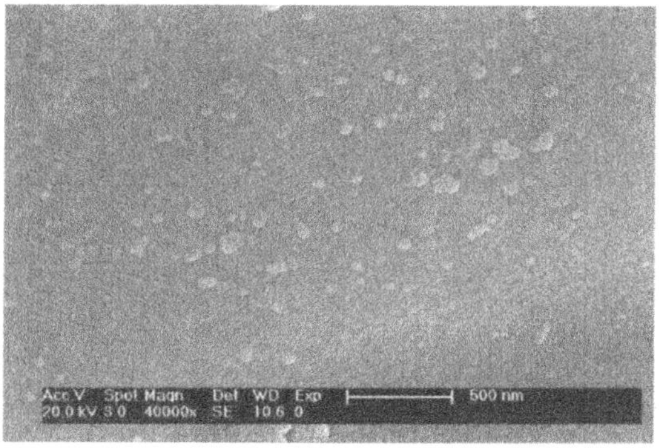

Fig. 5 SEM image for LZS glass-ceramic

CONCLUSION

In conclusion LZS glass-ceramic were prepared by normal melting method. For the LZS glass-ceramic the major phase are $Li_8Zn_{10}Si_7O_{28}$ and cristobalite. The transparency was investigated which can get 90%. And the LZS glass-ceramic has a good optical character. When the Nd ion was doped, four strong emission bands at 878nm, 1054nm, 1330nm and 1536nm were observed. The emission bands of the glass-ceramic are stronger than the parent glass. It shows that the LZS glass-ceramic is a better laser material than parent glass.

REFERENCES

[1]B. Sharma, Study on some thermo-physical properties in Li_2O-ZnO-SiO_2 glass-ceramics, *Material Letters*, **58**, 2423-28(2004)

[2] E. Demirkesen, Effect of Al_2O_3 additions on the acid durability of a Li_2O-ZnO-SiO_2 glass and its lass-ceramic, *Ceramics International*, **29**, 463-69(2003)

[3] I. W. Donald, B. L. Metcalfe, D. J. Wood, J. R. Copley, The preparation and properties of some lithium zinc silicate glass-ceramic, *J. Mater. Sci.*, **24**, 3892-903(1989)

[4] A. A. Maurer, A W A El-Shennavi, A R El-Ghannam, Thermal expansion of Li_2O-ZnO-SiO_2 glasses and corresponding glass-ceramics, *J.Mater. Sci.*, **26(5)**, 6049-56(1991)

[5] Zhou YANG, An-xian LU, Shu-jiang LIU, et al., Crystallization and thermal expansion behavior for Li_2O-ZnO-SiO_2 system glass-ceramics, *Materials Review*, **6(14)**, 296-301(2005)

[6] S. C. Clausbruch, M. Sehweiger, W. Holand, et al., The effect of P_2O_5 on the crystallization and microstructure of glass-ceramics in the Si_2O-Li_2O-K_2O-ZnO_2-P_2O_5 system, *Journal of Non-crystalline Solids*, **263/264**, 388-94(2000)

PREPARATION AND CHARACTERIZATION OF $ZnO-Al_2O_3-SiO_2$ TRANSPARENT GLASS-CERAMICS

Jing Shao[1],Guohui Feng[1,2], Hongbo Zhang[1],Guangyuan Ma[1], Chunhui Su[1,3,*],
[1]School of Chemistry and Environmental Engineering, Changchun University of Science and Technology
[2]Plastic Institute of Ji Lin Province
[3]Jilin Teacher's Institute of Engineering and Technology, Changchun, Ji Lin, China

ABSTRACT

Zinc aluminum silicate glass ($ZnO-Al_2O_3-SiO_2$) doped-Nd_2O_3 and TiO_2 as nucleus was prepared by conventional melt and quenching technique. It was melted in a platinum crucible at 1550°C for 2h to convert to transparent glass-ceramics. The characterizations were performed by DTA, XRD, SEM, UV–Vis-NIR etc. When heat-treated below 1000°C, main crystal phase was cubic gahnite $ZnAl_2O_4$ and the minor crystal phases were Zn_2SiO_4、SiO_2. When heat-treated temperature in excess of 1000°C, the main crystal phase was SiO_2, the minor crystal phases were Zn_2SiO_4 and $ZnAl_2O_4$. The lattice parameter was a =0.8087 nm(±0.0012).With the crystallization heat-treated temperature increased, the optical transmittances decreased. The fluorescence spectra of the glass and glass-ceramics were measured. The fluorescent characteristics of Nd^{3+} ion were stronger in the glass-ceramic than in the glass matrix.

INTRODUCTION

Glass-ceramics, discovered by S.D. Stookey in the mid-1950s[1-4], are polycrystalline ceramic materials consisting of at least one crystalline phase and a vitreous phase[5], formed through the controlled nucleation and crystallization of glass, where the amount of residual glassy phase is usually less than 50%. The precursor glass is melted, quenched and shape-processed, and then is thermally converted into a composite material formed by a crystalline phase dispersed within a glass matrix. The basis of controlled internal crystallization lies in efficient nucleation. The most frequently used nucleating agents are the TiO_2 and ZrO_2 oxides or their mixture which is more efficient according to Stewart[6]. During the controlled heat treatment, nuclei are formed and different crystalline phases are grown in the glass matrix depending on the heat treatment.

Glass-ceramics exhibit superior thermal, mechanical, electrical and other physicochemical properties compared to their counter base glasses. The various constituent oxides in the base glass have their own specific functions. The introduction of Na_2O and ZnO in the glass composition modifies thermo-physical properties, including reduction in viscosity and decrease in thermal expansion. Li_2O is partially substituted by MgO and ZnO in order to improve work properties of the parent glass, while lowering the cost of materials.

The transparent glass-ceramics, with almost zero thermal expansion, fine thermal conductivity, high transmittance in the laser wavelength, the superior chemical stability and high resistance to hot shock, have potential to substitute the single crystals and glasses in the near future. Nowadays, one of the most interesting fields of research is focused on the development of optical devices based on rare earth ions doped materials for their use in telecommunication systems, such as solid-state lasers, sensors, optical amplifiers, upconversion fibers and in other uses due to their optical and magnetic properties[7-10].

EXPERIMENTAL PROCEDURE

The raw materials used to produce Zinc aluminum silicate glasses were 58%SiO_2, 21%Al_2O_3, 10%ZnO, 3%Li_2O, 5%TiO_2, 1%Sb_2O_3, 1% Na_2O and rare earth oxides 1%Nd_2O_3 were as for the base glass preparation. TiO_2 was used as nucleating agents, and Sb_2O_3 was used as the clarifier. The raw materials were homogenized in the planetary ball mill (PM 4000 model of M/s. Retsch, Germany). The batch was melted in a Pt crucible at 1550°C for 2h duration in the $MoSi_2$ electrical resistance heating furnace having PID temperature controller. Then the melt was poured on a stainless steel plate at room temperature, subsequent obtained samples were annealed at 600℃ for 2h, and allowed to cool down to room temperature at a rate of 20°C /h to relieve internal stresses. Then two-step heat treatments for the nucleation and the crystal growth were adopted. The objective is to obtain transparent glass-ceramics based on $ZnO-Al_2O_3-SiO_2$ system.

The study has been carried out using Differential hermal analysis (DTA), X-ray diffraction (XRD), FEG-ESEM and UV-VIS-NIR spectrophotometer etc. The nucleation and crystallization temperatures were determined by differential thermal analysis (DTA). Specimens heat-treated at various durations were analyzed by the X-ray diffraction (XRD) to determine the optimum conditions for nucleation and the crystal growth. Scanning Electron Microscopy (SEM) was used to study the glass-ceramics morphology, the grain size and distribution in the residual glass matrix. The transmittance was measured by UV-VIS-NIR scanning spectrophotometer. The fluorescence spectra of the glass and glass-ceramics were measured. The fluorescent characteristics of Nd^{3+} ion were stronger in the glass-ceramic than in the glass matrix.

RESULTS AND DISCUSSION

The nucleation and crystallization temperatures of the glass were determined by differential thermal analysis (DTA) (TA Corp., SDT 2960, American), which was employed for recording crystallization behavior of the glass sample. The DTA curve was recorded over 30~1000°C at heating rate 10°C /min. Various heat-treatment schedules given in Table I were applied in order to study the nucleation and crystallization processes and the change of microstructure. X-ray powder diffraction system (D/max 2500V, Rigaku, Japan) with Cu Kα radiation (1.54056 Å) has been used for identification of the crystalline phase. Each pattern was scanned from 2θ= 5° to 80° at the rate of 4°/min.The resulting glass-ceramic samples have been investigated by UV-VIS-NIR scanning spectrophotometer (UV-3101PC, SHIMADZU,J apan). Scanning Electron Microscope (FEI Corp., XL30ESEM FEG, USA) has been used to study the glass-ceramics morphology, the grain size and distribution in the residual glass matrix.

Table I The Heat-treatment Schedules for Glass Samples

Samples	Temperature/°C ×Time/h for nucleation	Temperature/°C× Time/h for crystallization	Main Crystal phase	Minor Crystal phase	Fluorescence properties
Glass	660×1	730×1	-	-	Better
a#	660×2	730×1	$ZnAl_2O_4$	Zn_2SiO_4 SiO_2	Better
b#	660×3	730×1	$ZnAl_2O_4$	Zn_2SiO_4 SiO_2	Low
c#	660×2	730×2	$ZnAl_2O_4$	Zn_2SiO_4 SiO_2	Low
d#	700×1	880×1	$ZnAl_2O_4$	Zn_2SiO_4 SiO_2	Low
e#	700×1	1050×1	SiO_2	$ZnAl_2O_4$ Zn_2SiO	Bad

Typical DTA curve of the ZAS base glass is shown in Figure 1. The DTA curve is characterized by some endothermic peak. The endothermic peak is very obvious at 763°C, 827°C and 1067°C, which indicates there exists transformation of crystal configuration. The endothermic peak is not very obvious at 961°C and the endothermic base line shift at 600°C -700°C gives the beginning of glass nucleation temperature zone.

XRD patterns for some specimens under present investigations are shown in Figure 2. XRD patterns of the glass-ceramic indicate the evidence of crystallites formed. All the peaks were analyzed with PCPDF files. The diffraction peaks attributable to $ZnAl_2O_4$ are indicated by open circles, and observed for all of the heat-treated specimens. It is noted that the intensity of gahnite $ZnAl_2O_4$ crystalline phase becomes larger with an increase in heat-treatment temperature and time duration. It is clear that SiO_2 also precipitates as the main crystal phase in the heat-treated specimens when the temperature is above 1000°C.

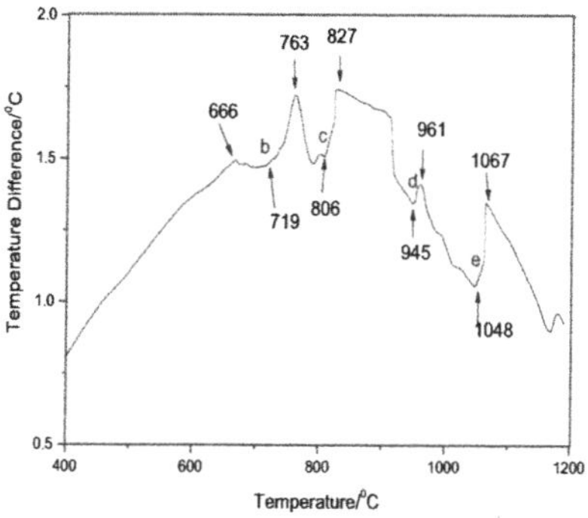

Figure 1. DTA curve for base glass

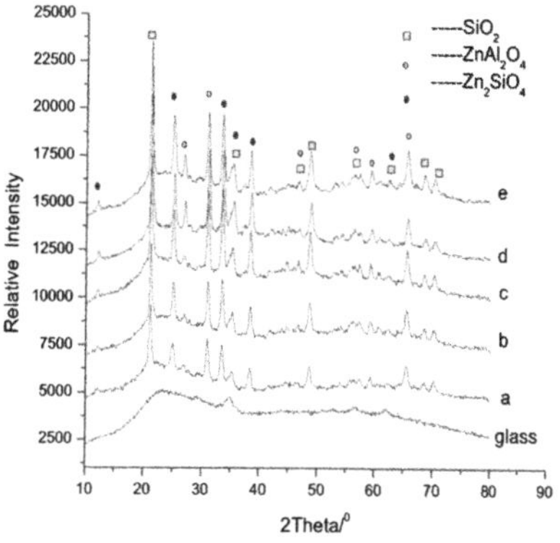

Figure 2. XRD patterns of the base glass and glass-ceramics of ZAS system

The d-spacing of standard sample (JCPDS 71-0968) signed by d_s. The lattice parameters a of the cubic gahnite have been calculated from the experimental lattice spacing with the expression relative to the cubic system:

$$d_{hkl}^2 = \frac{a^2}{(h^2 + k^2 + k^2)} \qquad (1)$$

The high intensity and well defined lines located namely (220), (311), (400), (331), (422), (511), (440), (620), and (533) lines. The a value has been calculated using d_s for these lines respectively. The mean value of the cell parameters of the crystalline phase of specimen b#, that is gahnite with cubic structure, is $\bar{a}$=8.087Å close to the standard value 8.088Å. The FWHM of diffraction peak around 2θ=37° corresponds to the reflection from the (311) plane of cubic gahnite. The mean grain sizes evaluated by Scherrer's equation were 10nm around.

SEM has been used to study the glass-ceramics morphology, the grain size and distribution in the residual glass matrix. Glass-ceramics b# is composed of crystallites of approximately 10nm in size as shown in Figure 3 (a), (b) and (c),which can also explains its transparency, nearly 70% in the infrared wave length.It can be verified that the grain size calculated by the Scherrer's formula are well in line with the practical sizes measured by SEM.

The UV-VIS-NIR transmittances of glass and glass-ceramic specimens have been measured in the range 240~2000nm. The transmission curves are presented in Figure 4. The transmittance of glass specimen reach nearly 80% within 700~2400nm, but the transmittance of the glass-ceramics decreases within 240~2000 compared with base glass. Heat treatment temperatures and time influence on the microstructure and the transmittance.

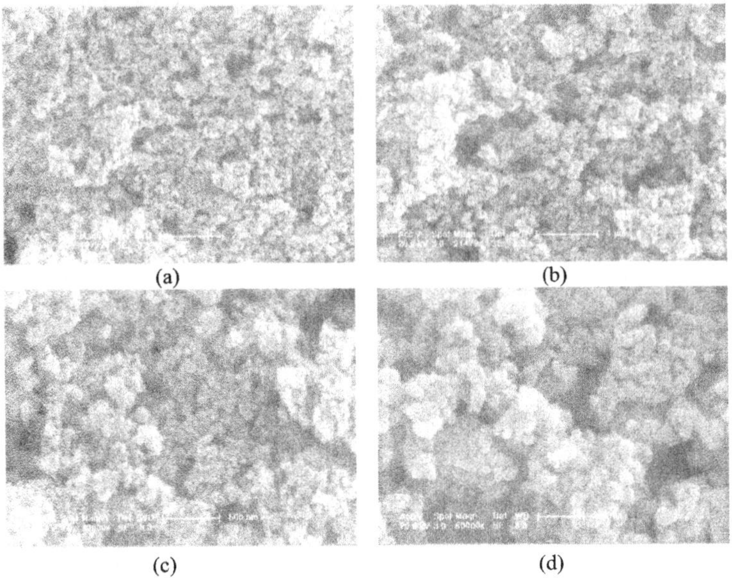

Figure 3. SEM micrograph of glass-ceramic samples

Fluorescence spectra of Nd^{3+} ions in ZAS system is presented in Figure 5. Emission measurements performed for the glass-ceramics samples with excitation between 800-1500nm.The emission spectra show three main fluorescence emission bands centered around 885nm,1064nm and 1337nm corresponding to the electronic transitions of ${}^4F_{3/2}\rightarrow{}^4I_{9/2}$, ${}^4F_{3/2}\rightarrow{}^4I_{11/2}$ and ${}^4F_{3/2}\rightarrow{}^4I_{13/2}$, respectively. It can be observed that the most intense emission band is located at 1064nm for the base glass and glass-ceramics, but peak intensity in glass-ceramics is higher compared to the base glass.

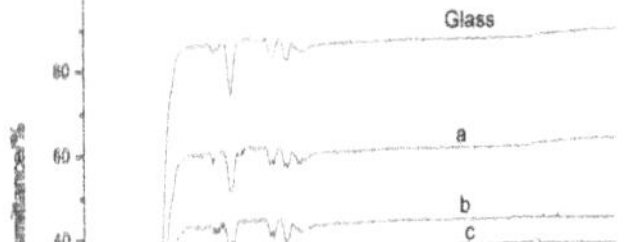

Figure 4. UV-VIS-NIR transmittance of glass and glass-ceramic samples

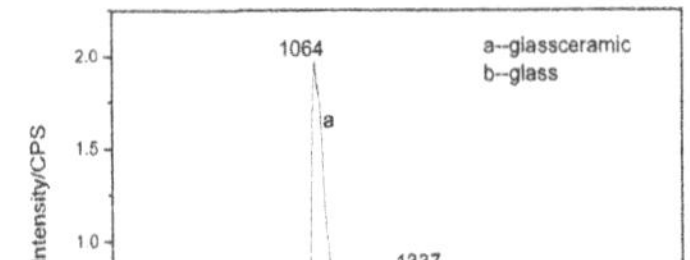

Figure5. Fluorescence spectra of Nd^{3+} ions in ZAS system between 800-1500 nm

CONCLUSIONS

Zinc aluminum silicate($ZnO-Al_2O_3-SiO_2$)doped-Nd_2O_3 glasses have been prepared by conventional melt and quenching technique, and subsequently converted to transparent glass-ceramics by controlled nucleation and crystallization. Results of XRD indicate that there is an obvious structure change in the glass-ceramics compared to the precursor glass. Major crystalline phase namely a cubic $ZnAl_2O_4$ is identified for all heat-treated samples. SiO_2 crystal phase becomes the main crystal phase when the heat-treatment temperature is above 1000°C, and the samples lose its transparency. The mean grain sizes, calculated by the Scherrer's equation using the results of XRD, keep well with the SEM results. Transparent glass-ceramics were obtained finally.

ACKNOWLEDGEMENT

This work was supported by the Chinese Education Ministry for financial support under Fund Item: KB20026.

REFERENCES

[1] S.D. Stookey, Catalyzed Crystallization of Glass in Theory and Practice, *J.Ind. Eng. Chem.*, **1**, 805-808 (1959).
[2] S.D. Stookey, Photosensitively Opacifiable Glass, *US Patent*, No: 2684911(1954).
[3] S.D. Stookey, Thermal Expansion of Some Sththetic Lithia Minerals, *J. Am. Soc.* **34**, 235-239 (1951)
[4] S.D. Stookey, Ceramic Body and Method of Making It, *Us Patent*, No:2971853,(1961).
[5] P. W.Micmillan, S. V. Phillips, G. Partridge, Glass-ceramics, *J. Mater. Sci.*, **1**, 269-273 (1966).
[6] D.R.Stewart, Advances in Nucleation and Crystallization in Glasses, *The American Ceramic. Society*, 83-90 (1971).
[7] M. Yamane, Y. Asahara. Glasses for Phtonics, Cambridge, UK: Cambridge University Press, 2002.
[8] M. Clara Goncalves, L.F. Santos, R.M. Almeida, *C.R.Chim.*, **5**, 845-852 (2002).
[9] Yuki Kishi,Setsuhisa Tanabe.Infrared-to-visible upconversion of rare-earth doped glass ceramics containing CaF_2 crystals, *J. Alloys and Compounds.***408-412**, 842-844 (2006).
[10] J. Marchi, D.S. Morais, J. Schneider, J.C. Bressiani, etc., Characterization of rare earth aluminosilicate glasses, *J. Non-Cryst. Solids*, **351**, 863-868 (2005).

LUMINESCENCE OF Yb^{3+}, Ho^{3+}: Lu_2O_3 NANOCRYSTALLINE POWDERS AND SINTERED CERAMIC

Liqiong An, Jian Zhang, Guohong Zhou, Shiwei Wang
Shanghai Institute of Ceramics, Chinese Academy of Sciences
1295 Dingxi Road, Shanghai 200050, P.R. China

ABSTRACT

The optical spectroscopy of $Lu_{1.896}Yb_{0.1}Ho_{0.004}O_3$ nanocrystalline powders and sintered ceramic were reported. Under the excitation of 360 nm, the nanocrystalline powders and sintered ceramic emitted green lights, which were centered at 548 nm while the dominant green, weak red and infrared emissions were observed in both samples when excited by 980 nm diode laser. The ultra-violet and blue upconversion emissions were also detected in the ranges of 381-394 nm and 409-428 nm. No noticeable shifting of these peaks between these two samples appeared. However, enhancement of luminescence in sintered ceramic was observed. Energy transfer and excited-state absorption may be responsible for the upconversion process.

INTRODUCTION

Recently, considerable attention has been given to lutetium-based materials for their potential applications, such as generation of artificial lights, detection of ionizing radiation, and medium of lasers[1-3]. Lu_2O_3, as a sesquioxide isostructural to Y_2O_3, crystallizes in a cubic bixbyite structure with space group Ia3. It possesses relatively low phonon energy (about 600 cm^{-1}) [4] and its powders can be sintered into transparent [5], which make it additionally attractive for practical applications. For example, J.A. Capobianco et al have studied the upconversion properties of nanocrystalline and bulk Lu_2O_3: Er^{3+} [4].

For Yb^{3+} ion, it is commonly used as sensitizer because of energy matching for the commercial high power laser diodes in NIR range. Ho3+ ion possesses several energy levels in the NIR portion of spectrum that can be pumped with NIR radiation or sensitized by Yb^{3+} ions along with several metastable energy levels. Moreover, Ho^{3+} is a suitable active ion for its energy levels with luminescence in visible wavelength, which has potential applications in visible solid-state lasers.

In this paper translucent Yb^{3+}, Ho^{3+}: Lu_2O_3 ceramics were fabricated at 1850^{0}C for 3 h in flowing H_2 atmosphere using co-precipitated nanocrystalline powders as starting materials. The luminescent properties of the nanocrystalline powders and sintered ceramic were investigated.

EXPERIMENTAL

$Lu_{1.896}Yb_{0.1}Ho_{0.004}O_3$ nanocrystalline powders were synthesized by co-precipitation method, which was described in detail before [6]. The as-prepared powders were pressed under 30 MPa into disk with 12-mm diameter, and then isostatically cold pressed under 200 MPa pressure. Finally, the disks were sintered at 1850°C for 3 h in flowing H_2 atmosphere. Therefore, the translucent ceramics were obtained.

The room temperature luminescent spectra of sintered ceramic were recorded by a spectrofluorometer (Fluorolog-3, Jobin Yvon, Edision, USA) equipped with Hamamatsu R928 photomultiplier and a 450 W Xenon lamp. The upconversion luminescent spectra were measured by the same equipment using a 980 nm continuous wave diode laser as excitation source. All the emission spectra were corrected for the setup characteristic.

RESULTS AND DISCUSSION

The FESEM image of the prepared powders was given in Fig. 1 (a). It can be evidently seen that the powders was about 70 nm in diameter and nearly spherical. Figure 1 (b) shows the photograph of the corresponding un-polished sintered ceramic. It is 1 mm in thickness and about 9 mm in diameter.

(a) (b)

Fig.1 (a) FESEM image of the prepared powders and (b) photograph of the un-polished corresponding sintered ceramic.

The excitation and emission spectra of the $Lu_{1.896}Yb_{0.1}Ho_{0.004}O_3$ sintered ceramic and nanocrystalline powders were measured at room temperature shown in Fig. 2. The excitation spectra of the sintered ceramic and nanocrystalline powders are similar, except that some peaks of nanocrystalline powders were absent for weak intensities compared to those of the sintered ceramic. Under a UV light (360 nm) excitation, the strong green emission band was detected in both samples, which was corresponding to the multiplets $^5F_4$, $^5S_2 \rightarrow {}^5I_8$ transition of the Ho^{3+} ions. Weak near-infrared emission band was also measured in the range of 740~775 nm, which is assigned to the $^5F_4$, $^5S_2 \rightarrow {}^5I_7$ transition. No noticeable peak shift was observed, which indicated the similar crystal field surrounding of Ho^{3+} in both samples.

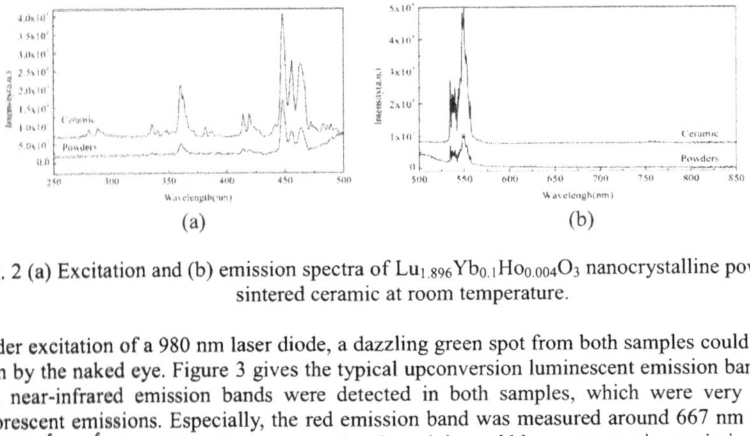

(a) (b)

Fig. 2 (a) Excitation and (b) emission spectra of $Lu_{1.896}Yb_{0.1}Ho_{0.004}O_3$ nanocrystalline powders and sintered ceramic at room temperature.

Under excitation of a 980 nm laser diode, a dazzling green spot from both samples could be clearly seen by the naked eye. Figure 3 gives the typical upconversion luminescent emission bands. Green and near-infrared emission bands were detected in both samples, which were very similar to fluorescent emissions. Especially, the red emission band was measured around 667 nm associated with the $^5F_5 \rightarrow {}^5I_8$ transition. Furthermore, the ultra-violet and blue upconversion emissions can also be measured. The ultra-violet emissions in the ranges of 381-394 nm and 409-428 nm are assigned to the $^5G_4 \rightarrow {}^5I_8$ and $^5G_5 \rightarrow {}^5I_8$ transitions, respectively. The blue emission between 473 and 500 nm is from the $^5F_3 \rightarrow ^5I_8$ transition. It should be mentioned that the intensity of the upconversion luminescence of the sintered ceramic is much stronger than that of the nanocrystalline powders. This behavior can be explained in two ways. One is from the density. The sintered ceramic apparently has a higher density than the nanocrystalline powders. The other is from nanocrystalline powders' aptness to affect by surroundings [7]. It is well-known that nanocrystalline powders preserve large surface areas, which easily adsorb water and CO_2 from air. Meanwhile, the preparing procedure also produces H_2O and CO_2, which enhanced the multi-phonon relaxation process and reduced the luminescent efficiency.

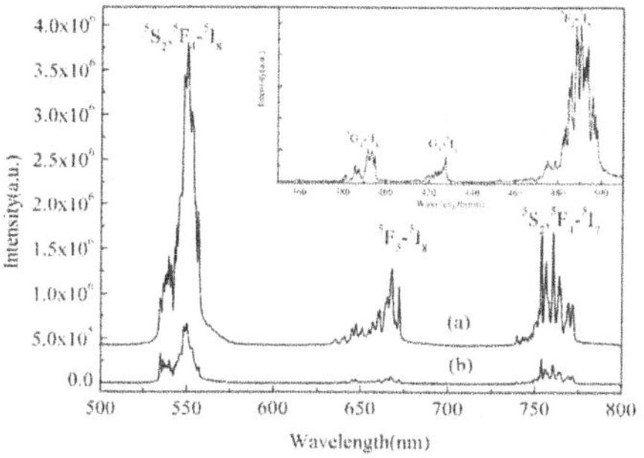

Fig. 3 Upconversion emission spectra of (a) sintered ceramic (b) and nanocrystalline powders $Lu_{1.896}Yb_{0.1}Ho_{0.004}O_3$. Inset : details of upconversion emissions in the range of 350-510 nm.

The mechanism of upconversion emission in Yb^{3+}-Ho^{3+} system has been extensively studied [8-10]. According to the energy match, the possible energy upconversion mechanism for the resulting emission bands is energy transfer from Yb^{3+} to Ho^{3+}. The $^5F_4$, $^5S_2$ levels can be populated by two successive energy transfers (SET) from Yb^{3+} ions and excited state absorption (ESA) after phonon-assisted energy transfer [11]. Subsequently, the Ho^{3+} ions in the $^5F_4$, $^5S_2$ levels can relax to lower excited state $^5F_5$, which can irradiate red emission to the ground state. Meanwhile, the Ho^{3+} ions in the $^5F_4$, $^5S_2$ levels can be excited to upper excited state by successive energy transfer and excited state absorption and then relaxed to lower excited states $^5G_4$, $^5G_5$ and $^5F_3$ by multiphonon relaxation process. The $^5F_3$ level can also be populated by stepwise phonon-assisted two-photon absorption followed by multiphonon relaxation [12]. Therefore the ultra-violet and relatively intense blue emissions can be detected.

CONCLUSION

Nanocrystalline powders and translucent sintered ceramic of $Lu_{1.896}Yb_{0.1}Ho_{0.004}O_3$ were synthesized by co-precipitation method and H_2 atmosphere sintering technology. Under UV excitation, the nanocrystalline powders and sintered ceramic emitted green lights, which were associated with the $^5F_4$, $^5S_2 \rightarrow ^5I_8$ transition of the Ho^{3+} ions. The dominant green, weak red and infrared emissions, together with ultra-violet and blue emissions were observed in both samples when excited by 980 nm diode laser. Both fluorescent and upconversion emission of the two samples were similar and luminescent upconversion in sintered ceramic was enhanced. It indicated that the crystal surroundings are similar and nanocrystalline powders are more easily affected by CO_2 and H_2O in air than sintered ceramic. Energy transfer and excited-state absorption may be responsible for the upconversion process.

ACKNOWLEDGEMENTS

This paper was financially supported by 863 Project (2006AA03Z535).

REFERENCES

[1]J. A. Capobianco, J. C. Boyer, F. Vetrone, A. Speghini and M. Bettinelli, Optical Spectroscopy of Bulk and Nanocrystalline Cubic Y_2O_3:Ho^{3+}, *Chem. Mater.*, **14**, 2915-2921 (2002).

[2]E. zych, Concentration Dependence of Energy Transfer Between Eu^{3+} Ions Oupying Two Symmetry Sites in Lu_2O_3, *J. Phys. Condens. Mater.*, **14,** 5637-50 (2002).

[3]J. Lu, K. Takaichi, T. Uematsu, A. Shriakawa, M. Musha, K. Ueda, H. Yagi, T. Yanagitani, A. A. Kaminskii, Promising ceramic laser material: highly transparent Nd^{3+}:Lu_2O_3 ceramic, *Appl. Phys. Lett.*, **81**, 4324-26 (2002).

[4]J.A. Capobianco, F. Vetrone, J.C. Boyer, A.Speghini, M. Bettinelli, Upconversion of Er^{3+} doped Nanocrystalline and Bulk Lu_2O_3, Opt. Mater., *Opt. Mater.*, **19**, 259-268 (2002).

[5]E. Zych, D. Hreniak, W. Stręk, L. Kepinski, K. Domagala, Sintering Properties of Urea-Derived Lu_2O_3-Based Phosphors, *J. Alloys Compd.*, **341**, 391-394 (2002).

[6]L.Q. An, J. Zhang, Min Liu, S.W. Wang, Preparation and Upconversion Properties of Yb^{3+}, Ho^{3+}: Lu_2O_3 Nanocrystalline Powders, *J. Am. Ceram. Soc.*, **88**, 1010-12 (2005).

[7]F. Vetrone, J.C. Boyer, J.A. Capobianco, NIR to Visible Upconversion in Nanocrystalline and Bulk Lu_2O_3:Er^{3+}, *J. Phys. Chem. B*, **106**, 5622-28 (2002).

[8]I.R. Martin, V.D. Rodriguez, V. Lavin, U.R. Rodriguez-Mendoza, Upconversion dynamics in Yb3+-Ho3+ doped fluoroindate glasses, *J. Alloys Compd.*, **275-277,** 345-348 (1998).

[9]J. Silver, E. Barrett, P.J. Marsh, et al., Yttrium oxide upconversion phosphors, *J. Phys. Chem. B*, **107**, 9236-42 (2003).

[10]A.M. Belovolov, M.I. Timoshechkin, M.J. Damzen, A. Minassian, Powerful visible (530-770 nm) luminescence in Yb,Ho:GGG with IR diode pumping, *Opt. Express*, **10,** 832-39 (2002).

[11]J. Li, J.Y. Wang, H. Tan, X.F. Cheng, F. Song, H.J. Zhang, S.R. Zhao, Growth and optical properties of Ho,Yb:$YAl_3(BO_3)_4$ crystal, *J. Cryst. Growth*, **256,** 324-327 (2003).

[12]A.S. Gouveia-Neto, E.B. da Costa, L.A. Bueno, S.J.L. Ribeiro, Intense red upconversion emission in infrared excited holmium-doped $PbGeO_3$-PbF_2-CdF_2 transparent glass ceramic, *J. Lumin.*, **110,** 79-84 (2004).

MIRRORLESS CONTINUOUS WAVE LASER EMISSION FROM Nd:YAG CERAMIC FEMTOSECOND-WRITTEN WAVEGUIDES

A. Benayas, D. Jaque, A. Ródenas, E. Cantelar
Departamento de Física de Materiales, Universidad Autónoma de Madrid, Madrid, 28049, Spain
L. Roso, G.A. Torchia
Grupo de Óptica, Departamento de Física Aplicada, Universidad de Salamanca, 37008, Spain

ABSTRACT

We have used a femtosecond-written Nd:YAG ceramic optical waveguide as an active media to achieve continuous wave 1.06 μm laser operation. We have obtained output laser power of 40 mW and with a laser slope efficiency in excess of 40%. Single mode and stable laser oscillation have been achieved by using the natural Fresnel reflection for optical feedback without the requirement of any kind of mirror or reflective component.

INTRODUCTION

From several years ago, ceramic materials are emerging as a valuable alternative to crystals for optical applications. Indeed, they are nowadays considered as advanced materials for obtaining high laser energies and intensities, due to their capability to incorporate higher doping levels without losses in their optical quality as well as to the possibility of composite fabrication. Among the different transparent ceramics, YAG host ceramic is at this moment the most consolidated laser material. Taking into account these two characteristics, it´s a well-established fact that the laser performance of Nd:YAG ceramics has been found to be equal or even superior to that corresponding to Nd:YAG crystals.

The outstanding properties of Nd:YAG ceramics as laser medium has been already demonstrated in different bulk configurations extending from continuous wave till pulse mode schemes[0]. Nevertheless, and despite of its interest in modern optoelectronics, the possibility of fabrication of Nd:YAG ceramic waveguide laser is still almost unexplored.

Among the different techniques used for the fabrication of waveguides (including thermal diffusion or ion implantation), ultrafast Direct Laser Writing (DLW) is merging as one of the most versatile ones. This technique is based on the micro-structural modifications that are permanently induced in a transparent material when a high-energy ultrafast laser pulse is focused in its volume. After the irradiation with this ultrafast high energy pulse, the material is modified at the focal volume so that a change in its refractive index is induced. This refractive index change is obtained at the micrometric scale and can be used for the fabrication of optoelectronic devices such as waveguides and photonic crystals[0-0]. This powerful technique has been successfully used for a wide range of glasses and crystals[0-0] and it opens a new and remarkable way to obtain control over the spatial and spectral properties of laser gain thank to micro structuring processes made on laser materials.

The fabrication of low loss Nd:YAG waveguides in Nd:YAG ceramics by ultrafast laser inscription has been recently demonstrated showing an outstanding laser behaviour. Nevertheless, previous results on Nd:YAG ceramic waveguide lasers were obtained by using dielectric mirrors for the achievement of optical feed-back[0]. Besides, the actual requirements on compactness would require reducing the number of optical elements down to minimum. In this sense, the achievement of laser gain in a mirrorless Nd:YAG ceramic waveguide constitutes a new step towards the development of high confined ultra compact laser source based on a transparent ceramic medium. Despite of its interest, mirrorless laser action from ultrafast DLW Nd:YAG ceramic waveguide has been not demonstrated until now.

In this work, we report on the fabrication of buried channel waveguide lasers in Nd:YAG ceramics by using a two line confinement approach. Light confinement has been achieved between two parallel tracks due to filamentation of the femtosecond laser pulses.

EXPERIMENTAL

The Nd:YAG ceramic sample used in this work was provided by Baikowski Ltd. (Japan). The sample was a 5×5×5 mm^3 cube with all its faces polished up to optical quality λ/4. The nominal Nd^{3+} concentration was 2 at. %. The waveguide laser was fabricated by using a Ti:Sapphire laser system providing 120 fs pulses at 796 nm and 1 kHz of repetition rate. The laser beam was focused with a 10× microscope objective (numerical aperture of 0.3). The translation of the sample was achived by a XYZ motorized stage with a spatial resolution of 0.8 μm. With the linear focus of the objective located 500 μm below surface, 29 μm long filaments were written with a pulse energy of 11 μJ, which corresponds to a laser power of 85 MW, well above the YAG threshold power for self-focusing (≈1 MW). Two parallel lines were written separated 12 μm by translating the simple with a speed of 50 μms. More details about the fabrication procedure can be found elsewhere. Figure 1 shows an optical microscope image of the end-face of the obtained waveguide. Note that the waveguide mode is located between the two ultrafast inscribed filaments that appear as the black parallel lines in the optical microscope image. We should remark at this point that the obtained channel waveguides were found stable up to annealing temperatures as high as 1400 °C without any observable deterioration in its guiding properties.

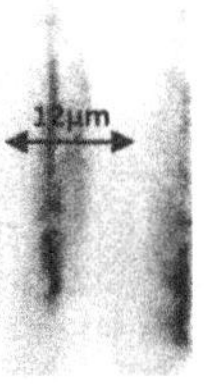

Figure 1.- Micro-photo of cut-in-section of ceramic sample obtained to illustrate the walls of filaments walls conforming buried waveguide structure.

For laser experiments the as fabricated waveguide was used as the laser gain medium without any coupling of external reflectors (mirrors). The experimental set-up used for laser gain experiments is schematically shown in Fig. 2. Optical excitation of the channel waveguide was performed by end-fired coupling a Ti:sapphire continuous wave laser operating at 748 nm ($^4I_{9/2}\rightarrow{}^4F_{7/2}$ transition). The absorption coefficient at this pump wavelength has been measured to be 9.6 cm^{-1}, so that the 5-mm-long waveguide absorbs virtually all the pump light. The excitation beam at 748 nm was coupled into the buried channel waveguide by using a 10X microscope objective. Under these conditions the coupling efficiency was estimated to be above 30 %. The out coming laser emission generated from the waveguide was collected by a second 20X objective microscope. Then the spectral and spatial properties of the oscillating radiation were measured by using a fibre-coupled spectrometer and a CCD camera, respectively. For the measurement of the laser curves the pump power was varied by varying the Ti:Sapphire power (while keeping the spatial characteristics of the pumping beam) and the output power was measured by a calibrated detector.

In absence of laser mirrors, optical feedback at the both faces was only provided by the Fresnel reflection. Taking into account the refractive index of Nd:YAG ceramics (n_0≈1.8), and using Fresnel equations,

$$R = \left(\frac{n_1 - n_2}{n_1 + n_2}\right)^2 \quad T = 1 - R = \frac{4n_1 n_2}{(n_1 + n_2)^2} \qquad (1)$$

we have estimated an effective output transmittance and reflectance of $T \approx 92\%$ and $R \approx 8\%$, respectively. The use of this relatively low reflectance ensures high extraction efficiencies at expenses of moderate laser thresholds but with the clear advantage of compactness reduction.

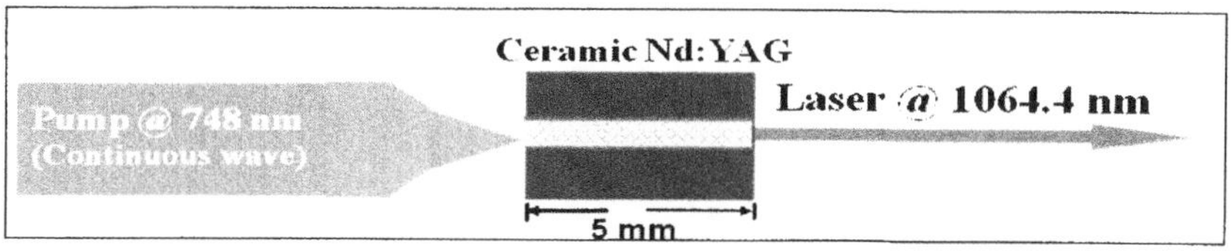

Figure 2.- Experimental set-up used.

RESULTS AND DISCUSSION

Stable laser radiation at 1.06 μm with good spectral and spatial quality was easily achieved once a minimum alignment of the set-up was done. For all the pump powers used in this work laser oscillation was found to be TM polarized. The spectral quality of the laser mode is plotted in Fig. 3.

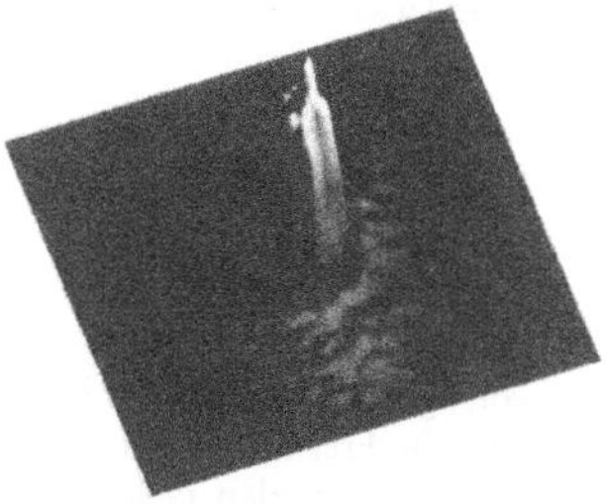

Figure 3.- 3D plot of near field TM_{00} mode intensity distribution of the laser radiation. The total spatial scale in both dimensions was 10 microns.

Note the presence of only one spot revealing the absence of parasitic reflections between parallel faces of the Nd:YAG ceramic sample. The spectral distribution of the laser radiation was also measured. From the laser spectrum (not shown in this work for the sake of brevity) we have corroborated single mode laser oscillation with a linewidth of 0.25 nm., centered around 1064.4 nm peak. And, in this way, we have experimental proofs about spectral and spatial quality of continuous wave laser emission from Nd:YAG ceramic waveguide structure.

Figure 4 shows the laser power generated from our ultrafast inscribed Nd:YAG ceramic waveguide as a function of the 748 nm pump power. We should remark here that the linear behaviour observed was achieved without any requirement of re-alignment during the measurement. The time instabilities in the laser power for any pump power were always below 10%. In Figure 3 we have also included the best linear fit to the laser curve. From this linear fit we have estimated a laser slope efficiency with respect to coupled power as high as 45 %, and a laser threshold of 35 mW. Taking into account the coupling efficiency (30%) this gives a total optical conversion efficiency close to 15%.

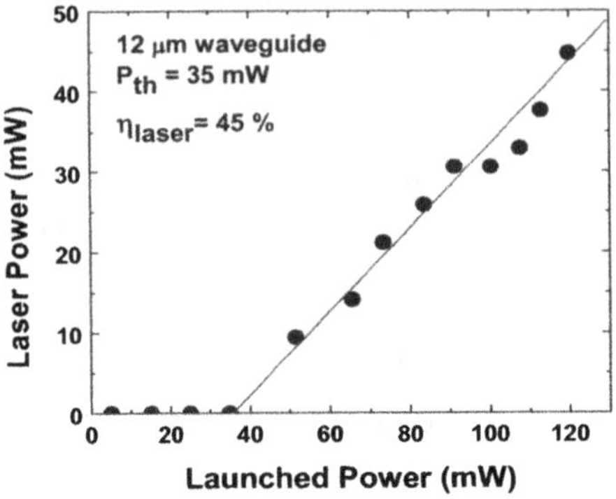

Figure 4.- Laser power as a function of the launched pump power

From the laser curve of Figure 4, it is also possible to get a rough estimation of the optical looses of the fabricated waveguide. Assuming a complete absorption of the launched pump power and a 100% pumping efficiency, the laser slope efficiency (η_{laser}) can be approximately written as[0]

$$\eta_{laser} = \frac{\lambda_{pump}}{\lambda_{laser}} \eta_Q \frac{\ln\left(\frac{1}{R}\right)}{2\alpha l + \ln\left(\frac{1}{R}\right)} \frac{dS}{dF} \tag{2}$$

where λ_{pump}=748 nm is the pumping wavelength, λ_{laser}=1064 nm is the laser wavelength, $R \approx 0.08$ is the output reflectance (given, in our case, by the Fresnel reflection), α is the loss coefficient, l=5 mm is the waveguide length, and dS/dF is the mode-overlap factor[0]. We are making the calculations, after a deep considerations about our experimental conditions, using the value of dSdF≈1, based on the following points: (a) in our experimental conditions pump beam waist is sligthly smaller than laser beam waist; (b) the population in the terminal laser level due to the advantageous four level scheme; (c) absence of reabsorption loss. Finally, we have over passed the usual uncertainty related to the quantitative value of η_Q (values ranging from 0.85 up to 0.95) reported from different authors[0,0]. We have used in our calculations a tentative value of 0.9. By substituting the experimental value found for the laser slope efficiency in expression (2) we have found α =1 cm^{-1} (4.3 dB/cm).
It is important to remark here that the laser slope efficiency exceeds 40% it is the highest even reported for a mirrorless waveguide laser and it is somehow comparable to that laser slope efficiency reported fro Nd:YAG waveguide lasers using mirrors for optical[5]. As a matter of fact, it is close 4 times the laser slope efficiency previously reported for a femtosecond written waveguide laser fabricated in a Nd:YAG crystal, slightly larger than the laser slope efficiency achieved with an epitaxial grown Nd:YAG waveguide laser and comparable to the laser slope efficiencies reported from Nd:YAG direct bonded waveguides.[0-0]

The value of the optical losses in our waveguide has been found to be higher than that reported for a femtosecond written depressed cladding waveguide laser formed in Nd:YAG crystals (0.3 dB/cm),[0] but comparable to that found in femtosecond laser written surface channel waveguides fabricated in Nd:YAG ceramics (1 dB/cm)[0] According to expression (2), the laser slope efficiency of our waveguide laser can be even improved by using longer pump wavelengths, since they would lead to a significant reduction in the quantum defect between pump and laser photons which, in turn, will be accompanied by a reduction in the pump induced thermal loading and, therefore, in the undesirable thermal effects[0].

Our results are mainly oriented to confirm that DLW technique is perfectly suitable to create microstructured laser systems. But, is equally relevant to point out that when we have used Nd:YAG ceramic sample were possible to achieve high doping levels and, then directly linked with it, to reach high laser emission powers. In addition, certain kind of high energy applications requires to employ ceramic material as active media because it high pump-damage resistance.

CONCLUSIONS

In summary, mirrorless laser waveguides have been fabricated in Nd:YAG ceramics by the ultrafast laser inscription technique. By using the Fresnel reflections for optical feedback at both faces of the cavity, laser action at 1064 nm has been demonstrated from the obtained waveguides with laser slope efficiencies as high as 45%, and output laser powers in excess of 45 mW. This laser outputs show a very high stability which clearly confirms the excellent mechanical and thermal properties of the Nd:YAG ceramic host besides the excellent optical properties of this matrix, and the versatility of the direct laser writing technique to construct highly efficient miniaturized three-dimensional devices. Finally, this buried channel waveguides are high thermally stable.

ACKNOWLEDGMENTS

This work has been supported by the Spanish Ministerio de Educación y Ciencia (MAT2004-03347, TEC2004-05260-C02-02, and MAT2005-05950) by FEDER founds (FIS2005-01351), by the Universidad Autónoma de Madrid and Comunidad Autonoma de Madrid (projects MICROSERES-CM and CCG06-UAM/MAT-0347), and by the Junta de Castilla y León (Grant No. SA026A05). G.A.T. wishes to thank to the Spanish Ministerio de Educación y Ciencia (Project No. FIS2006-04151), to the Agencia de Promoción Científica y Tecnológica de Argentina (Project No. PICT 15210) and to the Conicet for the financial support received.

REFERENCES

[1]G. Zhou and M. Gu, *Opt. Lett.*, **31**, 2783 (2006).

[2]K. Kawamura, T. Ogawa, N. Sarukura, M. Hirano, and H. Hosono, *Appl. Phys. B: Lasers Opt.* **71**, 119 (2000).

[3]K. Miura, J. Qiu, S. Fujiwara, S. Sakaguchi, and K. Hirao, *Appl. Phys. Lett.* **80**, 2263 (2002).

[4]M. Hughes, W. Yang, and D. Hewak, *Appl. Phys. Lett.* **90**, 131113 (2007).

[5]G. A. Torchia, A. Ródenas, A. Benayas, E. Cantelar L. Roso and D. Jaque, *App Phys.Lett.* **11**,1103 (2008).

[6]G. Della Valle, R. Osellame, N. Chiodo, S. Taccheo, G. Cerullo, P.Laporta, A. Killi, U. Morgner, M. Lederer, and D. Kopf, *Opt. Express* **13**,5976 (2005).

[7]J. Lu, M. Prabhu, J. Xu, K. Ueda, H. Yagi, T. Yanagitani, and A. A. Kaminskii, *Appl. Phys. Lett.* **77**, 3707 (2000).

[8]G. A. Torchia, P. Meilan, A. Rodenas, D. Jaque, C. Mendez, and L. Roso, *Opt. Express* **15**, 13266 (2007).

[9]A. G. Okhrimchuk, A. V. Shestakov, I. Khrushchev, and J. Mitchell, *Opt.Lett.* **30**, 2248 (2005).

[10]I. Chartier, B. Ferrand, D. Pelenc, S. J. Field, D. C. Hanna, A. C. Lage, D.P. Sheperd, and A. C.

Tropper, *Opt. Lett.* **17**, 810 (1992).
[11]J. I. Mackenzie, C. Li, and D. P. Shepherd, *IEEE J. Quantum Electron.* **39**,493 (2003)
[12]N. P. Barnes and B. M. Walsh, *OSA Proc. Adv. Solid-State Lasers* **68**, 284 (2002). W.P. Risk, J. Opt. Soc. Am. B **5**, 1412 (1988).
[13]W.P. Risk, *J. Opt. Soc. Am. B* **5**, 1412 (1988).
[14]W. Koechner, *Solid State Laser Engineering* (Springer, Berlin,1999).
[15]Z. D. Luo, Y. D. Huang, M. Montes, and D. Jaque, *Appl. Phys. Lett.* **85**,715 (2004)
[16]A. W. Snyder and J. D. Love, *Optical Waveguide Theory* (Chapman and Hall, London, (1983).

Author Index

Printed and bound by CPI Group (UK) Ltd, Croydon, CR0 4YY
17/09/2026
14964177-0001